Graduate Texts in Mathematics 249

Graduate Texts in Mathematics

Graduate Texts in Mathematics bridge the gap between passive study and creative understanding, offering graduate-level introductions to advanced topics in mathematics. The volumes are carefully written as teaching aids and highlight characteristic features of the theory. Although these books are frequently used as textbooks in graduate courses, they are also suitable for individual study.

For further volumes:
http://www.springer.com/series/136

Loukas Grafakos

Classical Fourier Analysis

Third Edition

Loukas Grafakos
Department of Mathematics
University of Missouri
Columbia, MO, USA

ISSN 0072-5285 ISSN 2197-5612 (electronic)
ISBN 978-1-4939-3916-9 ISBN 978-1-4939-1194-3 (eBook)
DOI 10.1007/978-1-4939-1194-3
Springer New York Heidelberg Dordrecht London

Mathematics Subject Classification (2010): 42Axx, 42Bxx

Printed on acid-free paper

Springer is part of Springer Science+Business Media (www.springer.com)

To Suzanne

Preface

The great response to the publication of my book *Classical and Modern Fourier Analysis* in 2004 has been especially gratifying to me. I was delighted when Springer offered to publish the second edition in 2008 in two volumes: *Classical Fourier Analysis, 2nd Edition,* and *Modern Fourier Analysis, 2nd Edition*. I am now elated to have the opportunity to write the present third edition of these books, which Springer has also kindly offered to publish. The third edition was born from my desire to improve the exposition in several places, fix a few inaccuracies, and add some new material. I have been very fortunate to receive several hundred e-mail messages that helped me improve the proofs and locate mistakes and misprints in the previous editions.

In this edition, I maintain the same style as in the previous ones. The proofs contain details that unavoidably make the reading more cumbersome. Although it will behoove many readers to skim through the more technical aspects of the presentation and concentrate on the flow of ideas, the fact that details are present will be comforting to some. (This last sentence is based on my experience as a graduate student.) Readers familiar with the second edition will notice that the chapter on weights has been moved from the second volume to the first.

This first volume *Classical Fourier Analysis* is intended to serve as a text for a one-semester course with prerequisites of measure theory, Lebesgue integration, and complex variables. I am aware that this book contains significantly more material than can be taught in a semester course; however, I hope that this additional information will be useful to researchers. Based on my experience, the following list of sections (or parts of them) could be taught in a semester without affecting the logical coherence of the book: Sections 1.1, 1.2, 1.3, 2.1, 2.2., 2.3, 3.1, 3.2, 3.3, 4.4, 4.5, 5.1, 5.2, 5.3, 5.5, 5.6, 6.1, 6.2.

A long list of people have assisted me in the preparation of this book, but I remain solely responsible for any misprints, mistakes, and omissions contained therein. Please contact me directly (grafakosl@missouri.edu) if you have corrections or comments. Any corrections to this edition will be posted to the website

`http://math.missouri.edu/~loukas/FourierAnalysis.html`

which I plan to update regularly. I have prepared solutions to all of the exercises for the present edition which will be available to instructors who teach a course out of this book.

Athens, Greece, *Loukas Grafakos*
March 2014

Acknowledgments

I am extremely fortunate that several people have pointed out errors, misprints, and omissions in the previous editions of the books in this series. All these individuals have provided me with invaluable help that resulted in the improved exposition of the text. For these reasons, I would like to express my deep appreciation and sincere gratitude to the all of the following people.

First edition acknowledgements: Georgios Alexopoulos, Nakhlé Asmar, Bruno Calado, Carmen Chicone, David Cramer, Geoffrey Diestel, Jakub Duda, Brenda Frazier, Derrick Hart, Mark Hoffmann, Steven Hofmann, Helge Holden, Brian Hollenbeck, Petr Honzík, Alexander Iosevich, Tunde Jakab, Svante Janson, Ana Jiménez del Toro, Gregory Jones, Nigel Kalton, Emmanouil Katsoprinakis, Dennis Kletzing, Steven Krantz, Douglas Kurtz, George Lobell, Xiaochun Li, José María Martell, Antonios Melas, Keith Mersman, Stephen Montgomery-Smith, Andrea Nahmod, Nguyen Cong Phuc, Krzysztof Oleszkiewicz, Cristina Pereyra, Carlos Pérez, Daniel Redmond, Jorge Rivera-Noriega, Dmitriy Ryabogin, Christopher Sansing, Lynn Savino Wendel, Shih-Chi Shen, Roman Shvidkoy, Elias M. Stein, Atanas Stefanov, Terence Tao, Erin Terwilleger, Christoph Thiele, Rodolfo Torres, Deanie Tourville, Nikolaos Tzirakis, Don Vaught, Igor Verbitsky, Brett Wick, James Wright, and Linqiao Zhao.

Second edition acknowledgements: Marco Annoni, Pascal Auscher, Andrew Bailey, Dmitriy Bilyk, Marcin Bownik, Juan Cavero de Carondelet Fiscowich, Leonardo Colzani, Simon Cowell, Mita Das, Geoffrey Diestel, Yong Ding, Jacek Dziubanski, Frank Ganz, Frank McGuckin, Wei He, Petr Honzík, Heidi Hulsizer, Philippe Jaming, Svante Janson, Ana Jiménez del Toro, John Kahl, Cornelia Kaiser, Nigel Kalton, Kim Jin Myong, Doowon Koh, Elena Koutcherik, David Kramer, Enrico Laeng, Sungyun Lee, Qifan Li, Chin-Cheng Lin, Liguang Liu, Stig-Olof Londen, Diego Maldonado, José María Martell, Mieczysław Mastyło, Parasar Mohanty, Carlo Morpurgo, Andrew Morris, Mihail Mourgoglou, Virginia Naibo, Tadahiro Oh, Marco Peloso, Maria Cristina Pereyra, Carlos Pérez, Humberto Rafeiro, Maria Carmen Reguera Rodríguez, Alexander Samborskiy, Andreas Seeger, Steven Senger, Sumi Seo, Christopher Shane, Shu Shen, Yoshihiro Sawano, Mark Spencer, Vladimir Stepanov, Erin Terwilleger, Rodolfo H. Torres, Suzanne Tourville,

Ignacio Uriarte-Tuero, Kunyang Wang, Huoxiong Wu, Kôzô Yabuta, Takashi Yamamoto, and Dachun Yang.

Third edition acknowledgments: Marco Annoni, Mark Ashbaugh, Daniel Azagra, Andrew Bailey, Árpad Bényi, Dmitriy Bilyk, Nicholas Boros, Almut Burchard, María Carro, Jameson Cahill, Juan Cavero de Carondelet Fiscowich, Xuemei Chen, Andrea Fraser, Shai Dekel, Fausto Di Biase, Zeev Ditzian, Jianfeng Dong, Oliver Dragičević, Sivaji Ganesh, Friedrich Gesztesy, Zhenyu Guo, Piotr Hajłasz, Danqing He, Andreas Heinecke, Steven Hofmann, Takahisa Inui, Junxiong Jia, Kasinathan Kamalavasanthi, Hans Koelsch, Richard Laugesen, Kaitlin Leach, Andrei Lerner, Yiyu Liang, Calvin Lin, Liguang Liu, Elizabeth Loew, Chao Lu, Richard Lynch, Diego Maldonado, Lech Maligranda, Richard Marcum, Mieczysław Mastyło, Mariusz Mirek, Carlo Morpurgo, Virginia Naibo, Hanh Van Nguyen, Seungly Oh, Tadahiro Oh, Yusuke Oi, Lucas da Silva Oliveira, Kevin O'Neil, Hesam Oveys, Manos Papadakis, Marco Peloso, Carlos Pérez, Jesse Peterson, Dmitry Prokhorov, Amina Ravi, Maria Carmen Reguera Rodríguez, Yoshihiro Sawano, Mirye Shin, Javier Soria, Patrick Spencer, Marc Strauss, Krystal Taylor, Naohito Tomita, Suzanne Tourville, Rodolfo H. Torres, Fujioka Tsubasa, Ignacio Uriarte-Tuero, Brian Tuomanen, Shibi Vasudevan, Michael Wilson, Dachun Yang, Kai Yang, Yandan Zhang, Fayou Zhao, and Lifeng Zhao.

Among all these people, I would like to give special thanks to an individual who has studied extensively the two books in the series and has helped me more than anyone else in the preparation of the third edition: Danqing He. I am indebted to him for all the valuable corrections, suggestions, and constructive help he has provided me with in this work. Without him, these books would have been a lot poorer.

Finally, I would also like to thank the University of Missouri for granting me a research leave during the academic year 2013-2014. This time off enabled me to finish the third edition of this book on time. I spent my leave in Greece.

Contents

Chapter 1
L^p Spaces and Interpolation

Many quantitative properties of functions are expressed in terms of their integrability to a power. For this reason it is desirable to acquire a good understanding of spaces of functions whose modulus to a power p is integrable. These are called Lebesgue spaces and are denoted by L^p. Although an in-depth study of Lebesgue spaces falls outside the scope of this book, it seems appropriate to devote a chapter to reviewing some of their fundamental properties.

The emphasis of this review is basic interpolation between Lebesgue spaces. Many problems in Fourier analysis concern boundedness of operators on Lebesgue spaces, and interpolation provides a framework that often simplifies this study. For instance, in order to show that a linear operator maps L^p to itself for all $1 < p < \infty$, it is sufficient to show that it maps the (smaller) Lorentz space $L^{p,1}$ into the (larger) Lorentz space $L^{p,\infty}$ for the same range of p's. Moreover, some further reductions can be made in terms of the Lorentz space $L^{p,1}$. This and other considerations indicate that interpolation is a powerful tool in the study of boundedness of operators.

Although we are mainly concerned with L^p subspaces of Euclidean spaces, we discuss in this chapter L^p spaces of arbitrary measure spaces, since they represent a useful general setting. Many results in the text require working with general measures instead of Lebesgue measure.

1.1 L^p and Weak L^p

A *measure space* is a set X equipped with a σ-algebra of subsets of it and a function μ from the σ-algebra to $[0,\infty]$ that satisfies $\mu(\emptyset) = 0$ and

$$\mu\Big(\bigcup_{j=1}^{\infty} B_j\Big) = \sum_{j=1}^{\infty} \mu(B_j)$$

for any sequence B_j of pairwise disjoint elements of the σ-algebra. The function μ is called a (positive) measure on X and elements of the σ-algebra of X are called

L. Grafakos, *Classical Fourier Analysis*, Graduate Texts in Mathematics 249,
DOI 10.1007/978-1-4939-1194-3_1,

measurable sets. Measure spaces will be assumed to be complete, i.e., subsets of the σ-algebra of measure zero also belong to the σ-algebra. A measure space X is called *σ-finite* if there is a sequence of measurable subsets X_n of it such that

$$X = \bigcup_{n=1}^{\infty} X_n$$

and $\mu(X_n) < \infty$. A real-valued function f on a measure space is called *measurable* if the set $\{x \in X : f(x) > \lambda\}$ is measurable for all real numbers λ. A complex-valued function is measurable if and only if its real and imaginary parts are measurable. A *simple function* is a finite linear combination of characteristic functions of measurable subsets of X; these subsets may have infinite measure. A *finitely simple* function has the form

$$\sum_{j=1}^{N} c_j \chi_{B_j}$$

where $N < \infty$, $c_j \in \mathbf{C}$, and B_j are pairwise disjoint measurable sets with $\mu(B_j) < \infty$. If $N = \infty$, this function will be called *countably simple*. Finitely simple functions are exactly the integrable simple functions. Every nonnegative measurable function is the pointwise limit of an increasing sequence of simple functions; if the space is σ-finite, these simple functions can be chosen to be finitely simple.

For $0 < p < \infty$, $L^p(X,\mu)$ denotes the set of all complex-valued μ-measurable functions on X whose modulus to the pth power is integrable. $L^\infty(X,\mu)$ is the set of all complex-valued μ-measurable functions f on X such that for some $B > 0$, the set $\{x\colon |f(x)| > B\}$ has μ-measure zero. Two functions in $L^p(X,\mu)$ are considered equal if they are equal μ-almost everywhere. When $0 < p < \infty$ finitely simple functions are dense in $L^p(X,\mu)$. Within context and in the absence of ambiguity, $L^p(X,\mu)$ is simply written as L^p.

The notation $L^p(\mathbf{R}^n)$ is reserved for the space $L^p(\mathbf{R}^n, |\cdot|)$, where $|\cdot|$ denotes n-dimensional Lebesgue measure. Lebesgue measure on $\mathbf{R}^n$ is also denoted by dx. Other measures will be considered on the *Borel σ-algebra* of $\mathbf{R}^n$, i.e., is the smallest σ-algebra that contains the closed subsets of $\mathbf{R}^n$. Measures on the σ-algebra of Borel measurable subsets are called *Borel measures*; such measures will be assumed to be finite on compact subsets of $\mathbf{R}^n$. A Borel measure μ with $\mu(\mathbf{R}^n) < \infty$ is called a *finite Borel measure*. A Borel measure on $\mathbf{R}^n$ is called *regular* for all Borel measurable sets E we have

$$\mu(E) = \inf\{\mu(O) : E \subseteqq O,\ O \text{ open}\} = \sup\{\mu(K) : K \subseteqq E,\ K \text{ compact}\}.$$

The space $L^p(\mathbf{Z})$ equipped with counting measure is denoted by $\ell^p(\mathbf{Z})$ or simply ℓ^p.

For $0 < p < \infty$, we define the L^p norm of a function f (or quasi-norm if $p < 1$) by

$$\|f\|_{L^p(X,\mu)} = \left(\int_X |f(x)|^p \, d\mu(x) \right)^{\frac{1}{p}} \tag{1.1.1}$$

and for $p = \infty$ by

$$\|f\|_{L^\infty(X,\mu)} = \text{ess.sup}|f| = \inf\{B > 0 : \mu(\{x : |f(x)| > B\}) = 0\}. \tag{1.1.2}$$

It is well known that Minkowski's (or the triangle) inequality

$$\|f+g\|_{L^p(X,\mu)} \le \|f\|_{L^p(X,\mu)} + \|g\|_{L^p(X,\mu)} \tag{1.1.3}$$

holds for all f, g in $L^p = L^p(X,\mu)$, whenever $1 \le p \le \infty$. Since in addition $\|f\|_{L^p(X,\mu)} = 0$ implies that $f = 0$ (μ-a.e.), the L^p spaces are normed linear spaces for $1 \le p \le \infty$. For $0 < p < 1$, inequality (1.1.3) is reversed when $f, g \ge 0$. However, the following substitute of (1.1.3) holds:

$$\|f+g\|_{L^p(X,\mu)} \le 2^{\frac{1-p}{p}} \left(\|f\|_{L^p(X,\mu)} + \|g\|_{L^p(X,\mu)}\right), \tag{1.1.4}$$

and thus $L^p(X,\mu)$ is a quasi-normed linear space. See also Exercise 1.1.5. For all $0 < p \le \infty$, it can be shown that every Cauchy sequence in $L^p(X,\mu)$ is convergent, and hence the spaces $L^p(X,\mu)$ are complete. For the case $0 < p < 1$ we refer to Exercise 1.1.8. Therefore, the L^p spaces are Banach spaces for $1 \le p \le \infty$ and quasi-Banach spaces for $0 < p < 1$. For any $p \in (0,\infty) \setminus \{1\}$ we use the notation $p' = \frac{p}{p-1}$. Moreover, we set $1' = \infty$ and $\infty' = 1$, so that $p'' = p$ for all $p \in (0,\infty]$. Hölder's inequality says that for all $p \in [1,\infty]$ and all measurable functions f, g on (X,μ) we have

$$\|fg\|_{L^1} \le \|f\|_{L^p} \|g\|_{L^{p'}}.$$

It is a well-known fact that the dual $(L^p)^*$ of L^p is isometric to $L^{p'}$ for all $1 \le p < \infty$. Furthermore, the L^p norm of a function can be obtained via duality when $1 \le p \le \infty$ as follows:

$$\|f\|_{L^p} = \sup_{\|g\|_{L^{p'}}=1} \left| \int_X f g \, d\mu \right|.$$

For the endpoint cases $p = 1$, $p = \infty$, see Exercise 1.4.12 (a), (b).

1.1.1 The Distribution Function

Definition 1.1.1. For f a measurable function on X, the *distribution function* of f is the function d_f defined on $[0,\infty)$ as follows:

$$d_f(\alpha) = \mu(\{x \in X : |f(x)| > \alpha\}). \tag{1.1.5}$$

The distribution function d_f provides information about the size of f but not about the behavior of f itself near any given point. For instance, a function on $\mathbf{R}^n$ and each of its translates have the same distribution function. It follows from Definition 1.1.1 that d_f is a decreasing function of α (not necessarily strictly).

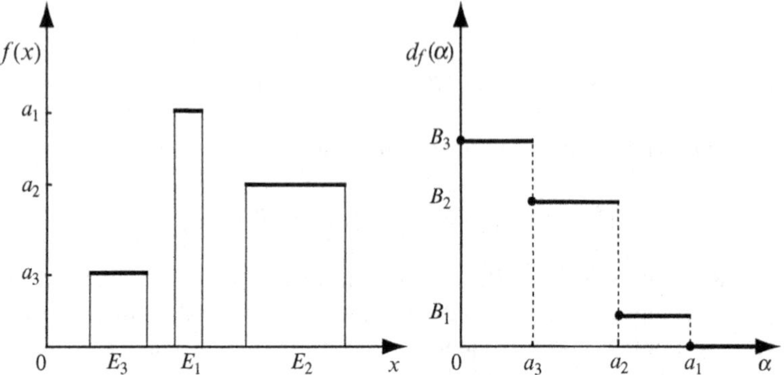

Fig. 1.1 The graph of a simple function $f=\sum_{k=1}^{3} a_k\chi_{E_k}$ and its distribution function $d_f(\alpha)$. Here $B_j=\sum_{k=1}^{j}\mu(E_k)$.

Example 1.1.2. For pedagogical reasons we compute the distribution function d_f of a nonnegative simple function

$$f(x)=\sum_{j=1}^{N} a_j\chi_{E_j}(x),$$

where the sets E_j are pairwise disjoint and $a_1>\cdots>a_N>0$. If $\alpha\geq a_1$, then clearly $d_f(\alpha)=0$. However, if $a_2\leq\alpha<a_1$ then $|f(x)|>\alpha$ precisely when $x\in E_1$, and in general, if $a_{j+1}\leq\alpha<a_j$, then $|f(x)|>\alpha$ precisely when $x\in E_1\cup\cdots\cup E_j$. Setting

$$B_j=\sum_{k=1}^{j}\mu(E_k),$$

for $j\in\{1,\dots,N\}$, $B_0=a_{N+1}=0$, and $a_0=\infty$, we have

$$d_f(\alpha)=\sum_{j=0}^{N} B_j\chi_{[a_{j+1},a_j)}(\alpha).$$

Note that these formulas are valid even when $\mu(E_i)=\infty$ for some i. Figure 1.1 presents an illustration of this example when $N=3$ and $\mu(E_j)<\infty$ for all j.

Proposition 1.1.3. *Let f and g be measurable functions on (X,μ). Then for all $\alpha,\beta>0$ we have*

(1) $|g|\leq|f|$ μ-a.e. implies that $d_g\leq d_f$;

(2) $d_{cf}(\alpha)=d_f(\alpha/|c|)$, for all $c\in\mathbf{C}\setminus\{0\}$;

(3) $d_{f+g}(\alpha+\beta)\leq d_f(\alpha)+d_g(\beta)$;

(4) $d_{fg}(\alpha\beta)\leq d_f(\alpha)+d_g(\beta)$.

Proof. The simple proofs are left to the reader. □

Knowledge of the distribution function d_f provides sufficient information to evaluate the L^p norm of a function f precisely. We state and prove the following important description of the L^p norm in terms of the distribution function.

Proposition 1.1.4. *Let (X,μ) be a σ-finite measure space. Then for f in $L^p(X,\mu)$, $0<p<\infty$, we have*

$$\|f\|_{L^p}^p = p\int_0^\infty \alpha^{p-1} d_f(\alpha)\,d\alpha\,. \tag{1.1.6}$$

Moreover, for any increasing continuously differentiable function φ on $[0,\infty)$ with $\varphi(0)=0$ and every measurable function f on X with $\varphi(|f|)$ integrable on X, we have

$$\int_X \varphi(|f|)\,d\mu = \int_0^\infty \varphi'(\alpha) d_f(\alpha)\,d\alpha\,. \tag{1.1.7}$$

Proof. Indeed, we have

$$\begin{aligned} p\int_0^\infty \alpha^{p-1} d_f(\alpha)\,d\alpha &= p\int_0^\infty \alpha^{p-1}\int_X \chi_{\{x:\,|f(x)|>\alpha\}}\,d\mu(x)\,d\alpha \\ &= \int_X \int_0^{|f(x)|} p\alpha^{p-1}\,d\alpha\,d\mu(x) \\ &= \int_X |f(x)|^p\,d\mu(x) \\ &= \|f\|_{L^p}^p\,, \end{aligned}$$

where in the second equality we used Fubini's theorem, which requires the measure space to be σ-finite. This proves (1.1.6). Identity (1.1.7) follows similarly, replacing the function α^p by the more general function $\varphi(\alpha)$ which has similar properties. □

Definition 1.1.5. For $0<p<\infty$, the space *weak* $L^p(X,\mu)$ is defined as the set of all μ-measurable functions f such that

$$\begin{aligned} \|f\|_{L^{p,\infty}} &= \inf\Big\{C>0:\, d_f(\alpha)\le \frac{C^p}{\alpha^p} \quad \text{for all} \quad \alpha>0\Big\} && (1.1.8)\\ &= \sup\big\{\gamma d_f(\gamma)^{1/p}:\, \gamma>0\big\} && (1.1.9) \end{aligned}$$

is finite. The space *weak* $L^\infty(X,\mu)$ is by definition $L^\infty(X,\mu)$.

One should check that (1.1.9) and (1.1.8) are in fact equal. The weak L^p spaces are denoted by $L^{p,\infty}(X,\mu)$. Two functions in $L^{p,\infty}(X,\mu)$ are considered equal if they are equal μ-a.e. The notation $L^{p,\infty}(\mathbf{R}^n)$ is reserved for $L^{p,\infty}(\mathbf{R}^n,|\cdot|)$. Using Proposition 1.1.3 (2), we can easily show that

$$\|kf\|_{L^{p,\infty}} = |k|\,\|f\|_{L^{p,\infty}}\,, \tag{1.1.10}$$

for any complex constant k. The analogue of (1.1.3) is

$$\|f+g\|_{L^{p,\infty}} \le c_p\big(\|f\|_{L^{p,\infty}} + \|g\|_{L^{p,\infty}}\big), \tag{1.1.11}$$

where $c_p = \max(2, 2^{1/p})$, a fact that follows from Proposition 1.1.3 (3), taking both α and β equal to $\alpha/2$. We also have that

$$\|f\|_{L^{p,\infty}(X,\mu)} = 0 \Rightarrow f = 0 \qquad \mu\text{-a.e.} \tag{1.1.12}$$

In view of (1.1.10), (1.1.11), and (1.1.12), $L^{p,\infty}$ is a *quasi-normed linear space* for $0<p<\infty$.

The weak L^p spaces are larger than the usual L^p spaces. We have the following:

Proposition 1.1.6. *For any $0<p<\infty$ and any f in $L^p(X,\mu)$ we have*

$$\|f\|_{L^{p,\infty}} \le \|f\|_{L^p}\,.$$

Hence the embedding $L^p(X,\mu) \subseteq L^{p,\infty}(X,\mu)$ holds.

Proof. This is just a trivial consequence of Chebyshev's inequality:

$$\alpha^p d_f(\alpha) \le \int_{\{x:\,|f(x)|>\alpha\}} |f(x)|^p\, d\mu(x) \le \|f\|_{L^p}^p\,.$$

Using (1.1.9) we obtain that $\|f\|_{L^{p,\infty}} \le \|f\|_{L^p}$. □

The inclusion $L^p \subsetneqq L^{p,\infty}$ is strict. For example, on $\mathbf{R}^n$ with the usual Lebesgue measure, let $h(x) = |x|^{-\frac{n}{p}}$. Obviously, h is not in $L^p(\mathbf{R}^n)$ but h is in $L^{p,\infty}(\mathbf{R}^n)$ with $\|h\|_{L^{p,\infty}(\mathbf{R}^n)} = v_n^{1/p}$, where v_n is the measure of the unit ball of $\mathbf{R}^n$.

It is not immediate from their definition that the weak L^p spaces are complete with respect to the quasi-norm $\|\cdot\|_{L^{p,\infty}}$. The completeness of these spaces is proved in Theorem 1.4.11, but it is also a consequence of Theorem 1.1.13, proved in this section.

1.1.2 Convergence in Measure

Next we discuss some convergence notions. The following notion is important in probability theory.

Definition 1.1.7. Let f, f_n, $n=1,2,\dots$, be measurable functions on the measure space (X,μ). The sequence f_n is said to *converge in measure* to f if for all $\varepsilon>0$ there exists an $n_0 \in \mathbf{Z}^+$ such that

$$n>n_0 \implies \mu(\{x\in X:\ |f_n(x)-f(x)|>\varepsilon\}) < \varepsilon\,. \tag{1.1.13}$$

Remark 1.1.8. The preceding definition is equivalent to the following statement:

$$\text{For all } \varepsilon > 0 \quad \lim_{n\to\infty} \mu(\{x \in X : |f_n(x) - f(x)| > \varepsilon\}) = 0. \tag{1.1.14}$$

Clearly (1.1.14) implies (1.1.13). To see the converse given $\varepsilon > 0$, pick $0 < \delta < \varepsilon$ and apply (1.1.13) for this δ. There exists an $n_0 \in \mathbf{Z}^+$ such that

$$\mu(\{x \in X : |f_n(x) - f(x)| > \delta\}) < \delta$$

holds for $n > n_0$. Since

$$\mu(\{x \in X : |f_n(x) - f(x)| > \varepsilon\}) \le \mu(\{x \in X : |f_n(x) - f(x)| > \delta\}),$$

we conclude that

$$\mu(\{x \in X : |f_n(x) - f(x)| > \varepsilon\}) < \delta$$

for all $n > n_0$. Let $n \to \infty$ to deduce that

$$\limsup_{n\to\infty} \mu(\{x \in X : |f_n(x) - f(x)| > \varepsilon\}) \le \delta. \tag{1.1.15}$$

Since (1.1.15) holds for all $0 < \delta < \varepsilon$, (1.1.14) follows by letting $\delta \to 0$.

Convergence in measure is a weaker notion than convergence in either L^p or $L^{p,\infty}$, $0 < p \le \infty$, as the following proposition indicates:

Proposition 1.1.9. *Let $0 < p \le \infty$ and f_n, f be in $L^{p,\infty}(X,\mu)$.*

(1) If f_n, f are in L^p and $f_n \to f$ in L^p, then $f_n \to f$ in $L^{p,\infty}$.
(2) If $f_n \to f$ in $L^{p,\infty}$, then f_n converges to f in measure.

Proof. Fix $0 < p < \infty$. Proposition 1.1.6 gives that for all $\varepsilon > 0$ we have

$$\mu(\{x \in X : |f_n(x) - f(x)| > \varepsilon\}) \le \frac{1}{\varepsilon^p} \int_X |f_n - f|^p \, d\mu.$$

This shows that convergence in L^p implies convergence in weak L^p. The case $p = \infty$ is tautological.

Given $\varepsilon > 0$ find an n_0 such that for $n > n_0$, we have

$$\big\|f_n - f\big\|_{L^{p,\infty}} = \sup_{\alpha>0} \alpha \mu(\{x \in X : |f_n(x) - f(x)| > \alpha\})^{\frac{1}{p}} < \varepsilon^{\frac{1}{p}+1}.$$

Taking $\alpha = \varepsilon$, we conclude that convergence in $L^{p,\infty}$ implies convergence in measure. □

Example 1.1.10. Note that there is no general converse of statement (2) in the preceding proposition. Fix $0 < p < \infty$ and on $[0,1]$ define the functions

$$f_{k,j} = k^{1/p} \chi_{(\frac{j-1}{k}, \frac{j}{k})}, \qquad k \ge 1,\ 1 \le j \le k.$$

Consider the sequence $\{f_{1,1}, f_{2,1}, f_{2,2}, f_{3,1}, f_{3,2}, f_{3,3}, \dots\}$. Observe that

$$|\{x:\ f_{k,j}(x) > 0\}| = 1/k.$$

Therefore, $f_{k,j}$ converges to 0 in measure. Likewise, observe that

$$\big\|f_{k,j}\big\|_{L^{p,\infty}} = \sup_{\alpha>0} \alpha|\{x:\ f_{k,j}(x) > \alpha\}|^{1/p} \geq \sup_{k\geq 1} \frac{(k-1/k)^{1/p}}{k^{1/p}} = 1,$$

which implies that $f_{k,j}$ does not converge to 0 in $L^{p,\infty}$.

It turns out that every sequence convergent in $L^p(X,\mu)$ or in $L^{p,\infty}(X,\mu)$ has a subsequence that converges a.e. to the same limit.

Theorem 1.1.11. *Let f_n and f be complex-valued measurable functions on a measure space (X,μ) and suppose that f_n converges to f in measure. Then some subsequence of f_n converges to f μ-a.e.*

Proof. For all $k = 1,2,\dots$ choose inductively n_k such that

$$\mu(\{x \in X:\ |f_{n_k}(x) - f(x)| > 2^{-k}\}) < 2^{-k} \tag{1.1.16}$$

and such that $n_1 < n_2 < \cdots < n_k < \cdots$. Define the sets

$$A_k = \{x \in X:\ |f_{n_k}(x) - f(x)| > 2^{-k}\}.$$

Equation (1.1.16) implies that

$$\mu\bigg(\bigcup_{k=m}^{\infty} A_k\bigg) \leq \sum_{k=m}^{\infty} \mu(A_k) \leq \sum_{k=m}^{\infty} 2^{-k} = 2^{1-m} \tag{1.1.17}$$

for all $m = 1,2,3,\dots$. It follows from (1.1.17) that

$$\mu\bigg(\bigcup_{k=1}^{\infty} A_k\bigg) \leq 1 < \infty. \tag{1.1.18}$$

Using (1.1.17) and (1.1.18), we conclude that the sequence of the measures of the sets $\{\bigcup_{k=m}^{\infty} A_k\}_{m=1}^{\infty}$ converges as $m \to \infty$ to

$$\mu\bigg(\bigcap_{m=1}^{\infty}\bigcup_{k=m}^{\infty} A_k\bigg) = 0. \tag{1.1.19}$$

To finish the proof, observe that the null set in (1.1.19) contains the set of all $x \in X$ for which $f_{n_k}(x)$ does not converge to $f(x)$. □

In many situations we are given a sequence of functions and we would like to extract a convergent subsequence. One way to achieve this is via the next theorem, which is a useful variant of Theorem 1.1.11. We first give a relevant definition.

Definition 1.1.12. We say that a sequence of measurable functions $\{f_n\}$ on the measure space (X,μ) is *Cauchy in measure* if for every $\varepsilon > 0$, there exists an $n_0 \in \mathbf{Z}^+$ such that for $n,m > n_0$ we have

$$\mu(\{x \in X: \ |f_m(x) - f_n(x)| > \varepsilon\}) < \varepsilon.$$

Theorem 1.1.13. *Let (X,μ) be a measure space and let $\{f_n\}$ be a complex-valued sequence on X that is Cauchy in measure. Then some subsequence of f_n converges μ-a.e.*

Proof. The proof is very similar to that of Theorem 1.1.11. For all $k = 1,2,\dots$ choose n_k inductively such that

$$\mu(\{x \in X: \ |f_{n_k}(x) - f_{n_{k+1}}(x)| > 2^{-k}\}) < 2^{-k} \tag{1.1.20}$$

and such that $n_1 < n_2 < \cdots < n_k < n_{k+1} < \cdots$. Define

$$A_k = \{x \in X: \ |f_{n_k}(x) - f_{n_{k+1}}(x)| > 2^{-k}\}.$$

As shown in the proof of Theorem 1.1.11, (1.1.20) implies that

$$\mu\bigg(\bigcap_{m=1}^{\infty}\bigcup_{k=m}^{\infty} A_k\bigg) = 0. \tag{1.1.21}$$

For $x \notin \bigcup_{k=m}^{\infty} A_k$ and $i \ge j \ge j_0 \ge m$ (and j_0 large enough) we have

$$|f_{n_i}(x) - f_{n_j}(x)| \le \sum_{l=j}^{i-1} |f_{n_l}(x) - f_{n_{l+1}}(x)| \le \sum_{l=j}^{i-1} 2^{-l} \le 2^{1-j} \le 2^{1-j_0}.$$

This implies that the sequence $\{f_{n_i}(x)\}_i$ is Cauchy for every x in the set $(\bigcup_{k=m}^{\infty} A_k)^c$ and therefore converges for all such x. We define a function

$$f(x) = \begin{cases} \lim\limits_{j\to\infty} f_{n_j}(x) & \text{when } x \notin \bigcap_{m=1}^{\infty}\bigcup_{k=m}^{\infty} A_k, \\ 0 & \text{when } x \in \bigcap_{m=1}^{\infty}\bigcup_{k=m}^{\infty} A_k. \end{cases}$$

Then $f_{n_j} \to f$ almost everywhere. □

1.1.3 A First Glimpse at Interpolation

It is a useful fact that if a function f is in $L^p(X,\mu)$ and in $L^q(X,\mu)$, then it also lies in $L^r(X,\mu)$ for all $p < r < q$. The usefulness of the spaces $L^{p,\infty}$ can be seen from the following sharpening of this statement:

Proposition 1.1.14. *Let $0 < p < q \le \infty$ and let f in $L^{p,\infty}(X,\mu) \cap L^{q,\infty}(X,\mu)$, where X is a σ-finite measure space. Then f is in $L^r(X,\mu)$ for all $p < r < q$ and*

$$\|f\|_{L^r} \le \left(\frac{r}{r-p} + \frac{r}{q-r}\right)^{\frac{1}{r}} \|f\|_{L^{p,\infty}}^{\frac{\frac{1}{r}-\frac{1}{q}}{\frac{1}{p}-\frac{1}{q}}} \|f\|_{L^{q,\infty}}^{\frac{\frac{1}{p}-\frac{1}{r}}{\frac{1}{p}-\frac{1}{q}}}, \tag{1.1.22}$$

with the interpretation that $1/\infty = 0$.

Proof. Let us take first $q < \infty$. We know that

$$d_f(\alpha) \le \min\left(\frac{\|f\|_{L^{p,\infty}}^p}{\alpha^p}, \frac{\|f\|_{L^{q,\infty}}^q}{\alpha^q}\right). \tag{1.1.23}$$

Set

$$B = \left(\frac{\|f\|_{L^{q,\infty}}^q}{\|f\|_{L^{p,\infty}}^p}\right)^{\frac{1}{q-p}}. \tag{1.1.24}$$

We now estimate the L^r norm of f. By (1.1.23), (1.1.24), and Proposition 1.1.4 we have

$$\begin{aligned}
\|f\|_{L^r(X,\mu)}^r &= r\int_0^\infty \alpha^{r-1} d_f(\alpha)\,d\alpha \\
&\le r\int_0^\infty \alpha^{r-1} \min\left(\frac{\|f\|_{L^{p,\infty}}^p}{\alpha^p}, \frac{\|f\|_{L^{q,\infty}}^q}{\alpha^q}\right) d\alpha \\
&= r\int_0^B \alpha^{r-1-p}\|f\|_{L^{p,\infty}}^p\,d\alpha + r\int_B^\infty \alpha^{r-1-q}\|f\|_{L^{q,\infty}}^q\,d\alpha \\
&= \frac{r}{r-p}\|f\|_{L^{p,\infty}}^p B^{r-p} + \frac{r}{q-r}\|f\|_{L^{q,\infty}}^q B^{r-q} \\
&= \left(\frac{r}{r-p} + \frac{r}{q-r}\right)\left(\|f\|_{L^{p,\infty}}^p\right)^{\frac{q-r}{q-p}} \left(\|f\|_{L^{q,\infty}}^q\right)^{\frac{r-p}{q-p}}.
\end{aligned} \tag{1.1.25}$$

Observe that the integrals converge, since $r-p>0$ and $r-q<0$.

The case $q=\infty$ is easier. Since $d_f(\alpha)=0$ for $\alpha > \|f\|_{L^\infty}$ we need to use only the inequality $d_f(\alpha) \le \alpha^{-p}\|f\|_{L^{p,\infty}}^p$ for $\alpha \le \|f\|_{L^\infty}$ in estimating the first integral in (1.1.25). We obtain

$$\|f\|_{L^r}^r \le \frac{r}{r-p}\|f\|_{L^{p,\infty}}^p \|f\|_{L^\infty}^{r-p},$$

which is nothing other than (1.1.22) when $q=\infty$. This completes the proof. □

Note that (1.1.22) holds with constant 1 if $L^{p,\infty}$ and $L^{q,\infty}$ are replaced by L^p and L^q, respectively. It is often convenient to work with functions that are only locally in some L^p space. This leads to the following definition.

Definition 1.1.15. For $0 < p < \infty$, the space $L^p_{\mathrm{loc}}(\mathbf{R}^n, |\cdot|)$ or simply $L^p_{\mathrm{loc}}(\mathbf{R}^n)$ is the set of all Lebesgue-measurable functions f on $\mathbf{R}^n$ that satisfy

$$\int_K |f(x)|^p\,dx < \infty \tag{1.1.26}$$

for any compact subset K of $\mathbf{R}^n$. Functions that satisfy (1.1.26) with $p = 1$ are called *locally integrable* functions on $\mathbf{R}^n$.

The union of all $L^p(\mathbf{R}^n)$ spaces for $1 \le p \le \infty$ is contained in $L^1_{\mathrm{loc}}(\mathbf{R}^n)$. More generally, for $0 < p < q < \infty$ we have the following:

$$L^q(\mathbf{R}^n) \subseteqq L^q_{\mathrm{loc}}(\mathbf{R}^n) \subseteqq L^p_{\mathrm{loc}}(\mathbf{R}^n).$$

Functions in $L^p(\mathbf{R}^n)$ for $0 < p < 1$ may not be locally integrable. For example, take $f(x) = |x|^{-n-\alpha}\chi_{|x|\le 1}$, which is in $L^p(\mathbf{R}^n)$ when $\alpha > 0$ and $p < n/(n+\alpha)$, and observe that f is not integrable over any open set in $\mathbf{R}^n$ containing the origin.

Exercises

1.1.1. Suppose f and f_n are measurable functions on (X,μ). Prove that
(a) d_f is right continuous on $[0,\infty)$.
(b) If $|f| \le \liminf_{n\to\infty} |f_n|$ μ-a.e., then $d_f \le \liminf_{n\to\infty} d_{f_n}$.
(c) If $|f_n| \uparrow |f|$, then $d_{f_n} \uparrow d_f$.
[*Hint:* Part (a): Let t_n be a decreasing sequence of positive numbers that tends to zero. Show that $d_f(\alpha_0 + t_n) \uparrow d_f(\alpha_0)$ using a convergence theorem. Part (b): Let $E = \{x \in X : |f(x)| > \alpha\}$ and $E_n = \{x \in X : |f_n(x)| > \alpha\}$. Use that $\mu\big(\bigcap_{n=m}^\infty E_n\big) \le \liminf_{n\to\infty} \mu(E_n)$ and $E \subseteqq \bigcup_{m=1}^\infty \bigcap_{n=m}^\infty E_n$ μ-a.e.]

1.1.2. (*Hölder's inequality*) Let $0 < p, p_1, \dots, p_k \le \infty$, where $k \ge 2$, and let f_j be in $L^{p_j} = L^{p_j}(X,\mu)$. Assume that

$$\frac{1}{p} = \frac{1}{p_1} + \cdots + \frac{1}{p_k}.$$

(a) Show that the product $f_1 \cdots f_k$ is in L^p and that

$$\big\|f_1 \cdots f_k\big\|_{L^p} \le \big\|f_1\big\|_{L^{p_1}} \cdots \big\|f_k\big\|_{L^{p_k}}.$$

(b) When no p_j is infinite, show that if equality holds in part (a), then it must be the case that $c_1|f_1|^{p_1} = \cdots = c_k|f_k|^{p_k}$ μ-a.e. for some $c_j \ge 0$.
(c) Let $0 < q < 1$ and $q' = \frac{q}{q-1}$. For $r < 0$ and $g > 0$ almost everywhere, define $\|g\|_{L^r} = \|g^{-1}\|_{L^{|r|}}^{-1}$. Show that if g is strictly positive μ-a.e. and lies in $L^{q'}$ and f is measurable such that fg belongs to L^1, we have

$$\big\|fg\big\|_{L^1} \ge \big\|f\big\|_{L^q}\big\|g\big\|_{L^{q'}}.$$

1.1.3. Let (X,μ) be a measure space.
(a) If f is in $L^{p_0}(X,\mu)$ for some $p_0 < \infty$, prove that

$$\lim_{p\to\infty} \big\|f\big\|_{L^p} = \big\|f\big\|_{L^\infty}.$$

(b) (*Jensen's inequality*) Suppose that $\mu(X) = 1$. Show that

$$\|f\|_{L^p} \geq \exp\left(\int_X \log|f(x)|\,d\mu(x)\right)$$

for all $0 < p < \infty$.
(c) If $\mu(X) = 1$ and f is in some $L^{p_0}(X,\mu)$ for some $p_0 > 0$, then

$$\lim_{p\to 0}\|f\|_{L^p} = \exp\left(\int_X \log|f(x)|\,d\mu(x)\right)$$

with the interpretation $e^{-\infty} = 0$.
[*Hint:* Part (a): If $0 < \|f\|_{L^\infty} < \infty$, use that $\|f\|_{L^p} \leq \|f\|_{L^\infty}^{(p-p_0)/p}\|f\|_{L^{p_0}}^{p_0/p}$ to obtain $\limsup_{p\to\infty}\|f\|_{L^p} \leq \|f\|_{L^\infty}$. Conversely, let $E_\gamma = \{x \in X : |f(x)| > \gamma\|f\|_{L^\infty}\}$ for γ in $(0,1)$. Then $\mu(E_\gamma) > 0$, $\|f\|_{L^{p_0}(E_\gamma)} > 0$, and $\|f\|_{L^p} \geq \left(\gamma\|f\|_{L^\infty}\right)^{(p-p_0)/p}\|f\|_{L^{p_0}(E_\gamma)}^{p_0/p}$, hence $\liminf_{p\to\infty}\|f\|_{L^p} \geq \gamma\|f\|_{L^\infty}$. If $\|f\|_{L^\infty} = \infty$, set $G_n = \{|f| > n\}$ and use that $\|f\|_{L^p} \geq \|f\|_{L^p(G_n)} \geq n\mu(G_n)^{\frac{1}{p}}$ to obtain $\liminf_{p\to\infty}\|f\|_{L^p} \geq n$. Part (b) is a direct consequence of Jensen's inequality $\int_X \log|h|\,d\mu \leq \log\left(\int_X |h|\,d\mu\right)$. Part (c): Fix a sequence $0 < p_n < p_0$ such that $p_n \downarrow 0$ and define

$$h_n(x) = \frac{1}{p_0}(|f(x)|^{p_0} - 1) - \frac{1}{p_n}(|f(x)|^{p_n} - 1).$$

Use that $\frac{1}{p}(t^p - 1) \downarrow \log t$ as $p \downarrow 0$ for all $t > 0$. The Lebesgue monotone convergence theorem yields $\int_X h_n\,d\mu \uparrow \int_X h\,d\mu$, hence $\int_X \frac{1}{p_n}(|f|^{p_n} - 1)\,d\mu \downarrow \int_X \log|f|\,d\mu$, where the latter could be $-\infty$. Use

$$\exp\left(\int_X \log|f|\,d\mu\right) \leq \left(\int_X |f|^{p_n}\,d\mu\right)^{\frac{1}{p_n}} \leq \exp\left(\int_X \frac{1}{p_n}(|f|^{p_n} - 1)\,d\mu\right)$$

to complete the proof.]

1.1.4. Let a_j be a sequence of positive reals. Show that
(a) $\left(\sum_{j=1}^\infty a_j\right)^\theta \leq \sum_{j=1}^\infty a_j^\theta$, for any $0 \leq \theta \leq 1$.
(b) $\sum_{j=1}^\infty a_j^\theta \leq \left(\sum_{j=1}^\infty a_j\right)^\theta$, for any $1 \leq \theta < \infty$.
(c) $\left(\sum_{j=1}^N a_j\right)^\theta \leq N^{\theta-1}\sum_{j=1}^N a_j^\theta$, when $1 \leq \theta < \infty$.
(d) $\sum_{j=1}^N a_j^\theta \leq N^{1-\theta}\left(\sum_{j=1}^N a_j\right)^\theta$, when $0 \leq \theta \leq 1$.

1.1.5. Let $\{f_j\}_{j=1}^N$ be a sequence of $L^p(X,\mu)$ functions.
(a) (*Minkowski's inequality*) For $1 \leq p \leq \infty$ show that

$$\Big\|\sum_{j=1}^N f_j\Big\|_{L^p} \leq \sum_{j=1}^N \|f_j\|_{L^p}.$$

(b) (*Reverse Minkowski inequality*) For $0 < p < 1$ and $f_j \geq 0$ prove that

$$\sum_{j=1}^{N} \|f_j\|_{L^p} \leq \|\sum_{j=1}^{N} f_j\|_{L^p}.$$

(c) For $0 < p < 1$ show that

$$\|\sum_{j=1}^{N} f_j\|_{L^p} \leq N^{\frac{1-p}{p}} \sum_{j=1}^{N} \|f_j\|_{L^p}.$$

(d) The constant $N^{\frac{1-p}{p}}$ in part (c) is best possible.
[*Hint:* Part (c): Use Exercise 1.1.4 (c). Part (d): Take $\{f_j\}_{j=1}^N$ to be characteristic functions of disjoint sets with the same measure.]

1.1.6. (a) (*Minkowski's integral inequality*) Let (X,μ) and (T,ν) be two σ-finite measure spaces and let $1 \leq p < \infty$. Show that for every nonnegative measurable function F on the product space $(X,\mu)\times(T,\nu)$ we have

$$\left[\int_T \left(\int_X F(x,t)\,d\mu(x)\right)^p d\nu(t)\right]^{\frac{1}{p}} \leq \int_X \left[\int_T F(x,t)^p\,d\nu(t)\right]^{\frac{1}{p}} d\mu(x),$$

(b) State and prove an analogous inequality when $p = \infty$.
(c) Prove that when $0 < p < 1$, then the preceding inequality is reversed.
(d) (Y. Sawano) Consider the example $X = T = [0,1]$, μ is counting measure, ν is Lebesgue measure, $F(x,t) = 1$ when $x = t$ and zero otherwise. What is the relevance of this example with the inequalities in (a) and (b)?
[*Hint:* Part (a) Split the power p as $1+(p-1)$ and apply Hölder's inequality with exponents p and p'. Part (b) Let $p \to \infty$ on subsets of X with finite measure.]

1.1.7. Let $f_1,\dots,f_N$ be in $L^{p,\infty}(X,\mu)$.
(a) Prove that for $1 \leq p < \infty$ we have

$$\|\sum_{j=1}^{N} f_j\|_{L^{p,\infty}} \leq N \sum_{j=1}^{N} \|f_j\|_{L^{p,\infty}}.$$

(b) Show that for $0 < p < 1$ we have

$$\|\sum_{j=1}^{N} f_j\|_{L^{p,\infty}} \leq N^{\frac{1}{p}} \sum_{j=1}^{N} \|f_j\|_{L^{p,\infty}}.$$

[*Hint:* Use that $\mu(\{|f_1+\cdots+f_N| > \alpha\}) \leq \sum_{j=1}^N \mu(\{|f_j| > \alpha/N\})$ and Exercise 1.1.4 (a) and (c).]

1.1.8. Let $0 < p < \infty$. Prove that $L^p(X,\mu)$ is a complete quasi-normed space. This means that every quasi-norm Cauchy sequence is quasi-norm convergent.

[*Hint:* Let f_n be a Cauchy sequence in L^p. Pass to a subsequence $\{n_i\}_i$ such that $\|f_{n_{i+1}} - f_{n_i}\|_{L^p} \le 2^{-i}$. Then the series $f = f_{n_1} + \sum_{i=1}^{\infty}(f_{n_{i+1}} - f_{n_i})$ converges in L^p.]

1.1.9. Let (X,μ) be a measure space with $\mu(X) < \infty$. Suppose that a sequence of measurable functions f_n on X converges to f μ-a.e. Prove that f_n converges to f in measure.

[*Hint:* For $\varepsilon > 0$, $\{x \in X : f_n(x) \to f(x)\} \subseteq \bigcup\limits_{m=1}^{\infty}\bigcap\limits_{n=m}^{\infty}\{x \in X : |f_n(x) - f(x)| < \varepsilon\}$.]

1.1.10. Let f be a measurable function on (X,μ) such such $d_f(\alpha) < \infty$ for all $\alpha > 0$. Fix $\gamma > 0$ and define $f_\gamma = f\chi_{|f|>\gamma}$ and $f^\gamma = f - f_\gamma = f\chi_{|f|\le\gamma}$.

(a) Prove that

$$d_{f_\gamma}(\alpha) = \begin{cases} d_f(\alpha) & \text{when} \quad \alpha > \gamma, \\ d_f(\gamma) & \text{when} \quad \alpha \le \gamma, \end{cases}$$

$$d_{f^\gamma}(\alpha) = \begin{cases} 0 & \text{when} \quad \alpha \ge \gamma, \\ d_f(\alpha) - d_f(\gamma) & \text{when} \quad \alpha < \gamma. \end{cases}$$

(b) If $f \in L^p(X,\mu)$ then

$$\begin{aligned} \|f_\gamma\|_{L^p}^p &= p\int_\gamma^\infty \alpha^{p-1} d_f(\alpha)\,d\alpha + \gamma^p d_f(\gamma), \\ \|f^\gamma\|_{L^p}^p &= p\int_0^\gamma \alpha^{p-1} d_f(\alpha)\,d\alpha - \gamma^p d_f(\gamma), \\ \int_{\gamma<|f|\le\delta} |f|^p\,d\mu &= p\int_\gamma^\delta d_f(\alpha)\alpha^{p-1}\,d\alpha - \delta^p d_f(\delta) + \gamma^p d_f(\gamma). \end{aligned}$$

(c) If f is in $L^{p,\infty}(X,\mu)$ prove that f^γ is in $L^q(X,\mu)$ for any $q > p$ and f_γ is in $L^q(X,\mu)$ for any $q < p$. Thus $L^{p,\infty} \subseteq L^{p_0} + L^{p_1}$ when $0 < p_0 < p < p_1 \le \infty$.

1.1.11. Let (X,μ) be a measure space and let E be a subset of X with $\mu(E) < \infty$. Assume that f is in $L^{p,\infty}(X,\mu)$ for some $0 < p < \infty$.

(a) Show that for $0 < q < p$ we have

$$\int_E |f(x)|^q\,d\mu(x) \le \frac{p}{p-q}\mu(E)^{1-\frac{q}{p}}\|f\|_{L^{p,\infty}}^q.$$

(b) Conclude that if $\mu(X) < \infty$ and $0 < q < p$, then

$$L^p(X,\mu) \subseteq L^{p,\infty}(X,\mu) \subseteq L^q(X,\mu).$$

[*Hint:* Part (a): Use $\mu(E \cap \{|f| > \alpha\}) \le \min\left(\mu(E), \alpha^{-p}\|f\|_{L^{p,\infty}}^p\right)$.]

1.1.12. (*Normability of weak L^p for $p > 1$*) Let (X,μ) be a σ-finite measure space and let $0 < p < \infty$. Pick $0 < r < p$ and define

$$|||f|||_{L^{p,\infty}} = \sup_{0<\mu(E)<\infty} \mu(E)^{-\frac{1}{r}+\frac{1}{p}}\left(\int_E |f|^r\,d\mu\right)^{\frac{1}{r}},$$

where the supremum is taken over all measurable subsets E of X of finite measure.
(a) Use Exercise 1.1.11 with $q = r$ to conclude that

$$\left|\left|\left| f \right|\right|\right|_{L^{p,\infty}} \le \left(\frac{p}{p-r}\right)^{\frac{1}{r}} \left\|f\right\|_{L^{p,\infty}}$$

for all f in $L^{p,\infty}(X,\mu)$. (It is not needed that X be σ-finite here).
(b) Prove that for all f in $L^{p,\infty}(X,\mu)$ we have

$$\left\|f\right\|_{L^{p,\infty}} \le \left|\left|\left| f \right|\right|\right|_{L^{p,\infty}}.$$

(Y. Oi) Notice that if $X = \{1,2\}$, $\mu(\{1\}) = 1$, $\mu(\{2\}) = \infty$, then X is not σ-finite, and verify that for the function $f = 1$ the preceding inequality fails.
(c) Show that $L^{p,\infty}(X,\mu)$ is metrizable for all $0 < p < \infty$, i.e., there is a metric on the space that generates the same topology as the quasi-norm. Also show that $L^{p,\infty}(X,\mu)$ is *normable* when $p > 1$, i.e., there is a norm on the space equivalent to $\|\cdot\|_{L^{p,\infty}}$.
(d) Use the characterization of the weak L^p quasi-norm obtained in parts (a) and (b) to prove Fatou's lemma for this space: For all measurable functions g_n on X we have

$$\left\|\liminf_{n\to\infty} |g_n|\right\|_{L^{p,\infty}} \le C_p \liminf_{n\to\infty} \left\|g_n\right\|_{L^{p,\infty}}$$

for some constant C_p that depends only on $p \in (0,\infty)$.
[*Hint:* Part (b): Write $X = \bigcup_{k=1}^{\infty} X_k$ with $\mu(X_k) < \infty$ and take $E = \{|f| > \alpha\} \cap X_k$.]

1.1.13. Consider the $N!$ functions on the line

$$f_\sigma = \sum_{j=1}^{N} \frac{N}{\sigma(j)} \chi_{[\frac{j-1}{N},\frac{j}{N})},$$

where σ is a permutation of the set $\{1,2,\dots,N\}$.
(a) Show that each f_σ satisfies $\|f_\sigma\|_{L^{1,\infty}} = 1$.
(b) Show that $\|\sum_{\sigma\in S_N} f_\sigma\|_{L^{1,\infty}} = N!\big(1+\frac{1}{2}+\cdots+\frac{1}{N}\big)$.
(c) Conclude that the space $L^{1,\infty}(\mathbf{R})$ is not normable (this means that $\|\cdot\|_{L^{1,\infty}}$ is not equivalent to a norm).
(d) Use a similar argument to prove that $L^{1,\infty}(\mathbf{R}^n)$ is not normable by considering the functions

$$f_\sigma(x_1,\dots,x_n) = \sum_{j_1=1}^{N}\cdots\sum_{j_n=1}^{N} \frac{N^n}{\sigma(\tau(j_1,\dots,j_n))} \chi_{[\frac{j_1-1}{N},\frac{j_1}{N})}(x_1)\cdots\chi_{[\frac{j_n-1}{N},\frac{j_n}{N})}(x_n),$$

where σ is a permutation of the set $\{1,2,\dots,N^n\}$ and τ is a fixed injective map from the set of all n-tuples of integers with coordinates $1 \le j \le N$ onto the set $\{1,2,\dots,N^n\}$. One may take

$$\tau(j_1,\dots,j_n) = j_1 + N(j_2-1) + N^2(j_3-1) + \cdots + N^{n-1}(j_n-1),$$

for instance.

1.1.14. Let (X,μ) be a measure space and let $s>0$.
(a) Let f be a measurable function on X. Show that if $0<p<q<\infty$ we have

$$\int_{|f|\le s} |f|^q \, d\mu \le \frac{q}{q-p} s^{q-p} \big\|f\big\|_{L^{p,\infty}}^p .$$

(b) Let f_j, $1\le j\le m$, be measurable functions on X and let $0<p<\infty$. Show that

$$\Big\| \max_{1\le j\le m} |f_j| \Big\|_{L^{p,\infty}}^p \le \sum_{j=1}^m \big\|f_j\big\|_{L^{p,\infty}}^p .$$

(c) Conclude from part (b) that for $0<p<1$ we have

$$\big\|f_1+\cdots+f_m\big\|_{L^{p,\infty}}^p \le \frac{2-p}{1-p} \sum_{j=1}^m \big\|f_j\big\|_{L^{p,\infty}}^p .$$

The latter estimate is referred to as the *p-normability* of weak L^p.
[*Hint:* Part (a): Use the distribution function. Part (c): First obtain the estimate

$$d_{f_1+\cdots+f_m}(\alpha) \le \mu(\{|f_1+\cdots+f_m|>\alpha, \max|f_j|\le\alpha\}) + d_{\max_j |f_j|}(\alpha)$$

for all $\alpha>0$ and then use part (b).]

1.1.15. (*Hölder's inequality for weak spaces*) Let f_j be in $L^{p_j,\infty}$ of a measure space X where $0<p_j<\infty$ and $1\le j\le k$. Let

$$\frac{1}{p} = \frac{1}{p_1} + \cdots + \frac{1}{p_k} .$$

Prove that

$$\big\|f_1\cdots f_k\big\|_{L^{p,\infty}} \le p^{-\frac{1}{p}} \prod_{j=1}^k p_j^{\frac{1}{p_j}} \prod_{j=1}^k \big\|f_j\big\|_{L^{p_j,\infty}} .$$

[*Hint:* Take $\|f_j\|_{L^{p_j,\infty}}=1$ for all j. Control $d_{f_1\cdots f_k}(\alpha)$ by

$$\begin{aligned} &\mu(\{|f_1|>\alpha/s_1\}) + \cdots + \mu(\{|f_{k-1}|>s_{k-2}/s_{k-1}\}) + \mu(\{|f_k|>s_{k-1}\}) \\ &\quad \le (s_1/\alpha)^{p_1} + (s_2/s_1)^{p_2} + \cdots + (s_{k-1}/s_{k-2})^{p_{k-1}} + (1/s_{k-1})^{p_k} . \end{aligned}$$

Set $x_1 = s_1/\alpha$, $x_2 = s_2/s_1, \dots, x_k = 1/s_{k-1}$. Minimize $x_1^{p_1}+\cdots+x_k^{p_k}$ subject to the constraint $x_1\cdots x_k = 1/\alpha$.]

1.1.16. Let $0<p_0<p<p_1\le\infty$ and let $\frac{1}{p} = \frac{1-\theta}{p_0} + \frac{\theta}{p_1}$ for some $\theta\in[0,1]$. Prove the following:

$$\begin{aligned} \big\|f\big\|_{L^p} &\le \big\|f\big\|_{L^{p_0}}^{1-\theta} \big\|f\big\|_{L^{p_1}}^{\theta} , \\ \big\|f\big\|_{L^{p,\infty}} &\le \big\|f\big\|_{L^{p_0,\infty}}^{1-\theta} \big\|f\big\|_{L^{p_1,\infty}}^{\theta} . \end{aligned}$$

1.1.17. ([231]) Follow the steps below to prove the *isoperimetric inequality*. For $n \geq 2$ and $1 \leq j \leq n$ define the projection maps $\pi_j : \mathbf{R}^n \to \mathbf{R}^{n-1}$ by setting for $x = (x_1, \dots, x_n)$,

$$\pi_j(x) = (x_1, \dots, x_{j-1}, x_{j+1}, \dots, x_n),$$

with the obvious interpretations when $j = 1$ or $j = n$.
(a) For maps $f_j : \mathbf{R}^{n-1} \to \mathbf{C}$ prove that

$$\Lambda(f_1, \dots, f_n) = \int_{\mathbf{R}^n} \prod_{j=1}^n |f_j \circ \pi_j| \, dx \leq \prod_{j=1}^n \|f_j\|_{L^{n-1}(\mathbf{R}^{n-1})}.$$

(b) Let Ω be a compact set with a rectifiable boundary in $\mathbf{R}^n$ where $n \geq 2$. Show that there is a constant c_n independent of Ω such that

$$|\Omega| \leq c_n |\partial \Omega|^{\frac{n}{n-1}},$$

where the expression $|\partial \Omega|$ denotes the $(n-1)$-dimensional surface measure of the boundary of Ω.
[*Hint:* Part (a): Use induction starting with $n = 2$. For $n \geq 3$ write

$$\begin{aligned}\Lambda(f_1, \dots, f_n) &\leq \int_{\mathbf{R}^{n-1}} P(x_1, \dots, x_{n-1}) |f_n(\pi_n(x))| \, dx_1 \cdots dx_{n-1} \\ &\leq \|P\|_{L^{\frac{n-1}{n-2}}(\mathbf{R}^{n-1})} \big\| f_n \circ \pi_n \big\|_{L^{n-1}(\mathbf{R}^{n-1})},\end{aligned}$$

where $P(x_1, \dots, x_{n-1}) = \int_{\mathbf{R}} |f_1(\pi_1(x)) \cdots f_{n-1}(\pi_{n-1}(x))| \, dx_n$, and apply the induction hypothesis to the $n-1$ functions

$$\left[\int_{\mathbf{R}} f_j(\pi_j(x))^{n-1} \, dx_n \right]^{\frac{1}{n-2}},$$

for $j = 1, \dots, n-1$, to obtain the required conclusion. Part (b): Specialize part (a) to the case $f_j = \chi_{\pi_j[\Omega]}$ to obtain

$$|\Omega| \leq |\pi_1[\Omega]|^{\frac{1}{n-1}} \cdots |\pi_n[\Omega]|^{\frac{1}{n-1}}$$

and then use that $|\pi_j[\Omega]| \leq \frac{1}{2} |\partial \Omega|$.]

1.2 Convolution and Approximate Identities

The notion of convolution can be defined on measure spaces endowed with a group structure. It turns out that the most natural environment to define convolution is the context of topological groups. Although the focus of this book is harmonic analysis on Euclidean spaces, we develop the notion of convolution on general groups. This allows us to study this concept on $\mathbf{R}^n$, $\mathbf{Z}^n$, and $\mathbf{T}^n$, in a unified way. Moreover, since the basic properties of convolutions and approximate identities do not require commutativity of the group operation, we may assume that the underlying groups are not necessarily abelian. Thus, the results in this section can be also applied to nonabelian structures such as the Heisenberg group.

1.2.1 Examples of Topological Groups

A *topological group* G is a Hausdorff topological space that is also a group with law

$$(x,y) \mapsto xy \tag{1.2.1}$$

such that the maps $(x,y) \mapsto xy$ and $x \mapsto x^{-1}$ are continuous. The identity element of the group is the unique element e with the property $xe = ex = x$ for all $x \in G$. We adopt the standard notation

$$AB = \{ab : \ a \in A, b \in B\}, \qquad A^{-1} = \{a^{-1} : \ a \in A\}$$

for subsets A and B of G. Note that $(AB)^{-1} = B^{-1}A^{-1}$. Every topological group G has an open basis at e consisting of symmetric neighborhoods, i.e., open sets U satisfying $U = U^{-1}$. A topological group is called *locally compact* if there is an open set U containing the identity element such that $\overline{U}$ is compact. Then every point in the group has an open neighborhood with compact closure.

Let G be a locally compact group. It is known that G possesses a positive measure λ on the Borel sets that is nonzero on all nonempty open sets, finite on compact sets, and is left invariant, meaning that

$$\lambda(tA) = \lambda(A), \tag{1.2.2}$$

for all measurable sets A and all $t \in G$. Such a measure λ is called a (left) *Haar measure* on G. Similarly, G possesses a *right Haar measure* which is right invariant, i.e., $\lambda(At) = \lambda(A)$ for all measurable $A \subseteq G$ and all $t \in G$. For the existence of Haar measure we refer to [152, §15] or [213, §16.3]. Furthermore, Haar measure is unique up to positive multiplicative constants. If G is abelian then any left Haar measure on G is a constant multiple of any given right Haar measure on G. A locally compact group which is a countable union of compact subsets is a σ-finite measure space under left or right Haar measure. This is case for connected locally compact groups.

Example 1.2.1. The standard examples are provided by the spaces $\mathbf{R}^n$ and $\mathbf{Z}^n$ with the usual topology and the usual addition of n-tuples. Another example is the space $\mathbf{T}^n = \mathbf{R}^n/\mathbf{Z}^n$ defined as follows:

$$\mathbf{T}^n = \underbrace{[0,1) \times \cdots \times [0,1)}_{n \text{ times}}$$

with the usual topology and group law:

$$(x_1,\dots,x_n)+(y_1,\dots,y_n)=((x_1+y_1)\bmod 1,\dots,(x_n+y_n)\bmod 1).$$

Example 1.2.2. Let $G=\mathbf{R}^*=\mathbf{R}\setminus\{0\}$ with group law the usual multiplication. It is easy to verify that the measure $\lambda=dx/|x|$ is invariant under multiplicative translations, that is,

$$\int_{-\infty}^{\infty} f(tx)\,\frac{dx}{|x|}=\int_{-\infty}^{\infty} f(x)\,\frac{dx}{|x|},$$

for all f in $L^1(G,\mu)$ and all $t\in\mathbf{R}^*$. Therefore, $dx/|x|$ is a Haar measure. [Taking $f=\chi_A$ gives $\lambda(tA)=\lambda(A)$.]

Example 1.2.3. Similarly, on the multiplicative group $G=\mathbf{R}^+$, a Haar measure is dx/x.

Example 1.2.4. Counting measure is a Haar measure on the group $\mathbf{Z}^n$ with the usual addition as group operation.

Example 1.2.5. The *Heisenberg group* $\mathbf{H}^n$ is the set $\mathbf{C}^n\times\mathbf{R}$ with the group operation

$$(z_1,\dots,z_n,t)(w_1,\dots,w_n,s)=\Big(z_1+w_1,\dots,z_n+w_n,t+s+2\operatorname{Im}\sum_{j=1}^{n} z_j\overline{w_j}\Big).$$

It can easily be seen that the identity element e of this group is $0\in\mathbf{C}^n\times\mathbf{R}$ and $(z_1,\dots,z_n,t)^{-1}=(-z_1,\dots,-z_n,-t)$. Topologically the Heisenberg group is identified with $\mathbf{C}^n\times\mathbf{R}$, and both left and right Haar measure on $\mathbf{H}^n$ is Lebesgue measure. The norm

$$|(z_1,\dots,z_n,t)|=\bigg[\Big(\sum_{j=1}^{n}|z_j|^2\Big)^2+t^2\bigg]^{\frac{1}{4}}$$

introduces balls $B_r(x)=\{y\in\mathbf{H}^n:\,|y^{-1}x|<r\}$ on the Heisenberg group that are quite different from Euclidean balls. For x close to the origin, the balls $B_r(x)$ are not far from being Euclidean, but for x far away from $e=0$ they look like slanted truncated cylinders. The Heisenberg group can be naturally identified as the boundary of the unit ball in $\mathbf{C}^n$ and plays an important role in quantum mechanics.

1.2.2 Convolution

Throughout the rest of this section, we fix a locally compact group G and a left invariant Haar measure λ on G. We assume that G is a countable union of compact subsets, hence the pair (G,λ) forms a σ-finite measure space. The spaces $L^p(G,\lambda)$ and $L^{p,\infty}(G,\lambda)$ are simply denoted by $L^p(G)$ and $L^{p,\infty}(G)$.

Left invariance of λ is equivalent to the fact that for all $t \in G$ and all nonnegative measurable functions f on G we have

$$\int_G f(tx)\,d\lambda(x) = \int_G f(x)\,d\lambda(x)\,. \tag{1.2.3}$$

Equation (1.2.3) is a restatement of (1.2.2) if f is a characteristic function. Obviously (1.2.3) also holds for $f \in L^1(G)$ by linearity and approximation.

We are now ready to define the operation of convolution.

Definition 1.2.6. Let f, g be in $L^1(G)$. Define the *convolution* $f * g$ by

$$(f*g)(x) = \int_G f(y)g(y^{-1}x)\,d\lambda(y)\,. \tag{1.2.4}$$

For instance, if $G = \mathbf{R}^n$ with the usual additive structure, then $y^{-1} = -y$ and the integral in (1.2.4) is written as

$$(f*g)(x) = \int_{\mathbf{R}^n} f(y)g(x-y)\,dy\,.$$

Remark 1.2.7. The right-hand side of (1.2.4) is defined a.e., since the following double integral converges absolutely:

$$\begin{aligned}
&\int_G \int_G |f(y)||g(y^{-1}x)|\,d\lambda(y)\,d\lambda(x) \\
&\qquad = \int_G \int_G |f(y)||g(y^{-1}x)|\,d\lambda(x)\,d\lambda(y) \\
&\qquad = \int_G |f(y)| \int_G |g(y^{-1}x)|\,d\lambda(x)\,d\lambda(y) \\
&\qquad = \int_G |f(y)| \int_G |g(x)|\,d\lambda(x)\,d\lambda(y) \qquad\qquad \text{by (1.2.2)} \\
&\qquad = \big\|f\big\|_{L^1(G)}\big\|g\big\|_{L^1(G)} < +\infty\,.
\end{aligned}$$

The change of variables $z = x^{-1}y$ yields that (1.2.4) is in fact equal to

$$(f*g)(x) = \int_G f(xz)g(z^{-1})\,d\lambda(z)\,, \tag{1.2.5}$$

where the substitution of $d\lambda(y)$ by $d\lambda(z)$ is justified by left invariance.

Example 1.2.8. On $\mathbf{R}$ let $f(x) = 1$ when $-1 \le x \le 1$ and zero otherwise. We see that $(f*f)(x)$ is equal to the length of the intersection of the intervals $[-1,1]$ and $[x-1, x+1]$. It follows that $(f*f)(x) = 2 - |x|$ for $|x| \le 2$ and zero otherwise. Observe that $f*f$ is a smoother function than f. Similarly, we obtain that $f*f*f$ is a smoother function than $f*f$.

There is an analogous calculation when g is the characteristic function of the unit disk $B(0,1)$ in $\mathbf{R}^2$. A simple computation gives

$$\begin{aligned}(g*g)(x) &= \left|B(0,1)\cap B(x,1)\right| = \int_{-\sqrt{1-\frac{1}{4}|x|^2}}^{+\sqrt{1-\frac{1}{4}|x|^2}} \left(2\sqrt{1-t^2}-|x|\right)dt \\ &= 2\arcsin\left(\sqrt{1-\tfrac{1}{4}|x|^2}\right) - |x|\sqrt{1-\tfrac{1}{4}|x|^2}\end{aligned}$$

when $x=(x_1,x_2)$ in $\mathbf{R}^2$ satisfies $|x|\le 2$, while $(g*g)(x)=0$ if $|x|\ge 2$.

A calculation similar to that in Remark 1.2.7 yields that

$$\left\|f*g\right\|_{L^1(G)} \le \left\|f\right\|_{L^1(G)}\left\|g\right\|_{L^1(G)}, \tag{1.2.6}$$

that is, the convolution of two integrable functions is also an integrable function with L^1 norm less than or equal to the product of the L^1 norms.

Proposition 1.2.9. *For all f, g, h in $L^1(G)$, the following properties are valid:*

(1) $f*(g*h)=(f*g)*h$ (associativity)
(2) $f*(g+h)=f*g+f*h$ and $(f+g)*h=f*h+g*h$ (distributivity)

Proof. The easy proofs are omitted. □

Proposition 1.2.9 implies that $L^1(G)$ is a (not necessarily commutative) Banach algebra under the convolution product.

1.2.3 Basic Convolution Inequalities

The most fundamental inequality involving convolutions is the following.

Theorem 1.2.10. *(**Minkowski's inequality**) Let $1\le p\le\infty$. For f in $L^p(G)$ and g in $L^1(G)$ we have that $g*f$ exists λ-a.e. and satisfies*

$$\left\|g*f\right\|_{L^p(G)} \le \left\|g\right\|_{L^1(G)}\left\|f\right\|_{L^p(G)}. \tag{1.2.7}$$

Proof. Estimate (1.2.7) follows directly from Exercise 1.1.6. Here we give a direct proof. We may assume that $1<p<\infty$, since the cases $p=1$ and $p=\infty$ are simple. We first show that the convolution $|g|*|f|$ exists λ-a.e. Indeed,

$$(|g|*|f|)(x) = \int_G |f(y^{-1}x)|\,|g(y)|\,d\lambda(y). \tag{1.2.8}$$

Apply Hölder's inequality in (1.2.8) with respect to the measure $|g(y)|\,d\lambda(y)$ to the functions $y\mapsto f(y^{-1}x)$ and 1 with exponents p and $p'=p/(p-1)$, respectively. We obtain

$$(|g|*|f|)(x) \le \left(\int_G |f(y^{-1}x)|^p|g(y)|\,d\lambda(y)\right)^{\frac{1}{p}}\left(\int_G |g(y)|\,d\lambda(y)\right)^{\frac{1}{p'}}. \tag{1.2.9}$$

Taking L^p norms of both sides of (1.2.9) we deduce

$$\begin{aligned}
\big\| |g| * |f| \big\|_{L^p} &\le \left(\|g\|_{L^1}^{p-1} \int_G \int_G |f(y^{-1}x)|^p |g(y)| \, d\lambda(y) \, d\lambda(x) \right)^{\frac{1}{p}} \\
&= \left(\|g\|_{L^1}^{p-1} \int_G \int_G |f(y^{-1}x)|^p \, d\lambda(x) |g(y)| \, d\lambda(y) \right)^{\frac{1}{p}} \\
&= \left(\|g\|_{L^1}^{p-1} \int_G \int_G |f(x)|^p \, d\lambda(x) |g(y)| \, d\lambda(y) \right)^{\frac{1}{p}} \qquad \text{by (1.2.3)} \\
&= \left(\|f\|_{L^p}^p \|g\|_{L^1} \|g\|_{L^1}^{p-1} \right)^{\frac{1}{p}} \\
&= \|f\|_{L^p} \|g\|_{L^1} < \infty,
\end{aligned}$$

where the second equality follows by Fubini's theorem. This shows that $|g| * |f|$ is finite λ-a.e. and satisfies (1.2.7); then $g * f$ exists λ-a.e. and also satisfies (1.2.7), since $|g * f| \le |g| * |f|$. □

Remark 1.2.11. Theorem 1.2.10 may fail for nonabelian groups if $g * f$ is replaced by $f * g$ in (1.2.7). Note, however, that if for all $h \in L^1(G)$ we have

$$\|h\|_{L^1} = \|\widetilde{h}\|_{L^1}, \tag{1.2.10}$$

where $\widetilde{h}(x) = h(x^{-1})$, then (1.2.7) holds when the quantity $\|g * f\|_{L^p(G)}$ is replaced by $\|f * g\|_{L^p(G)}$. To see this, observe that if (1.2.10) holds, then we can use (1.2.5) to conclude that if f in $L^p(G)$ and g in $L^1(G)$, then

$$\|f * g\|_{L^p(G)} \le \|g\|_{L^1(G)} \|f\|_{L^p(G)}. \tag{1.2.11}$$

If the left Haar measure satisfies

$$\lambda(A) = \lambda(A^{-1}) \tag{1.2.12}$$

for all measurable $A \subseteq G$, then (1.2.10) holds and thus (1.2.11) is satisfied for all g in $L^1(G)$ and $f \in L^p(G)$. This is, for instance, the case for the Heisenberg group $\mathbf{H}^n$.

Minkowski's inequality (1.2.11) is only a special case of Young's inequality in which the function g can be in any space $L^r(G)$ for $1 \le r \le \infty$.

Theorem 1.2.12. ***(Young's inequality)*** *Let $1 \le p, q, r \le \infty$ satisfy*

$$\frac{1}{q} + 1 = \frac{1}{p} + \frac{1}{r}. \tag{1.2.13}$$

Then for all f in $L^p(G)$ and all g in $L^r(G)$ satisfying $\|g\|_{L^r(G)} = \|\widetilde{g}\|_{L^r(G)}$ we have $f * g$ exists λ-a.e. and satisfies

$$\|f * g\|_{L^q(G)} \le \|g\|_{L^r(G)} \|f\|_{L^p(G)}. \tag{1.2.14}$$

Proof. Young's inequality is proved in a way similar to Minkowski's inequality. We do a suitable splitting of the product $|f(y)||g(y^{-1}x)|$ and apply Hölder's inequality. Observe that when $r < \infty$, the hypotheses on the indices imply that

$$\frac{1}{r'}+\frac{1}{q}+\frac{1}{p'}=1\,, \qquad \frac{p}{q}+\frac{p}{r'}=1\,, \qquad \frac{r}{q}+\frac{r}{p'}=1\,.$$

Using Hölder's inequality with exponents r', q, and p', we obtain

$$\begin{aligned}
|(|f|*|g|)(x)| &\le \int_G |f(y)|\,|g(y^{-1}x)|\,d\lambda(y)\\
&= \int_G |f(y)|^{\frac{p}{r'}}\big(|f(y)|^{\frac{p}{q}}|g(y^{-1}x)|^{\frac{r}{q}}\big)|g(y^{-1}x)|^{\frac{r}{p'}}\,d\lambda(y)\\
&\le \|f\|_{L^p}^{\frac{p}{r'}}\Big(\int_G |f(y)|^p|g(y^{-1}x)|^r\,d\lambda(y)\Big)^{\frac{1}{q}}\Big(\int_G |g(y^{-1}x)|^r\,d\lambda(y)\Big)^{\frac{1}{p'}}\\
&= \|f\|_{L^p}^{\frac{p}{r'}}\Big(\int_G |f(y)|^p|g(y^{-1}x)|^r\,d\lambda(y)\Big)^{\frac{1}{q}}\Big(\int_G |\widetilde{g}(x^{-1}y)|^r\,d\lambda(y)\Big)^{\frac{1}{p'}}\\
&= \Big(\int_G |f(y)|^p|g(y^{-1}x)|^r\,d\lambda(y)\Big)^{\frac{1}{q}}\|f\|_{L^p}^{\frac{p}{r'}}\|\widetilde{g}\|_{L^r}^{\frac{r}{p'}},
\end{aligned}$$

where we used left invariance. Now take L^q norms (in x) and apply Fubini's theorem to deduce that

$$\begin{aligned}
\big\||f|*|g|\big\|_{L^q} &\le \|f\|_{L^p}^{\frac{p}{r'}}\|\widetilde{g}\|_{L^r}^{\frac{r}{p'}}\Big(\int_G\int_G |f(y)|^p|g(y^{-1}x)|^r\,d\lambda(x)\,d\lambda(y)\Big)^{\frac{1}{q}}\\
&= \|f\|_{L^p}^{\frac{p}{r'}}\|\widetilde{g}\|_{L^r}^{\frac{r}{p'}}\|f\|_{L^p}^{\frac{p}{q}}\|g\|_{L^r}^{\frac{r}{q}}\\
&= \|g\|_{L^r}\|f\|_{L^p} < \infty,
\end{aligned}$$

using the hypothesis on g. This implies that $|f|*|g|$ is finite λ-a.e. and satisfies (1.2.14); then $f*g$ exists λ-a.e. and also satisfies (1.2.14).

Finally, note that if $r=\infty$, the assumptions on p and q imply that $p=1$ and $q=\infty$, in which case the required inequality trivially holds. □

We now give a version of Theorem 1.2.12 for weak L^p spaces. Theorem 1.2.13 is improved in Section 1.4.

Theorem 1.2.13. ***(Young's inequality for weak type spaces)*** *Let G be a locally compact group with left Haar measure λ that satisfies (1.2.12). Let $1 \le p < \infty$ and $1 < q,r < \infty$ satisfy*

$$\frac{1}{q}+1=\frac{1}{p}+\frac{1}{r}. \tag{1.2.15}$$

Then there exists a constant $C_{p,q,r} > 0$ such that for all f in $L^p(G)$ and g in $L^{r,\infty}(G)$, the convolution $f*g$ exists λ-a.e. and satisfies

$$\|f*g\|_{L^{q,\infty}(G)} \le C_{p,q,r}\|g\|_{L^{r,\infty}(G)}\|f\|_{L^p(G)}. \tag{1.2.16}$$

Proof. As in the proofs of Theorems 1.2.10 and 1.2.12, we first obtain (1.2.16) for the convolution of the absolute values of the functions. This implies that $|f| * |g| < \infty$ λ-a.e., and thus $f * g$ exists λ-a.e. and satisfies $|f * g| \le |f| * |g|$. We may therefore assume that $f, g \ge 0$ λ-a.e. The proof is based on a suitable splitting of the function g. Let M be a positive real number to be chosen later. Define $g_1 = g\chi_{|g| \le M}$ and $g_2 = g\chi_{|g|>M}$. In view of Exercise 1.1.10 (a) we have

$$d_{g_1}(\alpha) = \begin{cases} 0 & \text{if } \alpha \ge M, \\ d_g(\alpha) - d_g(M) & \text{if } \alpha < M, \end{cases} \tag{1.2.17}$$

$$d_{g_2}(\alpha) = \begin{cases} d_g(\alpha) & \text{if } \alpha > M, \\ d_g(M) & \text{if } \alpha \le M. \end{cases} \tag{1.2.18}$$

Proposition 1.1.3 gives for all $\beta > 0$

$$d_{f*g}(\beta) \le d_{f*g_1}(\beta/2) + d_{f*g_2}(\beta/2), \tag{1.2.19}$$

and thus it suffices to estimate the distribution functions of $f * g_1$ and $f * g_2$. Since g_1 is the "small" part of g, it is in L^s for any $s > r$. In fact, we have

$$\begin{aligned} \int_G g_1(x)^s \, d\lambda(x) &= s \int_0^\infty \alpha^{s-1} d_{g_1}(\alpha)\, d\alpha \\ &= s \int_0^M \alpha^{s-1}(d_g(\alpha) - d_g(M))\, d\alpha \\ &\le s \int_0^M \alpha^{s-1-r} \|g\|_{L^{r,\infty}}^r \, d\alpha - s\int_0^M \alpha^{s-1} d_g(M)\, d\alpha \\ &= \frac{s}{s-r} M^{s-r} \|g\|_{L^{r,\infty}}^r - M^s d_g(M), \end{aligned} \tag{1.2.20}$$

when $s < \infty$.

Similarly, since g_2 is the "large" part of g, it is in L^t for any $t < r$, and

$$\begin{aligned} \int_G g_2(x)^t \, d\lambda(x) &= t \int_0^\infty \alpha^{t-1} d_{g_2}(\alpha)\, d\alpha \\ &= t \int_0^M \alpha^{t-1} d_g(M)\, d\alpha + t\int_M^\infty \alpha^{t-1} d_g(\alpha)\, d\alpha \\ &\le M^t d_g(M) + t \int_M^\infty \alpha^{t-1-r} \|g\|_{L^{r,\infty}}^r \, d\alpha \\ &\le M^{t-r} \|g\|_{L^{r,\infty}}^r + \frac{t}{r-t} M^{t-r} \|g\|_{L^{r,\infty}}^r \\ &= \frac{r}{r-t} M^{t-r} \|g\|_{L^{r,\infty}}^r . \end{aligned} \tag{1.2.21}$$

Since $1/r = 1/p' + 1/q$, it follows that $1 < r < p'$. Select $t = 1$ and $s = p'$. Hölder's inequality and (1.2.20) give when $p' < \infty$

$$|(f * g_1)(x)| \le \|f\|_{L^p} \|g_1\|_{L^{p'}} \le \|f\|_{L^p} \left(\frac{p'}{p'-r} M^{p'-r} \|g\|_{L^{r,\infty}}^r \right)^{\frac{1}{p'}} \tag{1.2.22}$$

and

$$|(f * g_1)(x)| \leq \|f\|_{L^p} M \tag{1.2.23}$$

when $p' = \infty$. If $p' < \infty$ choose an M such that the right-hand side of (1.2.22) is equal to $\beta/2$. If $p' = \infty$ choose M such that the right-hand side of (1.2.23) is also equal to $\beta/2$. That is, choose

$$M = (\beta^{p'} 2^{-p'} r q^{-1} \|f\|_{L^p}^{-p'} \|g\|_{L^{r,\infty}}^{-r})^{1/(p'-r)}$$

if $p' < \infty$ and $M = \beta/(2\|f\|_{L^1})$ if $p' = \infty$. For these choices of M we have that

$$d_{f*g_1}(\beta/2) = 0.$$

Next by Theorem 1.2.10 and (1.2.21) with $t = 1$ we obtain

$$\|f * g_2\|_{L^p} \leq \|f\|_{L^p} \|g_2\|_{L^1} \leq \|f\|_{L^p} \frac{r}{r-1} M^{1-r} \|g\|_{L^{r,\infty}}^r. \tag{1.2.24}$$

For the value of M chosen, using (1.2.24) and Chebyshev's inequality, we obtain

$$\begin{aligned} d_{f*g}(\beta) &\leq d_{f*g_2}(\beta/2) \\ &\leq (2\|f * g_2\|_{L^p} \beta^{-1})^p \\ &\leq (2r\|f\|_{L^p} M^{1-r} \|g\|_{L^{r,\infty}}^r (r-1)^{-1} \beta^{-1})^p \\ &= C_{p,q,r}^q \beta^{-q} \|f\|_{L^p}^q \|g\|_{L^{r,\infty}}^q, \end{aligned} \tag{1.2.25}$$

which is the required inequality. This proof gives that the constant $C_{p,q,r}$ blows up like $(r-1)^{-p/q}$ as $r \to 1$. □

Example 1.2.14. Theorem 1.2.13 may fail at some endpoints:

(1) $r = 1$ and $1 \leq p = q \leq \infty$. On $\mathbf{R}$ take $g(x) = 1/|x|$ and $f = \chi_{[0,1]}$. Clearly, g is in $L^{1,\infty}$ and f in L^p for all $1 \leq p \leq \infty$, but the convolution of f and g is identically equal to infinity on the interval $[0,1]$. Therefore, (1.2.16) fails in this case.

(2) $q = \infty$ and $1 < r = p' < \infty$. On $\mathbf{R}$ let $f(x) = (|x|^{1/p} \log|x|)^{-1}$ for $|x| \geq 2$ and zero otherwise, and also let $g(x) = |x|^{-1/r}$. We see that $(f * g)(x) = \infty$ for $|x| \leq 1$. Thus (1.2.16) fails in this case also.

(3) $r = q = \infty$ and $p = 1$. Then inequality (1.2.16) trivially holds.

1.2.4 Approximate Identities

We now introduce the notion of approximate identities. The Banach algebra $L^1(G)$ may not have a unit element, that is, an element f_0 such that

$$f_0 * f = f = f * f_0 \tag{1.2.26}$$

for all $f \in L^1(G)$. In particular, this is the case when $G = \mathbf{R}$; in fact, the only f_0 that satisfies (1.2.26) for all $f \in L^1(\mathbf{R})$ is not a function but the Dirac delta distribution, introduced in Chapter 2. It is reasonable therefore to introduce the notion of approximate unit or identity, a family of functions k_ε with the property $k_\varepsilon * f \to f$ in L^1 as $\varepsilon \to 0$.

Definition 1.2.15. An *approximate identity* (as $\varepsilon \to 0$) is a family of $L^1(G)$ functions k_ε with the following three properties:

(i) There exists a constant $c > 0$ such that $\|k_\varepsilon\|_{L^1(G)} \le c$ for all $\varepsilon > 0$.

(ii) $\int_G k_\varepsilon(x)\, d\lambda(x) = 1$ for all $\varepsilon > 0$.

(iii) For any neighborhood V of the identity element e of the group G we have $\int_{V^c} |k_\varepsilon(x)|\, d\lambda(x) \to 0$ as $\varepsilon \to 0$.

The construction of approximate identities on general locally compact groups G is beyond the scope of this book and is omitted; see [152] for details. In this book we are interested only in groups with Euclidean structure, where approximate identities exist in abundance.

Sometimes we think of approximate identities as sequences $\{k_n\}_n$. In this case property (iii) holds as $n \to \infty$. It is best to visualize approximate identities as sequences of positive functions k_n that spike near 0 in such a way that the signed area under the graph of each function remains constant (equal to one) but the support shrinks to zero. See Figure 1.2.

Example 1.2.16. On $\mathbf{R}$ let $P(x) = (\pi(x^2+1))^{-1}$ and $P_\varepsilon(x) = \varepsilon^{-1}P(\varepsilon^{-1}x)$ for $\varepsilon > 0$. Since P_ε and P have the same L^1 norm and

$$\int_{-\infty}^{+\infty} \frac{1}{x^2+1}\, dx = \lim_{x\to+\infty} \big[\arctan(x) - \arctan(-x)\big] = (\pi/2) - (-\pi/2) = \pi,$$

property (ii) is satisfied. Property (iii) follows from the fact that

$$\frac{1}{\pi}\int_{|x|\ge\delta} \frac{1}{\varepsilon}\frac{1}{(x/\varepsilon)^2+1}\, dx = 1 - \frac{2}{\pi}\arctan(\delta/\varepsilon) \to 0 \qquad \text{as } \varepsilon \to 0,$$

for all $\delta > 0$. The function P_ε is called the *Poisson kernel.*

The Poisson kernel may be replaced by any integrable function of integral 1 as the following example indicates.

Example 1.2.17. On $\mathbf{R}^n$ let $k(x)$ be an integrable function with integral one. Let $k_\varepsilon(x) = \varepsilon^{-n}k(\varepsilon^{-1}x)$. It is straightforward to see that $k_\varepsilon(x)$ is an approximate identity. Property (iii) follows from the fact that

$$\int_{|x|\ge\delta/\varepsilon} |k(x)|\, dx \to 0$$

as $\varepsilon \to 0$ for δ fixed.

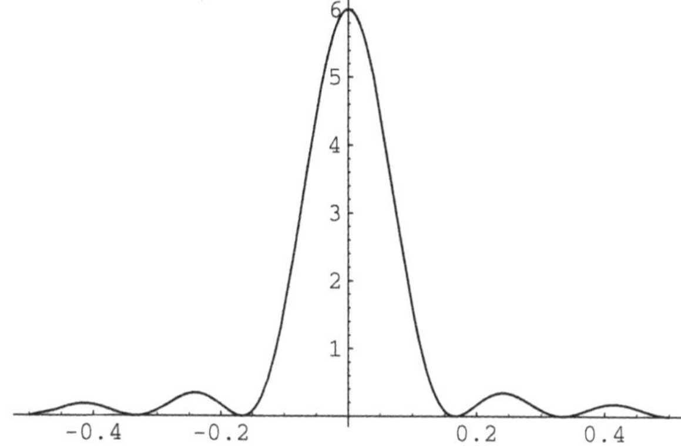

Fig. 1.2 The Fejér kernel F_5 plotted on the interval $[-\frac{1}{2}, \frac{1}{2}]$.

Example 1.2.18. On the circle group $\mathbf{T}^1$ let

$$F_N(t) = \sum_{j=-N}^{N} \left(1 - \frac{|j|}{N+1}\right) e^{2\pi i j t} = \frac{1}{N+1} \left(\frac{\sin(\pi(N+1)t)}{\sin(\pi t)}\right)^2. \tag{1.2.27}$$

To check the previous equality we use that

$$\sin^2(x) = (2 - e^{2ix} - e^{-2ix})/4,$$

and we carry out the calculation. F_N is called the *Fejér kernel*. See Figure 1.2. To see that the sequence $\{F_N\}_N$ is an approximate identity, we check conditions (i), (ii), and (iii) in Definition 1.2.15. Property (iii) follows from the expression giving F_N in terms of sines, while property (i) follows from the expression giving F_N in terms of exponentials. Property (ii) is identical to property (i), since F_N is nonnegative.

Next comes the basic theorem concerning approximate identities.

Theorem 1.2.19. *Let k_ε be an approximate identity on a locally compact group G with left Haar measure λ.*

(1) If f lies in $L^p(G)$ for $1 \le p < \infty$, then $\|k_\varepsilon * f - f\|_{L^p(G)} \to 0$ as $\varepsilon \to 0$.
(2) Let f be a function in $L^\infty(G)$ that is uniformly continuous on a subset K of G, in the sense that for all $\delta > 0$ there is a neighborhood V of the identity element such that for all $x \in K$ and $y \in V$ we have $|f(y^{-1}x) - f(x)| < \delta$. Then we have that $\|k_\varepsilon * f - f\|_{L^\infty(K)} \to 0$ as $\varepsilon \to 0$. In particular, if f is bounded and continuous at a point $x_0 \in G$, then $(k_\varepsilon * f)(x_0) \to f(x_0)$ as $\varepsilon \to 0$.

Proof. We start with the case $1 \le p < \infty$. We recall that continuous functions with compact support are dense in L^p of locally compact Hausdorff spaces equipped with measures arising from nonnegative linear functionals; see [152, Theorem 12.10]. For a continuous function g supported in a compact set L we have we have $|g(h^{-1}x) - g(x)|^p \le (2\|g\|_{L^\infty})^p \chi_{W^{-1}L}$ for h in a relatively compact neighborhood

W of the identity element e. By the Lebesgue dominated convergence theorem we obtain

$$\int_G |g(h^{-1}x) - g(x)|^p \, d\lambda(x) \to 0 \tag{1.2.28}$$

as $h \to e$. Now approximate a given f in $L^p(G)$ by a continuous function with compact support g to deduce that

$$\int_G |f(h^{-1}x) - f(x)|^p \, d\lambda(x) \to 0 \qquad \text{as} \qquad h \to e. \tag{1.2.29}$$

Because of (1.2.29), given a $\delta > 0$ there exists a neighborhood V of e such that

$$h \in V \implies \int_G |f(h^{-1}x) - f(x)|^p \, d\lambda(x) < \left(\frac{\delta}{2c}\right)^p, \tag{1.2.30}$$

where c is the constant that appears in Definition 1.2.15 (i). Since k_ε has integral one for all $\varepsilon > 0$, we have

$$\begin{aligned}(k_\varepsilon * f)(x) - f(x) &= (k_\varepsilon * f)(x) - f(x) \int_G k_\varepsilon(y) \, d\lambda(y) \\ &= \int_G (f(y^{-1}x) - f(x)) k_\varepsilon(y) \, d\lambda(y) \\ &= \int_V (f(y^{-1}x) - f(x)) k_\varepsilon(y) \, d\lambda(y) \\ &\qquad + \int_{V^c} (f(y^{-1}x) - f(x)) k_\varepsilon(y) \, d\lambda(y).\end{aligned} \tag{1.2.31}$$

Now take L^p norms in x in (1.2.31). In view of (1.2.30),

$$\begin{aligned}&\left\| \int_V (f(y^{-1}x) - f(x)) k_\varepsilon(y) \, d\lambda(y) \right\|_{L^p(G, d\lambda(x))} \\ &\qquad \le \int_V \|f(y^{-1}x) - f(x)\|_{L^p(G, d\lambda(x))} |k_\varepsilon(y)| \, d\lambda(y) \\ &\qquad \le \int_V \frac{\delta}{2c} |k_\varepsilon(y)| \, d\lambda(y) < \frac{\delta}{2},\end{aligned} \tag{1.2.32}$$

while

$$\begin{aligned}&\left\| \int_{V^c} (f(y^{-1}x) - f(x)) k_\varepsilon(y) \, d\lambda(y) \right\|_{L^p(G, d\lambda(x))} \\ &\qquad \le \int_{V^c} 2\|f\|_{L^p(G)} |k_\varepsilon(y)| \, d\lambda(y) < \frac{\delta}{2},\end{aligned} \tag{1.2.33}$$

provided we have that

$$\int_{V^c} |k_\varepsilon(x)| \, d\lambda(x) < \frac{\delta}{4(\|f\|_{L^p} + 1)}. \tag{1.2.34}$$

Choose $\varepsilon_0 > 0$ such that (1.2.34) is valid for $\varepsilon < \varepsilon_0$ by property (iii). Now (1.2.32) and (1.2.33) imply the required conclusion.

The case $p = \infty$ follows similarly. Let f be a bounded function on G that is uniformly continuous on K. Given $\delta > 0$, there is a neighborhood V of e such that, whenever $y \in V$ and $x \in K$ we have

$$|f(y^{-1}x) - f(x)| < \frac{\delta}{2c}, \tag{1.2.35}$$

where c is as in Definition 1.2.15 (i). By property (iii) in Definition 1.2.15, there is an $\varepsilon_0 > 0$ such that for $0 < \varepsilon < \varepsilon_0$ we have

$$\int_{V^c} |k_\varepsilon(y)|\,d\lambda(y) < \frac{\delta}{4\big(\|f\|_{L^\infty(G)} + 1\big)}. \tag{1.2.36}$$

Using (1.2.35) and (1.2.36), we deduce that

$$\begin{aligned}
&\sup_{x\in K} |(k_\varepsilon * f)(x) - f(x)| \\
&\le \int_V |k_\varepsilon(y)| \sup_{x\in K} |f(y^{-1}x) - f(x)|\,d\lambda(y) + \int_{V^c} |k_\varepsilon(y)| \sup_{x\in K} |f(y^{-1}x) - f(x)|\,d\lambda(y) \\
&\le c\,\frac{\delta}{2c} + \frac{\delta}{4\big(\|f\|_{L^\infty(G)} + 1\big)}\, 2\|f\|_{L^\infty(G)} \le \delta.
\end{aligned}$$

This shows that $k_\varepsilon * f$ converge uniformly to f on K as $\varepsilon \to 0$. In particular, if $K = \{x_0\}$ and f is bounded and continuous at x_0, we have $(k_\varepsilon * f)(x_0) \to f(x_0)$. □

Remark 1.2.20. Observe that if Haar measure satisfies (1.2.12), then the conclusion of Theorem 1.2.19 also holds for $f * k_\varepsilon$.

A simple modification in the proof of Theorem 1.2.19 yields the following variant, which presents a significant difference only when $a = 0$.

Theorem 1.2.21. *Let k_ε be a family of functions on a locally compact group G that satisfies properties (i) and (iii) of Definition 1.2.15 and also*

$$\int_G k_\varepsilon(x)\,d\lambda(x) = a$$

for some fixed $a \in \mathbf{C}$ and for all $\varepsilon > 0$. Let $f \in L^p(G)$ for some $1 \le p \le \infty$.

(a) If $1 \le p < \infty$, then $\|k_\varepsilon * f - af\|_{L^p(G)} \to 0$ as $\varepsilon \to 0$.
(b) If $p = \infty$ and f is uniformly continuous on a subset K of G, in the sense that for any $\delta > 0$ there is a neighborhood V of the identity element of G such that $\sup_{x\in G} \sup_{y\in V} |f(y^{-1}x) - f(x)| \le \delta$, then we have that $\|k_\varepsilon * f - af\|_{L^\infty(K)} \to 0$ as $\varepsilon \to 0$.

Exercises

1.2.1. Let G be a locally compact group and let f,g in $L^1(G)$ be supported in the subsets A and B of G, respectively. Prove that $f*g$ is supported in the algebraic product set AB.

1.2.2. For a function f on a locally compact group G and $t\in G$, let ${}^tf(x)=f(tx)$ and $f^t(x)=f(xt)$. Show that

$$ {}^tf*g={}^t(f*g) \qquad \text{and} \qquad f*g^t=(f*g)^t $$

whenever $f,g\in L^1(G)$, equipped with left Haar measure.

1.2.3. Let G be a locally compact group with left Haar measure. Let $f\in L^p(G)$ and $\tilde{g}\in L^{p'}(G)$, where $1<p<\infty$; recall that $\widetilde{g}(x)=g(x^{-1})$. For $t,x\in G$, let ${}^tg(x)=g(tx)$. Show that for any $\varepsilon>0$ there exists a relatively compact symmetric neighborhood of the origin U such that $u\in U$ implies $\|{}^u\widetilde{g}-\widetilde{g}\|_{L^{p'}(G)}<\varepsilon$ and therefore

$$ |(f*g)(v)-(f*g)(w)|<\|f\|_{L^p}\,\varepsilon $$

whenever $v^{-1}w\in U$.

1.2.4. (a) Prove that compactly supported functions are dense in $L^p(\mathbf{R}^n)$ for all $0<p<\infty$.
(b) Show that smooth functions with compact support are dense in $L^p(\mathbf{R}^n)$ for all $1\le p<\infty$.
[*Hint:* Part (b): Use Theorem 1.2.19 with $k_\varepsilon(x)=\varepsilon^{-n}k(\varepsilon^{-1}x)$ and k smooth and compactly supported function.]

1.2.5. Show that a Haar measure λ for the multiplicative group of all positive real numbers is

$$ \lambda(A)=\int_0^\infty \chi_A(t)\,\frac{dt}{t}\,. $$

1.2.6. Let $G=\mathbf{R}^2\setminus\{(0,y):\,y\in\mathbf{R}\}$ with group operation $(x,y)(z,w)=(xz,xw+y)$. [Think of G as the group of all 2×2 matrices with bottom row $(0,1)$ and nonzero top left entry.] Show that a left Haar measure on G is

$$ \lambda(A)=\int_{-\infty}^{+\infty}\int_{-\infty}^{+\infty}\chi_A(x,y)\,\frac{dx\,dy}{x^2}\,, $$

while a right Haar measure on G is

$$ \rho(A)=\int_{-\infty}^{+\infty}\int_{-\infty}^{+\infty}\chi_A(x,y)\,\frac{dx\,dy}{|x|}\,. $$

1.2.7. ([144], [145]) Use Theorem 1.2.10 to prove that

$$\left(\int_0^\infty \left(\frac{1}{x}\int_0^x |f(t)|\,dt\right)^p dx\right)^{\frac{1}{p}} \le \frac{p}{p-1}\|f\|_{L^p(0,\infty)},$$
$$\left(\int_0^\infty \left(\int_x^\infty |f(t)|\,dt\right)^p dx\right)^{\frac{1}{p}} \le p\left(\int_0^\infty |f(t)|^p t^p\,dt\right)^{\frac{1}{p}},$$

when $1 < p < \infty$.
[*Hint:* On the multiplicative group $(\mathbf{R}^+, \frac{dt}{t})$ consider the convolution of the function $|f(x)|x^{\frac{1}{p}}$ with the function $x^{-\frac{1}{p'}}\chi_{[1,\infty)}$ and the convolution of the function $|f(x)|x^{1+\frac{1}{p}}$ with $x^{\frac{1}{p}}\chi_{(0,1]}$.]

1.2.8. (*G. H. Hardy*) Let $0 < b < \infty$ and $1 \le p < \infty$. Prove that

$$\left(\int_0^\infty \left(\int_0^x |f(t)|\,dt\right)^p x^{-b-1}\,dx\right)^{\frac{1}{p}} \le \frac{p}{b}\left(\int_0^\infty |f(t)|^p t^{p-b-1}\,dt\right)^{\frac{1}{p}},$$
$$\left(\int_0^\infty \left(\int_x^\infty |f(t)|\,dt\right)^p x^{b-1}\,dx\right)^{\frac{1}{p}} \le \frac{p}{b}\left(\int_0^\infty |f(t)|^p t^{p+b-1}\,dt\right)^{\frac{1}{p}}.$$

[*Hint:* On the multiplicative group $(\mathbf{R}^+, \frac{dt}{t})$ consider the convolution of the function $|f(x)|x^{1-\frac{b}{p}}$ with $x^{-\frac{b}{p}}\chi_{[1,\infty)}$ and of the function $|f(x)|x^{1+\frac{b}{p}}$ with $x^{\frac{b}{p}}\chi_{(0,1]}$.]

1.2.9. On $\mathbf{R}^n$ let $T(f) = f * K$, where K is a positive L^1 function and f is in L^p, $1 \le p \le \infty$. Prove that the operator norm of $T: L^p \to L^p$ is equal to $\|K\|_{L^1}$.
[*Hint:* Clearly, $\|T\|_{L^p\to L^p} \le \|K\|_{L^1}$. Conversely, fix $0 < \varepsilon < 1$ and let N be a positive integer. Let $\chi_N = \chi_{B(0,N)}$ and for any $R > 0$ let $K_R = K\chi_{B(0,R)}$, where $B(x,R)$ is the ball of radius R centered at x. Observe that for $|x| \le (1-\varepsilon)N$, we have $B(0,N\varepsilon) \subseteq B(x,N)$; thus $\int_{\mathbf{R}^n} \chi_N(x-y)K_{N\varepsilon}(y)\,dy = \int_{\mathbf{R}^n} K_{N\varepsilon}(y)\,dy = \|K_{N\varepsilon}\|_{L^1}$. Then

$$\frac{\|K*\chi_N\|_{L^p}^p}{\|\chi_N\|_{L^p}^p} \ge \frac{\|K_{N\varepsilon}*\chi_N\|_{L^p(B(0,(1-\varepsilon)N)}^p}{\|\chi_N\|_{L^p}^p} \ge \|K_{N\varepsilon}\|_{L^1}^p(1-\varepsilon)^n.$$

Let $N \to \infty$ first and then $\varepsilon \to 0$.]

1.2.10. On the multiplicative group $(\mathbf{R}^+, \frac{dt}{t})$ let $T(f) = f * K$, where K is a positive L^1 function and f is in L^p, $1 \le p \le \infty$. Prove that the operator norm of $T: L^p \to L^p$ is equal to the L^1 norm of K. Deduce that the constants $p/(p-1)$ and p/b are sharp in Exercises 1.2.7 and 1.2.8.
[*Hint:* Adapt the idea of Exercise 1.2.9 to this setting.]

1.2.11. Let $Q_k(t) = c_k(1-t^2)^k$ for $t \in [-1,1]$ and zero elsewhere, where c_k is chosen such that $\int_{-1}^1 Q_k(t)\,dt = 1$ for all $k = 1, 2, \dots$.
(a) Show that $c_k < \sqrt{k}$.

(b) Use part (a) to show that $\{Q_k\}_k$ is an approximate identity on $\mathbf{R}$ as $k \to \infty$.
(c) Given a continuous function f on $\mathbf{R}$ that vanishes outside the interval $[-1,1]$, show that $f * Q_k$ converges to f uniformly on $[-1,1]$ as $k \to \infty$.
(d) (*Weierstrass*) Prove that every continuous function on $[-1,1]$ can be approximated uniformly by polynomials.
[*Hint:* Part (a): Estimate the integral $\int_{|t| \le k^{-1/2}} Q_k(t)\,dt$ from below using the inequality $(1-t^2)^k \ge 1 - kt^2$ for $|t| \le 1$. Part (d): Consider the function $g(t) = f(t) - f(-1) - \frac{t+1}{2}(f(1) - f(-1))$.]

1.2.12. Show that the *Laplace transform* $L(f)(x) = \int_0^\infty f(t) e^{-xt}\,dt$ maps $L^2(0,\infty)$ to itself with norm at most $\sqrt{\pi}$.
[*Hint:* Consider convolution with the kernel $\sqrt{t}\,e^{-t}$ on the group $L^2((0,\infty), \frac{dt}{t})$.]

1.2.13. ([62]) Let $F \ge 0$, $G \ge 0$ be measurable functions on the sphere $\mathbf{S}^{n-1}$ and let $K \ge 0$ be a measurable function on $[-1,1]$. Prove that

$$\int_{\mathbf{S}^{n-1}} \int_{\mathbf{S}^{n-1}} F(\theta) G(\varphi) K(\theta \cdot \varphi)\,d\varphi\,d\theta \le C \|F\|_{L^p(\mathbf{S}^{n-1})} \|G\|_{L^{p'}(\mathbf{S}^{n-1})},$$

where $1 \le p \le \infty$, $\theta \cdot \varphi = \sum_{j=1}^n \theta_j \varphi_j$ and $C = \int_{\mathbf{S}^{n-1}} K(\theta \cdot \varphi)\,d\varphi$, which is independent of θ. Moreover, show that C is the best possible constant in the preceding inequality. Using duality, compute the norm of the linear operator

$$F(\theta) \mapsto \int_{\mathbf{S}^{n-1}} F(\theta) K(\theta \cdot \varphi)\,d\varphi$$

from $L^p(\mathbf{S}^{n-1})$ to itself.
[*Hint:* Observe that $\int_{\mathbf{S}^{n-1}} \int_{\mathbf{S}^{n-1}} F(\theta) G(\varphi) K(\theta \cdot \varphi)\,d\varphi\,d\theta$ is bounded by the quantity

$$\left\{ \int_{\mathbf{S}^{n-1}} \left[\int_{\mathbf{S}^{n-1}} F(\theta) K(\theta \cdot \varphi)\,d\theta \right]^p d\varphi \right\}^{\frac{1}{p}} \|G\|_{L^{p'}(\mathbf{S}^{n-1})}.$$

Apply Hölder's inequality to the functions F and 1 with respect to the measure $K(\theta \cdot \varphi)\,d\theta$ to deduce that $\int_{\mathbf{S}^{n-1}} F(\theta) K(\theta \cdot \varphi)\,d\theta$ is controlled by

$$\left(\int_{\mathbf{S}^{n-1}} F(\theta)^p K(\theta \cdot \varphi)\,d\theta \right)^{1/p} \left(\int_{\mathbf{S}^{n-1}} K(\theta \cdot \varphi)\,d\theta \right)^{1/p'}.$$

Use Fubini's theorem to bound the latter by

$$\|F\|_{L^p(\mathbf{S}^{n-1})} \|G\|_{L^{p'}(\mathbf{S}^{n-1})} \int_{\mathbf{S}^{n-1}} K(\theta \cdot \varphi)\,d\varphi.$$

Note that equality is attained if and only if both F and G are constants.]

1.3 Interpolation

The theory of interpolation of operators is vast and extensive. In this section we are mainly concerned with a couple of basic interpolation results that appear in a variety of applications and constitute the foundation of the field. These results are the *Marcinkiewicz interpolation theorem* and the *Riesz–Thorin interpolation theorem*. These theorems are traditionally proved using real and complex variables techniques, respectively. A byproduct of the Riesz–Thorin interpolation theorem, *Stein's theorem on interpolation of analytic families of operators*, has also proved to be an important and useful tool in many applications and is presented at the end of the section.

We begin by setting up the background required to formulate the results of this section. Let (X,μ) and (Y,ν) be two measure spaces. Suppose we are given a linear operator T, initially defined on the set of simple functions on X, such that for all f simple on X, $T(f)$ is a ν-measurable function on Y. Let $0<p<\infty$ and $0<q<\infty$. If there exists a constant $C_{p,q}>0$ such that for all simple functions f on X we have

$$\big\|T(f)\big\|_{L^q(Y,\nu)} \le C_{p,q}\big\|f\big\|_{L^p(X,\mu)}, \tag{1.3.1}$$

then by density, T admits a unique bounded extension from $L^p(X,\mu)$ to $L^q(Y,\nu)$. This extension is also denoted by T. Operators that map L^p to L^q are called of *strong type* (p,q) and operators that map L^p to $L^{q,\infty}$ are called *weak type* (p,q).

1.3.1 Real Method: The Marcinkiewicz Interpolation Theorem

Definition 1.3.1. Let T be an operator defined on a linear space of complex-valued measurable functions on a measure space (X,μ) and taking values in the set of all complex-valued finite almost everywhere measurable functions on a measure space (Y,ν). Then T is called *linear* if for all f, g in the domain of T and all $\lambda\in\mathbf{C}$ we have

$$T(f+g)=T(f)+T(g) \qquad \text{and} \qquad T(\lambda f)=\lambda T(f). \tag{1.3.2}$$

T is called *sublinear* if for all f, g in the domain of T and all $\lambda\in\mathbf{C}$ we have

$$|T(f+g)|\le|T(f)|+|T(g)| \qquad \text{and} \qquad |T(\lambda f)|=|\lambda||T(f)|. \tag{1.3.3}$$

T is called *quasi-linear* if for all f, g in the domain of T and all $\lambda\in\mathbf{C}$ we have

$$|T(f+g)|\le K(|T(f)|+|T(g)|) \qquad \text{and} \qquad |T(\lambda f)|=|\lambda||T(f)| \tag{1.3.4}$$

for some constant $K>0$. Sublinearity is a special case of quasi-linearity.

For instance, T_1 and T_2 are linear operators, then $(|T_1|^p+|T_2|^p)^{1/p}$ is sublinear if $p\ge 1$ and quasi-linear if $0<p<1$.

Theorem 1.3.2. *Let (X,μ) be a σ-finite measure space, let (Y,ν) be another measure space, and let $0 < p_0 < p_1 \le \infty$. Let T be a sublinear operator defined on $L^{p_0}(X)+L^{p_1}(X)=\{f_0+f_1:\ f_j\in L^{p_j}(X_j), j=0,1\}$ and taking values in the space of measurable functions on Y. Assume that there exist $A_0, A_1<\infty$ such that*

$$\big\|T(f)\big\|_{L^{p_0,\infty}(Y)} \le A_0\big\|f\big\|_{L^{p_0}(X)} \qquad \text{for all } f\in L^{p_0}(X), \tag{1.3.5}$$

$$\big\|T(f)\big\|_{L^{p_1,\infty}(Y)} \le A_1\big\|f\big\|_{L^{p_1}(X)} \qquad \text{for all } f\in L^{p_1}(X). \tag{1.3.6}$$

Then for all $p_0<p<p_1$ and for all f in $L^p(X)$ we have the estimate

$$\big\|T(f)\big\|_{L^p(Y)} \le A\big\|f\big\|_{L^p(X)}, \tag{1.3.7}$$

where

$$A = 2\left(\frac{p}{p-p_0}+\frac{p}{p_1-p}\right)^{\frac{1}{p}} A_0^{\frac{\frac{1}{p}-\frac{1}{p_1}}{\frac{1}{p_0}-\frac{1}{p_1}}} A_1^{\frac{\frac{1}{p_0}-\frac{1}{p}}{\frac{1}{p_0}-\frac{1}{p_1}}}. \tag{1.3.8}$$

Proof. Assume first that $p_1<\infty$. Fix f a function in $L^p(X)$ and $\alpha>0$. We split $f=f_0^\alpha+f_1^\alpha$, where f_0^α is in L^{p_0} and f_1^α is in L^{p_1}. The splitting is obtained by cutting $|f|$ at height $\delta\alpha$ for some $\delta>0$ to be determined later. Set

$$f_0^\alpha(x) = \begin{cases} f(x) & \text{for } |f(x)|>\delta\alpha,\\ 0 & \text{for } |f(x)|\le\delta\alpha,\end{cases}$$

$$f_1^\alpha(x) = \begin{cases} f(x) & \text{for } |f(x)|\le\delta\alpha,\\ 0 & \text{for } |f(x)|>\delta\alpha.\end{cases}$$

It can be checked easily that f_0^α (the unbounded part of f) is an L^{p_0} function and that f_1^α (the bounded part of f) is an L^{p_1} function. Indeed, since $p_0<p$, we have

$$\big\|f_0^\alpha\big\|_{L^{p_0}}^{p_0} = \int_{|f|>\delta\alpha} |f(x)|^p|f(x)|^{p_0-p}\,d\mu(x) \le (\delta\alpha)^{p_0-p}\big\|f\big\|_{L^p}^p$$

and similarly, since $p<p_1$,

$$\big\|f_1^\alpha\big\|_{L^{p_1}}^{p_1} \le (\delta\alpha)^{p_1-p}\big\|f\big\|_{L^p}^p.$$

In view of the subadditivity property of T contained in (1.3.3) we obtain that

$$|T(f)| \le |T(f_0^\alpha)|+|T(f_1^\alpha)|,$$

which implies

$$\{y\in Y:\ |T(f)(y)|>\alpha\} \subseteq \{y\in Y:\ |T(f_0^\alpha)(y)|>\alpha/2\} \cup \{y\in Y:\ |T(f_1^\alpha)(y)|>\alpha/2\},$$

and therefore

$$d_{T(f)}(\alpha) \le d_{T(f_0^\alpha)}(\alpha/2) + d_{T(f_1^\alpha)}(\alpha/2). \tag{1.3.9}$$

Hypotheses (1.3.5) and (1.3.6) together with (1.3.9) now give

$$d_{T(f)}(\alpha) \le \frac{A_0^{p_0}}{(\alpha/2)^{p_0}} \int_{|f|>\delta\alpha} |f(x)|^{p_0}\, d\mu(x) + \frac{A_1^{p_1}}{(\alpha/2)^{p_1}} \int_{|f|\le\delta\alpha} |f(x)|^{p_1}\, d\mu(x).$$

In view of the last estimate and Proposition 1.1.4, we obtain that

$$\begin{aligned}
\big\|T(f)\big\|_{L^p}^p &\le p(2A_0)^{p_0} \int_0^\infty \alpha^{p-1}\alpha^{-p_0} \int_{|f|>\delta\alpha} |f(x)|^{p_0}\, d\mu(x)\, d\alpha \\
&\qquad + p(2A_1)^{p_1} \int_0^\infty \alpha^{p-1}\alpha^{-p_1} \int_{|f|\le\delta\alpha} |f(x)|^{p_1}\, d\mu(x)\, d\alpha \\
&= p(2A_0)^{p_0} \int_X |f(x)|^{p_0} \int_0^{\frac{1}{\delta}|f(x)|} \alpha^{p-1-p_0}\, d\alpha\, d\mu(x) \\
&\qquad + p(2A_1)^{p_1} \int_X |f(x)|^{p_1} \int_{\frac{1}{\delta}|f(x)|}^\infty \alpha^{p-1-p_1}\, d\alpha\, d\mu(x) \\
&= \frac{p(2A_0)^{p_0}}{p-p_0}\frac{1}{\delta^{p-p_0}} \int_X |f(x)|^{p_0}|f(x)|^{p-p_0}\, d\mu(x) \\
&\qquad + \frac{p(2A_1)^{p_1}}{p_1-p}\frac{1}{\delta^{p-p_1}} \int_X |f(x)|^{p_1}|f(x)|^{p-p_1}\, d\mu(x) \\
&= p\left(\frac{(2A_0)^{p_0}}{p-p_0}\frac{1}{\delta^{p-p_0}} + \frac{(2A_1)^{p_1}}{p_1-p}\delta^{p_1-p}\right)\big\|f\big\|_{L^p}^p,
\end{aligned}$$

and the convergence of the integrals in α is justified from $p_0 < p < p_1$, while the interchange of the integrals (Fubini's theorem) uses the hypothesis that (X,μ) is a σ-finite measure space. We pick $\delta > 0$ such that

$$(2A_0)^{p_0}\frac{1}{\delta^{p-p_0}} = (2A_1)^{p_1}\delta^{p_1-p},$$

and observe that the last displayed constant is equal to the pth power of the constant in (1.3.8). We have therefore proved the theorem when $p_1 < \infty$.

We now consider the case $p_1 = \infty$. Write $f = f_0^\alpha + f_1^\alpha$, where

$$f_0^\alpha(x) = \begin{cases} f(x) & \text{for } |f(x)| > \gamma\alpha, \\ 0 & \text{for } |f(x)| \le \gamma\alpha, \end{cases}$$

$$f_1^\alpha(x) = \begin{cases} f(x) & \text{for } |f(x)| \le \gamma\alpha, \\ 0 & \text{for } |f(x)| > \gamma\alpha. \end{cases}$$

We have

$$\big\|T(f_1^\alpha)\big\|_{L^\infty} \le A_1\big\|f_1^\alpha\big\|_{L^\infty} \le A_1\gamma\alpha = \alpha/2,$$

provided we choose $\gamma = (2A_1)^{-1}$. It follows that the set $\{y \in Y : |T(f_1^\alpha)(y)| > \alpha/2\}$ has measure zero. Therefore,

$$d_{T(f)}(\alpha) \leq d_{T(f_0^\alpha)}(\alpha/2).$$

Since T maps L^{p_0} to $L^{p_0,\infty}$ with norm at most A_0, it follows that

$$d_{T(f_0^\alpha)}(\alpha/2) \leq \frac{(2A_0)^{p_0} \big\|f_0^\alpha\big\|_{L^{p_0}}^{p_0}}{\alpha^{p_0}} = \frac{(2A_0)^{p_0}}{\alpha^{p_0}} \int_{|f|>\gamma\alpha} |f(x)|^{p_0}\, d\mu(x). \tag{1.3.10}$$

Using (1.3.10) and Proposition 1.1.4, we obtain

$$\begin{aligned}
\big\|T(f)\big\|_{L^p}^p &= p \int_0^\infty \alpha^{p-1} d_{T(f)}(\alpha)\, d\alpha \\
&\leq p \int_0^\infty \alpha^{p-1} d_{T(f_0^\alpha)}(\alpha/2)\, d\alpha \\
&\leq p \int_0^\infty \alpha^{p-1} \frac{(2A_0)^{p_0}}{\alpha^{p_0}} \int_{|f|>\alpha/(2A_1)} |f(x)|^{p_0}\, d\mu(x)\, d\alpha \\
&= p(2A_0)^{p_0} \int_X |f(x)|^{p_0} \int_0^{2A_1|f(x)|} \alpha^{p-p_0-1}\, d\alpha\, d\mu(x) \\
&= \frac{p(2A_1)^{p-p_0}(2A_0)^{p_0}}{p-p_0} \int_X |f(x)|^p\, d\mu(x).
\end{aligned}$$

This proves the theorem with constant

$$A = 2\left(\frac{p}{p-p_0}\right)^{\frac{1}{p}} A_1^{1-\frac{p_0}{p}} A_0^{\frac{p_0}{p}}. \tag{1.3.11}$$

Observe that when $p_1 = \infty$, the constant in (1.3.11) coincides with that in (1.3.8). □

Remark 1.3.3. Notice that the proof of Theorem 1.3.2 only makes use of the *subadditivity* property $|T(f+g)| \leq |T(f)|+|T(g)|$ of T in hypothesis (1.3.3).

If T is a linear operator (instead of sublinear), then we can relax the hypotheses of Theorem 1.3.2 by assuming that (1.3.5) and (1.3.6) hold for all simple functions f on X. Then the functions f_0^α and f_1^α constructed in the proof are also simple, and we conclude that (1.3.7) holds for all simple functions f on X. By density, T has a unique extension on $L^p(X)$ that also satisfies (1.3.7).

1.3.2 Complex Method: The Riesz–Thorin Interpolation Theorem

The next interpolation theorem assumes stronger endpoint estimates, but yields a more natural bound on the norm of the operator on the intermediate spaces. Unfortunately, it is mostly applicable for linear operators and in some cases for sublinear operators (often via a linearization process) but it does not apply to quasi-linear operators without some loss in the constant.

Recall that a simple function is called finitely simple if it is supported in a set of finite measure. Finitely simple functions are dense in $L^p(X,\mu)$ for $0 < p < \infty$, whenever (X,μ) is a σ-finite measure space.

Theorem 1.3.4. *Let (X,μ) and (Y,ν) be two σ-finite measure spaces. Let T be a linear operator defined on the set of all finitely simple functions on X and taking values in the set of measurable functions on Y. Let $1 \le p_0, p_1, q_0, q_1 \le \infty$ and assume that*

$$\begin{aligned} \big\|T(f)\big\|_{L^{q_0}} &\le M_0 \big\|f\big\|_{L^{p_0}}, \\ \big\|T(f)\big\|_{L^{q_1}} &\le M_1 \big\|f\big\|_{L^{p_1}}, \end{aligned} \tag{1.3.12}$$

for all finitely simple functions f on X. Then for all $0 < \theta < 1$ we have

$$\big\|T(f)\big\|_{L^q} \le M_0^{1-\theta} M_1^{\theta} \big\|f\big\|_{L^p} \tag{1.3.13}$$

for all finitely simple functions f on X, where

$$\frac{1}{p} = \frac{1-\theta}{p_0} + \frac{\theta}{p_1} \qquad \textit{and} \qquad \frac{1}{q} = \frac{1-\theta}{q_0} + \frac{\theta}{q_1}. \tag{1.3.14}$$

Consequently, when $p < \infty$, by density, T has a unique bounded extension from $L^p(X,\mu)$ to $L^q(Y,\nu)$ when p and q are as in (1.3.14).

Proof. Let

$$f = \sum_{k=1}^{m} a_k e^{i\alpha_k} \chi_{A_k}$$

be a finitely simple function on X, where $a_k > 0$, α_k are real, and A_k are pairwise disjoint subsets of X with finite measure.

We need to control

$$\big\|T(f)\big\|_{L^q(Y,\nu)} = \sup_g \left| \int_Y T(f)(y) g(y)\, d\nu(y) \right|,$$

where the supremum is taken over all finitely simple functions g on Y with $L^{q'}$ norm less than or equal to 1. Write

$$g = \sum_{j=1}^{n} b_j e^{i\beta_j} \chi_{B_j},$$

where $b_j > 0$, β_j are real, and B_j are pairwise disjoint subsets of Y with finite ν-measure. Let

$$P(z) = \frac{p}{p_0}(1-z) + \frac{p}{p_1} z \qquad \text{and} \qquad Q(z) = \frac{q'}{q_0'}(1-z) + \frac{q'}{q_1'} z. \tag{1.3.15}$$

For z in the closed strip $\overline{S} = \{z \in \mathbf{C} : 0 \le \operatorname{Re} z \le 1\}$, define

$$f_z = \sum_{k=1}^{m} a_k^{P(z)} e^{i\alpha_k} \chi_{A_k}, \quad g_z = \sum_{j=1}^{n} b_j^{Q(z)} e^{i\beta_j} \chi_{B_j}, \tag{1.3.16}$$

and

$$F(z) = \int_Y T(f_z)(y)\, g_z(y)\, d\nu(y).$$

Notice that $f_\theta = f$ and $g_\theta = f$. By linearity we have

$$F(z) = \sum_{k=1}^{m} \sum_{j=1}^{n} a_k^{P(z)} b_j^{Q(z)} e^{i\alpha_k} e^{i\beta_j} \int_Y T(\chi_{A_k})(y)\, \chi_{B_j}(y)\, d\nu(y).$$

Since $a_k, b_j > 0$, F is analytic in z, and the expression

$$\int_Y T(\chi_{A_k})(y)\, \chi_{B_j}(y)\, d\nu(y)$$

is a finite constant, being an absolutely convergent integral; this is seen by Hölder's inequality with exponents q_0 and q_0' (or q_1 and q_1') and (1.3.12).

By the disjointness of the sets A_k we have (even when $p_0 = \infty$)

$$\big\|f_{it}\big\|_{L^{p_0}} = \big\|f\big\|_{L^p}^{\frac{p}{p_0}},$$

since $|a_k^{P(it)}| = a_k^{\frac{p}{p_0}}$, and by the disjointness of the B_j's we have (even when $q_0 = 1$)

$$\big\|g_{it}\big\|_{L^{q_0'}} = \big\|g\big\|_{L^{q'}}^{\frac{q'}{q_0'}},$$

since $|b_j^{Q(it)}| = b_j^{\frac{q'}{q_0'}}$. Thus Hölder's inequality and the hypothesis give

$$\begin{aligned} |F(it)| &\le \big\|T(f_{it})\big\|_{L^{q_0}} \big\|g_{it}\big\|_{L^{q_0'}} \\ &\le M_0 \big\|f_{it}\big\|_{L^{p_0}} \big\|g_{it}\big\|_{L^{q_0'}} \\ &= M_0 \big\|f\big\|_{L^p}^{\frac{p}{p_0}} \big\|g\big\|_{L^{q'}}^{\frac{q'}{q_0'}}. \end{aligned} \tag{1.3.17}$$

By similar calculations, which are valid even when $p_1 = \infty$ and $q_1 = 1$, we have

$$\big\|f_{1+it}\big\|_{L^{p_1}} = \big\|f\big\|_{L^p}^{\frac{p}{p_1}}$$

and

$$\big\|g_{1+it}\big\|_{L^{q_1'}} = \big\|g\big\|_{L^{q'}}^{\frac{q'}{q_1'}}.$$

Also, in a way analogous to that we obtained (1.3.17) we deduce that

$$|F(1+it)| \le M_1 \big\|f\big\|_{L^p}^{\frac{p}{p_1}} \big\|g\big\|_{L^{q'}}^{\frac{q'}{q'_1}}. \tag{1.3.18}$$

To finish the proof we will need the following lemma, known as *Hadamard's three lines lemma.*

Lemma 1.3.5. *Let F be analytic in the open strip $S = \{z \in \mathbf{C} : 0 < \operatorname{Re} z < 1\}$, continuous and bounded on its closure, such that $|F(z)| \le B_0$ when $\operatorname{Re} z = 0$ and $|F(z)| \le B_1$ when $\operatorname{Re} z = 1$, for some $0 < B_0, B_1 < \infty$. Then $|F(z)| \le B_0^{1-\theta} B_1^{\theta}$ when $\operatorname{Re} z = \theta$, for any $0 \le \theta \le 1$.*

To prove the lemma we define analytic functions

$$G(z) = F(z)(B_0^{1-z}B_1^z)^{-1} \qquad \text{and} \qquad G_n(z) = G(z)e^{(z^2-1)/n}$$

for z in the unit strip S, for $n = 1,2,\dots$. Since F is bounded on $\overline{S}$ and

$$|B_0^{1-z}B_1^z| \ge \min(1,B_0)\min(1,B_1) > 0$$

for all $z \in \overline{S}$, we conclude that G is bounded by some constant M on $\overline{S}$. Since

$$|G_n(x+iy)| \le Me^{-y^2/n}e^{(x^2-1)/n} \le Me^{-y^2/n},$$

we deduce that $G_n(x+iy)$ converges to zero uniformly in $0 \le x \le 1$ as $|y| \to \infty$. Select $y(n) > 0$ such that for $|y| \ge y(n)$, we have $|G_n(x+iy)| \le 1$ for all $x \in [0,1]$. Also, the assumptions on F imply that G is bounded by one on the two lines forming the boundary of $\overline{S}$. By the maximum principle we obtain that $|G_n(z)| \le 1$ for all z in the rectangle $[0,1] \times [-y(n), y(n)]$; hence $|G_n(z)| \le 1$ everywhere in the closed strip. Letting $n \to \infty$, we conclude that $|G(z)| \le 1$ in the closed strip. Taking $z = \theta + it$ we deduce that

$$|F(\theta+it)| \le |B_0^{1-\theta-it}B_1^{\theta+it}| = B_0^{1-\theta}B_1^{\theta}$$

whenever t is real. This proves the required conclusion. □

Returning to the proof of Theorem 1.3.4, we observe that F is analytic in the open strip S and continuous on its closure. Also, F is bounded on the closed unit strip (by some constant that depends on f and g). Therefore, (1.3.17), (1.3.18), and Lemma 1.3.5 give

$$|F(z)| \le \Big(M_0 \big\|f\big\|_{L^p}^{\frac{p}{p_0}} \big\|g\big\|_{L^{q'}}^{\frac{q'}{q'_0}}\Big)^{1-\theta} \Big(M_1 \big\|f\big\|_{L^p}^{\frac{p}{p_1}} \big\|g\big\|_{L^{q'}}^{\frac{q'}{q'_1}}\Big)^{\theta} = M_0^{1-\theta}M_1^{\theta} \big\|f\big\|_{L^p} \big\|g\big\|_{L^{q'}},$$

when $\operatorname{Re} z = \theta$. Observe that $P(\theta) = Q(\theta) = 1$ and hence

$$F(\theta) = \int_Y T(f)\,g\,d\nu.$$

Taking the supremum over all finitely simple functions g on Y with $L^{q'}$ norm less than or equal to one, we conclude the proof of the theorem. □

We now give an application of Theorem 1.3.4.

Example 1.3.6. One may prove Young's inequality (Theorem 1.2.12) using the Riesz–Thorin interpolation theorem (Theorem 1.3.4). Fix a function g in L^r and let $T(f) = f * g$. Since $T : L^1 \to L^r$ with norm at most $\|g\|_{L^r}$ and $T : L^{r'} \to L^\infty$ with norm at most $\|g\|_{L^r}$, Theorem 1.3.4 gives that T maps L^p to L^q with norm at most the quantity $\|g\|_{L^r}^{\theta}\|g\|_{L^r}^{1-\theta} = \|g\|_{L^r}$, where

$$\frac{1}{p} = \frac{1-\theta}{1} + \frac{\theta}{r'} \quad \text{and} \quad \frac{1}{q} = \frac{1-\theta}{r} + \frac{\theta}{\infty}. \tag{1.3.19}$$

Finally, observe that equations (1.3.19) give (1.2.13).

1.3.3 Interpolation of Analytic Families of Operators

Theorem 1.3.4 can be extended to the case in which the interpolated operators are allowed to vary. In particular, if a family of operators depends analytically on a parameter z, then the proof of this theorem can be adapted to work in this setting.

We describe the setup for this theorem. Let (X,μ) and (Y,ν) be σ-finite measure spaces. Suppose that for every z in the closed strip $\overline{S} = \{z \in \mathbf{C} : 0 \le \operatorname{Re} z \le 1\}$ there is an associated linear operator T_z defined on the space of finitely simple functions on X and taking values in the space of measurable functions on Y such that

$$\int_Y |T_z(\chi_A)\,\chi_B|\,d\nu < \infty \tag{1.3.20}$$

whenever A and B are subsets of finite measure of X and Y, respectively. The family $\{T_z\}_z$ is said to be *analytic* if for all f,g finitely simple functions we have that the function

$$z \mapsto \int_Y T_z(f)\,g\,d\nu \tag{1.3.21}$$

is analytic in the open strip $S = \{z \in \mathbf{C} : 0 < \operatorname{Re} z < 1\}$ and continuous on its closure. The analytic family $\{T_z\}_z$ is called of *admissible growth* if there is a constant τ_0 with $0 \le \tau_0 < \pi$ such that for finitely simple functions f on X and g on Y there is constant $C(f,g)$ such that

$$\log\left|\int_Y T_z(f)\,g\,d\nu\right| \le C(f,g)\,e^{\tau_0|\operatorname{Im} z|} \tag{1.3.22}$$

for all z satisfying $0 \le \operatorname{Re} z \le 1$. Note that if there is $\tau_0 \in (0,\pi)$ such that for all measurable subsets A of X and B of Y of finite measure there is a constant $c(A,B)$ such that

$$\log\left|\int_B T_z(\chi_A)\,d\nu\right| \le c(A,B)\,e^{\tau_0|\operatorname{Im} z|}, \tag{1.3.23}$$

then (1.3.22) holds for $f = \sum_{k=1}^{M} a_k \chi_{A_k}$ and $g = \sum_{j=1}^{N} b_j \chi_{B_j}$ and

$$C(f,g) = \log(MN) + \sum_{k=1}^{M} \sum_{j=1}^{N} \left(c(A_k, B_j) + \left| \log |a_k b_j| \right| \right).$$

The extension of the Riesz–Thorin interpolation theorem is as follows.

Theorem 1.3.7. *Let T_z be an analytic family of linear operators of admissible growth defined on the space of finitely simple functions of a σ-finite measure space (X, μ) and taking values in the set of measurable functions of another σ-finite measure space (Y, ν). Let $1 \le p_0, p_1, q_0, q_1 \le \infty$ and suppose that M_0 and M_1 are positive functions on the real line such that for some τ_1 with $0 \le \tau_1 < \pi$ we have*

$$\sup_{-\infty < y < +\infty} e^{-\tau_1 |y|} \log M_j(y) < \infty \tag{1.3.24}$$

for $j = 0, 1$. Fix $0 < \theta < 1$ and define p, q by the equations

$$\frac{1}{p} = \frac{1-\theta}{p_0} + \frac{\theta}{p_1} \qquad \text{and} \qquad \frac{1}{q} = \frac{1-\theta}{q_0} + \frac{\theta}{q_1}. \tag{1.3.25}$$

Suppose that for all finitely simple functions f on X we have

$$\left\| T_{iy}(f) \right\|_{L^{q_0}} \le M_0(y) \left\| f \right\|_{L^{p_0}}, \tag{1.3.26}$$

$$\left\| T_{1+iy}(f) \right\|_{L^{q_1}} \le M_1(y) \left\| f \right\|_{L^{p_1}}. \tag{1.3.27}$$

Then for all finitely simple functions f on X we have

$$\left\| T_\theta(f) \right\|_{L^q} \le M(\theta) \left\| f \right\|_{L^p} \tag{1.3.28}$$

where for $0 < x < 1$

$$M(x) = \exp\left\{ \frac{\sin(\pi x)}{2} \int_{-\infty}^{\infty} \left[\frac{\log M_0(t)}{\cosh(\pi t) - \cos(\pi x)} + \frac{\log M_1(t)}{\cosh(\pi t) + \cos(\pi x)} \right] dt \right\}.$$

Thus, by density, T_θ has a unique bounded extension from $L^p(X, \mu)$ to $L^q(Y, \nu)$ when p and q are as in (1.3.25).

Note that in view of (1.3.24), the integral defining $M(t)$ converges absolutely. The proof of the previous theorem is based on an extension of Lemma 1.3.5.

Lemma 1.3.8. *Let F be analytic on the open strip $S = \{z \in \mathbf{C} : 0 < \operatorname{Re} z < 1\}$ and continuous on its closure such that for some $A < \infty$ and $0 \le \tau_0 < \pi$ we have*

$$\log |F(z)| \le A e^{\tau_0 |\operatorname{Im} z|} \tag{1.3.29}$$

for all $z \in \overline{S}$. *Then*

$$|F(x+iy)| \le \exp\left\{\frac{\sin(\pi x)}{2}\int_{-\infty}^{\infty}\left[\frac{\log|F(it+iy)|}{\cosh(\pi t)-\cos(\pi x)}+\frac{\log|F(1+it+iy)|}{\cosh(\pi t)+\cos(\pi x)}\right]dt\right\}$$

whenever $0 < x < 1$, *and* y *is real.*

Assuming Lemma 1.3.8, we prove Theorem 1.3.7.

Proof. Fix $0 < \theta < 1$ and finitely simple functions f on X and g on Y such that $\|f\|_{L^p} = \|g\|_{L^{q'}} = 1$. Note that since $0 < \theta < 1$ we must have $1 < p, q < \infty$. Let

$$f = \sum_{k=1}^{m} a_k e^{i\alpha_k} \chi_{A_k} \quad \text{and} \quad g = \sum_{j=1}^{n} b_j e^{i\beta_j} \chi_{B_j},$$

where $a_k > 0$, $b_j > 0$, α_k, β_j are real, A_k are pairwise disjoint subsets of X with finite measure, and B_j are pairwise disjoint subsets of Y with finite measure for all k, j. Let $P(z)$, $Q(z)$ be as in (1.3.15) and f_z, g_z as in (1.3.16). Define for $z \in \overline{S}$

$$F(z) = \int_Y T_z(f_z)\, g_z \, d\nu. \tag{1.3.30}$$

Linearity gives that

$$F(z) = \sum_{k=1}^{m}\sum_{j=1}^{n} a_k^{P(z)} b_j^{Q(z)} e^{i\alpha_k} e^{i\beta_j} \int_Y T_z(\chi_{A_k})(x)\, \chi_{B_j}(x)\, d\nu(x),$$

and conditions (1.3.20) together with the fact that $\{T_z\}_z$ is an analytic family imply that $F(z)$ is a well-defined analytic function on the unit strip that extends continuously to its boundary.

Since $\{T_z\}_z$ is a family of admissible growth, (1.3.23) holds for some $c(A_k, B_j)$ and $\tau_0 \in (0, \pi)$ and this combined with the facts that

$$|a_k^{P(z)}| \le a_k^{\frac{p}{p_0}+\frac{p}{p_1}} \qquad \text{and} \qquad |b_j^{Q(z)}| \le b_j^{\frac{q'}{q'_0}+\frac{q'}{q'_1}}$$

for all z with $0 < \operatorname{Re} z < 1$, implies (1.3.29) with τ_0 as in (1.3.23) and

$$A = \log(mn) + \sum_{k=1}^{m}\sum_{j=1}^{n}\left(c(A_k, B_j) + \left(\frac{p}{p_0}+\frac{p}{p_1}\right)\big|\log a_k\big| + \left(\frac{q'}{q'_0}+\frac{q'}{q'_1}\right)\big|\log b_j\big|\right).$$

Thus F satisfies the hypotheses of Lemma 1.3.8. Moreover, the calculations in the proof of Theorem 1.3.4 show that (even when $p_0 = \infty$, $q_0 = 1$, $p_1 = \infty$, $q_1 = 1$)

$$\big\|f_{iy}\big\|_{L^{p_0}} = \|f\|_{L^p}^{\frac{p}{p_0}} = 1 = \|g\|_{L^{q'}}^{\frac{q'}{q'_0}} = \big\|g_{iy}\big\|_{L^{q'_0}} \quad \text{when } y \in \mathbf{R}, \tag{1.3.31}$$

$$\big\|f_{1+iy}\big\|_{L^{p_1}} = \|f\|_{L^p}^{\frac{p}{p_1}} = 1 = \|g\|_{L^{q'}}^{\frac{q'}{q'_1}} = \big\|g_{1+iy}\big\|_{L^{q'_1}} \quad \text{when } y \in \mathbf{R}. \tag{1.3.32}$$

Hölder's inequality, (1.3.31), and the hypothesis (1.3.26) now give

$$|F(iy)| \le \big\|T_{iy}(f_{iy})\big\|_{L^{q_0}} \big\|g_{iy}\big\|_{L^{q_0'}} \le M_0(y)\big\|f_{iy}\big\|_{L^{p_0}} \big\|g_{iy}\big\|_{L^{q_0'}} = M_0(y)$$

for all y real. Similarly, (1.3.32), and (1.3.27) imply

$$|F(1+iy)| \le \big\|T_{1+iy}(f_{1+iy})\big\|_{L^{q_1}} \big\|g_{1+iy}\big\|_{L^{q_1'}} \le M_1(y)\big\|f_{1+iy}\big\|_{L^{p_1}} \big\|g_{1+iy}\big\|_{L^{q_1'}} = M_1(y)$$

for all $y \in \mathbf{R}$. These inequalities and the conclusion of Lemma 1.3.8 yield

$$|F(x)| \le \exp\left\{\frac{\sin(\pi x)}{2}\int_{-\infty}^{\infty}\left[\frac{\log M_0(t)}{\cosh(\pi t)-\cos(\pi x)}+\frac{\log M_1(t)}{\cosh(\pi t)+\cos(\pi x)}\right]dt\right\} = M(x)$$

for all $0 < x < 1$. But notice that

$$F(\theta) = \int_Y T_\theta(f)\, g\, d\nu. \tag{1.3.33}$$

Taking absolute values and the supremum over all finitely simple functions g on Y with $L^{q'}$ norm equal to one, we conclude the proof of (1.3.28) for finitely simple functions f with L^p norm one. Then (1.3.28) follows by replacing f by $f/\|f\|_{L^p}$. □

We end this section with the proof of Lemma 1.3.8.

Proof of Lemma 1.3.8. Recall the Poisson integral formula

$$U(z) = \frac{1}{2\pi}\int_{-\pi}^{+\pi} U(Re^{i\varphi})\frac{R^2-\rho^2}{|Re^{i\varphi}-\rho e^{i\theta}|^2}\,d\varphi, \qquad z = \rho e^{i\theta}, \tag{1.3.34}$$

which is valid for a harmonic function U defined on the unit disk $D = \{z : \, |z| < 1\}$ when $|z| < R < 1$. See [307, p. 258].

Consider now a subharmonic function u on D that is continuous on the circle $|\zeta| = R < 1$. When $U = u$, the right side of (1.3.34) defines a harmonic function on the set $\{z \in \mathbf{C} : \, |z| < R\}$ that coincides with u on the circle $|\zeta| = R$. The maximum principle for subharmonic functions ([307, p. 362]) implies that for $|z| < R < 1$ we have

$$u(z) \le \frac{1}{2\pi}\int_{-\pi}^{+\pi} u(Re^{i\varphi})\frac{R^2-\rho^2}{|Re^{i\varphi}-\rho e^{i\theta}|^2}\,d\varphi, \quad z = \rho e^{i\theta}. \tag{1.3.35}$$

This is valid for all subharmonic functions u on D that are continuous on the circle $|\zeta| = R$ when $\rho < R < 1$.

It is not difficult to verify that

$$h(\zeta) = \frac{1}{\pi i}\log\left(i\frac{1+\zeta}{1-\zeta}\right)$$

is a conformal map from D onto the strip $S = (0,1)\times\mathbf{R}$. Indeed, $i(1+\zeta)/(1-\zeta)$ lies in the upper half-plane and the preceding complex logarithm is a well defined holomorphic function that takes the upper half-plane onto the strip $\mathbf{R}\times(0,\pi)$. Since

$F \circ h$ is a holomorphic function on D, $\log|F \circ h|$ is a subharmonic function on D. Applying (1.3.35) to the function $z \mapsto \log|F(h(z))|$, we obtain

$$\log|F(h(z))| \le \frac{1}{2\pi}\int_{-\pi}^{+\pi} \log|F(h(Re^{i\varphi}))| \frac{R^2-\rho^2}{R^2-2\rho R\cos(\theta-\varphi)+\rho^2}\,d\varphi \quad (1.3.36)$$

when $z = \rho e^{i\theta}$ and $|z| = \rho < R$. Observe that when $|\zeta| = 1$ and $\zeta \neq \pm 1$, $h(\zeta)$ has real part zero or one. It follows from the hypothesis that

$$\log|F(h(\zeta))| \le Ae^{\tau_0|\mathrm{Im}\, h(\zeta)|} = Ae^{\tau_0\left|\mathrm{Im}\frac{1}{\pi i}\log\left(i\frac{1+\zeta}{1-\zeta}\right)\right|} = Ae^{\frac{\tau_0}{\pi}\left|\log\left|\frac{1+\zeta}{1-\zeta}\right|\right|}.$$

Therefore, $\log|F(h(\zeta))|$ is bounded by a multiple of $|1+\zeta|^{-\tau_0/\pi}|1-\zeta|^{-\tau_0/\pi}$, which is integrable over the set $|\zeta| = 1$, since $\tau_0 < \pi$. Fix now $z = \rho e^{i\theta}$ with $\rho < R$ and let $R \to 1$ in (1.3.36). The Lebesgue dominated convergence theorem gives that

$$\log|F(h(\rho e^{i\theta}))| \le \frac{1}{2\pi}\int_{-\pi}^{+\pi} \log|F(h(e^{i\varphi}))| \frac{1-\rho^2}{1-2\rho\cos(\theta-\varphi)+\rho^2}\,d\varphi. \quad (1.3.37)$$

Setting $x = h(\rho e^{i\theta})$, we obtain that

$$\rho e^{i\theta} = h^{-1}(x) = \frac{e^{\pi i x}-i}{e^{\pi i x}+i} = -i\frac{\cos(\pi x)}{1+\sin(\pi x)} = \left(\frac{\cos(\pi x)}{1+\sin(\pi x)}\right)e^{-i(\pi/2)},$$

from which it follows that $\rho = (\cos(\pi x))/(1+\sin(\pi x))$ and $\theta = -\pi/2$ when $0 < x \le \frac{1}{2}$, while $\rho = -(\cos(\pi x))/(1+\sin(\pi x))$ and $\theta = \pi/2$ when $\frac{1}{2} \le x < 1$. In either case we easily deduce that

$$\frac{1-\rho^2}{1-2\rho\cos(\theta-\varphi)+\rho^2} = \frac{\sin(\pi x)}{1+\cos(\pi x)\sin(\varphi)}.$$

Using this we write (1.3.37) as

$$\log|F(x)| \le \frac{1}{2\pi}\int_{-\pi}^{\pi} \frac{\sin(\pi x)}{1+\cos(\pi x)\sin(\varphi)} \log|F(h(e^{i\varphi}))|\,d\varphi. \quad (1.3.38)$$

We now change variables. On the interval $[-\pi, 0)$ we use the change of variables $it = h(e^{i\varphi})$ or, equivalently, $e^{i\varphi} = -\tanh(\pi t) - i\,\mathrm{sech}(\pi t)$. Observe that as φ ranges from $-\pi$ to 0, t ranges from $+\infty$ to $-\infty$. Furthermore, $d\varphi = -\pi\,\mathrm{sech}(\pi t)\,dt$. We have

$$\begin{aligned} &\frac{1}{2\pi}\int_{-\pi}^{0} \frac{\sin(\pi x)}{1+\cos(\pi x)\sin(\varphi)} \log|F(h(e^{i\varphi}))|\,d\varphi \\ &\quad = \frac{1}{2}\int_{-\infty}^{\infty} \frac{\sin(\pi x)}{\cosh(\pi t)-\cos(\pi x)} \log|F(it)|\,dt. \end{aligned} \quad (1.3.39)$$

On the interval $(0,\pi]$ we use the change of variables $1+it = h(e^{i\varphi})$ or, equivalently, $e^{i\varphi} = -\tanh(\pi t) + i\,\mathrm{sech}(\pi t)$. Observe that as φ ranges from 0 to π, t ranges from $-\infty$ to $+\infty$. Furthermore, $d\varphi = \pi\,\mathrm{sech}(\pi t)\,dt$. Similarly, we obtain

$$\begin{aligned}&\frac{1}{2\pi}\int_0^{\pi} \frac{\sin(\pi t)}{1+\cos(\pi t)\sin(\varphi)} \log|F(h(e^{i\varphi}))|\,d\varphi \\ &\qquad = \frac{1}{2}\int_{-\infty}^{+\infty} \frac{\sin(\pi x)}{\cosh(\pi t)+\cos(\pi x)} \log|F(1+it)|\,dt.\end{aligned} \tag{1.3.40}$$

Adding (1.3.39) and (1.3.40) and using (1.3.38) we conclude the proof when $y=0$.

We now consider the case where $y \neq 0$. Fix $y \neq 0$ and define the function $G(z) = F(z+iy)$. Then G is analytic on the open strip $S = \{z \in \mathbf{C} : 0 < \mathrm{Re}\, z < 1\}$ and continuous on its closure. Moreover, for some $A < \infty$ and $0 \le \tau_0 < \pi$ we have

$$\log|G(z)| = \log|F(z+iy)| \le A\, e^{\tau_0|\mathrm{Im}\, z+y|} \le A e^{\tau_0|y|}\, e^{\tau_0|\mathrm{Im}\, z|}$$

for all $z \in \overline{S}$. Then the case $y=0$ for G (with A replaced by $Ae^{\tau_0|y|}$) yields

$$|G(x)| \le \exp\left\{\frac{\sin(\pi x)}{2}\int_{-\infty}^{\infty}\left[\frac{\log|G(it)|}{\cosh(\pi t)-\cos(\pi x)} + \frac{\log|G(1+it)|}{\cosh(\pi t)+\cos(\pi x)}\right]dt\right\},$$

which yields the required conclusion for any real y, since $G(x) = F(x+iy)$, $G(it) = F(it+iy)$, and $G(1+it) = F(1+it+iy)$. □

Exercises

1.3.1. Generalize Theorem 1.3.2 to the situation in which T is *quasi-subadditive*, that is, it satisfies for some $K>0$,

$$|T(f+g)| \le K(|T(f)|+|T(g)|),$$

for all f, g in the domain of T. Prove that in this case, the constant A in (1.3.7) can be taken to be K times the constant in (1.3.8).

1.3.2. Let (X,μ), (Y,ν) be two σ-finite measure spaces. Let $1 < p < r \le \infty$ and suppose that T be a sublinear operator defined on the space $L^{p_0}(X) + L^{p_1}(X)$ and taking values in the space of measurable functions on Y. Assume that T maps $L^1(X)$ to $L^{1,\infty}(Y)$ with norm A_0 and $L^r(X)$ to $L^r(Y)$ with norm A_1. Let $0 < p_0 < p_1 \le \infty$. Prove that T maps L^p to L^p with norm at most

$$8\,(p-1)^{-\frac{1}{p}} A_0^{\frac{\frac{1}{p}-\frac{1}{r}}{1-\frac{1}{r}}} A_1^{\frac{1-\frac{1}{p}}{1-\frac{1}{r}}}.$$

[*Hint:* First interpolate between L^1 and L^r using Theorem 1.3.2 and then interpolate between $L^{\frac{p+1}{2}}$ and L^r using Theorem 1.3.4.]

1.3.3. Let $0 < p_0 < p < p_1 \le \infty$ and let T be an operator as in Theorem 1.3.2 that also satisfies

$$|T(f)| \le T(|f|),$$

for all $f \in L^{p_0} + L^{p_1}$.
(a) If $p_0 = 1$ and $p_1 = \infty$, prove that T maps L^p to L^p with norm at most

$$\frac{p}{p-1} A_0^{\frac{1}{p}} A_1^{1-\frac{1}{p}}.$$

(b) More generally, if $p_0 < p < \infty$, prove that the norm of T from L^p to L^p is at most

$$p^{1+\frac{1}{p}} \left[\frac{B(p_0+1, p-p_0)}{p_0^{p_0}(p-p_0)^{p-p_0}} \right]^{\frac{1}{p}} A_0^{\frac{p_0}{p}} A_1^{1-\frac{p_0}{p}},$$

where $B(s,t) = \int_0^1 x^{s-1}(1-x)^{t-1}\,dx$ is the usual Beta function.
(c) When $0 < p_0 < p_1 < \infty$, then the norm of T from L^p to L^p is at most

$$\min_{0<\lambda<1} p^{\frac{1}{p}} \left(\frac{B(p-p_0, p_0+1)}{(1-\lambda)^{p_0}} + \frac{\frac{p_1-p+1}{p_1-p}}{\lambda^{p_1}} \right)^{\frac{1}{p}} A_0^{\frac{\frac{1}{p}-\frac{1}{p_1}}{\frac{1}{p_0}-\frac{1}{p_1}}} A_1^{\frac{\frac{1}{p_0}-\frac{1}{p}}{\frac{1}{p_0}-\frac{1}{p_1}}}.$$

[*Hint:* The hypothesis $|T(f)| \le T(|f|)$ reduces matters to nonnegative functions. Parts (a), (b): Given $f \ge 0$ and $\alpha > 0$ write $f = f_0 + f_1$, where $f_0 = f - \lambda\alpha/A_1$ when $f \ge \lambda\alpha/A_1$ and zero otherwise. Here $0 < \lambda < 1$ to be chosen later. Then we have that $|\{|T(f)| > \alpha\}| \le |\{|T(f_0)| > (1-\lambda)\alpha\}|$. Part (c): Write $f = f_0 + f_1$, where $f_0 = f - \delta\alpha$ when $f \ge \delta\alpha$ and zero otherwise. Use that

$$|\{|T(f)| > \alpha\}| \le |\{|T(f_0)| > (1-\lambda)\alpha\}| + |\{|T(f_1)| > \lambda\alpha\}|$$

and optimize over $\delta > 0$.]

1.3.4. Let $0 \le \gamma, \delta < \pi$. For every $z \in S_{a,b} = \{z \in \mathbf{C} : a < \operatorname{Re} z < b\}$, let T_z be a family of linear operators defined on finetely simple functions on a σ-finite measure space (X,μ) and taking values in another σ-finite measure space (Y,ν). Assume that $\{T_z\}_z$ is an analytic on of $S_{a,b}$, in the sense of (1.3.21), continuous on its closure, and that for all simple functions f on X and g on Y there is a constant $C_{f,g} < \infty$ such that for all $z \in S_{a,b}$,

$$\log \left| \int_Y T_z(f) g\,d\nu \right| \le C_{f,g}\, e^{\gamma|\operatorname{Im} z|/(b-a)}.$$

Let $1 \le p_0, q_0, p_1, q_1 \le \infty$. Suppose that T_{a+iy} maps $L^{p_0}(X)$ to $L^{q_0}(Y)$ with bound $M_0(y)$ and T_{b+iy} maps $L^{p_1}(X)$ to $L^{q_1}(Y)$ with bound $M_1(y)$, where

$$\sup_{-\infty<y<\infty} e^{-\delta|y|/(b-a)} \log M_j(y) < \infty, \quad j = 0, 1.$$

Then for $a < t < b$, T_t maps $L^p(X)$ to $L^q(Y)$, where

$$\frac{1}{p} = \frac{\frac{b-t}{b-a}}{p_0} + \frac{\frac{t-a}{b-a}}{p_1} \quad \text{and} \quad \frac{1}{q} = \frac{\frac{b-t}{b-a}}{q_0} + \frac{\frac{t-a}{b-a}}{q_1}.$$

1.3.5. ([331]) On $\mathbf{R}^n$ let $x = (x_1, \dots, x_n)$ and $|x| = (x_1^2 + \cdots + x_n^2)^{1/2}$. Let

$$K_\lambda(x) = \frac{\pi^{\frac{n-1}{2}}\Gamma(\lambda+1)}{\Gamma(\lambda+\frac{n+1}{2})}\int_{-1}^{+1} e^{2\pi i s|x|}(1-s^2)^{\lambda+\frac{n-1}{2}}\,ds = \frac{\Gamma(\lambda+1)}{\pi^\lambda |x|^{\lambda+\frac{n}{2}}} J_{\lambda+\frac{n}{2}}(2\pi|x|),$$

where λ is a complex number and $J_{\lambda+\frac{n}{2}}$ is the Bessel function of order $\lambda+\frac{n}{2}$. Let T_λ be the operator given by convolution with K_λ. Show that T_λ maps $L^p(\mathbf{R}^n)$ to itself for $\operatorname{Re}\lambda > (n-1)|\frac{1}{2}-\frac{1}{p}|$.
[*Hint:* In view of the calculation of the Fourier transform of K_λ contained in Appendix B.5, we have that when $\operatorname{Re}\lambda = 0$, T_λ maps $L^2(\mathbf{R}^n)$ to itself with norm 1. Using the estimates in Appendices B.6 and B.7, conclude that K_λ is integrable and thus T_λ maps $L^1(\mathbf{R}^n)$ to itself with an appropriate constant when $\operatorname{Re}\lambda = (n-1)/2+\delta$ (for $\delta > 0$). Then use Exercise 1.3.4.]

1.3.6. Observe that Theorem 1.3.7 yields the stronger conclusion

$$\big\|T_z(f)\big\|_{L^q} \le M(z)\,\big\|f\big\|_{L^p}$$

for $z \in S = \{z \in \mathbf{C} : 0 < \operatorname{Re} z < 1\}$, where for $z = x + iy$

$$M(z) = \exp\left\{\frac{\sin(\pi x)}{2}\int_{-\infty}^{\infty}\left[\frac{\log M_0(t+y)}{\cosh(\pi t)-\cos(\pi x)} + \frac{\log M_1(t+y)}{\cosh(\pi t)+\cos(\pi x)}\right]dt\right\}.$$

1.3.7. ([380]) Let (X,μ) and (Y,ν) be two measure spaces with $\mu(X) < \infty$ and $\nu(Y) < \infty$. Let T be a *countably subadditive* operator that maps $L^p(X)$ to $L^p(Y)$ for every $1 < p \le 2$ with norm $\|T\|_{L^p \to L^p} \le A(p-1)^{-\alpha}$ for some fixed $A, \alpha > 0$. (Countably subadditive means that $|T(\sum_j f_j)| \le \sum_j |T(f_j)|$ for all f_j in $L^p(X)$ with $\sum_j f_j \in L^p$.) Prove that for all f measurable on X we have

$$\int_Y |T(f)|\,d\nu \le 6A(1+\nu(Y))^{\frac{1}{2}}\left[\int_X |f|(\log_2^+|f|)^\alpha\,d\mu + C_\alpha + \mu(X)^{\frac{1}{2}}\right],$$

where $C_\alpha = \sum_{k=1}^\infty k^\alpha (2/3)^k$. This result provides an example of *extrapolation*.
[*Hint:* Write

$$f = \sum_{k=0}^{\infty} f\chi_{S_k},$$

where $S_k = \{2^k \le |f| < 2^{k+1}\}$ when $k \ge 1$ and $S_0 = \{|f| < 2\}$. Using Hölder's inequality and the hypotheses on T, obtain that

$$\int_Y |T(f\chi_{S_k})|\,d\nu \le 2A\nu(Y)^{\frac{1}{k+1}} 2^k k^\alpha \mu(S_k)^{\frac{k}{k+1}}$$

for $k \ge 1$. Note that for $k \ge 1$ we have $\nu(Y)^{\frac{1}{k+1}} \le \max(1,\nu(Y))^{\frac{1}{2}}$ and consider the cases $\mu(S_k) \ge 3^{-k-1}$ and $\mu(S_k) \le 3^{-k-1}$ when summing in $k \ge 1$. The term with $k = 0$ is easier.]

1.3.8. Prove that for $0 < x < 1$ we have

$$\frac{\sin(\pi x)}{2}\int_{-\infty}^{+\infty} \frac{1}{\cosh(\pi t)+\cos(\pi x)}\,dt = x,$$
$$\frac{\sin(\pi x)}{2}\int_{-\infty}^{+\infty} \frac{1}{\cosh(\pi t)-\cos(\pi x)}\,dt = 1-x,$$

and conclude that Lemma 1.3.8 reduces to Lemma 1.3.5 when the functions $M_0(y)$ and $M_1(y)$ are constant and assumption (1.3.29) is replaced by the stronger assumption that F is bounded on $\overline{S}$.
[*Hint:* In the first integral write $\cosh(\pi t) = \frac{1}{2}(e^{\pi t} + e^{-\pi t})$. Then use the change of variables $s = e^{\pi t}$.]

1.3.9. Let (X,μ), (Y,ν) be σ-finite measure spaces, and let $0 < p_0 < p_1 \le \infty$. Let T be a sublinear operator defined on the space $L^{p_0}(X) + L^{p_1}(X)$ and taking values in the space of measurable functions on Y. Suppose T is a sublinear operator such that maps L^{p_0} to L^∞ with constant A_0 and L^{p_1} to L^∞ with constant A_1. Prove T maps L^p to L^∞ with constant $2A_0^{1-\theta}A_1^{\theta}$ where

$$\frac{1-\theta}{p_0} + \frac{\theta}{p_1} = \frac{1}{p}.$$

1.4 Lorentz Spaces

Suppose that f is a measurable function on a measure space (X,μ). It would be desirable to have another function f^* defined on $[0,\infty)$ that is decreasing and *equidistributed* with f. By this we mean

$$d_f(\alpha) = d_{f^*}(\alpha) \tag{1.4.1}$$

for all $\alpha \ge 0$. This is achieved via a simple construction discussed in this section.

1.4.1 Decreasing Rearrangements

Definition 1.4.1. Let f be a complex-valued function defined on X. The *decreasing rearrangement* of f is the function f^* defined on $[0,\infty)$ by

$$f^*(t) = \inf\{s > 0 : \, d_f(s) \le t\} = \inf\{s \ge 0 : \, d_f(s) \le t\}. \tag{1.4.2}$$

We adopt the convention $\inf\emptyset = \infty$, thus having $f^*(t) = \infty$ whenever $d_f(\alpha) > t$ for all $\alpha \geq 0$. Observe that f^* is decreasing and supported in $[0, \mu(X)]$.

Before we proceed with properties of the function f^*, we work out three examples.

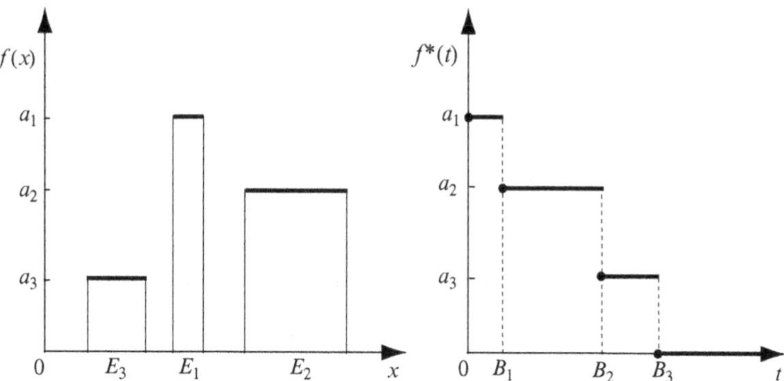

Fig. 1.3 The graph of a simple function $f(x)$ and its decreasing rearrangement $f^*(t)$.

Example 1.4.2. Consider the simple function of Example 1.1.2,

$$f(x) = \sum_{j=1}^{N} a_j \chi_{E_j}(x),$$

where E_j are pairwise disjoint sets of finite measure and $a_1 > \cdots > a_N > 0$. We saw in Example 1.1.2 that

$$d_f(\alpha) = \sum_{j=0}^{N} B_j \chi_{[a_{j+1}, a_j)}(\alpha),$$

where

$$B_j = \sum_{i=1}^{j} \mu(E_i)$$

and $a_{N+1} = B_0 = 0$ and $a_0 = \infty$. Observe that for $B_0 \leq t < B_1$, the smallest $s > 0$ with $d_f(s) \leq t$ is a_1. Similarly, for $B_1 \leq t < B_2$, the smallest $s > 0$ with $d_f(s) \leq t$ is a_2. Arguing this way, it is not difficult to see that

$$f^*(t) = \sum_{j=1}^{N} a_j \chi_{[B_{j-1}, B_j)}(t).$$

See Figure 1.3.

Example 1.4.3. On $(\mathbf{R}^n, dx)$ let

$$f(x) = \frac{1}{1 + |x|^p}, \qquad 0 < p < \infty.$$

A computation shows that

$$d_f(\alpha) = \begin{cases} v_n(\frac{1}{\alpha} - 1)^{\frac{n}{p}} & \text{if } \alpha < 1, \\ 0 & \text{if } \alpha \geq 1, \end{cases}$$

and therefore

$$f^*(t) = \frac{1}{(t/v_n)^{p/n} + 1},$$

where v_n is the volume of the unit ball in $\mathbf{R}^n$.

Example 1.4.4. Again on $(\mathbf{R}^n, dx)$ let $g(x) = 1 - e^{-|x|^2}$. We can easily see that $d_g(\alpha) = 0$ if $\alpha \geq 1$ and $d_g(\alpha) = \infty$ if $\alpha < 1$. We conclude that $g^*(t) = 1$ for all $t \geq 0$. This example indicates that although quantitative information is preserved, significant qualitative information is lost in passing from a function to its decreasing rearrangement.

It is clear from the previous examples that f^* is continuous from the right and decreasing. The following are some properties of the function f^*.

Proposition 1.4.5. *For f, g, f_n μ-measurable, $k \in \mathbf{C}$, and $0 \leq t, s, t_1, t_2 < \infty$ we have*

(1) $f^*(d_f(\alpha)) \leq \alpha$ *whenever* $\alpha > 0$.

(2) $d_f(f^*(t)) \leq t$.

(3) $f^*(t) > s$ *if and only if* $t < d_f(s)$*; that is,* $\{t \geq 0 : f^*(t) > s\} = [0, d_f(s))$.

(4) $|g| \leq |f|$ μ*-a.e. implies that* $g^* \leq f^*$ *and* $|f|^* = f^*$.

(5) $(kf)^* = |k| f^*$.

(6) $(f+g)^*(t_1+t_2) \leq f^*(t_1) + g^*(t_2)$.

(7) $(fg)^*(t_1+t_2) \leq f^*(t_1) g^*(t_2)$.

(8) $|f_n| \uparrow |f|$ μ*-a.e. implies* $f_n^* \uparrow f^*$.

(9) $|f| \leq \liminf_{n\to\infty} |f_n|$ μ*-a.e. implies* $f^* \leq \liminf_{n\to\infty} f_n^*$.

(10) f^* *is right continuous on* $[0, \infty)$.

(11) *If* $f^*(t) < \infty$, $c > 0$, *and* $\mu(\{|f| \geq f^*(t) - c\}) < \infty$, *then* $t \leq \mu(\{|f| \geq f^*(t)\})$.

(12) $d_f = d_{f^*}$.

(13) $(|f|^p)^* = (f^*)^p$ *when* $0 < p < \infty$.

(14) $\displaystyle\int_X |f|^p \, d\mu = \int_0^\infty f^*(t)^p \, dt$ *when* $0 < p < \infty$.

(15) $\big\|f\big\|_{L^\infty} = f^*(0)$.

(16) $\sup_{t>0} t^q f^*(t) = \sup_{\alpha>0} \alpha \, (d_f(\alpha))^q$ *for* $0 < q < \infty$.

Proof. Property (1): The set $A = \{s > 0 : d_f(s) \le d_f(\alpha)\}$ contains α and thus $f^*(d_f(\alpha)) = \inf A \le \alpha$.

Property (2): Let $s_n \in \{s > 0 : d_f(s) \le t\}$ be such that $s_n \downarrow f^*(t)$. Then $d_f(s_n) \le t$, and the right continuity of d_f (Exercise 1.1.1 (a)) implies that $d_f(f^*(t)) \le t$.

Property (3): If $s < f^*(t) = \inf\{u > 0 : d_f(u) \le t\}$, then $s \notin \{u > 0 : d_f(u) \le t\}$ which gives $d_f(s) > t$. Conversely, if for some $t < d_f(s)$ we had $f^*(t) \le s$, applying d_f and using property (2) would yield the contradiction $d_f(s) \le d_f(f^*(t)) \le t$.

Properties (4) and (5) are left to the reader.

Properties (6) and (7): Let $A = \{s_1 > 0 : d_f(s_1) \le t_1\}$, $B = \{s_2 > 0 : d_g(s_2) \le t_2\}$, $P = \{s > 0 : d_{fg}(s) \le t_1 + t_2\}$, and $S = \{s > 0 : d_{f+g}(s) \le t_1 + t_2\}$. Then $A + B \subseteq S$ and $A \cdot B \subseteq P$; thus $(f+g)^*(t_1+t_2) = \inf S \le s_1 + s_2$ and $(fg)^*(t_1+t_2) = \inf P \le s_1 s_2$ are valid for all $s_1 \in A$ and $s_2 \in B$. Taking the infimum over all $s_1 \in A$ and $s_2 \in B$ yields the conclusions.

Property (8): It follows from the definition of decreasing rearrangements that $f_n^* \le f_{n+1}^* \le f^*$ for all n. Let $h = \lim_{n\to\infty} f_n^*$; then obviously $h \le f^*$. Since $f_n^* \le h$, we have $d_{f_n}(h(t)) \le d_{f_n}(f_n^*(t)) \le t$, which implies, in view of Exercise 1.1.1 (c), that $d_f(h(t)) \le t$ by letting $n \to \infty$. It follows that $f^* \le h$, hence $h = f^*$.

Property (9): Set $F_n = \inf_{m \ge n} |f_m|$ and $h = \liminf_{n\to\infty} |f_n| = \sup_{n\ge 1} F_n$. Since $F_n \uparrow h$, property (8) yields that $F_n^* \uparrow h^*$ as $n \to \infty$. By hypothesis we have $|f| \le h$, hence $f^* \le h^* = \sup_n F_n^*$. Since $F_n \le |f_m|$ for $m \ge n$, it follows that $F_n^* \le f_m^*$ for $m \ge n$; thus $F_n^* \le \inf_{m\ge n} f_m^*$. Putting these facts together, we obtain $f^* \le h^* \le \sup_n \inf_{m\ge n} f_m^* = \liminf_{n\to\infty} f_n^*$.

Property (10): If $f^*(t_0) = 0$, then $f^*(t) = 0$ for all $t > t_0$ and thus f^* is right continuous at t_0. Suppose $f^*(t_0) > 0$. Pick α such that $0 < \alpha < f^*(t_0)$ and let $\{t_n\}_{n=1}^\infty$ be a sequence of real numbers decreasing to zero. The definition of f^* yields that $d_f(f^*(t_0) - \alpha) > t_0$. Since $t_n \downarrow 0$, there is an $n_0 \in \mathbf{Z}^+$ such that $d_f(f^*(t_0) - \alpha) > t_0 + t_n$ for all $n \ge n_0$. Property (3) yields that for all $n \ge n_0$ we have $f^*(t_0) - \alpha < f^*(t_0 + t_n)$, and since the latter is at most $f^*(t_0)$, the right continuity of f^* follows.

Property (11): The definition of f^* yields that the set $A_n = \{|f| > f^*(t) - c/n\}$ has measure $\mu(A_n) > t$. The sets A_n form a decreasing sequence as n increases and $\mu(A_1) < \infty$ by assumption. Consequently, $\{|f| \ge f^*(t)\} = \bigcap_{n=1}^\infty A_n$ has measure greater than or equal to t.

Property (12): This is immediate for nonnegative simple functions in view of Examples 1.1.2 and 1.4.2. For an arbitrary measurable function f, find a sequence of nonnegative simple functions f_n such that $f_n \uparrow |f|$ and apply (9).

Property (13): It follows from $d_{|f|^p}(\alpha) = d_f(\alpha^{1/p}) = d_{f^*}(\alpha^{1/p}) = d_{(f^*)^p}(\alpha)$ for all $\alpha > 0$.

Property (14): This is a consequence of property (12) and of Proposition 1.1.4.

Property (15): This is a restatement of (1.1.2).

Property (16): Given $\alpha > 0$, without loss of generality we may assume $d_f(\alpha) > 0$. Pick ε satisfying $0 < \varepsilon < d_f(\alpha)$. Property (3) yields $f^*(d_f(\alpha) - \varepsilon) > \alpha$, which implies that

$$\sup_{t>0} t^q f^*(t) \ge (d_f(\alpha) - \varepsilon)^q f^*(d_f(\alpha) - \varepsilon) > (d_f(\alpha) - \varepsilon)^q \alpha .$$

We first let $\varepsilon \to 0$ and then take the supremum over all $\alpha > 0$ to obtain one direction. Conversely, given $t > 0$, assume without loss of generality that $f^*(t) > 0$, and pick ε such that $0 < \varepsilon < f^*(t)$. Property (3) yields $d_f(f^*(t) - \varepsilon) > t$. This implies that $\sup_{\alpha>0} \alpha(d_f(\alpha))^q \geq (f^*(t) - \varepsilon)(d_f(f^*(t) - \varepsilon))^q > (f^*(t) - \varepsilon)t^q$. We first let $\varepsilon \to 0$ and then take the supremum over all $t > 0$ to obtain the opposite direction of the claimed equality. □

1.4.2 Lorentz Spaces

Having disposed of the basic properties of decreasing rearrangements of functions, we proceed with the definition of the Lorentz spaces.

Definition 1.4.6. Given f a measurable function on a measure space (X, μ) and $0 < p, q \leq \infty$, define

$$\|f\|_{L^{p,q}} = \begin{cases} \left(\int_0^\infty \left(t^{\frac{1}{p}} f^*(t) \right)^q \frac{dt}{t} \right)^{\frac{1}{q}} & \text{if } q < \infty, \\ \sup_{t>0} t^{\frac{1}{p}} f^*(t) & \text{if } q = \infty. \end{cases}$$

The set of all f with $\|f\|_{L^{p,q}} < \infty$ is denoted by $L^{p,q}(X, \mu)$ and is called the *Lorentz space* with indices p and q.

As in L^p and in weak L^p, two functions in $L^{p,q}(X, \mu)$ are considered equal if they are equal μ-almost everywhere. Observe that the previous definition implies that $L^{\infty,\infty} = L^\infty$, $L^{p,\infty} =$ weak L^p in view of Proposition 1.4.5 (16) and that $L^{p,p} = L^p$.

Remark 1.4.7. Observe that for all $0 < p, r < \infty$ and $0 < q \leq \infty$ we have

$$\big\| |g|^r \big\|_{L^{p,q}} = \|g\|_{L^{pr,qr}}^r . \tag{1.4.3}$$

On $\mathbf{R}^n$ let $\delta^\varepsilon(f)(x) = f(\varepsilon x)$, $\varepsilon > 0$, be the dilation operator. It is straightforward that $d_{\delta^\varepsilon(f)}(\alpha) = \varepsilon^{-n} d_f(\alpha)$ and $(\delta^\varepsilon(f))^*(t) = f^*(\varepsilon^n t)$. It follows that Lorentz norms satisfy the following dilation identity:

$$\big\| \delta^\varepsilon(f) \big\|_{L^{p,q}} = \varepsilon^{-n/p} \big\| f \big\|_{L^{p,q}} . \tag{1.4.4}$$

Next, we calculate the Lorentz norms of a finitely simple function.

Example 1.4.8. Using the notation of Example 1.4.2, when $0 < p, q < \infty$ we have

$$\|f\|_{L^{p,q}} = \left(\frac{p}{q} \right)^{\frac{1}{q}} \left[a_1^q B_1^{\frac{q}{p}} + a_2^q \left(B_2^{\frac{q}{p}} - B_1^{\frac{q}{p}} \right) + \cdots + a_N^q \left(B_N^{\frac{q}{p}} - B_{N-1}^{\frac{q}{p}} \right) \right]^{\frac{1}{q}},$$

and also

$$\|f\|_{L^{p,\infty}} = \sup_{1\le j\le N} a_j B_j^{\frac{1}{p}}.$$

Next, we calculate $\|f\|_{L^{\infty,q}}$ for the simple function f of Example 1.4.2. when $q<\infty$. It turns out that

$$\|f\|_{L^{\infty,q}} = \left[a_1^q \log\left(\frac{B_1}{B_0}\right) + a_2^q \log\left(\frac{B_2}{B_1}\right) + \cdots + a_N^q \log\left(\frac{B_N}{B_{N-1}}\right)\right]^{\frac{1}{q}} = \infty,$$

since $B_0 = 0$. We conclude that the only nonnegative simple function with finite $L^{\infty,q}$ norm is the zero function. Given a general nonzero function $g\in L^{\infty,q}$ with $0<q<\infty$, there is a nonzero simple function s with $0\le s\le g$. Then s has infinite norm, and therefore so does g. We deduce that $L^{\infty,q}(X) = \{0\}$ when $0<q<\infty$.

Proposition 1.4.9. *For $0<p<\infty$ and $0<q\le\infty$, we have the identity*

$$\|f\|_{L^{p,q}} = \begin{cases} p^{\frac{1}{q}}\left(\int_0^\infty \left[d_f(s)^{\frac{1}{p}} s\right]^q \frac{ds}{s}\right)^{\frac{1}{q}} & \text{when } q<\infty \\ \sup_{s>0} s\, d_f(s)^{\frac{1}{p}} & \text{when } q=\infty. \end{cases} \tag{1.4.5}$$

Proof. The case $q=\infty$ is statement (16) in Proposition 1.4.5, and we may therefore concentrate on the case $q<\infty$. If f is the simple function of Example 1.1.2, then

$$d_f(s) = \sum_{j=1}^N B_j \chi_{[a_{j+1},a_j)}(s)$$

with the understanding that $a_{N+1}=0$. Using the this formula and identity in Example 1.4.8, we obtain the validity of (1.4.5) for simple functions. In general, given a measurable function f, find a sequence of nonnegative simple functions such that $f_n \uparrow |f|$ a.e. Then $d_{f_n}\uparrow d_f$ (Exercise 1.1.1 (c)) and $f_n^*\uparrow f^*$ (Proposition 1.4.5 (8)). Using the Lebesgue monotone convergence theorem we deduce (1.4.5). □

Since $L^{p,p}\subsetneq L^{p,\infty}$, one may wonder whether these spaces are nested. The next result shows that for any fixed p, the Lorentz spaces $L^{p,q}$ increase as the exponent q increases.

Proposition 1.4.10. *Suppose $0<p\le\infty$ and $0<q<r\le\infty$. Then there exists a constant $c_{p,q,r}$ (which depends on p, q, and r) such that*

$$\|f\|_{L^{p,r}} \le c_{p,q,r}\|f\|_{L^{p,q}}. \tag{1.4.6}$$

In other words, $L^{p,q}$ is a subspace of $L^{p,r}$.

Proof. We may assume $p < \infty$, since the case $p = \infty$ is trivial. We have

$$\begin{aligned} t^{1/p} f^*(t) &= \left\{ \frac{q}{p} \int_0^t [s^{1/p} f^*(t)]^q \frac{ds}{s} \right\}^{1/q} \\ &\le \left\{ \frac{q}{p} \int_0^t [s^{1/p} f^*(s)]^q \frac{ds}{s} \right\}^{1/q} \qquad \text{since } f^* \text{ is decreasing,} \\ &\le \left(\frac{q}{p} \right)^{1/q} \big\| f \big\|_{L^{p,q}}. \end{aligned}$$

Hence, taking the supremum over all $t > 0$, we obtain

$$\big\| f \big\|_{L^{p,\infty}} \le \left(\frac{q}{p} \right)^{1/q} \big\| f \big\|_{L^{p,q}}. \tag{1.4.7}$$

This establishes (1.4.6) in the case $r = \infty$. Finally, when $r < \infty$, we have

$$\big\| f \big\|_{L^{p,r}} = \left\{ \int_0^\infty [t^{1/p} f^*(t)]^{r-q+q} \frac{dt}{t} \right\}^{1/r} \le \big\| f \big\|_{L^{p,\infty}}^{(r-q)/r} \big\| f \big\|_{L^{p,q}}^{q/r}. \tag{1.4.8}$$

Inequality (1.4.7) combined with (1.4.8) gives (1.4.6) with $c_{p,q,r} = (q/p)^{(r-q)/rq}$. □

Unfortunately, the functionals $\| \cdot \|_{L^{p,q}}$ do not satisfy the triangle inequality. For instance, consider the functions $f(t) = t$ and $g(t) = 1 - t$ defined on $[0,1]$. Then $f^*(\alpha) = g^*(\alpha) = (1-\alpha)\chi_{[0,1]}(\alpha)$. A simple calculation shows that the inequality $\|f + g\|_{L^{p,q}} \le \|f\|_{L^{p,q}} + \|g\|_{L^{p,q}}$ would be equivalent to

$$\frac{p}{q} \le 2^q \frac{\Gamma(q+1)\Gamma(q/p)}{\Gamma(q+1+q/p)},$$

which fails in general. However, since for all $t > 0$ we have

$$(f+g)^*(t) \le f^*(t/2) + g^*(t/2),$$

the estimate

$$\big\| f + g \big\|_{L^{p,q}} \le c_{p,q} \big(\big\| f \big\|_{L^{p,q}} + \big\| g \big\|_{L^{p,q}} \big), \tag{1.4.9}$$

where $c_{p,q} = 2^{1/p} \max(1, 2^{(1-q)/q})$, is a consequence of (1.1.4). Also, if $\|f\|_{L^{p,q}} = 0$ then we must have $f = 0$ μ-a.e. Therefore, $L^{p,q}$ is a quasi-normed space for all p, q with $0 < p, q \le \infty$. Is this space complete with respect to its quasi-norm? The next theorem answers this question.

Theorem 1.4.11. *Let (X, μ) be a measure space. Then for all $0 < p, q \le \infty$, the spaces $L^{p,q}(X, \mu)$ are complete with respect to their quasi-norm and they are therefore quasi-Banach spaces.*

Proof. We consider only the case $p < \infty$. First we note that convergence in $L^{p,q}$ implies convergence in measure. When $q = \infty$, this is proved in Proposition 1.1.9. When $q < \infty$, in view of Proposition 1.4.5 (16) and (1.4.7), it follows that

$$\sup_{t>0} t^{1/p} f^*(t) = \sup_{\alpha>0} \alpha d_f(\alpha)^{1/p} \le \left(\frac{q}{p}\right)^{1/q} \|f\|_{L^{p,q}}$$

for all $f \in L^{p,q}$, and from this it follows that convergence in $L^{p,q}$ implies convergence in measure.

Now let $\{f_n\}$ be a Cauchy sequence in $L^{p,q}$. Then $\{f_n\}$ is Cauchy in measure, and hence it has a subsequence $\{f_{n_k}\}$ that converges almost everywhere to some f by Theorem 1.1.13. Fix k_0 and apply property (9) in Proposition 1.4.5. Since $|f - f_{n_{k_0}}| = \lim_{k\to\infty} |f_{n_k} - f_{n_{k_0}}|$, it follows that

$$(f - f_{n_{k_0}})^*(t) \le \liminf_{k\to\infty} (f_{n_k} - f_{n_{k_0}})^*(t). \tag{1.4.10}$$

Raise (1.4.10) to the power q, multiply by $t^{q/p}$, integrate with respect to dt/t over $(0,\infty)$, and apply Fatou's lemma to obtain

$$\big\|f - f_{n_{k_0}}\big\|_{L^{p,q}}^q \le \liminf_{k\to\infty} \big\|f_{n_k} - f_{n_{k_0}}\big\|_{L^{p,q}}^q. \tag{1.4.11}$$

Now let $k_0 \to \infty$ in (1.4.11) and use the fact that $\{f_n\}$ is Cauchy to conclude that f_{n_k} converges to f in $L^{p,q}$. It is a general fact that if a Cauchy sequence has a convergent subsequence in a quasi-normed space, then the sequence is convergent to the same limit. It follows that f_n converges to f in $L^{p,q}$. □

Remark 1.4.12. It can be shown that the spaces $L^{p,q}$ are normable when p, q are bigger than 1; see Exercise 1.4.3. Therefore, these spaces can be normed to become Banach spaces.

It is well known that finitely simple functions are dense in L^p of any measure space, when $0 < p < \infty$. It is natural to ask whether finitely simple functions are also dense in $L^{p,q}$. This is in fact the case when $q \ne \infty$.

Theorem 1.4.13. *Finitely simple functions are dense in $L^{p,q}(X,\mu)$ when $0 < q < \infty$.*

Proof. Let $f \in L^{p,q}(X,\mu)$. Assume without loss of generality that $f \ge 0$. Since f lies in $L^{p,q} \subsetneq L^{p,\infty}$ we have $\mu(\{f > \varepsilon\})^{1/p}\varepsilon \le \|f\|_{L^{p,q}} < \infty$ for every $\varepsilon > 0$ and consequently for any $A > 0$, $\mu(\{f > A\})$ is finite and tends to zero as $A \to \infty$. Thus for every $n = 1,2,3,\dots$, there is an $A_n > 0$ such that $\mu(\{f > A_n\}) < 2^{-n}$.

For each $n = 1,2,3,\dots$ define the function

$$\varphi_n(x) = \sum_{k=0}^{1+2^n A_n} \frac{k}{2^n} \chi_{\{k2^{-n} < f \le (k+1)2^{-n}\}} \chi_{\{2^{-n} < f \le A_n\}}.$$

Then φ_n is supported in the set $\{2^{-n} < f \le A_n\}$ which has finite μ measure, thus φ_n is finitely simple and satisfies

$$f(x) - 2^{-n} \le \varphi_n(x) \le f(x)$$

for every $x \in \{x \in X : 2^{-n} < f(x) \le A_n\}$. It follows that

$$\mu(\{x \in X : |f(x) - \varphi_n(x)| > 2^{-n}\}) < 2^{-n}$$

which implies that $(f - \varphi_n)^*(t) \le 2^{-n}$ for $t \ge 2^{-n}$. Thus

$$(f - \varphi_n)^*(t) \to 0 \quad \text{as } n \to \infty \text{ and } \quad \varphi_n^*(t) \le f^*(t) \quad \text{for all } t > 0.$$

Since $(f - \varphi_n)^*(t) \le f^*(t)$, an application of the Lebesgue dominated convergence theorem gives that $\|\varphi_n - f\|_{L^{p,q}} \to 0$ as $n \to \infty$. □

Remark 1.4.14. One may wonder whether simple functions are dense in $L^{p,\infty}$. This turns out to be false for all $0 < p \le \infty$. However, countable linear combinations of characteristic functions of sets with finite measure are dense in $L^{p,\infty}(X,\mu)$. We call such functions *countably simple*. See Exercise 1.4.4 for details.

1.4.3 Duals of Lorentz Spaces

Given a quasi-Banach space Z with norm $\|\cdot\|_Z$, its dual Z^* is defined as the space of all continuous linear functionals T on Z equipped with the norm

$$\big\|T\big\|_{Z^*} = \sup_{\|x\|_Z=1} |T(x)|.$$

Observe that the dual of a quasi-Banach space is always a Banach space.

We are now considering the following question: What are the dual spaces $(L^{p,q})^*$ of $L^{p,q}$? The answer to this question presents some technical difficulties for general measure spaces. In this exposition we restrict our attention to σ-finite nonatomic measure spaces, where the situation is simpler.

Definition 1.4.15. A measurable subset A of a measure space (X,μ) is called an *atom* if $\mu(A) > 0$ and every measurable subset B of A has measure either equal to zero or equal to $\mu(A)$. A measure space (X,μ) is called *nonatomic* if it contains no atoms. In other words, X is nonatomic if and only if for any $A \subseteq X$ with $\mu(A) > 0$, there exists a proper subset $B \subsetneqq A$ with $\mu(B) > 0$ and $\mu(A \setminus B) > 0$.

For instance, $\mathbf{R}$ with Lebesgue measure is nonatomic, but any measure space with counting measure is atomic. Nonatomic spaces have the property that every measurable subset of them with strictly positive measure contains subsets of any given measure smaller than the measure of the original subset. See Exercise 1.4.5.

Theorem 1.4.16. *Suppose that (X,μ) is a nonatomic σ-finite measure space. Then*

(i)	$(L^{p,q})^* = \{0\}$,	*when* $0<p<1,\ 0<q\le\infty$,
(ii)	$(L^{p,q})^* = L^\infty$,	*when* $p=1,\ 0<q\le 1$,
(iii)	$(L^{p,q})^* = \{0\}$,	*when* $p=1,\ 1<q<\infty$,
(iv)	$(L^{p,q})^* \neq \{0\}$,	*when* $p=1,\ q=\infty$,
(v)	$(L^{p,q})^* = L^{p',\infty}$,	*when* $1<p<\infty,\ 0<q\le 1$,
(vi)	$(L^{p,q})^* = L^{p',q'}$,	*when* $1<p<\infty,\ 1<q<\infty$,
(vii)	$(L^{p,q})^* \neq \{0\}$,	*when* $1<p<\infty,\ q=\infty$,
(viii)	$(L^{p,q})^* \neq \{0\}$,	*when* $p=q=\infty$.

Proof. Since X is σ-finite, we have $X=\bigcup_{N=1}^\infty K_N$, where K_N is an increasing sequence of sets with $\mu(K_N)<\infty$. Let $\mathscr{A}$ be the σ-algebra on which μ is defined and define $\mathscr{A}_N=\{A\cap K_N : A\in\mathscr{A}\}$. Given $T\in(L^{p,q})^*$, where $0<p,q<\infty$, for each $N=1,2,\dots$, consider the measure $\sigma_N(E)=T(\chi_E)$ defined on $\mathscr{A}_N$. Since σ_N satisfies $|\sigma_N(E)|\le(p/q)^{1/q}\|T\|\mu(E)^{1/p}$, it follows that σ_N is absolutely continuous with respect to μ restricted on $\mathscr{A}_N$. By the Radon–Nikodym theorem (see [153] (19.36)), there exists a unique (up to a set of μ-measure zero) complex-valued measurable function g_N which satisfies $\int_{K_N}|g_N|\,d\mu<\infty$ such that

$$\int_{K_N} f\,d\sigma_N=\int_{K_N} g_N f\,d\mu \tag{1.4.12}$$

for all f in $L^1(K_N,\mathscr{A}_N,\sigma_N)$. Since $\sigma_N=\sigma_{N+1}$ on $\mathscr{A}_N$, it follows that $g_N=g_{N+1}$ μ-a.e. on K_N and hence there is a well-defined measurable function g on X that coincides with each g_N on K_N. But the linear functionals $f\mapsto T(f)$ and $f\mapsto\int_{K_N} f\,d\sigma_N$ coincide on simple functions supported in K_N and therefore they must be equal on $L^1(K_N,\mathscr{A}_N,\sigma_N)\cap L^{p,q}(X,\mu)$ by density; consequently, (1.4.12) is also equal to $T(f)$ for f in $L^1(K_N,\mathscr{A}_N,\sigma_N)\cap L^{p,q}(X,\mu)$.

Note that if $f\in L^\infty(K_N,\mu)$, then $f\in L^{p,q}(K_N,\mu)$ and also in $L^\infty(K_N,\sigma_N)$, which is contained in $L^1(K_N,\mathscr{A}_N,\sigma_N)$. It follows from (1.4.12) and the preceding discussion that

$$T(f)=\int_X g f\,d\mu \tag{1.4.13}$$

for every $f\in L^\infty(K_N)$. We have now proved that for every linear functional T on $L^{p,q}(X,\mu)$ with $0<p,q<\infty$ there is a function g satisfying $\int_{K_N}|g|\,d\mu<\infty$ for all $N=1,2,\dots$ such that (1.4.13) holds for all $f\in L^\infty(K_N)$.

We now examine each case (i)–(viii) separately.

(i) We consider the case $0<p<1$. Let $f=\sum_n a_n\chi_{E_n}$ be a finitely simple function on X (which is taken to be countably simple when $q=\infty$). Since X is nonatomic, we split each E_n as a union of m disjoint sets $E_{j,n}$, $j=1,2,\dots,m$, each having measure $m^{-1}\mu(E_n)$. Let $f_j=\sum_n a_n\chi_{E_{j,n}}$. We see that $\|f_j\|_{L^{p,q}}=m^{-1/p}\|f\|_{L^{p,q}}$. Now if $T\in(L^{p,q})^*$, it follows that

$$|T(f)| \le \sum_{j=1}^{m} |T(f_j)| \le \|T\| \sum_{j=1}^{m} \|f_j\|_{L^{p,q}} = \|T\| m^{1-1/p} \|f\|_{L^{p,q}}.$$

Let $m \to \infty$ and use that $p < 1$ to obtain that $T = 0$.

(ii) We now consider the case $p = 1$ and $0 < q \le 1$. Clearly, every $h \in L^\infty$ gives a bounded linear functional on $L^{1,q}$, since

$$\left| \int_X f h \, d\mu \right| \le \|h\|_{L^\infty} \|f\|_{L^1} \le C_q \|h\|_{L^\infty} \|f\|_{L^{1,q}}.$$

Conversely, suppose that $T \in (L^{1,q})^*$ where $q \le 1$. The function g given in (1.4.13) satisfies

$$\left| \int_E g \, d\mu \right| = |T(\chi_E)| \le \|T\| q^{-1/q} \mu(E)$$

for all $E \subsetneqq K_N$, and hence $|g| \le q^{-1/q} \|T\|$ μ-a.e. on every K_N; see [307, Theorem 1.40 on p. 31] for a proof of this fact. It follows that $\|g\|_{L^\infty} \le q^{-1/q} \|T\|$ and hence $(L^{1,q})^* = L^\infty$.

(iii) Let us now take $p = 1$, $1 < q < \infty$, and suppose that $T \in (L^{1,q})^*$. Then

$$\left| \int_X f g \, d\mu \right| \le \|T\| \, \|f\|_{L^{1,q}}, \tag{1.4.14}$$

where g is the function in (1.4.13) and $f \in L^\infty(K_N)$. We will show that $g = 0$ a.e. Suppose that $|g| \ge \delta$ on some set E_0 with $\mu(E_0) > 0$. Then there exists N such that $\mu(E_0 \cap K_N) > 0$. Let $f = \overline{g}|g|^{-2} \chi_{E_0 \cap K_N} h \chi_{h \le M}$, where h is a nonnegative function. Then (1.4.14) implies for all $h \ge 0$ that

$$\|h\chi_{h\le M}\|_{L^1(E_0 \cap K_N)} \le \|T\| \, \|h\chi_{h\le M}\|_{L^{1,q}(E_0 \cap K_N)}.$$

Letting $M \to \infty$, we obtain that $L^{1,q}(E_0 \cap K_N)$ is contained in $L^1(E_0 \cap K_N)$, but since the reverse inclusion is always valid, these spaces must be equal. Since X is nonatomic, this can't happen; see Exercise 1.4.8 (d). Thus $g = 0$ μ-a.e. and $T = 0$.

(iv) In the case $p = 1$, $q = \infty$ an interesting phenomenon appears. Since every continuous linear functional on $L^{1,\infty}$ extends to a continuous linear functional on $L^{1,q}$ for $1 < q < \infty$, it must necessarily vanish on all simple functions by part (iii). However, $(L^{1,\infty})^*$ contains nontrivial linear functionals; see [84], [85].

(v) We now take up the case $1 < p < \infty$ and $0 < q \le 1$. Using Exercise 1.4.1 (b) and Proposition 1.4.10, we see that if $f \in L^{p,q}$ and $h \in L^{p',\infty}$, then

$$\begin{aligned}
\int_X |fh| \, d\mu &\le \int_0^\infty t^{\frac{1}{p}} f^*(t) t^{\frac{1}{p'}} h^*(t) \frac{dt}{t} \\
&\le \|f\|_{L^{p,1}} \|h\|_{L^{p',\infty}} \\
&\le C_{p,q} \|f\|_{L^{p,q}} \|h\|_{L^{p',\infty}};
\end{aligned}$$

thus every $h \in L^{p',\infty}$ gives rise to a bounded linear functional $f \mapsto \int h f \, d\mu$ on $L^{p,q}$ with norm at most $C_{p,q}\|h\|_{L^{p',\infty}}$. Conversely, let $T \in (L^{p,q})^*$ where $1 < p < \infty$ and $0 < q \le 1$. Let g satisfy (1.4.13) for all $f \in L^\infty(K_N)$. Taking $f = \overline{g}|g|^{-1}\chi_{K_N \cap \{|g|>\alpha\}}$ for $\alpha > 0$ and using that

$$\left| \int_X f g \, d\mu \right| \le \|T\| \|f\|_{L^{p,q}} ,$$

we obtain that

$$\alpha \mu(K_N \cap \{|g| > \alpha\}) \le (p/q)^{1/q} \|T\| \, \mu(K_N \cap \{|g| > \alpha\})^{\frac{1}{p}} .$$

Divide by $\mu(K_N \cap \{|g| > \alpha\})^{\frac{1}{p}}$, let $N \to \infty$, and take the supremum over $\alpha > 0$ to obtain that $\|g\|_{L^{p',\infty}} \le (p/q)^{1/q}\|T\|$.

(vi) Using Exercise 1.4.1 (b) and Hölder's inequality, we obtain

$$\left| \int_X f \varphi \, d\mu \right| \le \int_0^\infty t^{\frac{1}{p}} f^*(t) t^{\frac{1}{p'}} \varphi^*(t) \frac{dt}{t} \le \|f\|_{L^{p,q}} \|\varphi\|_{L^{p',q'}} ;$$

thus every $\varphi \in L^{p',q'}$ gives a bounded linear functional on $L^{p,q}$ with norm at most $\|\varphi\|_{L^{p',q'}}$. Conversely, let T be in $(L^{p,q})^*$. By (1.4.13), T is given by integration against a locally integrable function g. It remains to prove that $g \in L^{p',q'}$. We let $g_{N,M} = g \chi_{K_N} \chi_{|g| \le M}$. Then $(g_{N,M})^* \le g^*$ for all $M, N = 1, 2, \dots$ and $(g_{N,M})^* \uparrow g^*$ as $M, N \to \infty$ by Proposition 1.4.5 (4), (8).

For a bounded function f in $L^{p,q}(X)$ we have

$$\begin{aligned}
\int_0^\infty f^*(t)(g_{N,M})^*(t)\, dt &= \sup_{h:\, d_h = d_f} \left| \int_X h g_{N,M} \, d\mu \right| \\
&= \sup_{h:\, d_h = d_f} \left| \int_{K_N} h \chi_{|g| \le M} \, g \, d\mu \right| \\
&= \sup_{h:\, d_h = d_f} \left| T(h \chi_{K_N} \chi_{|g| \le M}) \right| \\
&\le \sup_{h:\, d_h = d_f} \|T\| \, \|h \chi_{K_N} \chi_{|g| \le M}\|_{L^{p,q}} \\
&\le \sup_{h:\, d_h = d_f} \|T\| \, \|h\|_{L^{p,q}} \\
&= \|T\| \, \|f\|_{L^{p,q}} ,
\end{aligned} \tag{1.4.15}$$

where the first equality is a consequence of the fact that X is nonatomic (see Exercise 1.4.5 (d)). Using the result of Exercise 1.4.5 (b), pick a function f on X such that

$$f^*(t) = \int_{t/2}^\infty s^{\frac{q'}{p'} - 1} (g_{N,M})^*(s)^{q'-1} \frac{ds}{s} , \tag{1.4.16}$$

noting that the preceding integral converges since $(g_{N,M})^*(s) \le M\chi_{[0,\mu(K_N)]}(s)$. It follows that $f^* \le c_{p,q} M^{q'-1}$, which implies that f is bounded, and also that $f^*(t)=0$ when $t > 2\mu(K_N)$, which implies that f is supported in a set of measure at most $2\mu(K_N)$; thus the function f defined in (1.4.16) is bounded and lies in $L^{p,q}(X)$.

We have the following calculation regarding the $L^{p,q}$ norm of f:

$$\begin{aligned} \|f\|_{L^{p,q}} &= \left(\int_0^\infty t^{\frac{q}{p}} \left[\int_{t/2}^\infty s^{\frac{q'}{p'}-1}(g_{N,M})^*(s)^{q'-1}\frac{ds}{s}\right]^q \frac{dt}{t}\right)^{\frac{1}{q}} \\ &\le C_1(p,q)\left(\int_0^\infty (t^{\frac{1}{p'}}(g_{N,M})^*(t))^{q'}\frac{dt}{t}\right)^{\frac{1}{q}} \\ &= C_1(p,q)\|g_{N,M}\|_{L^{p',q'}}^{q'/q} < \infty, \end{aligned} \tag{1.4.17}$$

which is a consequence of Hardy's second inequality in Exercise 1.2.8 with $b = q/p$.

Using (1.4.15) and (1.4.17) we deduce that

$$\int_0^\infty f^*(t)(g_{N,M})^*(t)\,dt \le \|T\|\,\|f\|_{L^{p,q}} \le C_1(p,q)\|T\|\,\|g_{N,M}\|_{L^{p',q'}}^{q'-1}. \tag{1.4.18}$$

On the other hand, we have

$$\begin{aligned} \int_0^\infty f^*(t)(g_{N,M})^*(t)\,dt &\ge \int_0^\infty \int_{t/2}^t s^{\frac{q'}{p'}-1}(g_{N,M})^*(s)^{q'-1}\frac{ds}{s}(g_{N,M})^*(t)\,dt \\ &\ge \int_0^\infty (g_{N,M})^*(t)^{q'}\int_{t/2}^t s^{\frac{q'}{p'}-1}\frac{ds}{s}\,dt \\ &= C_2(p,q)\|g_{N,M}\|_{L^{p',q'}}^{q'}. \end{aligned} \tag{1.4.19}$$

Combining (1.4.18) and (1.4.19), and using the fact that $\|g_{N,M}\|_{L^{p',q'}} < \infty$, we obtain $\|g_{N,M}\|_{L^{p',q'}} \le C(p,q)\|T\|$. Letting $N,M\to\infty$ we deduce $\|g\|_{L^{p',q'}} \le C(p,q)\|T\|$ and this proves the reverse inequality required to complete case (vi).

(vii) For a complete characterization of this space, we refer to [83].

(viii) The dual of $L^\infty = L^{\infty,\infty}$ can be identified with the set of all bounded finitely additive set functions; see [99]. □

Remark 1.4.17. Some parts of Theorem 1.4.16 are false if X is atomic. For instance, the dual of $\ell^p(\mathbf{Z})$ contains ℓ^∞ when $0<p<1$ and thus it is not equal to $\{0\}$.

1.4.4 The Off-Diagonal Marcinkiewicz Interpolation Theorem

We now present the main result of this section, the off-diagonal extension of Marcinkiewicz's interpolation theorem (Theorem 1.3.2). For a measure space (X,μ), let $S(X)$ be the space of finitely simple functions on X and $S_0^+(X)$ be the subset of $S(X)$ of functions of the form

$$\sum_{i=m}^{n} 2^{-i}\chi_{A_i}$$

where $m < n$ are integers and A_i are subsets of X of finite measure. The sets A_i are not required to be different nor disjoint; consequently, the sum of two elements in $S_0^+(X)$ also belong to $S_0^+(X)$. We define $S_0^{real}(X) = S_0^+(X) - S_0^+(X)$ be the space of all functions of the form $f_1 - f_2$, where f_1, f_2 lie in $S_0^+(X)$ and $S_0(X)$ be the space of functions of the form $h_1 + ih_2$, where h_1, h_2 lie in $S_0^{real}(X)$.

An operator T defined on $S_0(X)$ is called *quasi-linear* if there is a $K \geq 1$ such that

$$|T(\lambda f)| = |\lambda|\,|T(f)| \qquad \text{and} \qquad |T(f+g)| \leq K(|T(f)| + |T(g)|),$$

for all $\lambda \in \mathbf{C}$ and all functions f, g in $S_0(X)$. If $K = 1$, then T is called *sublinear*.

Definition 1.4.18. Let T be a linear operator defined on the space of finitely simple functions $S(X)$ on a measure space (X, μ) and let $0 < p, q \leq \infty$. We say that T is of *restricted weak type* (p,q) if

$$\big\|T(\chi_A)\big\|_{L^{q,\infty}} \leq C\mu(A)^{1/p} \tag{1.4.20}$$

for all measurable subsets A of X with finite measure. Estimates of the form (1.4.20) are called *restricted weak type* estimates.

It is important to observe that if an operator is of restricted weak type (p_0, q_0) and of restricted weak type (p_1, q_1), then it is of restricted weak type (p, q), where the indices are as in (1.4.23). It will be a considerable effort to extend the latter estimate to all functions in $S_0(X)$. The next theorem addresses this extension.

Theorem 1.4.19. *Let $0 < r \leq \infty$, $0 < p_0 \neq p_1 \leq \infty$, and $0 < q_0 \neq q_1 \leq \infty$ and let (X, μ), (Y, ν) be σ-finite measure spaces. Let T be a quasi-linear operator defined on the space of simple functions on X and taking values in the set of measurable functions on Y. Assume that for some $M_0, M_1 < \infty$ the following restricted weak type estimates hold:*

$$\big\|T(\chi_A)\big\|_{L^{q_0,\infty}} \leq M_0\,\mu(A)^{1/p_0}, \tag{1.4.21}$$
$$\big\|T(\chi_A)\big\|_{L^{q_1,\infty}} \leq M_1\,\mu(A)^{1/p_1}, \tag{1.4.22}$$

for all measurable subsets A of X with $\mu(A) < \infty$. Fix $0 < \theta < 1$ and let

$$\frac{1}{p} = \frac{1-\theta}{p_0} + \frac{\theta}{p_1} \qquad \text{and} \qquad \frac{1}{q} = \frac{1-\theta}{q_0} + \frac{\theta}{q_1}. \tag{1.4.23}$$

Then there exists a constant $C_*(p_0, q_0, p_1, q_1, K, r, \theta) < \infty$ such that for all functions f in $S_0(X)$ we have

$$\big\|T(f)\big\|_{L^{q,r}} \leq C_*(p_0, q_0, p_1, q_1, K, r, \theta) M_0^{1-\theta} M_1^{\theta} \big\|f\big\|_{L^{p,r}}. \tag{1.4.24}$$

Additionally, if $0 < p, r < \infty$ and if T is linear (or sublinear with nonnegative values), then it admits a unique bounded extension from $L^{p,r}(X)$ to $L^{q,r}(Y,\nu)$ such that (1.4.24) holds for all f in $L^{p,r}$.

Before we give the proof of Theorem 1.4.19, we state and prove a lemma that is interesting on its own.

Lemma 1.4.20. *Let $0 < p < \infty$ and $0 < q \le \infty$ and let (X,μ), (Y,ν) be σ-finite measure spaces. Let T be a quasi-linear operator defined on $S(X)$ and taking values in the set of measurable functions on Y. Suppose that there exists a constant $M > 0$ such that for all measurable subsets A of X of finite measure we have*

$$\big\|T(\chi_A)\big\|_{L^{q,\infty}} \le M\,\mu(A)^{\frac{1}{p}}. \tag{1.4.25}$$

Then for all α with $0 < \alpha < \min(q, \frac{\log 2}{\log 2K})$ there exists a constant $C(p,q,K,\alpha) > 0$ such that for all functions f in $S_0(X)$ we have the estimate

$$\big\|T(f)\big\|_{L^{q,\infty}} \le C(p,q,K,\alpha) M \big\|f\big\|_{L^{p,\alpha}} \tag{1.4.26}$$

where

$$C(p,q,K,\alpha) = 2^{8+\frac{2}{p}+\frac{2}{q}} K^3 \left(\frac{q}{q-\alpha}\right)^{\frac{2}{\alpha}} (1-2^{-\alpha})^{-\frac{1}{\alpha}} (\log 2)^{-\frac{1}{\alpha}}.$$

Proof. A function f in $S_0(X)$ can be written as $f = h_1 - h_2 + i(h_3 - h_4)$, where h_j are in $S_0^+(X)$. We write $f = f_1 - f_2 + i(f_3 - f_4)$, where $f_1 = \max(h_1 - h_2, 0)$, $f_2 = \max(-(h_1 - h_2), 0)$, $f_3 = \max(h_3 - h_4, 0)$, and $f_4 = \max(-(h_3 - h_4), 0)$. We note that f_j lie in $S_0^+(X)$; indeed, if $h_1 = \sum_\ell 2^{-\ell}\chi_{A_\ell}$ and $h_2 = \sum_k 2^{-k}\chi_{B_k}$, where both sums are finite, then

$$f_1 = \sum_{\ell:\, A_\ell \cap (\cup_k B_k) = \emptyset} 2^{-\ell}\chi_{A_\ell} + \sum_{(\ell,k):\, \ell<k,\, A_\ell\cap B_k \ne \emptyset} (2^{-\ell} - 2^{-k})\chi_{A_\ell \cap B_k}.$$

Since the second sum is equal to $\sum_{s=\ell+1}^{k} 2^{-s}\chi_{A_\ell\cap B_k}$, we obtain that $h_1 \in S_0^+(X)$. Likewise we can show that f_2, f_3, f_4 lie in $S_0^+(X)$. Moreover, we have $f_j \le |f|$ and Proposition 1.4.5(4), yields

$$\|f_j\|_{L^{p,\alpha}(X)} \le \|f\|_{L^{p,\alpha}(X)}$$

for all $j = 1,2,3,4$. Suppose now that (1.4.26) holds for functions in $S_0^+(X)$ with constant $C'(p,q,\alpha)$ in place of $C(p,q,K,\alpha)$. By the quasi-linearity of T we have

$$
\begin{aligned}
\|T(f)\|_{L^{q,\infty}(Y)} &\le K^3 \left\| \sum_{j=1}^{4} |T(f_j)| \right\|_{L^{q,\infty}(Y)} \\
&\le K^3 4^{1+\frac{1}{q}} \sum_{j=1}^{4} \|T(f_j)\|_{L^{q,\infty}(Y)} \\
&\le K^3 4^{2+\frac{2}{q}} M C'(p,q,K,\alpha) \|f\|_{L^{p,\alpha}(X)}
\end{aligned}
$$

which proves (1.4.26) for all f in $S_0(X)$ with constant

$$
C(p,q,K,\alpha) = 2^{4+\frac{2}{q}} K^3 C'(p,q,\alpha).
$$

We now prove (1.4.26) for functions in $S_0^+(X)$ with constant $C'(p,q,\alpha)$ in place of $C(p,q,K,\alpha)$. It follows from the Aoki–Rolewicz theorem (Exercise 1.4.6) that for all $N \in \mathbf{Z}^+$ and for all $f_1,\dots,f_N$ in $S_0^+(X)$ we have the pointwise inequality

$$
|T(f_1+\dots+f_N)| \le 4\left(\sum_{j=1}^{N} |T(f_j)|^{\alpha_1}\right)^{\frac{1}{\alpha_1}} \le 4\left(\sum_{j=1}^{N} |T(f_j)|^{\alpha}\right)^{\frac{1}{\alpha}}, \tag{1.4.27}
$$

where $0<\alpha\le\alpha_1$ and α_1 satisfies the equation $(2K)^{\alpha_1}=2$. The second inequality in (1.4.27) is a simple consequence of the fact that $\alpha\le\alpha_1$. Fix α_0 with

$$
0<\alpha_0\le\alpha_1=\frac{\log 2}{\log 2K} \qquad \text{and} \qquad \alpha_0<q.
$$

This ensures that the quasi-normed space $L^{q/\alpha,\infty}$ is normable when $\alpha\le\alpha_0$. In fact, since Y is σ-finite, Exercise 1.1.12 gives that the space $L^{s,\infty}$ is normable as long as $s>1$ and there is an equivalent norm $|||f|||_{L^{s,\infty}}$ such that

$$
\|f\|_{L^{s,\infty}} \le |||f|||_{L^{s,\infty}} \le \frac{s}{s-1}\|f\|_{L^{s,\infty}}.
$$

Next we claim that for any nonnegative function f in $S_0^+(X)$ we have

$$
\big\|T(f\chi_A)\big\|_{L^{q,\infty}} \le 4\left(\frac{q}{q-\alpha}\right)^{\frac{1}{\alpha}} (1-2^{-\alpha})^{-\frac{1}{\alpha}} M\mu(A)^{\frac{1}{p}} \big\|f\chi_A\big\|_{L^\infty}. \tag{1.4.28}
$$

To show this, we write $f=\sum_{j=m}^{n} 2^{-j}\chi_{S_j}$, where $m<n$ are integers, S_j are subsets of X of finite measure for all $j\in\{m,m+1,\dots,n\}$, $\mu(S_m)\neq 0$ and $\mu(S_n)\neq 0$. Setting $B_j=S_j\cap A$ we have

$$
f\chi_A = \sum_{j=m}^{n} 2^{-j}\chi_{B_j}
$$

and $2^{-m}\le\|f\chi_A\|_{L^\infty(X)}$.

We use (1.4.27) once and (1.4.3) twice in the following argument. We have

$$
\begin{aligned}
\big\|T(f\chi_A)\big\|_{L^{q,\infty}} &\le 4\Big\|\Big(\sum_{j=m}^{n} 2^{-j\alpha}|T(\chi_{B_j})|^{\alpha}\Big)^{\frac{1}{\alpha}}\Big\|_{L^{q,\infty}} \\
&= 4\Big\|\sum_{j=m}^{n} 2^{-j\alpha}|T(\chi_{B_j})|^{\alpha}\Big\|_{L^{q/\alpha,\infty}}^{\frac{1}{\alpha}} \\
&\le 4\,\Big|\Big|\Big|\sum_{j=m}^{n} 2^{-j\alpha}|T(\chi_{B_j})|^{\alpha}\Big|\Big|\Big|_{L^{q/\alpha,\infty}}^{\frac{1}{\alpha}} \\
&\le 4\Big(\sum_{j=m}^{n} 2^{-j\alpha}\Big|\Big|\Big||T(\chi_{B_j})|^{\alpha}\Big|\Big|\Big|_{L^{q/\alpha,\infty}}\Big)^{\frac{1}{\alpha}} \\
&\le 4\Big(\frac{q}{q-\alpha}\Big)^{\frac{1}{\alpha}}\Big(\sum_{j=m}^{n} 2^{-j\alpha}\Big\||T(\chi_{B_j})|^{\alpha}\Big\|_{L^{q/\alpha,\infty}}\Big)^{\frac{1}{\alpha}} \\
&= 4\Big(\frac{q}{q-\alpha}\Big)^{\frac{1}{\alpha}}\Big(\sum_{j=m}^{n} 2^{-j\alpha}\big\|T(\chi_{B_j})\big\|_{L^{q,\infty}}^{\alpha}\Big)^{\frac{1}{\alpha}} \\
&\le 4\Big(\frac{q}{q-\alpha}\Big)^{\frac{1}{\alpha}} M\Big(\sum_{j=m}^{n} 2^{-j\alpha}\mu(B_j)^{\frac{\alpha}{p}}\Big)^{\frac{1}{\alpha}} \\
&\le 4\Big(\frac{q}{q-\alpha}\Big)^{\frac{1}{\alpha}}(1-2^{-\alpha})^{-\frac{1}{\alpha}} M\,\mu(A)^{\frac{1}{p}}2^{-m},
\end{aligned}
$$

using $B_j \subseteqq A$. Using that $2^{-m} \le \|f\chi_A\|_{L^\infty}$ establishes (1.4.28).

We now apply (1.4.28) to obtain (1.4.26). For any $f \in S_0^+(X)$ we define measurable sets

$$
A_k = \{x \in X:\ f^*(2^{k+1}) < |f(x)| \le f^*(2^k)\} \tag{1.4.29}
$$

and we note that these sets are pairwise disjoint. We may write the finitely simple function f as $\sum_{j=1}^n a_j\chi_{E_j}$, where $0 < a_j < \infty$, $E_1 \subseteqq E_2 \subseteqq \cdots \subseteqq E_n$ and $0 < \mu(E_j) < \infty$ for $j \in \{1,2,\dots,n\}$. Clearly, we have

$$
f^* = \sum_{j=1}^{n} a_j \chi_{[0,\mu(E_j))}.
$$

Thus, when $t \in (\mu(E_n),\infty)$, $f^*(t)$ vanishes, and when $t \in (0,\mu(E_1))$, $f^*(t) = \sum_{j=1}^n a_j$ is a positive constant. So there exists $N \in \mathbf{Z}^+$ such that $f^*(2^k) = 0$ when $k > N$, and that $f^*(2^k)$ is a positive constant when $k < -N$. This also implies that $A_k = \emptyset$ if $|k| > N$ and thus we express

$$
f = \sum_{k=-N}^{N} f\chi_{A_k}.
$$

Proposition 1.4.5(2) implies $\mu(A_k) \le d_f(f^*(2^{k+1})) \le 2^{k+1}$. Using (1.4.27) we obtain

$$\begin{aligned}
\big\|T(f)\big\|_{L^{q,\infty}(Y)} &\le 4\bigg\|\Big(\sum_{k=-N}^{N} |T(f\chi_{A_k})|^\alpha\Big)^{\frac{1}{\alpha}}\bigg\|_{L^{q,\infty}(Y)} \\
&= 4\bigg\|\sum_{k=-N}^{N} |T(f\chi_{A_k})|^\alpha\bigg\|_{L^{q/\alpha,\infty}(Y)}^{\frac{1}{\alpha}} \\
&\le 4\,\bigg|\!\bigg|\!\bigg|\sum_{k=-N}^{N} |T(f\chi_{A_k})|^\alpha\bigg|\!\bigg|\!\bigg|_{L^{q/\alpha,\infty}(Y)}^{\frac{1}{\alpha}} \\
&\le 4\bigg(\sum_{k=-N}^{N} \Big|\!\Big|\!\Big| |T(f\chi_{A_k})|^\alpha \Big|\!\Big|\!\Big|_{L^{q/\alpha,\infty}(Y)}\bigg)^{\frac{1}{\alpha}} \\
&\le 4\Big(\frac{q}{q-\alpha}\Big)^{\frac{1}{\alpha}}\bigg(\sum_{k=-N}^{N} \Big\||T(f\chi_{A_k})|^\alpha\Big\|_{L^{q/\alpha,\infty}(Y)}\bigg)^{\frac{1}{\alpha}} \\
&\le 4\Big(\frac{q}{q-\alpha}\Big)^{\frac{1}{\alpha}}\bigg(\sum_{k=-N}^{N} \|T(f\chi_{A_k})\|_{L^{q,\infty}(Y)}^\alpha\bigg)^{\frac{1}{\alpha}} \\
&\le 16\Big(\frac{q}{q-\alpha}\Big)^{\frac{2}{\alpha}}(1-2^{-\alpha})^{-\frac{1}{\alpha}}M\bigg(\sum_{k=-N}^{N} \mu(A_k)^{\frac{\alpha}{p}}\|f\chi_{A_k}\|_{L^\infty}^\alpha\bigg)^{\frac{1}{\alpha}} \\
&\le 16\Big(\frac{q}{q-\alpha}\Big)^{\frac{2}{\alpha}}(1-2^{-\alpha})^{-\frac{1}{\alpha}}2^{\frac{1}{p}}M\bigg(\sum_{k=-\infty}^{\infty} [f^*(2^k)]^\alpha 2^{\frac{k\alpha}{p}}\bigg)^{\frac{1}{\alpha}} \\
&\le 16\Big(\frac{q}{q-\alpha}\Big)^{\frac{2}{\alpha}}(1-2^{-\alpha})^{-\frac{1}{\alpha}}2^{\frac{2}{p}}(\log 2)^{-\frac{1}{\alpha}}M\|f\|_{L^{p,\alpha}(X)},
\end{aligned}$$

where we made use of (1.4.28) and in the last inequality, we used

$$\begin{aligned}
\|f\|_{L^{p,\alpha}(X)}^\alpha &= \sum_{k=-\infty}^{\infty}\int_{2^{k-1}}^{2^k} t^{\frac{\alpha}{p}}[f^*(t)]^\alpha\frac{dt}{t} \\
&\ge \sum_{k=-\infty}^{\infty}(2^{k-1})^{\frac{\alpha}{p}}[f^*(2^k)]^\alpha\int_{2^{k-1}}^{2^k}\frac{dt}{t} \\
&= 2^{-\frac{\alpha}{p}}\log 2\sum_{k=-\infty}^{\infty}[f^*(2^k)]^\alpha 2^{\frac{k\alpha}{p}}.
\end{aligned}$$

This completes the proof of the required inequality for nonnegative functions in $S_0^+(X)$ with constant

$$C'(p,q,\alpha) = 16\Big(\frac{q}{q-\alpha}\Big)^{\frac{2}{\alpha}}(1-2^{-\alpha})^{-\frac{1}{\alpha}}2^{\frac{2}{p}}(\log 2)^{-\frac{1}{\alpha}}$$

As noted, the constant in general is $C(p,q,K,\alpha) = 2^{4+\frac{2}{p}}K^3C'(p,q,\alpha)$. □

We now proceed with the proof of Theorem 1.4.19.

Proof. We assume that $p_0 < p_1$, since if $p_0 > p_1$ we may simply reverse the roles of p_0 and p_1. We first consider the case $p_1, r < \infty$. Lemma 1.4.20 implies that

$$\begin{aligned} \big\|T(f)\big\|_{L^{q_0,\infty}} &\le M_0' \big\|f\big\|_{L^{p_0,m}}, \\ \big\|T(f)\big\|_{L^{q_1,\infty}} &\le M_1' \big\|f\big\|_{L^{p_1,m}}, \end{aligned} \tag{1.4.30}$$

for all f in $S_0(X)$, where $m = \frac{1}{2}\min\big(q_0, q_1, \frac{\log 2}{\log 2K}, 2r\big)$, $M_0' = C(p_0, q_0, K, m) M_0$, $M_1' = C(p_1, q_1, K, m) M_1$, and $C(p, q, K, \alpha)$ is as in (1.4.26).

Fix a function f in $S_0(X)$. Split $f = f^t + f_t$ as follows:

$$f^t(x) = \begin{cases} f(x) & \text{if } |f(x)| > f^*(\delta t^\gamma), \\ 0 & \text{if } |f(x)| \le f^*(\delta t^\gamma), \end{cases}$$

$$f_t(x) = \begin{cases} 0 & \text{if } |f(x)| > f^*(\delta t^\gamma), \\ f(x) & \text{if } |f(x)| \le f^*(\delta t^\gamma), \end{cases}$$

where δ is to be determined later and γ is the following nonzero real number:

$$\gamma = \frac{\frac{1}{q_0} - \frac{1}{q}}{\frac{1}{p_0} - \frac{1}{p}} = \frac{\frac{1}{q} - \frac{1}{q_1}}{\frac{1}{p} - \frac{1}{p_1}}.$$

Using Exercise 1.1.10 we write

$$d_{f^t}(v) = \begin{cases} d_f(v) & \text{when } v > f^*(\delta t^\gamma) \\ d_f(f^*(\delta t^\gamma)) & \text{when } v \le f^*(\delta t^\gamma) \end{cases}$$

$$d_{f_t}(v) = \begin{cases} 0 & \text{when } v \ge f^*(\delta t^\gamma) \\ d_f(v) - d_f(f^*(\delta t^\gamma)) & \text{when } v < f^*(\delta t^\gamma). \end{cases}$$

Observe the following facts

$$\begin{aligned} v \ge \delta t^\gamma \implies (f^t)^*(v) &\le \inf\big\{s \in (0, f^*(\delta t^\gamma)] : d_{f^t}(s) \le v\big\} \\ &= \inf\big\{s \in (0, f^*(\delta t^\gamma)] : d_f(f^*(\delta t^\gamma)) \le v\big\} \\ &= \inf(0, f^*(\delta t^\gamma)] \\ &= 0, \end{aligned}$$

$$\begin{aligned} v < \delta t^\gamma \implies (f^t)^*(v) &\le \inf\big\{s > f^*(\delta t^\gamma) : d_{f^t}(s) \le v\big\} \\ &= \inf\big\{s > f^*(\delta t^\gamma) : d_f(s) \le v\big\} \\ &= \inf\Big\{\{s > 0 : d_f(s) \le v\} \cap \big(f^*(\delta t^\gamma), \infty\big)\Big\} \\ &= f^*(v), \qquad\qquad \text{since } f^*(v) \ge f^*(\delta t^\gamma), \end{aligned}$$

$$\begin{aligned} v \geq \delta t^{\gamma} \implies (f_t)^*(v) &= \inf\{s>0:\ d_{f_t}(s) \leq v\} \\ &\leq \inf\{s>0:\ d_f(s) \leq v\} && \text{since } d_{f_t} \leq d_f \\ &= f^*(v), \end{aligned}$$

$$\begin{aligned} v < \delta t^{\gamma} \implies (f_t)^*(v) &= \inf\{s>0:\ d_{f_t}(s) \leq v\} \\ &\leq f^*(\delta t^{\gamma}), \qquad \text{since } f^*(\delta t^{\gamma}) \in \{s>0:\ d_{f_t}(s) \leq v\}. \end{aligned}$$

We summarize these observations in a couple of inequalities:

$$(f^t)^*(s) \leq \begin{cases} f^*(s) & \text{if } 0 < s < \delta t^{\gamma}, \\ 0 & \text{if } s \geq \delta t^{\gamma}, \end{cases}$$

$$(f_t)^*(s) \leq \begin{cases} f^*(\delta t^{\gamma}) & \text{if } 0 < s < \delta t^{\gamma}, \\ f^*(s) & \text{if } s \geq \delta t^{\gamma}. \end{cases}$$

It follows from these inequalities that f^t lies in $L^{p_0,m}$ and f_t lies in $L^{p_1,m}$ for all $t>0$.

The quasi-linearity of the operator T and (1.4.9) imply

$$\begin{aligned} &\big\|T(f)\big\|_{L^{q,r}} \\ &= \big\|t^{\frac{1}{q}} T(f)^*(t)\big\|_{L^r(\frac{dt}{t})} \\ &\leq K\big\|t^{\frac{1}{q}}(|T(f_t)|+|T(f^t)|)^*(t)\big\|_{L^r(\frac{dt}{t})} \\ &\leq K\big\|t^{\frac{1}{q}}T(f_t)^*(\tfrac{t}{2})+t^{\frac{1}{q}}T(f^t)^*(\tfrac{t}{2})\big\|_{L^r(\frac{dt}{t})} \\ &\leq Ka_r\Big(\big\|t^{\frac{1}{q}}T(f_t)^*(\tfrac{t}{2})\big\|_{L^r(\frac{dt}{t})}+\big\|t^{\frac{1}{q}}T(f^t)^*(\tfrac{t}{2})\big\|_{L^r(\frac{dt}{t})}\Big) \\ &\leq K\max\{1,2^{\frac{1}{r}-1}\}\Big(\big\|t^{\frac{1}{q}}T(f_t)^*(\tfrac{t}{2})\big\|_{L^r(\frac{dt}{t})}+\big\|t^{\frac{1}{q}}T(f^t)^*(\tfrac{t}{2})\big\|_{L^r(\frac{dt}{t})}\Big). \end{aligned} \tag{1.4.31}$$

It follows from (1.4.30) that

$$t^{\frac{1}{q_0}}T(f^t)^*(\tfrac{t}{2}) \leq 2^{\frac{1}{q_0}}\sup_{s>0}s^{\frac{1}{q_0}}T(f^t)^*(s) \leq 2^{\frac{1}{q_0}}M_0'\big\|f^t\big\|_{L^{p_0,m}}, \tag{1.4.32}$$

$$t^{\frac{1}{q_1}}T(f_t)^*(\tfrac{t}{2}) \leq 2^{\frac{1}{q_1}}\sup_{s>0}s^{\frac{1}{q_1}}T(f_t)^*(s) \leq 2^{\frac{1}{q_1}}M_1'\big\|f_t\big\|_{L^{p_1,m}}, \tag{1.4.33}$$

for all $t>0$. Now use (1.4.32), (1.4.33), and the facts that

$$t^{\frac{1}{q}}T(f^t)^*(\tfrac{t}{2}) = t^{\frac{1}{q}-\frac{1}{q_0}}t^{\frac{1}{q_0}}T(f^t)^*(\tfrac{t}{2}) \leq t^{\frac{1}{q}-\frac{1}{q_0}}2^{\frac{1}{q_0}}M_0'\big\|f^t\big\|_{L^{p_0,m}}$$

$$t^{\frac{1}{q}}T(f^t)^*(\tfrac{t}{2}) = t^{\frac{1}{q}-\frac{1}{q_1}}t^{\frac{1}{q_1}}T(f^t)^*(\tfrac{t}{2}) \leq t^{\frac{1}{q}-\frac{1}{q_1}}2^{\frac{1}{q_1}}M_1'\big\|f^t\big\|_{L^{p_1,m}},$$

to estimate (1.4.31) by

$$K\max\{1,2^{\frac{1}{r}-1}\}\left[2^{\frac{1}{q_0}}M_0'\left\|t^{\frac{1}{q}-\frac{1}{q_0}}\|f^t\|_{L^{p_0,m}}\right\|_{L^r(\frac{dt}{t})}+2^{\frac{1}{q_1}}M_1'\left\|t^{\frac{1}{q}-\frac{1}{q_1}}\|f_t\|_{L^{p_1,m}}\right\|_{L^r(\frac{dt}{t})}\right],$$

which is the same as

$$K\max\{1,2^{\frac{1}{r}-1}\}2^{\frac{1}{q_0}}M_0'\left\|t^{-\gamma(\frac{1}{p_0}-\frac{1}{p})}\|f^t\|_{L^{p_0,m}}\right\|_{L^r(\frac{dt}{t})} \tag{1.4.34}$$

$$+K\max\{1,2^{\frac{1}{r}-1}\}2^{\frac{1}{q_1}}M_1'\left\|t^{\gamma(\frac{1}{p}-\frac{1}{p_1})}\|f_t\|_{L^{p_1,m}}\right\|_{L^r(\frac{dt}{t})}. \tag{1.4.35}$$

Next, we change variables $u=\delta t^\gamma$ in the L^r quasi-norm in (1.4.34) to obtain

$$\begin{aligned}
&\left\|t^{-\gamma(\frac{1}{p_0}-\frac{1}{p})}\|f^t\|_{L^{p_0,m}}\right\|_{L^r(\frac{dt}{t})}\\
&\le \frac{\delta^{\frac{1}{p_0}-\frac{1}{p}}}{|\gamma|^{\frac{1}{r}}}\left\|u^{-(\frac{1}{p_0}-\frac{1}{p})}\left(\int_0^u f^*(s)^m s^{\frac{m}{p_0}}\frac{ds}{s}\right)^{\frac{1}{m}}\right\|_{L^r(\frac{du}{u})}\\
&\le \frac{\delta^{\frac{1}{p_0}-\frac{1}{p}}}{|\gamma|^{\frac{1}{r}}}\left[\frac{\frac{r}{m}}{r(\frac{1}{p_0}-\frac{1}{p})}\right]^{\frac{1}{m}}\left(\int_0^\infty (s^{\frac{1}{p_0}}f^*(s))^r s^{-r(\frac{1}{p_0}-\frac{1}{p})}\frac{ds}{s}\right)^{\frac{1}{r}}\\
&= \frac{\delta^{\frac{1}{p_0}-\frac{1}{p}}}{m^{\frac{1}{m}}|\gamma|^{\frac{1}{r}}(\frac{1}{p_0}-\frac{1}{p})^{\frac{1}{m}}}\|f\|_{L^{p,r}},
\end{aligned}$$

where the last inequality is a consequence of Hardy's inequality:

$$\left(\int_0^\infty\left(\int_0^u g(s)\frac{ds}{s}\right)^p u^{-b}\frac{du}{u}\right)^{\frac{1}{p}}\le\frac{p}{b}\left(\int_0^\infty g(u)^p u^{-b}\frac{du}{u}\right)^{\frac{1}{p}} \tag{1.4.36}$$

with $g(s)=f^*(s)^m s^{m/p_0}\ge 0$, $p=r/m\ge 1$ and $b=r/p_0-r/p>0$. See Exercise 1.2.8 for the proof of (1.4.36).

Likewise, change variables $u=\delta t^\gamma$ in the L^r quasi-norm of (1.4.35) to obtain

$$\begin{aligned}
&\left\|t^{\gamma(\frac{1}{p}-\frac{1}{p_1})}\|f_t\|_{L^{p_1,m}}\right\|_{L^r(\frac{dt}{t})}\\
&\le \frac{\delta^{-(\frac{1}{p}-\frac{1}{p_1})}}{|\gamma|^{\frac{1}{r}}}\left\|u^{\frac{1}{p}-\frac{1}{p_1}}\left[\int_0^u f^*(u)^m s^{\frac{m}{p_1}}\frac{ds}{s}+\int_u^\infty f^*(s)^m s^{\frac{m}{p_1}}\frac{ds}{s}\right]^{\frac{1}{m}}\right\|_{L^r(\frac{du}{u})}\\
&= \frac{\delta^{-(\frac{1}{p}-\frac{1}{p_1})}}{|\gamma|^{\frac{1}{r}}}\left\|u^{\frac{m}{p}-\frac{m}{p_1}}\int_0^u f^*(u)^m s^{\frac{m}{p_1}}\frac{ds}{s}+\int_u^\infty f^*(s)^m s^{\frac{m}{p_1}}\frac{ds}{s}\right\|_{L^{r/m}(\frac{du}{u})}^{\frac{1}{m}}
\end{aligned}$$

$$\leq \frac{\delta^{-(\frac{1}{p}-\frac{1}{p_1})}}{|\gamma|^{\frac{1}{r}}} \left\{ \left\| u^{\frac{m}{p}-\frac{m}{p_1}} f^*(u)^m \int_0^u s^{\frac{m}{p_1}} \frac{ds}{s} \right\|_{L^{r/m}(\frac{du}{u})} \right.$$

$$\left. + \left\| u^{\frac{m}{p}-\frac{m}{p_1}} \int_u^\infty f^*(s)^m s^{\frac{m}{p_1}} \frac{ds}{s} \right\|_{L^{r/m}(\frac{du}{u})} \right\}^{\frac{1}{m}}$$

$$\leq \frac{\delta^{-(\frac{1}{p}-\frac{1}{p_1})}}{|\gamma|^{\frac{1}{r}}} \left\{ \frac{p_1}{m} \|f\|_{L^{p,r}}^m + \frac{\frac{r}{m}}{r(\frac{1}{p}-\frac{1}{p_1})} \left(\int_0^\infty (f^*(u)^m u^{\frac{m}{p_1}})^{\frac{r}{m}} u^{\frac{r}{p}-\frac{r}{p_1}} \frac{du}{u} \right)^{\frac{m}{r}} \right\}^{\frac{1}{m}}$$

$$= \frac{\delta^{-(\frac{1}{p}-\frac{1}{p_1})}}{m^{\frac{1}{m}} |\gamma|^{\frac{1}{r}}} \left\{ \frac{\frac{p_1}{p}}{\frac{1}{p}-\frac{1}{p_1}} \right\}^{\frac{1}{m}} \|f\|_{L^{p,r}},$$

where the last inequality above is Hardy's inequality:

$$\left(\int_0^\infty \left(\int_u^\infty g(s) \frac{ds}{s} \right)^p u^b \frac{du}{u} \right)^{\frac{1}{p}} \leq \frac{p}{b} \left(\int_0^\infty g(u)^p u^b \frac{du}{u} \right)^{\frac{1}{p}} \tag{1.4.37}$$

with $g(s) = f^*(s)^m s^{m/p_1} \geq 0$, $p = r/m \geq 1$ and $b = r/p - r/p_1 > 0$. See Exercise 1.2.8 for the proof of (1.4.37).

Combining these elements we deduce that given f in $S_0(X)$, we have that the expression in (1.4.34) plus the expression in (1.4.35) is at most

$$\frac{K \max\{1, 2^{\frac{1}{r}-1}\}}{m^{\frac{1}{m}} |\gamma|^{\frac{1}{r}}} \left\{ \frac{2^{\frac{1}{q_0}} M_0' \delta^{\frac{1}{p_0}-\frac{1}{p}}}{(\frac{1}{p_0}-\frac{1}{p})^{\frac{1}{m}}} + \frac{2^{\frac{1}{q_1}} (\frac{p_1}{p})^{\frac{1}{m}} M_1' \delta^{-(\frac{1}{p}-\frac{1}{p_1})}}{(\frac{1}{p}-\frac{1}{p_1})^{\frac{1}{m}}} \right\} \|f\|_{L^{p,r}}$$

We choose $\delta > 0$ such that the two terms in the curly brackets above are equal. We deduce that

$$\|T(f)\|_{L^{q,r}} \leq \frac{2K \max\{1, 2^{\frac{1}{r}-1}\}}{m^{\frac{1}{m}} |\gamma|^{\frac{1}{r}}} \left\{ \frac{2^{\frac{1-\theta}{q_0}} (M_0')^{1-\theta}}{(\frac{1}{p_0}-\frac{1}{p})^{\frac{1-\theta}{m}}} \frac{2^{\frac{\theta}{q_1}} (\frac{p_1}{p})^{\frac{\theta}{m}} (M_1')^\theta}{(\frac{1}{p}-\frac{1}{p_1})^{\frac{\theta}{m}}} \right\} \|f\|_{L^{p,r}}$$

where θ is as in (1.4.23), i.e.,

$$\theta = \frac{\frac{1}{p_0}-\frac{1}{p}}{\frac{1}{p_0}-\frac{1}{p_1}}.$$

This proves (1.4.24) in the case $p_1, r < \infty$ with constant $C_*(p_0, q_0, p_1, q_1, K, r, \theta)$ equal to

$$\frac{2K \max\{1, 2^{\frac{1}{r}-1}\}}{m^{\frac{1}{m}} |\gamma|^{\frac{1}{r}}} \left\{ \frac{2^{\frac{1-\theta}{q_0}} C(p_0, q_0, K, m)^{1-\theta} 2^{\frac{\theta}{q_1}} (\frac{p_1}{p})^{\frac{\theta}{m}} C(p_1, q_1, K, m)^\theta}{(\frac{1}{p_0}-\frac{1}{p})^{\frac{1-\theta}{m}} (\frac{1}{p}-\frac{1}{p_1})^{\frac{\theta}{m}}} \right\},$$

where we recall that $m = \frac{1}{2} \min\left(q_0, q_1, \frac{\log 2}{\log 2K}, 2r\right)$ and $C(p_j, q_j, K, m)$ is as in Lemma 1.4.20.

We now turn to the remaining cases $p=\infty$ or $r=\infty$. The restriction $r<\infty$ can be removed since $C_*(p_0,q_0,p_1,q_1,K,r,\theta)$ has a finite limit as $r\to\infty$ and, moreover, $\|f\|_{L^{p,r}}=\|t^{1/p}f^*(t)\|_{L^r(dt/t)}\to\|t^{1/p}f^*(t)\|_{L^\infty(dt/t)}=\|f\|_{L^{p,\infty}}$ as $r\to\infty$ and likewise $\|T(f)\|_{L^{p,r}}\to\|T(f)\|_{L^{p,\infty}}$ as $r\to\infty$; see Exercise 1.1.3 (a). The restriction $p_1<\infty$ can be removed as follows. Suppose that $p_1=\infty$. Then, since $\theta\in(0,1)$ it follows that $p<\infty$ and we pick $p_2>p$ and $p_2<\infty$. It is easy to see T satisfies the restricted weak type (p_2,q_2) estimate

$$\sup_{\alpha>0}\alpha\nu(\{|T(\chi_A)|>\alpha\})^{\frac{1}{q_2}}\le M_0^{1-\varphi}M_1^{\varphi}\mu(A)^{\frac{1}{p_2}},$$

where

$$\frac{1-\varphi}{p_0}+\frac{\varphi}{\infty}=\frac{1}{p_2},\qquad \frac{1-\varphi}{q_0}+\frac{\varphi}{q_1}=\frac{1}{q_2}. \tag{1.4.38}$$

Using the result obtained when $p_1<\infty$ with p_2 in place of p_1 we obtain that

$$\big\|T(f)\big\|_{L^{q,r}}\le C_*(p_0,q_0,p_2,q_2,K,r,\rho)M_0^{1-\rho}(M_0^{1-\varphi}M_1^{\varphi})^{\rho}\big\|f\big\|_{L^{p,r}} \tag{1.4.39}$$

for all functions f in $S_0(X)$, where

$$\frac{1-\rho}{p_0}+\frac{\rho}{p_2}=\frac{1}{p},\qquad \frac{1-\rho}{q_0}+\frac{\rho}{q_2}=\frac{1}{q}. \tag{1.4.40}$$

Combining (1.4.38) and (1.4.40) and using (1.4.23) we deduce that $\theta=\rho\varphi$ and hence (1.4.39) yields (1.4.24) in the case where $p_1=\infty$. In this case we have

$$C_*(p_0,q_0,\infty,q_1,K,r,\theta)=C_*(p_0,q_0,\tfrac{p_0}{1-\varphi},(\tfrac{1-\varphi}{q_0}+\tfrac{\varphi}{q_1})^{-1},K,r,\tfrac{\theta}{\varphi}),$$

where φ is any number satisfying $1>\varphi>1-\frac{p_0}{p}$.

Finally, we address the last assertion of the theorem which claims that when $p,r<\infty$ and $K=1$, the linear (or sublinear with nonnegative values) operator T initially defined on finitely simple functions has a unique bounded extension from $L^{p,r}(X)$ to $L^{q,r}(Y)$, which also satisfies (1.4.24) (with the same constant). To obtain this conclusion, we will need to know that the space $S_0(X)$ is dense in $L^{p,r}(X)$ whenever $0<p,r<\infty$. This is proved in Proposition 1.4.21 below. Assuming this proposition, we define the extension of T on $L^{p,r}(X)$ as follows:

Given f in $L^{p,r}(X)$ a sequence of functions f_j in $S_0(X)$ that converge to f in $L^{p,r}(X)$, notice that the linearity (or the sublinearity and the fact that $T(f)\ge 0$ for all f in $S_0(X)$) implies

$$|T(f_j)-T(f_k)|\le|T(f_j-f_k)|.$$

Using the boundedness of T from $L^{p,r}(X)$ to $L^{q,r}(Y)$ we obtain that the sequence $\{T(f_j)\}_j$ is Cauchy in $L^{q,r}(Y)$ and by the completeness of this space, it must converge to a limit which we call $\overline{T}(f)$. We observe that $\overline{T}(f)$ is independent of the choice of the sequence $\{f_j\}_j$ that converges to f in $L^{p,r}$. Moreover, one can show that $\overline{T}$ is linear (or sublinear with nonnegative values), $\overline{T}(f)$ coincides with $T(f)$

on $S_0(X)$ and $\overline{T}$ is bounded from $L^{p,r}(X)$ to $L^{q,r}(Y)$. Thus $\overline{T}$ is the unique bounded extension of T on the entire space $L^{p,r}(X)$. For details, see Exercise 1.4.17. $\square$

Proposition 1.4.21. *For all $0 < p,r < \infty$ the space $S_0(X)$ is dense in $L^{p,r}(X)$.*

Proof. Let $f \in L^{p,r}(X)$ and assume first that $f \geq 0$. Using (1.4.5) and the fact that d_f is decreasing on $[0,\infty)$, we obtain for any $n \in \mathbf{Z}^+$,

$$\begin{aligned}\|f\|_{L^{p,r}(X)}^r &= p\int_0^\infty \left[d_f(s)^{\frac{1}{p}}s\right]^r \frac{ds}{s}\\ &\geq p\int_0^{2^{-n}} \left[d_f(2^{-n})\right]^{\frac{r}{p}} s^{r-1}\,ds\\ &= \frac{p2^{-nr}}{r}\left[d_f(2^{-n})\right]^{\frac{r}{p}},\end{aligned}$$

which implies that $d_f(2^{-n}) < \infty$. Likewise, again in view of (1.4.5), we have

$$\|f\|_{L^{p,r}(X)}^r \geq p\int_0^{2^n} \left[d_f(s)\right]^{\frac{r}{p}} s^{r-1}\,ds = \frac{p2^{nr}}{r}\left[d_f(2^n)\right]^{\frac{r}{p}},$$

which implies that $\lim_{n\to+\infty} d_f(2^n) = 0$. Thus, for any $n \in \mathbf{Z}^+$, there exists $k_n \in \mathbb{N}$ such that

$$d_f(2^{k_n}) = \mu(\{x \in X : f(x) > 2^{k_n}\}) < 2^{-n}.$$

Let $E_n = \{x \in X : 2^{-n} < f(x) \leq 2^{k_n}\}$ and note that $\mu(E_n) \leq d_f(2^{-n}) < \infty$ for each $n \in \mathbf{Z}^+$. We write $f\chi_{E_n}$ in binary expansion, that is, $f\chi_{E_n}(x) = \sum_{j=-k_n}^{\infty} d_j(x)2^{-j}$, where $d_j(x) = 0$ or 1. Let $B_j = \{x \in E_n : d_j(x) = 1\}$. Then, $\mu(B_j) \leq \mu(E_n)$ and $f\chi_{E_n}$ can be expressed as $f\chi_{E_n} = \sum_{j=-k_n}^{\infty} 2^{-j}\chi_{B_j}$.

Set $f_n = \sum_{j=-k_n}^{n} 2^{-j}\chi_{B_j}$. It is obvious that $f_n \in S_0^+(X)$ and $f_n \leq f\chi_{E_n} \leq f$. Observe that when $x \in E_n$, we have

$$f(x) - f_n(x) = \sum_{j=n+1}^{\infty} 2^{-j}\chi_{B_j} \leq 2^{-n},$$

and that when $x \notin E_n$, we have $f_n(x) = 0$ and $f(x) > 2^{k_n}$ or $f(x) \leq 2^{-n}$. It follows from these facts that

$$d_{f-f_n}(2^{-n}) = \mu\left(E_n \cap \{f - f_n > 2^{-n}\}\right) + \mu\left(E_n^c \cap \{f - f_n > 2^{-n}\}\right) < 2^{-n}.$$

Hence, for $2^{-n} \leq t < \infty$ one has

$$(f - f_n)^*(t) \leq (f - f_n)^*(2^{-n}) = \inf\{s > 0 : d_{f-f_n}(s) \leq 2^{-n}\} \leq 2^{-n}.$$

This implies that $\lim_{n\to\infty}(f - f_n)^*(t) = 0$ for all $t \in (0,\infty)$. By Proposition 1.4.5 (5), (6), we obtain for all $t \in (0,\infty)$

$$(f - f_n)^*(t) \leq f^*(t/2) + f_n^*(t/2) \leq 2f^*(t/2).$$

The Lebesgue dominated convergence theorem gives $\|f_n - f\|_{L^{p,r}(X)} \to 0$ as $n \to \infty$ which yields the required conclusion for nonnegative functions f in $L^{p,r}(X)$.

For a complex-valued function $f \in L^{p,r}(X)$, we write $f = f_1 - f_2 + i(f_3 - f_4)$, where f_j are nonnegative functions in $L^{p,r}(X)$. By the preceding conclusion, there exist sequences $\{f_n^j\}_{n\in\mathbf{Z}^+}$, $j = 1,2,3,4$, in $S_0^+(X)$ such that $f_n^j \to f_j$ in $L^{p,r}(X)$ as $n \to \infty$. Set $f_n = f_n^1 - f_n^2 + i(f_n^3 - f_n^4)$. Using the fact that $\|\cdot\|_{L^{p,r}(X)}$ is a quasi-norm we obtain

$$\big\|f - f_n\big\|_{L^{p,r}(X)} \le C(p,r)\sum_{j=1}^{4}\big\|f_j - f_n^j\big\|_{L^{p,r}(X)}$$

which tends to zero as $n \to \infty$. This completes the proof. □

Corollary 1.4.22. *Let T be as in the statement of Theorem 1.4.19 and let $0 < p_0 \neq p_1 \le \infty$ and $0 < q_0 \neq q_1 \le \infty$. If T is restricted weak type (p_0,q_0) and (p_1,q_1) with constants M_0 and M_1, respectively, and for some $0 < \theta < 1$ we have*

$$\frac{1}{p} = \frac{1-\theta}{p_0} + \frac{\theta}{p_1}, \qquad \frac{1}{q} = \frac{1-\theta}{q_0} + \frac{\theta}{q_1},$$

and $p \le q$, then T satisfies the strong type estimate

$$\big\|T(f)\big\|_{L^q} \le C(p_0,q_0,p_1,q_1,\theta)M_0^{1-\theta}M_1^{\theta}\big\|f\big\|_{L^p} \tag{1.4.41}$$

for all f in $S_0(X)$. Moreover, if T is linear (or sublinear with nonnegative values), then it has a unique bounded extension from $L^p(X,\mu)$ to $L^q(Y,\nu)$ that satisfies estimate (1.4.41) *for all $f \in L^p(X)$ with the constant $C(p_0,q_0,p_1,q_1,\theta)$ replaced by $C(p_0,q_0,p_1,q_1,\theta)2^{2/p}\max(1,2^{1/p-1})^2$.*

Proof. Since $\theta \in (0,1)$ we must have $p,q < \infty$. Take $r = q$ in Theorem 1.4.19 and note that $\|f\|_{L^{p,r}} \le \|f\|_{L^p}$ since $p \le q = r$; see Proposition 1.4.10. The last assertion follows using Exercise 1.4.17. □

We now give examples to indicate why the assumptions $p_0 \neq p_1$ and $q_0 \neq q_1$ cannot be dropped in Theorem 1.4.19.

Example 1.4.23. Let $X = Y = \mathbf{R}$ and

$$T(f)(x) = |x|^{-1/2}\int_0^1 f(t)\,dt\,.$$

Then $\alpha|\{x:\ |T(\chi_A)(x)| > \alpha\}|^{1/2} = 2^{1/2}|A \cap [0,1]|$ and thus T is of restricted weak types $(1,2)$ and $(3,2)$. But observe that T does not map $L^2 = L^{2,2}$ to $L^{q,2}$. Thus Theorem 1.4.19 fails if the assumption $q_0 \neq q_1$ is dropped. The dual operator

$$S(f)(x) = \chi_{[0,1]}(x)\int_{-\infty}^{+\infty} f(t)|t|^{-1/2}\,dt$$

satisfies $\alpha|\{x:\ |S(\chi_A)(x)|>\alpha\}|^{1/q}\le c|A|^{1/2}$ when $q=1$ or 3, and thus it furnishes an example of an operator of restricted weak types $(2,1)$ and $(2,3)$ that is not L^2 bounded. Thus Theorem 1.4.19 fails if the assumption $p_0\neq p_1$ is dropped.

As an application of Theorem 1.4.19, we give the following strengthening of Theorem 1.2.13.

We end this chapter with a corollary of the proof of Theorem 1.4.19.

Corollary 1.4.24. *Let $1\le r<\infty$, $1\le p_0\neq p_1<\infty$, and $0<q_0\neq q_1\le\infty$ and let (X,μ) and (Y,ν) be σ-finite measure spaces. Let T be a quasi-linear operator defined on $L^{p_0}(X)+L^{p_1}(X)$ and taking values in the set of measurable functions on Y. Assume that for some $M_0',M_1'<\infty$ the following estimates hold for $j=0,1$*

$$\big\|T(f)\big\|_{L^{q_j,\infty}(Y)}\le M_j'\big\|f\big\|_{L^{p_j}(X)},\tag{1.4.42}$$

for all functions $f\in L^{p_j}(X)$. Fix $0<\theta<1$ and let

$$\frac{1}{p}=\frac{1-\theta}{p_0}+\frac{\theta}{p_1}\qquad\textit{and}\qquad\frac{1}{q}=\frac{1-\theta}{q_0}+\frac{\theta}{q_1}.\tag{1.4.43}$$

Then there exists a constant $C_*(p_0,q_0,p_1,q_1,K,r,\theta)<\infty$ such that for all functions f in $L^p(X)$ we have

$$\big\|T(f)\big\|_{L^{q,p}}\le C_*(p_0,q_0,p_1,q_1,K,r,\theta)(M_0')^{1-\theta}(M_1')^{\theta}\big\|f\big\|_{L^p}.\tag{1.4.44}$$

Proof. Since $L^p(X)$ is contained in the sum $L^{p_0}(X)+L^{p_1}(X)$, the operator T is well defined on $L^p(X)$. Hypothesis (1.4.42) implies that (1.4.30) holds for all $f\in L^{p_j,1}$. Repeat the proof of Theorem 1.4.19 starting from (1.4.30) fixing a function f in $L^p(X)$, $m=1$ and $r=p$. We obtain the required conclusion.

Theorem 1.4.25. ***(Young's inequality for weak type spaces)*** *Let G be a locally compact group with left Haar measure λ that satisfies (1.2.12) for all measurable subsets A of G. Let $1<p,q,r<\infty$ satisfy*

$$\frac{1}{q}+1=\frac{1}{p}+\frac{1}{r}.\tag{1.4.45}$$

Then there exists a constant $B_{p,q,r}>0$ such that for all f in $L^p(G)$ and g in $L^{r,\infty}(G)$ we have

$$\big\|f*g\big\|_{L^q(G)}\le B_{p,q,r}\big\|g\big\|_{L^{r,\infty}(G)}\big\|f\big\|_{L^p(G)}.\tag{1.4.46}$$

Proof. We fix $1<p,q<\infty$. Since p and q range in an open interval, we can find $p_0<p<p_1$, $q_0<q<q_1$, and $0<\theta<1$ such that (1.4.23) and (1.4.45) hold. Let $T(f)=f*g$, defined for all functions f on G. By Theorem 1.2.13, T extends to a bounded operator from L^{p_0} to $L^{q_0,\infty}$ and from L^{p_1} to $L^{q_1,\infty}$. It follows from the Corollary 1.4.24 that T extends to a bounded operator from $L^p(G)$ to $L^q(G)$. Notice that since G is locally compact, (G,λ) is a σ-finite measure space and for this reason, we were able to apply Corollary 1.4.24. □

Exercises

1.4.1. (a) Let g be a nonnegative integrable function on a measure space (X,μ) and let A be a measurable subset of X. Prove that

$$\int_A g\,d\mu \le \int_0^{\mu(A)} g^*(t)\,dt.$$

(b) (*G. H. Hardy and J. E. Littlewood*) For f and g measurable on a σ-finite measure space (X,μ), prove that

$$\int_X |f(x)g(x)|\,d\mu(x) \le \int_0^\infty f^*(t)g^*(t)\,dt.$$

Compare this result to the classical Hardy–Littlewood result asserting that for $a_j, b_j > 0$, the sum $\sum_j a_j b_j$ is greatest when both a_j and b_j are rearranged in decreasing order (for this see [148, p. 261]).

1.4.2. Let (X,μ) be a measure space. Prove that if $f \in L^{q_0,\infty}(X) \cap L^{q_1,\infty}(X)$ for some $0 < q_0 < q_1 \le \infty$, then $f \in L^{q,s}(X)$ for all $0 < s \le \infty$ and $q_0 < q < q_1$.

1.4.3. ([164]) Given $0 < p, q < \infty$, fix an $r = r(p,q) > 0$ such that $r \le 1$, $r \le q$ and $r < p$. Let (X,μ) be a measure space. For $t < \mu(X)$ define

$$f^{**}(t) = \sup_{\mu(E)\ge t} \left(\frac{1}{\mu(E)} \int_E |f|^r\,d\mu \right)^{1/r},$$

while for $t \ge \mu(X)$ (if $\mu(X) < \infty$) let

$$f^{**}(t) = \left(\frac{1}{t} \int_X |f|^r\,d\mu \right)^{1/r}.$$

Also define

$$|||\, f \,|||_{L^{p,q}} = \left(\int_0^\infty \left(t^{\frac{1}{p}} f^{**}(t)\right)^q \frac{dt}{t} \right)^{\frac{1}{q}}.$$

(The function f^{**} and the functional $f \to |||f|||_{L^{p,q}}$ depend on r.)
(a) Prove that the inequality $(((f+g)^{**})(t))^r \le (f^{**}(t))^r + (g^{**}(t))^r$ is valid for all $t \ge 0$. Since $r \le q$, conclude that the functional $f \to |||f|||_{L^{p,q}}^r$ is subadditive and hence it is a norm when $r = 1$ (this is possible only if $p > 1$).
(b) Show that for all f we have

$$\|f\|_{L^{p,q}} \le |||\, f \,|||_{L^{p,q}} \le \left(\frac{p}{p-r} \right)^{1/r} \|f\|_{L^{p,q}}.$$

(c) Conclude that $L^{p,q}(X)$ is metrizable and normable when $1 < p, q < \infty$.

1.4.4. Show that on a measure space (X,μ) the set of countable linear combinations of simple functions is dense in $L^{p,\infty}(X)$.
(b) Prove that finitely simple functions are not dense in $L^{p,\infty}(\mathbf{R})$ for any $0<p\le\infty$.
[*Hint:* Part (b): Show that the function $h(x)=x^{-1/p}\chi_{x>0}$ cannot be approximated in $L^{p,\infty}$ by a sequence of finitely simple functions. Given a finitely simple function s which is nonzero on a set A with $|A|>0$, show that $\|s-h\|_{L^{p,\infty}}\ge\sup_{0<\lambda<|A|^{1/p}}\lambda(\lambda^{-p}-|A|)^{1/p}=1$.]

1.4.5. Let (X,μ) be a nonatomic measure space.
(a) If $A_0\subseteq A_1\subseteq X$, $0<\mu(A_1)<\infty$, and $\mu(A_0)\le t\le\mu(A_1)$, show that there exists an $E_t\subseteq A_1$ with $\mu(E_t)=t$.
(b) Given a nonnegative continuous and decreasing function φ on $[0,\infty)$ such that $\varphi(t)=0$ whenever $t\ge\mu(X)$, prove that there exists a measurable function f on X with $f^*(t)=\varphi(t)$ for all $t>0$.
(c) Given $A\subseteq X$ with $0<\mu(A)<\infty$ and g an integrable function on X, show that there exists a subset $\widetilde{A}$ of X with $\mu(\widetilde{A})=\mu(A)$ such that

$$\int_{\widetilde{A}}|g|\,d\mu=\int_0^{\mu(A)}g^*(s)\,ds.$$

(d) If X is σ-finite, $f\in L^\infty(X)$, and $g\in L^1(X)$, prove that

$$\sup_{h:\,d_h=d_f}\left|\int_X hg\,d\mu\right|=\int_0^\infty f^*(s)g^*(s)\,ds,$$

where the supremum is taken over all functions h on X equidistributed with f.
[*Hint:* Part (a): Reduce matters to the situation in which $A_0=\emptyset$. Consider first the case that for all $A\subseteq X$ there exists a subset B of X satisfying $\frac{1}{10}\mu(A)\le\mu(B)\le\frac{9}{10}\mu(A)$. Then we can find subsets of A_1 of measure in any arbitrarily small interval, and by continuity the required conclusion follows. Next consider the case in which there is a subset A_1 of X such that every $B\subseteq A_1$ satisfies $\mu(B)<\frac{1}{10}\mu(A_1)$ or $\mu(B)>\frac{9}{10}\mu(A_1)$. Without loss of generality, normalize μ so that $\mu(A_1)=1$. Let $\mu_1=\sup\{\mu(C):\ C\subseteq A_1,\ \mu(C)<\frac{1}{10}\}$ and pick $B_1\subseteq A_1$ such that $\frac{1}{2}\mu_1\le\mu(B_1)\le\mu_1$. Set $A_2=A_1\setminus B_1$ and define $\mu_2=\sup\{\mu(C):\ C\subseteq A_2,\ \mu(C)<\frac{1}{10}\}$. Continue in this way and define sets $A_1\supseteq A_2\supseteq A_3\supseteq\cdots$ and numbers $\frac{1}{10}\ge\mu_1\ge\mu_2\ge\mu_3\ge\cdots$. If $C\subseteq A_{n+1}$ with $\mu(C)<\frac{1}{10}$, then $C\cup B_n\subseteq A_n$ with $\mu(C\cup B_n)<\frac{1}{5}<\frac{9}{10}$, and hence by assumption we must have $\mu(C\cup B_n)<\frac{1}{10}$. Conclude that $\mu_{n+1}\le\frac{1}{2}\mu_n$ and that $\mu(A_n)\ge\frac{4}{5}$ for all $n=1,2,\dots$. Then the set $\bigcap_{n=1}^\infty A_n$ must be an atom. Part (b): First show that when d is a simple right continuous decreasing function on $[0,\infty)$ there exists a measurable f on X such that $f^*=d$. For general continuous functions, use approximation. Part (c): Let $t=\mu(A)$ and define $A_1=\{x:\ |g(x)|>g^*(t)\}$ and $A_2=\{x:\ |g(x)|\ge g^*(t)\}$. Then $A_1\subseteq A_2$ and $\mu(A_1)\le t\le\mu(A_2)$. Pick $\widetilde{A}$ such that $A_1\subseteq\widetilde{A}\subseteq A_2$ and $\mu(\widetilde{A})=t=\mu(A)$ by part (a). Then $\int_{\widetilde{A}}g\,d\mu=\int_X g\chi_{\widetilde{A}}\,d\mu=\int_0^\infty(g\chi_{\widetilde{A}})^*\,ds=\int_0^{\mu(\widetilde{A})}g^*(s)\,ds$. Part (d): Reduce matters to functions $f,g\ge0$. Let

$f = \sum_{j=1}^N a_j \chi_{A_j}$ where $a_1 > a_2 > \cdots > a_N > 0$ and the A_j are pairwise disjoint. Write f as $\sum_{j=1}^N b_j \chi_{B_j}$, where $b_j = (a_j - a_{j+1})$ and $B_j = A_1 \cup \cdots \cup A_j$. Pick $\widetilde{B}_j$ as in part (c). Then $\widetilde{B}_1 \subsetneqq \cdots \subsetneqq \widetilde{B}_N$ and the function $f_1 = \sum_{j=1}^N b_j \chi_{\widetilde{B}_j}$ has the same distribution function as f. It follows from part (c) that $\int_X f_1 g\, d\mu = \int_0^\infty f^*(s) g^*(s)\, ds$. The case of a general $f \in L^\infty(X)$ follows by approximation by finitely simple functions.]

1.4.6. ([7], [297]) Let $K \geq 1$ and let $\|\cdot\|$ be a nonnegative functional on a vector space X that satisfies

$$\|x+y\| \leq K\big(\|x\| + \|y\|\big)$$

for all $x, y \in X$. For a fixed $\alpha \leq 1$ satisfying $(2K)^\alpha = 2$ show that

$$\|x_1 + \cdots + x_n\|^\alpha \leq 4\,(\|x_1\|^\alpha + \cdots + \|x_n\|^\alpha)$$

for all $n = 1, 2, \ldots$ and all $x_1, x_2, \ldots, x_n$ in X. This inequality is referred to as the *Aoki-Rolewicz theorem*.

[*Hint:* Quasi-linearity implies that $\|x_1 + \cdots + x_n\| \leq \max_{1\leq j\leq n}[(2K)^j \|x_j\|]$ for all $x_1, \ldots, x_n$ in X (use that $K \geq 1$). Define $H : X \to \mathbf{R}$ by setting $H(0) = 0$ and $H(x) = 2^{j/\alpha}$ if $2^{j-1} < \|x\|^\alpha \leq 2^j$. Then $\|x\| \leq H(x) \leq 2^{1/\alpha}\|x\|$ for all $x \in X$. Prove by induction that $\|x_1 + \cdots + x_n\|^\alpha \leq 2(H(x_1)^\alpha + \cdots + H(x_n)^\alpha)$. Suppose that this statement is true when $n = m$. To show its validity for $n = m+1$, without loss of generality assume that $\|x_1\| \geq \|x_2\| \geq \cdots \geq \|x_{m+1}\|$. Then $H(x_1) \geq H(x_2) \geq \cdots \geq H(x_{m+1})$. Assume that all the $H(x_j)$'s are distinct. Then since $H(x_j)^\alpha$ are distinct powers of 2, they must satisfy $H(x_j)^\alpha \leq 2^{-j+1} H(x_1)^\alpha$. Then

$$\begin{aligned}
\|x_1 + \cdots + x_{m+1}\|^\alpha &\leq \Big[\max_{1\leq j\leq m+1} (2K)^j \|x_j\|\Big]^\alpha \\
&\leq \Big[\max_{1\leq j\leq m+1} (2K)^j H(x_j)\Big]^\alpha \\
&\leq \Big[\max_{1\leq j\leq m+1} (2K)^j 2^{1/\alpha} 2^{-j/\alpha} H(x_1)\Big]^\alpha \\
&= 2H(x_1)^\alpha \\
&\leq 2(H(x_1)^\alpha + \cdots + H(x_{m+1})^\alpha).
\end{aligned}$$

We now consider the case that $H(x_j) = H(x_{j+1})$ for some $1 \leq j \leq m$. Then for some integer r we must have $2^{r-1} < \|x_{j+1}\|^\alpha \leq \|x_j\|^\alpha \leq 2^r$ and $H(x_j) = 2^{r/\alpha}$. Next note that

$$\|x_j + x_{j+1}\|^\alpha \leq K^\alpha(\|x_j\| + \|x_{j+1}\|)^\alpha \leq K^\alpha(2\, 2^{r/\alpha})^\alpha = 2^{r+1}.$$

This implies

$$H(x_j + x_{j+1})^\alpha \leq 2^{r+1} = 2^r + 2^r = H(x_j)^\alpha + H(x_{j+1})^\alpha.$$

Now apply the inductive hypothesis to $x_1, \ldots, x_{j-1}, x_j + x_{j+1}, x_{j+1}, \ldots, x_m$ and use the previous inequality to obtain the required conclusion.]

1.4.7. (a) ([347]) Let (X, μ) and (Y, ν) be measure spaces. Let Z be a Banach space of complex-valued measurable functions on Y. Assume that Z is closed under absolute values and satisfies $\|f\|_Z = \| |f| \|_Z$. Suppose that T is a linear operator defined on the space of finitely simple functions on (X,μ) and taking values in Z. Suppose that for some constant $A > 0$ the following restricted weak type estimate

$$\big\|T(\chi_E)\big\|_Z \le A\mu(E)^{1/p}$$

holds for some $0 < p < \infty$ and for all E measurable subsets of X of finite measure. Show that for all finitely simply functions f on X we have

$$\big\|T(f)\big\|_Z \le p^{-1}A\big\|f\big\|_{L^{p,1}}.$$

Consequently T has a bounded extension from $L^{p,1}(X)$ to Z.
(b) ([172]) As an application of part (a) prove that for any U, V measurable subsets of $\mathbf{R}^n$ with $|U|,|V| < \infty$ and any f measurable on $U \times V$ we have

$$\bigg(\int_U \big\|f(u,\cdot)\big\|^2_{L^{2,1}(V)}\,du\bigg)^{\frac{1}{2}} \le \frac{1}{2}\big\|f\big\|_{L^{2,1}(U\times V)}.$$

[*Hint:* Part (a): Let $f = \sum_{j=1}^N a_j\chi_{E_j} \ge 0$, where $a_1 > a_2 > \cdots > a_N > 0$, $\mu(E_j) < \infty$ pairwise disjoint. Let $F_j = E_1 \cup \cdots \cup E_j$, $B_0 = 0$, and $B_j = \mu(F_j)$ for $j \ge 1$. Write $f = \sum_{j=1}^N (a_j - a_{j+1})\chi_{F_j}$, where $a_{N+1} = 0$. Then

$$\begin{aligned}
\big\|T(f)\big\|_Z &= \big\| |T(f)| \big\|_Z \\
&\le \sum_{j=1}^N (a_j - a_{j+1})\big\|T(\chi_{F_j})\big\|_Z \\
&\le A\sum_{j=1}^N (a_j - a_{j+1})(\mu(F_j))^{1/p} \\
&= A\sum_{j=0}^{N-1} a_{j+1}(B_{j+1}^{1/p} - B_j^{1/p}) \\
&= p^{-1}A\big\|f\big\|_{L^{p,1}},
\end{aligned}$$

where the penultimate equality follows by a summation by parts; see Appendix F.]

1.4.8. Let $0 < p,q,\alpha,\beta < \infty$. Also let $0 < q_1 < q_2 < \infty$.
(a) Show that the function $f_{\alpha,\beta}(t) = t^{-\alpha}(\log t^{-1})^{-\beta}\chi_{[0,e^{-\beta/\alpha})}(t)$ lies in $L^{p,q}(\mathbf{R})$ if and only if either $p < 1/\alpha$ or both $p = 1/\alpha$ and $q > 1/\beta$ hold. Conclude that the function $t \mapsto t^{-1/p}(\log t^{-1})^{-1/q_1}\chi_{[0,e^{-p/q_1})}(t)$ lies in $L^{p,q_2}(\mathbf{R})$ but not in $L^{p,q_1}(\mathbf{R})$.
(b) Find a necessary and sufficient condition in terms of p,α,β for the function $g_{\alpha,\beta}(t) = (1+t)^{-\alpha}(\log(2+t))^{-\beta}\chi_{[0,\infty)}$ to lie in $L^{p,q}(\mathbf{R})$.

(c) Let $\psi(t)$ be smooth decreasing function on $[0,\infty)$ and let $F(x) = \psi(|x|)$ for x in $\mathbf{R}^n$, where $|x|$ is the modulus of x. Show that $F^*(t) = f((t/v_n)^{1/n})$, where v_n is the volume of the unit ball. Use this formula to construct examples showing that $L^{p,q_1}(\mathbf{R}^n) \subsetneqq L^{p,q_2}(\mathbf{R}^n)$.
(d) On a general nonatomic measure space (X,μ) prove that there *does not* exist a constant $C(p,q_1,q_2) > 0$ such that for all f in $L^{p,q_2}(X)$ the following is valid:

$$\big\|f\big\|_{L^{p,q_1}} \le C(p,q_1,q_2)\big\|f\big\|_{L^{p,q_2}}\,.$$

[*Hint:* Parts (a), (b): Use that $f_{\alpha,\beta}$ and $g_{\alpha,\beta}$ are equal to their decreasing rearrangements. Part (d): Use Exercise 1.4.5 (b) with $\varphi(t) = g_{1/p,1/q_1}(t)$.]

1.4.9. ([346]) Let $L^p(\omega)$ denote the space of all measurable functions f on $\mathbf{R}^n$ such that $\|f\|_{L^p(\omega)}^p = \int_{\mathbf{R}^n} |f(x)|^p\,\omega(x)\,dx < \infty$, where $0 < \omega < \infty$ a.e. Let T be a sublinear operator that maps $L^{p_0}(\omega_0)$ to $L^{q_0,\infty}(\omega)$ and $L^{p_1}(\omega_1)$ to $L^{q_1,\infty}(\omega)$, where ω_0,ω_1,ω are positive functions and $1 \le p_0 < p_1 < \infty$, $0 < q_0,q_1 < \infty$. Suppose that

$$\frac{1}{p_\theta} = \frac{1-\theta}{p_0} + \frac{\theta}{p_1}, \qquad \frac{1}{q_\theta} = \frac{1-\theta}{q_0} + \frac{\theta}{q_1}.$$

Let $\Omega_\theta = \omega_0^{\frac{1-\theta}{p_0}p_\theta}\,\omega_1^{\frac{\theta}{p_1}p_\theta}$. Show that T maps $L^{p_\theta}(\Omega_\theta) \to L^{q_\theta,p_\theta}(\omega)$.
[*Hint:* Define $L(f) = (\omega_1/\omega_0)^{\frac{1}{p_1-p_0}} f$ and observe that for each $\theta \in [0,1]$, L maps $L^{p_\theta}(\Omega_\theta) \to L^{p_\theta}\big((\omega_0^{p_1}\omega_1^{-p_0})^{\frac{1}{p_1-p_0}}\big)$ isometrically. Then apply Corollary 1.4.24 to the sublinear operator $T \circ L^{-1}$.]

1.4.10. ([185], [349]) Let λ_n be a sequence of positive numbers with $\sum_n \lambda_n \le 1$ and $\sum_n \lambda_n \log(\frac{1}{\lambda_n}) = K < \infty$. Suppose all sequences are indexed by a fixed countable set.
(a) Let f_n be a sequence of complex-valued functions in $L^{1,\infty}(X)$ with $\|f_n\|_{L^{1,\infty}} \le 1$ uniformly in n. Prove that $\sum_n \lambda_n f_n$ lies in $L^{1,\infty}(X)$ with norm at most $2(K+2)$. (This property is referred to as the *logconvexity* of $L^{1,\infty}$.)
(b) Let T_n be a sequence of sublinear operators that map $L^1(X)$ to $L^{1,\infty}(Y)$ with norms $\|T_n\|_{L^1 \to L^{1,\infty}} \le B$ uniformly in n. Use part (a) to prove that $\sum_n \lambda_n T_n$ maps $L^1(X)$ to $L^{1,\infty}(Y)$ with norm at most $2B(K+2)$.
(c) Given $\delta > 0$ pick $0 < \varepsilon < \delta$ and use the simple estimate

$$\mu(\{\sum_{n=1}^{\infty} 2^{-\delta n} f_n > \alpha\}) \le \sum_{n=1}^{\infty} \mu(\{2^{-\delta n} f_n > (2^\varepsilon - 1)2^{-\varepsilon n}\alpha\})$$

to obtain a simple proof of the statement in part (a) when $\lambda_n = 2^{-\delta n}$, $n = 1,2,\dots$.
[*Hint:* Part (a): For fixed $\alpha > 0$, write $f_n = u_n + v_n + w_n$, where $u_n = f_n\chi_{|f_n|\le\frac{\alpha}{2}}$, $v_n = f_n\chi_{|f_n|>\frac{\alpha}{2\lambda_n}}$, and $w_n = f_n\chi_{\frac{\alpha}{2}<|f_n|\le\frac{\alpha}{2\lambda_n}}$. Let $u = \sum_n \lambda_n u_n$, $v = \sum_n \lambda_n v_n$, and $w = \sum_n \lambda_n w_n$. Clearly $|u| \le \frac{\alpha}{2}$. Also $\{v \ne 0\} \subseteq \bigcup_n \{|f_n| > \frac{\alpha}{2\lambda_n}\}$; hence $\mu(\{v \ne 0\}) \le \frac{2}{\alpha}$. Finally,

$$\begin{aligned}\int_X |w|\,d\mu &\leq \sum_n \lambda_n \int_X |f_n| \chi_{\frac{\alpha}{2}<|f_n|\leq \frac{\alpha}{2\lambda_n}}\,d\mu \\ &\leq \sum_n \lambda_n \left[\int_{\alpha/2}^{\alpha/(2\lambda_n)} d_{f_n}(\beta)\,d\beta + \int_0^{\alpha/2} d_{f_n}(\alpha/2)\,d\beta\right] \\ &\leq K+1\,.\end{aligned}$$

Using $\mu(\{|u+v+w|>\alpha\}) \leq \mu(\{|u|>\alpha/2\}) + \mu(\{|v|\neq 0\}) + \mu(\{|w|>\alpha/2\})$, deduce the conclusion.]

1.4.11. Let $\{f_n\}_n$ be a sequence of measurable functions on a measure space (X,μ). Let $0<q,s\leq\infty$.
(a) Suppose that $f_n \geq 0$ for all n. Show that

$$\big\|\liminf_{n\to\infty} f_n\big\|_{L^{q,s}} \leq \liminf_{n\to\infty} \big\|f_n\big\|_{L^{q,s}}\,.$$

(b) Let $g_n \to g$ in $L^{q,s}$ as $n\to\infty$. Show that $\|g_n\|_{L^{q,s}} \to \|g\|_{L^{q,s}}$ as $n\to\infty$.

1.4.12. (a) Suppose that X is a quasi-Banach space and let X^* be its dual (which is always a Banach space). Prove that for all $T\in X^*$ we have

$$\big\|T\big\|_{X^*} = \sup_{\substack{x\in X\\ \|x\|_X\leq 1}} |T(x)|\,.$$

(b) Now suppose that X is a Banach space. Use the Hahn–Banach theorem to prove that for every $x\in X$ we have

$$\|x\|_X = \sup_{\substack{T\in X^*\\ \|T\|_{X^*}\leq 1}} |T(x)|\,.$$

Observe that this result may fail for quasi-Banach spaces. For example, if $X = L^{1,\infty}$, every linear functional on X^* vanishes on the set of simple functions.
(c) Let $1<p<\infty$, $X = L^{p,1}(Y)$, and $X^* = L^{p',\infty}(Y)$, where (Y,μ) is nonatomic σ-finite measure space. Conclude that

$$\begin{aligned}\big\|f\big\|_{L^{p,1}} &\approx \sup_{\|g\|_{L^{p',\infty}}\leq 1} \left|\int_Y fg\,d\mu\right|, \\ \big\|f\big\|_{L^{p,\infty}} &\approx \sup_{\|g\|_{L^{p',1}}\leq 1} \left|\int_Y fg\,d\mu\right|.\end{aligned}$$

1.4.13. Let $0<p,q<\infty$. Prove that any function in $L^{p,q}(X,\mu)$ can be written as

$$f = \sum_{n=-\infty}^{+\infty} c_n f_n,$$

where f_n is a function bounded by $2^{-n/p}$, supported on a set of measure 2^n, and the sequence $\{c_k\}_k$ lies in ℓ^q and satisfies

$$2^{-\frac{1}{p}}(\log 2)^{\frac{1}{q}}\big\|\{c_k\}_k\big\|_{\ell^q} \le \big\|f\big\|_{L^{p,q}} \le \big\|\{c_k\}_k\big\|_{\ell^q} 2^{\frac{1}{p}}(\log 2)^{\frac{1}{q}}.$$

[*Hint:* Let $c_n = 2^{n/p} f^*(2^n)$, $A_n = \{x : f^*(2^{n+1}) < |f(x)| \le f^*(2^n))\}$, and $f_n = c_n^{-1} f \chi_{A_n}$.]

1.4.14. (*T. Tao*) Let $0<p<\infty$, $0<\gamma<1$, $A,B>0$, and let f be a measurable function on a measure space (X,μ).
(a) Suppose that $\|f\|_{L^{p,\infty}} \le A$. Then for every measurable set E of finite measure there exists a measurable subset E' of E with $\mu(E') \ge \gamma\mu(E)$ such that f is integrable on E' and

$$\left|\int_{E'} f\,d\mu\right| \le (1-\gamma)^{-1/p} A\,\mu(E)^{1-\frac{1}{p}}.$$

(b) Suppose that (X,μ) is a σ-finite measure space and that f has the property that for any measurable subset E of X with $\mu(E)<\infty$ there is a measurable subset E' of E with $\mu(E') \ge \gamma\mu(E)$ such that f is integrable on E' and

$$\left|\int_{E'} f\,d\mu\right| \le B\,\mu(E)^{1-\frac{1}{p}}.$$

Then we have that $\|f\|_{L^{p,\infty}} \le B 4^{1/p}\gamma^{-1}\sqrt{2}$.
(c) Conclude that if (X,μ) is a σ-finite measure space then

$$\big\|f\big\|_{L^{p,\infty}} \approx \sup_{\substack{E\subseteq X\\ 0<\mu(E)<\infty}} \inf_{\substack{E'\subseteq E\\ \mu(E')\ge\frac{1}{2}\mu(E)\\ f\in L^1(E')}} \mu(E)^{-1+\frac{1}{p}}\left|\int_{E'} f\,d\mu\right|.$$

[*Hint:* Part (a): Take $E' = E\setminus\{|f| > A(1-\gamma)^{-\frac{1}{p}}\mu(E)^{-\frac{1}{p}}\}$. Part (b): Write $X = \bigcup_{n=1}^\infty X_n$ with $\mu(X_n)<\infty$. Given $\alpha>0$, note that the set $\{|f|>\alpha\}$ is contained in

$$\{\operatorname{Re} f > \tfrac{\alpha}{\sqrt{2}}\} \cup \{\operatorname{Im} f > \tfrac{\alpha}{\sqrt{2}}\} \cup \{\operatorname{Re} f < -\tfrac{\alpha}{\sqrt{2}}\} \cup \{\operatorname{Im} f < -\tfrac{\alpha}{\sqrt{2}}\}.$$

Let E_n be any of the preceding four sets intersected with X_n, let E_n' be a subset of it with measure at least $\gamma\mu(E_n)$ as in the hypothesis. Then $\left|\int_{E_n'} f\,d\mu\right| \ge \frac{\alpha}{\sqrt{2}}\gamma\mu(E_n)$, from which it follows that $\alpha\mu(E_n)^{1/p} \le B\sqrt{2}\gamma^{-1}$, and let $n\to\infty$.]

1.4.15. Let T be a linear operator defined on the set of finitely simple functions on a σ-finite measure space (X,μ) and taking values in the set of measurable functions on a σ-finite measure space (Y,ν) and T^t be a linear operator defined on the set of finitely simple functions on (Y,ν) and taking values in the set of measurable functions of (X,μ). Suppose that for all A subsets of X and B subsets of Y of finite measure we have

$$\int_B |T(\chi_A)|\,d\nu + \int_A |T^t(\chi_B)|\,d\mu < \infty$$

and that T and T^t are related via the "transpose identity"

$$\int_Y T(\chi_A)\,\chi_B\,d\nu = \int_X T^t(\chi_B)\,\chi_A\,d\mu = \Lambda(A,B)\,.$$

Assume that whenever $\mu(A_n)+\nu(B_n)\to 0$ as $n\to\infty$, we have $\Lambda(A_n,B_n)\to 0$. Suppose that T and T^t are *restricted weak type* $(1,1)$ operators, with constants C_1 and C_2, respectively. Show that, for all $1<p<\infty$, T is of *restricted weak type* (p,p). Precisely, show that there exists a constant K_p such that

$$\big\|T(\chi_A)\big\|_{L^p(Y)} \le K_p C_1^{\frac{1}{p}} C_2^{1-\frac{1}{p}} \mu(A)^{\frac{1}{p}}$$

for all measurable subsets A of X with $\mu(A)<\infty$.
[*Hint:* Suppose that $C_1\mu(F) > C_2\nu(E)$ and pick m so that $C_1\mu(F)\sim 2^m C_2\nu(E)$. Since T^t is restricted weak type $(1,1)$ there is an $F'\subseteqq F$ such that $\mu(F')\ge\frac{1}{2}\mu(F)$ and $|\Lambda(F',E)|\le 2C_2\nu(E)$. Find by induction sets $F^{(j)}\subseteqq F\setminus(F'\cup\cdots\cup F^{(j-1)})$ such that $\mu(F^{(j)})\ge\frac{1}{2}\mu(F\setminus(F'\cup\cdots\cup F^{(j-1)}))$ and $|\Lambda(F^{(j)},E)|\le 2C_2\nu(E)$, $j=1,2,\dots,m$. Stop when $F^{(m)}=F\setminus(F'\cup\cdots\cup F^{(m-1)})$ satisfies $C_1\mu(F^{(m)})\le C_2\nu(E)$. Since T is restricted weak type $(1,1)$ there is a subset E' of E such that $\nu(E')\ge\frac{1}{2}\nu(E)$ and $|\Lambda(F^{(m)},E')|\le 2C_1\mu(F^{(m)})\le 2C_2\nu(E)$. Now write

$$\Lambda(F,E)=\sum_{j=1}^{m-1}\Lambda(F^{(j)},E)+\Lambda(F^{(m)},E')+\Lambda(F^{(m)},E\setminus E')$$

from which it follows that

$$|\Lambda(F,E)|\le 2C_2\nu(E)\left(1+\log_2\frac{C_1\mu(F)}{C_2\nu(E)}\right)+|\Lambda(F_1,E_1)|$$

where $F_1=F^{(m)}$ and $E_1=E\setminus E'$. Note that the first term in the sum above is at most $K_p'(C_1\mu(F))^{1/p}(C_2\nu(E))^{1/p'}$ and that the identical estimate holds if the roles of E and F are reversed. Also observe that $\mu(F_1)\le\frac{1}{2}\mu(F)$ and $\nu(E_1)\le\frac{1}{2}\nu(E)$. Continuing this process we find sets (F_n,E_n) with $\mu(F_{n+1})\le\frac{1}{2}\mu(F_n)$ and $\nu(E_{n+1})\le\frac{1}{2}\nu(E_n)$.Using $\Lambda(F_n,E_n)\to 0$ as $n\to\infty$ we deduce that $|\Lambda(F,E)|\le 2K_p'(C_1\mu(F))^{1/p}(C_2\nu(E))^{1/p'}$. Considering the sets $E_+=E\cap\{T(\chi_F)>0\}$ and $E_-=E\cap\{T(\chi_F)<0\}$, obtain that $\int_E\big|T(\chi_F)\big|\,d\nu\le 4K_p'(C_1\mu(F))^{\frac{1}{p}}(C_2\nu(E))^{\frac{1}{p'}}$ for all F and E measurable sets of finite measure. Exercise 1.1.12 (a) with $r=1$ yields that $\|T(\chi_F)\|_{L^{p,\infty}}\le 4K_p C_1^{1/p}C_2^{1/p'}\mu(F)^{1/p}$.]

1.4.16. ([35]) Let $0<p_0<p_1<\infty$ and $0<\alpha,\beta,A,B<\infty$. Suppose that a family of sublinear operators T_k is of restricted weak type (p_0,p_0) with constant $A\,2^{-k\alpha}$ and of restricted weak type (p_1,p_1) with constant $B\,2^{k\beta}$ for all $k\in\mathbf{Z}$. Show that there is a constant $C = C(\alpha,\beta,p_0,p_1)$ such that $\sum_{k\in\mathbf{Z}} T_k$ is of restricted weak type (p,p) with constant $CA^{1-\theta}B^{\theta}$, where $\theta = \alpha/(\alpha+\beta)$ and

$$\frac{1}{p} = \frac{1-\theta}{p_0} + \frac{\theta}{p_1}.$$

[*Hint:* Estimate $\mu(\{|T(\chi_E)| > \lambda\})$ by the sum $\sum_{k\geq k_0} \mu(\{|T_k(\chi_E)| > c\lambda 2^{\alpha'(k_0-k)}\})+$ $\sum_{k\leq k_0} \mu(\{|T_k(\chi_E)| > c\lambda 2^{\beta'(k-k_0)}\})$, where c is a suitable constant and $0<\alpha'<\alpha$, $0<\beta'<\beta$. Apply the restricted weak type (p_0,p_0) hypothesis on each term of the first sum, the restricted weak type (p_1,p_1) hypothesis on each term of the second sum, and choose k_0 to optimize the resulting expression.]

1.4.17. Let (X,μ), (Y,ν) be measure spaces, $0<p,r,q,s\leq\infty$ and $0<B<\infty$. Suppose that a sublinear operator T is defined on a dense subspace $\mathscr{D}$ of $L^{p,r}(X)$, takes values in the space of measurable functions of another measure space Y, and satisfies $T(f)\geq 0$ for all f in $\mathscr{D}$. Assume that

$$\big\|T(\varphi)\big\|_{L^{q,s}} \leq B\big\|\varphi\big\|_{L^{p,r}}$$

for all φ in $\mathscr{D}$. Prove that T admits a unique sublinear extension $\overline{T}$ on $L^{p,r}(X)$ such that

$$\big\|\overline{T}(f)\big\|_{L^{q,s}} \leq B\big\|f\big\|_{L^{p,r}}$$

for all $f\in L^{p,r}(X)$.
[*Hint:* Given $f\in L^{p,r}(X)$ find a sequence of functions φ_j in $\mathscr{D}$ such that $\varphi_j\to f$ in $L^{p,r}$. Use the inequality $|T(\varphi_j)-T(\varphi_k)|\leq|T(\varphi_j-\varphi_k)|$, to obtain that the sequence $\{T(\varphi_j)\}_j$ is Cauchy in $L^{q,s}$ and thus it has a unique limit $\overline{T}(f)$ which is independent of the choice of sequence φ_j. Boundedness of $\overline{T}$ follows by density. To prove that $\overline{T}$ is sublinear use that convergence in $L^{q,s}$ implies convergence in measure and thus a subsequence of $T(\varphi_j)$ converges ν-a.e. to $\overline{T}(f)$. Also use Exercise 1.4.11.]

HISTORICAL NOTES

The modern theory of measure and integration was founded with the publication of Lebesgue's dissertation [214]; see also [215]. The theory of the Lebesgue integral reshaped the course of integration. The spaces $L^p([a,b])$, $1<p<\infty$, were first investigated by Riesz [290], who obtained many important properties of them. A rigorous treatise of harmonic analysis on general groups can be found in the book of Hewitt and Ross [152]. The best possible constant C_{pqr} in Young's inequality $\|f*g\|_{L^r(\mathbf{R}^n)}\leq C_{pqr}\|f\|_{L^p(\mathbf{R}^n)}\|g\|_{L^q(\mathbf{R}^n)}$, $\frac{1}{p}+\frac{1}{q}=\frac{1}{r}+1$, $1<p,q,r<\infty$, was shown by Beckner [21] to be $C_{pqr}=(B_pB_qB_{r'})^n$, where $B_p^2=p^{1/p}(p')^{-1/p'}$.

Theorem 1.3.2 first appeared without proof in Marcinkiewicz's brief note [240]. After his death in World War II, this theorem seemed to have escaped attention until Zygmund reintroduced it in [387]. This reference presents the more difficult off-diagonal version of the theorem, derived by Zygmund. Stein and Weiss [347] strengthened Zygmund's theorem by assuming that the initial estimates are of restricted weak type whenever $1\leq p_0,p_1,q_0,q_1\leq\infty$. The extension of this result to the case $0<p_0,p_1,q_0,q_1<1$ in Theorem 1.4.19 is due to the author. The critical Lemma 1.4.20 was suggested by Kalton. Improvements of these results, in particular, the appearance of the space $S_0(X)$ and the presence of the factor $M_0^{1-\theta}M_1^{\theta}$ in (1.4.24) appeared in Liang, Liu, and Yang [224]. Equivalence of restricted weak type $(1,1)$ and weak type $(1,1)$ properties for certain maximal multipliers was obtained by Moon [257]. The following partial converse of Theorem 1.2.13 is due to Stepanov [351]: If a convolution operator maps $L^1(\mathbf{R}^n)$ to $L^{q,\infty}(\mathbf{R}^n)$ for some $1 < q < \infty$ then its kernel must be in $L^{q,\infty}$.

The extrapolation result of Exercise 1.3.7 is due to Yano [380]; see also Zygmund [389, pp. 119–120] and the related work of Carro [56], Soria [330], and Tao [356].

The original version of Theorem 1.3.4 was proved by Riesz [293] in the context of bilinear forms. This version is called the Riesz convexity theorem, since it says that the logarithm of the function $M(\alpha,\beta) = \inf_{x,y} \left|\sum_{j=1}^n \sum_{k=1}^m a_{jk} x_j y_k\right| \|x\|_{\ell^{1/\alpha}}^{-1} \|y\|_{\ell^{1/\beta}}^{-1}$ (where the infimum is taken over all sequences $\{x_j\}_{j=1}^n$ in $\ell^{1/\alpha}$ and $\{y_k\}_{k=1}^m$ in $\ell^{1/\beta}$) is a convex function of (α,β) in the triangle $0 \le \alpha,\beta \le 1$, $\alpha+\beta \ge 1$. Riesz's student Thorin [360] extended this triangle to the unit square $0 \le \alpha,\beta \le 1$ and generalized this theorem by replacing the maximum of a bilinear form with the maximum of the modulus of an entire function in many variables. After the end of World War II, Thorin published his thesis [361], building the subject and giving a variety of applications. The original proof of Thorin was rather long, but a few years later, Tamarkin and Zygmund [354] gave a very elegant short proof using the maximum modulus principle in a more efficient way. Today, this theorem is referred to as the Riesz–Thorin interpolation theorem.

Calderón [42] elaborated the complex-variables proof of the Riesz–Thorin theorem into a general method of interpolation between Banach spaces. The complex interpolation method can also be defined for pairs of quasi-Banach spaces, although certain complications arise in this setting; however, the Riesz–Thorin theorem is true for pairs of L^p spaces (with the "correct" geometric mean constant) for all $0 < p \le \infty$ and also for Lorentz spaces. In this setting, duality cannot be used, but a well-developed theory of analytic functions with values in quasi-Banach spaces is crucial. We refer to the articles of Kalton [186] and [187] for details. Complex interpolation for sublinear maps is also possible; see the article of Calderón and Zygmund [47]. Interpolation for analytic families of operators (Theorem 1.3.7) is due to Stein [331]. The critical Lemma 1.3.8 used in the proof was previously obtained by Hirschman [154].

The fact that nonatomic measure spaces contain subsets of all possible measures is classical. An extension of this result to countably additive vector measures with values in finite-dimensional Banach spaces was obtained by Lyapunov [236]; for a proof of this fact, see Diestel and Uhl [95, p. 264]. The Aoki–Rolewicz theorem (Exercise 1.4.6) was proved independently by Aoki [7] and Rolewicz [297]. For a proof of this fact and a variety of its uses in the context of quasi-Banach spaces we refer to the book of Kalton, Peck, and Roberts [188].

Decreasing rearrangements of functions were introduced by Hardy and Littlewood [146]; the authors attribute their motivation to understanding cricket averages. The $L^{p,q}$ spaces were introduced by Lorentz in [232] and in [233]. A general treatment of Lorentz spaces is given in the article of Hunt [164]. The normability of the spaces $L^{p,q}$ (which holds exactly when $1 < p \le \infty$ and $1 \le q \le \infty$) can be traced back to general principles obtained by Kolmogorov [199]. The introduction of the function f^{**}, which was used in Exercise 1.4.3, to explicitly define a norm on the normable spaces $L^{p,q}$ is due to Calderón [42]. These spaces appear as intermediate spaces in the general interpolation theory of Calderón [42] and in that of Lions and Peetre [225]. The latter was pointed out by Peetre [275]. For a systematic study of the duals of Lorentz spaces we refer to Cwikel [83] and Cwikel and Fefferman [84], [85]. An extension of the Marcinkiewicz interpolation theorem to Lorentz spaces was obtained by Hunt [163]. Carro, Raposo, and Soria [57] provide a comprehensive presentation of the theory of Lorentz spaces in the context of weighted inequalities. For further topics on interpolation one may consult the books of Bennett and Sharpley [24], Bergh and Löfström [25], Sadosky [309], Kislyakov and Kruglyak [194], and Chapter 5 in Stein and Weiss [348].

Chapter 2
Maximal Functions, Fourier Transform, and Distributions

We have already seen that the convolution of a function with a fixed density is a smoothing operation that produces a certain average of the function. Averaging is an important operation in analysis and naturally arises in many situations. The study of averages of functions is better understood by the introduction of the maximal function which is defined as the largest average of a function over all balls containing a fixed point. Maximal functions are used to obtain almost everywhere convergence for certain integral averages and play an important role in this area, which is called differentiation theory. Although maximal functions do not preserve qualitative information about the given functions, they maintain crucial quantitative information, a fact of great importance in the subject of Fourier analysis.

Another important operation we study in this chapter is the Fourier transform, the father of all oscillatory integrals. This is as fundamental to Fourier analysis as marrow is to the human bone. It is a powerful transformation that carries a function from its spatial domain to its frequency domain. By doing this, it inverts the function's localization properties. If applied one more time, then magically reproduces the function composed with a reflection. It changes convolution to multiplication, translation to modulation, and expanding dilation to shrinking dilation. Its decay at infinity encodes information about the local smoothness of the function. The study of the Fourier transform also motivates the launch of a thorough study of general oscillatory integrals. We take a quick look at this topic with emphasis on one-dimensional results.

Distributions suppy a mathematical framework for many operations that do not exactly qualify to be called functions. These operations found their mathematical place in the world of functionals applied to smooth functions (called test functions). These functionals also introduced the correct interpretation for many physical objects, such as the Dirac delta function. Distributions have become an indispensable tool in analysis and have enhanced our perspective.

L. Grafakos, *Classical Fourier Analysis*, Graduate Texts in Mathematics 249,
DOI 10.1007/978-1-4939-1194-3_2,

2.1 Maximal Functions

Given a Lebesgue measurable subset A of $\mathbf{R}^n$, we denote by $|A|$ its Lebesgue measure. For $x \in \mathbf{R}^n$ and $r > 0$, we denote by $B(x,r)$ the open ball of radius r centered at x. We also use the notation $aB(x,\delta) = B(x,a\delta)$, for $a > 0$, for the ball with the same center and radius $a\delta$. Given $\delta > 0$ and f a locally integrable function on $\mathbf{R}^n$, let

$$\operatorname*{Avg}_{B(x,\delta)} |f| = \frac{1}{|B(x,\delta)|} \int_{B(x,\delta)} |f(y)|\,dy$$

denote the average of $|f|$ over the ball of radius δ centered at x.

2.1.1 The Hardy–Littlewood Maximal Operator

Definition 2.1.1. Let f be a locally integrable function on $\mathbf{R}^n$. The function

$$\mathcal{M}(f)(x) = \sup_{\delta>0} \operatorname*{Avg}_{B(x,\delta)} |f| = \sup_{\delta>0} \frac{1}{v_n \delta^n} \int_{|y|<\delta} |f(x-y)|\,dy$$

is called the *centered Hardy–Littlewood maximal function* of f.

Obviously we have $\mathcal{M}(f) = \mathcal{M}(|f|) \ge 0$; thus the maximal function is a positive operator. Information concerning cancellation of the function f is lost by passing to $\mathcal{M}(f)$. We show later that $\mathcal{M}(f)$ pointwise controls f (i.e., $\mathcal{M}(f) \ge |f|$ almost everywhere). Note that $\mathcal{M}$ maps L^∞ to itself, that is, we have

$$\big\|\mathcal{M}(f)\big\|_{L^\infty} \le \big\|f\big\|_{L^\infty}.$$

Let us compute the Hardy–Littlewood maximal function of a specific function.

Example 2.1.2. On $\mathbf{R}$, let f be the characteristic function of the interval $[a,b]$. For $x \in (a,b)$, clearly $\mathcal{M}(f) = 1$. For $x \ge b$, a simple calculation shows that the largest average of f over all intervals $(x-\delta, x+\delta)$ is obtained when $\delta = x-a$. Similarly, when $x \le a$, the largest average is obtained when $\delta = b-x$. Therefore,

$$\mathcal{M}(f)(x) = \begin{cases} (b-a)/2|x-b| & \text{when } x \le a\,, \\ 1 & \text{when } x \in (a,b)\,, \\ (b-a)/2|x-a| & \text{when } x \ge b\,. \end{cases}$$

Observe that $\mathcal{M}(f)$ has a jump at $x = a$ and $x = b$ equal to one-half that of f.

$\mathcal{M}$ is a sublinear operator, i.e., it satisfies $\mathcal{M}(f+g) \le \mathcal{M}(f) + \mathcal{M}(g)$ and $\mathcal{M}(\lambda f) = |\lambda|\mathcal{M}(f)$ for all locally integrable functions f and g and all complex constants λ. It also has some interesting properties:

If f is locally integrable, then by considering the average of f over the ball $B(x,|x|+R)$, which contains the ball $B(0,R)$, we obtain

$$\mathcal{M}(f)(x) \geq \frac{\int_{B(0,R)} |f(y)|\,dy}{v_n(|x|+R)^n}, \tag{2.1.1}$$

for all $x \in \mathbf{R}^n$, where v_n is the volume of the unit ball in $\mathbf{R}^n$. An interesting consequence of (2.1.1) is the following: suppose that $f \neq 0$ on a set of positive measure E, then $\mathcal{M}(f)$ is not in $L^1(\mathbf{R}^n)$. In other words, if f is in $L^1_{\mathrm{loc}}(\mathbf{R}^n)$ and $\mathcal{M}(f)$ is in $L^1(\mathbf{R}^n)$, then $f = 0$ a.e. To see this, integrate (2.1.1) over the ball $\mathbf{R}^n$ to deduce that $\|f\chi_{B(0,R)}\|_{L^1} = 0$ and thus $f(x) = 0$ for almost all x in the ball $B(0,R)$. Since this is valid for all $R = 1,2,3,\dots$, it follows that $f = 0$ a.e. in $\mathbf{R}^n$.

Another remarkable locality property of $\mathcal{M}$ is that if $\mathcal{M}(f)(x_0) = 0$ for some x_0 in $\mathbf{R}^n$, then $f = 0$ a.e. To see we take $x = x_0$ in (2.1.1) to deduce that $\|f\chi_{B(0,R)}\|_{L^1} = 0$ and as before we have that $f = 0$ a.e. on every ball centered at the origin, i.e., $f = 0$ a.e. in $\mathbf{R}^n$.

A related analogue of $\mathcal{M}(f)$ is its uncentered version $M(f)$, defined as the supremum of all averages of f over all open balls containing a given point.

Definition 2.1.3. The *uncentered Hardy–Littlewood maximal function* of f,

$$M(f)(x) = \sup_{\substack{\delta>0 \\ |y-x|<\delta}} \operatorname*{Avg}_{B(y,\delta)} |f|,$$

is defined as the supremum of the averages of $|f|$ over all open balls $B(y,\delta)$ that contain the point x.

Clearly $\mathcal{M}(f) \leq M(f)$; in other words, M is a larger operator than $\mathcal{M}$. However, $M(f) \leq 2^n\mathcal{M}(f)$ and the boundedness properties of M are identical to those of $\mathcal{M}$.

Example 2.1.4. On $\mathbf{R}$, let f be the characteristic function of the interval $I = [a,b]$. For $x \in (a,b)$, clearly $M(f)(x) = 1$. For $x > b$, a calculation shows that the largest average of f over all intervals $(y-\delta, y+\delta)$ that contain x is obtained when $\delta = \frac{1}{2}(x-a)$ and $y = \frac{1}{2}(x+a)$. Similarly, when $x < a$, the largest average is obtained when $\delta = \frac{1}{2}(b-x)$ and $y = \frac{1}{2}(b+x)$. We conclude that

$$M(f)(x) = \begin{cases} (b-a)/|x-b| & \text{when } x \leq a\,, \\ 1 & \text{when } x \in (a,b)\,, \\ (b-a)/|x-a| & \text{when } x \geq b\,. \end{cases}$$

Observe that M does not have a jump at $x = a$ and $x = b$ and is in fact equal to the function $\left(1 + \frac{\operatorname{dist}(x,I)}{|I|}\right)^{-1}$.

We are now ready to obtain some basic properties of maximal functions. We need the following simple covering lemma.

Lemma 2.1.5. *Let $\{B_1,B_2,\dots,B_k\}$ be a finite collection of open balls in $\mathbf{R}^n$. Then there exists a finite subcollection $\{B_{j_1},\dots,B_{j_l}\}$ of pairwise disjoint balls such that*

$$\sum_{r=1}^{l} |B_{j_r}| \ge 3^{-n}\Big|\bigcup_{i=1}^{k} B_i\Big|. \tag{2.1.2}$$

Proof. Let us reindex the balls so that

$$|B_1| \ge |B_2| \ge \cdots \ge |B_k|.$$

Let $j_1 = 1$. Having chosen $j_1, j_2, \dots, j_i$, let j_{i+1} be the least index $s > j_i$ such that $\bigcup_{m=1}^{i} B_{j_m}$ is disjoint from B_s. Since we have a finite number of balls, this process will terminate, say after l steps. We have now selected pairwise disjoint balls $B_{j_1},\dots,B_{j_l}$. If some B_m was not selected, that is, $m \notin \{j_1,\dots,j_l\}$, then B_m must intersect a selected ball B_{j_r} for some $j_r < m$. Then B_m has smaller size than B_{j_r} and we must have $B_m \subsetneqq 3B_{j_r}$. This shows that the union of the unselected balls is contained in the union of the triples of the selected balls. Therefore, the union of all balls is contained in the union of the triples of the selected balls. Thus

$$\Big|\bigcup_{i=1}^{k} B_i\Big| \le \Big|\bigcup_{r=1}^{l} 3B_{j_r}\Big| \le \sum_{r=1}^{l} |3B_{j_r}| = 3^n \sum_{r=1}^{l} |B_{j_r}|,$$

and the required conclusion follows. □

It was noted earlier that $\mathcal{M}(f)$ and $M(f)$ never map into L^1. However, it is true that these functions are in $L^{1,\infty}$ when f is in L^1. Operators that map L^1 to $L^{1,\infty}$ are said to be *weak type* $(1,1)$. The centered and uncentered maximal functions $\mathcal{M}$ and M are of weak type $(1,1)$ as shown in the next theorem.

Theorem 2.1.6. *The uncentered and centered Hardy–Littlewood maximal operators M and $\mathcal{M}$ map $L^1(\mathbf{R}^n)$ to $L^{1,\infty}(\mathbf{R}^n)$ with constant at most 3^n and also $L^p(\mathbf{R}^n)$ to $L^p(\mathbf{R}^n)$ for $1 < p < \infty$ with constant at most $3^{n/p}p(p-1)^{-1}$. For any $f \in L^1(\mathbf{R}^n)$ we also have*

$$\big|\{M(f) > \alpha\}\big| \le \frac{3^n}{\alpha}\int_{\{M(f)>\alpha\}} |f(y)|\,dy. \tag{2.1.3}$$

Proof. We claim that the set $E_\alpha = \{x \in \mathbf{R}^n : M(f)(x) > \alpha\}$ is open. Indeed, for $x \in E_\alpha$, there is an open ball B_x that contains x such that the average of $|f|$ over B_x is strictly bigger than α. Then the uncentered maximal function of any other point in B_x is also bigger than α, and thus B_x is contained in E_α. This proves that E_α is open.

Let K be a compact subset of E_α. For each $x \in K$ there exists an open ball B_x containing the point x such that

$$\int_{B_x} |f(y)|\,dy > \alpha|B_x|. \tag{2.1.4}$$

Observe that $B_x \subset E_\alpha$ for all x. By compactness there exists a finite subcover $\{B_{x_1}, \dots, B_{x_k}\}$ of K. Using Lemma 2.1.5 we find a subcollection of pairwise disjoint balls $B_{x_{j_1}}, \dots, B_{x_{j_l}}$ such that (2.1.2) holds. Using (2.1.4) and (2.1.2) we obtain

$$|K| \le \Big|\bigcup_{i=1}^k B_{x_i}\Big| \le 3^n \sum_{i=1}^l |B_{x_{j_i}}| \le \frac{3^n}{\alpha} \sum_{i=1}^l \int_{B_{x_{j_i}}} |f(y)|\,dy \le \frac{3^n}{\alpha}\int_{E_\alpha} |f(y)|\,dy,$$

since all the balls $B_{x_{j_i}}$ are disjoint and contained in E_α. Taking the supremum over all compact $K \subseteq E_\alpha$ and using the inner regularity of Lebesgue measure, we deduce (2.1.3). We have now proved that M maps $L^1 \to L^{1,\infty}$ with constant 3^n. It is a trivial fact that M maps $L^\infty \to L^\infty$ with constant 1. Since M is well defined and finite a.e. on $L^1 + L^\infty$, it is also on $L^p(\mathbf{R}^n)$ for $1 < p < \infty$. The Marcinkiewicz interpolation theorem (Theorem 1.3.2) implies that M maps $L^p(\mathbf{R}^n)$ to $L^p(\mathbf{R}^n)$ for all $1 < p < \infty$. Using Exercise 1.3.3, we obtain the following estimate for the operator norm of M on $L^p(\mathbf{R}^n)$:

$$\big\|M\big\|_{L^p \to L^p} \le \frac{p\,3^{\frac{n}{p}}}{p-1}. \tag{2.1.5}$$

Observe that a direct application of Theorem 1.3.2 would give the slightly worse bound of $2\big(\frac{p}{p-1}\big)^{\frac{1}{p}} 3^{\frac{n}{p}}$. Finally the boundedness of $\mathcal{M}$ follows from that of M. □

Remark 2.1.7. The previous proof gives a bound on the operator norm of M on $L^p(\mathbf{R}^n)$ that grows exponentially with the dimension. One may wonder whether this bound could be improved to a better one that does not grow exponentially in the dimension n, as $n \to \infty$. This is not possible; see Exercise 2.1.8.

Example 2.1.8. Let $R > 0$. Then we have

$$\frac{R^n}{(|x|+R)^n} \le M(\chi_{B(0,R)})(x) \le \frac{6^n R^n}{(|x|+R)^n}. \tag{2.1.6}$$

The lower estimate in (2.1.6), is an easy consequence of the fact that the ball $B(x, |x|+R)$ contains the ball $B(0,R)$. For the upper estimate, we first consider the case where $|x| \le 2R$, when clearly $M(\chi_{B(0,R)})(x) \le 1 \le \frac{3^n R^n}{(|x|+R)^n}$. In the case where $|x| > 2R$, if the balls $B(x,r)$ and $B(0,R)$ intersect, we must have that $r > |x| - R$. But note that $|x| - R > \frac{1}{3}(|x|+R)$, since $|x| > 2R$. We conclude that for $|x| > 2R$ we have

$$\mathcal{M}(\chi_{B(0,R)})(x) \le \sup_{r>0} \frac{|B(x,r)\cap B(0,R)|}{|B(x,r)|} \le \sup_{r>|x|-R} \frac{v_n R^n}{v_n r^n} \le \frac{R^n}{\big(\frac{1}{3}(|x|+R)\big)^n}$$

and thus the upper estimate in (2.1.6) holds since $M(\chi_{B(0,R)}) \le 2^n \mathcal{M}(\chi_{B(0,R)})$. Thus in both cases the upper estimate in (2.1.6) is valid.

Next we estimate $M(M(\chi_{B(0,R)}))(x)$. First we write

$$\frac{R^n}{(|x|+R)^n} \le \chi_{B(0,R)} + \sum_{k=0}^\infty \frac{R^n}{(R+2^k R)^n} \chi_{B(0,2^{k+1}R)\setminus B(0,2^k R)}.$$

Using the upper estimate in (2.1.6) and the sublinearity of M, we obtain

$$\begin{aligned} M\Big(\frac{R^n}{(|\cdot|+R)^n}\Big)(x) &\le M(\chi_{B(0,R)})(x)+\sum_{k=0}^{\infty}\frac{1}{(1+2^k)^n}M(\chi_{B(0,2^{k+1}R)})(x)\\ &\le \frac{6^n R^n}{(|x|+R)^n}+\sum_{k=0}^{\infty}\frac{1}{2^{nk}}\frac{6^n(2^{k+1}R)^n}{(|x|+2^{k+1}R)^n}\\ &\le \frac{C_n\log(e+|x|/R)}{(1+|x|/R)^n}, \end{aligned}$$

where the last estimate follows by summing separately over k satisfying $2^{k+1}\le|x|/R$ and $2^{k+1}\ge|x|/R$. Note that the presence of the logarithm does not affect the L^p boundedness of this function when $p>1$.

2.1.2 Control of Other Maximal Operators

We now study some properties of the Hardy–Littlewood maximal function. We begin with a notational definition that we plan to use throughout this book.

Definition 2.1.9. Given a function g on $\mathbf{R}^n$ and $\varepsilon>0$, we denote by g_ε the following function:

$$g_\varepsilon(x)=\varepsilon^{-n}g(\varepsilon^{-1}x). \tag{2.1.7}$$

As observed in Example 1.2.17, if g is an integrable function with integral equal to 1, then the family defined by (2.1.7) is an approximate identity. Therefore, convolution with g_ε is an averaging operation. The Hardy–Littlewood maximal function $\mathcal{M}(f)$ is obtained as the supremum of the averages of a function f with respect to the dilates of the kernel $k=v_n^{-1}\chi_{B(0,1)}$ in $\mathbf{R}^n$; here v_n is the volume of the unit ball $B(0,1)$. Indeed, we have

$$\begin{aligned} \mathcal{M}(f)(x) &= \sup_{\varepsilon>0}\frac{1}{v_n\varepsilon^n}\int_{\mathbf{R}^n}|f(x-y)|\,\chi_{B(0,1)}\Big(\frac{y}{\varepsilon}\Big)\,dy\\ &= \sup_{\varepsilon>0}(|f|*k_\varepsilon)(x). \end{aligned}$$

Note that the function $k=v_n^{-1}\chi_{B(0,1)}$ has integral equal to 1, and convolving with k_ε is an averaging operation.

It turns out that the Hardy–Littlewood maximal function controls the averages of a function with respect to any radially decreasing L^1 function. Recall that a function f on $\mathbf{R}^n$ is called *radial* if $f(x)=f(y)$ whenever $|x|=|y|$. Note that a radial function f on $\mathbf{R}^n$ has the form $f(x)=\varphi(|x|)$ for some function φ on $\mathbf{R}^+$. We have the following result.

Theorem 2.1.10. *Let $k \geq 0$ be a function on $[0,\infty)$ that is continuous except at a finite number of points. Suppose that $K(x) = k(|x|)$ is an integrable function on $\mathbf{R}^n$ that satisfies*

$$K(x) \geq K(y), \quad \text{whenever } |x| \leq |y|, \tag{2.1.8}$$

i.e., k is decreasing. Then the following estimate is true:

$$\sup_{\varepsilon>0}(|f| * K_\varepsilon)(x) \leq \big\|K\big\|_{L^1}\mathcal{M}(f)(x) \tag{2.1.9}$$

for all locally integrable functions f on $\mathbf{R}^n$.

Proof. We prove (2.1.9) when K is radial, satisfies (2.1.8), and is compactly supported and continuous. When this case is established, select a sequence K_j of radial, compactly supported, continuous functions that increase to K as $j \to \infty$. This is possible, since the function k is continuous except at a finite number of points. If (2.1.9) holds for each K_j, passing to the limit implies that (2.1.9) also holds for K. Next, we observe that it suffices to prove (2.1.9) for $x = 0$. When this case is established, replacing $f(t)$ by $f(t+x)$ implies that (2.1.9) holds for all x.

Let us now fix a radial, continuous, and compactly supported function K with support in the ball $B(0,R)$, satisfying (2.1.8). Also fix an $f \in L^1_{\mathrm{loc}}$ and take $x = 0$. Let e_1 be the vector $(1,0,0,\dots,0)$ on the unit sphere $\mathbf{S}^{n-1}$. Polar coordinates give

$$\int_{\mathbf{R}^n}|f(y)|K_\varepsilon(-y)\,dy = \int_0^\infty\int_{\mathbf{S}^{n-1}}|f(r\theta)|K_\varepsilon(re_1)r^{n-1}\,d\theta\,dr. \tag{2.1.10}$$

Define functions

$$\begin{aligned} F(r) &= \int_{\mathbf{S}^{n-1}}|f(r\theta)|\,d\theta\,, \\ G(r) &= \int_0^r F(s)s^{n-1}\,ds\,, \end{aligned}$$

where $d\theta$ denotes surface measure on $\mathbf{S}^{n-1}$. Using these functions, (2.1.10), and integration by parts, we obtain

$$\begin{aligned} \int_{\mathbf{R}^n}|f(y)|K_\varepsilon(y)\,dy &= \int_0^{\varepsilon R}F(r)r^{n-1}K_\varepsilon(re_1)\,dr \\ &= G(\varepsilon R)K_\varepsilon(\varepsilon Re_1) - G(0)K_\varepsilon(0) - \int_0^{\varepsilon R}G(r)\,dK_\varepsilon(re_1) \\ &= \int_0^\infty G(r)\,d(-K_\varepsilon(re_1))\,, \end{aligned} \tag{2.1.11}$$

where two of the integrals are of Lebesgue–Stieltjes type and we used our assumptions that $G(0) = 0$, $K_\varepsilon(0) < \infty$, $G(\varepsilon R) < \infty$, and $K_\varepsilon(\varepsilon Re_1) = 0$. Let v_n be the volume of the unit ball in $\mathbf{R}^n$. Since

$$G(r) = \int_0^r F(s)s^{n-1}\,ds = \int_{|y|\leq r}|f(y)|\,dy \leq \mathcal{M}(f)(0)v_nr^n\,,$$

it follows that the expression in (2.1.11) is dominated by

$$\begin{aligned}\mathcal{M}(f)(0)v_n\int_0^\infty r^n d(-K_\varepsilon(re_1)) &= \mathcal{M}(f)(0)\int_0^\infty nv_n r^{n-1}K_\varepsilon(re_1)\,dr\\ &= \mathcal{M}(f)(0)\big\|K\big\|_{L^1}.\end{aligned}$$

Here we used integration by parts and the fact that the surface measure of the unit sphere $\mathbf{S}^{n-1}$ is equal to nv_n. See Appendix A.3. The theorem is now proved. □

Remark 2.1.11. Theorem 2.1.10 can be generalized as follows. If K is an L^1 function on $\mathbf{R}^n$ such that $|K(x)| \le k_0(|x|) = K_0(x)$, where k_0 is a nonnegative decreasing function on $[0,\infty)$ that is continuous except at a finite number of points, then (2.1.9) holds with $\|K\|_{L^1}$ replaced by $\|K_0\|_{L^1}$. Such a K_0 is called a *radial decreasing majorant* of K. This observation is formulated as the following corollary.

Corollary 2.1.12. *If a function φ has an integrable radially decreasing majorant Φ, then the estimate*

$$\sup_{t>0}|(f*\varphi_t)(x)| \le \big\|\Phi\big\|_{L^1}\mathcal{M}(f)(x)$$

is valid for all locally integrable functions f on $\mathbf{R}^n$.

Example 2.1.13. Let

$$P(x) = \frac{c_n}{(1+|x|^2)^{\frac{n+1}{2}}},$$

where c_n is a constant such that

$$\int_{\mathbf{R}^n} P(x)\,dx = 1\,.$$

The function P is called the *Poisson kernel.* We define L^1 dilates P_t of the Poisson kernel P by setting

$$P_t(x) = t^{-n}P(t^{-1}x)$$

for $t > 0$. It is straightforward to verify that when $n \ge 2$,

$$\frac{d^2}{dt^2}P_t + \sum_{j=1}^n \partial_j^2 P_t = 0,$$

that is, $P_t(x_1,\dots,x_n)$ is a *harmonic function* of the variables $(x_1,\dots,x_n,t)$. Therefore, for $f \in L^p(\mathbf{R}^n)$, $1 \le p < \infty$, the function

$$u(x,t) = (f*P_t)(x)$$

is harmonic in $\mathbf{R}^{n+1}_+$ and converges to $f(x)$ in $L^p(dx)$ as $t \to 0$, since $\{P_t\}_{t>0}$ is an approximate identity. If we knew that $f*P_t$ converged to f a.e. as $t \to 0$, then we could say that $u(x,t)$ solves the *Dirichlet problem*

$$
\begin{aligned}
\partial_t^2 u + \sum_{j=1}^{n} \partial_j^2 u &= 0 \qquad\qquad \text{on } \mathbf{R}_+^{n+1}, \\
u(x,0) &= f(x) \qquad \text{a.e. on } \mathbf{R}^n.
\end{aligned}
\tag{2.1.12}
$$

Solving the Dirichlet problem (2.1.12) motivates the study of the almost everywhere convergence of the expressions $f * P_t$.

Let us now compute the value of the constant c_n. Denote by ω_{n-1} the surface area of $\mathbf{S}^{n-1}$. Using polar coordinates, we obtain

$$
\begin{aligned}
\frac{1}{c_n} &= \int_{\mathbf{R}^n} \frac{dx}{(1+|x|^2)^{\frac{n+1}{2}}} \\
&= \omega_{n-1} \int_0^\infty \frac{r^{n-1}}{(1+r^2)^{\frac{n+1}{2}}} dr \\
&= \omega_{n-1} \int_0^{\pi/2} (\sin\varphi)^{n-1} \, d\varphi \qquad\qquad (r = \tan\varphi) \\
&= \frac{2\pi^{\frac{n}{2}}}{\Gamma(\frac{n}{2})} \frac{1}{2} \frac{\Gamma(\frac{n}{2})\Gamma(\frac{1}{2})}{\Gamma(\frac{n+1}{2})} \\
&= \frac{\pi^{\frac{n+1}{2}}}{\Gamma(\frac{n+1}{2})},
\end{aligned}
$$

where we used the formula for ω_{n-1} in Appendix A.3 and an identity in Appendix A.4. We conclude that

$$
c_n = \frac{\Gamma(\frac{n+1}{2})}{\pi^{\frac{n+1}{2}}}
$$

and that the Poisson kernel on $\mathbf{R}^n$ is given by

$$
P(x) = \frac{\Gamma(\frac{n+1}{2})}{\pi^{\frac{n+1}{2}}} \frac{1}{(1+|x|^2)^{\frac{n+1}{2}}} . \tag{2.1.13}
$$

Theorem 2.1.10 implies that the solution of the Dirichlet problem (2.1.12) is pointwise bounded by the Hardy–Littlewood maximal function of f.

2.1.3 Applications to Differentiation Theory

We continue this section by obtaining some applications of the boundedness of the Hardy–Littlewood maximal function in differentiation theory.

We now show that the weak type $(1,1)$ property of the Hardy–Littlewood maximal function implies almost everywhere convergence for a variety of families of functions. We deduce this from the more general fact that a certain weak type property for the supremum of a family of linear operators implies almost everywhere convergence.

Here is our setup. Let (X,μ), (Y,ν) be measure spaces and let $0 < p \le \infty$, $0 < q < \infty$. Suppose that D is a dense subspace of $L^p(X,\mu)$. This means that for all $f \in L^p$ and all $\delta > 0$ there exists a $g \in D$ such that $\|f-g\|_{L^p} < \delta$. Suppose that for every $\varepsilon > 0$, T_ε is a linear operator that maps $L^p(X,\mu)$ into a subspace of measurable functions, which are defined everywhere on Y. For $y \in Y$, define a sublinear operator

$$T_*(f)(y) = \sup_{\varepsilon>0} |T_\varepsilon(f)(y)| \tag{2.1.14}$$

and assume that $T_*(f)$ is ν- measurable for any $f \in L^p(X,\mu)$. We have the following.

Theorem 2.1.14. *Let $0 < p < \infty$, $0 < q < \infty$, and T_ε and T_* as previously. Suppose that for some $B > 0$ and all $f \in L^p(X)$ we have*

$$\big\|T_*(f)\big\|_{L^{q,\infty}} \le B\big\|f\big\|_{L^p} \tag{2.1.15}$$

and that for all $f \in D$,

$$\lim_{\varepsilon\to 0} T_\varepsilon(f) = T(f) \tag{2.1.16}$$

exists and is finite ν-a.e. (and defines a linear operator on D). Then for all functions f in $L^p(X,\mu)$ the limit (2.1.16) exists and is finite ν-a.e., and defines a linear operator T on $L^p(X)$ (uniquely extending T defined on D) that satisfies

$$\big\|T(f)\big\|_{L^{q,\infty}} \le B\big\|f\big\|_{L^p} \tag{2.1.17}$$

for all functions f in $L^p(X)$.

Proof. Given f in L^p, we define the *oscillation* of f:

$$O_f(y) = \limsup_{\varepsilon\to 0}\limsup_{\theta\to 0} |T_\varepsilon(f)(y) - T_\theta(f)(y)|\,.$$

We would like to show that for all $f \in L^p$ and $\delta > 0$,

$$\nu(\{y \in Y:\ O_f(y) > \delta\}) = 0\,. \tag{2.1.18}$$

Once (2.1.18) is established, given $f \in L^p(X)$, we obtain that $O_f(y) = 0$ for ν-almost all y, which implies that $T_\varepsilon(f)(y)$ is Cauchy for ν-almost all y, and it therefore converges ν-a.e. to some $T(f)(y)$ as $\varepsilon \to 0$. The operator T defined this way on $L^p(X)$ is linear and extends T defined on D.

To approximate O_f we use density. Given $\eta > 0$, find a function $g \in D$ such that $\|f-g\|_{L^p} < \eta$. Since $T_\varepsilon(g) \to T(g)$ ν-a.e, it follows that $O_g = 0$ ν-a.e. Using this fact and the linearity of the T_ε's, we conclude that

$$O_f(y) \le O_g(y) + O_{f-g}(y) = O_{f-g}(y) \qquad \nu\text{-a.e.}$$

Now for any $\delta > 0$ we have

$$\begin{aligned} \nu(\{y \in Y:\, O_f(y) > \delta\}) &\le \nu(\{y \in Y:\, O_{f-g}(y) > \delta\}) \\ &\le \nu(\{y \in Y:\, 2T_*(f-g)(y) > \delta\}) \\ &\le \big(2B\|f-g\|_{L^p}/\delta\big)^q \\ &\le (2B\eta/\delta)^q\,. \end{aligned}$$

Letting $\eta \to 0$, we deduce (2.1.18). We conclude that $T_\varepsilon(f)$ is a Cauchy sequence, and hence it converges ν-a.e. to some $T(f)$. Since $|T(f)| \le |T_*(f)|$, the conclusion (2.1.17) of the theorem follows easily. □

We now derive some applications. First we return to the issue of almost everywhere convergence of the expressions $f * P_y$, where P is the Poisson kernel.

Example 2.1.15. Fix $1 \le p < \infty$ and $f \in L^p(\mathbf{R}^n)$. Let

$$P(x) = \frac{\Gamma(\frac{n+1}{2})}{\pi^{\frac{n+1}{2}}} \frac{1}{(1+|x|^2)^{\frac{n+1}{2}}}$$

be the Poisson kernel on $\mathbf{R}^n$ and let $P_\varepsilon(x) = \varepsilon^{-n}P(\varepsilon^{-1}x)$. We deduce from the previous theorem that the family $f * P_\varepsilon$ converges to f a.e. Let D be the set of all continuous functions with compact support on $\mathbf{R}^n$. Since the family $(P_\varepsilon)_{\varepsilon>0}$ is an approximate identity, Theorem 1.2.19 (2) implies that for f in D we have that $f * P_\varepsilon \to f$ uniformly on compact subsets of $\mathbf{R}^n$ and hence pointwise everywhere. In view of Theorem 2.1.10, the supremum of the family of linear operators $T_\varepsilon(f) = f * P_\varepsilon$ is controlled by the Hardy–Littlewood maximal function, and thus it maps L^p to $L^{p,\infty}$ for $1 \le p < \infty$. Theorem 2.1.14 now gives that $f * P_\varepsilon$ converges to f a.e. for all $f \in L^p$.

Here is another application of Theorem 2.1.14. Exercise 2.1.10 contains other applications.

Corollary 2.1.16. ***(Lebesgue's differentiation theorem)*** *For any locally integrable function f on $\mathbf{R}^n$ we have*

$$\lim_{r\to 0} \frac{1}{|B(x,r)|} \int_{B(x,r)} f(y)\,dy = f(x) \tag{2.1.19}$$

for almost all x in $\mathbf{R}^n$. Consequently we have $|f| \le \mathcal{M}(f)$ a.e. There is also an analogous statement to (2.1.19) *in which balls are replaced by cubes centered at x. Precisely, for any locally integrable function f on $\mathbf{R}^n$ we have*

$$\lim_{r\to 0} \frac{1}{(2r)^n} \int_{x+[-r,r]^n} f(y)\,dy = f(x) \tag{2.1.20}$$

for almost all x in $\mathbf{R}^n$.

Proof. Since $\mathbf{R}^n$ is the union of the balls $B(0,N)$ for $N = 1,2,3\dots$, it suffices to prove the required conclusion for almost all x inside a fixed ball $B(0,N)$. Given a locally integrable function f on $\mathbf{R}^n$, consider the function $f_N = f\chi_{B(0,N+1)}$. Then f_N lies in $L^1(\mathbf{R}^n)$. Let T_ε be the operator given with convolution with k_ε, where $k = v_n^{-1}\chi_{B(0,1)}$ and $0 < \varepsilon < 1$. We know that the corresponding maximal operator T_* is controlled by the centered Hardy–Littlewood maximal function $\mathcal{M}$, which maps L^1 to $L^{1,\infty}$. It is straightforward to verify that (2.1.19) holds for all continuous functions f with compact support. Since this set of functions is dense in L^1, and T_* maps L^1 to $L^{1,\infty}$, Theorem 2.1.14 implies that (2.1.19) holds for all integrable functions on $\mathbf{R}^n$, in particular for f_N. But for $0 < \varepsilon < 1$ and $x \in B(0,N)$ we have $f\chi_{B(x,\varepsilon)} = f_N\chi_{B(x,\varepsilon)}$, so it follows that

$$\lim_{\varepsilon\to 0}\frac{1}{|B(x,\varepsilon)|}\int_{B(x,\varepsilon)} f(y)\,dy = \lim_{\varepsilon\to 0}\frac{1}{|B(x,\varepsilon)|}\int_{B(x,\varepsilon)} f_N(y)\,dy = f_N(x)$$

for almost all $x \in \mathbf{R}^n$, in particular for almost all x in $B(0,N)$. But on this set $f_N = f$, so the required conclusion follows. The assertion that $|f| \le \mathcal{M}(f)$ a.e. is an easy consequence of (2.1.19) when the limit is replaced by a supremum.

Finally, with minor modifications, the proof can be adjusted to work for cubes in place of balls. To prove (2.1.20), for $f \in L^1_{\text{loc}}(\mathbf{R}^n)$ we introduce the maximal operator

$$\mathcal{M}_c(f)(x) = \sup_{r>0}\frac{1}{(2r)^n}\int_{x+[-r,r]^n} |f(y)|\,dy.$$

Then Exercise 2.1.3 yields that $\mathcal{M}_c$ maps $L^1(\mathbf{R}^n)$ to weak $L^1(\mathbf{R}^n)$ and the preceding proof with $\mathcal{M}_c$ in place of $\mathcal{M}$ yields (2.1.20). □

The following corollaries were inspired by Example 2.1.15.

Corollary 2.1.17. ***(Differentiation theorem for approximate identities)*** *Let K be an L^1 function on $\mathbf{R}^n$ with integral 1 that has a continuous integrable radially decreasing majorant. Then $f * K_\varepsilon \to f$ a.e. as $\varepsilon \to 0$ for all $f \in L^p(\mathbf{R}^n)$, $1 \le p < \infty$.*

Proof. It follows from Example 1.2.17 that K_ε is an approximate identity. Theorem 1.2.19 now implies that $f * K_\varepsilon \to f$ uniformly on compact sets when f is continuous. Let D be the space of all continuous functions with compact support. Then $f * K_\varepsilon \to f$ a.e. for $f \in D$. It follows from Corollary 2.1.12 that $T_*(f) = \sup_{\varepsilon>0}|f * K_\varepsilon|$ maps L^p to $L^{p,\infty}$ for $1 \le p < \infty$. Using Theorem 2.1.14, we conclude the proof of the corollary.

□

Remark 2.1.18. Fix $f \in L^p(\mathbf{R}^n)$ for some $1 \le p < \infty$. Theorem 1.2.19 implies that $f * K_\varepsilon$ converges to f in L^p and hence some subsequence $f * K_{\varepsilon_n}$ of $f * K_\varepsilon$ converges to f a.e. as $n \to \infty$, ($\varepsilon_n \to 0$). Compare this result with Corollary 2.1.17, which gives a.e. convergence for the whole family $f * K_\varepsilon$ as $\varepsilon \to 0$.

Corollary 2.1.19. ***(Differentiation theorem for multiples of approximate identities)*** *Let K be a function on $\mathbf{R}^n$ that has an integrable radially decreasing majorant.*

Let $a = \int_{\mathbf{R}^n} K(x)\,dx$. Then for all $f \in L^p(\mathbf{R}^n)$ and $1 \le p < \infty$, $(f * K_\varepsilon)(x) \to af(x)$ for almost all $x \in \mathbf{R}^n$ as $\varepsilon \to 0$.

Proof. Use Theorem 1.2.21 instead of Theorem 1.2.19 in the proof of Corollary 2.1.17. □

The following application of the Lebesgue differentiation theorem uses a simple *stopping-time argument*. This is the sort of argument in which a selection procedure stops when it is exhausted at a certain scale and is then repeated at the next scale. A certain refinement of the following proposition is of fundamental importance in the study of singular integrals given in Chapter 4.

Proposition 2.1.20. *Given a nonnegative integrable function f on $\mathbf{R}^n$ and $\alpha > 0$, there exists a collection of disjoint (possibly empty) open cubes Q_j such that for almost all $x \in \big(\bigcup_j Q_j\big)^c$ we have $f(x) \le \alpha$ and*

$$\alpha < \frac{1}{|Q_j|}\int_{Q_j} f(t)\,dt \le 2^n \alpha\,. \tag{2.1.21}$$

Proof. The proof provides an excellent paradigm of a stopping-time argument. Start by decomposing $\mathbf{R}^n$ as a union of cubes of equal size, whose interiors are disjoint, and whose diameter is so large that $|Q|^{-1}\int_Q f(x)\,dx \le \alpha$ for every Q in this mesh. This is possible since f is integrable and $|Q|^{-1}\int_Q f(x)\,dx \to 0$ as $|Q| \to \infty$. Call the union of these cubes $\mathscr{E}_0$.

Divide each cube in the mesh into 2^n congruent cubes by bisecting each of the sides. Call the new collection of cubes $\mathscr{E}_1$. Select a cube Q in $\mathscr{E}_1$ if

$$\frac{1}{|Q|}\int_Q f(x)\,dx > \alpha \tag{2.1.22}$$

and call the set of all selected cubes $\mathscr{S}_1$. Now subdivide each cube in $\mathscr{E}_1 \setminus \mathscr{S}_1$ into 2^n congruent cubes by bisecting each of the sides as before. Call this new collection of cubes $\mathscr{E}_2$. Repeat the same procedure and select a family of cubes $\mathscr{S}_2$ that satisfy (2.1.22). Continue this way ad infinitum and call the cubes in $\bigcup_{m=1}^{\infty} \mathscr{S}_m$ "selected." If Q was selected, then there exists Q_1 in $\mathscr{E}_{m-1}$ containing Q that was not selected at the $(m-1)th$ step for some $m \ge 1$. Therefore,

$$\alpha < \frac{1}{|Q|}\int_Q f(x)\,dx \le 2^n \frac{1}{|Q_1|}\int_{Q_1} f(x)\,dx \le 2^n \alpha\,.$$

Now call F the closure of the complement of the union of all selected cubes. If $x \in F$, then there exists a sequence of cubes containing x whose diameter shrinks down to zero such that the average of f over these cubes is less than or equal to α. By Corollary 2.1.16, it follows that $f(x) \le \alpha$ almost everywhere in F. This proves the proposition. □

In the proof of Proposition 2.1.20 it was not crucial to assume that f was defined on all $\mathbf{R}^n$, but only on a cube. We now give a local version of this result.

Corollary 2.1.21. *Let $f \geq 0$ be an integrable function over a cube Q in $\mathbf{R}^n$ and let $\alpha \geq \frac{1}{|Q|}\int_Q f\,dx$. Then there exist disjoint (possibly empty) open subcubes Q_j of Q such that for almost all $x \in Q \setminus \bigcup_j Q_j$ we have $f \leq \alpha$ and (2.1.21) holds for all Q_j.*

Proof. The proof easily follows by a simple modification of Proposition 2.1.20 in which $\mathbf{R}^n$ is replaced by the fixed cube Q. To apply Corollary 2.1.16, we extend f to be zero outside the cube Q. □

See Exercise 2.1.4 for an application of Proposition 2.1.20 involving maximal functions.

Exercises

2.1.1. A positive Borel measure μ on $\mathbf{R}^n$ is called *inner regular* if for any open subset U of $\mathbf{R}^n$ we have $\mu(U) = \sup\{\mu(K) : K \subsetneqq U,\ K \text{ compact}\}$ and μ is called *locally finite* if $\mu(B) < \infty$ for all balls B.
(a) Let μ be a positive inner regular locally finite measure on $\mathbf{R}^n$ that satisfies the following *doubling condition*: There exists a constant $D(\mu) > 0$ such that for all $x \in \mathbf{R}^n$ and $r > 0$ we have

$$\mu(3B(x,r)) \leq D(\mu)\,\mu(B(x,r)).$$

For $f \in L^1_{\mathrm{loc}}(\mathbf{R}^n, \mu)$ define the uncentered maximal function $M_\mu(f)$ with respect to μ by

$$M_\mu(f)(x) = \sup_{r>0} \sup_{\substack{z:\,|z-x|<r \\ \mu(B(z,r))\neq 0}} \frac{1}{\mu(B(z,r))}\int_{B(z,r)} f(y)\,d\mu(y).$$

Show that M_μ maps $L^1(\mathbf{R}^n, \mu)$ to $L^{1,\infty}(\mathbf{R}^n, \mu)$ with constant at most $D(\mu)$ and $L^p(\mathbf{R}^n, \mu)$ to itself with constant at most $2(\frac{p}{p-1})^{\frac{1}{p}} D(\mu)^{\frac{1}{p}}$.

(b) Obtain as a consequence a differentiation theorem analogous to Corollary 2.1.16.
[*Hint:* Part (a): For $f \in L^1(\mathbf{R}^n, \mu)$ show that the set $E_\alpha = \{M_\mu(f) > \alpha\}$ is open. Then use the argument of the proof of Theorem 2.1.6 and the inner regularity of μ.]

2.1.2. On $\mathbf{R}$ consider the maximal function M_μ of Exercise 2.1.1.
(a) (*W. H. Young*) Prove the following covering lemma. Given a finite set $\mathcal{F}$ of open intervals in $\mathbf{R}$, prove that there exist two subfamilies each consisting of pairwise disjoint intervals such that the union of the intervals in the original family is equal to the union of the intervals of both subfamilies. Use this result to show that the maximal function M_μ of Exercise 2.1.1 maps $L^1(\mu) \to L^{1,\infty}(\mu)$ with constant at most 2.

(b) ([134]) Prove that for any σ-finite positive measure μ on $\mathbf{R}$, $\alpha > 0$, and $f \in L^1_{\mathrm{loc}}(\mathbf{R},\mu)$ we have

$$\frac{1}{\alpha}\int_A |f|\,d\mu - \mu(A) \le \frac{1}{\alpha}\int_{\{|f|>\alpha\}} |f|\,d\mu - \mu(\{|f|>\alpha\}).$$

Use this result and part (a) to prove that for all $\alpha > 0$ and all locally integrable f we have

$$\mu(\{|f|>\alpha\}) + \mu(\{M_\mu(f)>\alpha\}) \le \frac{1}{\alpha}\int_{\{|f|>\alpha\}} |f|\,d\mu + \frac{1}{\alpha}\int_{\{M_\mu(f)>\alpha\}} |f|\,d\mu$$

and note that equality is obtained when $\alpha = 1$ and $f(x) = |x|^{-1/p}$.
(c) Conclude that M_μ maps $L^p(\mu)$ to $L^p(\mu)$, $1 < p < \infty$, with bound at most the unique positive solution A_p of the equation

$$(p-1)x^p - px^{p-1} - 1 = 0.$$

(d) ([136]) If μ is the Lebesgue measure show that for $1 < p < \infty$ we have

$$\big\|M\big\|_{L^p\to L^p} = A_p,$$

where A_p is the unique positive solution of the equation in part (c).
[*Hint:* Part (a): Select a subset $\mathcal{G}$ of $\mathcal{F}$ with minimal cardinality such that $\bigcup_{J\in\mathcal{G}} J = \bigcup_{I\in\mathcal{F}} I$. Part (d): One direction follows from part (c). Conversely, $M(|x|^{-1/p})(1) = \frac{p}{p-1}\frac{\gamma^{1/p'}+1}{\gamma+1}$, where γ is the unique positive solution of the equation $\frac{p}{p-1}\frac{\gamma^{1/p'}+1}{\gamma+1} = \gamma^{-1/p}$. Conclude that $M(|x|^{-1/p})(1) = A_p$ and that $M(|x|^{-1/p}) = A_p|x|^{-1/p}$. Since this function is not in L^p, consider the family $f_\varepsilon(x) = |x|^{-1/p}\min(|x|^{-\varepsilon},|x|^{\varepsilon})$, $\varepsilon > 0$, and show that $M(f_\varepsilon)(x) \ge (1+\gamma^{\frac{1}{p'}+\varepsilon})(1+\gamma)^{-1}(\frac{1}{p'}+\varepsilon)^{-1} f_\varepsilon(x)$ for $0 < \varepsilon < p'$.]

2.1.3. Define the centered Hardy–Littlewood maximal function $\mathcal{M}_c$ and the uncentered Hardy–Littlewood maximal function M_c using cubes with sides parallel to the axes instead of balls in $\mathbf{R}^n$. Prove that

$$1 \le \frac{M(f)}{\mathcal{M}(f)} \le 2^n, \quad \frac{1}{n^{\frac{n}{2}}}\frac{2^n}{v_n} \le \frac{M(f)}{M_c(f)} \le \frac{2^n}{v_n}, \quad \frac{1}{n^{\frac{n}{2}}}\frac{2^n}{v_n} \le \frac{\mathcal{M}(f)}{\mathcal{M}_c(f)} \le \frac{2^n}{v_n},$$

where v_n is the volume of the unit ball in $\mathbf{R}^n$. Conclude that $\mathcal{M}_c$ and M_c are weak type $(1,1)$ and they map $L^p(\mathbf{R}^n)$ to $L^p(\mathbf{R}^n)$ for $1 < p \le \infty$.

2.1.4. (a) Prove the estimate:

$$|\{x\in\mathbf{R}^n : M(f)(x) > 2\alpha\}| \le \frac{3^n}{\alpha}\int_{\{|f|>\alpha\}} |f(y)|\,dy$$

and conclude that M maps L^p to $L^{p,\infty}$ with norm at most $2\cdot 3^{n/p}$ for $1 \le p < \infty$.

(b) Deduce that if $f\log^+(2|f|)$ is integrable over a ball B, then $M(f)$ is integrable over the same ball B.
(c) ([375], [336]) Apply Proposition 2.1.20 to $|f|$ and $\alpha>0$ and Exercise 2.1.3 to show that with $c_n=2^n(n^{n/2}v_n)^{-1}$ we have

$$|\{x\in\mathbf{R}^n:\,M(f)(x)>c_n\alpha\}|\geq\frac{2^{-n}}{\alpha}\int_{\{|f|>\alpha\}}|f(y)|\,dy.$$

(d) Suppose that f is integrable and supported in a ball $B(0,\rho)$. Show that for x in $B(0,2\rho)\setminus B(0,\rho)$ we have $\mathcal{M}(f)(x)\leq\mathcal{M}(f)(\rho^2|x|^{-2}x)$. Conclude that

$$\int_{B(0,2\rho)}\mathcal{M}(f)\,dx\leq(4^n+1)\int_{B(0,\rho)}\mathcal{M}(f)\,dx$$

and from this deduce a similar inequality for $M(f)$.
(e) Suppose that f is integrable and supported in a ball B and that $M(f)$ is integrable over B. Let $\lambda_0=2^n|B|^{-1}\|f\|_{L^1}$. Use part (b) to prove that $f\log^+(\lambda_0^{-1}c_n|f|)$ is integrable over B.
[*Hint:* Part (a): Write $f=f\chi_{|f|>\alpha}+f\chi_{|f|\leq\alpha}$. Part (b): Show that $M(f\chi_E)$ is integrable over B, where $E=\{|f|\geq 1/2\}$. Part (c): Use Proposition 2.1.20. Part (d): Let $x'=\rho^2|x|^{-2}x$ for some $\rho<|x|<2\rho$. Show that for $R>|x|-\rho$, we have that

$$\int_{B(x,R)}|f(z)|\,dz\leq\int_{B(x',R)}|f(z)|\,dz$$

by showing that $B(x,R)\cap B(0,\rho)\subset B(x',R)$. Part (e): For $x\notin 2B$ we have $M(f)(x)\leq\lambda_0$, hence $\int_{2B}M(f)(x)\,dx\geq\int_{\lambda_0}^\infty|\{x\in 2B:\,M(f)(x)>\alpha\}|\,d\alpha$.]

2.1.5. (*A. Kolmogorov*) Let S be a sublinear operator that maps $L^1(\mathbf{R}^n)$ to $L^{1,\infty}(\mathbf{R}^n)$ with norm B. Suppose that $f\in L^1(\mathbf{R}^n)$. Prove that for any set A of finite Lebesgue measure and for all $0<q<1$ we have

$$\int_A|S(f)(x)|^q\,dx\leq(1-q)^{-1}B^q|A|^{1-q}\|f\|_{L^1}^q,$$

and in particular, for the Hardy–Littlewood maximal operator,

$$\int_A M(f)(x)^q\,dx\leq(1-q)^{-1}3^{nq}|A|^{1-q}\|f\|_{L^1}^q.$$

[*Hint:* Use the identity

$$\int_A|S(f)(x)|^q\,dx=\int_0^\infty q\alpha^{q-1}|\{x{\in}A:\,S(f)(x){>}\alpha\}|\,d\alpha$$

and estimate the last measure by $\min(|A|,\frac{B}{\alpha}\|f\|_{L^1})$.]

2.1.6. Let $M_s(f)(x)$ be the supremum of the averages of $|f|$ over all rectangles with sides parallel to the axes containing x. The operator M_s is called the *strong maximal function*.
(a) Prove that M_s maps $L^p(\mathbf{R}^n)$ to itself.
(b) Show that the operator norm of M_s is A_p^n, where A_p is as in Exercise 2.1.2 (c).
(c) Prove that M_s is not weak type (1,1).

2.1.7. Prove that if

$$|\varphi(x_1,\dots,x_n)| \le A(1+|x_1|)^{-1-\varepsilon}\cdots(1+|x_n|)^{-1-\varepsilon}$$

for some $A,\varepsilon>0$, and $\varphi_{t_1,\dots,t_n}(x)=t_1^{-1}\cdots t_n^{-1}\varphi(t_1^{-1}x_1,\dots,t_n^{-1}x_n)$, then the maximal operator

$$f \mapsto \sup_{t_1,\dots,t_n>0} |f*\varphi_{t_1,\dots,t_n}|$$

is pointwise controlled by the strong maximal function.

2.1.8. Prove that for any fixed $1<p<\infty$, the operator norm of M on $L^p(\mathbf{R}^n)$ tends to infinity as $n\to\infty$.
[*Hint:* Let f_0 be the characteristic function of the unit ball in $\mathbf{R}^n$. Consider the averages $|B_x|^{-1}\int_{B_x} f_0\,dy$, where $B_x = B\big(\frac{1}{2}(|x|-|x|^{-1})\frac{x}{|x|},\frac{1}{2}(|x|+|x|^{-1})\big)$ for $|x|>1$.]

2.1.9. (a) In $\mathbf{R}^2$ let $M_0(f)(x)$ be the maximal function obtained by taking the supremum of the averages of $|f|$ over all rectangles (of arbitrary orientation) containing x. Prove that M_0 is not bounded on $L^p(\mathbf{R}^n)$ for $p\le 2$ and conclude that M_0 is not weak type $(1,1)$.
(b) Let $M_{00}(f)(x)$ be the maximal function obtained by taking the supremum of the averages of $|f|$ over all rectangles in $\mathbf{R}^2$ of arbitrary orientation but fixed eccentricity containing x. (The eccentricity of a rectangle is the ratio of its longer side to its shorter side.) Using a covering lemma, show that M_{00} is weak type $(1,1)$ with a bound proportional to the square of the eccentricity.
(c) On $\mathbf{R}^n$ define a maximal function by taking the supremum of the averages of $|f|$ over all products of intervals $I_1\times\cdots\times I_n$ containing a point x with $|I_2| = a_2|I_1|,\dots,|I_n|=a_n|I_1|$ and $a_2,\dots,a_n>0$ fixed. Show that this maximal function is of weak type $(1,1)$ with bound independent of the numbers $a_2,\dots,a_n$.
[*Hint:* Part (b): Let b be the eccentricity. If two rectangles with the same eccentricity intersect, then the smaller one is contained in the bigger one scaled $4b$ times. Then use an argument similar to that in Lemma 2.1.5.]

2.1.10. (a) Let $0<p,q<\infty$ and let X,Y be measure spaces. Suppose that T_ε are maps from $L^p(X)$ to $L^{q,\infty}(Y)$ satisfy $|T_\varepsilon(f+g)|\le K(|T_\varepsilon(f)|+|T_\varepsilon(g)|)$ for all $\varepsilon>0$ and all $f,g\in L^p(X)$, and also $\lim_{\varepsilon\to 0}T_\varepsilon(f)=0$ a.e. for all f in some dense subspace D of $L^p(X)$. Assume furthermore that the maximal operator $T_*(f)=\sup_{\varepsilon>0}|T_\varepsilon(f)|$ maps $L^p(X)$ to $L^{q,\infty}(Y)$. Prove that $\lim_{\varepsilon\to 0}T_\varepsilon(f)=0$ a.e. for all f in $L^p(X)$.
(b) Use the result in part (a) to prove the following version of the Lebesgue differentiation theorem: Let $f\in L^p(\mathbf{R}^n)$ for some $0<p<\infty$. Then for almost all $x\in\mathbf{R}^n$ we have

$$\lim_{\substack{|B|\to 0\\ B\ni x}} \frac{1}{|B|}\int_B |g(y)-g(x)|^p\,dy = 0\,,$$

where the limit is taken over all open balls B containing x and shrinking to $\{x\}$.
(c) Conclude that for any f in $L^1_{\mathrm{loc}}(\mathbf{R}^n)$ and for almost all $x \in \mathbf{R}^n$ we have

$$\lim_{\substack{|B|\to 0\\ B\ni x}} \frac{1}{|B|}\int_B f(y)\,dy = f(x)\,,$$

where the limit is taken over all open balls B containing x and shrinking to $\{x\}$.
[*Hint:* (a) Define an oscillation $O_f(y) = \limsup_{\varepsilon\to 0}|T_\varepsilon(f)(y)|$. For all f in $L^p(X)$ and $g \in D$ we have that $O_f(y) \le K O_{f-g}(y)$. Then use the argument in the proof of Theorem 2.1.14. (b) Apply part (a) with

$$T_\varepsilon(f)(x) = \sup_{B(z,\varepsilon)\ni x}\left(\frac{1}{|B(z,\varepsilon)|}\int_{B(z,\varepsilon)} |f(y)-f(x)|^p\,dy\right)^{1/p},$$

observing that $T_*(f) = \sup_{\varepsilon>0} T_\varepsilon(f) \le \max(1, 2^{\frac{1-p}{p}})\big(|f| + M(|f|^p)^{\frac{1}{p}}\big)$. (c) Follows from part (b) with $p=1$. Note that part (b) can be proved without part (a) but using part (c) as follows: for every rational number a there is a set E_a of Lebesgue measure zero such that for $x \in \mathbf{R}^n \setminus E_a$ we have $\lim_{B\ni x, |B|\to 0}\frac{1}{|B|}\int_B |g(y)-a|^p\,dy = |g(x)-a|^p$, since the function $y \mapsto |f(y)-a|^p$ is in $L^1_{\mathrm{loc}}(\mathbf{R}^n)$. By considering an enumeration of the rationals, find a set of measure zero E such for $x \notin E$ the preceding limit exists for all rationals a and by continuity for all real numbers a, in particular for $a = g(x)$.]

2.1.11. Let f be in $L^1(\mathbf{R})$. Define the right maximal function $M_R(f)$ and the left maximal function $M_L(f)$ as follows:

$$\begin{aligned}
M_L(f)(x) &= \sup_{r>0}\frac{1}{r}\int_{x-r}^{x} |f(t)|\,dt\,,\\
M_R(f)(x) &= \sup_{r>0}\frac{1}{r}\int_{x}^{x+r} |f(t)|\,dt\,.
\end{aligned}$$

(a) Show that for all $\alpha > 0$ and $f \in L^1(\mathbf{R})$ we have

$$\begin{aligned}
|\{x\in\mathbf{R}:\, M_L(f)(x) > \alpha\}| &= \frac{1}{\alpha}\int_{\{M_L(f)>\alpha\}} |f(t)|\,dt\,,\\
|\{x\in\mathbf{R}:\, M_R(f)(x) > \alpha\}| &= \frac{1}{\alpha}\int_{\{M_R(f)>\alpha\}} |f(t)|\,dt\,.
\end{aligned}$$

(b) Extend the definition of $M_L(f)$ and $M_R(f)$ for $f \in L^p(\mathbf{R})$ for $1 \le p \le \infty$. Show that M_L and M_R map L^p to L^p with norm at most $p/(p-1)$ for all p with $1 < p < \infty$.
(c) Construct examples to show that the operator norms of M_L and M_R on $L^p(\mathbf{R})$ are exactly $p/(p-1)$ for $1 < p < \infty$.

(d) Prove that $M = \max(M_R, M_L)$.
(e) Let $N = \min(M_R, M_L)$. Obtain the following consequence of part (a),

$$\int_{\mathbf{R}} M(f)^p + N(f)^p \, dx = \frac{p}{p-1} \int_{\mathbf{R}} |f| \left(M(f)^{p-1} + N(f)^{p-1}\right) dx,$$

(f) Use part (e) to prove that

$$(p-1)\big\|M(f)\big\|_{L^p}^p - p\big\|f\big\|_{L^p}\big\|M(f)\big\|_{L^p}^{p-1} - \big\|f\big\|_{L^p}^p \le 0.$$

[*Hint:* (a) Write the set $E_\alpha = \{M_R(f) > \alpha\}$ as a union of open intervals (a_j, b_j). For each x in (a_j, b_j), let $N_x = \{s \in \mathbf{R} : \int_x^s |f| > \alpha(s-x)\} \cap (x, b_j]$. Show that N_x is nonempty and that $\sup N_x = b_j$ for every $x \in (a_j, b_j)$. Conclude that $\int_{a_j}^{b_j} |f(t)|\, dt \ge \alpha(b_j - a_j)$, which implies that each a_j is finite. For the reverse inequality use that $a_j \notin E_\alpha$. Part (d) is due to K. L. Phillips. (e) First obtain a version of the equality with M_R in the place of M and M_L in the place of N. Then use that $M(f)^q + N(f)^q = M_L(f)^q + M_R(f)^q$ for all q. (f) Use that $|f|N(f)^{p-1} \le \frac{1}{p}|f|^p + \frac{1}{p'}N(f)^p$. This alternative proof of the result in Exercise 2.1.2(c) was suggested by J. Duoandikoetxea.]

2.1.12. A cube $Q = [a_1 2^k, (a_1+1)2^k) \times \cdots \times [a_n 2^k, (a_n+1)2^k)$ on $\mathbf{R}^n$ is called *dyadic* if $k, a_1, \dots, a_n \in \mathbf{Z}$. Observe that either two dyadic cubes are disjoint or one contains the other. Define the *dyadic maximal function*

$$M_d(f)(x) = \sup_{Q \ni x} \frac{1}{|Q|} \int_Q f(y)\, dy,$$

where the supremum is taken over all dyadic cubes Q containing x.
(a) Prove that M_d maps L^1 to $L^{1,\infty}$ with constant at most one. Presicely, show that for all $\alpha > 0$ and $f \in L^1(\mathbf{R}^n)$ we have

$$\big|\{x \in \mathbf{R}^n : M_d(f)(x) > \alpha\}\big| \le \frac{1}{\alpha} \int_{\{M_d(f) > \alpha\}} f(t)\, dt.$$

(b) Conclude that M_d maps $L^p(\mathbf{R}^n)$ to itself with constant at most $p/(p-1)$.

2.1.13. Observe that the proof of Theorem 2.1.6 yields the estimate

$$\lambda \big|\{M(f) > \lambda\}\big|^{\frac{1}{p}} \le 3^n \big|\{M(f) > \lambda\}\big|^{-1+\frac{1}{p}} \int_{\{M(f) > \lambda\}} |f(y)|\, dy$$

for $\lambda > 0$ and f locally integrable. Use the result of Exercise 1.1.12(a) to prove that the Hardy–Littlewood maximal operator M maps the space $L^{p,\infty}(\mathbf{R}^n)$ to itself for $1 < p < \infty$.

2.1.14. Let $K(x) = (1+|x|)^{-n-\delta}$ be defined on $\mathbf{R}^n$. Prove that there exists a constant $C_{n,\delta}$ such that for all $\varepsilon_0 > 0$ we have the estimate

$$\sup_{\varepsilon>\varepsilon_0} (|f| * K_\varepsilon)(x) \le C_{n,\delta} \sup_{\varepsilon>\varepsilon_0} \frac{1}{\varepsilon^n} \int_{|y-x|\le\varepsilon} |f(y)|\, dy,$$

for all f locally integrable on $\mathbf{R}^n$.
[*Hint:* Apply only a minor modification to the proof of Theorem 2.1.10.]

2.2 The Schwartz Class and the Fourier Transform

In this section we introduce the single most important tool in harmonic analysis, the Fourier transform. It is often the case that the Fourier transform is introduced as an operation on L^1 functions. In this exposition we first define the Fourier transform on a smaller class, the space of Schwartz functions, which turns out to be a very natural environment. Once the basic properties of the Fourier transform are derived, we extend its definition to other spaces of functions.

We begin with some preliminaries. Given $x = (x_1,\dots,x_n) \in \mathbf{R}^n$, we set $|x| = (x_1^2 + \cdots + x_n^2)^{1/2}$. The partial derivative of a function f on $\mathbf{R}^n$ with respect to the jth variable x_j is denoted by $\partial_j f$ while the mth partial derivative with respect to the jth variable is denoted by $\partial_j^m f$. The *gradient* of a function f is the vector $\nabla f = (\partial_1 f,\dots,\partial_n f)$. A *multi-index* α is an ordered n-tuple of nonnegative integers. For a multi-index $\alpha = (\alpha_1,\dots,\alpha_n)$, $\partial^\alpha f$ denotes the derivative $\partial_1^{\alpha_1}\cdots\partial_n^{\alpha_n} f$. If $\alpha = (\alpha_1,\dots,\alpha_n)$ is a multi-index, $|\alpha| = \alpha_1 + \cdots + \alpha_n$ denotes its size and $\alpha! = \alpha_1!\cdots\alpha_n!$ denotes the product of the factorials of its entries. The number $|\alpha|$ indicates the *total order of differentiation* of $\partial^\alpha f$. The space of functions in $\mathbf{R}^n$ all of whose derivatives of order at most $N \in \mathbf{Z}^+$ are continuous is denoted by $\mathscr{C}^N(\mathbf{R}^n)$ and the space of all *infinitely differentiable functions* on $\mathbf{R}^n$ by $\mathscr{C}^\infty(\mathbf{R}^n)$. The space of $\mathscr{C}^\infty$ functions with compact support on $\mathbf{R}^n$ is denoted by $\mathscr{C}_0^\infty(\mathbf{R}^n)$. This space is nonempty; see Exercise 2.2.1(a).

For $x \in \mathbf{R}^n$ and $\alpha = (\alpha_1,\dots,\alpha_n)$ a multi-index, we set $x^\alpha = x_1^{\alpha_1}\cdots x_n^{\alpha_n}$. Multi-indices will be denoted by the letters $\alpha,\beta,\gamma,\delta,\dots$. It is a simple fact to verify that

$$|x^\alpha| \le c_{n,\alpha} |x|^{|\alpha|}, \tag{2.2.1}$$

for some constant that depends on the dimension n and on α. In fact, $c_{n,\alpha}$ is the maximum of the continuous function $(x_1,\dots,x_n) \mapsto |x_1^{\alpha_1}\cdots x_n^{\alpha_n}|$ on the sphere $\mathbf{S}^{n-1} = \{x \in \mathbf{R}^n : |x| = 1\}$. The converse inequality in (2.2.1) fails. However, the following substitute of the converse of (2.2.1) is of great use: for $k \in \mathbf{Z}^+$ we have

$$|x|^k \le C_{n,k} \sum_{|\beta|=k} |x^\beta| \tag{2.2.2}$$

for all $x \in \mathbf{R}^n \setminus \{0\}$. To prove (2.2.2), take $1/C_{n,k}$ to be the minimum of the function

$$x \mapsto \sum_{|\beta|=k} |x^\beta|$$

on $\mathbf{S}^{n-1}$; this minimum is positive since this function has no zeros on $\mathbf{S}^{n-1}$. A related inequality is

$$(1+|x|)^k \le 2^k(1+C_{n,k}) \sum_{|\beta|\le k} |x^\beta| . \tag{2.2.3}$$

This follows from (2.2.2) for $|x|\ge 1$, while for $|x|<1$ we note that the sum in (2.2.3) is at least one since $|x^{(0,\dots,0)}|=1$.

We end the preliminaries by noting the validity of the one-dimensional Leibniz rule

$$\frac{d^m}{dt^m}(fg) = \sum_{k=0}^{m} \binom{m}{k} \frac{d^k f}{dt^k}\frac{d^{m-k} g}{dt^{m-k}}, \tag{2.2.4}$$

for all $\mathscr{C}^m$ functions f,g on $\mathbf{R}$, and its multidimensional analogue

$$\partial^\alpha(fg) = \sum_{\beta\le\alpha} \binom{\alpha_1}{\beta_1}\cdots\binom{\alpha_n}{\beta_n}(\partial^\beta f)(\partial^{\alpha-\beta} g), \tag{2.2.5}$$

for f,g in $\mathscr{C}^{|\alpha|}(\mathbf{R}^n)$ for some multi-index α, where the notation $\beta\le\alpha$ in (2.2.5) means that β ranges over all multi-indices satisfying $0\le\beta_j\le\alpha_j$ for all $1\le j\le n$. We observe that identity (2.2.5) is easily deduced by repeated application of (2.2.4), which in turn is obtained by induction.

2.2.1 The Class of Schwartz Functions

We now introduce the class of *Schwartz functions* on $\mathbf{R}^n$. Roughly speaking, a function is Schwartz if it is smooth and all of its derivatives decay faster than the reciprocal of any polynomial at infinity. More precisely, we give the following definition.

Definition 2.2.1. A $\mathscr{C}^\infty$ complex-valued function f on $\mathbf{R}^n$ is called a Schwartz function if for every pair of multi-indices α and β there exists a positive constant $C_{\alpha,\beta}$ such that

$$\rho_{\alpha,\beta}(f) = \sup_{x\in\mathbf{R}^n} |x^\alpha \partial^\beta f(x)| = C_{\alpha,\beta} < \infty. \tag{2.2.6}$$

The quantities $\rho_{\alpha,\beta}(f)$ are called the *Schwartz seminorms* of f. The set of all Schwartz functions on $\mathbf{R}^n$ is denoted by $\mathscr{S}(\mathbf{R}^n)$.

Example 2.2.2. The function $e^{-|x|^2}$ is in $\mathscr{S}(\mathbf{R}^n)$ but $e^{-|x|}$ is not, since it fails to be differentiable at the origin. The $\mathscr{C}^\infty$ function $g(x)=(1+|x|^4)^{-a}$, $a>0$, is not in $\mathscr{S}$ since it decays only like the reciprocal of a fixed polynomial at infinity. The set of all smooth functions with compact support, $\mathscr{C}_0^\infty(\mathbf{R}^n)$, is contained in $\mathscr{S}(\mathbf{R}^n)$.

Remark 2.2.3. If f_1 is in $\mathscr{S}(\mathbf{R}^n)$ and f_2 is in $\mathscr{S}(\mathbf{R}^m)$, then the function of $m+n$ variables $f_1(x_1,\dots,x_n)f_2(x_{n+1},\dots,x_{n+m})$ is in $\mathscr{S}(\mathbf{R}^{n+m})$. If f is in $\mathscr{S}(\mathbf{R}^n)$ and $P(x)$ is a polynomial of n variables, then $P(x)f(x)$ is also in $\mathscr{S}(\mathbf{R}^n)$. If α is a multi-index and f is in $\mathscr{S}(\mathbf{R}^n)$, then $\partial^\alpha f$ is in $\mathscr{S}(\mathbf{R}^n)$. Also note that

$$f \in \mathscr{S}(\mathbf{R}^n) \iff \sup_{x\in\mathbf{R}^n} |\partial^\alpha(x^\beta f(x))| < \infty \qquad \text{for all multi-indices } \alpha,\, \beta.$$

Remark 2.2.4. The following alternative characterization of Schwartz functions is very useful. A $\mathscr{C}^\infty$ function f is in $\mathscr{S}(\mathbf{R}^n)$ if and only if for all positive integers N and all multi-indices α there exists a positive constant $C_{\alpha,N}$ such that

$$|(\partial^\alpha f)(x)| \le C_{\alpha,N}(1+|x|)^{-N}. \tag{2.2.7}$$

The simple proofs are omitted. We now discuss convergence in $\mathscr{S}(\mathbf{R}^n)$.

Definition 2.2.5. Let f_k, f be in $\mathscr{S}(\mathbf{R}^n)$ for $k = 1,2,\dots$. We say that the sequence f_k converges to f in $\mathscr{S}(\mathbf{R}^n)$ if for all multi-indices α and β we have

$$\rho_{\alpha,\beta}(f_k - f) = \sup_{x\in\mathbf{R}^n} |x^\alpha(\partial^\beta(f_k - f))(x)| \to 0 \qquad \text{as} \quad k\to\infty.$$

For instance, for any fixed $x_0 \in \mathbf{R}^n$, $f(x+x_0/k) \to f(x)$ in $\mathscr{S}(\mathbf{R}^n)$ for any f in $\mathscr{S}(\mathbf{R}^n)$ as $k\to\infty$.

This notion of convergence is compatible with a topology on $\mathscr{S}(\mathbf{R}^n)$ under which the operations $(f,g)\mapsto f+g$, $(a,f)\to af$, and $f\mapsto \partial^\alpha f$ are continuous for all complex scalars a and multi-indices α ($f,g\in\mathscr{S}(\mathbf{R}^n)$). A subbasis for open sets containing 0 in this topology is

$$\{f\in\mathscr{S} :\ \rho_{\alpha,\beta}(f) < r\},$$

for all α, β multi-indices and all $r\in\mathbf{Q}^+$. Observe the following: If $\rho_{\alpha,\beta}(f) = 0$, then $f = 0$. This means that $\mathscr{S}(\mathbf{R}^n)$ is a locally convex topological vector space equipped with the family of seminorms $\rho_{\alpha,\beta}$ that separate points. We refer to Reed and Simon [286] for the pertinent definitions. Since the origin in $\mathscr{S}(\mathbf{R}^n)$ has a countable base, this space is metrizable. In fact, the following is a metric on $\mathscr{S}(\mathbf{R}^n)$:

$$d(f,g) = \sum_{j=1}^\infty 2^{-j}\frac{\rho_j(f-g)}{1+\rho_j(f-g)},$$

where ρ_j is an enumeration of all the seminorms $\rho_{\alpha,\beta}$, α and β multi-indices. One may easily verify that $\mathscr{S}$ is complete with respect to the metric d. Indeed, a Cauchy sequence $\{h_j\}_j$ in $\mathscr{S}$ would have to be Cauchy in L^∞ and therefore it would converge uniformly to some function h. The same is true for the sequences $\{\partial^\beta h_j\}_j$ and $\{x^\alpha h_j(x)\}_j$, and the limits of these sequences can be shown to be the functions $\partial^\beta h$ and $x^\alpha h(x)$, respectively. It follows that the sequence $\{h_j\}$ converges to h in $\mathscr{S}$. Therefore, $\mathscr{S}(\mathbf{R}^n)$ is a *Fréchet space (complete metrizable locally convex space)*.

We note that convergence in $\mathscr{S}$ is stronger than convergence in all L^p. We have the following.

Proposition 2.2.6. *Let f, f_k, $k = 1,2,3,\dots$, be in $\mathscr{S}(\mathbf{R}^n)$. If $f_k\to f$ in $\mathscr{S}$ then $f_k\to f$ in L^p for all $0<p\le\infty$. Moreover, there exists a $C_{p,n}>0$ such that*

$$\left\|\partial^{\beta} f\right\|_{L^p} \leq C_{p,n} \sum_{|\alpha| \leq [\frac{n+1}{p}]+1} \rho_{\alpha,\beta}(f) \tag{2.2.8}$$

for all f for which the right-hand side is finite.

Proof. Observe that when $p < \infty$ we have

$$\begin{aligned}
\left\|\partial^{\beta} f\right\|_{L^p} &\leq \left[\int_{|x|\leq 1} |\partial^{\beta} f(x)|^p \, dx + \int_{|x|\geq 1} |x|^{n+1} |\partial^{\beta} f(x)|^p |x|^{-(n+1)} \, dx\right]^{1/p} \\
&\leq \left[v_n \left\|\partial^{\beta} f\right\|_{L^\infty}^p + \left(\sup_{|x|\geq 1} |x|^{n+1} |\partial^{\beta} f(x)|^p\right) \int_{|x|\geq 1} |x|^{-(n+1)} \, dx\right]^{1/p} \\
&\leq C_{p,n}\left(\left\|\partial^{\beta} f\right\|_{L^\infty} + \sup_{|x|\geq 1} \left(|x|^{[\frac{n+1}{p}]+1} |\partial^{\beta} f(x)|\right)\right).
\end{aligned}$$

The preceding inequality is also trivially valid when $p = \infty$. Now set $m = \left[\frac{n+1}{p}\right] + 1$ and use (2.2.2) to obtain

$$\sup_{|x|\geq 1} |x|^m |\partial^{\beta} f(x)| \leq \sup_{|x|\geq 1} C_{n,m} \sum_{|\alpha|=m} |x^{\alpha} \partial^{\beta} f(x)| \leq C_{n,m} \sum_{|\alpha|\leq m} \rho_{\alpha,\beta}(f).$$

Conclusion (2.2.8) now follows immediately. This shows that convergence in $\mathscr{S}$ implies convergence in L^p. □

We now show that the Schwartz class is closed under certain operations.

Proposition 2.2.7. *Let f, g be in $\mathscr{S}(\mathbf{R}^n)$. Then fg and $f * g$ are in $\mathscr{S}(\mathbf{R}^n)$. Moreover,*

$$\partial^{\alpha}(f * g) = (\partial^{\alpha} f) * g = f * (\partial^{\alpha} g) \tag{2.2.9}$$

for all multi-indices α.

Proof. Fix f and g in $\mathscr{S}(\mathbf{R}^n)$. Let e_j be the unit vector $(0,\dots,1,\dots,0)$ with 1 in the jth entry and zeros in all the other entries. Since

$$\frac{f(y+he_j) - f(y)}{h} - (\partial_j f)(y) \to 0 \tag{2.2.10}$$

as $h \to 0$, and since the expression in (2.2.10) is pointwise bounded by some constant depending on f, the integral of the expression in (2.2.10) with respect to the measure $g(x-y)\,dy$ converges to zero as $h \to 0$ by the Lebesgue dominated convergence theorem. This proves (2.2.9) when $\alpha = (0,\dots,1,\dots,0)$. The general case follows by repeating the previous argument and using induction.

We now show that the convolution of two functions in $\mathscr{S}$ is also in $\mathscr{S}$. For each $N > 0$ there is a constant C_N such that

$$\left|\int_{\mathbf{R}^n} f(x-y) g(y) \, dy\right| \leq C_N \int_{\mathbf{R}^n} (1+|x-y|)^{-N} (1+|y|)^{-N-n-1} dy. \tag{2.2.11}$$

Inserting the simple estimate

$$(1+|x-y|)^{-N} \le (1+|y|)^{N}(1+|x|)^{-N}$$

in (2.2.11) we obtain that

$$|(f*g)(x)| \le C_N(1+|x|)^{-N} \int_{\mathbf{R}^n} (1+|y|)^{-n-1} dy = C_N'(1+|x|)^{-N}.$$

This shows that $f*g$ decays like $(1+|x|)^{-N}$ at infinity, but since $N>0$ is arbitrary it follows that $f*g$ decays faster than the reciprocal of any polynomial.

Since $\partial^\alpha(f*g) = (\partial^\alpha f)*g$, replacing f by $\partial^\alpha f$ in the previous argument, we also conclude that all the derivatives of $f*g$ decay faster than the reciprocal of any polynomial at infinity. Using (2.2.7), we conclude that $f*g$ is in $\mathscr{S}$. Finally, the fact that fg is in $\mathscr{S}$ follows directly from Leibniz's rule (2.2.5) and (2.2.7). □

2.2.2 The Fourier Transform of a Schwartz Function

The Fourier transform is often introduced as an operation on L^1. In that setting, problems of convergence arise when certain manipulations of functions are performed. Also, Fourier inversion requires the additional assumption that the Fourier transform is in L^1. Here we initially introduce the Fourier transform on the space of Schwartz functions. The rapid decay of Schwartz functions at infinity allows us to develop its fundamental properties without encountering any convergence problems. The Fourier transform is a homeomorphism of the Schwartz class and Fourier inversion holds in it. For these reasons, this class is a natural environment for it.

For $x=(x_1,\dots,x_n)$, $y=(y_1,\dots,y_n)$ in $\mathbf{R}^n$ we use the notation

$$x\cdot y = \sum_{j=1}^n x_j y_j.$$

Definition 2.2.8. Given f in $\mathscr{S}(\mathbf{R}^n)$ we define

$$\widehat{f}(\xi) = \int_{\mathbf{R}^n} f(x) e^{-2\pi i x\cdot\xi}\, dx.$$

We call $\widehat{f}$ the Fourier transform of f.

Example 2.2.9. If $f(x) = e^{-\pi|x|^2}$ defined on $\mathbf{R}^n$, then $\widehat{f}(\xi) = f(\xi)$. To prove this, observe that the function

$$s \mapsto \int_{-\infty}^{+\infty} e^{-\pi(t+is)^2}\, dt, \qquad s\in\mathbf{R},$$

defined on the line is constant (and thus equal to $\int_{-\infty}^{+\infty} e^{-\pi t^2}\,dt$), since its derivative is

$$\int_{-\infty}^{+\infty} -2\pi i(t+is)e^{-\pi(t+is)^2}\,dt = \int_{-\infty}^{+\infty} i\frac{d}{dt}(e^{-\pi(t+is)^2})\,dt = 0.$$

Using this fact, we calculate the Fourier transform of the function $t \mapsto e^{-\pi t^2}$ on $\mathbf{R}$ by completing the squares as follows:

$$\int_{\mathbf{R}} e^{-\pi t^2} e^{-2\pi i t\tau}\,dt = \int_{\mathbf{R}} e^{-\pi(t+i\tau)^2} e^{\pi(i\tau)^2}\,dt = \left(\int_{-\infty}^{+\infty} e^{-\pi t^2}\,dt\right) e^{-\pi\tau^2} = e^{-\pi\tau^2},$$

where $\tau \in \mathbf{R}$, and we used that

$$\int_{-\infty}^{+\infty} e^{-t^2}\,dt = \sqrt{\pi}, \tag{2.2.12}$$

a fact that can be found in Appendix A.1.

Remark 2.2.10. It follows from the definition of the Fourier transform that if f is in $\mathscr{S}(\mathbf{R}^n)$ and g is in $\mathscr{S}(\mathbf{R}^m)$, then

$$[f(x_1,\dots,x_n)g(x_{n+1},\dots,x_{n+m})]\widehat{\ } = \widehat{f}(\xi_1,\dots,\xi_n)\widehat{g}(\xi_{n+1},\dots,\xi_{n+m}),$$

where the first $\widehat{\ }$ denotes the Fourier transform on $\mathbf{R}^{n+m}$. In other words, the Fourier transform preserves separation of variables. Combining this observation with the result in Example 2.2.9, we conclude that the function $f(x) = e^{-\pi|x|^2}$ defined on $\mathbf{R}^n$ is equal to its Fourier transform.

We now continue with some properties of the Fourier transform. Before we do this we introduce some notation. For a measurable function f on $\mathbf{R}^n$, $x \in \mathbf{R}^n$, and $a > 0$ we define the *translation*, *dilation*, and *reflection* of f by

$$\begin{aligned} (\tau^y f)(x) &= f(x-y) \\ (\delta^a f)(x) &= f(ax) \\ \widetilde{f}(x) &= f(-x). \end{aligned} \tag{2.2.13}$$

Also recall the notation $f_a = a^{-n}\delta^{1/a}(f)$ introduced in Definition 2.1.9.

Proposition 2.2.11. *Given f, g in $\mathscr{S}(\mathbf{R}^n)$, $y \in \mathbf{R}^n$, $b \in \mathbf{C}$, α a multi-index, and $t > 0$, we have*

(1) $\|\widehat{f}\|_{L^\infty} \le \|f\|_{L^1}$,

(2) $\widehat{f+g} = \widehat{f} + \widehat{g}$,

(3) $\widehat{bf} = b\widehat{f}$,

(4) $\widehat{\widetilde{f}} = \widetilde{\widehat{f}}$,

(5) $\widehat{\overline{f}} = \overline{\widetilde{\widehat{f}}}$,

(6) $\widehat{\tau^y f}(\xi) = e^{-2\pi i y\cdot\xi}\widehat{f}(\xi)$,

(7) $(e^{2\pi i x\cdot y}f(x))\widehat{\ }(\xi) = \tau^y(\widehat{f})(\xi)$,

(8) $(\delta^t f)\widehat{\ } = t^{-n}\delta^{t^{-1}}\widehat{f} = (\widehat{f})_t$,

(9) $(\partial^\alpha f)\widehat{\ }(\xi) = (2\pi i\xi)^\alpha \widehat{f}(\xi)$,

(10) $(\partial^\alpha \widehat{f})(\xi) = ((-2\pi i x)^\alpha f(x))\widehat{\ }(\xi)$,

(11) $\widehat{f} \in \mathscr{S}$,

(12) $\widehat{f*g} = \widehat{f}\widehat{g}$,

(13) $\widehat{f\circ A}(\xi) = \widehat{f}(A\xi)$, *where A is an orthogonal matrix and ξ is a column vector.*

Proof. Property (1) follows directly from Definition 2.2.8. Properties (2)–(5) are trivial. Properties (6)–(8) require a suitable change of variables but they are omitted. Property (9) is proved by integration by parts (which is justified by the rapid decay of the integrands):

$$\begin{aligned}(\partial^\alpha f)\widehat{\ }(\xi) &= \int_{\mathbf{R}^n}(\partial^\alpha f)(x)e^{-2\pi i x\cdot\xi}\,dx\\ &= (-1)^{|\alpha|}\int_{\mathbf{R}^n} f(x)(-2\pi i\xi)^\alpha e^{-2\pi i x\cdot\xi}\,dx\\ &= (2\pi i\xi)^\alpha \widehat{f}(\xi).\end{aligned}$$

To prove (10), let $\alpha = e_j = (0,\dots,1,\dots,0)$, where all entries are zero except for the jth entry, which is 1. Since

$$\frac{e^{-2\pi i x\cdot(\xi+he_j)} - e^{-2\pi i x\cdot\xi}}{h} - (-2\pi i x_j)e^{-2\pi i x\cdot\xi} \to 0 \tag{2.2.14}$$

as $h\to 0$ and the preceding function is bounded by $C|x|$ for all h and ξ, the Lebesgue dominated convergence theorem implies that the integral of the function in (2.2.14) with respect to the measure $f(x)dx$ converges to zero. This proves (10) for $\alpha = e_j$. For other α's use induction. To prove (11) we use (9), (10), and (1) in the following way:

$$\big\|x^\alpha(\partial^\beta \widehat{f})(x)\big\|_{L^\infty} = \frac{(2\pi)^{|\beta|}}{(2\pi)^{|\alpha|}}\big\|(\partial^\alpha(x^\beta f(x)))\widehat{\ }\big\|_{L^\infty} \le \frac{(2\pi)^{|\beta|}}{(2\pi)^{|\alpha|}}\big\|\partial^\alpha(x^\beta f(x))\big\|_{L^1} < \infty.$$

Identity (12) follows from the following calculation:

$$\begin{aligned}
\widehat{f*g}(\xi) &= \int_{\mathbf{R}^n}\int_{\mathbf{R}^n} f(x-y)g(y)e^{-2\pi i x\cdot\xi}\,dy\,dx \\
&= \int_{\mathbf{R}^n}\int_{\mathbf{R}^n} f(x-y)g(y)e^{-2\pi i(x-y)\cdot\xi}e^{-2\pi i y\cdot\xi}\,dy\,dx \\
&= \int_{\mathbf{R}^n} g(y)\int_{\mathbf{R}^n} f(x-y)e^{-2\pi i(x-y)\cdot\xi}dx\, e^{-2\pi i y\cdot\xi}\,dy \\
&= \widehat{f}(\xi)\widehat{g}(\xi),
\end{aligned}$$

where the application of Fubini's theorem is justified by the absolute convergence of the integrals. Finally, we prove (13). We have

$$\begin{aligned}
\widehat{f\circ A}(\xi) &= \int_{\mathbf{R}^n} f(Ax)e^{-2\pi i x\cdot\xi}\,dx \\
&= \int_{\mathbf{R}^n} f(y)e^{-2\pi i A^{-1}y\cdot\xi}\,dy \\
&= \int_{\mathbf{R}^n} f(y)e^{-2\pi i A^t y\cdot\xi}\,dy \\
&= \int_{\mathbf{R}^n} f(y)e^{-2\pi i y\cdot A\xi}\,dy \\
&= \widehat{f}(A\xi),
\end{aligned}$$

where we used the change of variables $y = Ax$ and the fact that $|\det A| = 1$. □

Corollary 2.2.12. *The Fourier transform of a radial function is radial. Products and convolutions of radial functions are radial.*

Proof. Let ξ_1, ξ_2 in $\mathbf{R}^n$ with $|\xi_1| = |\xi_2|$. Then for some orthogonal matrix A we have $A\xi_1 = \xi_2$. Since f is radial, we have $f = f\circ A$. Then

$$\widehat{f}(\xi_2) = \widehat{f}(A\xi_1) = \widehat{f\circ A}(\xi_1) = \widehat{f}(\xi_1),$$

where we used (13) in Proposition 2.2.11 to justify the second equality. Products and convolutions of radial functions are easily seen to be radial. □

2.2.3 The Inverse Fourier Transform and Fourier Inversion

We now define the inverse Fourier transform.

Definition 2.2.13. Given a Schwartz function f, we define

$$f^\vee(x) = \widehat{f}(-x),$$

for all $x \in \mathbf{R}^n$. The operation

$$f \mapsto f^\vee$$

is called the *inverse Fourier transform.*

It is straightforward that the inverse Fourier transform shares the same properties as the Fourier transform. One may want to list (and prove) properties for the inverse Fourier transform analogous to those in Proposition 2.2.11.

We now investigate the relation between the Fourier transform and the inverse Fourier transform. In the next theorem, we prove that one is the inverse operation of the other. This property is referred to as *Fourier inversion.*

Theorem 2.2.14. *Given f, g, and h in $\mathscr{S}(\mathbf{R}^n)$, we have*

(1) $\displaystyle\int_{\mathbf{R}^n} f(x)\widehat{g}(x)\,dx = \int_{\mathbf{R}^n} \widehat{f}(x)g(x)\,dx,$

(2) ***(Fourier Inversion)*** $\quad (\widehat{f})^\vee = f = (f^\vee)^\wedge\,,$

(3) ***(Parseval's relation)*** $\displaystyle\int_{\mathbf{R}^n} f(x)\overline{h}(x)\,dx = \int_{\mathbf{R}^n} \widehat{f}(\xi)\overline{\widehat{h}(\xi)}\,d\xi\,,$

(4) ***(Plancherel's identity)*** $\quad \big\|f\big\|_{L^2} = \big\|\widehat{f}\big\|_{L^2} = \big\|f^\vee\big\|_{L^2}\,,$

(5) $\displaystyle\int_{\mathbf{R}^n} f(x)h(x)\,dx = \int_{\mathbf{R}^n} \widehat{f}(x)h^\vee(x)\,dx.$

Proof. (1) follows immediately from the definition of the Fourier transform and Fubini's theorem. To prove (2) we use (1) with

$$g(\xi) = e^{2\pi i\xi\cdot t}e^{-\pi|\varepsilon\xi|^2}\,.$$

By Proposition 2.2.11 (7) and (8) and Example 2.2.9, we have that

$$\widehat{g}(x) = \frac{1}{\varepsilon^n}e^{-\pi|(x-t)/\varepsilon|^2},$$

which is an approximate identity. Now (1) gives

$$\int_{\mathbf{R}^n} f(x)\varepsilon^{-n}e^{-\pi\varepsilon^{-2}|x-t|^2}\,dx = \int_{\mathbf{R}^n} \widehat{f}(\xi)e^{2\pi i\xi\cdot t}e^{-\pi|\varepsilon\xi|^2}\,d\xi\,. \tag{2.2.15}$$

Now let $\varepsilon \to 0$ in (2.2.15). The left-hand side of (2.2.15) converges to $f(t)$ uniformly on compact sets by Theorem 1.2.19. The right-hand side of (2.2.15) converges to $(\widehat{f})^\vee(t)$ as $\varepsilon \to 0$ by the Lebesgue dominated convergence theorem. We conclude that $(\widehat{f})^\vee = f$ on $\mathbf{R}^n$. Replacing f by $\widetilde{f}$ and using the result just proved, we conclude that $(f^\vee)^\wedge = f$.

Note that if $g = \overline{\widehat{h}}$, then Proposition 2.2.11 (5) and identity (2) imply that $\widehat{g} = \overline{h}$. Then (3) follows from (1) by expressing h in terms of g. Identity (4) is a trivial consequence of (3). (Sometimes the polarized identity (3) is also referred to as Plancherel's identity.) Finally, (5) easily follows from (1) and (2) with $\widehat{g} = h$. □

Next we have the following simple corollary of Theorem 2.2.14.

Corollary 2.2.15. *The Fourier transform is a homeomorphism from $\mathscr{S}(\mathbf{R}^n)$ onto itself.*

Proof. The continuity of the Fourier transform (and its inverse) follows from Exercise 2.2.2, while Fourier inversion yields that this map is bijective. □

2.2.4 The Fourier Transform on $L^1 + L^2$

We have defined the Fourier transform on $\mathscr{S}(\mathbf{R}^n)$. We now extend this definition to the space $L^1(\mathbf{R}^n) + L^2(\mathbf{R}^n)$.

We begin by observing that the Fourier transform given in Definition 2.2.8,

$$\widehat{f}(\xi) = \int_{\mathbf{R}^n} f(x) e^{-2\pi i x \cdot \xi} \, dx,$$

makes sense as a convergent integral for functions $f \in L^1(\mathbf{R}^n)$. This allows us to extend the definition of the Fourier transform on L^1. Moreover, this operator satisfies properties (1)–(8) as well as (12) and (13) in Proposition 2.2.11, with f, g integrable. We also define the inverse Fourier transform on L^1 by setting $f^\vee(x) = \widehat{f}(-x)$ for $f \in L^1(\mathbf{R}^n)$ and we note that analogous properties hold for it. One problem in this generality is that when f is integrable, one may not necessarily have $(\widehat{f})^\vee = f$ a.e. This inversion is possible when $\widehat{f}$ is also integrable; see Exercise 2.2.6.

The integral defining the Fourier transform does not converge absolutely for functions in $L^2(\mathbf{R}^n)$; however, the Fourier transform has a natural definition in this space accompanied by an elegant theory. In view of the result in Exercise 2.2.8, the Fourier transform is an L^2 isometry on $L^1 \cap L^2$, which is a dense subspace of L^2. By density, there is a unique bounded extension of the Fourier transform on L^2. Let us denote this extension by $\mathscr{F}$. Then $\mathscr{F}$ is also an isometry on L^2, i.e.,

$$\big\|\mathscr{F}(f)\big\|_{L^2} = \big\|f\big\|_{L^2}$$

for all $f \in L^2(\mathbf{R}^n)$, and any sequence of functions $f_N \in L^1(\mathbf{R}^n) \cap L^2(\mathbf{R}^n)$ converging to a given f in $L^2(\mathbf{R}^n)$ satisfies

$$\big\|\widehat{f_N} - \mathscr{F}(f)\big\|_{L^2} \to 0, \tag{2.2.16}$$

as $N \to \infty$. In particular, the sequence of functions $f_N(x) = f(x)\chi_{|x| \le N}$ yields that

$$\widehat{f_N}(\xi) = \int_{|x| \le N} f(x) e^{-2\pi i x \cdot \xi} \, dx \tag{2.2.17}$$

converges to $\mathscr{F}(f)(\xi)$ in L^2 as $N \to \infty$. If f is both integrable and square integrable, the expressions in (2.2.17) also converge to $\widehat{f}(\xi)$ pointwise. Also, in view of Theorem 1.1.11 and (2.2.16), there is a subsequence of $\widehat{f_N}$ that converges to $\mathscr{F}(f)$ pointwise a.e. Consequently, for f in $L^1(\mathbf{R}^n) \cap L^2(\mathbf{R}^n)$ the expressions $\widehat{f}$ and $\mathscr{F}(f)$ coincide pointwise a.e. For this reason we often adopt the notation $\widehat{f}$ to denote the Fourier transform of functions f in L^2 as well.

In a similar fashion, we let $\mathscr{F}'$ be the isometry on $L^2(\mathbf{R}^n)$ that extends the operator $f \mapsto f^\vee$, which is an L^2 isometry on $L^1 \cap L^2$; the last statement follows by adapting the result of Exercise 2.2.8 to the inverse Fourier transform. Since $\varphi^\vee(x) = \widehat{\varphi}(-x)$ for φ in the Schwartz class, which is dense in L^2 (Exercise 2.2.5), it follows that $\mathscr{F}'(f)(x) = \mathscr{F}(f)(-x)$ for all $f \in L^2$ and almost all $x \in \mathbf{R}^n$. The operators $\mathscr{F}$ and $\mathscr{F}'$ are L^2-isometries that satisfy $\mathscr{F}' \circ \mathscr{F} = \mathscr{F} \circ \mathscr{F}' = \mathrm{Id}$ on the Schwartz space. By density this identity also holds for L^2 functions and implies that $\mathscr{F}$ and $\mathscr{F}'$ are injective and surjective mappings from L^2 to itself; consequently, $\mathscr{F}'$ coincides with the inverse operator $\mathscr{F}^{-1}$ of $\mathscr{F} : L^2 \to L^2$, and Fourier inversion

$$f = \mathscr{F}^{-1} \circ \mathscr{F}(f) = \mathscr{F} \circ \mathscr{F}^{-1}(f) \qquad \text{a.e.}$$

holds on L^2.

Having set down the basic facts concerning the action of the Fourier transform on L^1 and L^2, we extend its definition on L^p for $1 < p < 2$. Given a function f in $L^p(\mathbf{R}^n)$, with $1 < p < 2$, we define $\widehat{f} = \widehat{f_1} + \widehat{f_2}$, where $f_1 \in L^1(\mathbf{R}^n)$, $f_2 \in L^2(\mathbf{R}^n)$, and $f = f_1 + f_2$; we may take, for instance, $f_1 = f\chi_{|f|>1}$ and $f_2 = f\chi_{|f|\le 1}$. The definition of $\widehat{f}$ is independent of the choice of f_1 and f_2, for if $f_1 + f_2 = h_1 + h_2$ for $f_1, h_1 \in L^1(\mathbf{R}^n)$ and $f_2, h_2 \in L^2(\mathbf{R}^n)$, we have $f_1 - h_1 = h_2 - f_2 \in L^1(\mathbf{R}^n) \cap L^2(\mathbf{R}^n)$. Since these functions are equal on $L^1(\mathbf{R}^n) \cap L^2(\mathbf{R}^n)$, their Fourier transforms are also equal, and we obtain $\widehat{f_1} - \widehat{h_1} = \widehat{h_2} - \widehat{f_2}$, which yields $\widehat{f_1 + f_2} = \widehat{h_1 + h_2}$. We have the following result concerning the action of the Fourier transform on L^p.

Proposition 2.2.16. ***(Hausdorff–Young inequality)*** *For every function f in $L^p(\mathbf{R}^n)$ we have the estimate*

$$\|\widehat{f}\|_{L^{p'}} \le \|f\|_{L^p}$$

whenever $1 \le p \le 2$.

Proof. This follows easily from Theorem 1.3.4. Interpolate between the estimates $\|\widehat{f}\|_{L^\infty} \le \|f\|_{L^1}$ (Proposition 2.2.11 (1)) and $\|\widehat{f}\|_{L^2} \le \|f\|_{L^2}$ to obtain $\|\widehat{f}\|_{L^{p'}} \le \|f\|_{L^p}$. We conclude that the Fourier transform is a bounded operator from $L^p(\mathbf{R}^n)$ to $L^{p'}(\mathbf{R}^n)$ with norm at most 1 when $1 \le p \le 2$. □

Next, we are concerned with the behavior of the Fourier transform at infinity.

Proposition 2.2.17. ***(Riemann–Lebesgue lemma)*** *For a function f in $L^1(\mathbf{R}^n)$ we have that*

$$|\widehat{f}(\xi)| \to 0 \qquad \textit{as} \qquad |\xi| \to \infty.$$

Proof. Consider the function $\chi_{[a,b]}$ on $\mathbf{R}$. A simple computation gives

$$\widehat{\chi_{[a,b]}}(\xi) = \int_a^b e^{-2\pi i x\xi}\,dx = \frac{e^{-2\pi i\xi a} - e^{-2\pi i\xi b}}{2\pi i\xi},$$

which tends to zero as $|\xi| \to \infty$. Likewise, if $g = \prod_{j=1}^n \chi_{[a_j,b_j]}$ on $\mathbf{R}^n$, then

$$\widehat{g}(\xi) = \prod_{j=1}^n \frac{e^{-2\pi i\xi_j a_j} - e^{-2\pi i\xi_j b_j}}{2\pi i\xi_j}.$$

Given a $\xi = (\xi_1,\dots,\xi_n) \neq 0$, there is j_0 such that $|\xi_{j_0}| > |\xi|/\sqrt{n}$. Then

$$\left|\prod_{j=1}^n \frac{e^{-2\pi i\xi_j a_j} - e^{-2\pi i\xi_j b_j}}{2\pi i\xi_j}\right| \le \frac{2\sqrt{n}}{2\pi|\xi|} \sup_{1\le j_0\le n} \prod_{j\neq j_0} (b_j - a_j)$$

which also tends to zero as $|\xi| \to \infty$ in $\mathbf{R}^n$.

Given a general integrable function f on $\mathbf{R}^n$ and $\varepsilon > 0$, there is a simple function h, which is a finite linear combination of characteristic functions of rectangles (like g), such that $\|f - h\|_{L^1} < \frac{\varepsilon}{2}$. Then there is an M is such that for $|\xi| > M$ we have $|\widehat{h}(\xi)| < \frac{\varepsilon}{2}$. It follows that

$$|\widehat{f}(\xi)| \le |\widehat{f}(\xi) - \widehat{h}(\xi)| + |\widehat{h}(\xi)| \le \|f - h\|_{L^1} + |\widehat{h}(\xi)| < \frac{\varepsilon}{2} + \frac{\varepsilon}{2},$$

provided $|\xi| > M$. This implies that $|\widehat{f}(\xi)| \to 0$ as $|\xi| \to \infty$.

A different proof can be given by taking the function h in the preceding paragraph to be a Schwartz function and using that Schwartz functions are dense in $L^1(\mathbf{R}^n)$; see Exercise 2.2.5 about the last assertion. □

We end this section with an example that illustrates some of the practical uses of the Fourier transform.

Example 2.2.18. We would like to find a Schwartz function $f(x_1,x_2,x_3)$ on $\mathbf{R}^3$ that satisfies the partial differential equation

$$f(x) + \partial_1^2\partial_2^2\partial_3^4 f(x) + 4i\partial_1^2 f(x) + \partial_2^7 f(x) = e^{-\pi|x|^2}.$$

Taking the Fourier transform on both sides of this identity and using Proposition 2.2.11 (2), (9) and the result of Example 2.2.9, we obtain

$$\widehat{f}(\xi)\Big[1 + (2\pi i\xi_1)^2(2\pi i\xi_2)^2(2\pi i\xi_3)^4 + 4i(2\pi i\xi_1)^2 + (2\pi i\xi_2)^7\Big] = e^{-\pi|\xi|^2}.$$

Let $p(\xi) = p(\xi_1,\xi_2,\xi_3)$ be the polynomial inside the square brackets. We observe that $p(\xi)$ has no real zeros and we may therefore write

$$\widehat{f}(\xi) = e^{-\pi|\xi|^2} p(\xi)^{-1} \implies f(x) = \big(e^{-\pi|\xi|^2} p(\xi)^{-1}\big)^{\vee}(x).$$

In general, let

$$P(\xi) = \sum_{|\alpha|\le N} C_\alpha \xi^\alpha$$

be a polynomial in $\mathbf{R}^n$ with constant complex coefficients C_α indexed by multi-indices α. If $P(2\pi i\xi)$ has no real zeros, and u is in $\mathscr{S}(\mathbf{R}^n)$, then the partial differential equation

$$P(\partial)f = \sum_{|\alpha|\le N} C_\alpha \partial^\alpha f = u$$

is solved as before to give

$$f = \big(\widehat{u}(\xi)P(2\pi i\xi)^{-1}\big)^\vee .$$

Since $P(2\pi i\xi)$ has no real zeros and $u \in \mathscr{S}(\mathbf{R}^n)$, the function

$$\widehat{u}(\xi)P(2\pi i\xi)^{-1}$$

is smooth and therefore a Schwartz function. Then f is also in $\mathscr{S}(\mathbf{R}^n)$ by Proposition 2.2.11 (11).

Exercises

2.2.1. (a) Construct a Schwartz function supported in the unit ball of $\mathbf{R}^n$.
(b) Construct a $\mathscr{C}_0^\infty(\mathbf{R}^n)$ function equal to 1 on the annulus $1 \le |x| \le 2$ and vanishing off the annulus $1/2 \le |x| \le 4$.
(c) Construct a nonnegative nonzero Schwartz function f on $\mathbf{R}^n$ whose Fourier transform is nonnegative and compactly supported.
$\big[$*Hint:* Part (a): Try the construction in dimension one first using the $\mathscr{C}^\infty$ function $\eta(x) = e^{-1/x}$ for $x > 0$ and $\eta(x) = 0$ for $x < 0$. Part (c): Take $f = |\phi * \widetilde{\phi}|^2$, where $\widehat{\phi}$ is odd, real-valued, and compactly supported; here $\widetilde{\phi}(x) = \phi(-x)$.$\big]$

2.2.2. If $f_k, f \in \mathscr{S}(\mathbf{R}^n)$ and $f_k \to f$ in $\mathscr{S}(\mathbf{R}^n)$, then $\widehat{f_k} \to \widehat{f}$ and $f_k^\vee \to f^\vee$ in $\mathscr{S}(\mathbf{R}^n)$.

2.2.3. Find the *spectrum* (i.e., the set of all *eigenvalues* of the Fourier transform), that is, all complex numbers λ for which there exist nonzero functions f such that

$$\widehat{f} = \lambda f .$$

$\big[$*Hint:* Apply the Fourier transform three times to the preceding identity. Consider the functions $xe^{-\pi x^2}$, $(a+bx^2)e^{-\pi x^2}$, and $(cx+dx^3)e^{-\pi x^2}$ for suitable a,b,c,d to show that all fourth roots of unity are indeed eigenvalues of the Fourier transform.$\big]$

2.2.4. Use the idea of the proof of Proposition 2.2.7 to show that if the functions f, g defined on $\mathbf{R}^n$ satisfy $|f(x)| \le A(1+|x|)^{-M}$ and $|g(x)| \le B(1+|x|)^{-N}$ for some $M,N > n$, then

$$|(f*g)(x)| \le ABC(1+|x|)^{-L},$$

where $L = \min(N,M)$ and $C = C(N,M) > 0$.

2.2.5. Show that $\mathscr{C}_0^\infty(\mathbf{R}^n)$ is dense on $L^p(\mathbf{R}^n)$ for $0 < p < \infty$ but not for $p = \infty$.
[*Hint:* Use a smooth approximate identity when $p \ge 1$. Reduce the case $p < 1$ to $p = 1$.]

2.2.6. (a) Prove that if $f \in L^1$, then $\widehat{f}$ is uniformly continuous on $\mathbf{R}^n$.
(b) Prove that for $f, g \in L^1(\mathbf{R}^n)$ we have

$$\int_{\mathbf{R}^n} f(x)\widehat{g}(x)\,dx = \int_{\mathbf{R}^n} \widehat{f}(y)g(y)\,dy.$$

(c) Take $\widehat{g}(x) = \varepsilon^{-n}e^{-\pi\varepsilon^{-2}|x-t|^2}$ in (b) and let $\varepsilon \to 0$ to prove that if f and $\widehat{f}$ are both in L^1, then $(\widehat{f})^\vee = f$ a.e. This fact is called *Fourier inversion on* L^1.

2.2.7. (a) Prove that if a function f in $L^1(\mathbf{R}^n) \cap L^\infty(\mathbf{R}^n)$ is continuous at 0, then

$$\lim_{\varepsilon\to 0}\int_{\mathbf{R}^n} \widehat{f}(x)e^{-\pi|\varepsilon x|^2}\,dx = f(0).$$

(b) Let $f \in L^1(\mathbf{R}^n) \cap L^\infty(\mathbf{R}^n)$ be continuous at zero and satisfy $\widehat{f} \ge 0$. Show that $\widehat{f}$ is in L^1 and conclude that Fourier inversion holds at zero $f(0) = \|\widehat{f}\|_{L^1}$, and also $f = (\widehat{f})^\vee$ a.e. in general.
[*Hint:* Part (a): Let $g(x) = e^{-\pi|\varepsilon x|^2}$ in Exercise 2.2.6(b) and use Theorem 1.2.19 (2).]

2.2.8. Given f in $L^1(\mathbf{R}^n) \cap L^2(\mathbf{R}^n)$, without appealing to density, prove that

$$\|\widehat{f}\|_{L^2} = \|f\|_{L^2}.$$

[*Hint:* Let $h = f * \overline{\widetilde{f}}$, where $\widetilde{f}(x) = f(-x)$ and the bar indicates complex conjugation. Then $h \in L^1(\mathbf{R}^n) \cap L^\infty(\mathbf{R}^n)$, $\widehat{h} = |\widehat{f}|^2 \ge 0$, and h is continuous at zero. Exercise 2.2.7(b) yields $\|\widehat{f}\|_{L^2}^2 = \|\widehat{h}\|_{L^1} = h(0) = \int_{\mathbf{R}^n} f(x)\overline{\widetilde{f}(-x)}\,dx = \|f\|_{L^2}^2$.]

2.2.9. (a) Prove that for all $0 < \varepsilon < t < \infty$ we have

$$\left|\int_\varepsilon^t \frac{\sin(\xi)}{\xi}\,d\xi\right| \le 4.$$

(b) If f is an odd L^1 function on the line, conclude that for all $t > \varepsilon > 0$ we have

$$\left|\int_\varepsilon^t \frac{\widehat{f}(\xi)}{\xi}\,d\xi\right| \le 4\|f\|_{L^1}.$$

(c) Let $g(\xi)$ be a continuous odd function that is equal to $1/\log(\xi)$ for $\xi \ge 2$. Show that there does not exist an L^1 function whose Fourier transform is g.

2.2.10. Let f be in $L^1(\mathbf{R})$. Prove that

$$\int_{-\infty}^{+\infty} f\Big(x-\frac{1}{x}\Big)\,dx = \int_{-\infty}^{+\infty} f(u)\,du.$$

[*Hint:* For $x \in (-\infty,0)$ use the change of variables $u = x - \frac{1}{x}$ or $x = \frac{1}{2}\big(u - \sqrt{4+u^2}\big)$. For $x \in (0,\infty)$ use the change of variables $u = x - \frac{1}{x}$ or $x = \frac{1}{2}\big(u + \sqrt{4+u^2}\big)$.]

2.2.11. (a) Use Exercise 2.2.10 with $f(x) = e^{-tx^2}$ to obtain the *subordination* identity

$$e^{-2t} = \frac{1}{\sqrt{\pi}}\int_0^\infty e^{-y-t^2/y}\,\frac{dy}{\sqrt{y}}, \qquad \text{where } t>0.$$

(b) Set $t = \pi|x|$ and integrate with respect to $e^{-2\pi i\xi\cdot x}dx$ to prove that

$$(e^{-2\pi|x|})\widehat{\ }(\xi) = \frac{\Gamma(\frac{n+1}{2})}{\pi^{\frac{n+1}{2}}}\frac{1}{(1+|\xi|^2)^{\frac{n+1}{2}}}.$$

This calculation gives the Fourier transform of the Poisson kernel.

2.2.12. Let $1 \le p \le \infty$ and let p' be its dual index.
(a) Prove that Schwartz functions f on the line satisfy the estimate

$$\|f\|_{L^\infty}^2 \le 2\|f\|_{L^p}\|f'\|_{L^{p'}}.$$

(b) Prove that all Schwartz functions f on $\mathbf{R}^n$ satisfy the estimate

$$\|f\|_{L^\infty}^2 \le \sum_{|\alpha+\beta|=n} \|\partial^\alpha f\|_{L^p}\|\partial^\beta f\|_{L^{p'}},$$

where the sum is taken over all pairs of multi-indices α and β whose sum has size n.
[*Hint:* Part (a): Write $f(x)^2 = \int_{-\infty}^x \frac{d}{dt} f(t)^2\,dt$.]

2.2.13. The *uncertainty principle* says that the position and the momentum of a particle cannot be simultaneously localized. Prove the following inequality, which presents a quantitative version of this principle:

$$\|f\|_{L^2(\mathbf{R}^n)}^2 \le \frac{4\pi}{n}\inf_{y\in\mathbf{R}^n}\Big[\int_{\mathbf{R}^n}|x-y|^2|f(x)|^2\,dx\Big]^{\frac12}\inf_{z\in\mathbf{R}^n}\Big[\int_{\mathbf{R}^n}|\xi-z|^2|\widehat{f}(\xi)|^2\,d\xi\Big]^{\frac12},$$

where f is a Schwartz function on $\mathbf{R}^n$ (or an L^2 function with sufficient decay at infinity).
[*Hint:* Let y be in $\mathbf{R}^n$. Start with

$$\|f\|_{L^2}^2 = \frac{1}{n}\int_{\mathbf{R}^n} f(x)\overline{f(x)}\sum_{j=1}^n \frac{\partial}{\partial x_j}(x_j-y_j)\,dx,$$

integrate by parts, apply the Cauchy–Schwarz inequality, Plancherel's identity, and the identity $\sum_{j=1}^n |\widehat{\partial_j f}(\xi)|^2 = 4\pi^2|\xi|^2|\widehat{f}(\xi)|^2$ for all $\xi \in \mathbf{R}^n$. Then replace $f(x)$ by $f(x)e^{2\pi i x\cdot z}$.]

2.2.14. Let $-\infty < \alpha < \frac{n}{2} < \beta < +\infty$. Prove the validity of the following inequality:

$$\big\|g\big\|_{L^1(\mathbf{R}^n)} \le C\big\||x|^\alpha g(x)\big\|_{L^2(\mathbf{R}^n)}^{\frac{\beta-n/2}{\beta-\alpha}}\big\||x|^\beta g(x)\big\|_{L^2(\mathbf{R}^n)}^{\frac{n/2-\alpha}{\beta-\alpha}}$$

for some constant $C = C(n,\alpha,\beta)$ independent of g.
[*Hint:* First prove $\|g\|_{L^1} \le C\|\,|x|^\alpha g(x)\|_{L^2} + \|\,|x|^\beta g(x)\|_{L^2}$ and then replace $g(x)$ by $g(\lambda x)$ for some suitable $\lambda > 0$.]

2.3 The Class of Tempered Distributions

The fundamental idea of the theory of distributions is that it is generally easier to work with linear functionals acting on spaces of "nice" functions than to work with "bad" functions directly. The set of "nice" functions we consider is closed under the basic operations in analysis, and these operations are extended to distributions by duality. This wonderful interpretation has proved to be an indispensable tool that has clarified many situations in analysis.

2.3.1 Spaces of Test Functions

We recall the space $\mathscr{C}_0^\infty(\mathbf{R}^n)$ of all smooth functions with compact support, and $\mathscr{C}^\infty(\mathbf{R}^n)$ of all smooth functions on $\mathbf{R}^n$. We are mainly interested in the three spaces of "nice" functions on $\mathbf{R}^n$ that are nested as follows:

$$\mathscr{C}_0^\infty(\mathbf{R}^n) \subseteq \mathscr{S}(\mathbf{R}^n) \subseteq \mathscr{C}^\infty(\mathbf{R}^n).$$

Here $\mathscr{S}(\mathbf{R}^n)$ is the space of Schwartz functions introduced in Section 2.2.

Definition 2.3.1. We define convergence of sequences in these spaces. We say that

$$f_k \to f \text{ in } \mathscr{C}^\infty \iff f_k, f \in \mathscr{C}^\infty \text{ and } \lim_{k\to\infty} \sup_{|x|\le N} |\partial^\alpha(f_k - f)(x)| = 0$$
$$\forall\, \alpha \text{ multi-indices and all } N = 1,2,\dots.$$

$$f_k \to f \text{ in } \mathscr{S} \iff f_k, f \in \mathscr{S} \text{ and } \lim_{k\to\infty} \sup_{x\in\mathbf{R}^n} |x^\alpha \partial^\beta(f_k - f)(x)| = 0$$
$$\forall\, \alpha,\beta \text{ multi-indices.}$$

$$f_k \to f \text{ in } \mathscr{C}_0^\infty \iff f_k, f \in \mathscr{C}_0^\infty,\ \text{support}(f_k) \subseteq B \text{ for all } k,\ B \text{ compact,}$$
$$\text{and } \lim_{k\to\infty} \big\|\partial^\alpha(f_k - f)\big\|_{L^\infty} = 0\ \forall\, \alpha \text{ multi-indices.}$$

It follows that convergence in $\mathscr{C}_0^\infty(\mathbf{R}^n)$ implies convergence in $\mathscr{S}(\mathbf{R}^n)$, which in turn implies convergence in $\mathscr{C}^\infty(\mathbf{R}^n)$.

Example 2.3.2. Let φ be a nonzero $\mathscr{C}_0^\infty$ function on $\mathbf{R}$. We call such functions *smooth bumps*. Define the sequence of smooth bumps $\varphi_k(x) = \varphi(x-k)/k$. Then $\varphi_k(x)$ does not converge to zero in $\mathscr{C}_0^\infty(\mathbf{R})$, even though φ_k (and all of its derivatives) converge to zero uniformly. Furthermore, we see that φ_k does not converge to any function in $\mathscr{S}(\mathbf{R})$. Clearly $\varphi_k \to 0$ in $\mathscr{C}^\infty(\mathbf{R})$.

The space $\mathscr{C}^\infty(\mathbf{R}^n)$ is equipped with the family of seminorms

$$\widetilde{\rho}_{\alpha,N}(f) = \sup_{|x|\le N} |(\partial^\alpha f)(x)|, \tag{2.3.1}$$

where α ranges over all multi-indices and N ranges over $\mathbf{Z}^+$. It can be shown that $\mathscr{C}^\infty(\mathbf{R}^n)$ is complete with respect to this countable family of seminorms, i.e., it is a Fréchet space. However, it is true that $\mathscr{C}_0^\infty(\mathbf{R}^n)$ is not complete with respect to the topology generated by this family of seminorms.

The topology of $\mathscr{C}_0^\infty$ given in Definition 2.3.1 is the *inductive limit topology,* and under this topology it is complete. Indeed, letting $\mathscr{C}_0^\infty(B(0,k))$ be the space of all smooth functions with support in $B(0,k)$, then $\mathscr{C}_0^\infty(\mathbf{R}^n)$ is equal to $\bigcup_{k=1}^\infty \mathscr{C}_0^\infty(B(0,k))$ and each space $\mathscr{C}_0^\infty(B(0,k))$ is complete with respect to the topology generated by the family of seminorms $\widetilde{\rho}_{\alpha,N}$; hence so is $\mathscr{C}_0^\infty(\mathbf{R}^n)$. Nevertheless, $\mathscr{C}_0^\infty(\mathbf{R}^n)$ is not metrizable. Details on the topologies of these spaces can be found in [286].

2.3.2 Spaces of Functionals on Test Functions

The dual spaces (i.e., the spaces of continuous linear functionals on the sets of test functions) we introduced is denoted by

$$\begin{aligned}
(\mathscr{C}_0^\infty(\mathbf{R}^n))' &= \mathscr{D}'(\mathbf{R}^n),\\
(\mathscr{S}(\mathbf{R}^n))' &= \mathscr{S}'(\mathbf{R}^n),\\
(\mathscr{C}^\infty(\mathbf{R}^n))' &= \mathscr{E}'(\mathbf{R}^n).
\end{aligned}$$

By definition of the topologies on the dual spaces, we have

$$\begin{aligned}
T_k \to T \quad \text{in } \mathscr{D}' \quad &\iff \quad T_k, T \in \mathscr{D}' \text{ and } T_k(f) \to T(f) \text{ for all } f \in \mathscr{C}_0^\infty.\\
T_k \to T \quad \text{in } \mathscr{S}' \quad &\iff \quad T_k, T \in \mathscr{S}' \text{ and } T_k(f) \to T(f) \text{ for all } f \in \mathscr{S}.\\
T_k \to T \quad \text{in } \mathscr{E}' \quad &\iff \quad T_k, T \in \mathscr{E}' \text{ and } T_k(f) \to T(f) \text{ for all } f \in \mathscr{C}^\infty.
\end{aligned}$$

The dual spaces are nested as follows:

$$\mathscr{E}'(\mathbf{R}^n) \subseteq \mathscr{S}'(\mathbf{R}^n) \subseteq \mathscr{D}'(\mathbf{R}^n).$$

Definition 2.3.3. Elements of the space $\mathscr{D}'(\mathbf{R}^n)$ are called *distributions*. Elements of $\mathscr{S}'(\mathbf{R}^n)$ are called *tempered distributions*. Elements of the space $\mathscr{E}'(\mathbf{R}^n)$ are called *distributions with compact support*.

Before we discuss some examples, we give alternative characterizations of distributions, which are very useful from the practical point of view. The action of a distribution u on a test function f is represented in either one of the following two ways:

$$\langle u, f \rangle = u(f).$$

Proposition 2.3.4. *(a) A linear functional u on $\mathscr{C}_0^\infty(\mathbf{R}^n)$ is a distribution if and only if for every compact $K \subseteq \mathbf{R}^n$, there exist $C > 0$ and an integer m such that*

$$\left|\langle u, f \rangle\right| \le C \sum_{|\alpha| \le m} \left\|\partial^\alpha f\right\|_{L^\infty}, \quad \textit{for all } f \in \mathscr{C}^\infty \textit{ with support in } K. \tag{2.3.2}$$

(b) A linear functional u on $\mathscr{S}(\mathbf{R}^n)$ is a tempered distribution if and only if there exist $C > 0$ and k, m integers such that

$$\left|\langle u, f \rangle\right| \le C \sum_{\substack{|\alpha| \le m \\ |\beta| \le k}} \rho_{\alpha,\beta}(f), \qquad \textit{for all } f \in \mathscr{S}(\mathbf{R}^n). \tag{2.3.3}$$

(c) A linear functional u on $\mathscr{C}^\infty(\mathbf{R}^n)$ is a distribution with compact support if and only if there exist $C > 0$ and N, m integers such that

$$\left|\langle u, f \rangle\right| \le C \sum_{|\alpha| \le m} \widetilde{\rho}_{\alpha,N}(f), \qquad \textit{for all } f \in \mathscr{C}^\infty(\mathbf{R}^n). \tag{2.3.4}$$

The seminorms $\rho_{\alpha,\beta}$ and $\widetilde{\rho}_{\alpha,N}$ are defined in (2.2.6) and (2.3.1), respectively.

Proof. We prove only (2.3.3), since the proofs of (2.3.2) and (2.3.4) are similar. It is clear that (2.3.3) implies continuity of u. Conversely, it was pointed out in Section 2.2 that the family of sets $\{f \in \mathscr{S}(\mathbf{R}^n) : \rho_{\alpha,\beta}(f) < \delta\}$, where α, β are multi-indices and $\delta > 0$, forms a subbasis for the topology of $\mathscr{S}$. Thus if u is a continuous functional on $\mathscr{S}$, there exist integers k, m and a $\delta > 0$ such that

$$|\alpha| \le m,\ |\beta| \le k, \quad \text{and } \rho_{\alpha,\beta}(f) < \delta \implies \left|\langle u, f \rangle\right| \le 1. \tag{2.3.5}$$

We see that (2.3.3) follows from (2.3.5) with $C = 1/\delta$. □

Examples 2.3.5. We now discuss some important examples.

1. The *Dirac mass* at the origin δ_0. This is defined for $\varphi \in \mathscr{C}^\infty(\mathbf{R}^n)$ by

$$\langle \delta_0, \varphi \rangle = \varphi(0).$$

We claim that δ_0 is in $\mathscr{E}'$. To see this we observe that if $\varphi_k \to \varphi$ in $\mathscr{C}^\infty$ then $\langle \delta_0, \varphi_k \rangle \to \langle \delta_0, \varphi \rangle$. The Dirac mass at a point $a \in \mathbf{R}^n$ is defined similarly by

$$\langle \delta_a, \varphi \rangle = \varphi(a).$$

2. Some functions g can be thought of as distributions via the identification $g \mapsto L_g$, where L_g is the functional

$$L_g(\varphi) = \int_{\mathbf{R}^n} \varphi(x) g(x)\, dx.$$

Here are some examples: The function 1 is in $\mathscr{S}'$ but not in $\mathscr{E}'$. Compactly supported integrable functions are in $\mathscr{E}'$. The function $e^{|x|^2}$ is in $\mathscr{D}'$ but not in $\mathscr{S}'$.
3. Functions in L^1_{loc} are distributions. To see this, first observe that if $g \in L^1_{\mathrm{loc}}$, then the integral

$$L_g(\varphi) = \int_{\mathbf{R}^n} \varphi(x) g(x)\, dx$$

is well defined for all $\varphi \in \mathscr{D}$ and satisfies $|L_g(\varphi)| \le \big(\int_K |g(x)|\, dx\big) \|\varphi\|_{L^\infty}$ for all smooth functions φ supported in the compact set K.
4. Functions in L^p, $1 \le p \le \infty$, are tempered distributions, but may not in $\mathscr{E}'$ unless they have compact support.
5. Any finite Borel measure μ is a tempered distribution via the identification

$$L_\mu(\varphi) = \int_{\mathbf{R}^n} \varphi(x)\, d\mu(x).$$

To see this, observe that $\varphi_k \to \varphi$ in $\mathscr{S}$ implies that $L_\mu(\varphi_k) \to L_\mu(\varphi)$. Finite Borel measures may not be distributions with compact support.
6. Every function g that satisfies $|g(x)| \le C(1+|x|)^k$, for some real number k, is a tempered distribution. To see this, observe that

$$\big|L_g(\varphi)\big| \le \sup_{x \in \mathbf{R}^n} (1+|x|)^m |\varphi(x)| \int_{\mathbf{R}^n} (1+|x|)^{k-m} dx,$$

where $m > n+k$ and the expression $\sup_{x\in\mathbf{R}^n}(1+|x|)^m|\varphi(x)|$ is bounded by a finite sum of Schwartz seminorms $\rho_{\alpha,\beta}(\varphi)$.
7. The function $\log|x|$ is a tempered distribution; indeed for any $\varphi \in \mathscr{S}(\mathbf{R}^n)$, the integral of $\varphi(x)\log|x|$ is bounded by a finite number of Schwartz seminorms of φ. More generally, any function that is integrable on a ball $|x| \le M$ and for some $C > 0$ satisfies $|g(x)| \le C(1+|x|)^k$ for $|x| \ge M$, is a tempered distribution.
8. Here is an example of a compactly supported distribution on $\mathbf{R}$ that is neither a locally integrable function nor a finite Borel measure:

$$\langle u, \varphi \rangle = \lim_{\varepsilon \to 0} \int_{\varepsilon \le |x| \le 1} \varphi(x) \frac{dx}{x} = \lim_{\varepsilon \to 0} \int_{\varepsilon \le |x| \le 1} (\varphi(x) - \varphi(0)) \frac{dx}{x}.$$

We have that $|\langle u, \varphi \rangle| \le 2\|\varphi'\|_{L^\infty([-1,1])}$ and notice that $\|\varphi'\|_{L^\infty([-1,1])}$ is a $\widetilde{\rho}_{\alpha,N}$ seminorm of φ.

2.3.3 The Space of Tempered Distributions

Having set down the basic definitions of distributions, we now focus our study on the space of tempered distributions. These distributions are the most useful in harmonic analysis. The main reason for this is that the subject is concerned with boundedness of translation-invariant operators, and every such bounded operator from $L^p(\mathbf{R}^n)$ to $L^q(\mathbf{R}^n)$ is given by convolution with a tempered distribution. This fact is shown in Section 2.5.

Suppose that f and g are Schwartz functions and α a multi-index. Integrating by parts $|\alpha|$ times, we obtain

$$\int_{\mathbf{R}^n}(\partial^\alpha f)(x)g(x)\,dx=(-1)^{|\alpha|}\int_{\mathbf{R}^n}f(x)(\partial^\alpha g)(x)\,dx. \tag{2.3.6}$$

If we wanted to define the derivative of a tempered distribution u, we would have to give a definition that extends the definition of the derivative of the function and that satisfies (2.3.6) for g in $\mathscr{S}'$ and $f\in\mathscr{S}$ if the integrals in (2.3.6) are interpreted as actions of distributions on functions. We simply use equation (2.3.6) to define the derivative of a distribution.

Definition 2.3.6. Let $u\in\mathscr{S}'$ and α a multi-index. Define

$$\left\langle\partial^\alpha u,f\right\rangle=(-1)^{|\alpha|}\left\langle u,\partial^\alpha f\right\rangle. \tag{2.3.7}$$

If u is a function, the derivatives of u in the sense of distributions are called *distributional derivatives*.

In view of Theorem 2.2.14, it is natural to give the following:

Definition 2.3.7. Let $u\in\mathscr{S}'$. We define the Fourier transform $\widehat{u}$ and the inverse Fourier transform $u^\vee$ of a tempered distribution u by

$$\left\langle\widehat{u},f\right\rangle=\left\langle u,\widehat{f}\right\rangle\qquad\text{and}\qquad\left\langle u^\vee,f\right\rangle=\left\langle u,f^\vee\right\rangle, \tag{2.3.8}$$

for all f in $\mathscr{S}$.

Example 2.3.8. We observe that $\widehat{\delta_0}=1$. More generally, for any multi-index α we have

$$(\partial^\alpha\delta_0)^\wedge=(2\pi i x)^\alpha.$$

To see this, observe that for all $f\in\mathscr{S}$ we have

$$\begin{aligned}\left\langle(\partial^\alpha\delta_0)^\wedge,f\right\rangle&=\left\langle\partial^\alpha\delta_0,\widehat{f}\right\rangle\\&=(-1)^{|\alpha|}\left\langle\delta_0,\partial^\alpha\widehat{f}\right\rangle\\&=(-1)^{|\alpha|}\left\langle\delta_0,((-2\pi i x)^\alpha f(x))^\wedge\right\rangle\\&=(-1)^{|\alpha|}((-2\pi i x)^\alpha f(x))^\wedge(0)\end{aligned}$$

$$= (-1)^{|\alpha|} \int_{\mathbf{R}^n} (-2\pi i x)^\alpha f(x)\, dx$$
$$= \int_{\mathbf{R}^n} (2\pi i x)^\alpha f(x)\, dx.$$

This calculation indicates that $(\partial^\alpha \delta_0)^\wedge$ can be identified with the function $(2\pi i x)^\alpha$.

Example 2.3.9. Recall that for $x_0 \in \mathbf{R}^n$, $\delta_{x_0}(f) = \langle \delta_{x_0}, f \rangle = f(x_0)$. Then

$$\langle \widehat{\delta_{x_0}}, h \rangle = \langle \delta_{x_0}, \widehat{h} \rangle = \widehat{h}(x_0) = \int_{\mathbf{R}^n} h(x) e^{-2\pi i x \cdot x_0}\, dx, \qquad h \in \mathscr{S}(\mathbf{R}^n),$$

that is, $\widehat{\delta_{x_0}}$ can be identified with the function $x \mapsto e^{-2\pi i x \cdot x_0}$. In particular, $\widehat{\delta_0} = 1$.

Example 2.3.10. The function $e^{|x|^2}$ is not in $\mathscr{S}'(\mathbf{R}^n)$ and therefore its Fourier transform is not defined as a distribution. However, the Fourier transform of any locally integrable function with polynomial growth at infinity is defined as a tempered distribution.

Now observe that the following are true whenever f, g are in $\mathscr{S}$.

$$\begin{aligned} \int_{\mathbf{R}^n} g(x) f(x-t)\, dx &= \int_{\mathbf{R}^n} g(x+t) f(x)\, dx, \\ \int_{\mathbf{R}^n} g(ax) f(x)\, dx &= \int_{\mathbf{R}^n} g(x) a^{-n} f(a^{-1}x)\, dx, \\ \int_{\mathbf{R}^n} \widetilde{g}(x) f(x)\, dx &= \int_{\mathbf{R}^n} g(x) \widetilde{f}(x)\, dx, \end{aligned} \tag{2.3.9}$$

for all $t \in \mathbf{R}^n$ and $a > 0$. Recall now the definitions of τ^t, δ^a, and $\widetilde{\ }$ given in (2.2.13). Motivated by (2.3.9), we give the following:

Definition 2.3.11. The *translation* $\tau^t u$, the *dilation* $\delta^a u$, and the *reflection* $\widetilde{u}$ of a tempered distribution u are defined as follows:

$$\langle \tau^t u, f \rangle = \langle u, \tau^{-t} f \rangle, \tag{2.3.10}$$
$$\langle \delta^a u, f \rangle = \langle u, a^{-n} \delta^{1/a} f \rangle, \tag{2.3.11}$$
$$\langle \widetilde{u}, f \rangle = \langle u, \widetilde{f} \rangle, \tag{2.3.12}$$

for all $t \in \mathbf{R}^n$ and $a > 0$. Let A be an invertible matrix. The composition of a distribution u with an invertible matrix A is the distribution

$$\langle u^A, \varphi \rangle = |\det A|^{-1} \langle u, \varphi^{A^{-1}} \rangle, \tag{2.3.13}$$

where $\varphi^{A^{-1}}(x) = \varphi(A^{-1}x)$.

It is easy to see that the operations of translation, dilation, reflection, and differentiation are continuous on tempered distributions.

Example 2.3.12. The Dirac mass at the origin δ_0 is equal to its reflection, while $\delta^a\delta_0 = a^{-n}\delta_0$. Also, $\tau^x\delta_0 = \delta_x$ for any $x \in \mathbf{R}^n$.

Now observe that for f, g, and h in $\mathscr{S}$ we have

$$\int_{\mathbf{R}^n} (h*g)(x)f(x)\,dx = \int_{\mathbf{R}^n} g(x)(\widetilde{h}*f)(x)\,dx. \tag{2.3.14}$$

Motivated by (2.3.14), we define the convolution of a function with a tempered distribution as follows:

Definition 2.3.13. Let $u \in \mathscr{S}'$ and $h \in \mathscr{S}$. Define the convolution $h*u$ by

$$\langle h*u, f\rangle = \langle u, \widetilde{h}*f\rangle, \qquad f \in \mathscr{S}. \tag{2.3.15}$$

Example 2.3.14. Let $u = \delta_{x_0}$ and $f \in \mathscr{S}$. Then $f*\delta_{x_0}$ is the function $x \mapsto f(x-x_0)$, for when $h \in \mathscr{S}$, we have

$$\langle f*\delta_{x_0}, h\rangle = \langle \delta_{x_0}, \widetilde{f}*h\rangle = (\widetilde{f}*h)(x_0) = \int_{\mathbf{R}^n} f(x-x_0)h(x)\,dx.$$

It follows that convolution with δ_0 is the identity operator.

We now define the product of a function and a distribution.

Definition 2.3.15. Let $u \in \mathscr{S}'$ and let h be a $\mathscr{C}^\infty$ function that has at most polynomial growth at infinity and the same is true for all of its derivatives. This means that for all α it satisfies $|(\partial^\alpha h)(x)| \le C_\alpha(1+|x|)^{k_\alpha}$ for some $C_\alpha, k_\alpha > 0$. Then define the product hu of h and u by

$$\langle hu, f\rangle = \langle u, hf\rangle, \qquad f \in \mathscr{S}. \tag{2.3.16}$$

Note that hf is in $\mathscr{S}$ and thus (2.3.16) is well defined. The product of an arbitrary $\mathscr{C}^\infty$ function with a tempered distribution is not defined.

We observe that if a function g is supported in a set K, then for all $f \in \mathscr{C}_0^\infty(K^c)$ we have

$$\int_{\mathbf{R}^n} f(x)g(x)\,dx = 0. \tag{2.3.17}$$

Moreover, the support of g is the intersection of all closed sets K with the property (2.3.17) for all f in $\mathscr{C}_0^\infty(K^c)$. Motivated by the preceding observation we give the following:

Definition 2.3.16. Let u be in $\mathscr{D}'(\mathbf{R}^n)$. The *support* of u ($\operatorname{supp} u$) is the intersection of all closed sets K with the property

$$\varphi \in \mathscr{C}_0^\infty(\mathbf{R}^n), \qquad \operatorname{supp}\varphi \subseteq \mathbf{R}^n \setminus K \implies \langle u, \varphi\rangle = 0. \tag{2.3.18}$$

Distributions with compact support are exactly those whose support (as defined in the previous definition) is a compact set. To prove this assertion, we start with a distribution u with compact support as defined in Definition 2.3.3. Then there exist $C,N,m>0$ such that (2.3.4) holds. For a $\mathscr{C}^\infty$ function f whose support is contained in $B(0,N)^c$, the expression on the right in (2.3.4) vanishes and we must therefore have $\langle u,f\rangle = 0$. This shows that the support of u is contained in $\overline{B(0,N)}$ hence it is bounded, and since it is already closed (as an intersection of closed sets), it must be compact. Conversely, if the support of u as defined in Definition 2.3.16 is a compact set, then there exists an $N>0$ such that $\operatorname{supp} u$ is contained in $B(0,N)$. We take a smooth function η that is equal to 1 on $B(0,N)$ and vanishes off $B(0,N+1)$. Then for $h\in\mathscr{C}_0^\infty$ the support of $h(1-\eta)$ does not meet the support of u, and we must have

$$\langle u,h\rangle = \langle u,h\eta\rangle + \langle u,h(1-\eta)\rangle = \langle u,h\eta\rangle .$$

The distribution u can be thought of as an element of $\mathscr{E}'$ by defining for $f\in\mathscr{C}^\infty(\mathbf{R}^n)$

$$\langle u,f\rangle = \langle u,f\eta\rangle .$$

Taking m to be the integer that corresponds to the compact set $K=\overline{B(0,N+1)}$ in (2.3.2), and using that the L^∞ norm of $\partial^\alpha(f\eta)$ is controlled by a finite sum of seminorms $\widetilde{\rho}_{\alpha,N+1}(f)$ with $|\alpha|\le m$, we obtain the validity of (2.3.4) for $f\in\mathscr{C}^\infty$.

Example 2.3.17. The support of the Dirac mass at x_0 is the set $\{x_0\}$.

Along the same lines, we give the following definition:

Definition 2.3.18. We say that a distribution u in $\mathscr{D}'(\mathbf{R}^n)$ coincides with the function h on an open set Ω if

$$\langle u,f\rangle = \int_{\mathbf{R}^n} f(x)h(x)\,dx \qquad \text{for all } f \text{ in } \mathscr{C}_0^\infty(\Omega). \tag{2.3.19}$$

When (2.3.19) occurs we often say that u agrees with h away from Ω^c.

This definition implies that the support of the distribution $u-h$ is contained in the set Ω^c.

Example 2.3.19. The distribution $|x|^2+\delta_{a_1}+\delta_{a_2}$, where a_1, a_2 are in $\mathbf{R}^n$, coincides with the function $|x|^2$ on any open set not containing the points a_1 and a_2. Also, the distribution in Example 2.3.5 (8) coincides with the function $x^{-1}\chi_{|x|\le 1}$ away from the origin in the real line.

Having ended the streak of definitions regarding operations with distributions, we now discuss properties of convolutions and Fourier transforms.

Theorem 2.3.20. *If $u \in \mathscr{S}'$ and $\varphi \in \mathscr{S}$, then $\varphi * u$ is a $\mathscr{C}^\infty$ function and*

$$(\varphi * u)(x) = \langle u, \tau^x \widetilde{\varphi} \rangle$$

for all $x \in \mathbf{R}^n$. Moreover, for all multi-indices α there exist constants $C_\alpha, k_\alpha > 0$ such that

$$|\partial^\alpha (\varphi * u)(x)| \le C_\alpha (1+|x|)^{k_\alpha} .$$

Furthermore, if u has compact support, then $\varphi * u$ is a Schwartz function.

Proof. Let ψ be in $\mathscr{S}(\mathbf{R}^n)$. We have

$$\begin{aligned}
\big\langle \varphi * u, \psi \big\rangle &= \big\langle u, \widetilde{\varphi} * \psi \big\rangle \\
&= u\left(\int_{\mathbf{R}^n} \widetilde{\varphi}(\cdot - y) \psi(y)\, dy \right) \\
&= u\left(\int_{\mathbf{R}^n} (\tau^y \widetilde{\varphi})(\cdot) \psi(y)\, dy \right) \\
&= \int_{\mathbf{R}^n} \big\langle u, \tau^y \widetilde{\varphi} \big\rangle \psi(y)\, dy,
\end{aligned} \tag{2.3.20}$$

where the last step is justified by the continuity of u and by the fact that the Riemann sums of the inner integral in (2.3.20) converge to that integral in the topology of $\mathscr{S}$, a fact that will be justified later. This calculation identifies the function $\varphi * u$ as

$$(\varphi * u)(x) = \big\langle u, \tau^x \widetilde{\varphi} \big\rangle . \tag{2.3.21}$$

We now show that $(\varphi * u)(x)$ is a $\mathscr{C}^\infty$ function. Let $e_j = (0, \dots, 1, \dots, 0)$ with 1 in the jth entry and zero elsewhere. Then

$$\frac{\tau^{-he_j}(\varphi * u)(x) - (\varphi * u)(x)}{h} = u\left(\frac{\tau^{-he_j}(\tau^x \widetilde{\varphi}) - \tau^x \widetilde{\varphi}}{h} \right) \to \langle u, \tau^x(\partial_j \widetilde{\varphi}) \rangle$$

by the continuity of u and the fact that $\big(\tau^{-he_j}(\tau^x \widetilde{\varphi}) - \tau^x \widetilde{\varphi}\big)/h$ tends to $\partial_j \tau^x \widetilde{\varphi} = \tau^x(\partial_j \widetilde{\varphi})$ in $\mathscr{S}$ as $h \to 0$; see Exercise 2.3.5 (a). The same calculation for higher-order derivatives shows that $\varphi * u \in \mathscr{C}^\infty$ and that $\partial^\gamma(\varphi * u) = (\partial^\gamma \varphi) * u$ for all multi-indices γ. It follows from (2.3.3) that for some C, m, and k we have

$$\begin{aligned}
|\partial^\alpha (\varphi * u)(x)| &\le C \sum_{\substack{|\gamma| \le m \\ |\beta| \le k}} \sup_{y \in \mathbf{R}^n} |y^\gamma \tau^x (\partial^{\alpha+\beta} \widetilde{\varphi})(y)| \\
&= C \sum_{\substack{|\gamma| \le m \\ |\beta| \le k}} \sup_{y \in \mathbf{R}^n} |(x+y)^\gamma (\partial^{\alpha+\beta} \widetilde{\varphi})(y)| \\
&\le C_m \sum_{|\beta| \le k} \sup_{y \in \mathbf{R}^n} (1+|x|^m + |y|^m) |(\partial^{\alpha+\beta} \widetilde{\varphi})(y)| ,
\end{aligned} \tag{2.3.22}$$

and this clearly implies that $\partial^\alpha(\varphi * u)$ grows at most polynomially at infinity.

We now indicate why $\varphi * u$ is Schwartz whenever u has compact support. Applying estimate (2.3.4) to the function $y \mapsto \varphi(x-y)$ yields that

$$\big|\langle u, \varphi(x-\cdot)\rangle\big| = |(\varphi * u)(x)| \le C \sum_{|\alpha|\le m} \sup_{|y|\le N} |\partial_y^\alpha \varphi(x-y)|$$

for some constants C, m, N. Since for $|x| \ge 2N$ we have

$$|\partial_y^\alpha \varphi(x-y)| \le C_{\alpha,M}(1+|x-y|)^{-M} \le C_{\alpha,M,N}(1+|x|)^{-M},$$

it follows that $\varphi * u$ decays rapidly at infinity. Since $\partial^\gamma(\varphi * u) = (\partial^\gamma \varphi) * u$, the same argument yields that all the derivatives of $\varphi * u$ decay rapidly at infinity; hence $\varphi * u$ is a Schwartz function. Incidentally, this argument actually shows that any Schwartz seminorm of $\varphi * u$ is controlled by a finite sum of Schwartz seminorms of φ.

We now return to the point left open concerning the convergence of the Riemann sums in (2.3.20) in the topology of $\mathscr{S}(\mathbf{R}^n)$. For each $N = 1, 2, \dots$, consider a partition of $[-N,N]^n$ into $(2N^2)^n$ cubes Q_m of side length $1/N$ and let y_m be the center of each Q_m. For multi-indices α, β, we must show that

$$D_N(x) = \sum_{m=1}^{(2N^2)^n} x^\alpha \partial_x^\beta \widetilde{\varphi}(x-y_m)\psi(y_m)|Q_m| - \int_{\mathbf{R}^n} x^\alpha \partial_x^\beta \widetilde{\varphi}(x-y)\psi(y)\,dy$$

converges to zero in $L^\infty(\mathbf{R}^n)$ as $N \to \infty$. We have

$$\begin{aligned} x^\alpha \partial_x^\beta \widetilde{\varphi}(x-y_m)\psi(y_m)|Q_m| &- \int_{Q_m} x^\alpha \partial_x^\beta \widetilde{\varphi}(x-y)\psi(y)\,dy \\ &= \int_{Q_m} x^\alpha (y-y_m)\cdot \nabla\big(\partial_x^\beta \widetilde{\varphi}(x-\cdot)\psi\big)(\xi)\,dy \end{aligned}$$

for some $\xi = y + \theta(y_m - y)$, where $\theta \in [0,1]$. Distributing the gradient to both factors, we see that the last integrand is at most

$$C|x|^{|\alpha|}\frac{\sqrt{n}}{N}\frac{1}{(1+|x-\xi|)^{M/2}}\frac{1}{(2+|\xi|)^M}$$

for M large (pick $M > 2|\alpha|$), which in turn is at most

$$C'|x|^{|\alpha|}\frac{\sqrt{n}}{N}\frac{1}{(1+|x|)^{M/2}}\frac{1}{(2+|\xi|)^{M/2}} \le C'|x|^{|\alpha|}\frac{\sqrt{n}}{N}\frac{1}{(1+|x|)^{M/2}}\frac{1}{(1+|y|)^{M/2}},$$

since $|y| \le |\xi| + \theta|y-y_m| \le |\xi| + \sqrt{n}/N \le |\xi|+1$ for $N \ge \sqrt{n}$. Inserting the estimate obtained for the integrand in the last displayed integral, we obtain

$$|D_N(x)| \le \frac{C''}{N}\frac{|x|^{|\alpha|}}{(1+|x|)^{M/2}}\int_{[-N,N]^n}\frac{dy}{(1+|y|)^{M/2}} + \int_{([-N,N]^n)^c} |x^\alpha \partial_x^\beta \widetilde{\varphi}(x-y)\psi(y)|\,dy.$$

But the second integral in the preceding expression is bounded by

$$\int_{([-N,N]^n)^c} \frac{C'''|x|^{|\alpha|}}{(1+|x-y|)^M}\frac{dy}{(1+|y|)^M} \le \frac{C'''|x|^{|\alpha|}}{(1+|x|)^{M/2}} \int_{([-N,N]^n)^c} \frac{dy}{(1+|y|)^{M/2}}.$$

Using these estimates it is now easy to see that $\lim_{N\to\infty}\sup_{x\in\mathbf{R}^n}|D_N(x)|=0$. □

Next we have the following important result regarding distributions with compact support:

Theorem 2.3.21. *If u is in $\mathscr{E}'(\mathbf{R}^n)$, then $\widehat{u}$ is a real analytic function on $\mathbf{R}^n$. In particular, $\widehat{u}$ is a $\mathscr{C}^\infty$ function. Furthermore, $\widehat{u}$ and all of its derivatives have polynomial growth at infinity. Moreover, $\widehat{u}$ has a holomorphic extension on $\mathbf{C}^n$.*

Proof. Given a distribution u with compact support and a polynomial $p(\xi)$, the action of u on the $\mathscr{C}^\infty$ function $\xi \mapsto p(\xi)e^{-2\pi i x\cdot\xi}$ is a well defined function of x, which we denote by $u(p(\cdot)e^{-2\pi i x\cdot(\cdot)})$. Here x is an element of $\mathbf{R}^n$ but the same assertion is valid if $x=(x_1,\dots,x_n)\in\mathbf{R}^n$ is replaced by $z=(z_1,\dots,z_n)\in\mathbf{C}^n$. In this case we define the dot product of ξ and z via $\xi\cdot z=\sum_{k=1}^n \xi_k z_k$.

It is straightforward to verify that the function of $z=(z_1,\dots,z_n)$

$$F(z)=u\big(e^{-2\pi i(\cdot)\cdot z}\big)$$

defined on $\mathbf{C}^n$ is holomorphic, in fact entire. Indeed, the continuity and linearity of u and the fact that $(e^{-2\pi i\xi_j h}-1)/h \to -2\pi i\xi_j$ in $\mathscr{C}^\infty(\mathbf{R}^n)$ as $h\to 0$, $h\in\mathbf{C}$, imply that F is holomorphic in every variable and its derivative with respect to z_j is the action of the distribution u to the $\mathscr{C}^\infty$ function

$$\xi \mapsto (-2\pi i\xi_j)e^{-2\pi i\sum_{j=1}^n \xi_j z_j}.$$

By induction it follows that for all multi-indices α we have

$$\partial_{z_1}^{\alpha_1}\cdots\partial_{z_n}^{\alpha_n}F = u\big((-2\pi i(\cdot))^\alpha e^{-2\pi i\sum_{j=1}^n(\cdot)z_j}\big).$$

Since F is entire, its restriction on $\mathbf{R}^n$, i.e., $F(x_1,\dots,x_n)$, where $x_j=\operatorname{Re} z_j$, is real analytic. Also, an easy calculation using (2.3.4) and Leibniz's rule yield that the restriction of F on $\mathbf{R}^n$ and all of its derivatives have polynomial growth at infinity.

Now for f in $\mathscr{S}(\mathbf{R}^n)$ we have

$$\langle\widehat{u},f\rangle=\langle u,\widehat{f}\rangle=u\bigg(\int_{\mathbf{R}^n} f(x)e^{-2\pi i x\cdot\xi}\,dx\bigg)=\int_{\mathbf{R}^n} f(x)u(e^{-2\pi i x\cdot(\cdot)})\,dx,$$

provided we can justify the passage of u inside the integral. The reason for this is that the Riemann sums of the integral of $f(x)e^{-2\pi i x\cdot\xi}$ over $\mathbf{R}^n$ converge to it in the topology of $\mathscr{C}^\infty$, and thus the linear functional u can be interchanged with the integral. We conclude that the tempered distribution $\widehat{u}$ can be identified with the real analytic function $x\mapsto F(x)$ whose derivatives have polynomial growth at infinity.

To justify the fact concerning the convergence of the Riemann sums, we argue as in the proof of the previous theorem. For each $N = 1,2,\dots$, consider a partition of $[-N,N]^n$ into $(2N^2)^n$ cubes Q_m of side length $1/N$ and let y_m be the center of each Q_m. For a multi-index α let

$$D_N(\xi) = \sum_{m=1}^{(2N^2)^n} f(y_m)(-2\pi i y_m)^\alpha e^{-2\pi i y_m\cdot\xi}|Q_m| - \int_{\mathbf{R}^n} f(x)(-2\pi i x)^\alpha e^{-2\pi i x\cdot\xi}\,dx.$$

We must show that for every $M>0$, $\sup_{|\xi|\le M}|D_N(\xi)|$ converges to zero as $N\to\infty$. Setting $g(x) = f(x)(-2\pi i x)^\alpha$, we write

$$D_N(\xi) = \sum_{m=1}^{(2N^2)^n}\int_{Q_m}\left[g(y_m)e^{-2\pi i y_m\cdot\xi} - g(x)e^{-2\pi i x\cdot\xi}\right]dx + \int_{([-N,N]^n)^c} g(x)e^{-2\pi i x\cdot\xi}\,dx.$$

Using the mean value theorem, we bound the absolute value of the expression inside the square brackets by

$$\big(|\nabla g(z_m)| + 2\pi|\xi|\,|g(z_m)|\big)\frac{\sqrt{n}}{N} \le \frac{C_K(1+|\xi|)}{(1+|z_m|)^K}\frac{\sqrt{n}}{N},$$

for some point z_m in the cube Q_m. Since

$$\sum_{m=1}^{(2N^2)^n}\int_{Q_m}\frac{C_K(1+|\xi|)}{(1+|z_m|)^K}\,dx \le C_K'(1+M) < \infty$$

for $|\xi|\le M$, it follows that $\sup_{|\xi|\le M}|D_N(\xi)|\to 0$ as $N\to\infty$. □

Next we give a proposition that extends the properties of the Fourier transform to tempered distributions.

Proposition 2.3.22. *Given u, v in $\mathscr{S}'(\mathbf{R}^n)$, $f_j, f\in\mathscr{S}$, $y\in\mathbf{R}^n$, b a complex scalar, α a multi-index, and $a>0$, we have*

(1) $\widehat{u+v} = \widehat{u}+\widehat{v}$,

(2) $\widehat{bu} = b\widehat{u}$,

(3) *If $f_j\to f$ in $\mathscr{S}$, then $\widehat{f_j}\to\widehat{f}$ in $\mathscr{S}$ and if $u_j\to u$ in $\mathscr{S}'$, then $\widehat{u}_j\to\widehat{u}$ in $\mathscr{S}'$,*

(4) $(\widetilde{u})\widehat{\ } = (\widehat{u})\widetilde{\ }$,

(5) $(\tau^y u)\widehat{\ } = e^{-2\pi i y\cdot\xi}\widehat{u}$,

(6) $(e^{2\pi i x\cdot y}u)\widehat{\ } = \tau^y\widehat{u}$,

(7) $(\delta^a u)\widehat{\ } = (\widehat{u})_a = a^{-n}\delta^{a^{-1}}\widehat{u}$,

(8) $(\partial^\alpha u)\widehat{\ } = (2\pi i\xi)^\alpha\widehat{u}$,

(9) $\partial^\alpha\widehat{u} = ((-2\pi i x)^\alpha u)\widehat{\ }$,

(10) $(\widehat{u})^\vee = u$,

(11) $\widehat{f * u} = \widehat{f}\,\widehat{u}$,

(12) $\widehat{fu} = \widehat{f} * \widehat{u}$,

(13) **(Leibniz's rule)** $\partial_j^m(fu) = \sum_{k=0}^m \binom{m}{k}(\partial_j^k f)(\partial_j^{m-k}u)$, $m \in \mathbf{Z}^+$,

(14) **(Leibniz's rule)** $\partial^\alpha(fu) = \sum_{\gamma_1=0}^{\alpha_1} \cdots \sum_{\gamma_n=0}^{\alpha_n} \binom{\alpha_1}{\gamma_1} \cdots \binom{\alpha_n}{\gamma_n} (\partial^\gamma f)(\partial^{\alpha-\gamma}u)$,

(15) *If u_k, $u \in L^p(\mathbf{R}^n)$ and $u_k \to u$ in L^p $(1 \le p \le \infty)$, then $u_k \to u$ in $\mathscr{S}'(\mathbf{R}^n)$. Therefore, convergence in $\mathscr{S}$ implies convergence in L^p, which in turn implies convergence in $\mathscr{S}'(\mathbf{R}^n)$.*

Proof. All the statements can be proved easily using duality and the corresponding statements for Schwartz functions. □

We continue with an application of Theorem 2.3.21.

Proposition 2.3.23. *Given $u \in \mathscr{S}'(\mathbf{R}^n)$, there exists a sequence of $\mathscr{C}_0^\infty$ functions f_k such that $f_k \to u$ in the sense of tempered distributions; in particular, $\mathscr{C}_0^\infty(\mathbf{R}^n)$ is dense in $\mathscr{S}'(\mathbf{R}^n)$.*

Proof. Fix a function in $\mathscr{C}_0^\infty(\mathbf{R}^n)$ with $\varphi(x) = 1$ in a neighborhood of the origin. Let $\varphi_k(x) = \delta^{1/k}(\varphi)(x) = \varphi(x/k)$. It follows from Exercise 2.3.5 (b) that for $u \in \mathscr{S}'(\mathbf{R}^n)$, $\varphi_k u \to u$ in $\mathscr{S}'$. By Proposition 2.3.22 (3), we have that the map $u \mapsto (\varphi_k \widehat{u})^\vee$ is continuous on $\mathscr{S}'(\mathbf{R}^n)$. Now Theorem 2.3.21 gives that $(\varphi_k \widehat{u})^\vee$ is a $\mathscr{C}^\infty$ function and therefore $\varphi_j(\varphi_k \widehat{u})^\vee$ is in $\mathscr{C}_0^\infty(\mathbf{R}^n)$. As observed, $\varphi_j(\varphi_k \widehat{u})^\vee \to (\varphi_k \widehat{u})^\vee$ in $\mathscr{S}'$ when k is fixed and $j \to \infty$. Exercise 2.3.5 (c) gives that the diagonal sequence $\varphi_k(\varphi_k \widehat{f})^\wedge$ converges to f in $\mathscr{S}$ as $k \to \infty$ for all $f \in \mathscr{S}$. Using duality and Exercise 2.2.2, we conclude that the sequence of $\mathscr{C}_0^\infty$ functions $\varphi_k(\varphi_k \widehat{u})^\vee$ converges to u in $\mathscr{S}'$ as $k \to \infty$. □

Exercises

2.3.1. Show that a positive measure μ that satisfies

$$\int_{\mathbf{R}^n} \frac{d\mu(x)}{(1+|x|)^k} < +\infty,$$

for some $k > 0$, can be identified with a tempered distribution. Show that if we think of Lebesgue measure as a tempered distribution, then it coincides with the constant function 1 also interpreted as a tempered distribution.

2.3.2. Let $\varphi, f \in \mathscr{S}(\mathbf{R}^n)$, and for $\varepsilon > 0$ let $\varphi_\varepsilon(x) = \varepsilon^{-n}\varphi(\varepsilon^{-1}x)$. Prove that $\varphi_\varepsilon * f \to bf$ in $\mathscr{S}$, where b is the integral of φ.

2.3.3. Prove that for all $a>0$, $u\in\mathscr{S}'(\mathbf{R}^n)$, and $f\in\mathscr{S}(\mathbf{R}^n)$ we have

$$(\delta^a f)*(\delta^a u)=a^{-n}\delta^a(f*u)\,.$$

2.3.4. (a) Prove that the derivative of $\chi_{[a,b]}$ is $\delta_a-\delta_b$.
(b) Compute $\partial_j\chi_{B(0,1)}$ on $\mathbf{R}^2$.
(c) Compute the Fourier transforms of the locally integrable functions $\sin x$ and $\cos x$.
(d) Prove that the derivative of the distribution $\log|x|\in\mathscr{S}'(\mathbf{R})$ is the distribution

$$u(\varphi)=\lim_{\varepsilon\to 0}\int_{\varepsilon\le|x|}\varphi(x)\frac{dx}{x}\,.$$

2.3.5. Let $f\in\mathscr{S}(\mathbf{R}^n)$ and let $\varphi\in\mathscr{C}_0^\infty$ be identically equal to 1 in a neighborhood of the origin. Define $\varphi_k(x)=\varphi(x/k)$ as in the proof of Proposition 2.3.23.
(a) Prove that $(\tau^{-he_j}f-f)/h\to\partial_j f$ in $\mathscr{S}$ as $h\to 0$.
(b) Prove that $\varphi_k f\to f$ in $\mathscr{S}$ as $k\to\infty$.
(c) Prove that the sequence $\varphi_k(\varphi_k f)^\wedge$ converges to $\widehat{f}$ in $\mathscr{S}$ as $k\to\infty$.

2.3.6. Use Theorem 2.3.21 to show that there does not exist a nonzero $\mathscr{C}_0^\infty$ function whose Fourier transform is also a $\mathscr{C}_0^\infty$ function.

2.3.7. Let $f\in L^p(\mathbf{R}^n)$ for some $1\le p<\infty$. Show that the sequence of functions

$$g_N(\xi)=\int_{B(0,N)}f(x)e^{-2\pi i x\cdot\xi}\,dx$$

converges to $\widehat{f}$ in $\mathscr{S}'$.

2.3.8. Let $(c_k)_{k\in\mathbf{Z}^n}$ be a sequence that satisfies $|c_k|\le A(1+|k|)^M$ for all k and some fixed M and $A>0$. Let δ_k denote Dirac mass at the integer k. Show that the sequence of distributions

$$\sum_{|k|\le N}c_k\delta_k$$

converges to some tempered distribution u in $\mathscr{S}'(\mathbf{R}^n)$ as $N\to\infty$. Also show that $\widehat{u}$ is the $\mathscr{S}'$ limit of the sequence of functions

$$h_N(\xi)=\sum_{|k|\le N}c_k e^{-2\pi i\xi\cdot k}\,.$$

2.3.9. A distribution in $\mathscr{S}'(\mathbf{R}^n)$ is called *homogeneous of degree* $\gamma\in\mathbf{C}$ if for all $\lambda>0$ and for all $\varphi\in\mathscr{S}(\mathbf{R}^n)$ we have

$$\langle u,\delta^\lambda\varphi\rangle=\lambda^{-n-\gamma}\langle u,\varphi\rangle\,.$$

(a) Prove that this definition agrees with the usual definition for functions.
(b) Show that δ_0 is homogeneous of degree $-n$.
(c) Prove that if u is homogeneous of degree γ, then $\partial^\alpha u$ is homogeneous of degree $\gamma-|\alpha|$.

(d) Show that u is homogeneous of degree γ if and only if $\widehat{u}$ is homogeneous of degree $-n-\gamma$.

2.3.10. (a) Show that the functions e^{inx} and e^{-inx} converge to zero in $\mathscr{S}'$ and $\mathscr{D}'$ as $n \to \infty$. Conclude that multiplication of distributions is not a continuous operation even when it is defined.
(b) What is the limit of $\sqrt{n}(1+n|x|^2)^{-1}$ in $\mathscr{D}'(\mathbf{R})$ as $n \to \infty$?

2.3.11. (*S. Bernstein*) Let f be a bounded function on $\mathbf{R}^n$ with $\widehat{f}$ supported in the ball $B(0,R)$. Prove that for all multi-indices α there exist constants $C_{\alpha,n}$ (depending only on α and on the dimension n) such that

$$\big\|\partial^\alpha f\big\|_{L^\infty} \le C_{\alpha,n} R^{|\alpha|} \big\|f\big\|_{L^\infty}.$$

[*Hint:* Write $f = f * h_{1/R}$, where h is a Schwartz function h in $\mathbf{R}^n$ whose Fourier transform is equal to one on the ball $B(0,1)$ and vanishes outside the ball $B(0,2)$.]

2.3.12. Let $\widehat{\Phi}$ be a $\mathscr{C}_0^\infty$ function that is equal to 1 in $B(0,1)$ and let $\widehat{\Theta}$ be a $\mathscr{C}^\infty$ function that is equal to 1 in a neighborhood of infinity and equal to zero in a neighborhood of the origin. Prove the following.
(a) For all u in $\mathscr{S}'(\mathbf{R}^n)$ we have

$$\Big(\widehat{\Phi}(\xi/2^N)\widehat{u}\Big)^\vee \to u \quad \text{in } \mathscr{S}'(\mathbf{R}^n) \text{ as } N \to \infty.$$

(b) For all u in $\mathscr{S}'(\mathbf{R}^n)$ we have

$$\Big(\widehat{\Theta}(\xi/2^N)\widehat{u}\Big)^\vee \to 0 \quad \text{in } \mathscr{S}'(\mathbf{R}^n) \text{ as } N \to \infty.$$

2.3.13. Prove that there exists a function in L^p for $2 < p < \infty$ whose distributional Fourier transform is not a locally integrable function.
[*Hint:* Assume the converse. Then for all $f \in L^p(\mathbf{R}^n)$, $\widehat{f}$ is locally integrable and hence the map $f \mapsto \widehat{f}$ is a well defined linear operator from $L^p(\mathbf{R}^n)$ to $L^1(B(0,M))$ for all $M > 0$ (i.e. $\|\widehat{f}\|_{L^1(B(0,M))} < \infty$ for all $f \in L^p(\mathbf{R}^n)$). Use the closed graph theorem to deduce that $\|\widehat{f}\|_{L^1(B(0,M))} \le C_M \|f\|_{L^p(\mathbf{R}^n)}$ for some $C_M < \infty$. To violate this inequality whenever $p > 2$, take $f_N(x) = (1+iN)^{-n/2} e^{-\pi(1+iN)^{-1}|x|^2}$ and let $N \to \infty$, noting that $\widehat{f_N}(\xi) = e^{-\pi|\xi|^2(1+iN)}$.]

2.4 More About Distributions and the Fourier Transform

In this section we discuss further properties of distributions and Fourier transforms and bring up certain connections that arise between harmonic analysis and partial differential equations.

2.4.1 Distributions Supported at a Point

We begin with the following characterization of distributions supported at a single point.

Proposition 2.4.1. *If $u \in \mathscr{S}'(\mathbf{R}^n)$ is supported in the singleton $\{x_0\}$, then there exists an integer k and complex numbers a_α such that*

$$u = \sum_{|\alpha|\le k} a_\alpha \partial^\alpha \delta_{x_0}.$$

Proof. Without loss of generality we may assume that $x_0 = 0$. By (2.3.3) we have that for some C, m, and k,

$$\big|\langle u, f\rangle\big| \le C \sum_{\substack{|\alpha|\le m \\ |\beta|\le k}} \sup_{x\in\mathbf{R}^n} |x^\alpha(\partial^\beta f)(x)| \qquad \text{for all } f \in \mathscr{S}(\mathbf{R}^n).$$

We now prove that if $\varphi \in \mathscr{S}$ satisfies

$$(\partial^\alpha \varphi)(0) = 0 \qquad \text{for all } |\alpha| \le k, \tag{2.4.1}$$

then $\langle u, \varphi\rangle = 0$. To see this, fix a φ satisfying (2.4.1) and let $\zeta(x)$ be a smooth function on $\mathbf{R}^n$ that is equal to 1 when $|x| \ge 2$ and equal to zero for $|x| \le 1$. Let $\zeta^\varepsilon(x) = \zeta(x/\varepsilon)$. Then, using (2.4.1) and the continuity of the derivatives of φ at the origin, it is not hard to show that $\rho_{\alpha,\beta}(\zeta^\varepsilon\varphi - \varphi) \to 0$ as $\varepsilon \to 0$ for all $|\alpha| \le m$ and $|\beta| \le k$. Then

$$\big|\langle u, \varphi\rangle\big| \le \big|\langle u, \zeta^\varepsilon\varphi\rangle\big| + \big|\langle u, \zeta^\varepsilon\varphi - \varphi\rangle\big| \le 0 + C \sum_{\substack{|\alpha|\le m \\ |\beta|\le k}} \rho_{\alpha,\beta}(\zeta^\varepsilon\varphi - \varphi) \to 0$$

as $\varepsilon \to 0$. This proves our assertion.

Now let $f \in \mathscr{S}(\mathbf{R}^n)$. Let η be a $\mathscr{C}_0^\infty$ function on $\mathbf{R}^n$ that is equal to 1 in a neighborhood of the origin. Write

$$f(x) = \eta(x)\Big(\sum_{|\alpha|\le k} \frac{(\partial^\alpha f)(0)}{\alpha!}x^\alpha + h(x)\Big) + (1-\eta(x))f(x), \tag{2.4.2}$$

where $h(x) = O(x^{k+1})$ as $|x| \to 0$. Then ηh satisfies (2.4.1) and hence $\langle u, \eta h\rangle = 0$ by the claim. Also,

$$\big\langle u, ((1-\eta)f)\big\rangle = 0$$

by our hypothesis. Applying u to both sides of (2.4.2), we obtain

$$\langle u, f\rangle = \sum_{|\alpha|\le k} \frac{(\partial^\alpha f)(0)}{\alpha!} u(x^\alpha \eta(x)) = \sum_{|\alpha|\le k} a_\alpha (\partial^\alpha \delta_0)(f),$$

with $a_\alpha = (-1)^{|\alpha|} u(x^\alpha \eta(x))/\alpha!$. This proves the proposition. □

An immediate consequence is the following result.

Corollary 2.4.2. *Let $u \in \mathscr{S}'(\mathbf{R}^n)$. If $\widehat{u}$ is supported in the singleton $\{\xi_0\}$, then u is a finite linear combination of functions $(-2\pi i\xi)^\alpha e^{2\pi i\xi\cdot\xi_0}$, where α is a multi-index. In particular, if $\widehat{u}$ is supported at the origin, then u is a polynomial.*

Proof. Proposition 2.4.1 gives that $\widehat{u}$ is a linear combination of derivatives of Dirac masses at ξ_0. Then Proposition 2.3.22 (8) yields the required conclusion. □

2.4.2 The Laplacian

The *Laplacian* Δ is a partial differential operator acting on tempered distributions on $\mathbf{R}^n$ as follows:

$$\Delta(u) = \sum_{j=1}^{n} \partial_j^2 u.$$

Solutions of Laplace's equation $\Delta(u) = 0$ are called *harmonic* distributions. We have the following:

Corollary 2.4.3. *Let $u \in \mathscr{S}'(\mathbf{R}^n)$ satisfy $\Delta(u) = 0$. Then u is a polynomial.*

Proof. Taking Fourier transforms, we obtain that $\widehat{\Delta(u)} = 0$. Therefore,

$$-4\pi^2|\xi|^2\widehat{u} = 0 \qquad \text{in } \mathscr{S}'.$$

This implies that $\widehat{u}$ is supported at the origin, and by Corollary 2.4.2 it follows that u must be polynomial. □

Liouville's classical theorem that every bounded harmonic function must be constant is a consequence of Corollary 2.4.3. See Exercise 2.4.2.

Next we would like to compute the fundamental solutions of Laplace's equation in $\mathbf{R}^n$. A distribution is called a *fundamental solution* of a partial differential operator L if we have $L(u) = \delta_0$. The following result gives the fundamental solution of the Laplacian.

Proposition 2.4.4. *For $n \geq 3$ we have*

$$\Delta(|x|^{2-n}) = -(n-2)\frac{2\pi^{n/2}}{\Gamma(n/2)}\delta_0, \tag{2.4.3}$$

while for $n = 2$,

$$\Delta(\log|x|) = 2\pi\delta_0. \tag{2.4.4}$$

Proof. We use Green's identity

$$\int_\Omega (v\Delta(u) - u\Delta(v))\,dx = \int_{\partial\Omega}\left(v\frac{\partial u}{\partial \nu} - u\frac{\partial v}{\partial \nu}\right)ds,$$

where Ω is an open set in $\mathbf{R}^n$ with smooth boundary and $\partial v/\partial \nu$ denotes the derivative of v with respect to the outer unit normal vector. Take $\Omega = \mathbf{R}^n \setminus \overline{B(0,\varepsilon)}$, $v = |x|^{2-n}$, and $u = f$ a $\mathscr{C}_0^\infty(\mathbf{R}^n)$ function in the previous identity. The normal derivative of $f(r\theta)$ is the derivative with respect to the radial variable r. Observe that $\Delta(|x|^{2-n}) = 0$ for $x \neq 0$. We obtain

$$\int_{|x|>\varepsilon} \Delta(f)(x)|x|^{2-n}\,dx = -\int\limits_{|\theta|=\varepsilon} \left(\varepsilon^{2-n}\frac{\partial f}{\partial r} - f(r\theta)\frac{\partial r^{2-n}}{\partial r}\right)d\theta, \tag{2.4.5}$$

where $d\theta$ denotes surface measure on the sphere $|\theta| = \varepsilon$. Now observe two things: first, that for some $C = C(f)$ we have

$$\left|\int_{|\theta|=\varepsilon} \frac{\partial f}{\partial r}\,d\theta\right| \le C\varepsilon^{n-1};$$

second, that

$$\int_{|\theta|=\varepsilon} f(r\theta)\varepsilon^{1-n}\,d\theta \to \omega_{n-1} f(0)$$

as $\varepsilon \to 0$. Letting $\varepsilon \to 0$ in (2.4.5), we obtain that

$$\lim_{\varepsilon\to 0}\int_{|x|>\varepsilon} \Delta(f)(x)|x|^{2-n}\,dx = -(n-2)\omega_{n-1} f(0),$$

which implies (2.4.3) in view of the formula for ω_{n-1} given in Appendix A.3.

The proof of (2.4.4) is identical. The only difference is that the quantity $\partial r^{2-n}/\partial r$ in (2.4.5) is replaced by $\partial \log r/\partial r$. □

2.4.3 Homogeneous Distributions

The fundamental solutions of the Laplacian are locally integrable functions on $\mathbf{R}^n$ and also homogeneous of degree $2-n$ when $n \ge 3$. Since homogeneous distributions often arise in applications, it is desirable to pursue their study. Here we do not undertake such a study in depth, but we discuss a few important examples.

Our first goal is to understand the action of the distribution $|t|^z$ on $\mathbf{R}^n$ when $\operatorname{Re} z \le -n$. Let us consider first the case $n = 1$. The tempered distribution

$$\langle w_z, \varphi\rangle = \int_{-1}^{1} |t|^z \varphi(t)\,dt$$

is well-defined when $\operatorname{Re} z > -1$. But we can extend the definition for all z with $\operatorname{Re} z > -3$ and $z \neq -1$ by rewriting it as

$$\langle w_z, \varphi\rangle = \int_{-1}^{1} |t|^z \big(\varphi(t) - \varphi(0) - t\varphi'(0)\big)\,dt + \frac{2}{z+1}\varphi(0), \tag{2.4.6}$$

and noting that for all $\varphi \in \mathscr{S}(\mathbf{R})$ we have

$$\left|\langle w_z, \varphi\rangle\right| \le \frac{1}{z+3}\|\varphi''\|_{L^\infty} + \frac{2}{z+1}\|\varphi\|_{L^\infty},$$

thus $w_z \in \mathscr{S}'(\mathbf{R})$. Subtracting the Taylor polynomial of degree 3 centered at zero from $\varphi(t)$ instead of the linear one, as in (2.4.6), allows us to extend the definition for $\operatorname{Re} z > -5$ and $\operatorname{Re} z \notin \{-1,-3\}$. Subtracting higher order Taylor polynomials allows us to extend the definition of w_z for all $z \in \mathbf{C}$ except at the negative odd integers. To be able to include the points $z = -1,-3,-5,-7,\dots$ we need to multiply w_z by an entire function that has simple zeros at all the negative odd integers to be able to eliminate the simple poles at these points. Such a function is $\Gamma\big(\frac{z+1}{2}\big)^{-1}$. This discussion leads to the following definition.

Definition 2.4.5. For $z \in \mathbf{C}$ we define a distribution u_z as follows:

$$\langle u_z, f\rangle = \int_{\mathbf{R}^n} \frac{\pi^{\frac{z+n}{2}}}{\Gamma\big(\frac{z+n}{2}\big)} |x|^z f(x)\, dx. \tag{2.4.7}$$

Clearly the u_z's coincide with the locally integrable functions

$$\pi^{\frac{z+n}{2}}\Gamma\big(\tfrac{z+n}{2}\big)^{-1}|x|^z$$

when $\operatorname{Re} z > -n$ and the definition makes sense only for that range of z's. It follows from its definition that u_z is a homogeneous distribution of degree z.

We would like to extend the definition of u_z for $z \in \mathbf{C}$. Let $\operatorname{Re} z > -n$ first. Fix N to be a positive integer. Given $f \in \mathscr{S}(\mathbf{R}^n)$, write the integral in (2.4.7) as follows:

$$\int_{|x|<1} \frac{\pi^{\frac{z+n}{2}}}{\Gamma(\frac{z+n}{2})} \left\{ f(x) - \sum_{|\alpha|\le N} \frac{(\partial^\alpha f)(0)}{\alpha!} x^\alpha \right\} |x|^z\, dx$$
$$+ \int_{|x|>1} \frac{\pi^{\frac{z+n}{2}}}{\Gamma(\frac{z+n}{2})} f(x)|x|^z\, dx + \int_{|x|<1} \frac{\pi^{\frac{z+n}{2}}}{\Gamma(\frac{z+n}{2})} \sum_{|\alpha|\le N} \frac{(\partial^\alpha f)(0)}{\alpha!} x^\alpha |x|^z\, dx.$$

The preceding expression is equal to

$$\int_{|x|<1} \frac{\pi^{\frac{z+n}{2}}}{\Gamma(\frac{z+n}{2})} \left\{ f(x) - \sum_{|\alpha|\le N} \frac{(\partial^\alpha f)(0)}{\alpha!} x^\alpha \right\} |x|^z\, dx$$
$$+ \int_{|x|>1} \frac{\pi^{\frac{z+n}{2}}}{\Gamma(\frac{z+n}{2})} f(x)|x|^z\, dx$$
$$+ \sum_{|\alpha|\le N} \frac{(\partial^\alpha f)(0)}{\alpha!} \frac{\pi^{\frac{z+n}{2}}}{\Gamma(\frac{z+n}{2})} \int_{r=0}^{1} \int_{\mathbf{S}^{n-1}} (r\theta)^\alpha r^{z+n-1}\, dr\, d\theta,$$

where we switched to polar coordinates in the penultimate integral. Now set

$$\begin{aligned} b(n,\alpha,z) &= \frac{\pi^{\frac{z+n}{2}}}{\Gamma(\frac{z+n}{2})}\frac{1}{\alpha!}\left(\int_{\mathbf{S}^{n-1}}\theta^\alpha\,d\theta\right)\int_0^1 r^{|\alpha|+n+z-1}\,dr \\ &= \frac{\pi^{\frac{z+n}{2}}}{\Gamma(\frac{z+n}{2})}\frac{\frac{1}{\alpha!}\int_{\mathbf{S}^{n-1}}\theta^\alpha\,d\theta}{|\alpha|+z+n}, \end{aligned}$$

where $\alpha=(\alpha_1,\dots,\alpha_n)$ is a multi-index. These coefficients are zero when at least one α_j is odd. Consider now the case that all the α_j's are even; then $|\alpha|$ is also even. The function $\Gamma(\frac{z+n}{2})$ has simple poles at the points

$$z=-n,\;\; z=-(n+2),\;\; z=-(n+4), \qquad \text{and so on;}$$

see Appendix A.5. These poles cancel exactly the poles of the function

$$z\mapsto(|\alpha|+z+n)^{-1}$$

at $z=-n-|\alpha|$ when $|\alpha|$ is an even integer in $[0,N]$. We therefore have

$$\begin{aligned} \langle u_z,f\rangle &= \int_{|x|\ge 1}\frac{\pi^{\frac{z+n}{2}}}{\Gamma(\frac{z+n}{2})}f(x)|x|^z\,dx+\sum_{|\alpha|\le N}b(n,\alpha,z)\,(-1)^{|\alpha|}\langle\partial^\alpha\delta_0,f\rangle \\ &\quad+\int_{|x|<1}\frac{\pi^{\frac{z+n}{2}}}{\Gamma(\frac{z+n}{2})}\left\{f(x)-\sum_{|\alpha|\le N}\frac{(\partial^\alpha f)(0)}{\alpha!}x^\alpha\right\}|x|^z\,dx. \end{aligned} \tag{2.4.8}$$

Both integrals converge absolutely when $\operatorname{Re} z>-N-n-1$, since the expression inside the curly brackets above is bounded by a constant multiple of $|x|^{N+1}$, and the resulting function of z in (2.4.8) is a well defined analytic function in the range $\operatorname{Re} z>-N-n-1$.

Since N was arbitrary, $\langle u_z,f\rangle$ has an analytic extension to all of $\mathbf{C}$. Therefore, u_z is a *distribution-valued entire function* of z, i.e., for all $\varphi\in\mathscr{S}(\mathbf{R}^n)$, the function $z\mapsto\langle u_z,\varphi\rangle$ is entire.

Next we would like to calculate the Fourier transform of u_z. We know by Exercise 2.3.9 that $\widehat{u_z}$ is a homogeneous distribution of degree $-n-z$. The choice of constant in the definition of u_z was made to justify the following result:

Theorem 2.4.6. *For all $z\in\mathbf{C}$ we have $\widehat{u_z}=u_{-n-z}$.*

Proof. The idea of the proof is straightforward. First we show that for a certain range of z's we have

$$\int_{\mathbf{R}^n}|\xi|^z\widehat{\varphi}(\xi)\,d\xi=C(n,z)\int_{\mathbf{R}^n}|x|^{-n-z}\varphi(x)\,dx, \tag{2.4.9}$$

for some fixed constant $C(n,z)$ and all $\varphi \in \mathscr{S}(\mathbf{R}^n)$. Next we pick a specific φ to evaluate the constant $C(n,z)$. Then we use analytic continuation to extend the validity of (2.4.9) for all z's. Use polar coordinates by setting $\xi = \rho\varphi$ and $x = r\theta$ in (2.4.9). We have

$$\begin{aligned}
&\int_{\mathbf{R}^n} |\xi|^z \widehat{\varphi}(\xi)\,d\xi \\
&\quad = \int_0^\infty \rho^{z+n-1} \int_0^\infty \int_{\mathbf{S}^{n-1}} \varphi(r\theta) \left(\int_{\mathbf{S}^{n-1}} e^{-2\pi i r\rho(\theta\cdot\varphi)} d\varphi \right) d\theta\, r^{n-1}\, dr\, d\rho \\
&\quad = \int_0^\infty \left(\int_0^\infty \sigma_n(r\rho)\rho^{z+n-1}\, d\rho \right) \left(\int_{\mathbf{S}^{n-1}} \varphi(r\theta) d\theta \right) r^{n-1}\, dr \\
&\quad = C(n,z) \int_0^\infty r^{-z-n} \left(\int_{\mathbf{S}^{n-1}} \varphi(r\theta)\, d\theta \right) r^{n-1}\, dr \\
&\quad = C(n,z) \int_{\mathbf{R}^n} |x|^{-n-z} \varphi(x)\, dx,
\end{aligned}$$

where we set

$$\sigma_n(t) = \int_{\mathbf{S}^{n-1}} e^{-2\pi i t(\theta\cdot\varphi)}\, d\varphi = \int_{\mathbf{S}^{n-1}} e^{-2\pi i t(\varphi_1)}\, d\varphi, \tag{2.4.10}$$

$$C(n,z) = \int_0^\infty \sigma_n(t) t^{z+n-1}\, dt, \tag{2.4.11}$$

and the second equality in (2.4.10) is a consequence of rotational invariance. It remains to prove that the integral in (2.4.11) converges for some range of z's.

If $n = 1$, then

$$\sigma_1(t) = \int_{\mathbf{S}^0} e^{-2\pi i t\varphi} d\varphi = e^{-2\pi i t} + e^{2\pi i t} = 2\cos(2\pi t)$$

and the integral in (2.4.11) converges conditionally for $-1 < \operatorname{Re} z < 0$.

Let us therefore assume that $n \geq 2$. Since $|\sigma_n(t)| \leq \omega_{n-1}$, the integral converges near zero when $-n < \operatorname{Re} z$. Let us study the behavior of $\sigma_n(t)$ for t large. Using the formula in Appendix D.2 and the definition of Bessel functions in Appendix B.1, we write

$$\sigma_n(t) = \omega_{n-2} \int_{-1}^1 e^{2\pi i t s} \left(\sqrt{1-s^2}\right)^{n-2} \frac{ds}{\sqrt{1-s^2}} = c_n t^{-\frac{n-2}{2}} J_{\frac{n-2}{2}}(2\pi t),$$

for some constant c_n. Since $n \geq 2$ we have when $n-2 > -1/2$. Then the asymptotics for Bessel functions (Appendix B.7) apply and yield $|\sigma_n(t)| \leq c t^{-(n-1)/2}$ for $t \geq 1$. Splitting the integral in (2.4.11) in $t \leq 1$ and $t \geq 1$ and using the corresponding estimates, we notice that it converges absolutely on $[0,1]$ when $\operatorname{Re} z > -n$ and on $[1,\infty)$ when $\operatorname{Re} z + n - 1 - \frac{n-1}{2} < -1$.

We have now proved that when $-n < \operatorname{Re} z < -\frac{n+1}{2}$ and $n \geq 2$ we have

$$\widehat{u_z} = C(n,z) u_{-n-z}$$

for some constant $C(n,z)$ that we wish to compute. Insert the function $\varphi(x) = e^{-\pi|x|^2}$ in (2.4.9). Example 2.2.9 gives that this function is equal to its Fourier transform. Use polar coordinates to write

$$\omega_{n-1} \int_0^\infty r^{z+n-1} e^{-\pi r^2} dr = C(n,z)\, \omega_{n-1} \int_0^\infty r^{-z-n+n-1} e^{-\pi r^2} dr.$$

Change variables $s = \pi r^2$ and use the definition of the gamma function to obtain that

$$C(n,z) = \frac{\Gamma(\frac{z+n}{2})}{\Gamma(-\frac{z}{2})} \frac{\pi^{-\frac{z+n}{2}}}{\pi^{\frac{z}{2}}}.$$

It follows that $\widehat{u_z} = u_{-n-z}$ for the range of z's considered.

At this point observe that for every $f \in \mathscr{S}(\mathbf{R}^n)$, the function $z \mapsto \langle \widehat{u_z} - u_{-z-n}, f \rangle$ is entire and vanishes for $-n < \operatorname{Re} z < -n + 1/2$. Therefore, it must vanish everywhere and the theorem is proved. □

Homogeneous distributions were introduced in Exercise 2.3.9. We already saw that the Dirac mass on $\mathbf{R}^n$ is a homogeneous distribution of degree $-n$. There is another important example of a homogeneous distributions of degree $-n$, which we now discuss.

Let Ω be an integrable function on the sphere $\mathbf{S}^{n-1}$ with integral zero. Define a tempered distribution W_Ω on $\mathbf{R}^n$ by setting

$$\langle W_\Omega, f \rangle = \lim_{\varepsilon \to 0} \int_{|x| \ge \varepsilon} \frac{\Omega(x/|x|)}{|x|^n} f(x)\, dx. \tag{2.4.12}$$

We check that W_Ω is a well defined tempered distribution on $\mathbf{R}^n$. Indeed, since $\Omega(x/|x|)/|x|^n$ has integral zero over all annuli centered at the origin, we obtain

$$\begin{aligned} \left|\langle W_\Omega, \varphi \rangle\right| &= \left| \lim_{\varepsilon \to 0} \int_{\varepsilon \le |x| \le 1} \frac{\Omega(x/|x|)}{|x|^n} (\varphi(x) - \varphi(0))\, dx + \int_{|x| \ge 1} \frac{\Omega(x/|x|)}{|x|^n} \varphi(x)\, dx \right| \\ &\le \big\|\nabla \varphi\big\|_{L^\infty} \int_{|x| \le 1} \frac{|\Omega(x/|x|)|}{|x|^{n-1}}\, dx + \Big(\sup_{x \in \mathbf{R}^n} |x|\, |\varphi(x)| \Big) \int_{|x| \ge 1} \frac{|\Omega(x/|x|)|}{|x|^{n+1}}\, dx \\ &\le C_1 \big\|\nabla \varphi\big\|_{L^\infty} \big\|\Omega\big\|_{L^1(\mathbf{S}^{n-1})} + C_2 \sum_{|\alpha| \le 1} \big\|\varphi(x) x^\alpha\big\|_{L^\infty} \big\|\Omega\big\|_{L^1(\mathbf{S}^{n-1})}, \end{aligned}$$

for suitable constants C_1 and C_2 in view of (2.2.2).

One can verify that $W_\Omega \in \mathscr{S}'(\mathbf{R}^n)$ is a homogeneous distribution of degree $-n$ just like the Dirac mass at the origin. It is an interesting fact that all homogeneous distributions on $\mathbf{R}^n$ of degree $-n$ that coincide with a smooth function away from the origin arise in this way. We have the following result.

Proposition 2.4.7. *Suppose that m is a $\mathscr{C}^\infty$ function on $\mathbf{R}^n \setminus \{0\}$ that is homogeneous of degree zero. Then there exist a scalar b and a $\mathscr{C}^\infty$ function Ω on $\mathbf{S}^{n-1}$ with integral zero such that*

$$m^\vee = b\, \delta_0 + W_\Omega, \tag{2.4.13}$$

where W_Ω denotes the distribution defined in (2.4.12).

To prove this result we need the following proposition, whose proof we postpone until the end of this section.

Proposition 2.4.8. *Suppose that u is a $\mathscr{C}^\infty$ function on $\mathbf{R}^n \setminus \{0\}$ that is homogeneous of degree $z \in \mathbf{C}$. Then $\widehat{u}$ is a $\mathscr{C}^\infty$ function on $\mathbf{R}^n \setminus \{0\}$.*

We now prove Proposition 2.4.7 using Proposition 2.4.8.

Proof. Let a be the integral of the smooth function m over $\mathbf{S}^{n-1}$. The function $m-a$ is homogeneous of degree zero and thus locally integrable on $\mathbf{R}^n$; hence it can be thought of as a tempered distribution that we call $\widehat{u}$ (the Fourier transform of a tempered distribution u). Since $\widehat{u}$ is a $\mathscr{C}^\infty$ function on $\mathbf{R}^n \setminus \{0\}$, Proposition 2.4.8 implies that u is also a $\mathscr{C}^\infty$ function on $\mathbf{R}^n \setminus \{0\}$. Let Ω be the restriction of u on $\mathbf{S}^{n-1}$. Then Ω is a well defined $\mathscr{C}^\infty$ function on $\mathbf{S}^{n-1}$. Since u is a homogeneous function of degree $-n$ that coincides with the smooth function Ω on $\mathbf{S}^{n-1}$, it follows that $u(x) = \Omega(x/|x|)/|x|^n$ for x in $\mathbf{R}^n \setminus \{0\}$.

We show that Ω has mean value zero over $\mathbf{S}^{n-1}$. Pick a nonnegative, radial, smooth, and nonzero function ψ on $\mathbf{R}^n$ supported in the annulus $1 < |x| < 2$. Switching to polar coordinates, we write

$$\begin{aligned}
\langle u, \psi \rangle &= \int_{\mathbf{R}^n} \frac{\Omega(x/|x|)}{|x|^n} \psi(x)\, dx = c_\psi \int_{\mathbf{S}^{n-1}} \Omega(\theta)\, d\theta, \\
\langle u, \psi \rangle &= \langle \widehat{u}, \widehat{\psi} \rangle = \int_{\mathbf{R}^n} (m(\xi) - a)\widehat{\psi}(\xi)\, d\xi = c'_\psi \int_{\mathbf{S}^{n-1}} (m(\theta) - a)\, d\theta = 0,
\end{aligned}$$

and thus Ω has mean value zero over $\mathbf{S}^{n-1}$ (since $c_\psi \neq 0$).

We can now legitimately define the distribution W_Ω, which coincides with the function $\Omega(x/|x|)/|x|^n$ on $\mathbf{R}^n \setminus \{0\}$. But the distribution u also coincides with this function on $\mathbf{R}^n \setminus \{0\}$. It follows that $u - W_\Omega$ is supported at the origin. Proposition 2.4.1 now gives that $u - W_\Omega$ is a sum of derivatives of Dirac masses. Since both distributions are homogeneous of degree $-n$, it follows that

$$u - W_\Omega = c\delta_0 .$$

But $u = (m-a)^\vee = m^\vee - a\delta_0$, and thus $m^\vee = (c+a)\delta_0 + W_\Omega$. This proves the proposition. $\square$

We now turn to the proof of Proposition 2.4.8.

Proof. Let $u \in \mathscr{S}'$ be homogeneous of degree z and $\mathscr{C}^\infty$ on $\mathbf{R}^n \setminus \{0\}$. We need to show that $\widehat{u}$ is $\mathscr{C}^\infty$ away from the origin. We prove that $\widehat{u}$ is $\mathscr{C}^M$ for all M. Fix $M \in \mathbf{Z}^+$ and let α be any multi-index such that

$$|\alpha| > n + M + \operatorname{Re} z . \tag{2.4.14}$$

Pick a $\mathscr{C}^\infty$ function φ on $\mathbf{R}^n$ that is equal to 1 when $|x| \ge 2$ and equal to zero for $|x| \le 1$. Write $u_0 = (1-\varphi)u$ and $u_\infty = \varphi u$. Then

$$\partial^\alpha u = \partial^\alpha u_0 + \partial^\alpha u_\infty \qquad \text{and thus} \qquad \widehat{\partial^\alpha u} = \widehat{\partial^\alpha u_0} + \widehat{\partial^\alpha u_\infty},$$

where the operations are performed in the sense of distributions. Since u_0 is compactly supported, Theorem 2.3.21 implies that $\widehat{\partial^\alpha u_0}$ is $\mathscr{C}^\infty$. Now Leibniz's rule gives that

$$\partial^\alpha u_\infty = v + \varphi \partial^\alpha u,$$

where v is a smooth function supported in the annulus $1 \le |x| \le 2$. Then $\widehat{v}$ is $\mathscr{C}^\infty$ and we need to show only that $\widehat{\varphi \partial^\alpha u}$ is $\mathscr{C}^M$. The function $\varphi \partial^\alpha u$ is actually $\mathscr{C}^\infty$, and by the homogeneity of $\partial^\alpha u$ (Exercise 2.3.9 (c)) we obtain that $(\partial^\alpha u)(x) = |x|^{-|\alpha|+z}(\partial^\alpha u)(x/|x|)$. Since φ is supported away from zero, it follows that

$$|\varphi(x)(\partial^\alpha u)(x)| \le \frac{C_\alpha}{(1+|x|)^{|\alpha|-\operatorname{Re} z}} \tag{2.4.15}$$

for some $C_\alpha > 0$. It is now straightforward to see that if a function satisfies (2.4.15), then its Fourier transform is $\mathscr{C}^M$ whenever (2.4.14) is satisfied. See Exercise 2.4.1.

We conclude that $\widehat{\partial^\alpha u_\infty}$ is a $\mathscr{C}^M$ function whenever (2.4.14) is satisfied; thus so is $\widehat{\partial^\alpha u}$. Since $\widehat{\partial^\alpha u}(\xi) = (2\pi i \xi)^\alpha \widehat{u}(\xi)$, we deduce smoothness for $\widehat{u}$ away from the origin. Let $\xi \neq 0$. Pick a neighborhood V of ξ such that for η in V we have $\eta_j \neq 0$ for some $j \in \{1,\dots,n\}$. Consider the multi-index $(0,\dots,|\alpha|,\dots,0)$ with $|\alpha|$ in the jth coordinate and zeros elsewhere. Then $(2\pi i \eta_j)^{|\alpha|}\widehat{u}(\eta)$ is a $\mathscr{C}^M$ function on V, and thus so is $\widehat{u}(\eta)$, since we can divide by $\eta_j^{|\alpha|}$. We conclude that $\widehat{u}(\xi)$ is $\mathscr{C}^M$ on $\mathbf{R}^n \setminus \{0\}$. Since M is arbitrary, the conclusion follows. □

We end this section with an example that illustrates the usefulness of some of the ideas discussed in this section.

Example 2.4.9. Let η be a smooth radial function on $\mathbf{R}^n$ that is equal to 1 on the set $|x| \ge 1/2$ and vanishes on the set $|x| \le 1/4$. Fix $z \in \mathbf{C}$ satisfy $0 < \operatorname{Re} z < n$. Let $g = \big(\eta(x)|x|^{-z}\big)^\wedge$ be the distributional Fourier transform of $\eta(x)|x|^{-z}$. We show that g is a function that decays faster than $|\xi|^{-N}$ at infinity (for sufficiently large positive number N) and that

$$g(\xi) - \frac{\pi^{z-\frac{n}{2}}\Gamma(\frac{n-z}{2})}{\Gamma(\frac{z}{2})}|\xi|^{z-n} \tag{2.4.16}$$

is a $\mathscr{C}^\infty$ function on $\mathbf{R}^n$. This example indicates the interplay between the smoothness of a function and the decay of its Fourier transform. The smoothness of the function $\eta(x)|x|^{-z}$ near zero has as a consequence the rapid decay of g near infinity, while the slow decay of $\eta(x)|x|^{-z}$ at infinity reflects the lack of smoothness of $g(\xi)$ at zero, in view of the moderate blowup $|\xi|^{\operatorname{Re} z-n}$ as $|\xi| \to 0$.

To show that g is a function we write it as $g = (|x|^{-z})^\wedge + \big((\eta(x)-1)|x|^{-z}\big)^\wedge$ and we observe that the first term is a function, since $0 < \operatorname{Re} z < n$. Using Theorem 2.4.6 we write

$$g(\xi) = \frac{\pi^{z-\frac{n}{2}}\Gamma(\frac{n-z}{2})}{\Gamma(\frac{z}{2})}|\xi|^{z-n} + \widehat{\varphi}(\xi),$$

where $\widehat{\varphi}(\xi) = \big((\eta(x)-1)|x|^{-z}\big)^\wedge(\xi)$ is a $\mathscr{C}^\infty$ function, since it is the Fourier transform of a compactly supported integrable function. This proves that g is a function and that the difference in (2.4.16) is $\mathscr{C}^\infty$.

Finally, we assert that every derivative of g satisfies $|\partial^\gamma g(\xi)| \le C_{\gamma,N}|\xi|^{-N}$ for all sufficiently large positive integers N when $\xi \neq 0$. Indeed, fix a multi-index γ and write $\partial^\gamma g(\xi) = (|x|^{-z}\eta(x)(-2\pi i x)^\gamma)^\wedge(\xi)$. It follows that

$$(4\pi^2|\xi|^2)^N|\partial^\gamma g(\xi)| = \big|\big(\Delta^N(|x|^{-z}\eta(x)(-2\pi i x)^\gamma)\big)^\wedge(\xi)\big|$$

for all $N \in \mathbf{Z}^+$, where Δ is the Laplacian in the x variable. Using Leibniz's rule we distribute Δ^N to the product. If a derivative falls on η, we obtain a compactly supported smooth function, hence integrable. If all derivatives fall on $|x|^{-z}x^\gamma$, then we obtain a term that decays like $|x|^{-\operatorname{Re} z+|\gamma|-2N}$ at infinity, which is also integrable if N is sufficiently large. Thus the function $|\xi|^{2N}|\partial^\gamma g(\xi)|$ is equal to the Fourier transform of an L^1 function, hence it is bounded, when $2N > n - \operatorname{Re} z + |\gamma|$.

Exercises

2.4.1. Suppose that a function f satisfies the estimate

$$|f(x)| \le \frac{C}{(1+|x|)^N},$$

for some $C > 0$ and $N > n+1$. Then $\widehat{f}$ is $\mathscr{C}^M$ for all $M \in \mathbf{Z}^+$ with $1 \le M < N-n$.

2.4.2. Use Corollary 2.4.3 to prove Liouville's theorem that every bounded harmonic function on $\mathbf{R}^n$ must be a constant. Derive as a consequence the *fundamental theorem of algebra*, stating that every polynomial on $\mathbf{C}$ must have a complex root.

2.4.3. Prove that e^x is not in $\mathscr{S}'(\mathbf{R})$ but that $e^x e^{ie^x}$ is in $\mathscr{S}'(\mathbf{R})$.

2.4.4. Show that the Schwartz function $x \mapsto \operatorname{sech}(\pi x)$, $x \in \mathbf{R}$, coincides with its Fourier transform.
[*Hint:* Integrate the function e^{iaz} over the rectangular contour with corners $(-R,0)$, $(R,0)$, $(R,i\pi)$, and $(-R,i\pi)$.]

2.4.5. ([174]) Construct an uncountable family of linearly independent Schwartz functions f_a such that $|f_a| = |f_b|$ and $|\widehat{f_a}| = |\widehat{f_b}|$ for all f_a and f_b in the family.
[*Hint:* Let w be a smooth nonzero function whose Fourier transform is supported

in the interval $[-1/2,1/2]$ and let ϕ be a real-valued smooth nonconstant periodic function with period 1. Then take $f_a(x) = w(x)e^{i\phi(x-a)}$ for $a \in \mathbf{R}$.]

2.4.6. Let P_y be the Poisson kernel defined in (2.1.13). Prove that for $f \in L^p(\mathbf{R}^n)$, $1 \le p < \infty$, the function

$$(x,y) \mapsto (P_y * f)(x)$$

is a harmonic function on $\mathbf{R}^{n+1}_+$. Use the Fourier transform and Exercise 2.2.11 to prove that $(P_{y_1} * P_{y_2})(x) = P_{y_1+y_2}(x)$ for all $x \in \mathbf{R}^n$.

2.4.7. (a) For a fixed $x_0 \in \mathbf{S}^{n-1}$, show that the function

$$v(x;x_0) = \frac{1-|x|^2}{|x-x_0|^n}$$

is harmonic on $\mathbf{R}^n \setminus \{x_0\}$.
(b) For fixed $x_0 \in \mathbf{S}^{n-1}$, prove that the family of functions $\theta \mapsto v(rx_0;\theta)$, $0<r<1$, defined on the sphere satisfies

$$\lim_{r\uparrow 1} \int_{\substack{\theta \in \mathbf{S}^{n-1} \\ |\theta - x_0| > \delta}} v(rx_0;\theta)\,d\theta = 0$$

uniformly in x_0. The function $v(rx_0;\theta)$ is called the *Poisson kernel for the sphere*.
(c) Show that

$$\frac{1}{\omega_{n-1}}(1-|x|^2)\int_{\mathbf{S}^{n-1}} \frac{1}{|x-\theta|^n}\,d\theta = 1$$

for all $|x|<1$.
(d) Let f be a continuous function on $\mathbf{S}^{n-1}$. Prove that the function

$$u(x) = \frac{1}{\omega_{n-1}}(1-|x|^2)\int_{\mathbf{S}^{n-1}} \frac{f(\theta)}{|x-\theta|^n}\,d\theta$$

solves the Dirichlet problem $\Delta(u)=0$ on $|x|<1$ with boundary values $u=f$ on $\mathbf{S}^{n-1}$, in the sense $\lim_{r\uparrow 1} u(rx_0) = f(x_0)$ when $|x_0|=1$.
[*Hint:* Part (c): Apply the mean value property over spheres to the harmonic function $y \mapsto (1-|x|^2|y|^2)\big|\,|x|y - \frac{x}{|x|}\big|^{-n}$.]

2.4.8. Fix $n \in \mathbf{Z}^+$ with $n \ge 2$ and a real number λ, $0<\lambda<n$. Also fix $\eta \in \mathbf{S}^n$ and $y \in \mathbf{R}^n$.
(a) Prove that

$$\int_{\mathbf{S}^n} |\xi - \eta|^{-\lambda}\,d\xi = 2^{n-\lambda}\frac{\pi^{\frac{n}{2}}\Gamma(\frac{n-\lambda}{2})}{\Gamma(n-\frac{\lambda}{2})}.$$

(b) Prove that

$$\int_{\mathbf{R}^n} |x-y|^{-\lambda}(1+|x|^2)^{\frac{\lambda}{2}-n}\,dx = \frac{\pi^{\frac{n}{2}}\Gamma(\frac{n-\lambda}{2})}{\Gamma(n-\frac{\lambda}{2})}(1+|y|^2)^{-\frac{\lambda}{2}}.$$

[*Hint:* Part (a): See Appendix D.4 Part (b): Use the stereographic projection in Appendix D.6.]

2.4.9. Prove the following *beta integral identity:*

$$\int_{\mathbf{R}^n} \frac{dt}{|x-t|^{\alpha_1}|y-t|^{\alpha_2}} = \pi^{\frac{n}{2}} \frac{\Gamma\big(\frac{n-\alpha_1}{2}\big)\Gamma\big(\frac{n-\alpha_2}{2}\big)\Gamma\big(\frac{\alpha_1+\alpha_2-n}{2}\big)}{\Gamma\big(\frac{\alpha_1}{2}\big)\Gamma\big(\frac{\alpha_2}{2}\big)\Gamma\big(n-\frac{\alpha_1+\alpha_2}{2}\big)} |x-y|^{n-\alpha_1-\alpha_2},$$

where $0<\alpha_1,\alpha_2<n$, $\alpha_1+\alpha_2>n$.
[*Hint:* Reduce to the case $y=0$, interpret the integral as a convolution, and use Theorem 2.4.6.]

2.4.10. (a) Prove that if a continuous integrable function f on $\mathbf{R}^n$ $(n\ge 2)$ is constant on the spheres $r\mathbf{S}^{n-1}$ for all $r>0$, then so is its Fourier transform.
(b) If a continuous integrable function on $\mathbf{R}^n$ $(n\ge 3)$ is constant on all $(n-2)$–dimensional spheres orthogonal to $e_1=(1,0,\dots,0)$, then its Fourier transform has the same property.

2.4.11. ([137]) Suppose that $0<d_1,d_2,d_3<n$ satisfy $d_1+d_2+d_3=2n$. Prove that for any distinct $x,y,z\in\mathbf{R}^n$ we have the identity

$$\int_{\mathbf{R}^n} |x-t|^{-d_2}|y-t|^{-d_3}|z-t|^{-d_1}\,dt$$
$$= \pi^{\frac{n}{2}}\left(\prod_{j=1}^{3}\frac{\Gamma\big(n-\frac{d_j}{2}\big)}{\Gamma\big(\frac{d_j}{2}\big)}\right)|x-y|^{d_1-n}|y-z|^{d_2-n}|z-x|^{d_3-n}.$$

[*Hint:* Reduce matters to the case that $z=0$ and $y=e_1$. Then take the Fourier transform in x and use that the function $h(t)=|t-e_1|^{-d_3}|t|^{-d_1}$ satisfies $\widehat{h}(\xi)=\widehat{h}(A_\xi^{-2}\xi)$ for all $\xi\neq 0$, where A_ξ is an orthogonal matrix with $A_\xi e_1=\xi/|\xi|$.]

2.4.12. (a) Integrate the function e^{iz^2} over the contour consisting of the three pieces $P_1=\{x+i0:\ 0\le x\le R\}$, $P_2=\{Re^{i\theta}:\ 0\le\theta\le\frac{\pi}{4}\}$, and $P_3=\{re^{i\frac{\pi}{4}}:\ 0\le r\le R\}$ (with the proper orientation) to obtain the *Fresnel integral identity*:

$$\lim_{R\to\infty}\int_0^R e^{ix^2}\,dx = \tfrac{\sqrt{2\pi}}{4}(1+i)\,.$$

(b) Use the result in part (a) to show that the Fourier transform of the function $e^{i\pi|x|^2}$ in $\mathbf{R}^n$ is equal to $e^{i\frac{\pi n}{4}}e^{-i\pi|\xi|^2}$.
[*Hint:* Part (a): On P_2 we have $e^{-R^2\sin(2\theta)}\le e^{-\frac{4}{\pi}R^2\theta}$, and the integral over P_2 tends to 0. Part (b): Try first $n=1$.]

2.5 Convolution Operators on L^p Spaces and Multipliers

In this section we study the class of operators that commute with translations. We prove in this section that bounded operators that commute with translations must be of convolution type. Convolution operators arise in many situations, and we would like to know under what circumstances they are bounded between L^p spaces.

2.5.1 Operators That Commute with Translations

Definition 2.5.1. A vector space X of measurable functions on $\mathbf{R}^n$ is called *closed under translations* if for $f \in X$ we have $\tau^z(f) \in X$ for all $z \in \mathbf{R}^n$. Let X and Y be vector spaces of measurable functions on $\mathbf{R}^n$ that are closed under translations. Let also T be an operator from X to Y. We say that T *commutes with translations* or is *translation-invariant* if

$$T(\tau^y(f)) = \tau^y(T(f))$$

for all $f \in X$ and all $y \in \mathbf{R}^n$.

It is straightforward to see that convolution operators commute with translations, i.e., $\tau^y(f * g) = \tau^y(f) * g$ whenever the convolution is defined. One of the goals of this section is to prove the converse: every bounded linear operator that commutes with translations is of convolution type. We have the following:

Theorem 2.5.2. *Let $1 \le p,q \le \infty$ and suppose T is a bounded linear operator from $L^p(\mathbf{R}^n)$ to $L^q(\mathbf{R}^n)$ that commutes with translations. Then there exists a unique tempered distribution w such that*

$$T(f) = f * w \qquad \text{a.e. for all } f \in \mathscr{S}.$$

A very important point to make is that if $p = \infty$, the restriction of T on $\mathscr{S}$ does not uniquely determine T on the entire L^∞; see Example 2.5.9 and the comments preceding it about this. The theorem is a consequence of the following two results:

Lemma 2.5.3. *Under the hypotheses of Theorem 2.5.2 and for $f \in \mathscr{S}(\mathbf{R}^n)$, the distributional derivatives of $T(f)$ are L^q functions that satisfy*

$$\partial^\alpha\big(T(f)\big) = T(\partial^\alpha f), \qquad \text{for all multi-indices } \alpha. \tag{2.5.1}$$

Lemma 2.5.4. *Let $1 \le q \le \infty$ and let $h \in L^q(\mathbf{R}^n)$. If all distributional derivatives $\partial^\alpha h$ are also in L^q, then h is almost everywhere equal to a continuous function H satisfying*

$$|H(0)| \le C_{n,q} \sum_{|\alpha| \le n+1} \big\|\partial^\alpha h\big\|_{L^q}. \tag{2.5.2}$$

Proof. Assuming Lemmas 2.5.3 and 2.5.4, we prove Theorem 2.5.2.

Given $f \in \mathscr{S}(\mathbf{R}^n)$, by Lemmas 2.5.3 and 2.5.4, there is a continuous function H such that $T(f) = H$ a.e. and such that

$$|H(0)| \le C_{n,q} \sum_{|\alpha| \le n+1} \big\|\partial^\alpha T(f)\big\|_{L^q}$$

holds. Define a linear functional u on $\mathscr{S}$ by setting

$$\langle u, f \rangle = H(0).$$

This functional is well-defined, for, if there is another continuous function G such that $G = T(f)$ a.e., then $G = H$ a.e. and since both functions are continuous, it follows that $H = G$ everywhere and thus $H(0) = G(0)$.

By (2.5.1), (2.5.2), and the boundedness of T, we have

$$\begin{aligned} \big|\langle u, f \rangle\big| &\le C_{n,q} \sum_{|\alpha| \le n+1} \big\|\partial^\alpha T(f)\big\|_{L^q} \\ &\le C_{n,q} \sum_{|\alpha| \le n+1} \big\|T(\partial^\alpha f)\big\|_{L^q} \\ &\le C_{n,q} \big\|T\big\|_{L^p \to L^q} \sum_{|\alpha| \le n+1} \big\|\partial^\alpha f\big\|_{L^p} \\ &\le C'_{n,q} \big\|T\big\|_{L^p \to L^q} \sum_{\substack{|\gamma| \le [\frac{n+1}{p}]+1 \\ |\alpha| \le n+1}} \rho_{\gamma,\alpha}(f), \end{aligned}$$

where the last estimate uses (2.2.8). This implies that u is in $\mathscr{S}'$. We now set $w = \widetilde{u}$ and we claim that for all $x \in \mathbf{R}^n$ we have

$$\langle u, \tau^{-x} f \rangle = H(x). \tag{2.5.3}$$

Assuming (2.5.3) we prove that $T(f) = f * w$ for $f \in \mathscr{S}$. To see this, by Theorem 2.3.20 and by the translation invariance of T, for a given $f \in \mathscr{S}(\mathbf{R}^n)$ we have

$$(f * w)(x) = \langle \widetilde{u}, \tau^x \widetilde{f} \rangle = \langle u, \tau^{-x} f \rangle = H(x) = T(f)(x),$$

where the last equality holds for almost all x, by the definition of H. Thus $f * w = T(f)$ a.e., as claimed. The uniqueness of w follows from the simple observation that if $f * w = f * w'$ for all $f \in \mathscr{S}(\mathbf{R}^n)$, then $w = w'$.

We now turn to the proof of (2.5.3). Given $f \in \mathscr{S}(\mathbf{R}^n)$ and $x \in \mathbf{R}^n$ and let H_x be the continuous function such that $H_x = T(\tau^{-x} f)$. We show that $H_x(0) = H(x)$. Indeed, we have

$$H_x(y) = T(\tau^{-x} f)(y) = \tau^{-x} T(f)(y) = T(f)(x+y) = H(x+y) = \tau^{-x} H(y),$$

where the equality $T(f)(x+y) = H(x+y)$ holds a.e. in y. Thus the continuous functions H_x and $\tau^{-x}H$ are equal a.e. and thus they must be everywhere equal, in particular, when $y = 0$. This proves that $H_x(0) = H(x)$, which is a restatement of (2.5.3). □

We now return to Lemmas 2.5.3 and 2.5.4. We begin with Lemma 2.5.3.

Proof. Consider first the multi-index $\alpha = (0,\dots,1,\dots,0)$, where 1 is in the jth entry and 0 is elsewhere. Let $e_j = (0,\dots,1,\dots,0)$, where 1 is in the jth entry and zero elsewhere. We have

$$\int_{\mathbf{R}^n} T(f)(y)\frac{\varphi(y+he_j)-\varphi(y)}{h}\,dy = \int_{\mathbf{R}^n} \varphi(y)T\left(\frac{\tau^{he_j}(f)-f}{h}\right)(y)\,dy \tag{2.5.4}$$

since both of these expressions are equal to

$$\int_{\mathbf{R}^n} \varphi(y)\frac{T(f)(y-he_j)-T(f)(y)}{h}\,dy$$

and T commutes with translations. We will let $h \to 0$ in both sides of (2.5.4). We write

$$\frac{\varphi(y+he_j)-\varphi(y)}{h} = \int_0^1 \partial_j\varphi(y+hte_j)\,dt\,,$$

from which it follows that for $|h| < 1/2$ we have

$$\left|\frac{\varphi(y+he_j)-\varphi(y)}{h}\right| \le \int_0^1 \frac{C_M\,dt}{(1+|y+hte_j|)^M} \le \int_0^1 \frac{C_M\,dt}{(1+|y|-\frac{1}{2})^M} \le \frac{C'_M}{(|y|+1)^M}.$$

The integrand on the left-hand side of (2.5.4) is bounded by the integrable function $|T(f)(y)|C'_M(|y|+1)^{-M}$ and converges to $T(f)(y)\,\partial_j\varphi(y)$ as $h \to 0$. The Lebesgue dominated convergence theorem yields that the integral on the left-hand side of (2.5.4) converges to

$$\int_{\mathbf{R}^n} T(f)(y)\,\partial_j\varphi(y)\,dy. \tag{2.5.5}$$

Moreover, for a Schwartz function f we have

$$\frac{\tau^{he_j}(f)(y)-f(y)}{h} = \int_0^1 \partial_j f(y+hte_j)\,dt\,,$$

which converges to $\partial_j f(y)$ pointwise as $h \to 0$ and is bounded by $C'_M(1+|y|)^{-M}$ for $|h| < 1/2$ by an argument similar to the preceding one for φ in place of f. Thus

$$\frac{\tau^{he_j}(f)-f}{h} \to \partial_j f \quad \text{in } L^p \text{ as } h \to 0, \tag{2.5.6}$$

by the Lebesgue dominated convergence theorem. The boundedness of T from L^p to L^q yields that

$$T\left(\frac{\tau^{he_j}(f)-f}{h}\right) \to T(\partial_j f) \quad \text{in } L^q \text{ as } h\to 0. \tag{2.5.7}$$

Since $\varphi \in L^{q'}$, by Hölder's inequality, the right-hand side of (2.5.4) converges to

$$\int_{\mathbf{R}^n} \varphi(y) T(\partial_j f)(y)\, dy$$

as $h \to 0$. This limit is equal to (2.5.5) and the required conclusion follows for $\alpha = (0,\dots,0,1,0,\dots,0)$. The general case follows by induction on $|\alpha|$. □

We now prove Lemma 2.5.4.

Proof. Let $R \geq 1$. Fix a $\mathscr{C}_0^\infty$ function φ_R that is equal to 1 in the ball $|x| \leq R$ and equal to zero when $|x| \geq 2R$. Since h is in $L^q(\mathbf{R}^n)$, it follows that $\varphi_R h$ is in $L^1(\mathbf{R}^n)$. We show that $\widehat{\varphi_R h}$ is also in L^1. We begin with the inequality

$$1 \leq C_n(1+|x|)^{-(n+1)} \sum_{|\alpha|\leq n+1} |(-2\pi i x)^\alpha|, \tag{2.5.8}$$

which is just a restatement of (2.2.3). Now multiply (2.5.8) by $|\widehat{\varphi_R h}(x)|$ to obtain

$$\begin{aligned}
|\widehat{\varphi_R h}(x)| &\leq C_n(1+|x|)^{-(n+1)} \sum_{|\alpha|\leq n+1} |(-2\pi i x)^\alpha \widehat{\varphi_R h}(x)| \\
&\leq C_n(1+|x|)^{-(n+1)} \sum_{|\alpha|\leq n+1} \big\|(\partial^\alpha(\varphi_R h))^\wedge\big\|_{L^\infty} \\
&\leq C_n(1+|x|)^{-(n+1)} \sum_{|\alpha|\leq n+1} \big\|\partial^\alpha(\varphi_R h)\big\|_{L^1} \\
&\leq C_n(2^n R^n v_n)^{1/q'}(1+|x|)^{-(n+1)} \sum_{|\alpha|\leq n+1} \big\|\partial^\alpha(\varphi_R h)\big\|_{L^q} \\
&\leq C_{n,R}(1+|x|)^{-(n+1)} \sum_{|\alpha|\leq n+1} \big\|\partial^\alpha h\big\|_{L^q},
\end{aligned}$$

where we used Leibniz's rule (Proposition 2.3.22 (14)) and the fact that all derivatives of φ_R are pointwise bounded by constants depending on R.

Integrate the previously displayed inequality with respect to x to obtain

$$\big\|\widehat{\varphi_R h}\big\|_{L^1} \leq C_{R,n} \sum_{|\alpha|\leq n+1} \big\|\partial^\alpha h\big\|_{L^q} < \infty. \tag{2.5.9}$$

Therefore, Fourier inversion holds for $\varphi_R h$ (see Exercise 2.2.6). This implies that $\varphi_R h$ is equal a.e. to a continuous function, namely the inverse Fourier transform of its Fourier transform. Since $\varphi_R = 1$ on the ball $B(0,R)$, we conclude that h is a.e. equal to a continuous function in this ball. Since $R > 0$ was arbitrary, it follows that

h is a.e. equal to a continuous function on $\mathbf{R}^n$, which we denote by H. Finally, (2.5.2) is a direct consequence of (2.5.9) with $R = 1$, since $|H(0)| \le \|\widehat{\varphi_1 h}\|_{L^1}$. □

2.5.2 The Transpose and the Adjoint of a Linear Operator

We briefly discuss the notions of the transpose and the adjoint of a linear operator. We first recall real and complex inner products. For f, g measurable functions on $\mathbf{R}^n$, we define the *complex inner product*

$$\langle f \,|\, g \rangle = \int_{\mathbf{R}^n} f(x)\overline{g(x)}\,dx,$$

whenever the integral converges absolutely. We reserve the notation

$$\langle f, g \rangle = \int_{\mathbf{R}^n} f(x)g(x)\,dx$$

for the *real inner product* on $L^2(\mathbf{R}^n)$ and also for the action of a distribution f on a test function g. (This notation also makes sense when a distribution f coincides with a function.)

Let $1 \le p, q \le \infty$. For a bounded linear operator T from $L^p(X,\mu)$ to $L^q(Y,\nu)$ we denote by T^* its *adjoint operator* defined by

$$\langle T(f) \,|\, g \rangle = \int_Y T(f)\overline{g}\,d\nu = \int_X f\overline{T^*(g)}\,d\mu = \langle f \,|\, T^*(g) \rangle \tag{2.5.10}$$

for f in $L^p(X,\mu)$ and g in $L^{q'}(Y,\nu)$ (or in a dense subspace of it). We also define the *transpose* of T as the unique operator T^t that satisfies

$$\langle T(f), g \rangle = \int_Y T(f)\,g\,dx = \int_X f\,T^t(g)\,dx = \langle f, T^t(g) \rangle$$

for all $f \in L^p(X,\mu)$ and all $g \in L^{q'}(Y,\nu)$.

If T is an integral operator of the form

$$T(f)(x) = \int_X K(x,y)f(y)\,d\mu(y),$$

then T^* and T^t are also integral operators with kernels $K^*(x,y) = \overline{K(y,x)}$ and $K^t(x,y) = K(y,x)$, respectively. If T has the form $T(f) = (\widehat{f}m)^\vee$, that is, it is given by multiplication on the Fourier transform by a (complex-valued) function $m(\xi)$, then T^* is given by multiplication on the Fourier transform by the function $\overline{m(\xi)}$. Indeed for f, g in $\mathscr{S}(\mathbf{R}^n)$ we have

$$\begin{aligned}\int_{\mathbf{R}^n} f\,\overline{T^*(g)}\,dx &= \int_{\mathbf{R}^n} T(f)\,\overline{g}\,dx \\ &= \int_{\mathbf{R}^n} \widehat{T(f)}\,\overline{\widehat{g}}\,d\xi \\ &= \int_{\mathbf{R}^n} \widehat{f}\,\overline{\overline{m}\widehat{g}}\,d\xi \\ &= \int_{\mathbf{R}^n} f\,\overline{(\overline{m}\widehat{g})^\vee}\,dx\,.\end{aligned}$$

A similar argument (using Theorem 2.2.14 (5)) gives that if T is given by multiplication on the Fourier transform by the function $m(\xi)$, then T^t is given by multiplication on the Fourier transform by the function $m(-\xi)$. Since the complex-valued functions $\overline{m(\xi)}$ and $m(-\xi)$ may be different, the operators T^* and T^t may be different in general. Also, if $m(\xi)$ is real-valued, then T is *self-adjoint* (i.e., $T = T^*$) while if $m(\xi)$ is even, then T is *self-transpose* (i.e., $T = T^t$).

2.5.3 The Spaces $\mathscr{M}^{p,q}(\mathbf{R}^n)$

Definition 2.5.5. Given $1 \le p,q \le \infty$, we denote by $\mathscr{M}^{p,q}(\mathbf{R}^n)$ the set of all bounded linear operators from $L^p(\mathbf{R}^n)$ to $L^q(\mathbf{R}^n)$ that commute with translations.

By Theorem 2.5.2 we have that every T in $\mathscr{M}^{p,q}$ is given by convolution with a tempered distribution. We introduce a norm on $\mathscr{M}^{p,q}$ by setting

$$\big\|T\big\|_{\mathscr{M}^{p,q}} = \big\|T\big\|_{L^p\to L^q},$$

that is, the norm of T in $\mathscr{M}^{p,q}$ is the operator norm of T as an operator from L^p to L^q. It is a known fact that under this norm, $\mathscr{M}^{p,q}$ is a complete normed space (i.e., a Banach space).

Next we show that when $p > q$ the set $\mathscr{M}^{p,q}$ consists of only one element, namely the zero operator $T = 0$. This means that the only interesting classes of operators arise when $p \le q$.

Theorem 2.5.6. *$\mathscr{M}^{p,q} = \{0\}$ whenever $1 \le q < p < \infty$.*

Proof. Let f be a nonzero $\mathscr{C}_0^\infty$ function and let $h \in \mathbf{R}^n$. We have

$$\big\|\tau^h(T(f)) + T(f)\big\|_{L^q} = \big\|T(\tau^h(f)+f)\big\|_{L^q} \le \big\|T\big\|_{L^p\to L^q}\big\|\tau^h(f)+f\big\|_{L^p}.$$

Now let $|h| \to \infty$ and use Exercise 2.5.1. We conclude that

$$2^{\frac{1}{q}}\big\|T(f)\big\|_{L^q} \le \big\|T\big\|_{L^p\to L^q} 2^{\frac{1}{p}}\big\|f\big\|_{L^p},$$

which is impossible if $q < p$ unless T is the zero operator. □

Next we have a theorem concerning the duals of the spaces $\mathscr{M}^{p,q}(\mathbf{R}^n)$.

Theorem 2.5.7. *Let $1 < p \le q < \infty$ and $T \in \mathscr{M}^{p,q}(\mathbf{R}^n)$. Then T can be defined on $L^{q'}(\mathbf{R}^n)$, coinciding with its previous definition on the subspace $L^p(\mathbf{R}^n) \cap L^{q'}(\mathbf{R}^n)$ of $L^p(\mathbf{R}^n)$, so that it maps $L^{q'}(\mathbf{R}^n)$ to $L^{p'}(\mathbf{R}^n)$ with norm*

$$\big\|T\big\|_{L^{q'}\to L^{p'}} = \big\|T\big\|_{L^p\to L^q}. \tag{2.5.11}$$

In other words, we have the following isometric identification of spaces:

$$\mathscr{M}^{q',p'}(\mathbf{R}^n) = \mathscr{M}^{p,q}(\mathbf{R}^n).$$

Proof. We first observe that if $T : L^p \to L^q$ is given by convolution with $u \in \mathscr{S}'$, then the adjoint operator $T^* : L^{q'} \to L^{p'}$ is given by convolution with $\overline{\widetilde{u}} \in \mathscr{S}'$. Indeed, for $f, g \in \mathscr{S}(\mathbf{R}^n)$ we have

$$\begin{aligned}
\int_{\mathbf{R}^n} f\,\overline{T^*(g)}\,dx &= \int_{\mathbf{R}^n} T(f)\,\overline{g}\,dx \\
&= \int_{\mathbf{R}^n} (f*u)\,\overline{g}\,dx \\
&= \int_{\mathbf{R}^n} f\,(\overline{g}*\widetilde{u})\,dx \\
&= \int_{\mathbf{R}^n} f\,\overline{g*\overline{\widetilde{u}}}\,dx.
\end{aligned}$$

Therefore T^* is given by convolution with $\overline{\widetilde{u}}$ when applied to Schwartz functions.

Next we observe the validity of the identity

$$\overline{f*\overline{\widetilde{u}}} = (\overline{\widetilde{f}}*u)^{\sim}, \qquad f \in \mathscr{S}. \tag{2.5.12}$$

It remains to show that T (convolution with u) and T^* (convolution with $\overline{\widetilde{u}}$) map $L^{q'}$ to $L^{p'}$ with the same norm. But this easily follows from (2.5.12), which implies that

$$\frac{\big\|f*\overline{\widetilde{u}}\big\|_{L^{p'}}}{\big\|f\big\|_{L^{q'}}} = \frac{\big\|\overline{\widetilde{f}}*u\big\|_{L^{p'}}}{\big\|\overline{\widetilde{f}}\big\|_{L^{q'}}},$$

for all nonzero Schwartz functions f. We conclude that

$$\|T^*\|_{L^{q'}\to L^{p'}} = \|T\|_{L^{q'}\to L^{p'}}$$

and therefore

$$\|T\|_{L^p\to L^q} = \|T\|_{L^{q'}\to L^{p'}}.$$

This establishes the claimed assertion. □

We next focus attention on the spaces $\mathscr{M}^{p,q}(\mathbf{R}^n)$ whenever $p = q$. These spaces are of particular interest, since they include the singular integral operators, which we study in Chapter 5.

2.5.4 Characterizations of $\mathscr{M}^{1,1}(\mathbf{R}^n)$ and $\mathscr{M}^{2,2}(\mathbf{R}^n)$

It would be desirable to have a characterization of the spaces $\mathscr{M}^{p,p}$ in terms of properties of the convolving distribution. Unfortunately, this is unknown at present (it is not clear whether it is possible) except for certain cases.

Theorem 2.5.8. *An operator T is in $\mathscr{M}^{1,1}(\mathbf{R}^n)$ if and only if it is given by convolution with a finite Borel (complex-valued) measure. In this case, the norm of the operator is equal to the total variation of the measure.*

Proof. If T is given with convolution with a finite Borel measure μ, then clearly T maps L^1 to itself and $\|T\|_{L^1\to L^1} \le \|\mu\|_{\mathscr{M}}$, where $\|\mu\|_{\mathscr{M}}$ is the total variation of μ.

Conversely, let T be an operator bounded from L^1 to L^1 that commutes with translations. By Theorem 2.5.2, T is given by convolution with a tempered distribution u. Let

$$f_\varepsilon(x) = \varepsilon^{-n} e^{-\pi|x/\varepsilon|^2}.$$

Since the functions f_ε are uniformly bounded in L^1, it follows from the boundedness of T that $f_\varepsilon * u$ are also uniformly bounded in L^1. Since L^1 is naturally embedded in the space of finite Borel measures, which is the dual of the space $\mathscr{C}_{00}$ of continuous functions that tend to zero at infinity, we obtain that the family $f_\varepsilon * u$ lies in a fixed multiple of the unit ball of $\mathscr{C}_{00}^*$. By the Banach–Alaoglu theorem, this is a weak* compact set. Therefore, some subsequence of $f_\varepsilon * u$ converges in the weak* topology to a measure μ. That is, for some $\varepsilon_k \to 0$ and all $g \in \mathscr{C}_{00}(\mathbf{R}^n)$ we have

$$\lim_{k\to\infty} \int_{\mathbf{R}^n} g(x)(f_{\varepsilon_k} * u)(x)\,dx = \int_{\mathbf{R}^n} g(x)\,d\mu(x). \tag{2.5.13}$$

We claim that $u = \mu$. To see this, fix $g \in \mathscr{S}$. Equation (2.5.13) implies that

$$\langle u, \widetilde{f_{\varepsilon_k}} * g\rangle = \langle u, f_{\varepsilon_k} * g\rangle \to \langle \mu, g\rangle$$

as $k \to \infty$. Exercise 2.3.2 gives that $g * f_{\varepsilon_k}$ converges to g in $\mathscr{S}$. Therefore,

$$\langle u, f_{\varepsilon_k} * g\rangle \to \langle u, g\rangle.$$

It follows from (2.5.13) that $\langle u, g\rangle = \langle \mu, g\rangle$, and since g was arbitrary, $u = \mu$.

Next, (2.5.13) implies that for all $g \in \mathscr{C}_{00}$ we have

$$\left| \int_{\mathbf{R}^n} g(x)\,d\mu(x)\right| \le \|g\|_{L^\infty} \sup_k \|f_{\varepsilon_k} * u\|_{L^1} \le \|g\|_{L^\infty} \|T\|_{L^1\to L^1}. \tag{2.5.14}$$

The Riesz representation theorem gives that the norm of the functional

$$g \mapsto \int_{\mathbf{R}^n} g(x)\,d\mu(x)$$

on $\mathscr{C}_{00}$ is exactly $\|\mu\|_{\mathscr{M}}$. It follows from (2.5.14) that $\|T\|_{L^1\to L^1} \ge \|\mu\|_{\mathscr{M}}$. Since the reverse inequality is obvious, we conclude that $\|T\|_{L^1\to L^1} = \|\mu\|_{\mathscr{M}}$. □

Let μ be a finite Borel measure. The operator $h \mapsto h * \mu$ maps $L^p(\mathbf{R}^n)$ to itself for all $1 \le p \le \infty$; hence $\mathscr{M}^{1,1}(\mathbf{R}^n)$ can be identified with a subspace of $\mathscr{M}^{\infty,\infty}(\mathbf{R}^n)$. But there exist bounded linear operators Φ on L^∞ that commute with translations for which there does not exist a finite Borel measure μ such that $\Phi(h) = h * \mu$ for all $h \in L^\infty(\mathbf{R}^n)$. The following example captures such a behavior.

Example 2.5.9. Let $(X, \|\cdot\|_{L^\infty})$ be the space of all complex-valued bounded functions on the real line such that

$$\Phi(f) = \lim_{R \to +\infty} \frac{1}{R} \int_0^R f(t)\,dt$$

exists. Then Φ is a bounded linear functional on X with norm 1 and has a bounded extension $\widetilde{\Phi}$ on L^∞ with norm 1, by the Hahn–Banach theorem. We may view $\widetilde{\Phi}$ as a bounded linear operator from $L^\infty(\mathbf{R})$ to the space of constant functions, which is contained in $L^\infty(\mathbf{R})$. We note that $\widetilde{\Phi}$ commutes with translations, since for all $f \in L^\infty(\mathbf{R})$ and $x \in \mathbf{R}$ we have

$$\widetilde{\Phi}(\tau^x(f)) - \tau^x(\widetilde{\Phi}(f)) = \widetilde{\Phi}(\tau^x(f)) - \widetilde{\Phi}(f) = \widetilde{\Phi}(\tau^x(f) - f) = \Phi(\tau^x(f) - f) = 0,$$

where the last two equalities follow from the fact that for L^∞ functions f the expression $\frac{1}{R}\int_0^R (f(t-x) - f(t))\,dt$ is bounded by $\frac{|x|}{R}\|f\|_{L^\infty}$ when $R > |x|$ and thus tends to zero as $R \to \infty$. If $\Phi(\varphi) = \varphi * u$ for some $u \in \mathscr{S}'(\mathbf{R}^n)$ and all $\varphi \in \mathscr{S}(\mathbf{R}^n)$, since Φ vanishes on $\mathscr{S}$, the uniqueness in Theorem 2.5.2 yields that $u = 0$. Hence, if there existed a finite Borel measure μ such that $\widetilde{\Phi}(h) = h * \mu$ all $h \in L^\infty$, in particular we would have $0 = \Phi(\varphi) = \varphi * \mu$ for all $\varphi \in \mathscr{S}$, hence μ would be the zero measure. But obviously, this is not the case, since Φ is not the zero operator on X.

We now study the case $p = 2$. We have the following theorem:

Theorem 2.5.10. *An operator T is in $\mathscr{M}^{2,2}(\mathbf{R}^n)$ if and only if it is given by convolution with some $u \in \mathscr{S}'$ whose Fourier transform $\widehat{u}$ is an L^∞ function. In this case the norm of $T : L^2 \to L^2$ is equal to $\|\widehat{u}\|_{L^\infty}$.*

Proof. If $\widehat{u} \in L^\infty$, Plancherel's theorem gives

$$\int_{\mathbf{R}^n} |f * u|^2\,dx = \int_{\mathbf{R}^n} |\widehat{f}(\xi)\widehat{u}(\xi)|^2\,d\xi \le \|\widehat{u}\|_{L^\infty}^2 \|\widehat{f}\|_{L^2}^2;$$

therefore, $\|T\|_{L^2 \to L^2} \le \|\widehat{u}\|_{L^\infty}$, and hence T is in $\mathscr{M}^{2,2}(\mathbf{R}^n)$.

Now suppose that $T \in \mathscr{M}^{2,2}(\mathbf{R}^n)$ is given by convolution with a tempered distribution u. We show that $\widehat{u}$ is a bounded function. For $R > 0$ let φ_R be a $\mathscr{C}_0^\infty$ function supported inside the ball $B(0,2R)$ and equal to one on the ball $B(0,R)$. The product of the function φ_R with the distribution $\widehat{u}$ is $\varphi_R \widehat{u} = ((\varphi_R)^\vee * u)\widehat{\ } = T(\varphi_R^\vee)\widehat{\ }$, which is an L^2 function. Since the L^2 function $\varphi_R \widehat{u}$ coincides with the distribution $\widehat{u}$ on the set $B(0,R)$, it follows that $\widehat{u}$ is in $L^2(B(0,R))$ for all $R > 0$ and therefore it is in L^2_{loc}. If $f \in L^\infty(\mathbf{R}^n)$ has compact support, the function $f\widehat{u}$ is in L^2, and therefore Plancherel's theorem and the boundedness of T give

$$\int_{\mathbf{R}^n} |f(x)\widehat{u}(x)|^2\,dx = \int_{\mathbf{R}^n} |T(f^\vee)(x)|^2\,dx \le \|T\|^2_{L^2\to L^2} \int_{\mathbf{R}^n} |f(x)|^2\,dx.$$

We conclude that for all bounded functions with compact support f we have

$$\int_{\mathbf{R}^n} \left(\|T\|^2_{L^2\to L^2} - |\widehat{u}(x)|^2\right)|f(x)|^2\,dx \ge 0.$$

Taking $f(x_1,\dots,x_n) = (2r)^{-n/2}\prod_{j=1}^n \chi_{[-r,r]}(x_j)$ for $r>0$ and using Corollary 2.1.16, we obtain that $\|T\|^2_{L^2\to L^2} - |\widehat{u}(x)|^2 \ge 0$ for almost all x. Hence $\widehat{u}$ is in L^∞ and $\|\widehat{u}\|_{L^\infty} \le \|T\|_{L^2\to L^2}$. Combining this with the estimate $\|T\|_{L^2\to L^2} \le \|\widehat{u}\|_{L^\infty}$, which holds if $\widehat{u} \in L^\infty$, we deduce that $\|T\|_{L^2\to L^2} = \|\widehat{u}\|_{L^\infty}$. □

2.5.5 The Space of Fourier Multipliers $\mathscr{M}_p(\mathbf{R}^n)$

We have now characterized all convolution operators that map L^2 to L^2. Suppose now that T is in $\mathscr{M}^{p,p}$, where $1<p<2$. As discussed in Theorem 2.5.7, T also maps $L^{p'}$ to $L^{p'}$. Since $p<2<p'$, by Theorem 1.3.4, it follows that T also maps L^2 to L^2. Thus T is given by convolution with a tempered distribution whose Fourier transform is a bounded function.

Definition 2.5.11. Given $1\le p<\infty$, we denote by $\mathscr{M}_p(\mathbf{R}^n)$ the space of all bounded functions m on $\mathbf{R}^n$ such that the operator

$$T_m(f) = (\widehat{f}m)^\vee, \qquad f\in\mathscr{S},$$

is bounded on $L^p(\mathbf{R}^n)$ (or is initially defined in a dense subspace of $L^p(\mathbf{R}^n)$ and has a bounded extension on the whole space). The norm of m in $\mathscr{M}_p(\mathbf{R}^n)$ is defined by

$$\|m\|_{\mathscr{M}_p} = \|T_m\|_{L^p\to L^p}. \tag{2.5.15}$$

Definition 2.5.11 implies that $m\in\mathscr{M}_p$ if and only if $T_m \in \mathscr{M}^{p,p}$. Elements of the space $\mathscr{M}_p$ are called *L^p multipliers* or *L^p Fourier multipliers*. It follows from Theorem 2.5.10 that $\mathscr{M}_2$, the set of all L^2 multipliers, is L^∞. Theorem 2.5.8 implies that $\mathscr{M}_1(\mathbf{R}^n)$ is the set of the Fourier transforms of finite Borel measures that is usually denoted by $\mathscr{M}(\mathbf{R}^n)$. Theorem 2.5.7 states that a bounded function m is an L^p multiplier if and only if it is an $L^{p'}$ multiplier, and in this case

$$\|m\|_{\mathscr{M}_p} = \|m\|_{\mathscr{M}_{p'}}, \qquad 1<p<\infty.$$

It is a consequence of Theorem 1.3.4 that the normed spaces $\mathscr{M}_p$ are nested, that is, for $1 \le p \le q \le 2$ we have

$$\mathscr{M}_1 \subseteq \mathscr{M}_p \subseteq \mathscr{M}_q \subseteq \mathscr{M}_2 = L^\infty .$$

Moreover, if $m \in \mathscr{M}_p$ and $1 \le p \le 2 \le p'$, Theorem 1.3.4 gives

$$\big\| T_m \big\|_{L^2 \to L^2} \le \big\| T_m \big\|_{L^p \to L^p}^{\frac{1}{2}} \big\| T_m \big\|_{L^{p'} \to L^{p'}}^{\frac{1}{2}} = \big\| T_m \big\|_{L^p \to L^p} , \tag{2.5.16}$$

since $1/2 = (1/2)/p + (1/2)/p'$. Theorem 1.3.4 also gives that

$$\big\| m \big\|_{\mathscr{M}_p} \le \big\| m \big\|_{\mathscr{M}_q}$$

whenever $1 \le q \le p \le 2$. Thus the $\mathscr{M}_p$'s form an increasing family of spaces as p increases from 1 to 2.

Example 2.5.12. The function $m(\xi) = e^{2\pi i \xi \cdot b}$ is an L^p multiplier for all $b \in \mathbf{R}^n$, since the corresponding operator $T_m(f)(x) = f(x+b)$ is bounded on $L^p(\mathbf{R}^n)$. Clearly $\|m\|_{\mathscr{M}_p} = 1$.

Proposition 2.5.13. *For $1 \le p < \infty$, the normed space $\big(\mathscr{M}_p, \|\cdot\|_{\mathscr{M}_p}\big)$ is a Banach space. Furthermore, $\mathscr{M}_p$ is closed under pointwise multiplication and is a Banach algebra.*

Proof. It suffices to consider the case $1 \le p \le 2$. It is straightforward that if m_1, m_2 are in $\mathscr{M}_p$ and $b \in \mathbf{C}$ then $m_1 + m_2$ and $b m_1$ are also in $\mathscr{M}_p$. Observe that $m_1 m_2$ is the multiplier that corresponds to the operator $T_{m_1} T_{m_2} = T_{m_1 m_2}$ and thus

$$\big\| m_1 m_2 \big\|_{\mathscr{M}_p} = \big\| T_{m_1} T_{m_2} \big\|_{L^p \to L^p} \le \big\| m_1 \big\|_{\mathscr{M}_p} \big\| m_2 \big\|_{\mathscr{M}_p} .$$

This proves that $\mathscr{M}_p$ is an algebra. To show that $\mathscr{M}_p$ is a complete normed space, consider a Cauchy sequence m_j in $\mathscr{M}_p$. It follows from (2.5.16) that m_j is Cauchy in L^∞, and hence it converges to some bounded function m in the L^∞ norm; moreover all the m_j are a.e. bounded by some constant C uniformly in j. We have to show that $m \in \mathscr{M}_p$. Fix $f \in \mathscr{S}$. We have

$$T_{m_j}(f)(x) = \int_{\mathbf{R}^n} \widehat{f}(\xi) m_j(\xi) e^{2\pi i x \cdot \xi} \, d\xi \to \int_{\mathbf{R}^n} \widehat{f}(\xi) m(\xi) e^{2\pi i x \cdot \xi} \, d\xi = T_m(f)(x)$$

a.e. by the Lebesgue dominated convergence theorem, since $C \big| \widehat{f} \big|$ is an integrable upper bound of all integrands on the left in the preceding expression. Since $\{m_j\}_j$ is a Cauchy sequence in $\mathscr{M}_p$, it is bounded in $\mathscr{M}_p$, and thus $\sup_j \|m_j\|_{\mathscr{M}_p} < +\infty$. An application of Fatou's lemma yields that

$$\begin{aligned} \int_{\mathbf{R}^n} |T_m(f)|^p \, dx &= \int_{\mathbf{R}^n} \liminf_{j \to \infty} |T_{m_j}(f)|^p \, dx \\ &\le \liminf_{j \to \infty} \int_{\mathbf{R}^n} |T_{m_j}(f)|^p \, dx \\ &\le \liminf_{j \to \infty} \big\| m_j \big\|_{\mathscr{M}_p}^p \big\| f \big\|_{L^p}^p , \end{aligned}$$

which implies that $m \in \mathscr{M}_p$. This argument shows that if $m_j \in \mathscr{M}_p$ and $m_j \to m$ uniformly, then m is in $\mathscr{M}_p$ and satisfies

$$\|m\|_{\mathscr{M}_p} \le \liminf_{j\to\infty} \|m_j\|_{\mathscr{M}_p}.$$

Apply this inequality to $m_k - m_j$ in place of m_j and $m_k - m$ in place of m, for some fixed k. We obtain

$$\|m_k - m\|_{\mathscr{M}_p} \le \liminf_{j\to\infty} \|m_k - m_j\|_{\mathscr{M}_p} \tag{2.5.17}$$

for each k. Given $\varepsilon > 0$, by the Cauchy criterion, there is an N such that for $j,k > N$ we have $\|m_k - m_j\|_{\mathscr{M}_p} < \varepsilon$. Using (2.5.17) we conclude that $\|m_k - m\|_{\mathscr{M}_p} \le \varepsilon$ when $k > N$, thus m_k converges to m in $\mathscr{M}_p$.

This proves that $\mathscr{M}_p$ is a Banach space. □

The following proposition summarizes some simple properties of multipliers.

Proposition 2.5.14. *For all $m \in \mathscr{M}_p$, $1 \le p < \infty$, $x \in \mathbf{R}^n$, and $h > 0$ we have*

$$\big\|\tau^x(m)\big\|_{\mathscr{M}_p} = \big\|m\big\|_{\mathscr{M}_p}, \tag{2.5.18}$$

$$\big\|\delta^h(m)\big\|_{\mathscr{M}_p} = \big\|m\big\|_{\mathscr{M}_p}, \tag{2.5.19}$$

$$\big\|\widetilde{m}\big\|_{\mathscr{M}_p} = \big\|m\big\|_{\mathscr{M}_p},$$

$$\big\|e^{2\pi i(\cdot)\cdot x} m\big\|_{\mathscr{M}_p} = \big\|m\big\|_{\mathscr{M}_p},$$

$$\big\|m\circ A\big\|_{\mathscr{M}_p} = \big\|m\big\|_{\mathscr{M}_p}, \qquad \text{A is an orthogonal matrix.}$$

Proof. See Exercise 2.5.2. □

Example 2.5.15. We show that for $-\infty<a<b<\infty$ we have $\|\chi_{[a,b]}\|_{\mathscr{M}_p} = \|\chi_{[0,1]}\|_{\mathscr{M}_p}$. Indeed, using (2.5.18) we obtain that $\|\chi_{[a,b]}\|_{\mathscr{M}_p} = \|\chi_{[0,b-a]}\|_{\mathscr{M}_p}$, and the latter is equal to $\|\chi_{[0,1]}\|_{\mathscr{M}_p}$ in view of (2.5.19). The fact that we have $\|\chi_{[0,1]}\|_{\mathscr{M}_p} < \infty$ for all $1 < p < \infty$ is shown in Chapter 5.

We continue with the following interesting result.

Theorem 2.5.16. *Suppose that $m(\xi,\eta) \in \mathscr{M}_p(\mathbf{R}^{n+m})$, where $1 < p < \infty$. Then for almost every $\xi \in \mathbf{R}^n$ the function $\eta \mapsto m(\xi,\eta)$ is in $\mathscr{M}_p(\mathbf{R}^m)$, with*

$$\big\|m(\xi,\cdot)\big\|_{\mathscr{M}_p(\mathbf{R}^m)} \le \big\|m\big\|_{\mathscr{M}_p(\mathbf{R}^{n+m})}.$$

Proof. Since m lies in $L^\infty(\mathbf{R}^{n+m})$, it follows by Fubini's theorem that for almost all $\xi \in \mathbf{R}^n$, the function $\eta \mapsto m(\xi,\eta)$ lies in $L^\infty(\mathbf{R}^m)$ and

$$\big\|m(\xi,\cdot)\big\|_{L^\infty(\mathbf{R}^m)} \le \big\|m\big\|_{L^\infty(\mathbf{R}^{n+m})}. \tag{2.5.20}$$

Fix f_1, g_1 in $\mathscr{S}(\mathbf{R}^n)$ and f_2, g_2 in $\mathscr{S}(\mathbf{R}^m)$. Define the functions $(f_1 \otimes f_2)(x,y) = f_1(x)f_2(y)$ when $x \in \mathbf{R}^n$ and $y \in \mathbf{R}^m$. For all ξ for which (2.5.20) is satisfied define

$$M(\xi) = \int_{\mathbf{R}^m} (m(\xi,\cdot)\widehat{f_2})^\vee(y)\,g_2(y)\,dy = \int_{\mathbf{R}^m} m(\xi,\eta)\widehat{f_2}(\eta)g_2{}^\vee(\eta)\,d\eta$$

and observe that

$$\begin{aligned}
\left|\int_{\mathbf{R}^n} (M(\cdot)\widehat{f_1})^\vee(x)\,g_1(x)\,dx\right| &= \left|\int_{\mathbf{R}^n} M(\xi)\widehat{f_1}(\xi)g_1{}^\vee(\xi)\,d\xi\right| \\
&= \left|\iint_{\mathbf{R}^{n+m}} m(\xi,\eta)\widehat{f_1\otimes f_2}(\xi,\eta)(g_1\otimes g_2)^\vee(\xi,\eta)\,d\xi\,d\eta\right| \\
&= \left|\iint_{\mathbf{R}^{n+m}} (m\widehat{f_1\otimes f_2})^\vee(x,y)(g_1\otimes g_2)(x,y)\,dx\,dy\right| \\
&\le \big\|m\big\|_{\mathscr{M}_p(\mathbf{R}^{n+m})}\big\|f_1\big\|_{L^p}\big\|f_2\big\|_{L^p}\big\|g_1\big\|_{L^{p'}}\big\|g_2\big\|_{L^{p'}}.
\end{aligned}$$

In view of the identity

$$\big\|(M(\cdot)\widehat{f_1})^\vee\big\|_{L^p} = \sup_{\|g_1\|_{L^{p'}}\le 1}\left|\int_{\mathbf{R}^n} (M(\cdot)\widehat{f_1})^\vee(x)\,g_1(x)\,dx\right|,$$

it follows that, for the ξ that satisfy (2.5.20), $M(\xi)$ lies in $\mathscr{M}_p(\mathbf{R}^n)$ with

$$\big\|M\big\|_{\mathscr{M}_p(\mathbf{R}^n)} \le \big\|m\big\|_{\mathscr{M}_p(\mathbf{R}^{n+m})}\big\|f_2\big\|_{L^p}\big\|g_2\big\|_{L^{p'}}.$$

Since $\|M\|_{L^\infty} \le \|M\|_{\mathscr{M}_p}$ for almost all $\xi \in \mathbf{R}^n$, we obtain

$$\left|\int_{\mathbf{R}^m} (m(\xi,\cdot)\widehat{f_2})^\vee(y)\,g_2(y)\,dy\right| = |M(\xi)| \le \big\|m\big\|_{\mathscr{M}_p(\mathbf{R}^{n+m})}\big\|f_2\big\|_{L^p}\big\|g_2\big\|_{L^{p'}}, \tag{2.5.21}$$

which of course implies the required conclusion, by taking the supremum over all g_2 in $L^{p'}$ with norm at most 1. □

Example 2.5.17. (The cone multiplier) On $\mathbf{R}^{n+1}$ define the function

$$m_\lambda(\xi_1,\dots,\xi_{n+1}) = \left(1 - \frac{\xi_1^2+\dots+\xi_n^2}{\xi_{n+1}^2}\right)_+^\lambda, \qquad \lambda > 0,$$

where the plus sign indicates that $m_\lambda = 0$ if the expression inside the parentheses is negative. The multiplier m_λ is called the *cone multiplier with parameter* λ. If m_λ is in $\mathscr{M}_p(\mathbf{R}^{n+1})$, then the function $b_\lambda(\xi) = (1-|\xi|^2)_+^\lambda$ defined on $\mathbf{R}^n$ is in $\mathscr{M}_p(\mathbf{R}^n)$. Indeed, by Theorem 2.5.16 we have that for some $\xi_{n+1} = h$, $b_\lambda(\xi_1/h,\dots,\xi_n/h)$ is in $\mathscr{M}_p(\mathbf{R}^n)$ and hence so is b_λ by property (2.5.19).

Exercises

2.5.1. Prove that if $f \in L^q(\mathbf{R}^n)$ and $0 < q < \infty$, then

$$\big\|\tau^h(f) + f\big\|_{L^q} \to 2^{1/q}\big\|f\big\|_{L^q} \qquad \text{as } |h| \to \infty.$$

2.5.2. Prove Proposition 2.5.14. Also prove that if $\delta_j^{h_j}$ is a dilation operator in the jth variable (for instance $\delta_1^{h_1} f(x) = f(h_1x_1, x_2, \dots, x_n)$), then

$$\big\|\delta_1^{h_1} \cdots \delta_n^{h_n} m\big\|_{\mathscr{M}_p} = \big\|m\big\|_{\mathscr{M}_p}.$$

2.5.3. Let $m \in \mathscr{M}_p(\mathbf{R}^n)$ where $1 \le p < \infty$.
(a) If ψ is a function on $\mathbf{R}^n$ whose inverse Fourier transform is an integrable function, then prove that

$$\big\|\psi m\big\|_{\mathscr{M}_p} \le \big\|\psi^\vee\big\|_{L^1}\big\|m\big\|_{\mathscr{M}_p}.$$

(b) If ψ is in $L^1(\mathbf{R}^n)$, then prove that

$$\big\|\psi * m\big\|_{\mathscr{M}_p} \le \big\|\psi\big\|_{L^1}\big\|m\big\|_{\mathscr{M}_p}.$$

2.5.4. Fix a multi-index γ.
(a) Prove that the map $T(f) = f * \partial^\gamma \delta_0$ maps $\mathscr{S}$ continuously into $\mathscr{S}$.
(b) Prove that when $1/p - 1/q \neq |\gamma|/n$, T does not extend to an element of the space $\mathscr{M}^{p,q}$.

2.5.5. Let $K_\gamma(x) = |x|^{-n+\gamma}$, where $0 < \gamma < n$. Use Theorem 1.4.25 to show that the operator

$$T_\gamma(f) = f * K_\gamma, \qquad f \in \mathscr{S},$$

extends to a bounded operator in $\mathscr{M}^{p,q}(\mathbf{R}^n)$, where $1/p - 1/q = \gamma/n$, $1 < p < q < \infty$. This provides an example of a nontrivial operator in $\mathscr{M}^{p,q}(\mathbf{R}^n)$ when $p < q$.

2.5.6. (a) Use the ideas of the proof of Proposition 2.5.13 to show that if $m_j \in \mathscr{M}_p$, $1 \le p < \infty$, $\|m_j\|_{\mathscr{M}_p} \le C$ for all $j = 1, 2, \dots$, and $m_j \to m$ a.e., then $m \in \mathscr{M}_p$ and

$$\big\|m\big\|_{\mathscr{M}_p(\mathbf{R}^n)} \le \liminf_{j\to\infty} \big\|m_j\big\|_{\mathscr{M}_p(\mathbf{R}^n)} \le C.$$

(b) Prove that if $m \in \mathscr{M}_p$, $1 \le p < \infty$, and the limit $m_0(\xi) = \lim_{R\to\infty} m(\xi/R)$ exists for all $\xi \in \mathbf{R}^n$, then m_0 is a radial function in $\mathscr{M}_p(\mathbf{R}^n)$ and satisfies $\|m_0\|_{\mathscr{M}_p} \le \|m\|_{\mathscr{M}_p}$.
(c) If $m \in \mathscr{M}_p(\mathbf{R})$ has left and right limits at the origin, then prove that

$$\big\|m\big\|_{\mathscr{M}_p(\mathbf{R})} \ge \max(|m(0+)|, |m(0-)|).$$

(d) Suppose that for some $1 \le p < \infty$, $m_t \in \mathscr{M}_p(\mathbf{R}^n)$ for all $0 < t < \infty$. Prove that

$$\int_0^\infty \left\| m_t \right\|_{\mathscr{M}_p(\mathbf{R}^n)} \frac{dt}{t} < \infty \implies m(\xi) = \int_0^\infty m_t(\xi) \frac{dt}{t} \in \mathscr{M}_p .$$

2.5.7. Let $1 \le p < \infty$ and suppose that $m \in \mathscr{M}_p(\mathbf{R}^n)$ satisfies $|m(\xi)| \ge c(1+|\xi|)^{-N}$ for some $c, N > 0$. Prove that the operator $T(f) = (\widehat{f} m^{-1})^\vee$ satisfies $\|T(f)\|_{L^p} \ge c_p \|f\|_{L^p}$ for all $f \in \mathscr{S}(\mathbf{R}^n)$, where $c_p = \|m\|_{\mathscr{M}_p}^{-1}$.

2.5.8. (a) Prove that if $m \in L^\infty(\mathbf{R}^n)$ satisfies $m^\vee \ge 0$, then for all $1 \le p < \infty$ we have

$$\left\| m \right\|_{\mathscr{M}_p} = \left\| m^\vee \right\|_{L^1} .$$

(b) (*L. Colzani and E. Laeng*) On the real line let

$$m_1(\xi) = \begin{cases} -1 & \text{for } \xi > 0 \\ 1 & \text{for } \xi < 0, \end{cases} \qquad m_2(\xi) = \begin{cases} \min(\xi - 1, 0) & \text{for } \xi > 0 \\ \max(\xi + 1, 0) & \text{for } \xi < 0. \end{cases}$$

Prove that

$$\left\| m_1 \right\|_{\mathscr{M}_p} = \left\| m_2 \right\|_{\mathscr{M}_p}$$

for all $1 < p < \infty$.
[*Hint:* Part (a): Use Exercise 1.2.9. Part (b): Use part (a) to show that $\|m_2 m_1^{-1}\|_{\mathscr{M}_p} = 1$. Deduce that $\|m_2\|_{\mathscr{M}_p} \le \|m_1\|_{\mathscr{M}_p}$. For the converse use Exercise 2.5.6 (c).]

2.5.9. ([94]) Let $1 < p < \infty$ and $0 < A < \infty$. Prove that the following are equivalent:
(a) The operator $f \mapsto \sum_{m \in \mathbf{Z}^n} a_m f(x - m)$ is bounded on $L^p(\mathbf{R}^n)$ with norm A.
(b) The $\mathscr{M}_p$ norm of the function $\sum_{m \in \mathbf{Z}^n} a_m e^{-2\pi i m \cdot x}$ is exactly A.
(c) The operator given by convolution with the sequence $\{a_m\}$ is bounded on $\ell^p(\mathbf{Z}^n)$ with norm A.

2.5.10. ([177]) Let $m(\xi)$ in $\mathscr{M}_p(\mathbf{R}^n)$ be supported in $[0,1]^n$. Then the periodic extension of m in $\mathbf{R}^n$,

$$M(\xi) = \sum_{k \in \mathbf{Z}^n} m(\xi - k),$$

is also in $\mathscr{M}_p(\mathbf{R}^n)$.

2.5.11. Suppose that u is a $\mathscr{C}^\infty$ function on $\mathbf{R}^n \setminus \{0\}$ that is homogeneous of degree $-n + i\tau$, $\tau \in \mathbf{R}$. Prove that the operator given by convolution with u maps $L^2(\mathbf{R}^n)$ to $L^2(\mathbf{R}^n)$.

2.5.12. ([142]) Let $m_1 \in L^r(\mathbf{R}^n)$ and $m_2 \in L^{r'}(\mathbf{R}^n)$ for some $2 \le r \le \infty$. Prove that $m_1 * m_2 \in \mathscr{M}_p(\mathbf{R}^n)$ when $\frac{1}{p} - \frac{1}{2} = \frac{1}{r}$ and $1 \le p \le 2$.
[*Hint:* Prove that the trilinear operator $(m_1, m_2, f) \mapsto ((m_1 * m_2)\widehat{f})^\vee$ is bounded from $L^2 \times L^2 \times L^1 \to L^1$ and $L^\infty \times L^1 \times L^2 \to L^2$. Apply trilinear complex interpolation (Corollary 7.2.11 in [131]) to deduce the required conclusion for $1 \le p \le 2$.]

2.5.13. Show that the function $e^{i|\xi|^2}$ is an L^p Fourier multiplier on $\mathbf{R}^n$ if and only if $p=2$.
$\big[$*Hint:* By Exercise 2.4.12 the inverse Fourier transform of $e^{i|\xi|^2}$ is in L^∞, thus the operator $f \mapsto \big(\widehat{f}(\xi)e^{i\pi|\xi|^2}\big)^\vee$ maps L^1 to L^∞. Since this operator also maps L^2 to L^2, it should map L^p to $L^{p'}$ for all $1 \le p \le 2$.$\big]$

2.6 Oscillatory Integrals

Oscillatory integrals have played an important role in harmonic analysis from its outset. The Fourier transform is the prototype of oscillatory integrals and provides the simplest example of a nontrivial phase, a linear function of the variable of integration. More complicated phases naturally appear in the subject; for instance, Bessel functions provide examples of oscillatory integrals in which the phase is a sinusoidal function.

In this section we take a quick look at oscillatory integrals. We mostly concentrate on one-dimensional results, which already require some significant analysis. We examine only a very simple higher-dimensional situation. Our analysis here is far from adequate.

Definition 2.6.1. An *oscillatory integral* is an expression of the form

$$I(\lambda) = \int_{\mathbf{R}^n} e^{i\lambda\varphi(x)}\psi(x)\,dx, \tag{2.6.1}$$

where λ is a positive real number, φ is a real-valued function on $\mathbf{R}^n$ called the *phase*, and ψ is a complex-valued and smooth integrable function on $\mathbf{R}^n$, which is often taken to have compact support.

2.6.1 Phases with No Critical Points

We begin by studying the simplest possible one-dimensional case. Suppose that φ and ψ are smooth functions on the real line such that $\operatorname{supp}\psi$ is a closed interval and

$$\varphi'(x) \neq 0 \qquad \text{for all } x \in \operatorname{supp}\psi.$$

Since φ' has no zeros, it must be either strictly positive or strictly negative everywhere on the support of ψ. It follows that φ is monotonic on the support of ψ and we are allowed to change variables

$$u = \varphi(x)$$

in (2.6.1). Then $dx = (\varphi'(x))^{-1}du = (\varphi^{-1})'(u)\,du$, where φ^{-1} is the inverse function of φ. We transform the integral in (2.6.1) into

$$\int_{\mathbf{R}} e^{i\lambda u}\psi(\varphi^{-1}(u))(\varphi^{-1})'(u)\,du \tag{2.6.2}$$

and we note that the function $\theta(u) = \psi(\varphi^{-1}(u))(\varphi^{-1})'(u)$ is smooth and has compact support on $\mathbf{R}$. We therefore interpret the integral in (2.6.1) as $\widehat{\theta}(-\lambda/2\pi)$, where $\widehat{\theta}$ is the Fourier transform of θ. Since θ is a smooth function with compact support, it follows that the integral in (2.6.2) has rapid decay as $\lambda \to \infty$.

A quick way to see that the expression $\widehat{\theta}(-\lambda/2\pi)$ has decay of order λ^{-N} for all $N > 0$ as λ tends to ∞ is the following. Write

$$e^{i\lambda u} = \frac{1}{(i\lambda)^N}\frac{d^N}{du^N}(e^{i\lambda u})$$

and integrate by parts N times to express the integral in (2.6.2) as

$$\frac{(-1)^N}{(i\lambda)^N}\int_{\mathbf{R}} e^{i\lambda u}\frac{d^N\theta(u)}{du^N}\,du\,,$$

from which the assertion follows. Hence

$$|I(\lambda)| = |\widehat{\theta}(-\lambda/2\pi)| \le C_N\lambda^{-N}, \tag{2.6.3}$$

where $C_N = \|\theta^{(N)}\|_{L^1}$, which depends on derivatives of φ and ψ.

We now turn to a higher-dimensional analogue of this situation.

Definition 2.6.2. We say that a point x_0 is a *critical point* of a phase function φ if

$$\nabla\varphi(x_0) = \big(\partial_1\varphi(x_0),\dots,\partial_n\varphi(x_0)\big) = 0\,.$$

Example 2.6.3. Let $\xi \in \mathbf{R}^n \setminus \{0\}$. Then the phase functions $\varphi_1(x) = x\cdot\xi$, $\varphi_2(x) = e^{x\cdot\xi}$ have no critical points, while the phase function $\varphi_3(x) = |x|^2 - x\cdot\xi$ has one critical point at $x_0 = \frac{1}{2}\xi$.

The next result concerns the behavior of oscillatory integrals whose phase functions have no critical points.

Proposition 2.6.4. *Suppose that ψ is a compactly supported smooth function on $\mathbf{R}^n$ and that φ is a real-valued $\mathscr{C}^\infty$ function on $\mathbf{R}^n$ that has no critical points on the support of ψ. Then the oscillatory integral*

$$I(\lambda) = \int_{\mathbf{R}^n} e^{i\lambda\varphi(x)}\psi(x)\,dx \tag{2.6.4}$$

obeys a bound of the form $|I(\lambda)| \le C_N\lambda^{-N}$ for all $\lambda \ge 1$ and all $N > 0$, where C_N depends on N and on φ and ψ.

Proof. Since the case $n = 1$ has already been discussed, we concentrate on dimensions $n \geq 2$. For each y in the support of ψ there is a unit vector θ_y such that

$$\theta_y \cdot \nabla\varphi(y) = |\nabla\varphi(y)|.$$

By the continuity of $\nabla\varphi$ there is a small neighborhood $B(y, r_y)$ of y such that for all $x \in B(y, r_y)$ we have

$$\theta_y \cdot \nabla\varphi(x) \geq \frac{1}{2}|\nabla\varphi(y)| > 0.$$

Cover the support of ψ by a finite number of balls $B(y_j, r_{y_j})$, $j = 1, \dots, m$, and pick $c = \min_j \frac{1}{2}|\nabla\varphi(y_j)|$; we have

$$\theta_{y_j} \cdot \nabla\varphi(x) \geq c > 0 \tag{2.6.5}$$

for all $x \in B(y_j, r_{y_j})$ and $j = 1, \dots, m$.

Next we find a smooth partition of unity of $\mathbf{R}^n$ such that each member ζ_k of the partition is supported in some ball $B(y_j, r_{y_j})$ or lies outside the support of ψ. We therefore write

$$I(\lambda) = \sum_k \int_{\mathbf{R}^n} e^{i\lambda\varphi(x)} \psi(x)\zeta_k(x)\, dx, \tag{2.6.6}$$

where the sum contains only a finite number of indices, since only a finite number of the ζ_k's meet the support of ψ. It suffices to show that every term in the sum in (2.6.6) has rapid decay in λ as $\lambda \to \infty$.

To this end, we fix a k and we pick a j such that the support of $\psi\zeta_k$ is contained in some ball $B(y_j, r_{y_j})$. We find unit vectors $\theta_{y_j,2}, \dots, \theta_{y_j,n}$, such that the system $\{\theta_{y_j}, \theta_{y_j,2}, \dots, \theta_{y_j,n}\}$ is an orthonormal basis of $\mathbf{R}^n$. Let e_j be the unit (column) vector on $\mathbf{R}^n$ whose jth coordinate is one and whose remaining coordinates are zero. We find an orthogonal matrix R such that $R^t e_1 = \theta_{y_j}$ and we introduce the change of variables $u = y_j + R(x - y_j)$ in the integral

$$I_k(\lambda) = \int_{\mathbf{R}^n} e^{i\lambda\varphi(x)} \psi(x)\zeta_k(x)\, dx.$$

The map $x \mapsto u = (u_1, \dots, u_n)$ is a rotation that fixes y_j and preserves the ball $B(y_j, r_{y_j})$. Defining $\varphi(x) = \varphi^o(u)$, $\psi(x) = \psi^o(u)$, $\zeta_k(x) = \zeta_k^o(u)$, under this new coordinate system we write

$$I_k(\lambda) = \int_K \left\{ \int_{\mathbf{R}} e^{i\lambda\varphi^o(u)} \psi^o(u_1, \dots, u_n)\zeta_k^o(u_1, \dots, u_n)\, du_1 \right\} du_2 \cdots du_n, \tag{2.6.7}$$

where K is a compact subset of $\mathbf{R}^{n-1}$. Since R is an orthogonal matrix, $R^{-1} = R^t$, and the change of variables $x = y_j + R^t(u - y_j)$ implies that

$$\frac{\partial x}{\partial u_1} = \text{first column of } R^t = \text{first row of } R = R^t e_1 = \theta_{y_j}.$$

Thus for all $x \in B(y_j, r_j)$ we have

$$\frac{\partial \varphi^o(u)}{\partial u_1} = \frac{\partial \varphi(y_j + R^t(u - y_j))}{\partial u_1} = \nabla \varphi(x) \cdot \frac{\partial x}{\partial u_1} = \nabla \varphi(x) \cdot \theta_{y_j} \geq c > 0$$

in view of condition (2.6.5). This lower estimate is valid for all $u \in B(y_j, r_{y_j})$, and therefore the inner integral inside the curly brackets in (2.6.7) is at most $C_N \lambda^{-N}$ by estimate (2.6.3). Integrating over K results in the same conclusion for $I(\lambda)$ defined in (2.6.4). □

2.6.2 Sublevel Set Estimates and the Van der Corput Lemma

We discuss a sharp decay estimate for one-dimensional oscillatory integrals. This estimate is obtained as a consequence of delicate size estimates for the Lebesgue measures of the sublevel sets $\{|u| \leq \alpha\}$ for a function u. In what follows, $u^{(k)}$ denotes the kth derivative of a function $u(t)$ defined on $\mathbf{R}$, and $\mathscr{C}^k$ the space of all functions whose kth derivative exists and is continuous.

Lemma 2.6.5. *Let $k \geq 1$ and suppose that $a_0, \dots, a_k$ are distinct real numbers. Let $a = \min(a_j)$ and $b = \max(a_j)$ and let f be a real-valued $\mathscr{C}^{k-1}$ function on $[a,b]$ that is $\mathscr{C}^k$ on (a,b). Then there exists a point y in (a,b) such that*

$$\sum_{m=0}^{k} c_m f(a_m) = f^{(k)}(y),$$

where $c_m = (-1)^k k! \prod_{\substack{\ell=0 \\ \ell \neq m}}^{k} (a_\ell - a_m)^{-1}$.

Proof. Suppose we could find a polynomial $p_k(x) = \sum_{j=0}^{k} b_j x^j$ such that the function

$$\varphi(x) = f(x) - p_k(x) \tag{2.6.8}$$

satisfies $\varphi(a_m) = 0$ for all $0 \leq m \leq k$. Since the a_j are distinct, we apply Rolle's theorem k times to find a point y in (a,b) such that

$$f^{(k)}(y) = k!\, b_k .$$

The existence of a polynomial p_k such that (2.6.8) is satisfied is equivalent to the existence of a solution to the matrix equation

$$\begin{pmatrix} a_0^k & a_0^{k-1} & \dots & a_0 & 1 \\ a_1^k & a_1^{k-1} & \dots & a_1 & 1 \\ \vdots & \vdots & \vdots & \vdots & \vdots \\ a_{k-1}^k & a_{k-1}^{k-1} & \dots & a_{k-1} & 1 \\ a_k^k & a_k^{k-1} & \dots & a_k & 1 \end{pmatrix} \begin{pmatrix} b_k \\ b_{k-1} \\ \vdots \\ b_1 \\ b_0 \end{pmatrix} = \begin{pmatrix} f(a_0) \\ f(a_1) \\ \vdots \\ f(a_{k-1}) \\ f(a_k) \end{pmatrix}.$$

The determinant of the square matrix on the left is called the *Vandermonde determinant* and is equal to

$$\prod_{\ell=0}^{k-1} \prod_{j=\ell+1}^{k} (a_\ell - a_j) \neq 0.$$

Since the a_j are distinct, it follows that the system has a unique solution. Using Cramer's rule, we solve this system to obtain

$$\begin{aligned} b_k &= \sum_{m=0}^{k} (-1)^m f(a_m) \frac{\prod\limits_{\substack{\ell=0 \\ \ell\neq m}}^{k-1} \prod\limits_{\substack{j=\ell+1 \\ j\neq m}}^{k} (a_\ell - a_j)}{\prod\limits_{\ell=0}^{k-1} \prod\limits_{j=\ell+1}^{k} (a_\ell - a_j)} \\ &= \sum_{m=0}^{k} (-1)^m f(a_m) \prod_{\substack{\ell=0 \\ \ell\neq m}}^{k} (a_\ell - a_m)^{-1} (-1)^{k-m}. \end{aligned}$$

The required conclusion now follows with c_m as claimed. □

Lemma 2.6.6. *Let E be a measurable subset of $\mathbf{R}$ with finite nonzero Lebesgue measure and let $k \in \mathbf{Z}^+$. Then there exist $a_0, \dots, a_k$ in E such that for all $\ell = 0, 1, \dots, k$ we have*

$$\prod_{\substack{j=0 \\ j\neq\ell}}^{k} |a_j - a_\ell| \geq (|E|/2e)^k. \tag{2.6.9}$$

Proof. Given a measurable set E with finite measure, pick a compact subset E' of E such that $|E \setminus E'| < \delta$, for some $\delta > 0$. For $x \in \mathbf{R}$ define $T(x) = |(-\infty, x) \cap E'|$. Then T enjoys the distance-decreasing property

$$|T(x) - T(y)| \leq |x - y|$$

for all $x, y \in E'$; consequently, by the intermediate value theorem, T is a surjective map from E' to $[0, |E'|]$. Let a_j be points in E' such that $T(a_j) = \frac{j}{k}|E'|$ for $j = 0, 1, \dots, k$. For k an even integer, we have

$$\prod_{\substack{j=0 \\ j\neq\ell}}^{k} |a_j - a_\ell| \geq \prod_{\substack{j=0 \\ j\neq\ell}}^{k} \left| \frac{j}{k}|E'| - \frac{\ell}{k}|E'| \right| \geq \prod_{\substack{j=0 \\ j\neq\frac{k}{2}}}^{k} \left| \frac{j}{k} - \frac{1}{2} \right| |E'|^k = \prod_{r=0}^{\frac{k}{2}-1} \left(\frac{r - \frac{k}{2}}{k} \right)^2 |E'|^k,$$

and it is easily shown that $\left((k/2)!\right)^2 k^{-k} \geq (2e)^{-k}$.

For k an odd integer we have

$$\prod_{\substack{j=0\\ j\neq\ell}}^{k}|a_j-a_\ell|\geq\prod_{\substack{j=0\\ j\neq\ell}}^{k}\left|\frac{j}{k}|E'|-\frac{\ell}{k}|E'|\right|\geq\prod_{\substack{j=0\\ j\neq\frac{k+1}{2}}}^{k}\left|\frac{j}{k}-\frac{k+1}{2k}\right||E'|^k,$$

while the last product is at least

$$\left\{\frac{1}{k}\cdot\frac{2}{k}\cdots\frac{\frac{k-1}{2}}{k}\right\}^2\frac{k+1}{2k}\geq(2e)^{-k}.$$

We have therefore proved (2.6.9) with E' replacing E. Since $|E\setminus E'|<\delta$ and $\delta>0$ is arbitrarily small, the required conclusion follows. □

The following is the main result of this section.

Proposition 2.6.7. *(a) Let u be a real-valued $\mathscr{C}^k$ function, $k\in\mathbf{Z}^+$, that satisfies $u^{(k)}(t)\geq 1$ for all $t\in\mathbf{R}$. Then the following estimate is valid for all $\alpha>0$:*

$$\left|\{t\in\mathbf{R}:\ |u(t)|\leq\alpha\}\right|\leq(2e)((k+1)!)^{\frac{1}{k}}\alpha^{\frac{1}{k}}.\tag{2.6.10}$$

(b) Let $-\infty<a<b<\infty$. For all $k\geq 2$, for every real-valued $\mathscr{C}^k$ function u on the line that satisfies $u^{(k)}(t)\geq 1$ for all $t\in[a,b]$, and every $\lambda\in\mathbf{R}\setminus\{0\}$ we have:

$$\left|\int_a^b e^{i\lambda u(t)}\,dt\right|\leq 12k|\lambda|^{-\frac{1}{k}}.\tag{2.6.11}$$

(c) If $k=1$, $u'(t)$ is monotonic on (a,b), and $u'(t)\geq 1$ for all $t\in(a,b)$, then for all nonzero real numbers λ we have

$$\left|\int_a^b e^{i\lambda u(t)}\,dt\right|\leq 3|\lambda|^{-1}.\tag{2.6.12}$$

Proof. Part (a): Let $E=\{t\in\mathbf{R}:\ |u(t)|\leq\alpha\}$. If $|E|$ is nonzero, then by Lemma 2.6.6 there exist $a_0,a_1,\dots,a_k$ in E such that for all ℓ we have

$$|E|^k\leq(2e)^k\prod_{\substack{j=0\\ j\neq\ell}}^{k}|a_j-a_\ell|.\tag{2.6.13}$$

Lemma 2.6.5 implies that there exists $y\in\left(\min a_j,\max a_j\right)$ such that

$$u^{(k)}(y)=(-1)^k k!\sum_{m=0}^{k}u(a_m)\prod_{\substack{\ell=0\\ \ell\neq m}}^{k}(a_\ell-a_m)^{-1}.\tag{2.6.14}$$

Using (2.6.13), we obtain that the expression on the right in (2.6.14) is in absolute value at most

$$(k+1)! \max_{0\le j\le k} |u(a_j)|\,(2e)^k\,|E|^{-k} \le (k+1)!\,\alpha\,(2e)^k\,|E|^{-k},$$

since $a_j \in E$. The bound $u^{(k)}(t) \ge 1$ now implies

$$|E|^k \le (k+1)!\,(2e)^k\,\alpha$$

as claimed. This proves (2.6.10).

Part (b): We now take $k \ge 2$ and we split the interval (a,b) in (2.6.11) into the sets

$$\begin{aligned} R_1 &= \{t \in (a,b) : \ |u'(t)| \le \beta\},\\ R_2 &= \{t \in (a,b) : \ |u'(t)| > \beta\}, \end{aligned}$$

for some parameter β to be chosen momentarily. The function $v = u'$ satisfies $v^{(k-1)} \ge 1$ and $k-1 \ge 1$. It follows from part (a) that

$$\left| \int_{R_1} e^{i\lambda u(t)}\,dt \right| \le |R_1| \le 2e\,(k!)^{\frac{1}{k-1}}\,\beta^{\frac{1}{k-1}} \le 6k\,\beta^{\frac{1}{k-1}}.$$

To obtain the corresponding estimate over R_2, we note that if $u^{(k)} \ge 1$, then the set $\{|u'| > \beta\}$ is the union of at most $2k-2$ intervals on each of which u' is monotone. Let (c,d) be one of these intervals on which u' is monotone. Then u' has a fixed sign on (c,d) and we have

$$\begin{aligned} \left| \int_c^d e^{i\lambda u(t)}\,dt \right| &= \left| \int_c^d \big(e^{i\lambda u(t)}\big)' \frac{1}{\lambda u'(t)}\,dt \right| \\ &\le \left| \int_c^d e^{i\lambda u(t)} \Big(\frac{1}{\lambda u'(t)}\Big)'\,dt \right| + \frac{1}{|\lambda|}\left| \frac{e^{i\lambda u(d)}}{u'(d)} - \frac{e^{i\lambda u(c)}}{u'(c)} \right| \\ &\le \frac{1}{|\lambda|} \int_c^d \left| \Big(\frac{1}{u'(t)}\Big)' \right| dt + \frac{2}{|\lambda|\beta} \\ &= \frac{1}{|\lambda|} \left| \int_c^d \Big(\frac{1}{u'(t)}\Big)'\,dt \right| + \frac{2}{|\lambda|\beta} \\ &= \frac{1}{|\lambda|} \left| \frac{1}{u'(d)} - \frac{1}{u'(c)} \right| + \frac{2}{|\lambda|\beta} \\ &\le \frac{3}{|\lambda|\beta}, \end{aligned}$$

where we use the monotonicity of $1/u'(t)$ in moving the absolute value from inside the integral to outside. It follows that

$$\left| \int_{R_2} e^{i\lambda u(t)}\,dt \right| \le \frac{6k}{|\lambda|\beta}.$$

Choosing $\beta = |\lambda|^{-(k-1)/k}$ to optimize and adding the corresponding estimates for R_1 and R_2, we deduce the claimed estimate (2.6.11).

Part (c): Repeat the argument in part (b) setting $\beta = 1$ and replacing the interval (c,d) by (a,b). □

Corollary 2.6.8. *Let (a,b), $u(t)$, $\lambda > 0$, and k be as in Proposition 2.6.7. Then for any function ψ on (a,b) with an integrable derivative and $k \geq 2$, we have*

$$\left| \int_a^b e^{i\lambda u(t)} \psi(t)\, dt \right| \leq 12k\lambda^{-1/k} \left[|\psi(b)| + \int_a^b |\psi'(s)|\, ds \right].$$

We also have

$$\left| \int_a^b e^{i\lambda u(t)} \psi(t)\, dt \right| \leq 3\lambda^{-1} \left[|\psi(b)| + \int_a^b |\psi'(s)|\, ds \right],$$

when $k = 1$ and u' is monotonic on (a,b).

Proof. Set

$$F(x) = \int_a^x e^{i\lambda u(t)}\, dt$$

and use integration by parts to write

$$\int_a^b e^{i\lambda u(t)} \psi(t)\, dt = F(b)\psi(b) - \int_a^b F(t)\psi'(t)\, dt.$$

The conclusion easily follows. □

Example 2.6.9. The *Bessel function* of order m is defined as

$$J_m(r) = \frac{1}{2\pi} \int_0^{2\pi} e^{ir\sin\theta} e^{-im\theta}\, d\theta.$$

Here we take both r and m to be real numbers, and we suppose that $m > -\frac{1}{2}$; we refer to Appendix B for an introduction to Bessel functions and their basic properties.

We use Corollary 2.6.8 to calculate the decay of the Bessel function $J_m(r)$ as $r \to \infty$. Set

$$\varphi(\theta) = \sin(\theta)$$

and note that $\varphi'(\theta)$ vanishes only at $\theta = \pi/2$ and $3\pi/2$ inside the interval $[0, 2\pi]$ and that $\varphi''(\pi/2) = -1$, while $\varphi''(3\pi/2) = 1$. We now write $1 = \psi_1 + \psi_2 + \psi_3$, where ψ_1 is smooth and compactly supported in a small neighborhood of $\pi/2$, and ψ_2 is smooth and compactly supported in a small neighborhood of $3\pi/2$. For $j = 1, 2$, Corollary 2.6.8 yields

$$\left| \int_0^{2\pi} e^{ir\sin(\theta)} (\psi_j(\theta) e^{-im\theta})\, d\theta \right| \leq C m r^{-1/2}$$

for some constant C, while the corresponding integral containing ψ_3 has arbitrary decay in r in view of estimate (2.6.3) (or Proposition 2.6.4 when $n=1$).

Exercises

2.6.1. Suppose that u is a real-valued $\mathscr{C}^k$ function defined on the line that satisfies $|u^{(k)}(t)| \ge c_0 > 0$ for some $k \ge 2$ and all $t \in (a,b)$. Prove that for $\lambda \in \mathbf{R} \setminus \{0\}$ we have

$$\left| \int_a^b e^{i\lambda u(t)}\, dt \right| \le 12k\,(\lambda c_0)^{-1/k}$$

and that the same conclusion is valid when $k=1$, provided u' is monotonic.

2.6.2. Show that if u' is not monotonic in part (c) of Proposition 2.6.7, then the conclusion may fail.
[*Hint:* Let $\varphi(t)$ be a real-valued smooth function that is equal to $2t$ on intervals $[2\pi k + \varepsilon_k, 2\pi(k+\frac{1}{2}) - \varepsilon_k]$ and equal to t on intervals $[2\pi(k+\frac{1}{2}) + \varepsilon_k, 2\pi(k+1) - \varepsilon_k]$, where $0 \le k \le N$, for some $N \in \mathbf{Z}^+$. Show that the absolute value of the integral of $e^{i\varphi(t)}$ over the interval $[\varepsilon_0, 2\pi(N+1) - \varepsilon_N]$ tends to infinity as $N \to \infty$.]

2.6.3. Prove that the dependence on k of the constant in part (b) of Proposition 2.6.7 is indeed linear.
[*Hint:* Take $u(t) = t^k/k!$ over the interval $(0,k!)$.]

2.6.4. Follow the steps below to give an alternative proof of part (b) of Proposition 2.6.7. Assume that the statement is known for some $k \ge 2$ and some constant $C(k)$ for all intervals $[a,b]$ and all $\mathscr{C}^k$ functions satisfying $u^{(k)} \ge 1$ on $[a,b]$. Fix a $\mathscr{C}^{k+1}$ function u such that $u^{(k+1)} \ge 1$ on an interval $[a,b]$. Let c be the unique point at which the function $u^{(k)}$ attains its minimum in $[a,b]$.
(a) If $u^{(k)}(c) = 0$, then for all $\delta > 0$ we have $u^{(k)}(t) \ge \delta$ in the complement of the interval $(c-\delta, c+\delta)$ and derive the bound

$$\left| \int_a^b e^{i\lambda u(t)} dt \right| \le 2C(k)(\lambda\delta)^{-1/k} + 2\delta .$$

(b) If $u^{(k)}(c) \ne 0$, then we must have $c \in \{a,b\}$. Obtain the bound

$$\left| \int_a^b e^{i\lambda u(t)} dt \right| \le C(k)(\lambda\delta)^{-1/k} + \delta .$$

(c) Choose a suitable δ to optimize and deduce the validity of the statement for $k+1$ with $C(k+1) = 2C(k)+2$, hence $C(k) = 3 \cdot 2^{k-1} + 2^k - 2$, since $C(1) = 3$.

2.6.5. (a) Prove that for some constant C and all $\lambda \in \mathbf{R}$ and $\varepsilon \in (0,1)$ we have

$$\left| \int_{\varepsilon \le |t| \le 1} e^{i\lambda t} \frac{dt}{t} \right| \le C .$$

(b) Prove that for some $C' < \infty$, all $\lambda \in \mathbf{R}$, $k > 0$, and $\varepsilon \in (0,1)$ we have

$$\left| \int_{\varepsilon \le |t| \le 1} e^{i\lambda t \pm t^k} \frac{dt}{t} \right| \le C' .$$

(c) Show that there is a constant C'' such that for any $0 < \varepsilon < N < \infty$, for all ξ_1, ξ_2 in $\mathbf{R}$, and for all integers $k \ge 2$, we have

$$\left| \int_{\varepsilon \le |s| \le N} e^{i(\xi_1 s + \xi_2 s^k)} \frac{ds}{s} \right| \le C'' .$$

[*Hint:* Part (a): For $|\lambda|$ small use the inequality $|e^{i\lambda t} - 1| \le |\lambda t|$. If $|\lambda|$ is large, split the domains of integration into the regions $|t| \le |\lambda|^{-1}$ and $|t| \ge |\lambda|^{-1}$ and use integration by parts in the second case. Part (b): Write

$$\frac{e^{i(\lambda t \pm t^k)} - 1}{t} = e^{i\lambda t}\, \frac{e^{\pm i t^k} - 1}{t} + \frac{e^{i\lambda t}}{t}$$

and use part (a). Part (c): When $\xi_1 = \xi_2 = 0$ it is trivial. If $\xi_2 = 0$, $\xi_1 \neq 0$, change variables $t = \xi_1 s$ and then split the domain of integration into the sets $|t| \le 1$ and $|t| \ge 1$. In the interval over the set $|t| \le 1$ apply part (b) and over the set $|t| \ge 1$ use integration by parts. In the case $\xi_2 \neq 0$, change variables $t = |\xi_2|^{1/k} s$ and split the domain of integration into the sets $|t| \ge 1$ and $|t| \le 1$. When $|t| \le 1$ use part (b) and in the case $|t| \ge 1$ use Corollary 2.6.8, noting that $\frac{d^k(\xi_1 |\xi_2|^{-1/k} t \pm t^k)}{dt} = k! \ge 1$.]

2.6.6. (a) Show that for all $a > 0$ and $\lambda > 0$ the following is valid:

$$\left| \int_0^{a\lambda} e^{i\lambda \log t}\, dt \right| \le a .$$

(b) Prove that there is a constant $c > 0$ such that for all $b > \lambda > 10$ we have

$$\left| \int_0^b e^{i\lambda t \log t}\, dt \right| \le \frac{c}{\lambda \log \lambda} .$$

[*Hint:* Part (b): Consider the intervals $(0,\delta)$ and $[\delta, b)$ for some δ. Apply Proposition 2.6.7 with $k = 1$ on one of these intervals and with $k = 2$ on the other. Then choose a suitable δ.]

2.6.7. Show that there is a constant $C < \infty$ such that for all nonintegers $\gamma > 1$ and all $\lambda, b > 1$ we have

$$\left| \int_0^b e^{i\lambda t^\gamma} dt \right| \le \frac{C}{\lambda^\gamma} .$$

[*Hint:* On the interval $(0,\delta)$ apply Proposition 2.6.7 with $k = [\gamma] + 1$ and on the interval (δ, b) with $k = [\gamma]$. Then optimize by choosing $\delta = \lambda^{-1/\gamma}$.]

HISTORICAL NOTES

The one-dimensional maximal function originated in the work of Hardy and Littlewood [146]. Its n-dimensional analogue was introduced by Wiener [375], who used Lemma 2.1.5, a variant of the Vitali covering lemma, to derive its L^p boundedness. One may consult the books of de Guzmán [92], [93] for extensions and other variants of such covering lemmas. The actual covering lemma proved by Vitali [368] says that if a family of closed cubes in $\mathbf{R}^n$ has the property that for every point $x \in A \subseteq \mathbf{R}^n$ there exists a sequence of cubes in the family that tends to x, then it is always possible to extract a sequence of pairwise disjoint cubes E_j from the family such that $|A \setminus \bigcup_j E_j| = 0$. We refer to Saks [310] for details and extensions of this theorem.

The class $L\log L$ was introduced by Zygmund to give a sufficient condition on the local integrability of the Hardy–Littlewood maximal operator. The necessity of this condition was observed by Stein [336]. Stein [341] also showed that the $L^p(\mathbf{R}^n)$ norm of the centered Hardy–Littlewood maximal operator $\mathcal{M}$ is bounded above by some dimension-free constant; see also Stein and Strömberg [345]. Analogous results for maximal operators associated with convex bodies are contained in Bourgain [35], Carbery [51], and Müller [263]. Bourgain [37] showed the the Hardy-Littlewood maximal operator associated with cubes is bounded on $L^p(\mathbf{R}^n)$ with dimension-free bounds when $p > 1$. Aldaz [2] studied the corresponding weak type $(1,1)$ bounds and proved that they grow to infinity with the dimension; the constant was improved by Aubrun [15]. The situation for the uncentered maximal operator M on L^p is different, since given any $1 < p < \infty$ there exists $C_p > 1$ such that $\|M\|_{L^p(\mathbf{R}^n)\to L^p(\mathbf{R}^n)} \ge C_p^n$ (see Exercise 2.1.8 for a value of such a constant C_p and also the article of Grafakos and Montgomery-Smith [136] for a larger value).

The centered maximal function $\mathcal{M}_\mu$ with respect to a general inner regular locally finite positive measure μ on $\mathbf{R}^n$ is bounded on $L^p(\mathbf{R}^n,\mu)$ without the additional hypothesis that the measure is doubling; see Fefferman [117]. The proof of this result requires the following covering lemma, obtained by Besicovitch [27]: Given any family of closed balls whose centers form a bounded subset of $\mathbf{R}^n$, there exists an at most countable subfamily of balls that covers the set of centers and has bounded overlap, i.e., no point in $\mathbf{R}^n$ belongs to more than a finite number (depending on the dimension) of the balls in the subfamily. A similar version of this lemma was obtained independently by Morse [258]. See also Ziemer [385] for an alternative formulation. The uncentered maximal operator M_μ of Exercise 2.1.1 may not be weak type $(1,1)$ if the measure μ is nondoubling, as shown by Sjögren [323]; related positive weak type $(1,1)$ results are contained in the article of Vargas [365]. The precise value of the operator norm of the uncentered Hardy–Littlewood maximal function on $L^p(\mathbf{R})$ was shown by Grafakos and Montgomery-Smith [136] to be the unique positive solution of the equation $(p-1)x^p - px^{p-1} - 1 = 0$. This constant raised to the power n is the operator norm of the strong maximal function M_s on $L^p(\mathbf{R}^n)$ for $1 < p \le \infty$. The best weak type $(1,1)$ constant for the centered Hardy–Littlewood maximal operator was shown by Melas [248] to be the largest root of the quadratic equation $12x^2 - 22x + 5 = 0$. The strong maximal operator M_s is not weak type $(1,1)$, but it satisfies the substitute inequality $d_{M_s(f)}(\alpha) \le C\int_{\mathbf{R}^n} \frac{|f(x)|}{\alpha}(1+\log^+\frac{|f(x)|}{\alpha})^{n-1}\,dx$. This result is due to Jessen, Marcinkiewicz, and Zygmund [176], but a geometric proof of it was obtained by Córdoba and Fefferman [73].

The basic facts about the Fourier transform go back to Fourier [119]. The theory of distributions was developed by Schwartz [314], [315]. For a concise introduction to the theory of distributions we refer to Hörmander [160] and Yosida [382]. Homogeneous distributions were considered by Riesz [295] in the study of the Cauchy problem in partial differential equations, although some earlier accounts are found in the work of Hadamard. They were later systematically studied by Gelfand and Šilov [126], [127]. References on the uncertainty principle include the articles of Fefferman [114] and Folland and Sitaram [118]. The best possible constant B_p in the Hausdorff–Young inequality $\|\widehat{f}\|_{L^{p'}(\mathbf{R}^n)} \le B_p\|f\|_{L^p(\mathbf{R}^n)}$ when $1 \le p \le 2$ was shown by Beckner [21] to be $B_p = (p^{1/p}(p')^{-1/p'})^{n/2}$. This best constant was previously obtained by Babenko [16] in the case when p' is an even integer.

A nice treatise of the spaces $\mathscr{M}^{p,q}$ is found in Hörmander [159]. This reference also contains Theorem 2.5.6, which is due to him. Theorem 2.5.16 is due to de Leeuw [94], but the proof presented here is taken from Jodeit [178]. De Leeuw's result in Exercise 2.5.9 says that periodic elements of $\mathscr{M}_p(\mathbf{R}^n)$ can be isometrically identified with elements of $\mathscr{M}(\mathbf{T}^n)$, the latter being the space of all multipliers on $\ell^p(\mathbf{Z}^n)$. The hint in Exercise 2.5.13 was suggested by M. Peloso.

Parts (b) and (c) of Proposition 2.6.7 are due to van der Corput [364] and are referred to in the literature as van der Corput's lemma. The refinement in part (a) was subsequently obtained by Arhipov, Karachuba, and Čubarikov [8]. The treatment of these results in the text is based on the article of Carbery, Christ, and Wright [53], which also investigates higher-dimensional analogues of the theory. Precise asymptotics can be obtained for a variety of oscillatory integrals via the method of stationary phase; see Hörmander [160]. References on oscillatory integrals include the books of Titchmarsh [362], Erdélyi [107], Zygmund [388], [389], Stein [344], and Sogge [328]. The latter provides a treatment of Fourier integral operators.

Chapter 3
Fourier Series

Principles of Fourier series go back to ancient times. The attempts of the Pythagorean school to explain musical harmony in terms of whole numbers embrace early elements of a trigonometric nature. The theory of epicycles in the *Almagest* of Ptolemy, based on work related to the circles of Appolonius, contains ideas of astronomical periodicities that we would interpret today as harmonic analysis. Early studies of acoustical and optical phenomena, as well as periodic astronomical and geophysical occurrences, provided a stimulus in the physical sciences toward the rigorous study of expansions of periodic functions. This study is carefully pursued in this chapter.

The modern theory of Fourier series begins with attempts to solve boundary value problems using trigonometric functions. The work of d'Alembert, Bernoulli, Euler, and Clairaut on the vibrating string led to the belief that it might be possible to represent arbitrary periodic functions as sums of sines and cosines. Fourier announced belief in this possibility in his solution of the problem of heat distribution in spatial bodies (in particular, for the cube $\mathbf{T}^3$) by expanding an arbitrary function of three variables as a triple sine series. Fourier's approach, although heuristic, was appealing and eventually attracted attention. It was carefully studied and further developed by many scientists, but most notably by Laplace and Dirichlet, who were the first to investigate the validity of the representation of a function in terms of its Fourier series. This is the main topic of study in this chapter.

3.1 Fourier Coefficients

We discuss some basic facts of Fourier analysis on the torus $\mathbf{T}^n$. Throughout this chapter, n denotes the dimension, i.e., a fixed positive integer.

L. Grafakos, *Classical Fourier Analysis*, Graduate Texts in Mathematics 249, DOI 10.1007/978-1-4939-1194-3_3, © Springer Science+Business Media New York 2014

3.1.1 The n-Torus $\mathbf{T}^n$

The n-torus $\mathbf{T}^n$ is the cube $[0,1]^n$ with opposite sides identified. This means that the points $(x_1,\dots,0,\dots,x_n)$ and $(x_1,\dots,1,\dots,x_n)$ are identified whenever 0 and 1 appear in the same coordinate. A more precise definition can be given as follows: We say that x,y in $\mathbf{R}^n$ are equivalent and we write

$$x \equiv y \tag{3.1.1}$$

if $x-y \in \mathbf{Z}^n$. Here $\mathbf{Z}^n$ is the additive subgroup of all points in $\mathbf{R}^n$ with integer coordinates. It is a simple fact that $\equiv$ is an equivalence relation that partitions $\mathbf{R}^n$ into equivalence classes. The n-torus $\mathbf{T}^n$ is then defined as the set $\mathbf{R}^n/\mathbf{Z}^n$ of all such equivalence classes. When $n=1$, this set can be geometrically viewed as a circle by bending the line segment $[0,1]$ so that its endpoints are brought together. When $n=2$, the identification brings together the left and right sides of the unit square $[0,1]^2$ as well as the top and bottom sides. The resulting figure is a two-dimensional manifold embedded in $\mathbf{R}^3$ that looks like a donut. See Figure 3.1.

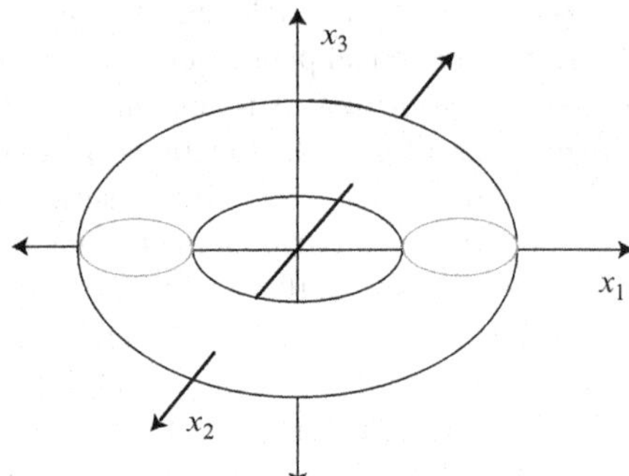

Fig. 3.1 The graph of the two-dimensional torus $\mathbf{T}^2$.

The n-torus is an additive group. The identity element of the group is 0, which of course coincides with every $e_j = (0,\dots,0,1,0,\dots,0)$. To avoid multiple appearances of the identity element in the group, we often think of the n-torus as the set $[-1/2,1/2]^n$. Since the group $\mathbf{T}^n$ is additive, the inverse of an element $x \in \mathbf{T}^n$ is denoted by $-x$. For example, $-(1/3,1/4) \equiv (2/3,3/4)$ on $\mathbf{T}^2$, or, equivalently, $-(1/3,1/4)-(2/3,3/4) \in \mathbf{Z}^2$.

The n-torus $\mathbf{T}^n$ can also be thought of as the following subset of $\mathbf{C}^n$,

$$\{(e^{2\pi i x_1},\dots,e^{2\pi i x_n}) \in \mathbf{C}^n \,:\, (x_1,\dots,x_n) \in [0,1]^n\}, \tag{3.1.2}$$

in a way analogous to which the unit interval $[0,1]$ can be thought of as the unit circle in $\mathbf{C}$ once 1 and 0 are identified.

Functions on $\mathbf{T}^n$ are functions f on $\mathbf{R}^n$ that satisfy $f(x+m)=f(x)$ for all $x \in \mathbf{R}^n$ and $m \in \mathbf{Z}^n$. Such functions are called 1-*periodic* in every coordinate. Haar measure on the n-torus is the restriction of n-dimensional Lebesgue measure to the set

$\mathbf{T}^n = [0,1]^n$. This measure is still denoted by dx, while the measure of a set $A \subseteq \mathbf{T}^n$ is denoted by $|A|$. Translation invariance of Lebesgue measure and the periodicity of functions on $\mathbf{T}^n$ imply that for all integrable functions f on $\mathbf{T}^n$, we have

$$\int_{\mathbf{T}^n} f(x)\,dx = \int_{[-1/2,1/2]^n} f(x)\,dx = \int_{[a_1,1+a_1]\times\cdots\times[a_n,1+a_n]} f(x)\,dx \tag{3.1.3}$$

for any real numbers $a_1,\dots,a_n$. In view of periodicity, integration by parts on the torus does not produce boundary terms; given f,g continuously differentiable functions on $\mathbf{T}^n$ we have

$$\int_{\mathbf{T}^n} \partial_j f(x)\,g(x)\,dx = -\int_{\mathbf{T}^n} \partial_j g(x)\,f(x)\,dx.$$

Elements of $\mathbf{Z}^n$ are denoted by $m = (m_1,\dots,m_n)$. For $m \in \mathbf{Z}^n$, we define the *total size* of m to be the number $|m| = (m_1^2+\cdots+m_n^2)^{1/2}$. Recall that for $x = (x_1,\dots,x_n)$ and $y = (y_1,\dots,y_n)$ in $\mathbf{R}^n$,

$$x \cdot y = x_1y_1 + \cdots + x_ny_n$$

denotes the usual dot product. Finally, for $x \in \mathbf{T}^n$, $|x|$ denotes the usual Euclidean norm of x. If we identify $\mathbf{T}^n$ with $[-1/2,1/2]^n$, then $|x|$ can be interpreted as the distance of the element x from the origin, and then we have $0 \le |x| \le \sqrt{n}/2$ for all $x \in \mathbf{T}^n$.

Multi-indices are elements of $(\mathbf{Z}^+ \cup \{0\})^n$. For a multi-index $\alpha = (\alpha_1,\dots,\alpha_n)$, we denote the partial derivative $\partial_1^{\alpha_1}\cdots\partial_n^{\alpha_n} f$ by $\partial^\alpha f$. The spaces $\mathscr{C}^k(\mathbf{T}^n)$ of *continuously differentiable functions of order k*, where $k \in \mathbf{Z}^+$, are defined as the sets of functions φ for which $\partial^\alpha\varphi$ exist and are continuous for all $|\alpha| \le k$. When $k=0$ we set $\mathscr{C}^0(\mathbf{T}^n) = \mathscr{C}(\mathbf{T}^n)$ to be the space of continuous functions on $\mathbf{T}^n$. The space $\mathscr{C}^\infty(\mathbf{T}^n)$ of infinitely differentiable functions on $\mathbf{T}^n$ is the union of all the $\mathscr{C}^k(\mathbf{T}^n)$. All of these spaces are contained in $L^p(\mathbf{T}^n)$, which are nested, with $L^1(\mathbf{T}^n)$ being the largest.

3.1.2 Fourier Coefficients

Definition 3.1.1. For a complex-valued function f in $L^1(\mathbf{T}^n)$ and m in $\mathbf{Z}^n$, we define

$$\widehat{f}(m) = \int_{\mathbf{T}^n} f(x)e^{-2\pi i m\cdot x}dx. \tag{3.1.4}$$

We call $\widehat{f}(m)$ the mth *Fourier coefficient* of f. We note that $\widehat{f}(\xi)$ is not defined for $\xi \in \mathbf{R}^n \setminus \mathbf{Z}^n$, since the function $x \mapsto e^{-2\pi i\xi\cdot x}$ is not 1-periodic in any coordinate and therefore not well defined on $\mathbf{T}^n$. For a finite Borel measure μ on $\mathbf{T}^n$ and $m \in \mathbf{Z}^n$ the expression

$$\widehat{\mu}(m) = \int_{\mathbf{T}^n} e^{-2\pi i m\cdot x}d\mu \tag{3.1.5}$$

is called the mth *Fourier coefficient* of μ.

The *Fourier series* of f at $x \in \mathbf{T}^n$ is the series

$$\sum_{m\in\mathbf{Z}^n} \widehat{f}(m)e^{2\pi i m\cdot x}. \tag{3.1.6}$$

It is not clear at present in which sense and for which $x \in \mathbf{T}^n$ (3.1.6) converges. The study of convergence of Fourier series is the main topic of study in this chapter.

We quickly recall the notation we introduced in Chapter 2. We denote by $\overline{f}$ the complex conjugate of the function f, by $\widetilde{f}$ the function $\widetilde{f}(x) = f(-x)$, and by $\tau^y(f)$ the function $\tau^y(f)(x) = f(x-y)$ for all $y \in \mathbf{T}^n$. We mention some elementary properties of Fourier coefficients.

Proposition 3.1.2. *Let f, g be in $L^1(\mathbf{T}^n)$. Then for all $m,k \in \mathbf{Z}^n$, $\lambda \in \mathbf{C}$, $y \in \mathbf{T}^n$, and all multi-indices α we have*

(1) $\widehat{f+g}(m) = \widehat{f}(m) + \widehat{g}(m)$,

(2) $\widehat{\lambda f}(m) = \lambda \widehat{f}(m)$,

(3) $\widehat{\overline{f}}(m) = \overline{\widehat{f}(-m)}$,

(4) $\widehat{\widetilde{f}}(m) = \widehat{f}(-m)$,

(5) $\widehat{\tau^y(f)}(m) = \widehat{f}(m)e^{-2\pi i m\cdot y}$,

(6) $(e^{2\pi i k(\cdot)}f)\widehat{\ }(m) = \widehat{f}(m-k)$,

(7) $\widehat{f}(0) = \int_{\mathbf{T}^n} f(x)\,dx$,

(8) $\sup_{m\in\mathbf{Z}^n} |\widehat{f}(m)| \le \|f\|_{L^1(\mathbf{T}^n)}$,

(9) $\widehat{f*g}(m) = \widehat{f}(m)\widehat{g}(m)$,

(10) $\widehat{\partial^\alpha f}(m) = (2\pi i m)^\alpha \widehat{f}(m)$, *whenever* $f \in \mathscr{C}^\alpha$.

Proof. The proof of properties (1)–(10) is rather easy and is left to the reader. We only sketch the proof of (9). We have

$$\widehat{f*g}(m) = \int_{\mathbf{T}^n}\int_{\mathbf{T}^n} f(x-y)g(y)e^{-2\pi i m\cdot(x-y)}e^{-2\pi i m\cdot y}\,dy\,dx = \widehat{f}(m)\widehat{g}(m),$$

where the interchange of integrals is justified by the absolute convergence of the integrals and Fubini's theorem. □

Remark 3.1.3. The Fourier coefficients have the following property. For a function f_1 on $\mathbf{T}^{n_1}$ and a function f_2 on $\mathbf{T}^{n_2}$, the tensor function

$$(f_1 \otimes f_2)(x_1,x_2) = f_1(x_1)f_2(x_2)$$

is a periodic function on $\mathbf{T}^{n_1+n_2}$ whose Fourier coefficients are

$$\widehat{f_1 \otimes f_2}(m_1, m_2) = \widehat{f_1}(m_1)\widehat{f_2}(m_2), \tag{3.1.7}$$

for all $m_1 \in \mathbf{Z}^{n_1}$ and $m_2 \in \mathbf{Z}^{n_2}$.

Definition 3.1.4. A *trigonometric polynomial* on $\mathbf{T}^n$ is a function of the form

$$P(x) = \sum_{m \in \mathbf{Z}^n} a_m e^{2\pi i m \cdot x}, \tag{3.1.8}$$

where $\{a_m\}_{m \in \mathbf{Z}^n}$ is a finitely supported sequence in $\mathbf{Z}^n$. The *degree* of P is the largest number $|q_1| + \cdots + |q_n|$ such that a_q is nonzero, where $q = (q_1, \dots, q_n)$. Observe that in view of the orthonormality of the exponentials we have for all $m \in \mathbf{Z}^n$

$$\widehat{P}(m) = a_m .$$

Example 3.1.5. If the sequence $\{a_m\}_m$ has only one nonzero term, then the trigonometric polynomial of Definition 3.1.4 reduces to a *trigonometric monomial*, which has the form

$$P(x) = a\, e^{2\pi i (q_1 x_1 + \cdots + q_n x_n)}$$

for some $q = (q_1, \dots, q_n) \in \mathbf{Z}^n$ and $a \in \mathbf{C}$.

Let

$$P(x) = \sum_{|m| \le N} a_m e^{2\pi i m \cdot x} = \sum_{|m| \le N} \widehat{P}(m) e^{2\pi i m \cdot x}$$

be a trigonometric polynomial on $\mathbf{T}^n$ and let μ be a finite Borel measure on $\mathbf{T}^n$. Then we have

$$(P * \mu)(x) = \int_{\mathbf{T}^n} \sum_{|m| \le N} \widehat{P}(m) e^{2\pi i m \cdot (x-y)} \, d\mu(y) = \sum_{|m| \le N} \widehat{P}(m) \widehat{\mu}(m) e^{2\pi i m \cdot x}. \tag{3.1.9}$$

In particular, if f is an integrable function on $\mathbf{T}^n$ we have

$$(P * f)(x) = \int_{\mathbf{T}^n} f(y) \sum_{|m| \le N} \widehat{P}(m) e^{2\pi i m \cdot (x-y)} \, dy = \sum_{|m| \le N} \widehat{P}(m) \widehat{f}(m) e^{2\pi i m \cdot x}. \tag{3.1.10}$$

This implies that the partial sums

$$\sum_{|m| \le N} \widehat{f}(m) e^{2\pi i m \cdot x}$$

of the Fourier series of f in (3.1.6) can be obtained by convolving f with the function

$$D_N(x) = \sum_{|m| \le N} e^{2\pi i m \cdot x}. \tag{3.1.11}$$

This function is called the *Dirichlet kernel*.

3.1.3 The Dirichlet and Fejér Kernels

Definition 3.1.6. Let $0 \le R < \infty$. The *square Dirichlet kernel* on $\mathbf{T}^n$ is the function

$$D_R^n(x) = \sum_{\substack{m \in \mathbf{Z}^n \\ |m_j| \le R}} e^{2\pi i m \cdot x}. \tag{3.1.12}$$

The *circular (or spherical) Dirichlet kernel* on $\mathbf{T}^n$ is the function

$$\overset{\circ}{D}{}_R^n(x) = \sum_{\substack{m \in \mathbf{Z}^n \\ |m| \le R}} e^{2\pi i m \cdot x}. \tag{3.1.13}$$

In dimension $n = 1$ these functions coincide and are denoted by

$$D_R(x) = D_R^1(x) = \overset{\circ}{D}{}_R^1(x).$$

This function is called the *Dirichlet kernel* and coincides with $D_N(x)$ in (3.1.11) when $N \le R < N+1$ and $N \in \mathbf{Z}^+ \cup \{0\}$; see Figure 3.2.

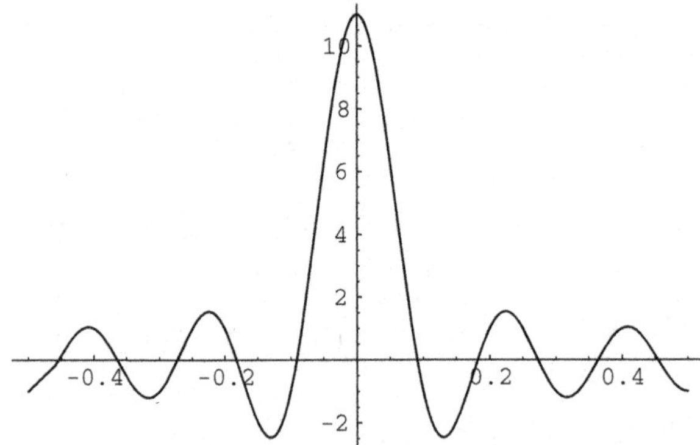

Fig. 3.2 The graph of the Dirichlet kernel D_5 plotted on the interval $[-1/2, 1/2]$.

Both the square and circular (or spherical) Dirichlet kernels are trigonometric polynomials. The square Dirichlet kernel on $\mathbf{T}^n$ is equal to a product of one-dimensional Dirichlet kernels, that is,

$$D_R^n(x_1, \dots, x_n) = D_R(x_1) \cdots D_R(x_n). \tag{3.1.14}$$

We have the following two equivalent ways to write the Dirichlet kernel D_N:

$$D_N(x) = \sum_{|m| \le N} e^{2\pi i m \cdot x} = \frac{\sin((2N+1)\pi x)}{\sin(\pi x)}, \tag{3.1.15}$$

for $x \in [0,1]$. To verify the validity of (3.1.15), we write

$$\sum_{|m|\le N} e^{2\pi i m\cdot x} = e^{-2\pi i N x}\frac{e^{2\pi i(2N+1)x}-1}{e^{2\pi i x}-1} = \frac{e^{2\pi i(N+1)x}-e^{-2\pi i N x}}{e^{\pi i x}(e^{\pi i x}-e^{-\pi i x})} = \frac{\sin((2N+1)\pi x)}{\sin(\pi x)}.$$

It follows that for $R \in \mathbf{R}^+ \cup \{0\}$ we have

$$D_R(x) = \frac{\sin(\pi x(2[R]+1))}{\sin(\pi x)}. \tag{3.1.16}$$

It is reasonable to ask whether the family $\{D_R\}_{R>0}$ forms an approximate identity as $R \to \infty$. Using (3.1.15) we see that each D_R is integrable over $[-1/2,1/2]$ and has integral equal to 1. But it follows from Exercise 3.1.5 that $\|D_R\|_{L^1} \approx \log R$ as $R \to \infty$, and therefore property (i) in Definition 1.2.15 fails for D_R. We conclude that the family $\{D_R\}_{R>0}$ is not an approximate identity on $\mathbf{T}^1$, which significantly complicates the study of Fourier series. Consequently, the family $\{D_R^n\}_{R>0}$ is not an approximate identity on $\mathbf{T}^n$, since $\|D_R^n\|_{L^1(\mathbf{T}^1)} \approx (\log R)^n$. The same is true for the family of circular (or spherical) Dirichlet kernels $\{\overset{\circ}{D}{}_R^n\}_{R>0}$. Although this is harder to prove, it will be a consequence of the results in Section 4.2.

A typical situation encountered in analysis is that the means of a sequence behave better than the original sequence. This fact led Cesàro and independently Fejér to consider the arithmetic means of the Dirichlet kernel in dimension 1, that is, the expressions

$$F_N(x) = \frac{1}{N+1}\big[D_0(x)+D_1(x)+D_2(x)+\cdots+D_N(x)\big]. \tag{3.1.17}$$

The expression in (3.1.17) is in fact equal to the Fejér kernel given in Example 1.2.18. We have the following identity concerning the kernel F_N.

Proposition 3.1.7. *For every nonnegative integer N the identity holds*

$$F_N(x) = \sum_{j=-N}^{N}\left(1-\frac{|j|}{N+1}\right)e^{2\pi i j x} = \frac{1}{N+1}\left(\frac{\sin(\pi(N+1)x)}{\sin(\pi x)}\right)^2 \tag{3.1.18}$$

for all $x \in \mathbf{T}^1$. Thus $\widehat{F_N}(m) = 1-\frac{|m|}{N+1}$ if $|m| \le N$ and zero otherwise.

Proof. The fact that the expression in (3.1.17) is equal to the middle term in (3.1.18) is a consequence of the trivial calculation:

$$\frac{1}{N+1}\sum_{k=0}^{N} D_k(x) = \frac{1}{N+1}\sum_{k=0}^{N}\sum_{|j|\le k} e^{2\pi i j x} = \sum_{|j|\le N}\frac{\#\{k\in\mathbf{Z}:\ |j|\le k\le N\}}{N+1}e^{2\pi i j x}.$$

To verify the second equality in (3.1.18) we use the simple geometric series identity $1+r+r^2+\cdots+r^N = \frac{1-r^{N+1}}{1-r}$ to write for $x \neq 0$

$$\sum_{j=1}^{N} e^{2\pi i j x} = \frac{e^{2\pi i (N+1)x} - 1}{e^{2\pi i x} - 1} - 1 = \frac{e^{i\pi(N+1)x}}{e^{i\pi x}} \frac{e^{\pi i (N+1)x} - e^{-\pi i (N+1)x}}{e^{\pi i x} - e^{-\pi i x}} - 1$$

from which it follows that

$$\sum_{j=1}^{N} |j| e^{2\pi i j x} = \frac{1}{2\pi i} \frac{d}{dx} \left(e^{i\pi N x} \frac{\sin(\pi(N+1)x)}{\sin(\pi x)} \right). \tag{3.1.19}$$

Likewise we prove that

$$\sum_{j=-N}^{-1} |j| e^{2\pi i j x} = -\frac{1}{2\pi i} \frac{d}{dx} \left(e^{-i\pi N x} \frac{\sin(\pi(N+1)x)}{\sin(\pi x)} \right). \tag{3.1.20}$$

Adding (3.1.19) and (3.1.20) we deduce

$$\sum_{|j| \le N} |j| e^{2\pi i j x} = \frac{1}{\pi} \frac{d}{dx} \left(\frac{\sin(\pi N x) \sin(\pi(N+1)x)}{\sin(\pi x)} \right). \tag{3.1.21}$$

Multiplying (3.1.21) by $-\frac{1}{N+1}$ and adding $D_N(x)$ we obtain

$$\sum_{j=-N}^{N} \left(1 - \frac{|j|}{N+1}\right) e^{2\pi i j x} = \frac{\sin(\pi(2N+1)x)}{\sin(\pi x)} - \frac{1}{\pi} \frac{d}{dx} \left(\sin(\pi N x) \frac{\sin(\pi(N+1)x)}{(N+1)\sin(\pi x)} \right).$$

Writing the preceding expression on the right as

$$\frac{(N+1)\sin(\pi(N+1)x)\cos(\pi N x)\sin(\pi x) + (N+1)\cos(\pi(N+1)x)\sin(\pi N x)\sin(\pi x)}{(N+1)\sin^2(\pi x)}$$

$$-\frac{1}{\pi} \frac{\frac{d}{dx}\{\sin(\pi N x)\sin(\pi(N+1)x)\}\sin(\pi x) - \{\sin(\pi N x)\sin(\pi(N+1)x)\}\pi\cos(\pi x)}{(N+1)\sin^2(\pi x)},$$

computing the derivative of the expression in the curly brackets, and simplifying, we finally obtain that

$$\sum_{j=-N}^{N} \left(1 - \frac{|j|}{N+1}\right) e^{2\pi i j x} = \frac{1}{N+1} \left(\frac{\sin(\pi(N+1)x)}{\sin(\pi x)} \right)^2. \tag{3.1.22}$$

This proves the second identity in (3.1.18). □

Definition 3.1.8. Let N be a nonnegative integer. The function F_N on $\mathbf{T}^1$ given by (3.1.22) is called the *Fejér kernel*.

The Fejér kernel F_N^n on $\mathbf{T}^n$ is defined as the product of the 1-dimensional Fejér kernels, or as the average of the product of the Dirichlet kernels in each variable, precisely, $F_N^1(x) = F_N(x)$ and

$$\begin{aligned} F_N^n(x_1,\dots,x_n) &= \prod_{j=1}^{n} F_N(x_j) \\ &= \prod_{j=1}^{n}\left(\frac{1}{N+1}\sum_{k=0}^{N} D_k(x_j)\right) \\ &= \frac{1}{(N+1)^n}\sum_{k_1=0}^{N}\cdots\sum_{k_n=0}^{N} D_{k_1}(x_1)\cdots D_{k_n}(x_n). \end{aligned}$$

Note that F_N^n is a trigonometric polynomial of degree nN.

Remark 3.1.9. Using the identities for F_N in (3.1.18), we may write for all $N \ge 0$

$$F_N^n(x_1,\dots,x_n) = \sum_{\substack{m\in\mathbf{Z}^n \\ |m_j|\le N}} \left(1-\frac{|m_1|}{N+1}\right)\cdots\left(1-\frac{|m_n|}{N+1}\right) e^{2\pi i m\cdot x} \tag{3.1.23}$$

$$= \frac{1}{(N+1)^n}\prod_{j=1}^{n}\left(\frac{\sin(\pi(N+1)x_j)}{\sin(\pi x_j)}\right)^2, \tag{3.1.24}$$

thus $F_N^n \ge 0$. Observe that $F_0^n(x) = 1$ for all $x \in \mathbf{T}^n$ and that $F_N^n(0) = (N+1)^n$.

Proposition 3.1.10. *The family of Fejér kernels $\{F_N^n\}_{N=0}^{\infty}$ is an approximate identity on $\mathbf{T}^n$.*

Proof. Since $F_N^n \ge 0$ we have that $\|F_N^n\|_{L^1} = \int_{\mathbf{T}^n} F_N^n\,dx$. Also $\int_{\mathbf{T}^n} F_N^n\,dx = 1$, in view of identity (3.1.23). Thus properties (i) and (ii) of approximate identities (according to Definition 1.2.15) hold. To prove property (iii) of the definition we make use of identity (3.1.24). Using the fact that $1 \le \frac{|t|}{|\sin t|} \le \frac{\pi}{2}$ when $|t| \le \frac{\pi}{2}$, we obtain

$$F_N(x) \le \frac{1}{N+1}\min\left(\frac{(N+1)|\pi x|}{|\sin(\pi x)|}, \frac{1}{|\sin(\pi x)|}\right)^2 \le \frac{1}{N+1}\frac{\pi^2}{4}\min\left(N+1, \frac{1}{|\pi x|}\right)^2$$

when $|x| \le \frac{1}{2}$. This implies that for $\delta > 0$ we have

$$\int_{\delta\le|x|\le\frac{1}{2}} F_N(x)\,dx \le \frac{1}{N+1}\frac{\pi^2}{4}\int_{\delta\le|x|\le\frac{1}{2}} \frac{dx}{|\pi\delta|^2} \le \frac{1}{4\delta^2}\frac{1}{N+1} \to 0$$

as $N \to \infty$. In higher dimensions, given $x = (x_1,\dots,x_n) \in [-1/2,1/2]^n$ with $|x| \ge \delta$, there is a $j \in \{1,\dots,n\}$ such that $|x_j| \ge \delta/\sqrt{n}$ and thus

$$\int_{\substack{x\in\mathbf{T}^n \\ |x|\ge\delta}} F_N^n(x)\,dx \le \sum_{j=1}^{n}\int_{|x_j|\ge\frac{\delta}{\sqrt{n}}} F_N(x_j)\,dx_j \prod_{k\ne j}\int_{\mathbf{T}^1} F_N(x_k)\,dx_k \le \frac{n}{4(\delta/\sqrt{n})^2}\frac{1}{N+1} \to 0.$$

This proves the claim. □

Exercises

3.1.1. Identify $\mathbf{T}^1$ with $[-1/2, 1/2]$ and let $h(t)$ be an integrable function on $\mathbf{T}^1$.
(a) If $h(t) \geq 0$ is even, show that $\widehat{h}(m)$ is real and $|\widehat{h}(m)| \leq \widehat{h}(0)$ for all $m \in \mathbf{Z}$.
(b) If $h(t)$ is odd and $h(t) \geq 0$ on $[0, 1/2)$, then $i\widehat{h}(m)$ is real and $|\widehat{h}(m)| \leq im\widehat{h}(1)$ for all $m \in \mathbf{Z}$.

3.1.2. Suppose that h is a periodic integrable function on $[-1/2, 1/2)$ with integral zero. Define another periodic function H on $\mathbf{T}^1$ by setting

$$H(x) = \int_{-1/2}^{x} h(t)\,dt\,.$$

Compute $\widehat{H}(m)$ in terms of $\widehat{h}(m)$ for $m \in \mathbf{Z}$.

3.1.3. Suppose that $\{g_\varepsilon\}_{\varepsilon>0}$ is an approximate identity on $\mathbf{R}^n$ as $\varepsilon \to 0$ and let

$$G_\varepsilon(x) = \sum_{\ell \in \mathbf{Z}^n} g_\varepsilon(x + \ell)\,.$$

Show that the family $\{G_\varepsilon\}_{\varepsilon>0}$ is an approximate identity on $\mathbf{T}^n$.

3.1.4. On $\mathbf{T}^1$ define the *de la Vallée Poussin kernel*

$$V_N(x) = 2F_{2N+1}(x) - F_N(x)\,.$$

(a) Show that the sequence V_N is an approximate identity.
(b) Prove that $\widehat{V_N}(m) = 1$ when $|m| \leq N+1$, and $\widehat{V_N}(m) = 0$ when $|m| \geq 2N+2$.

3.1.5. (a) Show that for all $|t| \leq \frac{\pi}{2}$ we have

$$\left|\frac{1}{\sin(t)} - \frac{1}{t}\right| \leq 1 - \frac{2}{\pi}\,.$$

(b) Let D_N be the Dirichlet kernel on $\mathbf{T}^1$. Prove that for $N \in \mathbf{Z}^+$ we have

$$\frac{4}{\pi^2} \sum_{k=1}^{N} \frac{1}{k} \leq \big\|D_N\big\|_{L^1} \leq 3 - \frac{2}{\pi} + \frac{4}{\pi^2} \sum_{k=1}^{N} \frac{1}{k}\,.$$

Conclude that the numbers $\|D_N\|_{L^1}$ grow logarithmically as $N \to \infty$ and therefore the family $\{D_N\}_{N=1}^{\infty}$ is not an approximate identity on $\mathbf{T}^1$. The numbers $\|D_N\|_{L^1}$, $N = 1, 2, \ldots$ are called the *Lebesgue constants*.
[*Hint:* Part (a): Show that the derivative of $\frac{1}{\sin(t)} - \frac{1}{t}$ is nonnegative on $(0, \frac{\pi}{2}]$, or equivalently prove that $\tan(t)\sin(t) \geq t^2$ on $(0, \frac{\pi}{2}]$; this is a consequence of the inequality $\sqrt{\sin(t)\tan(t)} \geq 2(\frac{1}{\sin(t)} + \frac{1}{\tan(t)})^{-1} = 2\tan(\frac{t}{2}) \geq t$. Part (b): Replace $D_N(t)$ by $\frac{\sin((2N+1)\pi t)}{\pi t}$ and estimate the difference using part (a).]

3.1.6. Let D_N be the Dirichlet kernel on $\mathbf{T}^1$. Prove that for all $1 < p < \infty$ there exist two constants $C_p, c_p > 0$ such that

$$c_p\,(2N+1)^{1/p'} \le \big\|D_N\big\|_{L^p} \le C_p\,(2N+1)^{1/p'}.$$

$\big[$*Hint:* Consider the two closest zeros of D_N near the origin and split the integral into the intervals thus obtained.$\big]$

3.1.7. The Poisson kernel on $\mathbf{T}^n$ is the function

$$P_{r_1,\dots,r_n}(x) = \sum_{m\in\mathbf{Z}^n} r_1^{|m_1|}\cdots r_n^{|m_n|} e^{2\pi i m\cdot x}$$

and is defined for $0 < r_1,\dots,r_n < 1$. Prove that $P_{r_1,\dots,r_n}$ can be written as

$$P_{r_1,\dots,r_n}(x_1,\dots,x_n) = \prod_{j=1}^{n} \mathrm{Re}\left(\frac{1+r_je^{2\pi i x_j}}{1-r_je^{2\pi i x_j}}\right) = \prod_{j=1}^{n} \frac{1-r_j^2}{1-2r_j\cos(2\pi x_j)+r_j^2},$$

and conclude that $P_{r,\dots,r}(x)$ is an approximate identity as $r\uparrow 1$.

3.2 Reproduction of Functions from Their Fourier Coefficients

We can obtain very interesting results using the Fejér kernel.

Proposition 3.2.1. *The set of trigonometric polynomials is dense in $L^p(\mathbf{T}^n)$ for $1\le p<\infty$.*

Proof. Given f in $L^p(\mathbf{T}^n)$ for $1\le p<\infty$, consider $f*F_N^n$. Clearly $f*F_N^n$ is also a trigonometric polynomial. In view of Theorem 1.2.19 (1), $f*F_N^n$ converges to f in L^p as $N\to\infty$. □

Corollary 3.2.2. (***Weierstrass approximation theorem for trigonometric polynomials***) *Every continuous function on the torus is a uniform limit of trigonometric polynomials.*

Proof. Since f is continuous on $\mathbf{T}^n$, which is a compact set, Theorem 1.2.19 (2) gives that $f*F_N^n$ converges uniformly to f as $N\to\infty$. But $f*F_N^n$ is a trigonometric polynomial, and so we conclude that every continuous function on $\mathbf{T}^n$ can be uniformly approximated by trigonometric polynomials. □

3.2.1 Partial sums and Fourier inversion

We now define the partial sums of Fourier series.

Definition 3.2.3. Let $R \geq 0$ and $N \in \mathbf{Z}^+ \cup \{0\}$. The expressions

$$(f * D_R^n)(x) = \sum_{\substack{m \in \mathbf{Z}^n \\ |m_j| \leq R}} \widehat{f}(m) e^{2\pi i m \cdot x}$$

are called the *square partial sums of the Fourier series* of f. Then the expressions

$$(f * \overset{\circ}{D}{}_R^n)(x) = \sum_{\substack{m \in \mathbf{Z}^n \\ |m| \leq R}} \widehat{f}(m) e^{2\pi i m \cdot x}$$

are called the *circular (or spherical) partial sums* of the Fourier series of f. The expressions

$$(f * F_N^n)(x) = \sum_{\substack{m \in \mathbf{Z}^n \\ |m_j| \leq N}} \left(1 - \frac{|m_1|}{N+1}\right) \cdots \left(1 - \frac{|m_n|}{N+1}\right) \widehat{f}(m) e^{2\pi i m \cdot x}$$

are called the *square Cesàro means* (or *square Fejér means*) of f.

Finally, for $R \geq 0$ the expressions

$$(f * \overset{\circ}{F}{}_R^n)(x) = \sum_{\substack{m \in \mathbf{Z}^n \\ |m| \leq R}} \left(1 - \frac{|m|}{R}\right) \widehat{f}(m) e^{2\pi i m \cdot x}$$

are called the *circular Cesàro means* (or *circular Fejér means*) of f.

Observe that $f * \overset{\circ}{F}{}_R^n$ is equal to the average of the expressions $f * \overset{\circ}{D}{}_r^n$ in the following sense:

$$(f * \overset{\circ}{F}{}_R^n)(x) = \frac{1}{R} \int_0^R (f * \overset{\circ}{D}{}_r^n)(x)\, dr.$$

This is analogous to the fact that the Fejér kernel F_N is the average of the Dirichlet kernels $D_0, D_1, \dots, D_N$.

A fundamental problem is in what sense the partial sums of the Fourier series converge back to the function as $R \to \infty$ or $N \to \infty$. This problem is of central importance in harmonic analysis and is in part investigated in this chapter.

We now ask the question whether the Fourier coefficients uniquely determine the function. The answer is affirmative and simple.

Proposition 3.2.4. *If $f, g \in L^1(\mathbf{T}^n)$ satisfy $\widehat{f}(m) = \widehat{g}(m)$ for all m in $\mathbf{Z}^n$, then $f = g$ a.e.*

Proof. By linearity of the problem, it suffices to assume that $g = 0$. If $\widehat{f}(m) = 0$ for all $m \in \mathbf{Z}^n$, Definition 3.2.3 implies that $F_N^n * f = 0$ for all $N \in \mathbf{Z}^+$. In view of Proposition 3.1.10, the sequence $\{F_N^n\}_{N \in \mathbf{Z}^+}$ is an approximate identity as $N \to \infty$. Therefore,

$$\left\| f - F_N^n * f \right\|_{L^1} \to 0$$

as $N \to \infty$; hence $\|f\|_{L^1} = 0$, from which we conclude that $f = 0$ a.e. □

A useful consequence of the result just proved is the following.

Proposition 3.2.5. *(**Fourier inversion**) Suppose that $f \in L^1(\mathbf{T}^n)$ and that*

$$\sum_{m\in\mathbf{Z}^n} |\widehat{f}(m)| < \infty .$$

Then

$$f(x) = \sum_{m\in\mathbf{Z}^n} \widehat{f}(m)e^{2\pi i m\cdot x} \qquad a.e., \tag{3.2.1}$$

and therefore f is almost everywhere equal to a continuous function.

Proof. It is straightforward to check that both functions in (3.2.1) are well defined and have the same Fourier coefficients. Therefore, they must be almost everywhere equal by Proposition 3.2.4. Moreover, the function on the right in (3.2.1) is everywhere continuous. □

3.2.2 Fourier series of square summable functions

Let H be a separable Hilbert space with complex inner product $\langle \cdot \,|\, \cdot \rangle$. Recall that a subset E of H is called *orthonormal* if $\langle f \,|\, g\rangle = 0$ for all f, g in E with $f \neq g$, while $\langle f \,|\, f\rangle = 1$ for all f in E. A *complete orthonormal system* is a subset of H having the additional property that the only vector orthogonal to all of its elements is the zero vector. We summarize basic properties about orthonormal systems in the proposition below (see [307]).

Proposition 3.2.6. *Let H be a separable Hilbert space and let $\{\varphi_k\}_{k\in\mathbf{Z}}$ be an orthonormal system in H. Then the following are equivalent:*
(1) $\{\varphi_k\}_{k\in\mathbf{Z}}$ is a complete orthonormal system.
(2) For every $f \in H$ we have

$$\|f\|_H^2 = \sum_{k\in\mathbf{Z}} |\langle f \,|\, \varphi_k\rangle|^2 .$$

(3) For every $f \in H$ we have

$$f = \lim_{N\to\infty} \sum_{|k|\le N} \langle f \,|\, \varphi_k\rangle \varphi_k ,$$

where the series converges in H.

Now consider the Hilbert space space $L^2(\mathbf{T}^n)$ with inner product

$$\langle f \,|\, g\rangle = \int_{\mathbf{T}^n} f(t)\overline{g(t)}\,dt .$$

Let φ_m be the sequence of functions $\xi \mapsto e^{2\pi i m\cdot\xi}$ indexed by $m \in \mathbf{Z}^n$. The orthonormality of the sequence $\{\varphi_m\}$ is a consequence of the following simple but crucial identity:

$$\int_{[0,1]^n} e^{2\pi i m\cdot x}\overline{e^{2\pi i k\cdot x}}\,dx = \begin{cases} 1 & \text{when } m = k, \\ 0 & \text{when } m \neq k. \end{cases}$$

The completeness of the sequence $\{\varphi_m\}$ is also evident. Since $\langle f \,|\, \varphi_m\rangle = \widehat{f}(m)$ for all $f \in L^2(\mathbf{T}^n)$, it follows from Proposition 3.2.4 that if $\langle f \,|\, \varphi_m\rangle = 0$ for all $m \in \mathbf{Z}^n$, then $f = 0$ a.e.

The next result is a consequence of Proposition 3.2.6.

Proposition 3.2.7. *The following are valid for $f, g \in L^2(\mathbf{T}^n)$:*
(1) ***(Plancherel's identity)***

$$\|f\|_{L^2}^2 = \sum_{m\in\mathbf{Z}^n} |\widehat{f}(m)|^2 .$$

(2) The function $f(t)$ is a.e. equal to the $L^2(\mathbf{T}^n)$ limit of the sequence

$$\lim_{M\to\infty} \sum_{|m|\le M} \widehat{f}(m) e^{2\pi i m\cdot t} .$$

(3) ***(Parseval's relation)***

$$\int_{\mathbf{T}^n} f(t)\overline{g(t)}\,dt = \sum_{m\in\mathbf{Z}^n} \widehat{f}(m)\overline{\widehat{g}(m)} .$$

(4) The map $f \mapsto \{\widehat{f}(m)\}_{m\in\mathbf{Z}^n}$ is an isometry from $L^2(\mathbf{T}^n)$ onto ℓ^2.
(5) For all $k \in \mathbf{Z}^n$ we have

$$\widehat{fg}(k) = \sum_{m\in\mathbf{Z}^n} \widehat{f}(m)\widehat{g}(k-m) = \sum_{m\in\mathbf{Z}^n} \widehat{f}(k-m)\widehat{g}(m) .$$

Proof. (1) and (2) follow from the corresponding statements in Proposition 3.2.6. Notice that both sides of (3) converge by the Cauchy-Schwarz inequality. Parseval's relation (3) follows from *polarization*. By this we mean the following procedure. First replace f by $f+g$ in (1) and expand the squares to obtain

$$\begin{aligned} \|f\|_{L^2}^2 + \|g\|_{L^2}^2 + 2\mathrm{Re}\,\langle f \,|\, g\rangle &= \|f+g\|_{L^2}^2 \\ &= \sum_{m\in\mathbf{Z}^n} |\widehat{f}(m) + \widehat{g}(m)|^2 \\ &= \sum_{m\in\mathbf{Z}^n} |\widehat{f}(m)|^2 + \sum_{m\in\mathbf{Z}^n} |\widehat{g}(m)|^2 + 2\mathrm{Re} \sum_{m\in\mathbf{Z}^n} \widehat{f}(m)\overline{\widehat{g}(m)} \end{aligned}$$

and from this it follows that the real parts of the expressions in (3) are equal. Next replace f by $f+ig$ in (1) and expand the squares. Using $\mathrm{Re}\,(-iw) = \mathrm{Im}\,w$ we obtain

$$
\begin{aligned}
\|f\|_{L^2}^2+\|g\|_{L^2}^2+2\mathrm{Im}\,\langle f\,|\,g\rangle &=\|f+ig\|_{L^2}^2 \\
&=\sum_{m\in\mathbf{Z}^n}|\widehat{f}(m)+i\widehat{g}(m)|^2 \\
&=\sum_{m\in\mathbf{Z}^n}|\widehat{f}(m)|^2+\sum_{m\in\mathbf{Z}^n}|\widehat{g}(m)|^2+2\mathrm{Im}\sum_{m\in\mathbf{Z}^n}\widehat{f}(m)\overline{\widehat{g}(m)},
\end{aligned}
$$

and thus the imaginary parts of the expressions in (3) are equal. Thus (3) holds. Next we prove (4). We already know that the map $f\mapsto\{\widehat{f}(m)\}_{m\in\mathbf{Z}^n}$ is an injective isometry. It remains to show that it is onto. Given a square summable sequence $\{a_m\}_{m\in\mathbf{Z}^n}$ of complex numbers, define

$$
f_N(t)=\sum_{|m|\le N}a_m e^{2\pi i m\cdot t}.
$$

Observe that f_N is a Cauchy sequence in $L^2(\mathbf{T}^n)$ and it therefore converges to some $f\in L^2(\mathbf{T}^n)$. Then we have $\widehat{f}(m)=a_m$ for all $m\in\mathbf{Z}^n$. Finally, (5) is a consequence of (3) and Proposition 3.1.2 (6) and (3). □

3.2.3 The Poisson Summation Formula

We end this section with an important result that connects Fourier analysis on the torus with Fourier analysis on $\mathbf{R}^n$. Suppose that f is an integrable function on $\mathbf{R}^n$ and let $\widehat{f}$ be its Fourier transform. Restrict $\widehat{f}$ on $\mathbf{Z}^n$ and form the "Fourier series" (assuming that it converges)

$$
\sum_{m\in\mathbf{Z}^n}\widehat{f}(m)e^{2\pi i m\cdot x}.
$$

What does this series represent? Since the preceding function is 1-periodic in every variable, it follows that it cannot be equal to f, unless f is itself periodic. However, it should not come as a surprise that it is in fact equal to the periodization of f on $\mathbf{R}^n$. In other words, the Fourier expansion of a function on $\mathbf{R}^n$ reproduces the periodization of the function.

Theorem 3.2.8. ***(Poisson summation formula)*** *Let f be a continuous function on $\mathbf{R}^n$ which satisfies for some $C,\delta>0$ and for all $x\in\mathbf{R}^n$*

$$
|f(x)|\le C(1+|x|)^{-n-\delta},
$$

and whose Fourier transform $\widehat{f}$ restricted on $\mathbf{Z}^n$ satisfies

$$
\sum_{m\in\mathbf{Z}^n}|\widehat{f}(m)|<\infty. \tag{3.2.2}
$$

Then for all $x \in \mathbf{R}^n$ we have

$$\sum_{m\in\mathbf{Z}^n} \widehat{f}(m)e^{2\pi i m\cdot x} = \sum_{k\in\mathbf{Z}^n} f(x+k), \tag{3.2.3}$$

and in particular

$$\sum_{m\in\mathbf{Z}^n} \widehat{f}(m) = \sum_{k\in\mathbf{Z}^n} f(k).$$

Proof. Define a 1-periodic function on $\mathbf{T}^n$ by setting

$$F(x) = \sum_{k\in\mathbf{Z}^n} f(x+k).$$

It is straightforward to verify that $\|F\|_{L^1([0,1]^n)} = \|f\|_{L^1(\mathbf{R}^n)}$, thus F lies in $L^1(\mathbf{T}^n)$. We prove that the sequence of the Fourier coefficients of F coincides with the restriction of the Fourier transform of f on $\mathbf{Z}^n$. This follows from the calculation

$$\begin{aligned}
\widehat{F}(m) &= \int_{\mathbf{T}^n} \sum_{k\in\mathbf{Z}^n} f(x+k)e^{-2\pi i m\cdot x}\,dx \\
&= \sum_{k\in\mathbf{Z}^n} \int_{\mathbf{T}^n} f(x+k)e^{-2\pi i m\cdot x}\,dx \\
&= \sum_{k\in\mathbf{Z}^n} \int_{[-\frac{1}{2},\frac{1}{2}]^n - k} f(x)e^{-2\pi i m\cdot x}\,dx \\
&= \int_{\mathbf{R}^n} f(x)e^{-2\pi i m\cdot x}\,dx \\
&= \widehat{f}(m),
\end{aligned}$$

in which the interchange of the sum and the integral is justified by the Weierstrass M-test of uniform convergence of series, since

$$\sum_{k\in\mathbf{Z}^n} \frac{1}{(1+|k+x|)^{n+\delta}} \le \sum_{k\in\mathbf{Z}^n} \frac{(1+\sqrt{n})^{n+\delta}}{(1+\sqrt{n}+|k+x|)^{n+\delta}} \le \sum_{k\in\mathbf{Z}^n} \frac{C_{n,\delta}}{(1+|k|)^{n+\delta}} < \infty,$$

where we used $|k+x| \ge |k| - |x| \ge |k| - \sqrt{n}$. This calculation also shows that F is the sum of a uniformly convergent series of continuous functions on $[0,1]^n$, thus it is itself continuous. It follows that (3.2.2) holds with $|\widehat{F}(m)|$ in place of $|\widehat{f}(m)|$. Hence, Proposition 3.2.5 applies, and given the fact that F is continuous, it yields conclusion (3.2.3) for all $x \in \mathbf{T}^n$ and, by periodicity, for all $x \in \mathbf{R}^n$. □

Example 3.2.9. We have seen earlier (Exercise 2.2.11) that the following identity gives the Fourier transform of the Poisson kernel in $\mathbf{R}^n$:

$$(e^{-2\pi|x|})^{\widehat{}}(\xi) = \frac{\Gamma(\frac{n+1}{2})}{\pi^{\frac{n+1}{2}}} \frac{1}{(1+|\xi|^2)^{\frac{n+1}{2}}}.$$

The Poisson summation formula yields the identity

$$\frac{\Gamma(\frac{n+1}{2})}{\pi^{\frac{n+1}{2}}}\sum_{k\in\mathbf{Z}^n}\frac{\varepsilon^{-n}}{(1+\frac{|k+x|^2}{\varepsilon^2})^{\frac{n+1}{2}}}=\sum_{m\in\mathbf{Z}^n}e^{-2\pi\varepsilon|m|}e^{-2\pi i m\cdot x} \tag{3.2.4}$$

which implies

$$\frac{\Gamma(\frac{n+1}{2})}{\pi^{\frac{n+1}{2}}}\sum_{k\in\mathbf{Z}^n}\frac{1}{(1+|k|^2)^{\frac{n+1}{2}}}=\sum_{m\in\mathbf{Z}^n}e^{-2\pi|m|}. \tag{3.2.5}$$

It follows from (3.2.4) that

$$\sum_{k\in\mathbf{Z}^n\setminus\{0\}}\frac{1}{(\varepsilon^2+|k|^2)^{\frac{n+1}{2}}}=\frac{1}{\varepsilon}\bigg(\frac{\pi^{\frac{n+1}{2}}}{\Gamma(\frac{n+1}{2})}\sum_{m\in\mathbf{Z}^n}e^{-2\pi\varepsilon|m|}-\frac{1}{\varepsilon^n}\bigg),$$

from which we obtain the identity

$$\sum_{k\in\mathbf{Z}^n\setminus\{0\}}\frac{1}{|k|^{n+1}}=\lim_{\varepsilon\to 0}\frac{1}{\varepsilon}\bigg(\frac{\pi^{\frac{n+1}{2}}}{\Gamma(\frac{n+1}{2})}\sum_{m\in\mathbf{Z}^n}e^{-2\pi\varepsilon|m|}-\frac{1}{\varepsilon^n}\bigg). \tag{3.2.6}$$

The limit in (3.2.6) can be easily calculated in dimension 1 using that

$$\lim_{\delta\to 0}\frac{\pi^2}{\delta}\bigg(\frac{1+e^{-2\delta}}{1-e^{-2\delta}}-\frac{1}{\delta}\bigg)=\frac{\pi^2}{3},$$

and this yields

$$\sum_{k\neq 0}\frac{1}{k^2}=\frac{\pi^2}{3}.$$

Also, in dimension 1, from (3.2.5) we obtain the related identity

$$\sum_{k\in\mathbf{Z}}\frac{1}{1+k^2}=\pi\sum_{m\in\mathbf{Z}}e^{-2\pi|m|}=\pi\,\frac{1+e^{-2\pi}}{1-e^{-2\pi}}.$$

Example 3.2.10. Let $0<\operatorname{Re}\alpha<n$. We introduce a smooth function $\widehat{\Phi}(\xi)$ which is equal to 1 on the ball $|\xi|\le 1$ and vanishes outside the ball $|\xi|\le 2$. We investigate the behavior as $x\to 0$ of the expression

$$\lim_{R\to\infty}\sum_{m\in\mathbf{Z}^n\setminus\{0\}}\frac{e^{2\pi i m\cdot x}}{|m|^\alpha}\widehat{\Phi}\Big(\frac{m}{R}\Big)$$

when $x\in[-\frac{1}{2},\frac{1}{2})^n\setminus\{0\}$. As in Example 2.4.9, let η be a smooth radial function on $\mathbf{R}^n$ that is equal to 1 outside the ball $B(0,1/2)$ and vanishes on the ball $B(0,1/4)$ and define

$$g=(\eta(\xi)|\xi|^{-\alpha})^{\widehat{}}.$$

Let $\Phi_\delta(x) = \delta^{-n}\Phi(x/\delta)$. The Poisson summation formula (Theorem 3.2.8) gives

$$\sum_{k\in\mathbf{Z}^n\setminus\{0\}} \frac{e^{2\pi i k\cdot x}}{|k|^\alpha}\widehat{\Phi}\left(\frac{k}{R}\right) = \sum_{k\in\mathbf{Z}^n} \frac{\eta(k)e^{2\pi i k\cdot x}}{|k|^\alpha}\widehat{\Phi}\left(\frac{k}{R}\right)$$
$$= (g*\Phi_{1/R})(x) + \sum_{m\in\mathbf{Z}^n\setminus\{0\}} (g*\Phi_{1/R})(x+m).$$

It was shown in Example 2.4.9 that $g(y)$ decays faster than the reciprocal of any polynomial at infinity and is equal to $\pi^{\alpha-\frac{n}{2}}\Gamma(\frac{n-\alpha}{2})\Gamma(\frac{\alpha}{2})^{-1}|y|^{\alpha-n} + h(y)$, where h is a smooth function on $\mathbf{R}^n$. Since $x \neq 0$, it follows that g is smooth in a small relatively compact neighborhood of x and, since $\{\Phi_{1/R}\}_{R>0}$ is an approximate identity, Theorem 1.2.19 (2) yields that $(g*\Phi_{1/R})(x) \to g(x)$ as $R\to\infty$. Assume for a moment that

$$\lim_{R\to\infty}\sum_{m\in\mathbf{Z}^n\setminus\{0\}} (g*\Phi_{1/R})(x+m) = \sum_{m\in\mathbf{Z}^n\setminus\{0\}} \lim_{R\to\infty}(g*\Phi_{1/R})(x+m). \tag{3.2.7}$$

Since $x+m$ does not vanish for any $m\in\mathbf{Z}^n\setminus\{0\}$, the function g is smooth in a relatively compact neighborhood of $x+m$, and thus $(g*\Phi_{1/R})(x+m)\to g(x+m)$ as $R\to\infty$. Consequently, the sum on the right in (3.2.7) is equal to

$$\sum_{m\in\mathbf{Z}^n\setminus\{0\}} g(x+m).$$

We conclude that

$$\lim_{R\to\infty}\sum_{m\in\mathbf{Z}^n\setminus\{0\}} \frac{e^{2\pi i m\cdot x}}{|m|^\alpha}\widehat{\Phi}\left(\frac{m}{R}\right) = \frac{\pi^{\alpha-\frac{n}{2}}\Gamma(\frac{n-\alpha}{2})}{\Gamma(\frac{\alpha}{2})}|x|^{\alpha-n} + h_1(x),$$

where h_1 is a $\mathscr{C}^\infty$ function on $[-\frac{1}{2},\frac{1}{2})^n$ given by

$$h_1(t) = h(t) + \sum_{m\in\mathbf{Z}^n\setminus\{0\}} g(t+m).$$

We now explain the passage of the limit inside the sum in (3.2.7). This is a consequence of the Lebesgue dominated convergence theorem, provided we know that

$$|(g*\Phi_{1/R})(x+m)| \le \frac{C}{|m|^{n+1}}, \qquad |m| > 5\sqrt{n} \tag{3.2.8}$$

for some constant C independent of $R\ge 1$ and of m. Indeed, the expression on the left of this inequality is bounded by $I+II$, where

$$I = C_{n,\alpha}\int_{|x+m-y|\le 2\sqrt{n}} |x+m-y|^{\alpha-n}\frac{R^n}{(1+R|y|)^{2n+2}}\,dy$$
$$II = C_{n,\alpha}\int_{|x+m-y|\ge 2\sqrt{n}} \frac{1}{(1+|x+m-y|)^{2n+2}}\frac{R^n}{(1+R|y|)^{2n+2}}\,dy.$$

In I we have

$$1+R|y| \geq R|x+m| - R|x+m-y| \geq R|m| - \tfrac{1}{2}R\sqrt{n} - 2R\sqrt{n} \geq \tfrac{1}{2}R|m|,$$

hence

$$I \leq C'_{n,\alpha} R^{-n-2} |m|^{-2n-2} \leq C'_{n,\alpha} |m|^{-2n-2}.$$

In II we use

$$(1+|x+m-y|)^{n+1}(1+R|y|)^{n+1} \geq (1+|m|)^{n+1}$$

while the term left produces a convergent integral, which is uniformly bounded in $R \geq 1$. This proves (3.2.8).

Exercises

3.2.1. On $\mathbf{T}^1$ let P be a trigonometric polynomial of degree $N > 0$. Show that P has at most $2N$ zeros. Construct a trigonometric polynomial with exactly $2N$ zeros.

3.2.2. (*Hausdorff–Young inequality*) Prove that when $f \in L^p(\mathbf{T}^n)$, $1 \leq p \leq 2$, the sequence of Fourier coefficients of f is in $\ell^{p'}(\mathbf{Z}^n)$ and

$$\Big(\sum_{m \in \mathbf{Z}^n} |\widehat{f}(m)|^{p'} \Big)^{1/p'} \leq \|f\|_{L^p(\mathbf{T}^n)}.$$

Also observe that 1 is the best constant in the preceding inequality.

3.2.3. Use without proof that there exists a constant $C > 0$ such that for all $t \in \mathbf{R}$ we have

$$\Big| \sum_{k=2}^{N} e^{ik\log k} e^{ikt} \Big| \leq C\sqrt{N}, \qquad N = 2,3,4,\dots,$$

to construct an example of a continuous function g on $\mathbf{T}^1$ with

$$\sum_{m \in \mathbf{Z}} |\widehat{g}(m)|^q = \infty$$

for all $q < 2$. Thus the Hausdorff–Young inequality of Exercise 3.2.2 fails for $p > 2$.
[*Hint:* Consider $g(x) = \sum_{k=2}^{\infty} \frac{e^{ik\log k}}{k^{1/2}(\log k)^2} e^{2\pi i k x}$. For a proof of the previous estimate, see Zygmund [388, Theorem (4.7) p. 199].]

3.2.4. (*S. Bernstein*) Let $P(x)$ be a trigonometric polynomial of degree N on $\mathbf{T}^1$. Prove that $\|P'\|_{L^\infty} \leq 4\pi N \|P\|_{L^\infty}$.
[*Hint:* Prove first that $P'(x)/2\pi i N$ is equal to

$$\big((e^{-2\pi i N(\cdot)}P) * F_{N-1}\big)(x)\, e^{2\pi i N x} - \big((e^{2\pi i N(\cdot)}P) * F_{N-1}\big)(x)\, e^{-2\pi i N x}$$

and then take L^∞ norms.]

3.2.5. (*Fejér and F. Riesz*) Let $P(\xi) = \sum_{k=-N}^{N} a_k e^{2\pi i k\xi}$ be a trigonometric polynomial on $\mathbf{T}^1$ of degree N such that $P(\xi) > 0$ for all ξ. Prove that there exists a trigonometric polynomial $Q(\xi)$ of the form $\sum_{k=0}^{N} b_k e^{2\pi i k\xi}$ such that $P(\xi) = |Q(\xi)|^2$.
[*Hint:* Since $P \geq 0$ the complex-variable polynomial $R(z) = \sum_{k=-N}^{N} a_k z^{k+N}$ must satisfy $R(z) = z^{2N}\overline{R(1/\overline{z})}$, and thus it must have N zeros inside the unit circle and the other N outside. Therefore we may write $R(z) = a_N \prod_{k=1}^{s} (z - z_k)^{r_k} (z - 1/\overline{z_k})^{r_k}$ for some $0 < |z_k| < 1$ and $r_k \geq 1$ with $\sum_{k=1}^{s} r_k = N$. Then take $z = e^{2\pi i \xi}$.]

3.2.6. Let g be a function on $\mathbf{R}^n$ that satisfies $|g(x)| + |\widehat{g}(x)| \leq C(1+|x|)^{-n-\delta}$ for some $C, \delta > 0$ and all $x \in \mathbf{R}^n$. Prove that

$$\lambda^n \sum_{m \in \mathbf{Z}^n} \widehat{g}(\lambda m + \alpha) e^{2\pi i x \cdot (m + \frac{\alpha}{\lambda})} = \sum_{k \in \mathbf{Z}^n} g\Big(\frac{x+k}{\lambda}\Big) e^{-2\pi i \frac{k \cdot \alpha}{\lambda}}$$

for any $x, \alpha \in \mathbf{R}^n$ and $\lambda > 0$.

3.2.7. Verify the following identity when $0 < r < 1$ and $x \in \mathbf{R}^n$

$$\frac{\Gamma(\frac{n+1}{2})}{\pi^{\frac{n+1}{2}}} \sum_{k \in \mathbf{Z}^n} \frac{\frac{1}{2\pi}\log\frac{1}{r}}{\big((\frac{1}{2\pi}\log\frac{1}{r})^2 + |x-k|^2\big)^{\frac{n+1}{2}}} = \sum_{m \in \mathbf{Z}^n} r^{|m|} e^{2\pi i m \cdot x}.$$

In the special case $n = 1$ and $x \in \mathbf{R}$ we have

$$\frac{1}{\pi} \sum_{k \in \mathbf{Z}} \frac{\frac{1}{2\pi}\log\frac{1}{r}}{(\frac{1}{2\pi}\log\frac{1}{r})^2 + |x-k|^2} = \frac{1-r^2}{1 - 2r\cos(2\pi x) + r^2}.$$

[*Hint:* Use identity (3.2.4) and Exercise 3.1.7 when $n = 1$.]

3.2.8. Let $\gamma \in \mathbf{R}$ and $\lambda > 0$. Show that

$$\sum_{k \in \mathbf{Z}} \frac{\cos(2\pi k\gamma)}{\lambda^2 + k^2} = \frac{\pi}{\lambda} \frac{\cosh(2\pi\lambda(\gamma - [\gamma] - \frac{1}{2}))}{\sinh(\pi\lambda)}.$$

[*Hint:* Use Exercise 3.2.6 when $n = 1$ with $x = 0$, $\alpha = -\gamma\lambda$, $g(x) = \frac{1}{\pi}\frac{1}{1+x^2}$ and sum in m.]

3.3 Decay of Fourier Coefficients

In this section we investigate the interplay between the smoothness of a function and the decay of its Fourier coefficients.

3.3.1 Decay of Fourier Coefficients of Arbitrary Integrable Functions

We begin with the classical result asserting that the Fourier coefficients of any integrable function tend to zero at infinity. One should compare the following proposition with Proposition 2.2.17.

Proposition 3.3.1. ***(Riemann–Lebesgue lemma)*** *Given a function f in $L^1(\mathbf{T}^n)$, we have that $|\widehat{f}(m)| \to 0$ as $|m| \to \infty$.*

Proof. Given $f \in L^1(\mathbf{T}^n)$ and $\varepsilon > 0$, let P be a trigonometric polynomial such that $\|f - P\|_{L^1} < \varepsilon$. If $|m| > \text{degree}\,(P)$, then $\widehat{P}(m) = 0$ and thus

$$|\widehat{f}(m)| = |\widehat{f}(m) - \widehat{P}(m)| \le \|f - P\|_{L^1} < \varepsilon.$$

This proves that $|\widehat{f}(m)| \to 0$ as $|m| \to \infty$. □

Several questions are naturally raised. How fast can the Fourier coefficients of an L^1 function tend to zero? Does additional smoothness of the function imply faster decay of the Fourier coefficients? Can such a decay be quantitatively expressed in terms of the smoothness of the function?

We answer the first question. Fourier coefficients of an L^1 function may tend to zero arbitrarily slowly, that is, slower than any given rate of decay. To achieve this, we need the following two lemmas.

Lemma 3.3.2. *Given a sequence of positive real numbers $\{a_m\}_{m=0}^{\infty}$ that tends to zero as $m \to \infty$, there exists a sequence $\{c_m\}_{m=0}^{\infty}$ that satisfies*

$$c_m \ge a_m, \qquad c_m \downarrow 0, \quad \text{and} \quad c_{m+2} + c_m \ge 2c_{m+1} \tag{3.3.1}$$

for all $m = 0, 1, \ldots$. A sequence $\{c_m\}_{m=0}^{\infty}$ that satisfies (3.3.1) *is called convex.*

Proof. Let $k_0 = 0$ and suppose that $a_m \le M$ for all $m \ge 0$. Find $k_1 > k_0$ such that for $m \ge k_1$ we have $a_m \le M/2$. Now find $k_2 > k_1 + \frac{k_1 - k_0}{2}$ such that for $m \ge k_2$ we have $a_m \le M/4$. Next find $k_3 > k_2 + \frac{k_2 - k_1}{2}$ such that for $m \ge k_3$ we have $a_m \le M/8$. Continue inductively in this way and construct a subsequence $k_0 < k_1 < k_2 < \cdots$ of the integers such that for $m \ge k_{j+1}$ we have $a_m \le 2^{-j-1}M$ and $k_{j+1} > k_j + \frac{k_j - k_{j-1}}{2}$ for $j \ge 1$. Join the points $(k_0, 2M)$, (k_1, M), $(k_2, M/2)$, $(k_3, M/4), \ldots$ by straight lines and note that by the choice of the sequence $\{k_j\}_{j=0}^{\infty}$ the resulting piecewise linear function h is convex on $[0, \infty)$. Define $c_m = h(m)$ and observe that the sequence $\{c_m\}_{m=0}^{\infty}$ satisfies the required properties. Exercise 3.3.1 contains an alternative proof.

□

Lemma 3.3.3. *Given a convex decreasing sequence $\{c_m\}_{m=0}^{\infty}$ of positive real numbers satisfying $\lim_{m \to \infty} c_m = 0$ and a fixed integer $s \ge 0$, we have that*

$$\sum_{r=0}^{\infty}(r+1)(c_{r+s}+c_{r+s+2}-2c_{r+s+1})=c_s\,. \tag{3.3.2}$$

Proof. We begin by observing the validity of the telescoping sum

$$\begin{aligned}&\sum_{r=0}^{N}(r+1)(c_{r+s}+c_{r+s+2}-2c_{r+s+1})\\ &\qquad\qquad =c_s-(N+1)(c_{s+N+1}-c_{s+N+2})-c_{s+N+1}\,.\end{aligned} \tag{3.3.3}$$

To show that the last expression tends to c_s as $N\to\infty$, we take $M=[\frac{N}{2}]$ and we use convexity $\big(c_{s+M+j}-c_{s+M+j+1}\ge c_{s+M+j+1}-c_{s+M+j+2}\big)$ to obtain

$$\begin{aligned}c_{s+M+1}-c_{s+N+2}&=c_{s+M+1}-c_{s+M+2}\\ &\quad+c_{s+M+2}-c_{s+M+3}\\ &\quad+\cdots\\ &\quad+c_{s+N+1}-c_{s+N+2}\\ &\ge(N-M+1)(c_{s+N+1}-c_{s+N+2})\\ &\ge\tfrac{N+1}{2}(c_{s+N+1}-c_{s+N+2})\ge 0\,.\end{aligned}$$

The preceding calculation implies that $(N+1)(c_{s+N+1}-c_{s+N+2})$ tends to zero as $N\to\infty$ and thus the expression in (3.3.3) converges to c_s as $N\to\infty$. □

The proof of Lemma 3.3.3 appears more natural after examining Exercise 3.3.3(a). We now state the theorem we alluded to earlier.

Theorem 3.3.4. *Let $(d_m)_{m\in\mathbf{Z}^n}$ be a sequence of positive real numbers with $d_m\to 0$ as $|m|\to\infty$. Then there exists a function $f\in L^1(\mathbf{T}^n)$ such that $\widehat{f}(m)\ge d_m$ for all $m\in\mathbf{Z}^n$. In other words, given any rate of decay, there exists an integrable function on the torus whose Fourier coefficients have slower rate of decay.*

Proof. We are given a sequence of positive numbers $\{a_m\}_{m\in\mathbf{Z}}$ that converges to zero as $|m|\to\infty$ and we would like to find an integrable function on $\mathbf{T}^1$ with $\widehat{f}(m)\ge a_m$ for all $m\in\mathbf{Z}$. Apply Lemma 3.3.2 to the sequence $\{a_m+a_{-m}\}_{m\ge0}$ to find a convex sequence $\{c_m\}_{m\ge0}$ that dominates $\{a_m+a_{-m}\}_{m\ge0}$ and decreases to zero as $m\to\infty$. Extend c_m for $m<0$ by setting $c_m=c_{|m|}$. Now define

$$f(x)=\sum_{j=0}^{\infty}(j+1)(c_j+c_{j+2}-2c_{j+1})F_j(x)\,, \tag{3.3.4}$$

where F_j is the (one-dimensional) Fejér kernel. The convexity of the sequence c_m and the positivity of the Fejér kernel imply that $f\ge 0$. Lemma 3.3.3 with $s=0$ gives that

$$\sum_{j=0}^{\infty}(j+1)(c_j+c_{j+2}-2c_{j+1})\big\|F_j\big\|_{L^1}=c_0<\infty\,, \tag{3.3.5}$$

since $\|F_j\|_{L^1} = 1$ for all j. Therefore (3.3.4) defines an integrable function f on $\mathbf{T}^1$. We now compute the Fourier coefficients of f. Since the series in (3.3.4) converges in L^1, for $m \in \mathbf{Z}$ we have

$$\begin{aligned} \widehat{f}(m) &= \sum_{j=0}^{\infty}(j+1)(c_j+c_{j+2}-2c_{j+1})\widehat{F_j}(m) \\ &= \sum_{j=|m|}^{\infty}(j+1)(c_j+c_{j+2}-2c_{j+1})\left(1-\frac{|m|}{j+1}\right) \\ &= \sum_{r=0}^{\infty}(r+1)(c_{r+|m|}+c_{r+|m|+2}-2c_{r+|m|+1}) \\ &= c_{|m|} = c_m\,, \end{aligned} \tag{3.3.6}$$

where we used Lemma 3.3.3 with $s = |m|$.

Let us now extend this result on $\mathbf{T}^n$. Let $(d_m)_{m\in\mathbf{Z}^n}$ be a positive sequence with $d_m \to 0$ as $|m| \to \infty$. By Exercise 3.3.2, there exists a positive sequence $(a_j)_{j\in\mathbf{Z}}$ with $a_{m_1}\cdots a_{m_n} \geq d_{(m_1,\dots,m_n)}$ and $a_j \to 0$ as $|j| \to \infty$. Let

$$\mathbf{f}(x_1,\dots,x_n) = f(x_1)\cdots f(x_n),$$

where f is the function previously constructed when $n=1$ so that $\widehat{f}(m) \geq a_m$. It can be seen easily using (3.1.7) that $\widehat{\mathbf{f}}(m) \geq d_m$. □

3.3.2 Decay of Fourier Coefficients of Smooth Functions

We next study the decay of the Fourier coefficients of functions that possess a certain amount of smoothness. In this section we see that the decay of the Fourier coefficients reflects the smoothness of the function in a rather precise quantitative way. Conversely, one can infer some information about the smoothness of a function from the decay of its Fourier coefficients.

Definition 3.3.5. Given $0<\gamma<1$ and f a function on $\mathbf{T}^n$, define the *homogeneous Lipschitz seminorm* of order γ of f by

$$\|f\|_{\dot{\Lambda}_\gamma} = \sup_{\substack{x,h\in\mathbf{T}^n \\ h\neq 0}} \frac{|f(x+h)-f(x)|}{|h|^\gamma}$$

and define the *homogeneous Lipschitz space* of order γ as

$$\dot{\Lambda}_\gamma(\mathbf{T}^n) = \{f : \mathbf{T}^n \to \mathbf{C} \text{ with } \|f\|_{\dot{\Lambda}_\gamma} < \infty\}.$$

Functions in $\dot{\Lambda}_\gamma(\mathbf{T}^n)$ are called *homogeneous Lipschitz* functions of order γ.

There is an analogous definition for the inhomogeneous norm.

Definition 3.3.6. For $0 < \gamma < 1$ and f a function on $\mathbf{T}^n$, define the *inhomogeneous Lipschitz norm* of order γ of f by

$$\big\|f\big\|_{\Lambda_\gamma} = \big\|f\big\|_{L^\infty} + \sup_{\substack{x,h\in\mathbf{T}^n\\ h\neq 0}} \frac{|f(x+h)-f(x)|}{|h|^\gamma} = \big\|f\big\|_{L^\infty} + \big\|f\big\|_{\dot{\Lambda}_\gamma}.$$

Also define the *inhomogeneous Lipschitz space* of order γ as

$$\Lambda_\gamma(\mathbf{T}^n) = \{f : \mathbf{T}^n \to \mathbf{C} \text{ with } \big\|f\big\|_{\Lambda_\gamma} < \infty\}.$$

Functions in $\Lambda_\gamma(\mathbf{T}^n)$ are called *inhomogeneous Lipschitz* functions of order γ.

Remark 3.3.7. Functions in both spaces $\Lambda_\gamma(\mathbf{T}^n)$ and $\dot{\Lambda}_\gamma(\mathbf{T}^n)$ are obviously continuous and therefore bounded. Moreover, the functional $\|\cdot\|_{\Lambda_\gamma}$ is a norm on $\Lambda_\gamma(\mathbf{T}^n)$. The positive functional $\|\cdot\|_{\dot{\Lambda}_\gamma}$ satisfies the triangle inequality, but it does not satisfy the property $\|f\|_{\dot{\Lambda}_\gamma} = 0 \implies f = 0$ required to be a norm. It is therefore a seminorm on $\dot{\Lambda}_\gamma(\mathbf{T}^n)$. However, if we identify functions whose difference is a constant, we form a space of the equivalence classes $\dot{\Lambda}_\gamma(\mathbf{T}^n)/\{\text{constants}\}$ on which $\|\cdot\|_{\dot{\Lambda}_\gamma}$ is a norm.

Remark 3.3.8. We already observed that elements of $\dot{\Lambda}_\gamma(\mathbf{T}^n)$ are continuous and thus bounded. Therefore, $\dot{\Lambda}_\gamma(\mathbf{T}^n) \subseteq L^\infty(\mathbf{T}^n)$ in the set-theoretic sense. However, the norm inequality $\|f\|_{L^\infty} \le C\|f\|_{\dot{\Lambda}_\gamma}$ for all $f \in \dot{\Lambda}_\gamma$ fails for all constants C. For example, take $f = N + \sin(2\pi x_1)$ on $\mathbf{T}^n$ and let $N \to \infty$ to see that this is the case.

The following theorem indicates how the smoothness of a function is reflected by the decay of its Fourier coefficients.

Theorem 3.3.9. *Let $s \in \mathbf{Z}$ with $s \ge 0$.*
(a) Suppose that $\partial^\alpha f$ exist and are integrable for all $|\alpha| \le s$. Then

$$|\widehat{f}(m)| \le \left(\frac{\sqrt{n}}{2\pi}\right)^s \frac{\max\limits_{|\alpha|=s} |\widehat{\partial^\alpha f}(m)|}{|m|^s}, \qquad m \neq 0, \tag{3.3.7}$$

and thus

$$|\widehat{f}(m)|(1+|m|^s) \to 0$$

as $|m| \to \infty$. In particular this holds when f lies in $\mathscr{C}^s(\mathbf{T}^n)$.
(b) Suppose that $\partial^\alpha f$ exist for all $|\alpha| \le s$ and whenever $|\alpha| = s$, $\partial^\alpha f$ are in $\dot{\Lambda}_\gamma(\mathbf{T}^n)$ for some $0 < \gamma < 1$. Then

$$|\widehat{f}(m)| \le \frac{(\sqrt{n})^{s+\gamma}}{(2\pi)^s 2^{\gamma+1}} \frac{\max\limits_{|\alpha|=s} \big\|\partial^\alpha f\big\|_{\dot{\Lambda}_\gamma}}{|m|^{s+\gamma}}, \qquad m \neq 0. \tag{3.3.8}$$

Proof. Fix $m \in \mathbf{Z}^n \setminus \{0\}$ and pick a j such that $|m_j| = \sup_{1\le k\le n} |m_k|$. Then clearly $m_j \neq 0$. Integrating by parts s times with respect to the variable x_j, we obtain

$$\widehat{f}(m) = \int_{\mathbf{T}^n} f(x) e^{-2\pi i x\cdot m}\, dx = (-1)^s \int_{\mathbf{T}^n} (\partial_j^s f)(x) \frac{e^{-2\pi i x\cdot m}}{(-2\pi i m_j)^s}\, dx, \tag{3.3.9}$$

where the boundary terms all vanish because of the periodicity of the integrand. Taking absolute values and using $|m| \le \sqrt{n}\,|m_j|$, we obtain assertion (3.3.7).

We now turn to the second part of the theorem. Let $e_j = (0,\dots,1,\dots,0)$ be the element of the torus $\mathbf{T}^n$ whose jth coordinate is one and all the others are zero. A simple change of variables together with the fact that $e^{\pi i} = -1$ gives that

$$\int_{\mathbf{T}^n} (\partial_j^s f)(x) e^{-2\pi i x\cdot m}\, dx = -\int_{\mathbf{T}^n} (\partial_j^s f)(x - \tfrac{e_j}{2m_j}) e^{-2\pi i x\cdot m}\, dx,$$

which implies that

$$\int_{\mathbf{T}^n} (\partial_j^s f)(x) e^{-2\pi i x\cdot m}\, dx = \frac{1}{2}\int_{\mathbf{T}^n} \big[(\partial_j^s f)(x) - (\partial_j^s f)(x - \tfrac{e_j}{2m_j})\big] e^{-2\pi i x\cdot m}\, dx.$$

Now use the estimate

$$|(\partial_j^s f)(x) - (\partial_j^s f)(x - \tfrac{e_j}{2m_j})| \le \frac{\|\partial_j^s f\|_{\dot\Lambda_\gamma}}{(2|m_j|)^\gamma}$$

and identity (3.3.9) to conclude the proof of (3.3.8). □

The following is an immediate consequence.

Corollary 3.3.10. *Let $s \in \mathbf{Z}$ with $s \ge 0$.*
(a) Suppose that $\partial^\alpha f$ exist and are integrable for all $|\alpha| \le s$. Then for some constant $c_{n,s}$ we have

$$|\widehat{f}(m)| \le c_{n,s} \frac{\max\big(\|f\|_{L^1}, \max_{|\alpha|=s} \|\partial^\alpha f\|_{L^1}\big)}{(1+|m|)^s}. \tag{3.3.10}$$

(b) Suppose that $\partial^\alpha f$ exist for all $|\alpha| \le s$ and whenever $|\alpha| = s$, $\partial^\alpha f$ are in $\dot\Lambda_\gamma(\mathbf{T}^n)$ for some $0<\gamma<1$. Then for some constant $c'_{n,s}$ we have

$$|\widehat{f}(m)| \le c'_{n,s} \frac{\max\big(\|f\|_{L^1}, \max_{|\alpha|=s} \|\partial^\alpha f\|_{\dot\Lambda_\gamma}\big)}{(1+|m|)^{s+\gamma}}. \tag{3.3.11}$$

Remark 3.3.11. The conclusions of Theorem 3.3.9 and Corollary 3.3.10 are also valid when $\gamma = 1$. In this case the spaces $\dot\Lambda_\gamma$ should be replaced by the space Lip 1 equipped with the seminorm

$$\|f\|_{\mathrm{Lip}\,1} = \sup_{\substack{x,h\in\mathbf{T}^n\\ h\neq 0}} \frac{|f(x+h)-f(x)|}{|h|}.$$

There is a slight lack of uniformity in the notation here, since in the theory of Lipschitz spaces the notation $\dot{\Lambda}_1$ is usually reserved for the space with seminorm

$$\|f\|_{\dot{\Lambda}_1} = \sup_{\substack{x,h\in\mathbf{T}^n\\ h\neq 0}} \frac{|f(x+h)+f(x-h)-2f(x)|}{|h|}.$$

The following proposition provides a partial converse to Theorem 3.3.9. We denote below by $[[s]]$ the largest integer strictly less than a given real number s. Then $[[s]]$ is equal to the integer part $[s]$ of s, unless s is an integer, in which case $[[s]] = [s]-1$.

Proposition 3.3.12. *Let $s>0$ and suppose that f is an integrable function on the torus with*

$$|\widehat{f}(m)| \le C(1+|m|)^{-s-n} \tag{3.3.12}$$

for all $m\in\mathbf{Z}^n$. Then f has partial derivatives of all orders $|\alpha|\le[[s]]$, and for $0<\gamma<s-[[s]]$, $\partial^\alpha f\in\dot{\Lambda}_\gamma$ for all multi-indices α satisfying $|\alpha|=[[s]]$.

Proof. Since f has an absolutely convergent Fourier series, Proposition 3.2.5 gives that

$$f(x) = \sum_{m\in\mathbf{Z}^n} \widehat{f}(m)e^{2\pi i x\cdot m}, \tag{3.3.13}$$

for almost all $x\in\mathbf{T}^n$.

Suppose that a series $g=\sum_m g_m$ satisfies $\sum_m \|\partial^\beta g_m\|_{L^\infty}<\infty$ for all $|\beta|\le M$. Then the function g is in $\mathscr{C}^M$ and $\partial^\beta g=\sum_m \partial^\beta g_m$; indeed this can be proved by induction on the degree of the multi-index, since for all $|\beta|\le M-1$ we have

$$\begin{aligned}\lim_{t\to 0}\frac{\partial^\beta g(x+te_j)-\partial^\beta g(x)}{t} &= \sum_m \lim_{t\to 0}\frac{\partial^\beta g_m(x+te_j)-\partial^\beta g_m(x)}{t}\\ &= \sum_m \partial_j\partial^\beta g(x),\end{aligned}$$

where the passage of the limit inside the sum is due to the Lebesgue dominated convergence theorem, which can be applied using the uniform convergence of $\sum_m \partial_j\partial^\beta g$ via the mean value theorem.

Using the preceding observation, the function f in (3.3.13) is $\mathscr{C}^{[[s]]}(\mathbf{T}^n)$ and

$$(\partial^\alpha f)(x) = \sum_{m\in\mathbf{Z}^n} \widehat{f}(m)(2\pi i m)^\alpha e^{2\pi i x\cdot m}$$

for all multi-indices $(\alpha_1,\dots,\alpha_n)$ with $|\alpha|\le[[s]]$, since

$$\sum_{m\in\mathbf{Z}^n} \widehat{f}(m)\sup_{x\in\mathbf{T}^n}\left|(2\pi i m)^\alpha e^{2\pi i x\cdot m}\right| < \infty,$$

which holds because of (3.3.12).

Now suppose that $|\alpha| = [[s]]$ and that $0 < \gamma < s - [[s]]$. Then

$$
\begin{aligned}
|(\partial^\alpha f)(x+h) - (\partial^\alpha f)(x)| &= \Big| \sum_{m \in \mathbf{Z}^n} \widehat{f}(m)(2\pi i m)^\alpha e^{2\pi i x \cdot m}\big(e^{2\pi i m \cdot h} - 1\big)\Big| \\
&\le (2\pi)^{[[s]]} \sum_{m \in \mathbf{Z}^n} |m|^{[[s]]} \frac{C 2^{1-\gamma}(2\pi)^\gamma |h|^\gamma |m|^\gamma}{(1+|m|)^{n+s}} \\
&\le C 2^{1-\gamma}(2\pi)^s |h|^\gamma,
\end{aligned}
$$

where we used the relation $[[s]] + \gamma - s - n < -n$ to conclude the convergence of the series and the fact that

$$
|e^{2\pi i m \cdot h} - 1| \le \min(2, 2\pi |m|\,|h|) \le 2^{1-\gamma}(2\pi)^\gamma |m|^\gamma |h|^\gamma .
$$

□

Next we recall the definition of functions of bounded variation.

Definition 3.3.13. A measurable function f on $\mathbf{T}^1$ is said to be of *bounded variation* if it is defined everywhere and

$$
\mathrm{Var}(f) = \sup\Big\{ \sum_{j=1}^{M} |f(x_j) - f(x_{j-1})| : \; 0 = x_0 < x_1 < \cdots < x_M = 1 \Big\} < \infty,
$$

where the supremum is taken over all partitions of the interval $[0,1]$. The expression $\mathrm{Var}(f)$ is called the *total variation* of f. The class of functions of bounded variation on $\mathbf{T}^1$ is denoted by $BV(\mathbf{T}^1)$.

Examples of functions of bounded variation can be constructed as follows: given f_1, f_2 nonnegative integrable functions on $[0,1]$ with

$$
\int_0^1 f_1(t)\,dt = \int_0^1 f_2(t)\,dt ,
$$

then the periodic function

$$
g(x) = \int_0^x f_1(t)\,dt - \int_0^x f_2(t)\,dt ,
$$

defined on $[0,1]$, is of bounded variation. Analogous examples can be constructed when f_1 and f_2 are replaced by nonnegative finite Borel measures on $[0,1]$.

Every function of bounded variation can be represented as the difference of two (not necessarily strictly) increasing functions and thus it has a finite derivative at almost every point. Moreover, for functions of bounded variation, the Lebesgue–Stieltjes integral with respect to df is well defined.

Proposition 3.3.14. *If f is in $BV(\mathbf{T}^1)$, then*

$$|\widehat{f}(m)| \le \frac{\mathrm{Var}(f)}{2\pi|m|}$$

whenever $m \neq 0$.

Proof. Integration by parts gives

$$\widehat{f}(m) = \int_{\mathbf{T}^1} f(x)e^{-2\pi imx}\,dx = \int_{\mathbf{T}^1} \frac{e^{-2\pi imx}}{-2\pi im}\,df,$$

where the boundary terms vanish because of periodicity. The conclusion follows from the fact that the norm of the measure df is the total variation of f. □

The following chart (Table 3.1) summarizes the decay of Fourier coefficients in terms of scales of spaces measuring the smoothness of the functions. Recall that for $q \ge 0$, $\widehat{f}(m) = o(|m|^{-q})$ means that $|\widehat{f}(m)|\,|m|^q \to 0$ as $|m| \to \infty$ and $\widehat{f}(m) = O(|m|^{-q})$ means that $|\widehat{f}(m)| \le C|m|^{-q}$ when $|m|$ is large. In this chart, we denote by $\mathscr{C}^{s,\gamma}(\mathbf{T}^n)$ the space of all $\mathscr{C}^s$ functions on $\mathbf{T}^n$, all of whose derivatives of total order s lie in $\Lambda_\gamma(\mathbf{T}^n)$, for some $0 < \gamma < 1$.

SPACE	*SEQUENCE OF FOURIER COEFFICIENTS*
$L^1(\mathbf{T}^n)$	$o(1)$
$L^p(\mathbf{T}^n)$	$\ell^{p'}(\mathbf{Z}^n)$
$L^2(\mathbf{T}^n)$	$\ell^2(\mathbf{Z}^n)$
$\dot{\Lambda}_\gamma(\mathbf{T}^n)$	$O(\|m\|^{-\gamma})$
$BV(\mathbf{T}^1)$	$O(\|m\|^{-1})$
$\mathscr{C}^1(\mathbf{T}^n)$	$o(\|m\|^{-1})$
$\mathscr{C}^{1,\gamma}(\mathbf{T}^n)$	$O(\|m\|^{-1-\gamma})$
$\mathscr{C}^2(\mathbf{T}^n)$	$o(\|m\|^{-2})$
$\mathscr{C}^{2,\gamma}(\mathbf{T}^n)$	$O(\|m\|^{-2-\gamma})$
$\mathscr{C}^3(\mathbf{T}^n)$	$o(\|m\|^{-3})$
...	...
$\mathscr{C}^\infty(\mathbf{T}^n)$	$o(\|m\|^{-N})$ for all $N > 0$

Table 3.1 Interconnection between smoothness of functions and decay of their Fourier coefficients. We take $0 < \gamma < 1$ and $1 < p < 2$.

3.3.3 Functions with Absolutely Summable Fourier Coefficients

Decay for the Fourier coefficients can also be indirectly deduced from knowledge about the summability of these coefficients. The simplest kind of summability is in the sense of ℓ^1. It is therefore natural to consider the class of functions on the torus whose Fourier coefficients form an absolutely summable series.

Definition 3.3.15. An integrable function f on the torus is said to have an *absolutely convergent* Fourier series if

$$\sum_{m\in\mathbf{Z}^n} |\widehat{f}(m)| < +\infty.$$

We denote by $A(\mathbf{T}^n)$ the space of all integrable functions on the torus $\mathbf{T}^n$ whose Fourier series are absolutely convergent. We then introduce a norm on $A(\mathbf{T}^n)$ by setting

$$\big\|f\big\|_{A(\mathbf{T}^n)} = \sum_{m\in\mathbf{Z}^n} |\widehat{f}(m)|.$$

In view of Proposition 3.2.5, every function f in $A(\mathbf{T}^n)$ can be changed on a set of measure zero to be made continuous and under this modification, Fourier inversion

$$f(x) = \sum_{m\in\mathbf{Z}^n} \widehat{f}(m)e^{2\pi i m\cdot x}$$

holds for all $x \in \mathbf{T}^n$. Thus functions in $A(\mathbf{T}^n)$ are continuous and bounded. Moreover, Theorem 3.3.9 yields that every function in $\mathscr{C}^n(\mathbf{T}^n)$ whose partial derivatives of order n are in $\dot{\Lambda}_\gamma$, $\gamma > 0$, must lie in $A(\mathbf{T}^n)$. The following theorem gives us a significantly better sufficient condition for a function to be in $A(\mathbf{T}^n)$.

Theorem 3.3.16. *Suppose f is a given function in $\mathscr{C}^{[n/2]}(\mathbf{T}^n)$ and that all partial derivatives of order $[\frac{n}{2}]$ of f lie in $\dot{\Lambda}_\gamma(\mathbf{T}^n)$ for some γ with $\frac{n}{2} - [\frac{n}{2}] < \gamma < 1$. Then f lies in $A(\mathbf{T}^n)$ and*

$$\big\|f\big\|_{A(\mathbf{T}^n)} \le |\widehat{f}(0)| + C(n,\gamma) \sup_{|\alpha|=[\frac{n}{2}]} \big\|\partial^\alpha f\big\|_{\dot{\Lambda}_\gamma(\mathbf{T}^n)},$$

where $C(n,\gamma)$ is a constant depending on n and γ.

Proof. For each $\ell = 0,1,2,\dots$, let

$$S_\ell = \bigg(\sum_{2^\ell \le |m| < 2^{\ell+1}} |\widehat{f}(m)|^2\bigg)^{1/2}.$$

We begin by writing

$$\big\|f\big\|_{A(\mathbf{T}^n)} = |\widehat{f}(0)| + \sum_{\ell=0}^\infty \sum_{2^\ell \le |m| < 2^{\ell+1}} |\widehat{f}(m)| \le |\widehat{f}(0)| + \sqrt{c_n} \sum_{\ell=0}^\infty 2^{\frac{\ell n}{2}} S_\ell, \tag{3.3.14}$$

where we used the Cauchy-Schwarz inequality and the fact that there are at most $c_n 2^{\ell n}$ points in $\mathbf{Z}^n$ inside the open ball $B(0,2^{\ell+1})$, for some dimensional constant c_n.

Notice that for a multi-index $m = (m_1,\dots,m_n)$ satisfying $2^\ell \le |m| \le 2^{\ell+1}$ and for j in $\{1,\dots,n\}$ such that $|m_j| = \sup_k |m_k|$ we have

$$\frac{|m_j|}{2^\ell} \ge \frac{|m|}{2^\ell\sqrt{n}} \ge \frac{1}{\sqrt{n}}. \tag{3.3.15}$$

For $1 \le j \le n$, let e_j be the element of $\mathbf{R}^n$ with zero entries except for the jth coordinate, which is 1, and define

$$h_j^\ell = 2^{-\ell-2} e_j . \tag{3.3.16}$$

Using the elementary fact that $|t| \le \pi \implies |e^{it} - 1| \ge 2|t|/\pi$, we obtain

$$|e^{2\pi i m\cdot h_j^\ell} - 1| = |e^{2\pi i m_j 2^{-\ell-2}} - 1| \ge \frac{2}{\pi}\frac{|2\pi m_j|}{2^{\ell+2}} = \frac{|m_j|}{2^\ell} \ge \frac{1}{\sqrt{n}}, \tag{3.3.17}$$

whenever $\frac{|2\pi m_j|}{2^{\ell+2}} \le \pi$, which is always true since $\frac{|2\pi m_j|}{2^{\ell+2}} \le \frac{2\pi 2^{\ell+1}}{2^{\ell+2}} \le \pi$.

We now have

$$\begin{aligned}
S_\ell^2 &= \sum_{j=1}^{n} \sum_{\substack{2^\ell \le |m| < 2^{\ell+1} \\ |m_j| = \sup_k |m_k|}} |\widehat{f}(m)|^2 \\
&\le n \sum_{j=1}^{n} \sum_{\substack{2^\ell \le |m| < 2^{\ell+1} \\ |m_j| = \sup_k |m_k|}} |e^{2\pi i m\cdot h_j^\ell} - 1|^2 |\widehat{f}(m)|^2 \frac{|2\pi i m_j|^{2[\frac{n}{2}]}}{|2\pi m_j|^{2[\frac{n}{2}]}} \\
&\le n \frac{n^{[\frac{n}{2}]}}{(2\pi 2^\ell)^{2[\frac{n}{2}]}} \sum_{j=1}^{n} \sum_{m\in\mathbf{Z}^n} |e^{2\pi i m\cdot h_j^\ell} - 1|^2 |\widehat{\partial_j^{[n/2]} f}(m)|^2 \\
&= C_n 2^{-2\ell[\frac{n}{2}]} \sum_{j=1}^{n} \big\|\partial_j^{[n/2]} f(\cdot + h_j^\ell) - \partial_j^{[n/2]} f\big\|_{L^2}^2 \\
&\le C_n 2^{-2\ell[\frac{n}{2}]} \sum_{j=1}^{n} \big\|\partial_j^{[n/2]} f(\cdot + h_j^\ell) - \partial_j^{[n/2]} f\big\|_{L^\infty}^2 \\
&\le C_n' 2^{-2\ell[\frac{n}{2}]} \sup_{|\alpha| = [\frac{n}{2}]} \big\|\partial^\alpha f\big\|_{\dot\Lambda_\gamma}^2 \sum_{j=1}^{n} |h_j^\ell|^{2\gamma} \\
&= C_{n,\gamma}' 2^{-2\ell[\frac{n}{2}] - 2\ell\gamma} \sup_{|\alpha| = [\frac{n}{2}]} \big\|\partial^\alpha f\big\|_{\dot\Lambda_\gamma}^2 ,
\end{aligned}$$

where we used (3.3.17), (3.3.15), and (3.3.16). We conclude that

$$S_\ell \le C_{n,\gamma}'' 2^{-\ell([\frac{n}{2}] + \gamma)} \sup_{|\alpha| = [\frac{n}{2}]} \big\|\partial^\alpha f\big\|_{\dot\Lambda_\gamma}$$

which inserted in (3.3.14) yields the desired conclusion since $\gamma > \frac{n}{2} - [\frac{n}{2}]$. □

Exercises

3.3.1. Given a sequence $\{a_n\}_{n=0}^\infty$ of positive numbers such that $a_n \to 0$ as $n \to \infty$, find a nonnegative integrable function h on $[0,1]$ such that

$$\int_0^1 h(t)t^m\,dt \geq a_m.$$

Use this result to deduce a different proof of Lemma 3.3.2.
[*Hint:* Try $h = e\sum_{k=0}^{\infty}(\sup_{j\geq k} a_j - \sup_{j\geq k+1} a_j)(k+2)\chi_{[\frac{k+1}{k+2},1]}$.]

3.3.2. Prove that given a positive sequence $\{d_m\}_{m\in\mathbf{Z}^n}$ with $d_m \to 0$ as $|m| \to \infty$, there exists a positive sequence $\{a_j\}_{j\in\mathbf{Z}}$ with $a_{m_1}\cdots a_{m_n} \geq d_{(m_1,\dots,m_n)}$ and $a_j \to 0$ as $|j| \to \infty$.

3.3.3. (a) Use the idea of the proof of Lemma 3.3.3 to prove that if a twice continuously differentiable function $f \geq 0$ is defined on $(0,\infty)$ and satisfies $f'(x) \leq 0$ and $f''(x) \geq 0$ for all $x > 0$, then $\lim_{x\to\infty} xf'(x) = 0$.
(b) Suppose that a continuously differentiable function g is defined on $(0,\infty)$ and satisfies $g \geq 0$, $g' \leq 0$, and $\int_1^\infty g(x)\,dx < +\infty$. Prove that

$$\lim_{x\to\infty} xg(x) = 0.$$

3.3.4. Prove that for $0 < \gamma < \delta < 1$ we have $\|f\|_{\dot{\Lambda}_\gamma} \leq C_{n,\gamma,\delta}\|f\|_{\dot{\Lambda}_\delta}$ for all functions f and thus $\dot{\Lambda}_\delta$ is a subspace of $\dot{\Lambda}_\gamma$.

3.3.5. Suppose that f is a differentiable function on $\mathbf{T}^1$ whose derivative f' is in $L^2(\mathbf{T}^1)$. Prove that $f \in A(\mathbf{T}^1)$ and that

$$\|f\|_{A(\mathbf{T}^1)} \leq \|f\|_{L^1} + \frac{1}{2\pi}\Big(\sum_{j\neq 0} j^{-2}\Big)^{1/2}\|f'\|_{L^2}.$$

3.3.6. (a) Prove that the product of two functions in $A(\mathbf{T}^n)$ is also in $A(\mathbf{T}^n)$ and that

$$\|fg\|_{A(\mathbf{T}^n)} \leq \|f\|_{A(\mathbf{T}^n)}\|g\|_{A(\mathbf{T}^n)}.$$

(b) Prove that the convolution of two square integrable functions on $\mathbf{T}^n$ always gives a function in $A(\mathbf{T}^n)$.

3.3.7. Fix $0 < \alpha < 1$ and define f on $\mathbf{T}^1$ by setting

$$f(x) = \sum_{k=0}^{\infty} 2^{-\alpha k} e^{2\pi i 2^k x}.$$

Prove that the function f lies in $\dot{\Lambda}_\alpha(\mathbf{T}^1)$. Conclude that there does not exist positive $\beta > \alpha$ such that for all f in $\dot{\Lambda}_\alpha(\mathbf{T}^1)$ we have $\sup_{m\in\mathbf{Z}} |m|^\beta |\widehat{f}(m)| < \infty$.
[*Hint:* For $h \neq 0$ pick $N \in \mathbf{Z}^+$ such that $2^N|h| > 1 \geq 2^{N-1}|h|$. To estimate the difference $|f(x+h) - f(x)|$, consider the cases $k \leq N$ and $k \geq N+1$ in the sum.]

3.3.8. Use without proof that there exists a constant $C > 0$ such that

$$\sup_{t\in\mathbf{R}}\left|\sum_{k=2}^{N} e^{ik\log k}e^{ikt}\right| \le C\sqrt{N}, \qquad N = 2,3,4,\dots,$$

to prove that the function

$$g(x) = \sum_{k=2}^{\infty}\frac{e^{ik\log k}}{k}e^{2\pi ikx}$$

is in $\dot{\Lambda}_{1/2}(\mathbf{T}^1)$ but not in $A(\mathbf{T}^1)$. Conclude that the restriction $s > 1/2$ in Theorem 3.3.16 is sharp.
[*Hint:* Estimate the difference $|g(x+h)-g(x)|$ using the summation by parts identity in Appendix F, taking sums of the sequence $e^{ik\log k}e^{2\pi ikx}$ and differences of the sequence $\frac{e^{2\pi ikh}-1}{k}$.]

3.3.9. Show that there exist sequences $\{a_m\}_{m\in\mathbf{Z}^n}$ that tend to zero as $|m| \to \infty$ for which there do not exist functions f in $L^1(\mathbf{T}^n)$ with $\widehat{f}(m) = a_m$ for all m.
[*Hint:* Suppose the contrary. Then the open mapping theorem would imply the inequality $\|f\|_{L^1(\mathbf{T}^n)} \le A\|\widehat{f}\|_{\ell^\infty(\mathbf{Z}^n)}$ for some $A > 0$. To contradict it, fix a smooth nonzero function h equal to 1 on $B(0,\frac{1}{4})$ and supported in $B(0,\frac{1}{2})$. For $b > 0$ define $g_b(x) = h(x)e^{-\pi(1+ib)|x|^2}$ and extend g_b to a 1-periodic function in each variable on $\mathbf{R}^n$. Use that $\widehat{g_b}(m) = \int_{\mathbf{R}^n}\widehat{h}(y)(1+ib)^{-n/2}e^{-\frac{\pi}{1+ib}|m-y|^2}dy$, and let $b\to\infty$ in the inequality $\|g_b\|_{L^1(\mathbf{T}^n)} \le A\|\widehat{g_b}\|_{\ell^\infty(\mathbf{Z}^n)}$ to obtain a contradiction.]

3.4 Pointwise Convergence of Fourier Series

In this section we are concerned with the pointwise convergence of the square partial sums and the Fejér means of a function defined on the torus.

3.4.1 Pointwise Convergence of the Fejér Means

We saw in Section 3.1 that the Fejér kernel is an approximate identity. This implies that the Fejér (or Cesàro) means of an L^p function f on $\mathbf{T}^n$ converge to it in L^p for any $1 \le p < \infty$. Moreover, if f is continuous at x_0, then the means $(F_N^n * f)(x_0)$ converge to $f(x_0)$ as $N\to\infty$ in view of Theorem 1.2.19 (2). Although this is a satisfactory result, it is natural to ask what happens for more general functions.

Using properties of the Fejér kernel, we obtain the following one-dimensional result regarding the convergence of the Fejér means:

Theorem 3.4.1. *(Fejér) If a function f in $L^1(\mathbf{T}^1)$ has left and right limits at a point x_0, denoted by $f(x_0-)$ and $f(x_0+)$, respectively, then*

$$(F_N * f)(x_0) \to \frac{1}{2}\big(f(x_0+)+f(x_0-)\big) \qquad \text{as} \qquad N\to\infty. \tag{3.4.1}$$

In particular, this is the case for functions of bounded variation.

Proof. Let us identify $\mathbf{T}^1$ with $[-1/2,1/2]$. Given $\varepsilon>0$, find $\delta\in(0,1/2)$ such that

$$0<t<\delta \implies \left|\frac{f(x_0+t)+f(x_0-t)}{2}-\frac{f(x_0+)+f(x_0-)}{2}\right|<\varepsilon. \tag{3.4.2}$$

Using the second expression for F_N in (3.1.18), we can find an $N_0>0$ such that for $N\ge N_0$ we have

$$\sup_{\delta\le t\le\frac12} F_N(t)=\frac{1}{N+1}\sup_{\delta\le t\le\frac12}\left(\frac{\sin(\pi(N+1)t)}{\sin(\pi t)}\right)^2\le\frac{1}{N+1}\frac{1}{\sin^2(\pi\delta)}<\varepsilon. \tag{3.4.3}$$

We now have

$$\begin{aligned}(F_N*f)(x_0)-f(x_0+) &= \int_{\mathbf{T}^1}F_N(t)\big(f(x_0+t)-f(x_0+)\big)\,dt,\\ (F_N*f)(x_0)-f(x_0-) &= \int_{\mathbf{T}^1}F_N(t)\big(f(x_0-t)-f(x_0-)\big)\,dt.\end{aligned}$$

Averaging these two identities and using that the integrand is even, we obtain

$$\begin{aligned}&(F_N*f)(x_0)-\frac{f(x_0+)+f(x_0-)}{2}\\ &=2\int_0^{1/2}F_N(t)\left(\frac{f(x_0+t)+f(x_0-t)}{2}-\frac{f(x_0+)+f(x_0-)}{2}\right)dt.\end{aligned} \tag{3.4.4}$$

We split the integral in (3.4.4) into two pieces, the integral over $[0,\delta)$ and the integral over $[\delta,1/2]$. By (3.4.2), the integral over $[0,\delta)$ is controlled by $\varepsilon\int_{\mathbf{T}^1}F_N(t)\,dt=\varepsilon$. Also (3.4.3) gives that for $N\ge N_0$

$$\begin{aligned}&\left|\int_\delta^{1/2}F_N(t)\left(\frac{f(x_0-t)+f(x_0+t)}{2}-\frac{f(x_0-)+f(x_0+)}{2}\right)dt\right|\\ &\qquad\le\frac{\varepsilon}{2}\left(\big\|f-f(x_0-)\big\|_{L^1}+\big\|f-f(x_0+)\big\|_{L^1}\right)=\varepsilon\,c(f,x_0),\end{aligned}$$

where $c(f,x_0)$ is a constant depending on f and x_0. We have now proved that given $\varepsilon>0$ there exists an N_0 such that for $N\ge N_0$ the second expression in (3.4.4) is bounded by $2\varepsilon\,(c(f,x_0)+1)$. This proves the required conclusion.

Functions of bounded variation can be written as differences of increasing functions, and since increasing functions have left and right limits everywhere, (3.4.1) holds for these functions. □

We continue with an elementary but very useful application of the preceding result.

Proposition 3.4.2. *Let $x_0 \in \mathbf{T}^1$ and let f be a complex-valued function on $\mathbf{T}^1$. Suppose that the left and right limits of f exist as $x \to x_0$ and that the partial sums (Dirichlet means) $(D_N * f)(x_0)$ converge. Then*

$$(D_N * f)(x_0) \to \frac{1}{2}\big(f(x_0+) + f(x_0-)\big)$$

as $N \to \infty$.

Proof. If $(D_N * f)(x_0) \to L(x_0)$ as $N \to \infty$, then

$$(F_N * f)(x_0) = \frac{(D_0 * f)(x_0) + (D_1 * f)(x_0) + \cdots + (D_N * f)(x_0)}{N+1} \to L(x_0)$$

as $N \to \infty$. But $(F_N * f)(x_0) \to \frac{1}{2}\big(f(x_0+) + f(x_0-)\big)$ as $N \to \infty$ in view of Theorem 3.4.1. We conclude that

$$L(x_0) = \frac{1}{2}\big(f(x_0+) + f(x_0-)\big),$$

Thus $(D_N * f)(x_0) \to \frac{1}{2}\big(f(x_0+) + f(x_0-)\big)$ as $N \to \infty$. □

This theorem is quite useful when we have a priori knowledge that the Fourier series converges. For instance, consider the following example.

Example 3.4.3. On $(-1/2, 1/2)$ let $f(t) = t$ and $f(1/2) = f(-1/2) = 1000$. Then f is discontinuous at the point $-1/2 \equiv 1/2$ but it has left and right limits at this point:

$$\lim_{t \to -\frac{1}{2}+} f(t) = -\frac{1}{2} \qquad \lim_{t \to \frac{1}{2}-} f(t) = \frac{1}{2}. \tag{3.4.5}$$

Moreover $\widehat{f}(m) = \frac{i(-1)^m}{2\pi m}$ when $m \neq 0$ and $\widehat{f}(0) = 0$ by Exercise 3.4.1 (a). It is not hard to see that the series

$$(D_N * f)(x) = \frac{i}{2\pi} \sum_{0<|m|\le N} \frac{(-1)^m}{m} e^{2\pi i m x} = \frac{i}{2\pi} \sum_{0<|m|\le N} \frac{e^{2\pi i m(x+\frac{1}{2})}}{m} \tag{3.4.6}$$

converges for every $x \in (-1/2, 1/2)$. Indeed, by Appendix F, (3.4.6) equals

$$\frac{i}{2\pi}\frac{1}{N}\sum_{m=1}^{N}\left(e^{2\pi i m(x+\frac{1}{2})} - e^{-2\pi i m(x+\frac{1}{2})}\right)$$
$$-\frac{i}{2\pi}\sum_{k=1}^{N-1}\left(\sum_{m=1}^{k}\left(e^{2\pi i m(x+\frac{1}{2})} - e^{-2\pi i m(x+\frac{1}{2})}\right)\right)\left(\frac{1}{k+1}-\frac{1}{k}\right)$$

which has a limit as $N \to \infty$, since the geometric sums

$$\sum_{m=1}^{N} e^{\pm 2\pi i m(x+\frac{1}{2})} = \frac{1 - e^{\pm 2\pi i (N+1)(x+\frac{1}{2})}}{1 - e^{\pm 2\pi i (x+\frac{1}{2})}} - 1$$

are bounded above independently of N when $x \in (-1/2, 1/2)$. We conclude that

$$f(x) = x = \lim_{N\to\infty} \frac{i}{2\pi} \sum_{0<|m|\le N} \frac{e^{2\pi i m(x+\frac{1}{2})}}{m} = -\lim_{N\to\infty} \sum_{0<|m|\le N} \frac{\sin(2\pi m(x+\frac{1}{2}))}{2\pi m}$$

whenever $|x| < 1/2$. Moreover, we have that

$$(D_N * f)(1/2) = \lim_{N\to\infty} \frac{i}{2\pi} \sum_{0<|m|\le N} \frac{0}{m} = 0,$$

which is the average of the left and right limits in (3.4.5) as Proposition 3.4.2 states. Exercise 3.4.2 contains other applications of this sort.

3.4.2 Almost Everywhere Convergence of the Fejér Means

We have seen that the Fejér means of a relatively nice function (such as of bounded variation) converge everywhere. What can we say about the Fejér means of a general integrable function? Since the Fejér kernel is an approximate identity that satisfies good estimates, the following result should not come as a surprise.

Theorem 3.4.4. *(a) For* $f \in L^1(\mathbf{T}^n)$, *let*

$$\mathscr{H}(f) = \sup_{N\in\mathbf{Z}^+} |f * F_N^n| .$$

Then $\mathscr{H}$ *maps* $L^1(\mathbf{T}^n)$ *to* $L^{1,\infty}(\mathbf{T}^n)$ *and* $L^p(\mathbf{T}^n)$ *to itself for* $1 < p \le \infty$.
(b) For any function $f \in L^1(\mathbf{T}^n)$, *we have as* $N \to \infty$

$$(F_N^n * f) \to f \qquad \text{a.e.}$$

Proof. It is an elementary fact that $|t| \le \frac{\pi}{2} \implies |\sin t| \ge \frac{2}{\pi}|t|$; see Appendix E. Using this fact and the expression (3.1.18) we obtain for all t in $[-\frac{1}{2}, \frac{1}{2}]$,

$$
\begin{aligned}
|F_N(t)| &= \frac{1}{N+1}\left|\frac{\sin(\pi(N+1)t)}{\sin(\pi t)}\right|^2 \\
&\leq \frac{N+1}{4}\left|\frac{\sin(\pi(N+1)t)}{(N+1)t}\right|^2 \\
&\leq \frac{N+1}{4}\min\left(\pi^2, \frac{1}{(N+1)^2t^2}\right) \\
&\leq \frac{\pi^2}{2}\frac{N+1}{1+(N+1)^2|t|^2}.
\end{aligned}
$$

For $t \in \mathbf{R}$ let us set $\varphi(t) = (1+|t|^2)^{-1}$ and $\varphi_\varepsilon(t) = \frac{1}{\varepsilon}\varphi(\frac{t}{\varepsilon})$ for $\varepsilon > 0$. For $x = (x_1,\dots,x_n) \in \mathbf{R}^n$ and $\varepsilon > 0$ we also set

$$
\Phi(x) = \varphi(x_1)\cdots\varphi(x_n)
$$

and $\Phi_\varepsilon(x) = \varepsilon^{-n}\Phi(\varepsilon^{-1}x)$. Then for $|t| \leq \frac{1}{2}$ we have $|F_N(t)| \leq \frac{\pi^2}{2}\varphi_\varepsilon(t)$ with $\varepsilon = (N+1)^{-1}$, and for $y \in [-\frac{1}{2},\frac{1}{2}]^n$ we have

$$
|F_N^n(y)| \leq (\tfrac{\pi^2}{2})^n\,\Phi_\varepsilon(y), \qquad \text{with } \varepsilon = (N+1)^{-1}.
$$

Now let f be an integrable function on $\mathbf{T}^n$ and let f_0 denote its periodic extension on $\mathbf{R}^n$. For $x \in [-\frac{1}{2},\frac{1}{2}]^n$ we have

$$
\begin{aligned}
\mathscr{H}(f)(x) &= \sup_{N>0}\left|\int_{\mathbf{T}^n} F_N^n(y)f(x-y)\,dy\right| \\
&\leq (\tfrac{\pi^2}{2})^n \sup_{\varepsilon>0}\int_{[-\frac{1}{2},\frac{1}{2}]^n} |\Phi_\varepsilon(y)|\,|f_0(x-y)|\,dy \\
&\leq 5^n \sup_{\varepsilon>0}\int_{\mathbf{R}^n} |\Phi_\varepsilon(y)|\,|(f_0\chi_Q)(x-y)|\,dy \\
&= 5^n\mathscr{G}(f_0\chi_Q)(x),
\end{aligned}
\tag{3.4.7}
$$

where Q is the cube $[-1,1]^n$ and $\mathscr{G}$ is the operator defined on integrable functions on $\mathbf{R}^n$ by

$$
\mathscr{G}(h) = \sup_{\varepsilon>0} |h| * \Phi_\varepsilon .
$$

If we can show that $\mathscr{G}$ maps $L^1(\mathbf{R}^n)$ to $L^{1,\infty}(\mathbf{R}^n)$, the corresponding conclusion for $\mathscr{H}$ on $\mathbf{T}^n$ would follow from the fact $\mathscr{H}(f) \leq 5^n\mathscr{G}(f_0\chi_Q)$ proved in (3.4.7) and the sequence of inequalities

$$
\big\|\mathscr{H}(f)\big\|_{L^{1,\infty}(\mathbf{T}^n)} \leq 5^n\big\|\mathscr{G}(f_0\chi_Q)\big\|_{L^{1,\infty}(\mathbf{R}^n)} \leq 5^nC\big\|f_0\chi_Q\big\|_{L^1(\mathbf{R}^n)} = C'\big\|f\big\|_{L^1(\mathbf{T}^n)}.
$$

Moreover, the L^p conclusion about $\mathscr{H}$ follows from the weak type $(1,1)$ result and the trivial L^∞ inequality, in view of the Marcinkiewicz interpolation theorem (Theorem 1.3.2). The required weak type $(1,1)$ estimate for $\mathscr{G}$ on $\mathbf{R}^n$ is a consequence of Lemma 3.4.5. Modulo the proof of this lemma, part (a) of the theorem is proved.

To prove the statement in part (b) observe that for $f \in \mathscr{C}^\infty(\mathbf{T}^n)$, which is a dense subspace of L^1, we have $F_N^n * f \to f$ uniformly on $\mathbf{T}^n$ as $N \to \infty$, since the sequence $\{F_N\}_N$ is an approximate identity. Since by part (a), $\mathscr{H}$ maps $L^1(\mathbf{T}^n)$ to $L^{1,\infty}(\mathbf{T}^n)$, Theorem 2.1.14 yields that for $f \in L^1(\mathbf{T}^n)$, $F_N^n * f \to f$ a.e. □

We now prove the weak type $(1,1)$ boundedness of $\mathscr{G}$ used earlier.

Lemma 3.4.5. *Let $\Phi(x_1,\dots,x_n) = (1+|x_1|^2)^{-1}\cdots(1+|x_n|^2)^{-1}$ and for $\varepsilon > 0$ let $\Phi_\varepsilon(x) = \varepsilon^{-n}\Phi(\varepsilon^{-1}x)$. Then the maximal operator*

$$\mathscr{G}(f) = \sup_{\varepsilon>0} |f| * \Phi_\varepsilon$$

maps $L^1(\mathbf{R}^n)$ to $L^{1,\infty}(\mathbf{R}^n)$.

Proof. Let $I_0 = [-1,1]$ and $I_k = \{t \in \mathbf{R} : 2^{k-1} \le |t| \le 2^k\}$ for $k = 1,2,\dots$. Also, let $\widetilde{I}_k$ be the convex hull of I_k, that is, the interval $[-2^k, 2^k]$. For $a_2,\dots,a_n$ fixed positive numbers, let $M_{a_2,\dots,a_n}$ be the maximal operator obtained by averaging a function on $\mathbf{R}^n$ over all products of closed intervals $J_1 \times \cdots \times J_n$ containing a given point with

$$|J_1| = 2^{a_2}|J_2| = \cdots = 2^{a_n}|J_n|.$$

In view of Exercise 2.1.9(c), we have that $M_{a_2,\dots,a_n}$ maps L^1 to $L^{1,\infty}$ with some constant independent of the a_j's. (This is due to the nice doubling property of this family of rectangles.) For a fixed $\varepsilon > 0$ we estimate the expression

$$(\Phi_\varepsilon * |f|)(0) = \int_{\mathbf{R}^n} \frac{|f(-\varepsilon y)|\,dy}{(1+y_1^2)\cdots(1+y_n^2)}.$$

Split $\mathbf{R}^n$ into $n!$ regions of the form $|y_{j_1}| \ge \cdots \ge |y_{j_n}|$, where $\{j_1,\dots,j_n\}$ is a permutation of the set $\{1,\dots,n\}$ and $y = (y_1,\dots,y_n)$. By symmetry, we examine the region $\mathscr{R}$ where $|y_1| \ge \cdots \ge |y_n|$. Then for some constant $C > 0$ we have

$$\int_{\mathscr{R}} \frac{|f(-\varepsilon y)|\,dy}{(1+y_1^2)\cdots(1+y_n^2)} \le C \sum_{k_1=0}^{\infty}\sum_{k_2=0}^{k_1}\cdots\sum_{k_n=0}^{k_{n-1}} 2^{-(2k_1+\cdots+2k_n)} \int_{I_{k_1}}\!\!\cdots\int_{I_{k_n}} |f(-\varepsilon y)|\,dy_{k_n}\cdots dy_1,$$

and the last expression is trivially controlled by the corresponding expression, where the I_k's are replaced by the $\widetilde{I}_k$'s. This, in turn, is controlled by

$$C' \sum_{k_1=0}^{\infty}\sum_{k_2=0}^{k_1}\cdots\sum_{k_n=0}^{k_{n-1}} 2^{-(k_1+\cdots+k_n)} M_{k_1-k_2,\dots,k_1-k_n}(f)(0). \tag{3.4.8}$$

Now set $s_2 = k_1 - k_2, \dots, s_n = k_1 - k_n$, observe that $s_j \geq 0$, use that

$$2^{-(k_1+\cdots+k_n)} \leq 2^{-\frac{k_1}{2}} 2^{-\frac{s_2}{2n}} \cdots 2^{-\frac{s_n}{2n}},$$

and change the indices of summation to estimate the expression in (3.4.8) by

$$C'' \sum_{k_1=0}^{\infty} \sum_{s_2=0}^{\infty} \cdots \sum_{s_n=0}^{\infty} 2^{-\frac{k_1}{2}} 2^{-\frac{s_2}{2n}} \cdots 2^{-\frac{s_n}{2n}} M_{s_2,\dots,s_n}(f)(0).$$

Argue similarly for the remaining regions $|y_{j_1}| \geq \cdots \geq |y_{j_n}|$. Finally, translate to an arbitrary point x to obtain the estimate

$$|(\Phi_\varepsilon * f)(x)| \leq C'' n! \sum_{s_2=0}^{\infty} \cdots \sum_{s_n=0}^{\infty} 2^{-\frac{s_2}{2n}} \cdots 2^{-\frac{s_n}{2n}} M_{s_2,\dots,s_n}(f)(x).$$

Now take the supremum over all $\varepsilon > 0$ and use the fact that the maximal functions $M_{s_2,\dots,s_n}$ map L^1 to $L^{1,\infty}$ uniformly in $s_2, \dots, s_n$ as well as the result of Exercise 1.4.10 to obtain the desired conclusion for $\mathscr{G}$. □

3.4.3 Pointwise Divergence of the Dirichlet Means

We now pass to the more difficult question of convergence of the square partial sums of a Fourier series. It is natural to start our investigation with the class of continuous functions. Do the partial sums of the Fourier series of continuous functions converge pointwise? The following simple proposition warns about the behavior of partial sums.

Proposition 3.4.6. *(a)* ***(duBois Reymond)*** *There exists a continuous function f on $\mathbf{T}^1$ whose partial sums diverge at a point. Precisely, for some point $x_0 \in \mathbf{T}^1$ we have*

$$\limsup_{N\to\infty} \left| \sum_{\substack{m\in\mathbf{Z} \\ |m_j|\leq N}} \widehat{f}(m) e^{2\pi i x_0 m} \right| = \infty.$$

(b) There exists a continuous function F on $\mathbf{T}^n$ and $x_0 \in \mathbf{T}^1$ such that the sequence

$$\limsup_{N\to\infty} \left| \sum_{\substack{m\in\mathbf{Z}^n \\ |m_j|\leq N}} \widehat{F}(m) e^{2\pi i (x_0 m_1 + x_2 m_2 + \cdots + x_n m_n)} \right| = \infty$$

for all $x_2, \dots, x_m$ in $\mathbf{T}^1$.

Proof. The proof of part (b) is obtained by considering the continuous function $F(x_1,\dots,x_n) = f(x_1)$, where f is as in part (a). Then we have

$$(F * D_N^n)(x_1,\dots,x_n) = (f * D_N)(x_1)$$

and thus the square partial sums of F diverge on the $(n-1)$-dimensional plane $\{(x_0,x_2,\dots,x_n) : x_2,\dots,x_n \in \mathbf{T}^1\}$.

We now prove part (a) using functional analysis. For a constructive proof, see Exercise 3.4.7. Let $C(\mathbf{T}^1)$ be the Banach space of all continuous functions on the circle equipped with the L^∞ norm. Consider the continuous linear functionals

$$f \to T_N(f) = (D_N * f)(0)$$

on $C(\mathbf{T}^1)$ for $N = 1,2,\dots$. We show that the norms of the T_N's on $C(\mathbf{T}^1)$ converge to infinity as $N \to \infty$. To see this, given any integer $N \geq 100$, let $\varphi_N(x)$ be a continuous even function on $[-\frac{1}{2},\frac{1}{2}]$ that is bounded by 1 and is equal to the sign of $D_N(x)$ except at small intervals of length $(2N+1)^{-2}$ around the $2N+1$ zeros of D_N. Call the union of all these intervals B_N and set $A_N = [-\frac{1}{2},\frac{1}{2}] \setminus B_N$. Then

$$\int_{B_N} |D_N(x)|\,dx + \left|\int_{B_N} \varphi_N(x) D_N(x)\,dx\right| \leq 2|B_N|(2N+1) = 2.$$

Using this estimate we obtain

$$\begin{aligned}
\|T_N\|_{C(\mathbf{T}^1)\to\mathbf{C}} &\geq |T_N(\varphi_N)| = \left|\int_{\mathbf{T}^1} D_N(-x)\varphi_N(x)\,dx\right| \\
&\geq \int_{A_N} |D_N(x)|\,dx - \left|\int_{B_N} D_N(x)\varphi_N(x)\,dx\right| \\
&= \int_{\mathbf{T}^1} |D_N(x)|\,dx - \left|\int_{B_N} D_N(x)\varphi_N(x)\,dx\right| - \int_{B_N} |D_N(x)|\,dx \\
&\geq \frac{4}{\pi^2}\sum_{k=1}^{N}\frac{1}{k} - 2.
\end{aligned}$$

It follows that the norms of the linear functionals T_N are not uniformly bounded. The uniform boundedness principle now implies the existence of a function $f \in C(\mathbf{T}^1)$ and of a sequence $N_j \to \infty$ such that

$$|T_{N_j}(f)| \to \infty$$

as $j \to \infty$. The Fourier series of this f diverges at $x_0 = 0$. □

3.4.4 Pointwise Convergence of the Dirichlet Means

We have seen that continuous functions may have divergent Fourier series. How about Lipschitz continuous functions? As it turns out, there is a more general condition that implies convergence for the Fourier series of functions that satisfy a certain integrability condition.

Theorem 3.4.7. ***(Dini)*** *Let f be an integrable function on $\mathbf{T}^1$, let t_0 be a point on $\mathbf{T}^1$ for which $f(t_0)$ is defined and assume that*

$$\int_{|t|\le\frac{1}{2}} \frac{|f(t+t_0)-f(t_0)|}{|t|}\,dt < \infty. \tag{3.4.9}$$

Then $(D_N^n * f)(t_0) \to f(t_0)$ as $N\to\infty$.
(Tonelli) *Let f be an integrable function on $\mathbf{T}^n$ and let $a=(a_1,\dots,a_n)\in\mathbf{T}^n$. If f is defined at a and*

$$\int_{|x_1|\le\frac{1}{2}} \cdots \int_{|x_n|\le\frac{1}{2}} \frac{|f(x+a)-f(a)|}{|x_1|\cdots|x_n|}\,dx_n\cdots dx_1 < \infty, \tag{3.4.10}$$

then we have $(D_N^n * f)(a) \to f(a)$ as $N\to\infty$.

Proof. Since the one-dimensional result is contained in the multidimensional one, we prove the latter. Replacing $f(x)$ by $f(x+a)-f(a)$, we may assume that $a=0$ and $f(a)=0$. Using identities (3.1.15) and (3.1.14), we can write

$$\begin{aligned}(D_N^n * f)(0) &= \int_{\mathbf{T}^n} f(-x)\prod_{j=1}^n \frac{\sin((2N+1)\pi x_j)}{\sin(\pi x_j)}\,dx_n\cdots dx_1 \\ &= \int_{\mathbf{T}^n} f(-x)\prod_{j=1}^n \left(\frac{\sin(2N\pi x_j)\cos(\pi x_j)}{\sin(\pi x_j)} + \cos(2N\pi x_j)\right)dx_n\cdots dx_1\,.\end{aligned} \tag{3.4.11}$$

Expand out the product to express the integrand as a sum of terms of the form

$$\left\{f(-x)\prod_{j\in I}\frac{\cos(\pi x_j)}{\sin(\pi x_j)}\right\}\prod_{j\in I}\sin(2N\pi x_j)\prod_{k\in\{1,2,\dots,n\}\setminus I}\cos(2N\pi x_k), \tag{3.4.12}$$

where I is a subset of $\{1,2,\dots,n\}$; here we use the convention that the product over an empty set of indices is 1. The function f_I inside the curly brackets in (3.4.12) is integrable on $[-\frac{1}{2},\frac{1}{2})^n$ except possibly in a neighborhood of the origin, since $|\sin(\pi x_j)|\ge 2|x_j|$ when $|x_j|\le\frac{1}{2}$. But condition (3.4.10) with $a=0$ and $f(a)=0$ guarantees that f_I is also integrable in a neighborhood of the origin. Expressing the sines and cosines in (3.4.12) in terms of exponentials, we obtain that the integral of (3.4.12) over $[-\frac{1}{2},\frac{1}{2})^n$ is a finite linear combination of Fourier coefficients of f_I at the points $(\pm N,\dots,\pm N)\in\mathbf{Z}^n$. Applying Lemma 3.3.1 yields that the expression in (3.4.11) tends to zero as $N\to\infty$. □

The following are consequences of this test.

Corollary 3.4.8. *(a) (**Riemann's principle of localization**) Let f be an integrable function on $\mathbf{T}^1$ that vanishes on an open interval I. Then $D_N * f$ converges to zero on the interval I.*
(b) Let $a = (a_1, \dots, a_n) \in \mathbf{T}^n$ and suppose that an integrable function f on $\mathbf{T}^n$ is constant on the cross

$$\{x = (x_1, \dots, x_n) \in \mathbf{T}^n : \ |x_j - a_j| < \delta_j \quad \textit{for some } j\},$$

where $0 < \delta_j < 1/2$ are fixed. Then $(D_N^n * f)(a) \to f(a)$ as $N \to \infty$.

Proof. (a) Let $t_0 \in I$. If f vanishes on I, condition (3.4.9) holds, since the function $t \mapsto f(t+t_0) - f(t_0)$ vanishes on $-t_0 + I$, which is an interval containing the origin, and is integrable outside $-t_0 + I$. Thus $(D_N * f)(t_0) \to f(t_0) = 0$ for every $t_0 \in I$.
(b) We need to show that the function

$$\frac{|f(x+a) - f(a)|}{|x_1| \cdots |x_n|}$$

is integrable over $\mathbf{T}^n = [-1/2, 1/2)^n$. The integral of this function over $\mathbf{T}^n$ is equal to its integral over the region

$$S = \{(x_1, \dots, x_n) \in \mathbf{T}^n : \ |x_k| \geq \delta_k \quad \text{for all } k\},$$

since $f(x+a) - f(a)$ vanishes whenever $|x_j| < \delta_j$ for some $j \in \{1, 2, \dots, n\}$. But on S we have that

$$\frac{|f(x+a) - f(a)|}{|x_1| \cdots |x_n|} \leq \frac{|f(x+a) - f(a)|}{\delta_1 \cdots \delta_n}$$

and this function is integrable over S, since f is. We deduce that (3.4.10) holds. □

Corollary 3.4.9. *Let $a \in \mathbf{T}^n$ and suppose that $f \in L^1(\mathbf{T}^n)$ satisfies*

$$|f(x) - f(a)| \leq C |x_1 - a_1|^{\varepsilon_1} \cdots |x_n - a_n|^{\varepsilon_n}$$

for some $C, \varepsilon_j > 0$ and for all $x \in \mathbf{T}^n$. Then the square partial sums $(D_N^n * f)(a)$ converge to $f(a)$.

Proof. Note that condition (3.4.10) holds. □

Corollary 3.4.10. *(**Dirichlet**) If f is defined on $\mathbf{T}^1$ and is a differentiable function at a point a in $\mathbf{T}^1$, then $(D_N * f)(a) \to f(a)$.*

Proof. There exists a $\delta > 0$ (say less than $1/2$) such that $|f(x) - f(a)|/|x-a|$ is bounded by $|f'(a)| + 1$ for $|x - a| \leq \delta$. Also $|f(x) - f(a)|/|x-a|$ is bounded by $|f(x) - f(a)|/\delta$ when $|x - a| > \delta$. It follows that condition (3.4.9) holds. □

Exercises

3.4.1. Identify $\mathbf{T}^1$ with $[-1/2,1/2)$ and fix $0<b<1/2$. Prove the following:

(a) The mth Fourier coefficient of the function x is $i\frac{(-1)^m}{2\pi m}$ when $m\neq 0$ and 0 when $m=0$.
(b) The mth Fourier coefficient of the function $\chi_{[-b,b]}$ is $\frac{\sin(2\pi bm)}{m\pi}$ when $m\neq 0$ and $2b$ when $m=0$.
(c) The mth Fourier coefficient of the function $\big(1-\frac{|x|}{b}\big)_+$ is $\frac{\sin^2(\pi bm)}{bm^2\pi^2}$ when $m\neq 0$ and b when $m=0$.
(d) The mth Fourier coefficient of the function $|x|$ is $-\frac{1}{2m^2\pi^2}+\frac{(-1)^m}{2m^2\pi^2}$ when $m\neq 0$ and $\frac{1}{4}$ when $m=0$.
(e) The mth Fourier coefficient of the function x^2 is $\frac{(-1)^m}{2m^2\pi^2}$ when $m\neq 0$ and $\frac{1}{12}$ when $m=0$.
(f) The mth Fourier coefficient of the function $\cosh(2\pi x)$ is $\frac{(-1)^m}{1+m^2}\frac{\sinh\pi}{\pi}$.
(g) The mth Fourier coefficient of the function $\sinh(2\pi x)$ is $\frac{im(-1)^m}{1+m^2}\frac{\sinh\pi}{\pi}$.

3.4.2. Use Exercise 3.4.1 and Proposition 3.4.2 to prove that

$$\sum_{k\in\mathbf{Z}}\frac{1}{(2k+1)^2}=\frac{\pi^2}{4}\qquad \sum_{k\in\mathbf{Z}\setminus\{0\}}\frac{1}{k^2}=\frac{\pi^2}{3}$$

$$\sum_{k\in\mathbf{Z}\setminus\{0\}}\frac{(-1)^{k+1}}{k^2}=\frac{\pi^2}{6}\qquad \sum_{k\in\mathbf{Z}}\frac{(-1)^k}{k^2+1}=\frac{2\pi}{e^\pi-e^{-\pi}}.$$

3.4.3. Let $M>N$ be given positive integers.
(a) For $f\in L^1(\mathbf{T}^1)$, prove the following identity:

$$\begin{aligned}(D_N*f)(x)&=\frac{M+1}{M-N}(F_M*f)(x)-\frac{N+1}{M-N}(F_N*f)(x)\\ &\quad-\frac{M+1}{M-N}\sum_{N<|j|\le M}\left(1-\frac{|j|}{M+1}\right)\widehat{f}(j)e^{2\pi ijx}.\end{aligned}$$

(b) (*G. H. Hardy*) Suppose that a function f on $\mathbf{T}^1$ satisfies the following condition: for any $\varepsilon>0$ there exists an $a>1$ and a $k_0>0$ such that for all $k\ge k_0$ we have

$$\sum_{k<|m|\le[ak]}|\widehat{f}(m)|<\varepsilon.$$

Use part (a) to prove that if $(F_N*f)(x)$ converges (uniformly) to $A(x)$ as $N\to\infty$, then $(D_N*f)(x)$ also converges (uniformly) to $A(x)$ as $N\to\infty$.

3.4.4. Use Proposition 3.4.2 to show that for $0 < b < \frac{1}{2}$ we have

$$\lim_{N\to\infty} \sum_{\substack{m=-N \\ m\neq 0}}^{N} \frac{\sin(2\pi bm)}{m\pi} e^{2\pi i bm} = \frac{1}{2} - 2b.$$

[*Hint:* Use Exercise 3.4.1(b).]

3.4.5. Let f be an integrable function on $\mathbf{T}^n$ and g be a bounded function on $\mathbf{T}^n$ and let K be a compact subset of $\mathbf{T}^n$. Consider the family $\mathscr{F} = \{f_w : w \in K\}$, where $f_w(x) = f(x-w)g(x)$ for all $x \in \mathbf{T}^n$. Prove that the Riemann–Lebesgue lemma holds uniformly for the family $\mathscr{F}$. This means that given $\varepsilon > 0$ there exists an $N_0(K) > 0$ such that for $|m| \geq N_0$ we have $|\widehat{f_w}(m)| \leq \varepsilon$ for all $w \in K$.

3.4.6. Prove the following version of Corollary 3.4.8 (b). Suppose that a function f on $\mathbf{T}^n$ is constant on the cross $U = \{(x_1,\dots,x_n) \in \mathbf{T}^n : |x_j - a_j| < \delta \text{ for some } j\}$, for some $\delta < 1/2$. Then $D_N^n * f$ converges to $f(a)$ uniformly on compact subsets of the box $W = \{(x_1,\dots,x_n) \in \mathbf{T}^n : |x_j| < \delta \text{ for all } j\}$.
[*Hint:* Use Exercise 3.4.5.]

3.4.7. Follow the steps given to obtain a constructive proof of the existence of a continuous function whose Fourier series diverges at a point. Identify $\mathbf{T}^1$ with $[0,1)$ and define

$$g(x) = -2\pi i(x - 1/2).$$

(a) Prove that $\widehat{g}(m) = 1/m$ when $m \neq 0$ and zero otherwise.
(b) Prove that for all nonnegative integers M and N we have

$$\big((e^{2\pi i N(\cdot)}(g * D_N)) * D_M\big)(x) = e^{2\pi i N x} \sum_{1\leq|r|\leq N} \frac{1}{r} e^{2\pi i r x}$$

when $M \geq 2N$ and

$$\big((e^{2\pi i N(\cdot)}(g * D_N)) * D_M\big)(x) = e^{2\pi i N x} \sum_{\substack{-N\leq r\leq M-N \\ r\neq 0}} \frac{1}{r} e^{2\pi i r x}$$

when $M < 2N$. Conclude that there exists a constant $C > 0$ such that for all M, N, and $x \neq 0$ we have

$$\big|(e^{2\pi i N(\cdot)}(g * D_N)) * D_M(x)\big| \leq \frac{C}{|x|}.$$

(c) Show that there exists a constant $C_1 > 0$ such that

$$\sup_{N>0} \sup_{x\in\mathbf{T}^1} \big|(g * D_N)(x)\big| = \sup_{N>0} \sup_{x\in\mathbf{T}^1} \left| \sum_{1\leq|r|\leq N} \frac{1}{r} e^{2\pi i r x} \right| \leq C_1 < \infty.$$

(d) Let $\lambda_k = 1 + e^{e^k}$. Define

$$f(x) = \sum_{k=1}^{\infty} \frac{1}{k^2} e^{2\pi i \lambda_k x} (g * D_{\lambda_k})(x)$$

and prove that f is continuous on $\mathbf{T}^1$ and that its Fourier series converges at every $x \neq 0$, but $\limsup_{M\to\infty} |(f * D_M)(0)| = \infty$.
$\big[$*Hint:* Take $M = e^{e^{e^m}}$ with $m \to \infty$. The inequality in part (b) follows by summation by parts.$\big]$

3.5 A Tauberian theorem and Functions of Bounded Variation

The relation between the partial sums of a Fourier series and the Fejér means is a particular situation of a relation between sequences of complex numbers and their arithmetic means. Given a sequence $\{a_k\}_{k=0}^{\infty}$ of complex numbers, we denote its *partial sums* by

$$s_N = a_1 + \cdots + a_N$$

for $N \geq 0$, and its arithmetic or *Cesàro means* by

$$\sigma_N = \frac{1}{N+1} \sum_{k=0}^{N} s_k = \frac{1}{N+1} \sum_{k=0}^{N} (N+1-k) a_k \,.$$

A classical result says that if $s_N \to L$ as $N \to \infty$, then $\sigma_N \to L$ as $N \to \infty$. The converse is not true, as the example $a_k = (-1)^k$ indicates. But in a particular situation the reverse implication holds.

3.5.1 A Tauberian theorem

We have the following result concerning the convergence of $\{s_k\}_{k=0}^{\infty}$ as a consequence of that of $\{\sigma_k\}_{k=0}^{\infty}$.

Theorem 3.5.1. *(a) Suppose that for a sequence $\{a_k\}_{k=0}^{\infty}$ of complex numbers we have that $\sigma_N \to L$ as $N \to \infty$ and that $|ka_k| \leq M < \infty$ for all $k = 0, 1, 2, \dots$. Then $s_k \to L$ as $k \to \infty$.*
(b) Let X be a nonempty set. Suppose that for a sequence $\{a_k(x)\}_{k=0}^{\infty}$ of complex-valued functions on X we have that $\sigma_N(x) \to L(x)$ uniformly in $x \in X$ as $N \to \infty$ and that $\sup_{k\geq 0} \sup_{x\in X} |ka_k(x)| \leq M < \infty$. Then $s_k(x) \to L(x)$ uniformly in $x \in X$ as $k \to \infty$.

Proof. We prove part (b), noting that the proof of part (a) is subsumed in that of (b).

For $0 \le k < m < \infty$ we have

$$\begin{aligned}
&(m+1)\sigma_m(x) - (k+1)\sigma_k(x) - \sum_{j=k+1}^{m} (m+1-j)a_j(x) \\
=&\sum_{j=0}^{m}(m+1-j)a_j(x) - \sum_{j=0}^{k}(k+1-j)a_j(x) - \sum_{j=k+1}^{m}(m+1-j)a_j(x) \\
=&(m-k)\sum_{j=0}^{k} a_j(x) \\
=&(m-k)s_k(x)\,.
\end{aligned}$$

Therefore we have

$$\frac{m+1}{m-k}\sigma_m(x) - \frac{k+1}{m-k}\sigma_k(x) - \frac{1}{m-k}\sum_{j=k+1}^{m}(m+1-j)a_j(x) = s_k(x)$$

and thus

$$s_k(x) - \sigma_k(x) = \frac{m+1}{m-k}(\sigma_m(x) - \sigma_k(x)) - \frac{m+1}{m-k}\sum_{j=k+1}^{m}\Big(\frac{1}{j} - \frac{1}{m+1}\Big) j a_j(x)\,. \tag{3.5.1}$$

Notice that

$$\sum_{j=k+1}^{m}\Big(\frac{1}{j} - \frac{1}{m+1}\Big) \le \int_k^m \frac{dt}{t} - \sum_{j=k+1}^{m}\frac{1}{m+1} = \log\frac{m}{k} - \frac{m-k}{m+1}\,. \tag{3.5.2}$$

Now fix $\varepsilon > 0$ such that $\varepsilon < 1$. For each $k \in \mathbf{Z}^+$ pick an $m_k \in \{k, k+1, \dots, 2k\}$ such that $\frac{m_k}{k} \to 1+\varepsilon$. Then $\frac{m_k+1}{m_k-k}$ converges to $\varepsilon^{-1}+1$ as $k \to \infty$, hence it is bounded by some constant C_ε. Then (3.5.1) and (3.5.2) with m_k in place of m yield

$$\sup_{x\in X}\big|s_k(x) - \sigma_k(x)\big| \le C_\varepsilon \sup_{x\in X}|\sigma_{m_k}(x) - \sigma_k(x)| + M\frac{m_k+1}{m_k-k}\Big[\log\frac{m_k}{k} - \frac{m_k-k}{m_k+1}\Big]\,.$$

Taking the $\limsup_{k\to\infty}$ in the preceding inequality and using that

$$\limsup_{k\to\infty}\sup_{x\in X}|\sigma_{m_k}(x) - \sigma_k(x)| = 0\,,$$

which is a consequence of the hypothesis that σ_k (and thus σ_{m_k}) converges to L uniformly, we obtain

$$\limsup_{k\to\infty}\sup_{x\in X}\big|s_k(x) - \sigma_k(x)\big| \le M\Big[\Big(1+\frac{1}{\varepsilon}\Big)\log(1+\varepsilon) - 1\Big]\,.$$

In view of the Taylor expansion

$$\log(1+\varepsilon) = \varepsilon - \frac{1}{2}\varepsilon^2 + \frac{1}{3}\varepsilon^3 - \cdots = \varepsilon + O(\varepsilon^2),$$

which is valid for $0 < \varepsilon < 1$, we conclude that

$$\limsup_{k\to\infty} \sup_{x\in X} \big|s_k(x) - \sigma_k(x)\big| \le M\, c\, \varepsilon$$

for some absolute constant $c > 0$. Since $\varepsilon > 0$ was arbitrary, we finally deduce that $s_k(x)$ converges uniformly to the same limit as $\sigma_k(x)$, which is $L(x)$. □

Corollary 3.5.2. *Suppose that a function f on $\mathbf{T}^1$ is continuous and there is a constant $M > 0$ such that $|\widehat{f}(m)| \le M|m|^{-1}$ for all $m \in \mathbf{Z}^+ \setminus \{0\}$. Then the Fourier series of f converges uniformly to f. In particular, if f is a continuous function of bounded variation on the circle, then $f * D_N \to f$ uniformly on $\mathbf{T}^1$ as $N \to \infty$.*

Proof. The Fejér means $\{F_N\}_{N=0}^{\infty}$ are an approximate identity on $\mathbf{T}^n$ (Proposition 3.1.10) and so $F_N * f$ converge uniformly to f on $\mathbf{T}^1$ as $N \to \infty$ in view of Theorem 1.2.19 (2). Moreover, we have $|m|\,|\widehat{f}(m)| \le M$ for all $m \in \mathbf{Z}$. It follows from Theorem 3.5.1 that $D_N * f$ converges uniformly to f.

If, additionally, f is a function of bounded variation, then $|m|\,|\widehat{f}(m)| \le \frac{1}{2\pi}\mathrm{Var}(f)$, as shown in Proposition 3.3.14. Then the claimed conclusion follows. □

3.5.2 The sine integral function

We examine a few useful properties of the antiderivative of $\sin(t)/t$.

Definition 3.5.3. For $0 \le x < \infty$ define the *sine integral function*

$$Si(x) = \int_0^x \frac{\sin(t)}{t}\, dt. \tag{3.5.3}$$

Integrating by parts we write

$$Si(x) = \int_0^1 \frac{\sin(t)}{t}\, dt + \frac{-\cos(x)}{x} + \cos(1) - \int_1^x \frac{\cos(t)}{t^2}\, dt,$$

from which it follows that the limit of $Si(x)$ as $x \to \infty$ exists and is equal to

$$\lim_{x\to\infty} Si(x) = \int_0^1 \frac{\sin(t)}{t}\, dt + \cos(1) - \int_1^\infty \frac{\cos(t)}{t^2}\, dt.$$

To precisely evaluate the limit of $Si(x)$ as $x \to \infty$ we write

$$\begin{aligned} Si((N+\tfrac{1}{2})\pi) =& \pi \int_0^{\frac{1}{2}} \frac{\sin((2N+1)\pi t)}{\pi t}\, dt \\ =& \pi \int_0^{\frac{1}{2}} \frac{\sin((2N+1)\pi t)}{\sin(\pi t)}\, dt + \pi \int_0^{\frac{1}{2}} \sin((2N+1)\pi t) \left\{ \frac{1}{\pi t} - \frac{1}{\sin(\pi t)} \right\} dt \\ =& \frac{\pi}{2} \int_{-\frac{1}{2}}^{\frac{1}{2}} D_N(t)\, dt + \frac{\pi}{2} \int_{-\frac{1}{2}}^{\frac{1}{2}} \frac{e^{(2N+1)\pi it} - e^{-(2N+1)\pi it}}{2i} \left\{ \frac{1}{\pi t} - \frac{1}{\sin(\pi t)} \right\} dt, \end{aligned}$$

which converges to $\pi/2 + 0$ as $N \to \infty$, in view of the Riemann-Lebesgue lemma (Proposition 3.3.1), since the function inside the curly brackets is integrable over the circle. We conclude that $\lim_{x\to\infty} Si(x) = \pi/2$.

Note that Si' vanishes at $n\pi$, $n = 0, 1, 2, \dots$ and $Si''(n\pi) = (-1)^n/n\pi$. Consequently, $Si(x)$ has local maxima at the points π, 3π, 5π, $\dots$ and local minima at the points 2π, 4π, 6π, $\dots$. Moreover, it is increasing on the intervals $[2k\pi, (2k+1)\pi]$ and decreasing on $[(2k+1)\pi, (2k+2)\pi]$, $k = 0, 1, 2, \dots$. Also, observe that

$$Si(3\pi) - Si(\pi) = \int_\pi^{2\pi} \frac{\sin(t)}{t}\, dt + \int_\pi^{2\pi} \frac{\sin(t+\pi)}{t+\pi}\, dt = \int_\pi^{2\pi} \sin(t) \left(\frac{1}{t} - \frac{1}{t+\pi} \right) dt < 0$$

and likewise we can prove the remaining inequalities in the sequence

$$Si(\pi) > Si(3\pi) > Si(5\pi) > Si(7\pi) > \cdots > \frac{\pi}{2}.$$

Similarly, one can show that

$$Si(2\pi) < Si(4\pi) < Si(6\pi) < \cdots < \frac{\pi}{2}.$$

Hence $Si(\pi)$ is the absolute maximum of $Si(x)$ on $[0, \infty)$, while 0 is the absolute minimum of $Si(x)$ on $[0, \infty)$; $Si(\pi)$ is the absolute minimum of $Si(x)$ on $[\pi, \infty)$.

3.5.3 Further properties of functions of bounded variation

Next we have the following theorem concerning functions of bounded variation. Recall that functions of bounded variation are differences of increasing functions and thus have left and right limits at every point.

Theorem 3.5.4. *Let $0 < \delta \le 1/2$. Suppose that f is an integrable function on $\mathbf{T}^1$ which is of bounded variation on the neighborhood $[t_0 - \delta, t_0 + \delta]$ of the point $t_0 \in \mathbf{T}^1$. Then*

$$\lim_{N\to\infty} (f * D_N)(t_0) = \frac{f(t_0+) + f(t_0-)}{2}.$$

Proof. We write W for the neighborhood $(-\delta,\delta)$ of 0, $F_{t_0}(t)=\frac{f(t_0-t)+f(t_0+t)}{2}$, and $L_{t_0}=\frac{f(t_0+)+f(t_0-)}{2}$. We have

$$(f*D_N)(t_0)=\int_{\mathbf{T}^1} f(t_0-t)D_N(t)\,dt=\int_{\mathbf{T}^1} f(t_0+t)D_N(t)\,dt\,,$$

hence, averaging yields

$$(f*D_N)(t_0)=\int_{\mathbf{T}^1}\frac{f(t_0-t)+f(t_0+t)}{2}D_N(t)\,dt=\int_{\mathbf{T}^1}F_{t_0}(t)D_N(t)\,dt\,.$$

Therefore we have

$$(f*D_N)(t_0)-L_{t_0}=\int_W \big(F_{t_0}(t)-L_{t_0}\big)D_N(t)\,dt+\int_{\mathbf{T}^1\setminus W}\big(F_{t_0}(t)-L_{t_0}\big)D_N(t)\,dt$$

and since in the second integral $|t|\geq\delta$, the Riemann-Lebesgue lemma shows that the second term is $o(1)$, i.e., it tends to zero as $N\to\infty$. We now show that the first integral also goes to zero. We write

$$\begin{aligned}\int_W \big(F_{t_0}(t)-L_{t_0}\big)D_N(t)\,dt&=\int_W \big(F_{t_0}(t)-L_{t_0}\big)\frac{\sin((2N+1)\pi t)}{\pi t}\,dt\\&\quad+\int_W \big(F_{t_0}(t)-L_{t_0}\big)\left(\frac{1}{\sin(\pi t)}-\frac{1}{\pi t}\right)\sin((2N+1)\pi t)\,dt\,,\end{aligned}\tag{3.5.4}$$

but since the function $\frac{1}{\pi t}-\frac{1}{\sin(\pi t)}$ remains bounded on $[-\frac{1}{2},\frac{1}{2}]$ (Exercise 3.1.5 (a)), it follows from the Riemann-Lebesgue lemma that the second term is $o(1)$ as $N\to\infty$. Consequently,

$$(f*D_N)(t_0)-L_{t_0}=\frac{1}{\pi}\int_W \big(F_{t_0}(t)-L_{t_0}\big)\frac{\sin((2N+1)\pi t)}{t}\,dt+o(1)$$

as $N\to\infty$. To prove the required conclusion, it will suffice to show that

$$\frac{2}{\pi}\int_0^\delta \big(F_{t_0}(t)-L_{t_0}\big)\frac{\sin((2N+1)\pi t)}{t}\,dt\to 0\tag{3.5.5}$$

as $N\to\infty$. Let $Si(t)$ be as defined in (3.5.3). We express the integral in (3.5.5) as

$$\int_0^\delta \big(F_{t_0}(t)-L_{t_0}\big)Si'((2N+1)\pi t)\,dt\tag{3.5.6}$$

Integrating by parts we obtain that (3.5.6) is equal to

$$(F_{t_0}(\delta-)-L_{t_0})Si((2N+1)\pi\delta)-\int_0^\delta Si((2N+1)\pi t)dF_{t_0}(t)\,.\tag{3.5.7}$$

Letting $N \to \infty$ and using the Lebesgue dominated convergence theorem, we conclude that (3.5.7) converges to

$$(F_{t_0}(\delta-) - L_{t_0})\frac{\pi}{2} - \int_0^\delta \frac{\pi}{2} dF_{t_0}(t) = (F_{t_0}(\delta-) - L_{t_0})\frac{\pi}{2} - (F_{t_0}(\delta-) - F_{t_0}(0+))\frac{\pi}{2} = 0$$

noticing that $L_{t_0} = F_{t_0}(0+)$. □

Next, we obtain an explicit bound for the partial sums of functions of bounded variation. Let $Si(t)$ be as in (3.5.3).

Theorem 3.5.5. *Suppose that f is a function of bounded variation on the circle $\mathbf{T}^1$. Then the partial sums of the Fourier series of f are uniformly bounded, in particular, we have*

$$\sup_{t_0 \in \mathbf{T}^1} \sup_{N \in \mathbf{Z}^+} |(f * D_N)(t_0)| \le \left(1 - \frac{2}{\pi} + Si(\pi)\right) \|f\|_{L^\infty} + Si(\pi)\mathrm{Var}(f). \tag{3.5.8}$$

Proof. We take $\delta = 1/2$ in the proof of the preceding theorem. For a point $t_0 \in \mathbf{T}^1$, let $F_{t_0}(t) = \frac{f(t_0-t)+f(t_0+t)}{2}$. We have that

$$\begin{aligned}(f * D_N)(t_0) = &\int_{\mathbf{T}^1} F_{t_0}(t) \frac{\sin((2N+1)\pi t)}{\pi t} dt \\ &+ \int_{\mathbf{T}^1} F_{t_0}(t) \left(\frac{1}{\sin(\pi t)} - \frac{1}{\pi t}\right) \sin((2N+1)\pi t)\, dt.\end{aligned} \tag{3.5.9}$$

Using that $\left|\frac{1}{\sin(\pi t)} - \frac{1}{\pi t}\right| \le 1 - \frac{2}{\pi}$ when $|t| \le \frac{1}{2}$ (Exercise 3.1.5 (a)), we obtain that the second integral in (3.5.9) is bounded by $(1 - \frac{2}{\pi})\|f\|_{L^\infty}$. Integrating by parts as in the proof of the preceding theorem, we express the first integral in (3.5.9) as

$$F_{t_0}(\tfrac{1}{2}-)Si((2N+1)\pi\tfrac{1}{2}) - \int_0^{\frac{1}{2}} Si((2N+1)\pi t)\, dF_{t_0}(t), \tag{3.5.10}$$

which is bounded (in absolute value) by $\|f\|_{L^\infty} Si(\pi) + Si(\pi)\mathrm{Var}(f)$. Assertion (3.5.8) now follows. □

3.5.4 Gibbs phenomenon

It is not reasonable to expect that the Fourier series of a discontinuous function converges uniformly in a neighborhood of a discontinuity. The lack of uniformity in the convergence can be measured in terms of the worst jump, called the *overshoot*. The exact form of nonuniform convergence is illustrated in the following example:

Example 3.5.6. Consider the function

$$h(t)=\begin{cases}\frac{1}{2}-t & \text{when } 0<t\le\frac{1}{2}\\ 0 & \text{when } t=0\\ -\frac{1}{2}-t & \text{when } -\frac{1}{2}<t<0.\end{cases} \tag{3.5.11}$$

Clearly $h(t)$ is a function of bounded variation and is continuous except at the point $t=0$ at which it has a jump discontinuity. Since h is an odd function, its Fourier coefficients are

$$\widehat{h}(m)=\int_{-1/2}^{1/2}h(t)e^{-2\pi imt}\,dt=-2i\int_0^{1/2}(\tfrac{1}{2}-t)\sin(2\pi mt)\,dt=-\frac{i}{2m\pi}$$

when $m\neq 0$ and $\widehat{h}(0)=0$. The partial sums of the Fourier series of h are

$$(h*D_N)(t)=-\frac{i}{2\pi}\sum_{\substack{|m|\le N\\ m\neq 0}}\frac{e^{2\pi imt}}{m}.$$

Notice that

$$\frac{d}{dt}(h*D_N)(t)=\sum_{\substack{|m|\le N\\ m\neq 0}}e^{2\pi imt}=D_N(t)-1.$$

Then, if we define $d(s)=\frac{1}{\sin(\pi s)}-\frac{1}{\pi s}$, we can write

$$\begin{aligned}(h*D_N)(t)&=\int_0^t D_N(s)-1\,ds\\ &=-t+\int_0^t\frac{\sin((2N+1)\pi s)}{\sin(\pi s)}\,ds\\ &=-t+\int_0^t d(s)\sin((2N+1)\pi s)\,ds+\int_0^t\frac{\sin((2N+1)\pi s)}{\pi s}\,ds.\end{aligned}$$

Notice that $d(s)$ is continuous at zero and $d(0)=0$, while $\lim_{s\to 0}\frac{d(s)}{s}=\frac{\pi}{6}$; thus d is a differentiable function on $[0,\frac{1}{2}]$ and $d'(0)=\frac{\pi}{6}$. Moreover, $\lim_{s\to 0}d'(s)=d'(0)$, thus d is continuously differentiable on $[0,\frac{1}{2}]$. Additionally both d and d' are nonnegative and increasing on $[0,\frac{1}{2}]$, thus $d'\le\frac{4}{\pi}$; see the hint of Exercise 3.1.5. It follows that

$$\int_0^t d(s)\sin((2N+1)\pi s)ds=-\frac{\cos((2N+1)\pi t)}{(2N+1)\pi}d(t)+\int_0^t d'(s)\frac{\cos((2N+1)\pi s)}{(2N+1)\pi}ds$$

and the preceding expression is bounded in absolute value by $(\frac{d(\frac{1}{2})}{\pi}+\frac{1}{2}\frac{d'(\frac{1}{2})}{\pi})\frac{1}{2N+1}$.

We deduce that

$$(h*D_N)(t)=-t+\frac{1}{\pi}\int_0^t\frac{\sin((2N+1)\pi s)}{s}\,ds+O\Big(\frac{1}{2N+1}\Big),$$

where $O\left(\frac{1}{2N+1}\right)$ is a function bounded by $\frac{1}{\pi}\frac{1}{2N+1}$. Consequently,

$$(h*D_N)(t) = -t + \frac{1}{\pi}\int_0^{(2N+1)\pi t} \frac{\sin(s)}{s}\,ds + O\Big(\frac{1}{2N+1}\Big).$$

Hence for $t \in (0, -\frac{1}{2}]$ we have $\lim_{N\to\infty}(h*D_N)(t) = -t + \frac{1}{\pi}\frac{\pi}{2} = \frac{1}{2} - t$ as expected. Analogously for $t \in [-\frac{1}{2}, 0)$ we have $\lim_{N\to\infty}(h*D_N)(t) = -\frac{1}{2} - t$. Also for $t = 0$, $\lim_{N\to\infty}(h*D_N)(0) = 0$. Thus the Fourier series of h at zero converges to the "fair" value of the average of $h(0+)$ and $h(0-)$ which happens to be $h(0) = 0$.

To quantitatively estimate the nonuniformity of the convergence of $(h*D_N)(t)$ we note that

$$(h*D_N)(t) - (\tfrac{1}{2} - t) = \frac{1}{\pi}\int_0^{(2N+1)\pi t} \frac{\sin(s)}{s}\,ds - \frac{1}{2} + O\Big(\frac{1}{2N+1}\Big).$$

Thus for all $N = 1, 2, \dots$ and $t \in (0, \frac{1}{2}]$ we have

$$(h*D_N)(t) - h(t) \le \frac{Si(\pi)}{\pi} - \frac{1}{2} + \frac{1}{\pi}\frac{1}{2N+1} \le .08949\cdots + \frac{\pi^{-1}}{2N+1}.$$

Also, for any sequence $t_N \to 0+$ we have

$$\limsup_{N\to\infty}\big[(h*D_N)(t_N) - h(t_N)\big] \le \frac{Si(\pi)}{\pi} - \frac{1}{2} = .08949\dots, \tag{3.5.12}$$

while if for each N we consider the value $t_N = 1/(2N+1)$, we obtain that

$$\limsup_{N\to\infty}\big[(h*D_N)(t_N) - h(t_N)\big] = \frac{Si(\pi)}{\pi} - \frac{1}{2} = .08949\dots. \tag{3.5.13}$$

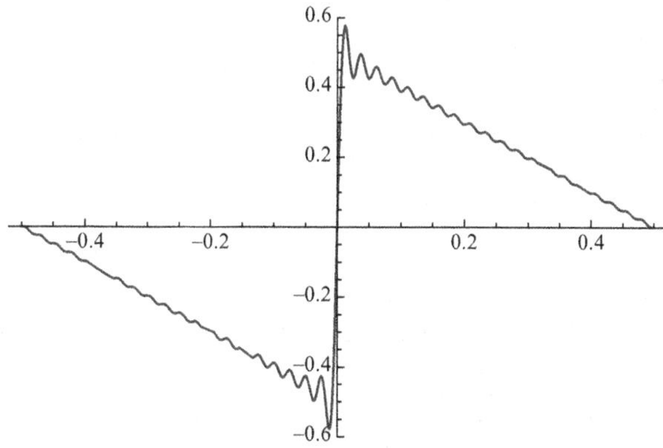

Fig. 3.3 The partial sums $(h*D_{40})(t)$ showing the overshoot of approximately 9% of the jump of h at zero.

The quantity .08949... is called the *overshoot* of the partial sums of the Fourier series of the function h in a neighborhood of zero. See Figure 3.3.

We now examine the preceding phenomenon in the setting of functions of bounded variation. These functions can be written as a differences of two increasing functions, so they have countable sets of discontinuities. Suppose we are given a function $f \in L^1(\mathbf{T}^1)$ of bounded variation, and for the sake of simplicity, let us consider the situation where it has exactly one discontinuity, say at the point $t_0 \in \mathbf{T}^1$. Consider the function h defined in (3.5.11) and define

$$f_0(t) = \begin{cases} \big(f(t_0+) - f(t_0-)\big)h(t-t_0) + \frac{f(t_0+)+f(t_0-)}{2} & \text{when } t \neq t_0, \\ f(t_0) & \text{when } t = t_0. \end{cases} \tag{3.5.14}$$

Now the function $f - f_0$ is of bounded variation and is also continuous and satisfies $(f - f_0)(t_0) = 0$. In view of Corollary 3.5.2, the Fourier series of $f - f_0$ converges uniformly to $f - f_0$ and so the lack of uniformity of the convergence of the partial sums of f is due to the presence of f_0.

We express these observations as a theorem.

Theorem 3.5.7. *(a) Let h be defined in* (3.5.11). *Then the set of accumulation points of sets of the form $\{(h * D_N)(t_N)\}_{N\in\mathbf{Z}^+}$, where $t_N \in [0,1/2]$, is the interval*

$$\Big[0, \tfrac{Si(\pi)}{\pi}\Big] = \big[0, 0.58949\ldots\big].$$

In particular if $t_N \to 0$ such that $Nt_N \to \frac{1}{2}$, then

$$\lim_{N\to\infty} (h * D_N)(t_N) = \frac{Si(\pi)}{\pi} = 0.58949\ldots.$$

(b) Let f be a function of bounded variation on the circle with a single discontinuity at the point t_0, such that $f(t_0+) - f(t_0-) > 0$. Then the set of accumulation points of sets of the form $\{(f * D_N)(t_N)\}_{N\in\mathbf{Z}^+}$, where $t_N \in [t_0, t_0+\delta]$, for some $\delta > 0$, is the interval

$$\Big[\tfrac{f(t_0+)+f(t_0-)}{2}, \tfrac{f(t_0+)+f(t_0-)}{2} + \tfrac{Si(\pi)}{\pi}\big(f(t_0+) - f(t_0-)\big)\Big].$$

In particular if $t_N \to t_0+$ such that $N(t_N - t_0) \to \frac{1}{2}$, then

$$\lim_{N\to\infty} (f * D_N)(t_N) = \frac{f(t_0+)+f(t_0-)}{2} + \frac{Si(\pi)}{\pi}\big(f(t_0+) - f(t_0-)\big).$$

Proof. (a) Since $h \geq 0$ on $(0, \frac{1}{2}]$ and $(h * D_N)(t) \to \frac{1}{2} - t$ for $0 < t \leq \frac{1}{2}$ we have that all accumulation points of sequences $(h * D_N)(t_N)$ are nonnegative. We showed in (3.5.12) that all accumulation points of sequences $(h * D_N)(t_N) - h(t_N)$ are at most $\frac{Si(\pi)}{\pi} - \frac{1}{2}$; but $h(t_N) \leq \frac{1}{2}$, when $t_N \in [0, \frac{1}{2}]$, hence all accumulation points of sequences $(h * D_N)(t_N)$ are at most $\frac{Si(\pi)}{\pi}$ and thus contained in $[0, \frac{Si(\pi)}{\pi}]$. Also, 0 is attained as the accumulation point of $(h * D_N)(0)$ and the number $\frac{Si(\pi)}{\pi}$ is attained as the accumulation point of the sequence $(h * D_N)(\frac{1}{2N+1})$ as shown in (3.5.13); notice that the same assertion is valid for any other sequence $t_N \to 0$ such that $Nt_N \to \frac{1}{2}$. Now, since the functions $(h * D_N)(t)$ are continuous, given any c in $\left[0, \frac{Si(\pi)}{\pi}\right]$, there is a t'_N between 0 and $\frac{1}{2N+1}$ such that $(h * D_N)(t'_N) = c$ for all N; this shows that the set of accumulation points of sequences of the form $\{(f * D_N)(t_N)\}_{N \in \mathbf{Z}^+}$ is the interval $\left[0, \frac{Si(\pi)}{\pi}\right]$.

(b) To examine the behavior of $f * D_N$ near the point of a single jump discontinuity t_0 of f, we reduce matters to the preceding situation as alluded earlier, by introducing the function f_0 defined in (3.5.14). Then for a sequence t_N converging to t_0 from the right we have

$$\begin{aligned}
&(f * D_N)(t_N) \\
&= (f - f_0) * D_N(t_N) + \big(f(t_0+) - f(t_0-)\big)(h * D_N)(t_N - t_0) + \tfrac{f(t_0+)+f(t_0-)}{2} \\
&= \big(f(t_0+) - f(t_0-)\big)\Big[(h * D_N)(t_N - t_0) - h(t_N - t_0)\Big] \\
&\qquad + \big(f(t_0+) - f(t_0-)\big)h(t_N - t_0) + (f - f_0) * D_N(t_N) + \tfrac{f(t_0+)+f(t_0-)}{2}.
\end{aligned}$$

Applying $\limsup_{N\to\infty}$ and using (3.5.12), we obtain as $N \to \infty$

$$\limsup_{N\to\infty} (f * D_N)(t_N) \le \big(f(t_0+) - f(t_0-)\big)\left(\tfrac{Si(\pi)}{\pi} - \tfrac{1}{2} + \tfrac{1}{2}\right) + \tfrac{f(t_0+)+f(t_0-)}{2},$$

where we used the following consequence of Theorem 3.5.4

$$\limsup_{N\to\infty} (f - f_0) * D_N(t_N) = (f - f_0)(t_0) = 0.$$

This shows that all accumulation points of sequences of the form $\{(f * D_N)(t_N)\}_{N \in \mathbf{Z}^+}$ are at most $\big(f(t_0+) - f(t_0-)\big)\frac{Si(\pi)}{\pi} + \frac{1}{2}\big(f(t_0+) + f(t_0-)\big)$. Also, since all accumulation points of sequences $(h * D_N)(t_N)$ are nonnegative, when t_N lies to the right of t_0, it follows from the identity $f = (f - f_0) + f_0$ and (3.5.14) that all accumulation points of $\{(f * D_N)(t_N)\}_{N \in \mathbf{Z}^+}$ are at least $\frac{1}{2}\big(f(t_0+) + f(t_0-)\big)$. As in part (a) the intermediate value theorem implies that every point in the interval having the aforementioned endpoints is also an accumulation point for the sequence at hand. □

Exercises

3.5.1. Let $Si(t)$ be the sine function as defined in (3.5.3).
(a) Prove that $|\frac{\pi}{2} - Si(t)| \le \frac{2}{t}$.
(b) Show that $Si(Nt) \to \frac{\pi}{2}$ uniformly in $t \in [\delta, \infty)$ for any $\delta > 0$.

3.5.2. Show that the sine integral function has the following expansion

$$Si(x) = \sum_{k=1}^{\infty} \frac{(-1)^k x^{2k+1}}{(2k+1)\,(2k+1)!}.$$

3.5.3. Let $L_1^1(\mathbf{T}^1)$ be the space of all differentiable functions on $\mathbf{T}^1$ whose derivatives are integrable. Obtain the inclusions $L_1^1(\mathbf{T}^1) \subseteq BV(\mathbf{T}^1) \subseteq L^\infty(\mathbf{T}^1)$ as follows:
(a) If $f \in L_1^1(\mathbf{T}^1)$, then $\mathrm{Var}(f) \le \|f'\|_{L^1}$.
(b) If $f \in BV(\mathbf{T}^1)$, then $\|f\|_{L^\infty} \le \mathrm{Var}(f) + |f(0)|$.

3.5.4. (a) Let $a_k \ge 0$, $s_N = \sum_{k=-N}^{N} a_k$, and $\sigma_N = \frac{1}{N+1}(\sigma_0 + \cdots + \sigma_N)$. Suppose that $\sigma_N \to L < \infty$ as $N \to \infty$. Prove that $s_N \to L$ as $N \to \infty$.
(b) Apply the preceding result to show that if a complex-valued function h on $\mathbf{T}^1$ is continuous in a neighborhood of 0 and $\widehat{h}(m) \ge 0$ for all $m \in \mathbf{Z}$, then $h(0) \ge 0$ and $\sum_{m \in \mathbf{Z}} \widehat{h}(m) = h(0) < \infty$; i.e., the partial sums of the Fourier series of h converge at zero.

3.5.5. Let $h \in L^1(\mathbf{T}^1)$, $t_0 \in \mathbf{T}^1$, and $0 < \delta < 1/2$.
(a) Show that $(h * D_N)(t_0) \to L$ as $N \to \infty$ if and only if

$$\lim_{M\to\infty} \int_0^\delta \left(\frac{h(t_0 - t) + h(t_0 + t)}{2} - L \right) \frac{\sin(Mt)}{t}\, dt = 0 .$$

(b) Conclude that if an integrable function h on $\mathbf{T}^1$ satisfies

$$\int_0^\delta \frac{|h(t_0 - t) + h(t_0 + t) - 2L|}{t}\, dt < \infty ,$$

then $(h * D_N)(t_0) \to L$ as $N \to \infty$.
(c) In particular, if there are constants $C, \beta > 0$ with $\beta < 1$ such that for all t with $0 < t < \delta$ we have

$$|h(t_0 - t) + h(t_0 + t) - 2h(t_0)| \le C t^\beta ,$$

then $(h * D_N)(t_0) \to h(t_0)$ as $N \to \infty$.
(d) If h is an odd function, then $(h * D_N)(0) \to 0$ as $N \to \infty$.

3.5.6. Let $f \in L^1(\mathbf{T}^1)$ and suppose that (a,b) is an interval in $\mathbf{T}^1$. Then we have

$$\lim_{N\to\infty} \int_a^b (f * D_N)(t)\, dt = \int_a^b f(t)\, dt .$$

[*Hint:* Use Theorems 3.5.4 and 3.5.5 and the fact that the operator $f \mapsto f * D_N$ is self-adjoint.]

3.6 Lacunary Series and Sidon Sets

Lacunary series provide examples of 1-periodic functions on the line that possess certain remarkable properties.

3.6.1 Definition and Basic Properties of Lacunary Series

We begin by defining lacunary sequences.

Definition 3.6.1. A sequence of positive integers $\Lambda = \{\lambda_k\}_{k=1}^{\infty}$ is called *lacunary* if there exists a constant $A > 1$ such that $\lambda_{k+1} \geq A\lambda_k$ for all $k \in \mathbf{Z}^+$.

Examples of lacunary sequences are provided by exponential sequences, such as $\lambda_k = 2^k, 3^k, 4^k, \dots$. Observe that polynomial sequences such as $\lambda_k = 1 + k^2$ are not lacunary. Note that lacunary sequences tend to infinity as $k \to \infty$.

An important observation about lacunary sequences is the following: for any $m, k_0 \in \mathbf{Z}^+$ we have

$$1 \leq |m - \lambda_{k_0}| < (1 - A^{-1})\lambda_{k_0} \implies m \notin \Lambda. \tag{3.6.1}$$

Indeed, to prove this assertion, notice that the closest numbers to λ_{k_0} among of the terms of the sequence $\{\lambda_k\}_{k=1}^{\infty}$ are λ_{k_0+1} and λ_{k_0-1} (the latter only if $k_0 > 1$) and thus if $j > k_0$ we have

$$|\lambda_j - \lambda_{k_0}| \geq \lambda_{k_0+1} - \lambda_{k_0} \geq A\lambda_{k_0} - \lambda_{k_0} = (A-1)\lambda_{k_0} \geq (1 - A^{-1})\lambda_{k_0},$$

while if $j < k_0$

$$|\lambda_j - \lambda_{k_0}| \geq \lambda_{k_0} - \lambda_{k_0-1} \geq \lambda_{k_0} - \frac{1}{A}\lambda_{k_0} = (1 - A^{-1})\lambda_{k_0}.$$

Thus (3.6.1) follows.

We begin with the following result.

Proposition 3.6.2. *Let $\{\lambda_k\}_{k=1}^{\infty}$ be a lacunary sequence and let f be an integrable function on the circle that is differentiable at a point and has Fourier coefficients*

$$\widehat{f}(m) = \begin{cases} a_m & \text{when } m = \lambda_k, \\ 0 & \text{when } m \neq \lambda_k. \end{cases} \tag{3.6.2}$$

Then we have

$$\lim_{k \to +\infty} \widehat{f}(\lambda_k)\lambda_k = 0.$$

Proof. Applying translation, we may assume that the point at which f is differentiable is the origin. Replacing f by the 1-periodic function

$$g(t) = f(t) - f(0)\cos(2\pi t) - f'(0)\frac{\sin(2\pi t)}{2\pi}$$

we may assume that $f(0) = f'(0) = 0$. (We have $\widehat{g}(m) = \widehat{f}(m)$ for $|m| \geq 2$ and thus the final conclusion for f is equivalent to that for g.)

Using (3.6.1) and (3.6.2), we obtain that for any $m \in \mathbf{Z}$ we have

$$1 \le |m-\lambda_k| < (1-A^{-1})\lambda_k \implies \widehat{f}(m) = 0. \tag{3.6.3}$$

Let $[t]$ denote the integer part of t. Given $\varepsilon > 0$, pick a positive integer k_0 such that if $[(1-A^{-1})\lambda_{k_0}] = 2N_0$, then $N_0^{-2} < \varepsilon$, and

$$\sup_{|x|<N_0^{-\frac{1}{4}}} \left| \frac{f(x)}{x} \right| < \varepsilon. \tag{3.6.4}$$

The expression in (3.6.4) can be made arbitrarily small, since f is differentiable at the origin. Now take an integer k with $k \ge k_0$ and set $2N = [\min(A-1, 1-A^{-1})\lambda_k]$, which is of course at least $2N_0$. Using (3.6.3), we obtain that for any trigonometric polynomial K_N of degree $2N$ with $\widehat{K_N}(0) = 1$ we have

$$\widehat{f}(\lambda_k) = \int_{|x|\le\frac{1}{2}} f(x)K_N(x)e^{-2\pi i\lambda_k x}\,dx. \tag{3.6.5}$$

We take $K_N = (F_N/\|F_N\|_{L^2})^2$, where F_N is the Fejér kernel. Using (3.1.18), we obtain first the identity

$$\|F_N\|_{L^2}^2 = \sum_{j=-N}^{N}\left(1-\frac{|j|}{N+1}\right)^2 = 1+\frac{1}{3}\frac{N(2N+1)}{N+1} > \frac{N}{3} \tag{3.6.6}$$

and also the estimate

$$F_N(x)^2 \le \left(\frac{1}{N+1}\frac{1}{4x^2}\right)^2, \tag{3.6.7}$$

which is valid for $|x| \le 1/2$. In view of (3.6.6) and (3.6.7), we have the estimate

$$K_N(x) \le \frac{3}{16}\frac{1}{N^3}\frac{1}{x^4}. \tag{3.6.8}$$

We now use (3.6.5) to obtain

$$\lambda_k\widehat{f}(\lambda_k) = \lambda_k\int_{|x|\le\frac{1}{2}} f(x)K_N(x)e^{-2\pi i\lambda_k x}\,dx = I_k^1 + I_k^2 + I_k^3,$$

where

$$\begin{aligned}
I_k^1 &= \lambda_k\int_{|x|\le N^{-1}} f(x)K_N(x)e^{-2\pi i\lambda_k x}\,dx,\\
I_k^2 &= \lambda_k\int_{N^{-1}<|x|\le N^{-\frac{1}{4}}} f(x)K_N(x)e^{-2\pi i\lambda_k x}\,dx,\\
I_k^3 &= \lambda_k\int_{N^{-\frac{1}{4}}<|x|\le\frac{1}{2}} f(x)K_N(x)e^{-2\pi i\lambda_k x}\,dx.
\end{aligned}$$

Since $\|K_N\|_{L^1} = 1$, it follows that

$$|I_k^1| \le \frac{\lambda_k}{N} \sup_{|x|<N^{-1}} \left|\frac{f(x)}{x}\right| \le \frac{(2N+1)\,\varepsilon}{\min(A-1, 1-A^{-1})N},$$

which can be made arbitrarily small if ε is small. Also, using (3.6.8), we obtain

$$|I_k^2| \le \frac{3\lambda_k}{16N^3} \sup_{|x|<N^{-\frac{1}{4}}} \left|\frac{f(x)}{x}\right| \int_{N^{-1}<|x|\le N^{-\frac{1}{4}}} \frac{dx}{|x|^3} \le \frac{3\lambda_k}{16N} \sup_{|x|<N^{-\frac{1}{4}}} \left|\frac{f(x)}{x}\right|,$$

which, as observed, is bounded by a constant multiple of ε. Finally, using again (3.6.8), we obtain

$$|I_k^3| \le \frac{3}{16N^3} N\lambda_k \int_{N^{-\frac{1}{4}}<|x|\le\frac{1}{2}} |f(x)|\,dx \le \frac{3}{16N^2}\|f\|_{L^1} < \frac{3\varepsilon}{16}\|f\|_{L^1}.$$

It follows that for all $k \ge k_0$ we have

$$|\lambda_k \widehat{f}(\lambda_k)| \le |I_k^1| + |I_k^2| + |I_k^3| \le C(f)\,\varepsilon$$

for some fixed constant $C(f)$. This proves the required conclusion. □

Corollary 3.6.3. *(Weierstrass) There exists a continuous function on the circle that is nowhere differentiable.*

Proof. Consider the 1-periodic function

$$f(t) = \sum_{k=0}^{\infty} 2^{-k} e^{2\pi i 3^k t}.$$

Since this series converges absolutely and uniformly, f is a continuous function. If f were differentiable at a point, then by Proposition 3.6.2 we would have that $3^k \widehat{f}(3^k)$ tends to zero as $k \to \infty$. Since $\widehat{f}(3^k) = 2^{-k}$ for $k \ge 0$, this is not the case. Therefore, f is nowhere differentiable. The real and imaginary parts of this function are displayed in Figure 3.4. □

3.6.2 Equivalence of L^p Norms of Lacunary Series

We now turn to one of the most important properties of lacunary series, equivalence of their norms. It is a remarkable result that lacunary Fourier series have comparable L^p norms for $1 \le p < \infty$. More precisely, we have the following theorem:

Theorem 3.6.4. *Let $1 \le \lambda_1 < \lambda_2 < \lambda_3 < \cdots$ be a lacunary sequence with constant $A > 1$. Set $\Lambda = \{\lambda_k : k \in \mathbf{Z}^+\}$. Then for all $1 \le p < \infty$, there exists a constant $C_p(A)$ such that for all $f \in L^1(\mathbf{T}^1)$, with $\widehat{f}(k) = 0$, when $k \in \mathbf{Z} \setminus \Lambda$ we have*

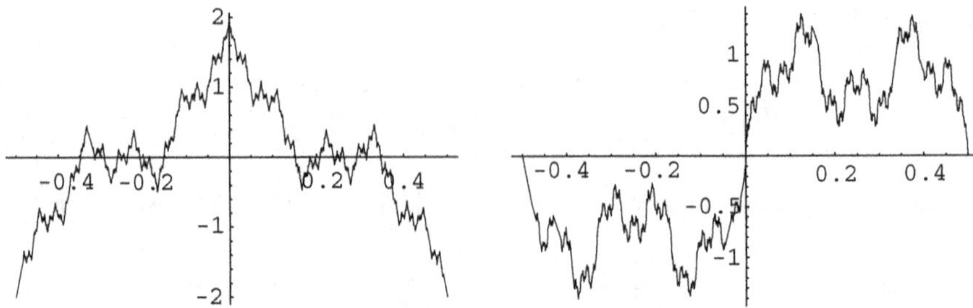

Fig. 3.4 The graph of the real and imaginary parts of the function $f(t) = \sum_{k=0}^{\infty} 2^{-k} e^{2\pi i 3^k t}$.

$$\big\|f\big\|_{L^p(\mathbf{T}^1)} \le C_p(A)\big\|f\big\|_{L^1(\mathbf{T}^1)}\,. \tag{3.6.9}$$

Moreover, the converse inequality to (3.6.9) is valid, and thus all L^p norms of lacunary Fourier series are equivalent for $1 \le p < \infty$.

Proof. We suppose initially that $f \in L^2(\mathbf{T}^1)$ and f is nonzero. We define

$$f_N(x) = \sum_{j=1}^{N} \widehat{f}(\lambda_j)e^{2\pi i \lambda_j x}\,. \tag{3.6.10}$$

Given $2 \le p < \infty$, we pick an integer m with $2m > p$ and we also pick a positive integer r such that $A^r > m$. Then we can write f_N as a sum of r functions φ_s, $s = 1,2,\dots,r$, where each φ_s has Fourier coefficients that vanish except possibly on the lacunary set

$$\{\lambda_{kr+s} :\, k \in \mathbf{Z}^+ \cup \{0\}\} = \{\mu_1, \mu_2, \mu_3, \dots\}\,.$$

It is a simple fact that the sequence $\{\mu_k\}_k$ is lacunary with constant A^r. Then we have

$$\int_0^1 |\varphi_s(x)|^{2m}\,dx = \sum_{\substack{1\le j_1,\dots,j_m,k_1,\dots,k_m \le N \\ \mu_{j_1}+\dots+\mu_{j_m}=\mu_{k_1}+\dots+\mu_{k_m}}} \widehat{\varphi_s}(\mu_{j_1})\cdots\widehat{\varphi_s}(\mu_{j_m})\overline{\widehat{\varphi_s}(\mu_{k_1})}\cdots\overline{\widehat{\varphi_s}(\mu_{k_m})}\,.$$

We claim that if $\mu_{j_1} + \dots + \mu_{j_m} = \mu_{k_1} + \dots + \mu_{k_m}$, then

$$\max(\mu_{j_1},\dots,\mu_{j_m}) = \max(\mu_{k_1},\dots,\mu_{k_m})\,.$$

Indeed, if $\max(\mu_{j_1},\dots,\mu_{j_m}) > \max(\mu_{k_1},\dots,\mu_{k_m})$, then

$$\max(\mu_{j_1},\dots,\mu_{j_m}) \le \mu_{k_1} + \dots + \mu_{k_m} \le m\max(\mu_{k_1},\dots,\mu_{k_m})\,.$$

But since

$$A^r\max(\mu_{k_1},\dots,\mu_{k_m}) \le \max(\mu_{j_1},\dots,\mu_{j_m})\,,$$

it would follow that $A^r \le m$, which contradicts our choice of r. Likewise, we eliminate the case $\max(\mu_{j_1},\dots,\mu_{j_m}) < \max(\mu_{k_1},\dots,\mu_{k_m})$. We conclude that these numbers are equal. We can now continue the same reasoning using induction to conclude that if $\mu_{j_1}+\cdots+\mu_{j_m} = \mu_{k_1}+\cdots+\mu_{k_m}$, then

$$\{\mu_{k_1},\dots,\mu_{k_m}\} = \{\mu_{j_1},\dots,\mu_{j_m}\}.$$

Using this fact in the evaluation of the previous multiple sum, we obtain

$$\int_0^1 |\varphi_s(x)|^{2m}\,dx = \sum_{j_1=1}^{N}\cdots\sum_{j_m=1}^{N} |\widehat{\varphi_s}(\mu_{j_1})|^2\cdots|\widehat{\varphi_s}(\mu_{j_m})|^2 = \big(\big\|\varphi_s\big\|_{L^2}^2\big)^m,$$

which implies that $\|\varphi_s\|_{L^{2m}} = \|\varphi_s\|_{L^2}$ for all $s\in\{1,2,\dots,r\}$. Then we have

$$\big\|f_N\big\|_{L^p} \le \big\|f_N\big\|_{L^{2m}} \le \sqrt{r}\Big(\sum_{s=1}^{r}\big\|\varphi_s\big\|_{L^{2m}}^2\Big)^{\frac12} = \sqrt{r}\Big(\sum_{s=1}^{r}\big\|\varphi_s\big\|_{L^2}^2\Big)^{\frac12} = \sqrt{r}\big\|f_N\big\|_{L^2},$$

since the functions φ_s are orthogonal in L^2. Since r can be chosen to be $[\log_A m]+1$ and m can be taken to be $[\frac{p}{2}]+1$, we have now established the inequality

$$\big\|f_N\big\|_{L^p(\mathbf{T}^1)} \le c_p(A)\big\|f_N\big\|_{L^2(\mathbf{T}^1)}, \qquad p\ge 2, \tag{3.6.11}$$

with $c_p(A) = \sqrt{1+\big[\log_A\big([\frac{p}{2}]+1\big)\big]}$ for every f_N of the form (3.6.10).

To replace f_N by f in (3.6.11), we recall our assumption that $f\in L^2(\mathbf{T}^1)$. We observe that $f_N\to f$ in L^2 and thus f_{N_j} tends to f a.e. for some subsequence. Then Fatou's lemma and (3.6.11) imply for $1<p<\infty$

$$\begin{aligned}\int_0^1 |f(x)|^p\,dx &= \int_0^1 \liminf_{j\to\infty}|f_{N_j}(x)|^p\,dx\\ &\le \liminf_{j\to\infty}\int_0^1 |f_{N_j}(x)|^p\,dx\\ &\le c_p(A)^p\liminf_{j\to\infty}\big\|f_{N_j}\big\|_{L^2}^p\\ &= c_p(A)^p\big\|f\big\|_{L^2}^p.\end{aligned}$$

We conclude that

$$\big\|f\big\|_{L^p(\mathbf{T}^1)} \le c_p(A)\big\|f\big\|_{L^2(\mathbf{T}^1)}, \qquad p\ge 2. \tag{3.6.12}$$

By interpolation we obtain

$$\big\|f\big\|_{L^2} \le \big\|f\big\|_{L^4}^{\frac23}\big\|f\big\|_{L^1}^{\frac13} \le \big([\log_A 3]+1\big)^{\frac12\cdot\frac23}\big\|f\big\|_{L^2}^{\frac23}\big\|f\big\|_{L^1}^{\frac13}.$$

We are assuming that $0 < \|f\|_{L^2} < \infty$ and the preceding inequality implies that

$$\|f\|_{L^2(\mathbf{T}^1)} \le \big([\log_A 3] + 1\big)\|f\|_{L^1(\mathbf{T}^1)}. \tag{3.6.13}$$

Finally an easy consequence of Hölder's inequality is that

$$\|f\|_{L^p(\mathbf{T}^1)} \le \|f\|_{L^2(\mathbf{T}^1)}, \qquad 1 \le p < 2. \tag{3.6.14}$$

Combining (3.6.12) and (3.6.14) with (3.6.13) yields (3.6.9) with

$$C_p(A) = c_p(A)\big([\log_A 3] + 1\big)$$

for all $1 \le p < \infty$ under the hypothesis that $\widehat{f}(k) = 0$ for all $k \in \mathbf{Z} \setminus \Lambda$ and the additional assumption that $f \in L^2$.

We now extend the result to $f \in L^1(\mathbf{T}^1)$. Given $f \in L^1(\mathbf{T}^1)$ with $\widehat{f}(k) = 0$ when $k \in \mathbf{Z} \setminus \Lambda$, consider the functions $f * F_M$, where F_M is the Fejér kernel and $M \in \mathbf{Z}^+$. Then $f * F_M$ lie in L^2, $f * F_M$ converge to f in L^1 and in L^p, and $\widehat{f * F_M}(k) = 0$ when $k \in \mathbf{Z} \setminus \Lambda$. The inequality

$$\|f * F_M\|_{L^p(\mathbf{T}^1)} \le C_p(A)\|f * F_M\|_{L^1(\mathbf{T}^1)} \tag{3.6.15}$$

holds since $f * F_M$ lie in L^2, so letting $M \to \infty$ yields (3.6.9). □

Theorem 3.6.4 describes the equivalence of the L^p norms of lacunary Fourier series for $p < \infty$. The question that remains is whether there is a similar characterization for the L^∞ norms of lacunary Fourier series. Such a characterization is investigated below. Before we state and prove this theorem, we need a classical tool, referred to as a Riesz product.

Definition 3.6.5. A *Riesz product* is a function of the form

$$P_N(x) = \prod_{j=1}^{N} \big(1 + a_j \cos(2\pi\lambda_j x + 2\pi\gamma_j)\big), \tag{3.6.16}$$

where N is a positive integer, $\lambda_1 < \lambda_2 < \cdots < \lambda_N$ is a lacunary sequence of positive integers, a_j are real numbers in $[-1, 1]$, and $\gamma_j \in [0, 1]$.

We make a few observations about Riesz products. A simple calculation gives that if $P_{N,j}(x) = 1 + a_j \cos(2\pi\lambda_j x + 2\pi\gamma_j)$, then

$$\widehat{P_{N,j}}(m) = \begin{cases} 1 & \text{when } m = 0, \\ \frac{1}{2}a_j e^{2\pi i \gamma_j} & \text{when } m = \lambda_j, \\ \frac{1}{2}a_j e^{-2\pi i \gamma_j} & \text{when } m = -\lambda_j, \\ 0 & \text{when } m \notin \{0\} \bigcup_{j=1}^{\infty}\{\lambda_j, -\lambda_j\}. \end{cases} \tag{3.6.17}$$

Assume that the constant A associated with the lacunary sequence $\lambda_1 < \lambda_2 < \cdots < \lambda_N$ satisfies $A \ge 3$. Then each integer m has *at most one* representation as a sum

$$m = \varepsilon_1\lambda_1 + \cdots + \varepsilon_N\lambda_N,$$

where $\varepsilon_j \in \{-1,1,0\}$; see Exercise 3.6.1. We now calculate the Fourier coefficients of the Riesz product defined in (3.6.16). For a fixed integer b, let us denote by δ_b the sequence of integers that is equal to 1 at b and zero otherwise. Then, using (3.6.17), we obtain that

$$\widehat{P_{N,j}} = \delta_0 + \tfrac{1}{2}a_j e^{2\pi i\gamma_j}\delta_{\lambda_j} + \tfrac{1}{2}a_j e^{-2\pi i\gamma_j}\delta_{-\lambda_j},$$

and thus $\widehat{P_N}$ is the N-fold convolution of these functions. Using that $\delta_a * \delta_b = \delta_{a+b}$, we obtain

$$\widehat{P_N}(m) = \begin{cases} 1 & \text{when } m = 0, \\ \prod_{j=1}^N \frac{1}{2}a_j e^{2\pi i\varepsilon_j\gamma_j} & \text{when } m = \sum_{j=1}^N \varepsilon_j\lambda_j \text{ and } \sum_{j=1}^N |\varepsilon_j| > 0, \\ 0 & \text{otherwise.} \end{cases}$$

It follows that $\widehat{P_N}(\lambda_j) = 0$ since for $j \ge N+1$, λ_j cannot be expressed as a linear combination of $\lambda_1, \dots, \lambda_N$ with coefficients in $\{\pm 1, 0\}$. Also $\widehat{P_N}(\lambda_k) = \frac{1}{2}a_j e^{2\pi i\gamma_k}$ for $1 \le k \le N$, since each λ_k is written uniquely as $0\cdot\lambda_1 + \cdots + 0\cdot\lambda_{k-1} + 1\cdot\lambda_k$. Hence when $A \ge 3$ we have that $\widehat{P_N}(\lambda_k) = \frac{1}{2}a_j e^{2\pi i\gamma_k}$ when $1 \le k \le N$ and $\widehat{P_N}(\lambda_k) = 0$ for $k \ge N+1$.

Next, we discuss an important property of Riesz products. Suppose that for some $m \in \mathbf{Z}$ we have $\widehat{P_N}(m) \ne 0$. We write $m = \sum_{j=1}^N \varepsilon_j\lambda_j$ uniquely with $\varepsilon_j \in \{-1,0,1\}$. Let k be the largest integer less than or equal to N such that $\varepsilon_k \ne 0$. Then we have

$$\big||m| - \lambda_k\big| \le \lambda_1 + \cdots + \lambda_{k-1} \le \frac{\lambda_k}{A^{k-1}} + \cdots + \frac{\lambda_k}{A} \le \frac{\lambda_k}{A}\frac{1}{1-\frac{1}{A}} = \frac{\lambda_k}{A-1}. \tag{3.6.18}$$

Another important property of the Riesz product is that since $P_N \ge 0$ we have

$$\|P_N\|_{L^1} = \int_{\mathbf{T}^1} P_N(t)\,dt = \widehat{P_N}(0) = 1.$$

We recall the space $A(\mathbf{T}^1)$ of all functions with absolutely summable Fourier coefficients normed with the ℓ^1 norm of the coefficients.

Theorem 3.6.6. *Let $1 < \lambda_1 < \lambda_2 < \lambda_3 < \cdots$ be a lacunary sequence of integers with constant $A > 1$. Set $\Lambda = \{\lambda_k : k \in \mathbf{Z}^+\}$. Then there exists a constant $C(A)$ such that for all $f \in L^\infty(\mathbf{T}^1)$ with $\widehat{f}(k) = 0$ when $k \in \mathbf{Z} \setminus \Lambda$ we have*

$$\big\|f\big\|_{A(\mathbf{T}^1)} = \sum_{k\in\Lambda} |\widehat{f}(k)| \le C(A)\big\|f\big\|_{L^\infty(\mathbf{T}^1)}. \tag{3.6.19}$$

Proof. Let us assume first that $A \ge 3$. Also fix $f \in L^\infty(\mathbf{T}^1)$. We consider the Riesz product

$$P_N(x) = \prod_{j=1}^N \big(1 + \cos(2\pi\lambda_j x + 2\pi\gamma_j)\big),$$

where γ_j is chosen to satisfy the identity $|\widehat{f}(\lambda_j)| = e^{2\pi i\gamma_j}\overline{\widehat{f}(\lambda_j)}$. In view of Parseval's relation and of the fact that $\|P_N\|_{L^1} = 1$ we obtain

$$\left|\sum_{m\in\mathbf{Z}} \widehat{P_N}(m)\overline{\widehat{f}(m)}\right| = \left|\int_0^1 P_N(x)\overline{f(x)}\,dx\right| \le \|f\|_{L^\infty}, \tag{3.6.20}$$

and the sum in (3.6.20) is finite, since the Fourier coefficients of $\widehat{P_N}$ form a finitely supported sequence. But $\widehat{f}(m) = 0$ for $m \notin \Lambda$, while $\widehat{P_N}(\lambda_j) = \frac{1}{2}e^{2\pi i\gamma_j}$ for $1 \le j \le N$ since $A \ge 3$, and moreover, $\widehat{P_N}(\lambda_j) = 0$ for $j \ge N+1$, as observed earlier. Thus (3.6.20) reduces to

$$\frac{1}{2}\sum_{j=1}^{N} |\widehat{f}(\lambda_j)| = \left|\sum_{j=1}^{N} \frac{1}{2}e^{2\pi i\gamma_j}\overline{\widehat{f}(\lambda_j)}\right| \le \|f\|_{L^\infty}.$$

Letting $N \to \infty$, we deduce that $\sum_{j=1}^{\infty} |\widehat{f}(\lambda_j)| \le 2\|f\|_{L^\infty}$, which proves (3.6.19) when $A \ge 3$.

We now consider the case $A < 3$. We fix $1 < A < 3$ and we pick a positive integer r such that

$$A^r > 3 \qquad \text{and} \qquad \frac{1}{A^r - 1} < 1 - \frac{1}{A}. \tag{3.6.21}$$

This is possible, since $(A^r - 1)^{-1} \to 0$ as $r \to \infty$.

For each $s \in \{1,\dots,r\}$, define the sequences $\lambda_k^s = \lambda_{s+(k-1)r}$ indexed by $k = 1,2,3,\dots$ and observe that $\lambda_{(k+1)}^s > A^r\lambda_k^s$ for all $k = 1,2,\dots$; i.e., each such sequence is lacunary with constant A^r. We consider the Riesz product

$$P_N^s(x) = \prod_{k=1}^{N} \left(1 + \cos(2\pi\lambda_k^s x + 2\pi\gamma_k^s)\right),$$

where γ_k^s is defined via the identity $|\widehat{f}(\lambda_k^s)| = e^{2\pi i\gamma_k^s}\overline{\widehat{f}(\lambda_k^s)}$.

Using (3.6.18) we obtain that, if $m \in \mathbf{Z}$ is such that $\widehat{P_N^s}(m) \neq 0$, then there exists a $k \in \{1,2\dots,N\}$ such that

$$\big||m| - \lambda_k^s\big| < \frac{\lambda_k^s}{A^r - 1}.$$

This combined with (3.6.21) yields

$$\big||m| - \lambda_k^s\big| < \left(1 - \frac{1}{A}\right)\lambda_k^s.$$

Using (3.6.1) we obtain that either $m = \pm\lambda_k^s$ or $|m| \notin \Lambda$. Thus we have

$$\{m \in \mathbf{Z}^+ : \widehat{P_N^s}(m) \neq 0\} \subseteq \{\lambda_1^s, \lambda_2^s, \dots, \lambda_N^s\} \cup \Lambda^c.$$

This observation, the fact that $\widehat{f}$ is supported in Λ, and Parseval's relation yield

$$\left|\sum_{k=1}^{N}\widehat{P_N^s}(\lambda_k^s)\overline{\widehat{f}(\lambda_k^s)}\right| = \left|\sum_{m\in\mathbf{Z}}\widehat{P_N^s}(m)\overline{\widehat{f}(m)}\right| = \left|\int_0^1 P_N^s(x)\overline{f(x)}\,dx\right| \le \big\|f\big\|_{L^\infty}. \quad (3.6.22)$$

Since $\widehat{P_N}(\lambda_k^s) = \frac{1}{2}e^{2\pi i\gamma_k^s}$ for $1\le k\le N$, (3.6.22) reduces to

$$\frac{1}{2}\sum_{k=1}^{N}\big|\widehat{f}(\lambda_k^s)\big| \le \big\|f\big\|_{L^\infty}.$$

Letting $N\to\infty$ gives

$$\sum_{j=1}^{\infty}\big|\widehat{f}(\lambda_j^s)\big| \le 2\big\|f\big\|_{L^\infty}.$$

Summing over s in the set $\{1,2,\dots,r\}$, we obtain the required conclusion with $C(A)=2r$ and note that r can be taken to be $[\max(\log_A\frac{2A-1}{A-1},\log_3 A)]+2$. □

Corollary 3.6.7. *Let $\Lambda=\{\lambda_k:\ k\in\mathbf{Z}^+\}$ be a lacunary set and let f be a bounded function on the circle that satisfies $\widehat{f}(k)=0$ when $k\in\mathbf{Z}\setminus\Lambda$. Then f is almost everywhere equal to the absolutely (and uniformly) convergent series*

$$f(x)=\sum_{k\in\Lambda}\widehat{f}(k)e^{2\pi ikx} \qquad \text{a.e.} \quad (3.6.23)$$

and thus it is almost everywhere equal to a continuous function.

Proof. It follows from Theorem 3.6.6 that if $\widehat{f}(k)=0$ when $k\in\mathbf{Z}\setminus\Lambda$, then we have that $f\in A(\mathbf{T}^1)$. Applying the inversion result in Proposition 3.2.5 we obtain that f is almost everywhere equal to a continuous function and that (3.6.23) holds for almost all $x\in\mathbf{T}^1$. □

3.6.3 Sidon sets

Given a subset E of the integers, we denote by $\mathscr{C}_E$ the space of all continuous functions on $\mathbf{T}^1$ such that

$$m\in\mathbf{Z}\setminus E \implies \widehat{f}(m)=0. \quad (3.6.24)$$

It is straightforward that $\mathscr{C}_E$ is a closed subspace of all bounded functions on the circle $\mathbf{T}^1$ with the standard L^∞ norm.

Definition 3.6.8. A set of integers E is called a *Sidon set* if every function in $\mathscr{C}_E$ has an absolutely convergent Fourier series.

There are several characterizations of Sidon sets. We state them below.

Proposition 3.6.9. *The following assertions are equivalent for a subset E of $\mathbf{Z}$.*

(1) There is a constant K such that for all trigonometric polynomials P with $\widehat{P}$ supported in E we have

$$\sum_{m\in\mathbf{Z}} |\widehat{P}(m)| \le K\|P\|_{L^\infty}$$

(2) There exists a constant K such that

$$\|\widehat{f}\|_{\ell^1(\mathbf{Z})} \le K\|f\|_{L^\infty(\mathbf{T}^1)}$$

for every bounded function f on $\mathbf{T}^1$ with $\widehat{f}$ supported in E.

(3) Every function f in $\mathscr{C}_E$ has an absolutely convergent Fourier series; i.e., E is a Sidon set.

(4) For every bounded function b on E there is a finite Borel measure μ on $\mathbf{T}^1$ such that $\widehat{\mu}(m) = b(m)$ for all $m \in E$.

(5) For every function b on $\mathbf{Z}$ with the property $b(m) \to 0$ as $m \to \infty$, there is a function $g \in L^1(\mathbf{T}^1)$ such that $\widehat{g}(m) = b(m)$ for all $m \in E$.

Proof. Suppose that (1) holds. Given f in $L^\infty(\mathbf{T}^1)$ with $\widehat{f}$ is supported in E, write

$$(f*F_N)(x) = \sum_{m=-N}^{N}\left(1-\frac{|m|}{N+1}\right)\widehat{f}(m)e^{2\pi imx},$$

where F_N is the Fejér kernel. These are trigonometric polynomials whose Fourier coefficients vanish on $\mathbf{Z}\setminus E$. Applying (1) we obtain

$$\sum_{k\in\mathbf{Z}}\left(1-\frac{|m|}{N+1}\right)|\widehat{f}(m)| \le K\|f*F_N\|_{L^\infty}.$$

Letting $N\to\infty$ we obtain (2).

It is trivial that (2) implies (3).

If (3) holds, then the map $f\mapsto\widehat{f}$ is a linear bijection from $\mathscr{C}_E$ to $\ell^1(E)$. Moreover its inverse mapping $\widehat{f}\mapsto f$ is continuous, since

$$\|f\|_{L^\infty(\mathbf{T}^1)} \le \sup_{t\in[0,1]}\left|\sum_{k\in\mathbf{Z}}\widehat{f}(k)e^{2\pi ikt}\right| \le \sum_{k\in\mathbf{Z}}|\widehat{f}(k)| = \|\widehat{f}\|_{\ell^1(\mathbf{Z})}.$$

By the open mapping theorem, it follows that $f\mapsto\widehat{f}$ is a continuous mapping, which proves the existence of a constant K such that (1) holds.

We have now proved the equivalence of (1), (2), and (3).

We show that (2) implies (4). If E is a Sidon set and if b is a bounded function on E, say $\|b\|_{\ell^\infty}\le 1$, then the mapping

$$f\mapsto\sum_{m\in E}\widehat{f}(m)\widehat{b}(m)$$

is a bounded linear functional on $\mathscr{C}_E$ with norm at most K. By the Hanh-Banach theorem this functional admits an extension to $\mathscr{C}(\mathbf{T}^1)$ with the same norm. Hence there is a measure μ, whose total variation $\|\mu\|$ does not exceed K, such that

$$\sum_{m\in E} \widehat{f}(m)\widehat{b}(m) = \int_{\mathbf{T}^1} f(t)d\mu(t).$$

Taking $f(t) = e^{2\pi imt}$ in (2) we obtain $\widehat{\mu}(m) = b(m)$ for all $m \in E$.

If (4) holds and $b(m) \to 0$ as $|m| \to \infty$, using Lemma 3.3.2 there is a convex sequence $c(m)$ such that $c(m) > 0$, $c(m) \to 0$ as $|m| \to \infty$, $c(-m) = c(m)$, and $|b(m)| \le c(m)$ for all $m \in \mathbf{Z}$. By (4), there is a finite Borel measure μ with $\widehat{\mu}(m) = b(m)/c(m)$ for all $m \in E$.

By Theorem 3.3.4, there is a function g in $L^1(\mathbf{T}^1)$ such that $\widehat{g}(m) = c(m)$ for all $m \in \mathbf{Z}$. Then $b(m) = \widehat{g}(m)\widehat{\mu}(m)$ for all $m \in E$. Since $f = g * \mu$ is in L^1, we have $b(m) = \widehat{f}(m)$ for all $m \in E$, and thus (4) implies (5).

Finally, if (5) holds, we show (3). Given $f \in \mathscr{C}_E$, we show that for an arbitrary sequence d_m tending to zero, we have $\sum_{m\in\mathbf{Z}} |\widehat{f}(m)d_m| < \infty$; this implies that $\sum_{m\in\mathbf{Z}} |\widehat{f}(m)| < \infty$. Given a sequence $d_m \to 0$, pick a function g in L^1 such that $\widehat{g}(m)\widehat{f}(m) = |\widehat{f}(m)|\,|d_m|$ for all $m \in E$ by assumption (5). Then the series

$$\sum_{m\in\mathbf{Z}} \widehat{g}(m)\widehat{f}(m) = \sum_{m\in\mathbf{Z}} \widehat{f * g}(m) \tag{3.6.25}$$

has nonnegative terms and the function $f * g$ is continuous, thus $F_N * (f * g)(0) \to (f * g)(0)$ as $N \to \infty$. It follows that $D_N * (f * g)(0) \to (f * g)(0)$, thus the series in (3.6.25) converges (see Exercise 3.5.4) and hence $\sum_{m\in\mathbf{Z}} |\widehat{f}(m)d_m| < \infty$. □

Example 3.6.10. Every lacunary set is a Sidon set. Indeed, suppose that E is a lacunary set with constant A. If f is a continuous function which satisfies (3.6.24), then Theorem 3.6.6 gives that

$$\sum_{m\in\Lambda} |\widehat{f}(m)| \le C(A)\|f\|_{L^\infty} < \infty;$$

hence f has an absolutely convergent Fourier series.

Example 3.6.11. There exist subsets of $\mathbf{Z}$ that are not Sidon. For example, $\mathbf{Z} \setminus \{0\}$ is not a Sidon set. See Exercise 3.6.7.

Exercises

3.6.1. Suppose that $0 < \lambda_1 < \lambda_2 < \cdots < \lambda_N$ is a lacunary sequence of integers with constant $A \ge 3$. Prove that for every integer m there exists at most one N-tuple $(\varepsilon_1, \dots, \varepsilon_N)$ with each $\varepsilon_j \in \{-1, 1, 0\}$ such that

$$m = \varepsilon_1\lambda_1 + \cdots + \varepsilon_N\lambda_N.$$

[*Hint:* Suppose there exist two such N-tuples. Pick the largest k such that the coefficients of λ_k are different.]

3.6.2. Is the sequence $\lambda_k = \left[e^{(\log k)^2}\right]$, $k = 2,3,4,\dots$ lacunary?

3.6.3. Let $a_k \geq 0$ for all $k \in \mathbf{Z}^+$ and $1 \leq p < \infty$. Show that there exist constants C_p, c_p such that for all $N \in \mathbf{Z}^+$ we have

$$c_p\Big(\sum_{k=1}^{N}|a_k|^2\Big)^{\frac{1}{2}} \leq \Big(\int_0^1 \Big|\sum_{k=1}^{N} a_k e^{2\pi i 2^k x}\Big|^p dx\Big)^{\frac{1}{p}} \leq C_p\Big(\sum_{k=1}^{N}|a_k|^2\Big)^{\frac{1}{2}},$$

while

$$\sup_{x\in[0,1]}\Big|\sum_{k=1}^{N} a_k e^{2\pi i 2^k x}\Big| = \sum_{k=1}^{N}|a_k|.$$

3.6.4. Suppose that $0 < \lambda_1 < \lambda_2 < \cdots$ is a lacunary sequence and let f be a bounded function on the circle that satisfies $\widehat{f}(m) = 0$ whenever $m \in \mathbf{Z}\setminus\{\lambda_1, \lambda_2, \dots\}$. Suppose also that

$$\sup_{t\neq 0}\frac{|f(t) - f(0)|}{|t|^\alpha} = B < \infty$$

for some $0 < \alpha < 1$.
(a) Prove that there is a constant C such that $|\widehat{f}(\lambda_k)| \leq CB\lambda_k^{-\alpha}$ for all $k \geq 1$.
(b) Prove that $f \in \dot{\Lambda}_\alpha(\mathbf{T}^1)$.
[*Hint:* Let $2N = [(1 - A^{-1})\lambda_k]$ and let K_N be as in the proof of Proposition 3.6.2. Write

$$\begin{aligned}\widehat{f}(\lambda_k) &= \int_{|x|\leq N^{-1}}(f(x) - f(0))e^{-2\pi i\lambda_k x}K_N(x)\,dx\\ &\quad + \int_{N^{-1}\leq|x|\leq\frac{1}{2}}(f(x) - f(0))e^{-2\pi i\lambda_k x}K_N(x)\,dx\,.\end{aligned}$$

Use that $\|K_N\|_{L^1} = 1$ and also the estimate (3.6.7). Part (b): Use the estimate in part (a).]

3.6.5. Let f be an integrable function on the circle whose Fourier coefficients vanish outside a lacunary set $\Lambda = \{\lambda_1, \lambda_2, \lambda_3, \dots\}$. Suppose that f vanishes identically in a small neighborhood of the origin. Show that f is in $\mathscr{C}^\infty(\mathbf{T}^1)$.
[*Hint:* Let $2N = [(1 - A^{-1})\lambda_k]$ and let K_N be as in the proof of Proposition 3.6.2. Write

$$\widehat{f}(\lambda_k) = \int_{|x|\leq\frac{1}{2}} f(x)e^{-2\pi i\lambda_k x}K_N(x)\,dx$$

and use estimate (3.6.7) to obtain that f is in $\mathscr{C}^2$. Continue by induction.]

3.6.6. Let $1 < a, b < \infty$. Consider the 1-periodic function

$$f(x) = \sum_{k=0}^{\infty} a^{-k}e^{2\pi i b^k x}.$$

Prove that the following statements are equivalent:
(a) f is differentiable at a point.
(b) $b < a$.
(c) f is differentiable everywhere.

3.6.7. Use the example in Proposition 3.4.6 (a) to show that $\mathbf{Z} \setminus \{q_1, \dots, q_L\}$ is not a Sidon set for any finite subset $\{q_1, \dots, q_L\}$ of the integers.

3.6.8. Let $0 < \delta < 1$. Let E be a subset of the integers such that for any sequence of complex numbers $\{d_m\}_{m \in E}$ with $|d_m| = 1$ there is a finite Borel measure μ on $\mathbf{T}^1$ such that

$$|\widehat{\mu}(m) - d_m| < 1 - \delta$$

for all $m \in E$. Show that E is a Sidon set.
[*Hint:* Given f be in $\mathscr{C}_E$ define d_m via the identity $d_m \widehat{f}(m) = |\widehat{f}(m)|$ if $\widehat{f}(m) \neq 0$, otherwise set $d_m = 1$. For the measure μ given by the hypothesis, notice that $\mathrm{Re}\,(\widehat{\mu}(m)\widehat{f}(m)) \geq \delta |\widehat{f}(m)|$ for all $m \in \mathbf{Z}$.]

HISTORICAL NOTES

Trigonometric series in one dimension were first considered in the study of the vibrating string problem and are implicitly contained in the work of d'Alembert, D. Bernoulli, Clairaut, and Euler. The analogous problem for vibrating higher-dimensional bodies naturally suggested the use of multiple trigonometric series. However, it was the work of Fourier on steady-state heat conduction that inspired the subsequent systematic development of such series. Fourier announced his results in 1811, although his classical book *Théorie de la chaleur* was published in 1822. This book contains several examples of heuristic use of trigonometric expansions and motivated other mathematicians to carefully study such expansions. The systematic development of the theory of Fourier series began by Dirichlet [96], who studied the pointwise convergence of the Fourier series of piecewise monotonic functions via the use of the kernel D_N, today called the Dirichlet kernel.

The fact that the Fourier series of a continuous function can diverge was first observed by DuBois Reymond in 1873. The Riemann–Lebesgue lemma was first proved by Riemann in his memoir on trigonometric series (appeared between 1850 and 1860). It carries Lebesgue's name today because Lebesgue later extended it to his notion of integral. The rebuilding of the theory of Fourier series based on Lebesgue's integral was mainly achieved by de la Vallée-Poussin and Fatou.

Theorem 3.3.16 was obtained by Bernstein [26] in dimension $n = 1$. Higher-dimensional analogues of the Hardy–Littlewood series of Exercise 3.3.8 were studied by Wainger [370]. These series can be used to produce examples indicating that the restriction $s > \alpha + n/2$ in Bernstein's theorem is sharp even in higher dimensions. Part (b) of Theorem 3.4.4 is due to Lebesgue when $n = 1$ and Marcinkiewicz and Zygmund [243] when $n = 2$. Marcinkiewicz and Zygmund's proof also extends to higher dimensions. The proof given here is based on Lemma 3.4.5 proved by Stein [342] in a different context. The proof of Lemma 3.4.5 presented here was suggested by T. Tao.

Abel proved that if an infinite series $\sum_{k=0}^{\infty} a_k$ converges and has sum L, then the power series $f(x) = \sum_{k=0}^{\infty} a_k x^k$ converges for $|x| < 1$ and tends to L as $x \to 1-$. The converse of this theorem under the additional assumption that $k a_k \to 0$ as $k \to \infty$ was proved by Tauber [358]. Hardy [143] extended Tauber's result (Theorem 3.5.1) for Cesàro summability under the weaker assumption that the sequence $k a_k$ is bounded. Jordan [180] studied of functions of bounded variation and proved Theorem 3.5.4. The existence of a continuous function which is nowhere differentiable (Corollary 3.6.3) was first published in 1872 by K. Weierstrass, although earlier findings of such functions were published later. The exposition on Sidon sets is taken from the classical article of Rudin [305], which also contains Exercise 3.6.8.

The Gibbs phenomenon (a version of Theorem 3.5.7) was discovered by Wilbraham [376] and rediscovered by Gibbs [128]; this phenomenon describes the particular way in which the Fourier sums of a piecewise continuously differentiable periodic function have large oscillations and overshoot at the jump discontinuity of the function. Bôcher [29] gave a detailed mathematical analysis of that overshoot, which he called the "Gibbs phenomenon".

The main references for trigonometric series are the books of Bary [20] and Zygmund [388], [389]. Other references for one-dimensional Fourier series include the books of Edwards [106], Dym and McKean [105], Katznelson [190], Körner [202], Pinsky [283], and the first eight chapters in Torchinsky [363]. The reader may also consult the book of Krantz [203] for a historical introduction to the subject of Fourier series. A review of the heritage and continuing significance of Fourier Analysis is written by Kahane [182].

A classical treatment of multiple Fourier series can be found in the last chapter of Bochner's book [32] and in parts of his other book [31]. Other references include the last chapter in Zygmund [389], the books of Yanushauskas [381] (in Russian) and Zhizhiashvili [384], the last chapter in Stein and Weiss [348], and the article of Alimov, Ashurov, and Pulatov in [3]. A brief survey article on the subject was written by Ash [11]. More extensive expositions were written by Shapiro [320], Igari [171], and Zhizhiashvili [383]. A short note on the history of Fourier series was written by Zygmund [390]. The book of Shapiro [321] contains a very detailed study of Fourier series in several variables as well as applications of this theory.

Chapter 4
Topics on Fourier Series

In this chapter we go deeper into the theory of Fourier series and we study topics such as convergence in norm and the conjugate function, divergence of Fourier series and Bochner–Riesz summability. We also study transference of multipliers on the torus and of maximal multipliers. This is a powerful technique that allows one to infer results concerning Fourier series from corresponding results about Fourier integrals and vice versa.

We also take a quick look at applications of Fourier series such as the isopertimetric inequality problem, the distribution of lattice points in a ball, and the heat equation. The power of Fourier series techniques manifests itself in the study of these problems which represent only a small part of the wide and vast range of applications of the subject known today.

4.1 Convergence in Norm, Conjugate Function, and Bochner–Riesz Means

In this section we address the following fundamental question: Do Fourier series converge in norm? We begin with some abstract necessary and sufficient conditions that guarantee such a convergence. In one dimension, we are able to reduce matters to the study of the so-called conjugate function on the circle, a sister operator of the Hilbert transform, which is the center of study of the next chapter. In higher dimensions the situation is more complicated, but we are able to give a positive answer in the case of square summability.

L. Grafakos, *Classical Fourier Analysis*, Graduate Texts in Mathematics 249,
DOI 10.1007/978-1-4939-1194-3_4,

4.1.1 Equivalent Formulations of Convergence in Norm

The question we pose is for which indices p, with $1 \le p < \infty$, we have

$$\big\| D_N^n * f - f \big\|_{L^p(\mathbf{T}^n)} \to 0 \qquad \text{as } N \to \infty, \tag{4.1.1}$$

and similarly for the circular Dirichlet kernel $\overset{\circ}{D}{}_N^n$. We tackle this question by looking at an equivalent formulation of it.

Theorem 4.1.1. *For $R > 0$ and $m \in \mathbf{Z}^n$, let $a(m,R)$ be complex numbers such that*

(i) For every $R > 0$ there is a q_R such that $a(m,R) = 0$ if $|m| > q_R$.
(ii) There is an $M_0 < \infty$ such that $|a(m,R)| \le M_0$ for all $m \in \mathbf{Z}^n$ and all $R > 0$.
(iii) For each $m \in \mathbf{Z}^n$, the limit of $a(m,R)$ exists as $R \to \infty$ and $\lim_{R\to\infty} a(m,R) = a_m$.

Let $1 \le p < \infty$. For $f \in L^p(\mathbf{T}^n)$ and $x \in \mathbf{T}^n$ define

$$S_R(f)(x) = \sum_{m \in \mathbf{Z}^n} a(m,R)\widehat{f}(m)e^{2\pi i m\cdot x}$$

noting that the sum is well defined because of (i). Also, for $h \in \mathscr{C}^\infty(\mathbf{T}^n)$ define

$$A(h)(x) = \sum_{m \in \mathbf{Z}^n} a_m \widehat{h}(m)e^{2\pi i m\cdot x}.$$

Then for all $f \in L^p(\mathbf{T}^n)$ the sequence $S_R(f)$ converges in L^p as $R \to \infty$ if and only if there exists a constant $K < \infty$ such that

$$\sup_{R>0} \big\| S_R \big\|_{L^p \to L^p} \le K. \tag{4.1.2}$$

Furthermore, if (4.1.2) holds, then for the same constant K we have

$$\sup_{\substack{h \in \mathscr{C}^\infty \\ h \neq 0}} \frac{\big\| A(h) \big\|_{L^p}}{\big\| h \big\|_{L^p}} \le K, \tag{4.1.3}$$

and then A extends to a bounded operator $\widetilde{A}$ from $L^p(\mathbf{T}^n)$ to itself; moreover, for every $f \in L^p(\mathbf{T}^n)$ we have that $S_R(f) \to \widetilde{A}(f)$ in L^p as $R \to \infty$.

Proof. If $S_R(f)$ converges in L^p, then $\|S_R(f)\|_{L^p} \le C_f$ for some constant C_f that depends on $f \in L^p(\mathbf{T}^n)$. Moreover, each S_R is a bounded operator from $L^p(\mathbf{T}^n)$ to itself with norm at most $\#\{m \in \mathbf{Z}^n : |m| \le q_R\} M_0$. Thus $\{S_R\}_{R>0}$ is a family of L^p-bounded linear operators that satisfy $\sup_{R>0} \|S_R(f)\|_{L^p} \le C_f$ for each $f \in L^p(\mathbf{T}^n)$. The uniform boundedness theorem applies and yields that the operator norms of S_R from L^p to L^p are bounded uniformly in R. This proves (4.1.2).

Conversely, assume (4.1.2). For $h \in \mathscr{C}^\infty(\mathbf{T}^n)$, we have that

$$\lim_{R\to\infty} \sum_{m \in \mathbf{Z}^n} a(m,R)\widehat{h}(m)e^{2\pi i m\cdot x} = \sum_{m \in \mathbf{Z}^n} a_m \widehat{h}(m)e^{2\pi i m\cdot x}$$

in view of property (ii) and of the Lebesgue dominated convergence theorem, since $\sum_{m\in\mathbf{Z}^n}|\widehat{h}(m)|<\infty$. Fatou's lemma now gives

$$\big\|A(h)\big\|_{L^p}=\big\|\lim_{R\to\infty}S_R(h)\big\|_{L^p}\le\liminf_{R\to\infty}\big\|S_R(h)\big\|_{L^p}\le K\big\|h\big\|_{L^p};$$

hence (4.1.3) holds. Thus A extends to a bounded operator $\widetilde{A}$ on $L^p(\mathbf{T}^n)$ by density.

We show that for all $f\in L^p(\mathbf{T}^n)$ we have $S_R(f)\to\widetilde{A}(f)$ in L^p as $R\to\infty$. Fix f in $L^p(\mathbf{T}^n)$ and let $\varepsilon>0$ be given. Pick a trigonometric polynomial P satisfying $\|f-P\|_{L^p}\le\varepsilon$. Let d be the degree of P. Then there is an $R_0>0$ such that for all $R>R_0$ we have

$$\sum_{|m_1|+\cdots+|m_n|\le d}|a(m,R)-a_m|\,|\widehat{P}(m)|\le\varepsilon$$

since $a(m,R)\to a_m$ for every m with $|m_1|+\cdots+|m_n|\le d$. We deduce that

$$\begin{aligned}\big\|S_R(P)-A(P)\big\|_{L^p}&\le\big\|S_R(P)-A(P)\big\|_{L^\infty}\\&\le\sum_{|m_1|+\cdots+|m_n|\le d}|a(m,R)-a_m|\,|\widehat{P}(m)|\\&\le\varepsilon,\end{aligned}$$

whenever $R>R_0$. Then

$$\begin{aligned}\big\|S_R(f)-\widetilde{A}(f)\big\|_{L^p}&\le\big\|S_R(f)-S_R(P)\big\|_{L^p}+\big\|S_R(P)-\widetilde{A}(P)\big\|_{L^p}+\big\|\widetilde{A}(P)-\widetilde{A}(f)\big\|_{L^p}\\&\le K\varepsilon+\varepsilon+K\varepsilon=(2K+1)\varepsilon\end{aligned}$$

for $R>R_0$. This proves that $S_R(f)$ converges to $\widetilde{A}(f)$ in L^p as $R\to\infty$. □

The most interesting situation arises, of course, when $a(m,R)\to a_m=1$ for all $m\in\mathbf{Z}^n$. In this case A (and $\widetilde{A}$) is the identity operator, and thus we expect the operators $S_R(f)$ to converge back to f as $R\to\infty$. We should keep in mind the following three examples:
(a) The sequence $a(m,R)=1$ when $|m_j|\le R$ for all $j\in\{1,2,\dots,n\}$ and zero otherwise, in which case the operator S_R of Theorem 4.1.1 is

$$S_R(f)=f*D_R^n;\tag{4.1.4}$$

(b) The sequence $a(m,R)=1$ when $|m|\le R$ and zero otherwise, in which case the S_R of Theorem 4.1.1 is

$$\mathring{S}_R(f)=f*\mathring{D}_R^n;\tag{4.1.5}$$

(c) The sequence $a(m,R)=\big(1-\frac{|m|^2}{R^2}\big)_+^\alpha$, for some $\alpha>0$, in which case we denote S_R by B_R^α.

Definition 4.1.2. The *Bochner–Riesz operator* or *Bochner–Riesz means* of order $\alpha \geq 0$ is the operator

$$B_R^\alpha(f)(x) = \sum_{\substack{m \in \mathbf{Z}^n \\ |m| \leq R}} \left(1 - \frac{|m|^2}{R^2}\right)^\alpha \widehat{f}(m) e^{2\pi i m \cdot x} \tag{4.1.6}$$

defined on integrable functions f on $\mathbf{T}^n$.

Corollary 4.1.3. *Let $1 \leq p < \infty$ and $\alpha \geq 0$. Let S_R and $\overset{\circ}{S}_R$ be as in (4.1.4) and (4.1.5), respectively, and let B_R^α be the Bochner–Riesz means as defined in (4.1.6). Then*

$$\forall f \in L^p(\mathbf{T}^n), \ \lim_{R\to\infty} \left\|D_R^n * f - f\right\|_{L^p} = 0 \qquad \Longleftrightarrow \qquad \sup_{R \geq 0} \left\|S_R\right\|_{L^p \to L^p} < \infty,$$

$$\forall f \in L^p(\mathbf{T}^n), \ \lim_{R\to\infty} \left\|\overset{\circ}{D}{}_R^n * f - f\right\|_{L^p} = 0 \qquad \Longleftrightarrow \qquad \sup_{R \geq 0} \left\|\overset{\circ}{S}_R\right\|_{L^p \to L^p} < \infty,$$

$$\forall f \in L^p(\mathbf{T}^n), \ \lim_{R\to\infty} \left\|B_R^\alpha * f - f\right\|_{L^p} = 0 \qquad \Longleftrightarrow \qquad \sup_{R \geq 0} \left\|B_R^\alpha\right\|_{L^p \to L^p} < \infty.$$

Example 4.1.4. We investigate the one-dimensional case in some detail. We take $n = 1$, and we define $a(m,N) = 1$ for all $-N \leq m \leq N$, and zero otherwise. Then $S_N(f) = \overset{\circ}{S}_N(f) = D_N * f$, where D_N is the Dirichlet kernel. Clearly, the expressions $\|S_N\|_{L^p \to L^p}$ are bounded above by the L^1 norm of D_N, but this estimation yields a bound that blows up as $N \to \infty$. We later show, via a more delicate argument, that the expressions $\|S_N\|_{L^p \to L^p}$ are uniformly bounded in N when $1 < p < \infty$.

This reasoning, however, allows us to deduce that for some function $g \in L^1(\mathbf{T}^1)$, $S_N(g)$ may not converge in L^1. This is also a consequence of the proof of Theorem 4.2.1; see (4.2.13). Note that since the Fejér kernel F_M has L^1 norm 1, we have

$$\left\|S_N\right\|_{L^1 \to L^1} \geq \lim_{M\to\infty} \left\|D_N * F_M\right\|_{L^1} = \left\|D_N\right\|_{L^1}.$$

This implies that the expressions $\|S_N\|_{L^1 \to L^1}$ are not uniformly bounded in N, and therefore Corollary 4.1.3 gives that for some $f_0 \in L^1(\mathbf{T}^1)$, $S_N(f_0)$ does not converge to f_0 in L^1.

Although the partial sums of Fourier series fail to convergence in $L^1(\mathbf{T}^n)$, it is a consequence of Plancherel's theorem that they converge in $L^2(\mathbf{T}^n)$. More precisely, if $f \in L^2(\mathbf{T}^n)$, then

$$\left\|\overset{\circ}{D}{}_N^n * f - f\right\|_{L^2}^2 = \sum_{|m| > N} |\widehat{f}(m)|^2 \to 0$$

as $N \to \infty$ and the same result is true for $D_N^n * f$ and for $B_R^\alpha * f$; for the latter, we apply Theorem 4.1.1, noting that

$$\left\|B_R^\alpha(f)\right\|_{L^2}^2 = \sum_{m \in \mathbf{Z}^n} \left|\left(1 - \frac{|m|^2}{R^2}\right)_+^\alpha\right|^2 |\widehat{f}(m)|^2 \leq \sum_{m \in \mathbf{Z}^n} |\widehat{f}(m)|^2 = \left\|f\right\|_{L^2}^2$$

and thus $\sup_{R>0} \|B_R^\alpha\|_{L^2 \to L^2} \leq 1$.

Motivated by the preceding discussion for $p = 2$, it is natural to pose the following question. Can 2 be replaced by $p \neq 2$ in the preceding results? This question has an affirmative answer in dimension one for D_N. In higher dimensions an interesting dichotomy appears. As a consequence of the one-dimensional result, the square partial sums $D_N^n * f$ converge in L^p to a given f in $L^p(\mathbf{T}^n)$, but for the circular partial sums this may not be the case.

We begin the discussion with the one-dimensional situation.

Definition 4.1.5. For $f \in \mathscr{C}^\infty(\mathbf{T}^1)$ define the *conjugate function* $\widetilde{f}$ by

$$\widetilde{f}(x) = -i \sum_{m \in \mathbf{Z}^1} \operatorname{sgn}(m) \widehat{f}(m) e^{2\pi i m x},$$

where $\operatorname{sgn}(m) = 1$ for $m > 0$, -1 for $m < 0$, and 0 for $m = 0$. Also define the *Riesz projections* P_+ and P_- by

$$P_+(f)(x) = \sum_{m=1}^{\infty} \widehat{f}(m) e^{2\pi i m x}, \tag{4.1.7}$$

$$P_-(f)(x) = \sum_{m=-\infty}^{-1} \widehat{f}(m) e^{2\pi i m x}. \tag{4.1.8}$$

Observe that $f = P_+(f) + P_-(f) + \widehat{f}(0)$, while $\widetilde{f} = -iP_+(f) + iP_-(f)$, when f is in $\mathscr{C}^\infty(\mathbf{T}^1)$. Consequently, one has

$$P_+(f) = \frac{1}{2}(f + i\widetilde{f}) - \frac{1}{2}\widehat{f}(0) \tag{4.1.9}$$

and therefore the L^p boundedness of the operator $f \mapsto \widetilde{f}$ is equivalent to that of the operator $f \mapsto P_+(f)$, since the identity and the operator $f \mapsto \widehat{f}(0)$ are obviously L^p bounded. Clearly, these statements are also valid for the other Riesz projection $f \mapsto P_-(f)$. The following is a consequence of Theorem 4.1.1.

Proposition 4.1.6. *Let $1 \le p < \infty$. Then the expressions $S_N(f) = D_N * f$ converge to f in $L^p(\mathbf{T}^1)$ as $N \to \infty$ if and only if there exists a constant $C_p > 0$ such that for all smooth functions f on $\mathbf{T}^n$ we have $\|\widetilde{f}\|_{L^p(\mathbf{T}^1)} \le C_p \|f\|_{L^p(\mathbf{T}^1)}$.*

Proof. In view of Corollary 4.1.3, the fact that for all $f \in L^p(\mathbf{T}^1)$, $S_N(f) \to f$ in L^p as $N \to \infty$ is equivalent to the uniform (in N) L^p boundedness of S_N.

We note the validity of the identity

$$e^{-2\pi i N x} \sum_{m=0}^{2N} \big(f(\cdot)e^{2\pi i N(\cdot)}\big)\widehat{\ }(m) e^{2\pi i m x} = \sum_{m=-N}^{N} \widehat{f}(m) e^{2\pi i m x}.$$

Since multiplication by exponentials does not affect L^p norms, this identity implies that the norm of the operator $S_N(f) = D_N * f$ from L^p to L^p is equal to that of the operator

$$S'_N(g)(x) = \sum_{m=0}^{2N} \widehat{g}(m)e^{2\pi imx}$$

from L^p to L^p. Therefore,

$$\sup_{N\geq 0} \|S_N\|_{L^p \to L^p} < \infty \iff \sup_{N\geq 0} \|S'_N\|_{L^p \to L^p} < \infty, \tag{4.1.10}$$

and both of these statements are equivalent to the fact that for all $f \in L^p(\mathbf{T}^1)$, $S_N(f) \to f$ in L^p as $N \to \infty$.

We have already observed that the L^p boundedness of the conjugate function is equivalent to that of P_+. Therefore, it suffices to show that the L^p boundedness of P_+ is equivalent to the uniform L^p boundedness of S'_N.

Suppose first that $\sup_{N\geq 0} \|S'_N\|_{L^p \to L^p} < \infty$. Theorem 4.1.1 applied to the sequence $a(m,R) = 1$ for $0 \leq m \leq R$ and $a(m,R) = 0$ otherwise gives that the operator $A(f) = P_+(f) + \widehat{f}(0)$ is bounded on $L^p(\mathbf{T}^1)$. Hence so is P_+.

Conversely, suppose that P_+ extends to a bounded operator from $L^p(\mathbf{T}^1)$ to itself. For all h in $\mathscr{C}^\infty(\mathbf{T}^n)$ we can write

$$\begin{aligned} S'_N(h)(x) &= \sum_{m=0}^{\infty} \widehat{h}(m)e^{2\pi imx} - \sum_{m=2N+1}^{\infty} \widehat{h}(m)e^{2\pi imx} \\ &= \sum_{m=1}^{\infty} \widehat{h}(m)e^{2\pi imx} + \widehat{h}(0) - e^{2\pi i(2N)x} \sum_{m=1}^{\infty} \widehat{h}(m+2N)e^{2\pi imx} \\ &= P_+(h)(x) - e^{2\pi i(2N)x} P_+(e^{-2\pi i(2N)(\cdot)}h) + \widehat{h}(0)\,. \end{aligned}$$

This identity implies that

$$\sup_{N\geq 0} \|S'_N(f)\|_{L^p} \leq \big(2\|P_+\|_{L^p \to L^p} + 1\big)\|f\|_{L^p} \tag{4.1.11}$$

for all f smooth, and by density for all $f \in L^p(\mathbf{T}^1)$. Note that S'_N is well defined on $L^p(\mathbf{T}^1)$. Thus the operators S'_N are uniformly bounded on $L^p(\mathbf{T}^n)$.

Thus the uniform L^p boundedness of S_N is equivalent to the uniform L^p boundedness of S'_N, which is equivalent to the L^p boundedness of P_+, which in turn is equivalent to the L^p boundedness of the conjugate function. □

4.1.2 The L^p Boundedness of the Conjugate Function

We know now that convergence of Fourier series in L^p is equivalent to the L^p boundedness of the conjugate function or either of the two Riesz projections. It is natural to ask whether these operators are L^p bounded.

Theorem 4.1.7. *Given $1 < p < \infty$, there is a constant $A_p > 0$ such that for all f in $\mathscr{C}^\infty(\mathbf{T}^1)$ we have*

$$\|\widetilde{f}\|_{L^p} \le A_p \|f\|_{L^p}. \tag{4.1.12}$$

Thus the operator $f \mapsto \widetilde{f}$ has a bounded extension on $L^p(\mathbf{T}^1)$ that also satisfies (4.1.12).

Consequently, the Fourier series of L^p functions on the circle converge back to the functions in the L^p norm for $1 < p < \infty$.

Proof. In proving the inequality (4.1.12), we make the following reductions:
(a) We assume that f is trigonometric polynomial.
(b) We assume that $\widehat{f}(0) = 0$.
(c) We assume that f is real valued.

Since f is a real-valued function, we have that $\widehat{f}(-m) = \overline{\widehat{f}(m)}$ for all m, and since $\widehat{f}(0) = 0$, we can write

$$\widetilde{f}(t) = -i \sum_{m=1}^{\infty} \widehat{f}(m) e^{2\pi i m t} + i \sum_{m=1}^{\infty} \widehat{f}(-m) e^{-2\pi i m t} = 2\mathrm{Re}\left[-i \sum_{m=1}^{\infty} \widehat{f}(m) e^{2\pi i m t} \right],$$

which implies that $\widetilde{f}$ is also real-valued (see also Exercise 4.1.4(b)). Therefore the polynomial $f + i\widetilde{f}$ contains only positive frequencies. Thus for $k \in \mathbf{Z}^+$ we have

$$\int_{\mathbf{T}^1} (f(t) + i\widetilde{f}(t))^{2k}\, dt = 0.$$

Expanding the $2k$ power and taking real parts, we obtain

$$\sum_{j=0}^{k} (-1)^{k-j} \binom{2k}{2j} \int_{\mathbf{T}^1} \widetilde{f}(t)^{2k-2j} f(t)^{2j}\, dt = 0,$$

where we used that f is real-valued. Therefore,

$$\begin{aligned} \|\widetilde{f}\|_{L^{2k}}^{2k} &\le \sum_{j=1}^{k} \binom{2k}{2j} \int_{\mathbf{T}^1} \widetilde{f}(t)^{2k-2j} f(t)^{2j}\, dt \\ &\le \sum_{j=1}^{k} \binom{2k}{2j} \|\widetilde{f}\|_{L^{2k}}^{2k-2j} \|f\|_{L^{2k}}^{2j}, \end{aligned}$$

by applying Hölder's inequality with exponents $2k/(2k-2j)$ and $2k/(2j)$ to the jth term of the sum. Dividing the last inequality by $\|f\|_{L^{2k}}^{2k}$, we obtain

$$R^{2k} \le \sum_{j=1}^{k} \binom{2k}{2j} R^{2k-2j}, \tag{4.1.13}$$

where $R = \|\widetilde{f}\|_{L^{2k}}/\|f\|_{L^{2k}}$. If $R > 0$ satisfies (4.1.13), then $R \le C_{2k}$, where C_{2k} is the largest real root of the polynomial $g(t) = t^{2k} - \sum_{j=1}^{k} \binom{2k}{2j} t^{2k-2j}$. (Since $g(0) < 0$ and $\lim_{t\to\infty} g(t) = \infty$, g has at least one real root.) We conclude that if f satisfies (a), (b), and (c), then we have for all $k = 1, 2, \dots$

$$\|\widetilde{f}\|_{L^{2k}} \le C_{2k}\|f\|_{L^{2k}}. \tag{4.1.14}$$

We now remove assumptions (a), (b), and (c). We first remove assumption (c). Given a complex-valued trigonometric polynomial f with $\widehat{f}(0) = 0$, we write

$$f(t) = \sum_{j=-N}^{N} c_j e^{2\pi i j t} = \left[\sum_{j=-N}^{N} \frac{c_j + \overline{c_{-j}}}{2} e^{2\pi i j t}\right] + i\left[\sum_{j=-N}^{N} \frac{c_j - \overline{c_{-j}}}{2i} e^{2\pi i j t}\right]$$

(with $c_0 = 0$) and we note that the expressions inside the square brackets are real-valued trigonometric polynomials. Thus we can express f as $P + iQ$, where P and Q are real-valued trigonometric polynomials, and applying (4.1.14) to P and Q we obtain the inequality

$$\|\widetilde{f}\|_{L^{2k}} \le 2C_{2k}\|f\|_{L^{2k}} \tag{4.1.15}$$

for all trigonometric polynomials f with $\widehat{f}(0) = 0$.

Next, we remove the assumption that $\widehat{f}(0) = 0$. We write $f = (f - \widehat{f}(0)) + \widehat{f}(0)$, we observe that the conjugate function of a constant is zero, and we apply (4.1.15) to obtain

$$\|\widetilde{f}\|_{L^{2k}} \le 2C_{2k}\|f - \widehat{f}(0)\|_{L^{2k}} \le 2C_{2k}\Big[\|f\|_{L^{2k}} + \|f\|_{L^1}\Big] \le 4C_{2k}\|f\|_{L^{2k}}.$$

Since trigonometric polynomials are dense in L^p, it follows that the operator $f \mapsto \widetilde{f}$ has a bounded extension on L^{2k} that satisfies (4.1.12) for all $f \in L^{2k}$, and in particular for all $f \in \mathscr{C}^\infty(\mathbf{T}^1)$.

Every real number $p \ge 2$ lies in an interval of the form $[2k, 2k+2]$, for some $k \in \mathbf{Z}^+$. Theorem 1.3.4 gives that for all $2 \le p < \infty$ there is a constant A_p such that

$$\|\widetilde{f}\|_{L^p} \le A_p\|f\|_{L^p} \tag{4.1.16}$$

when f is a simple function. Thus the conjugate function has a bounded extension on L^p that satisfies (4.1.16) when $p \ge 2$.

To extend this result for $p < 2$ we use duality. We observe that the adjoint operator of $f \mapsto \widetilde{f}$ is $f \mapsto -\widetilde{f}$. Indeed, for f, g in $\mathscr{C}^\infty(\mathbf{T}^1)$ we have

$$\langle \widetilde{f} \,|\, g\rangle = \sum_{m\in\mathbf{Z}} -i\,\mathrm{sgn}\,(m)\widehat{f}(m)\overline{\widehat{g}(m)} = -\sum_{m\in\mathbf{Z}} \widehat{f}(m)\overline{-i\,\mathrm{sgn}\,(m)\widehat{g}(m)} = -\langle f \,|\, \widetilde{g}\rangle.$$

By duality, estimate (4.1.16) is also valid for $1 < p \le 2$ with constant $A_{p'} = A_p$. □

We extend the preceding result to higher dimensions.

Theorem 4.1.8. *Let $1 < p < \infty$ and $f \in L^p(\mathbf{T}^n)$. Then $D_N^n * f$ converges to f in L^p as $N \to \infty$.*

Proof. As a consequence of Corollary 4.1.3, Proposition 4.1.6, and Theorem 4.1.7, it suffices to show that for all f trigonometric polynomials on $\mathbf{T}^n$ we have

$$\sup_{N>0} \int_0^1 \cdots \int_0^1 \left|(D_N^n * f)(x)\right|^p dx_1 \cdots dx_n \leq K^{np} \|f\|_{L^p(\mathbf{T}^n)}^p .$$

Obviously, this inequality is valid in dimension $n = 1$. We extend it by induction to all dimensions. We assume that it is valid in dimension $n-1$ and we prove it in dimension n.

Let $x' = (x_2, \dots, x_n) \in \mathbf{T}^{n-1}$. For a fixed trigonometric polynomial f, and for fixed $N \geq 0$ and $x' \in \mathbf{T}^{n-1}$, define a trigonometric polynomial $g_{N,x'}$ on $\mathbf{T}^1$ by setting

$$g_{N,x'}(x_1) = \sum_{m_1 \in \mathbf{Z}} \left[\sum_{|m_2|,\dots,|m_n| \leq N} e^{2\pi i m' \cdot x'} \widehat{f}(m_1, m') \right] e^{2\pi i m_1 x_1}$$

where $m' = (m_2, \dots, m_n)$. Then we have

$$\begin{aligned} g_{N,x'}(x_1) &= \sum_{|m_2|,\dots,|m_n| \leq N} e^{2\pi i m' \cdot x'} \left[\sum_{m_1 \in \mathbf{Z}} e^{2\pi i m_1 x_1} \widehat{f}(m_1, m') \right] \\ &= \sum_{|m_2|,\dots,|m_n| \leq N} e^{2\pi i m' \cdot x'} \left[\int_{\mathbf{T}^{n-1}} f(x_1, y') e^{-2\pi i m' \cdot y'} dy' \right] \\ &= \sum_{|m_2|,\dots,|m_n| \leq N} e^{2\pi i m' \cdot x'} \widehat{f_{x_1}}(m') \\ &= (D_N^{n-1} * f_{x_1})(x'), \end{aligned}$$

where f_{x_1} is the trigonometric polynomial of $n-1$ variables defined by $f_{x_1}(x') = f(x_1, x')$. We also have that

$$(D_N * g_{N,x'})(x_1) = (D_N^n * f)(x_1, x') .$$

Combining this information, we write

$$\begin{aligned} &\int_{\mathbf{T}^{n-1}} \int_0^1 \left|(D_N^n * f)(x_1, x')\right|^p dx_1 \, dx' \\ &\qquad = \int_{\mathbf{T}^{n-1}} \int_0^1 \left|(D_N * g_{N,x'})(x_1)\right|^p dx_1 \, dx' \\ &\qquad \leq K^p \int_{\mathbf{T}^{n-1}} \int_0^1 \left|g_{N,x'}(x_1)\right|^p dx_1 \, dx' \\ &\qquad = K^p \int_0^1 \int_{\mathbf{T}^{n-1}} \left|(D_N^{n-1} * f_{x_1})(x')\right|^p dx' \, dx_1 \end{aligned}$$

$$\leq K^p K^{(n-1)p} \int_{\mathbf{T}^{n-1}} \int_0^1 |f_{x_1}(x')|^p \, dx' \, dx_1$$
$$= K^{np} \|f\|_{L^p(\mathbf{T}^n)}^p ,$$

where the penultimate inequality follows from the induction hypothesis. □

4.1.3 Bochner–Riesz Summability

In dimension 1 the Fejér means of an integrable function are better behaved than the Dirichlet means. We investigate whether there is a similar phenomenon in higher dimensions. Recall that the circular (or spherical) partial sums of the Fourier series of f are given by

$$(f * \overset{\circ}{D}{}_R^n)(x) = \sum_{\substack{m\in\mathbf{Z}^n\\ |m|\leq R}} \widehat{f}(m) e^{2\pi i m\cdot x},$$

where $R \geq 0$. Taking the averages of these expressions, we obtain

$$\frac{1}{R}\int_0^R (f * \overset{\circ}{D}{}_r^n)(x)\, dr = \sum_{\substack{m\in\mathbf{Z}^n\\ |m|\leq R}} \big(1 - \tfrac{|m|}{R}\big)\widehat{f}(m) e^{2\pi i m\cdot x} = B_R^1(f)(x),$$

and we call these expressions the *circular Cesàro means* (or *circular Fejér means*) of f. It turns out that the circular Cesàro means of integrable functions on $\mathbf{T}^2$ always converge in L^1, but in dimension 3, this may fail. Theorem 4.2.5 gives an example of an integrable function f on $\mathbf{T}^3$ whose circular Cesàro means diverge a.e. However, we show below that this is not the case if the circular Cesàro means of a function f in $L^1(\mathbf{T}^3)$ are replaced by the only slightly different-looking means

$$\sum_{\substack{m\in\mathbf{Z}^n\\ |m|\leq R}} \Big(1 - \tfrac{|m|}{R}\Big)^{1+\varepsilon} \widehat{f}(m) e^{2\pi i m\cdot x},$$

for some $\varepsilon > 0$. This discussion suggests that the preceding expressions behave better as ε increases, but for a fixed ε they get worse as the dimension increases. The need to understand the behavior of these operators for different values of $\alpha \geq 0$ led to introduction of the operators B_R^α given in Definition 4.1.2.

The family of operators B_R^α forms a natural "spherical" analogue of the Cesàro–Fejér sums. It turns out that there is no significant difference in the behavior of these means if the expression $\big(1 - \frac{|m|^2}{R^2}\big)^\alpha$ in (4.1.6) is replaced by the expression $\big(1 - \frac{|m|}{R}\big)^\alpha$; see Exercise 4.3.1. The advantage of the quadratic expression in (4.1.6) is that it has an easily computable kernel and yields the elegant reproducing formula

$$B_R^\alpha(f) = \frac{2\Gamma(\alpha+1)}{\Gamma(\alpha-\beta)\Gamma(\beta+1)} \frac{1}{R}\int_0^R \Big(1 - \frac{r^2}{R^2}\Big)^{\alpha-\beta-1} \Big(\frac{r^2}{R^2}\Big)^{\beta+\frac{1}{2}} B_r^\beta(f)\, dr, \quad (4.1.17)$$

which precisely quantifies the way in which B_R^α is smoother than B_R^β when $\alpha > \beta$. Identity (4.1.17) also says that when $\alpha > \beta$, the operator $B_R^\alpha(f)$ is an average of the operators $B_r^\beta(f)$, $0 < r < R$, with respect to a certain density.

Note that the Bochner–Riesz means of order zero coincide with the circular (or spherical) Dirichlet means, and, as we have seen, these converge in $L^2(\mathbf{T}^n)$. We address an analogous question on $L^p(\mathbf{T}^n)$ for $p \neq 2$.

Proposition 4.1.9. *Let $1 \le p < \infty$ and $f \in L^p(\mathbf{T}^n)$. Then the Bochner–Riesz means $B_R^\alpha(f)$ converge to f in $L^p(\mathbf{T}^n)$ as $R \to \infty$ when $\alpha > (n-1)\left|\frac{1}{p} - \frac{1}{2}\right|$. Moreover, if f is continuous on $\mathbf{T}^n$ and $\alpha > \frac{n-1}{2}$, then $B_R^\alpha(f)$ converges to f uniformly as $R \to \infty$.*

Proof. For $z \in \mathbf{C}$ with $\operatorname{Re} z \ge 0$, consider the function

$$m_z(\xi) = (1 - |\xi|^2)_+^z$$

defined for ξ in $\mathbf{R}^n$. Note that $\|m_z\|_{L^\infty} = 1$. Using an identity proved in Appendix B.5, we have that

$$(m_z)^\vee(y) = K^z(y) = \frac{\Gamma(z+1)}{\pi^z} \frac{J_{\frac{n}{2}+z}(2\pi|y|)}{|y|^{\frac{n}{2}+z}}, \tag{4.1.18}$$

where $y \in \mathbf{R}^n$ and J_ν is the Bessel function of order ν. The estimates in Appendices B.6 and B.7 imply that there is a constant $C(\operatorname{Re}\nu)$ such that

$$|J_\nu(r)| \le C(\operatorname{Re}\nu) e^{10|\operatorname{Im}\nu|^2} (1+r)^{-\frac{1}{2}}$$

whenever $\operatorname{Re}\nu > 0$. This yields that if $\operatorname{Re} z > \frac{n-1}{2}$, then there is a constant $C'(t)$ such that the function K^z obeys the estimate

$$|K^z(y)| \le C'(\tfrac{n}{2} + \operatorname{Re} z)\, e^{10|\operatorname{Im} z|^2} (1+|y|)^{-n-(\operatorname{Re} z - \frac{n-1}{2})}, \tag{4.1.19}$$

and hence it lies in $L^1(\mathbf{R}^n)$. Using identity (3.1.10), whenever $\operatorname{Re} z > \frac{n-1}{2}$, we define for an integrable function f on $\mathbf{T}^n$ and $x \in \mathbf{T}^n$ the operator

$$B_R^z(f)(x) = \sum_{\ell \in \mathbf{Z}^n} m_z(\tfrac{\ell}{R}) \widehat{f}(\ell) e^{2\pi i \ell \cdot x} = (f * L^{z,R})(x),$$

where $L^{z,R}$ is a function whose sequence of Fourier coefficients is $\{m_z(\frac{\ell}{R})\}_{\ell \in \mathbf{Z}^n}$. But the function $L^{z,R}$ can be precisely identified. By the Poisson summation formula (Theorem 3.2.8), which applies since both $K_z(x)$ and $m_z(x)$ are bounded by a constant multiple of $(1+|x|)^{-n-\delta}$ for some $\delta > 0$, we have

$$L^{z,R}(x) = \sum_{k \in \mathbf{Z}^n} m_z(\tfrac{k}{R}) e^{2\pi i x \cdot k} = R^n \sum_{\ell \in \mathbf{Z}^n} K^z((x+\ell)R)$$

for all $x \in \mathbf{T}^n$. We show that the family $\{L^{z,R}\}_{R>0}$ is an approximate identity on $\mathbf{T}^n$ when $\mathrm{Re}\, z > \frac{n-1}{2}$; on this see the related Exercise 3.1.3. Obviously, using (4.1.19) we have that

$$\int_{\mathbf{T}^n} |L^{z,R}(x)|\, dx = \int_{\mathbf{R}^n} |K^z(y)|\, dy = C''(n, \mathrm{Re}\, z) e^{10|\mathrm{Im}\, z|^2} < \infty \tag{4.1.20}$$

for some constant $C''(n, \mathrm{Re}\, z)$, and also

$$\int_{\mathbf{T}^n} L^{z,R}(x)\, dx = \int_{\mathbf{R}^n} K^z(y)\, dy = m_z(0) = 1$$

for all $R > 0$ when $\mathrm{Re}\, z > \frac{n-1}{2}$. Moreover, for $\delta < \frac{1}{2}$ using (4.1.19) we have

$$\int_{\delta \le \sup_j |x_j| \le \frac{1}{2}} |L^{z,R}(x)|\, dx \le \frac{C_{n,z}}{R^{\mathrm{Re}\, z - \frac{n-1}{2}}} \int_{\delta \le \sup_j |x_j| \le \frac{1}{2}} \sum_{\ell \in \mathbf{Z}^n} \frac{1}{|x+\ell|^{n+\mathrm{Re}\, z - \frac{n-1}{2}}}\, dx \to 0,$$

thus the integral of $L^{z,R}$ over $[-1/2, 1/2]^n \setminus [-\delta, \delta]^n$ tends to zero as $R \to \infty$.

Using Theorem 1.2.19, we obtain these conclusions for $\mathrm{Re}\, z > \frac{n-1}{2}$:

(a) For $f \in L^1(\mathbf{T}^n)$, $B_R^z(f)$ converge to f in L^1 as $R \to \infty$.
(b) For f continuous on $\mathbf{T}^n$, $B_R^z(f)$ converge to f uniformly as $R \to \infty$.

We turn to the corresponding results for $1 < p < \infty$. We have that

$$\mathrm{Re}\, z > \tfrac{n-1}{2} \implies \sup_{R>0} \big\| B_R^z \big\|_{L^1(\mathbf{T}^n) \to L^1(\mathbf{T}^n)} = C''(n, \mathrm{Re}\, z) e^{10|\mathrm{Im}\, z|^2} \tag{4.1.21}$$

$$\mathrm{Re}\, z = 0 \implies \sup_{R>0} \big\| B_R^z \big\|_{L^2(\mathbf{T}^n) \to L^2(\mathbf{T}^n)} = \|m_z\|_{L^\infty} = 1. \tag{4.1.22}$$

The family of operators $f \mapsto B_R^z(f)$ is of admissible growth for all $\mathrm{Re}\, z \ge 0$, since for all measurable subsets A, B of $\mathbf{T}^n$ we have

$$\left| \int_{\mathbf{T}^n} B_R^z(\chi_A) \chi_B\, dx \right| = \left| \sum_{k \in \mathbf{Z}^n} \widehat{\chi_A}(k) m^z(k) \overline{\widehat{\chi_B}(k)} \right| \le \sum_{|k| \le R} 1 \le C_n R^n,$$

thus condition (1.3.23) holds. Moreover, hypothesis (1.3.24) of Theorem 1.3.7 holds in view of (4.1.21) and (4.1.22). Applying Theorem 1.3.7 (or rather Exercise 1.3.4 in which the strip $[0,1] \times \mathbf{R}$ is replaced by the more general strip $[a,b] \times \mathbf{R}$) we obtain that when $\alpha = \mathrm{Re}\, z > (n-1)|\frac{1}{p} - \frac{1}{2}|$, we have

$$\sup_{R>0} \big\| B_R^\alpha \big\|_{L^p(\mathbf{T}^n) \to L^p(\mathbf{T}^n)} < \infty.$$

Finally, using Corollary 4.1.3, we deduce that $B_R^\alpha(f) \to f$ in $L^p(\mathbf{T}^n)$ as $R \to \infty$ for all $f \in L^p(\mathbf{T}^n)$. □

The preceding result is sharp in the case $p = 1$ (Theorem 4.2.5). For this reason, the number $\alpha = (n-1)/2$ is referred to as the *critical index of Bochner–Riesz summability*.

Exercises

4.1.1. If $f \in \mathscr{C}^\infty(\mathbf{T}^n)$, then show that $D_N^n * f$ and $\overset{\circ}{D}{}_N^n * f$ converge to f uniformly and in L^p for $1 \le p \le \infty$.

4.1.2. Prove that

$$\|P_+\|_{L^2(\mathbf{T}^1)\to L^2(\mathbf{T}^1)} = \|P_-\|_{L^2(\mathbf{T}^1)\to L^2(\mathbf{T}^1)} = \|W\|_{L^2(\mathbf{T}^1)\to L^2(\mathbf{T}^1)} = 1,$$

where $W(f) = \widetilde{f}$ is the conjugate function on the circle. Moreover, show that the mappings $f \mapsto W(f) + \widehat{f}(0)$ and $f \mapsto W(f) - \widehat{f}(0)$ are isometries on $L^2(\mathbf{T}^1)$.

4.1.3. Let $-\infty \le a_j < b_j \le +\infty$ for $1 \le j \le n$. Consider the rectangular projection operator defined on $\mathscr{C}^\infty(\mathbf{T}^n)$ by

$$P(f)(x) = \sum_{a_j \le m_j \le b_j} \widehat{f}(m) e^{2\pi i(m_1x_1+\cdots+m_nx_n)}.$$

Prove that when $1 < p < \infty$, P extends to a bounded operator from $L^p(\mathbf{T}^n)$ to itself with bounds independent of the a_j, b_j.
[*Hint:* Express P in terms of the Riesz projection P_+.]

4.1.4. Let $P_r(t)$ be the Poisson kernel on $\mathbf{T}^1$ as defined in Exercise 3.1.7. For $0 < r < 1$, define the *conjugate Poisson kernel* $Q_r(t)$ on the circle by

$$Q_r(t) = -i \sum_{m=-\infty}^{+\infty} \operatorname{sgn}(m)\, r^{|m|} e^{2\pi i m t}.$$

(a) For $0 < r < 1$, prove the identity

$$Q_r(t) = \frac{2r\sin(2\pi t)}{1 - 2r\cos(2\pi t) + r^2}.$$

(b) Prove that $\widetilde{f}(t) = \lim_{r\to 1}(Q_r * f)(t)$ whenever f is smooth. Conclude that if f is real-valued, then so is $\widetilde{f}$.
(c) Let $f \in L^1(\mathbf{T}^1)$. Prove that the function

$$z \mapsto (P_r * f)(t) + i(Q_r * f)(t)$$

is analytic in $z = re^{2\pi i t}$ on the open unit disc $\{z \in \mathbf{C} : |z| < 1\}$.
(d) Let $f \in L^1(\mathbf{T}^1)$. Conclude that the functions $z \mapsto (P_r * f)(t)$ and $z \mapsto (Q_r * f)(t)$ are *conjugate harmonic functions* of $z = re^{2\pi i t}$ in the region $|z| < 1$. The term conjugate Poisson kernel stems from this property.

4.1.5. Let f be in $\dot{\Lambda}_\alpha(\mathbf{T}^1)$ for some $0<\alpha<1$. Prove that the conjugate function $\widetilde{f}$ is well defined and can be written as

$$\begin{aligned}\widetilde{f}(x) &= \lim_{\varepsilon\to 0}\int_{\varepsilon\le|t|\le 1/2} f(x-t)\cot(\pi t)\,dt\\ &= \int_{|t|\le 1/2}\big(f(x-t)-f(x)\big)\cot(\pi t)\,dt.\end{aligned}$$

[*Hint:* Use part (b) of Exercise 4.1.4 and the fact that Q_r has integral zero over the circle to write $(f*Q_r)(x)=\big((f-f(x))*Q_r\big)(x)$, allowing use of the Lebesgue dominated convergence theorem.]

4.1.6. Suppose that f is a real-valued function on $\mathbf{T}^1$ with $|f|\le 1$ and $0\le\lambda<\pi/2$.
(a) Prove that

$$\int_{\mathbf{T}^1} e^{\lambda\widetilde{f}(t)}\,dt \le \frac{1}{\cos(\lambda)}.$$

(b) Conclude that for $0\le\lambda<\pi/2$ we have

$$\int_{\mathbf{T}^1} e^{\lambda|\widetilde{f}(t)|}\,dt \le \frac{2}{\cos(\lambda)}.$$

[*Hint:* Part (a): Consider the analytic function $F(z)$ on the disk $|z|<1$ defined by $F(z)=-i(P_r*f)(\theta)+(Q_r*f)(\theta)$, where $z=re^{2\pi i\theta}$. Then Re $e^{\lambda F(z)}$ is harmonic and its average over the circle $|z|=r$ is equal to its value at the origin, which is $\cos(\lambda f(0))\le 1$. Let $r\uparrow 1$ and use that for $z=e^{2\pi it}$ on the circle we have Re $e^{\lambda F(z)}\ge e^{\lambda\widetilde{f}(t)}\cos(\lambda)$.]

4.1.7. Prove that for $0<\alpha<1$ there is a constant C_α such that

$$\big\|\widetilde{f}\,\big\|_{\dot{\Lambda}_\alpha(\mathbf{T}^1)} \le C_\alpha\big\|f\big\|_{\dot{\Lambda}_\alpha(\mathbf{T}^1)}.$$

[*Hint:* Using Exercise 4.1.5, for $|h|\le 1/10$ write $\widetilde{f}(x+h)-\widetilde{f}(x)$ as

$$\begin{aligned}&\int_{|t|\le 5|h|}\big(f(x-t)-f(x+h)\big)\cot(\pi(t+h))\,dt\\ &\quad-\int_{|t|\le 5|h|}\big(f(x-t)-f(x)\big)\cot(\pi t)\,dt\\ &\quad+\int_{5|h|\le|t|\le 1/2}\big(f(x-t)-f(x)\big)\big(\cot(\pi(t+h))-\cot(\pi t)\big)\,dt\\ &\quad+\big(f(x)-f(x+h)\big)\int_{5|h|\le|t|\le 1/2}\cot(\pi(t+h))\,dt.\end{aligned}$$

You may use the fact that $\cot(\pi t)=\frac{1}{\pi t}+b(t)$, where $b(t)$ is a bounded function when $|t|\le 1/2$. The case $|h|\ge 1/10$ is easy.]

4.1.8. The beta function is defined in Appendix A.2. Derive the identity

$$t^{\alpha} = \frac{1}{B(\alpha-\beta,\beta+1)} \int_0^t (t-s)^{\alpha-\beta-1} s^{\beta}\, ds$$

and show that the function $K_R^{\alpha}(x) = \sum_{|m|\le R} \big(1 - \frac{|m|^2}{R^2}\big)^{\alpha} e^{2\pi i m\cdot x}$ satisfies (4.1.17). $\big[$*Hint:* Take $t = 1 - \frac{|m|^2}{R^2}$ and change variables $s = \frac{r^2-|m|^2}{R^2}$ in the displayed identity.$\big]$

4.2 A. E. Divergence of Fourier Series and Bochner–Riesz means

We saw in Proposition 3.4.6 that the Fourier series of a continuous function may diverge at a point. As expected, the situation can only get worse as the functions get worse. In this section we present an example, due to A. N. Kolmogorov, of an integrable function on $\mathbf{T}^1$ whose Fourier series diverges almost everywhere. We also prove an analogous result for the Bochner–Riesz means at the critical index.

4.2.1 Divergence of Fourier Series of Integrable Functions

It is natural to start our investigation with the case $n = 1$. We begin with the following important result:

Theorem 4.2.1. *There exists an integrable function on the circle* $\mathbf{T}^1$ *whose Fourier series diverges almost everywhere.*

Proof. The proof of this theorem is a bit involved, and we need a sequence of lemmas, which we prove first.

Lemma 4.2.2. ***(Kronecker)*** *Suppose that* $N \in \mathbf{Z}^+$ *and*

$$\{x_1, x_2, \dots, x_N, 1\}$$

is a linearly independent set over the rationals. Then for any $\varepsilon > 0$ *and any complex numbers* $z_1, z_2, \dots, z_N$ *with* $|z_j| = 1$*, there exists an integer* $L \in \mathbf{Z}$ *such that*

$$|e^{2\pi i L x_j} - z_j| < \varepsilon \qquad \textit{for all} \quad 1 \le j \le N.$$

Proof. Suppose that the assertion claimed is false. Then there is an $\varepsilon > 0$ and complex numbers $z_j = e^{2\pi i \theta_j}$, $j = 1, \dots, N$, with $0 \le \theta_j < 1$, such that

$$\{m(x_1, \dots, x_N) : \, m \in \mathbf{Z}\} \cap B\big((\theta_1, \dots, \theta_N), \varepsilon\big) = \emptyset,$$

where $B\big((\theta_1,\dots,\theta_N),\varepsilon\big)$ denotes a neighborhood in $\mathbf{T}^N$ of radius ε centered at the point $(\theta_1,\dots,\theta_N)$. Pick a smooth, nonzero, and nonnegative function f on $\mathbf{T}^N$ supported in $B\big((\theta_1,\dots,\theta_N),\varepsilon\big)$. Then $f(m(x_1,\dots,x_N))=0$ for all $m\in\mathbf{Z}$, but

$$\widehat{f}(0)=\int_{\mathbf{T}^N} f(y)\,dy>0. \tag{4.2.1}$$

Set $x=(x_1,\dots,x_N)$. Then we have

$$\begin{aligned}
0=\frac{1}{M}\sum_{m=0}^{M-1} f(mx) &= \frac{1}{M}\sum_{m=0}^{M-1}\bigg(\sum_{\ell\in\mathbf{Z}^N}\widehat{f}(\ell)e^{2\pi i\ell\cdot mx}\bigg)\\
&=\sum_{\ell\in\mathbf{Z}^N}\widehat{f}(\ell)\bigg(\frac{1}{M}\sum_{m=0}^{M-1}e^{2\pi i m(\ell\cdot x)}\bigg)\\
&=\widehat{f}(0)+\sum_{\ell\in\mathbf{Z}^N\setminus\{0\}}\widehat{f}(\ell)\bigg(\frac{1}{M}\frac{e^{2\pi i M(\ell\cdot x)}-1}{e^{2\pi i(\ell\cdot x)}-1}\bigg).
\end{aligned}$$

Note that $e^{2\pi i(\ell\cdot x)}-1\neq 0$ because $\ell\cdot x=\ell_1x_1+\cdots+\ell_Nx_N\notin\mathbf{Z}$, since by assumption the set $\{x_1,x_2,\dots,x_N,1\}$ is linearly independent over the rationals. Observe that

$$\widehat{f}(\ell)\frac{1}{M}\frac{e^{2\pi i M(\ell\cdot x)}-1}{e^{2\pi i(\ell\cdot x)}-1}=\widehat{f}(\ell)\frac{1}{M}\sum_{m=0}^{M-1}e^{2\pi i m(\ell\cdot x)}$$

tends to 0 as $M\to\infty$ for every fixed $\ell\in\mathbf{Z}^N$ and is bounded uniformly in M by $|\widehat{f}(\ell)|$ which satisfies $\sum_{\ell\in\mathbf{Z}^N}|\widehat{f}(\ell)|<\infty$. Using the Lebesgue dominated convergence theorem, we obtain that

$$\begin{aligned}
0&=\widehat{f}(0)+\lim_{M\to\infty}\sum_{\ell\in\mathbf{Z}^N\setminus\{0\}}\widehat{f}(\ell)\bigg(\frac{1}{M}\frac{e^{2\pi i M(\ell\cdot x)}-1}{e^{2\pi i(\ell\cdot x)}-1}\bigg)\\
&=\widehat{f}(0)+\sum_{\ell\in\mathbf{Z}^N\setminus\{0\}}\widehat{f}(\ell)\lim_{M\to\infty}\bigg(\frac{1}{M}\frac{e^{2\pi i M(\ell\cdot x)}-1}{e^{2\pi i(\ell\cdot x)}-1}\bigg)\\
&=\widehat{f}(0)+0,
\end{aligned}$$

which contradicts (4.2.1). Therefore the claimed L exists. □

Lemma 4.2.3. *There exists a positive constant $c>0$ such that given any integer $N\geq 2$ there exists a positive measure μ_N on $\mathbf{T}^1$ with $\mu_N(\mathbf{T}^1)=1$ such that*

$$\sup_{L\geq 1}\big|(\mu_N*D_L)(x)\big|=\sup_{L\geq 1}\bigg|\sum_{k=-L}^{L}\widehat{\mu_N}(k)e^{2\pi ikx}\bigg|\geq c\log N \tag{4.2.2}$$

for almost all $x\in\mathbf{T}^1$ (c is a fixed constant).

Proof. Given irrational real numbers $x_1,\dots,x_N$ such that the set $\{x_1,\dots,x_N,1\}$ is linearly independent over the rationals, we define $\mathbf{Q}[x_1,\dots,x_N]$ to be the field extension of $\mathbf{Q}$ consisting of all linear combinations of the form $q_0+q_1x_1+\cdots+q_Nx_N$, where q_j are rational numbers. Obviously $\mathbf{Q}[x_1,\dots,x_N]$ is a countable set. Fix $N\ge 100$ and choose points x_j as follows:

$$0<x_1<\frac{1}{N}<x_2<\frac{2}{N}<x_3<\frac{3}{N}<\cdots<\frac{N-1}{N}<x_N<1 \tag{4.2.3}$$

and such that $x_1\notin\mathbf{Q}$, $x_2\notin\mathbf{Q}[x_1]$, $\dots$, $x_N\notin\mathbf{Q}[x_1,\dots,x_{N-1}]$. Then obviously the set $\{x_1,\dots,x_N,1\}$ is linearly independent over the rationals. Let

$$E_N=\big\{x\in[0,1]:\ \{x-x_1,\dots,x-x_N,1\}\quad\text{is linearly independent over }\mathbf{Q}\big\}$$

and observe that every x in $[0,1]\setminus\mathbf{Q}[x_1,\dots,x_N]$ belongs to E_N. Indeed, if $x\notin E_N$, then there are rational numbers q_j such that

$$q_0+q_1(x-x_1)+\cdots+q_N(x-x_N)=0.$$

Then $q=q_1+\cdots+q_N\neq 0$, since $\{x_1,\dots,x_N,1\}$ are linearly independent over $\mathbf{Q}$. It follows that

$$x=-q^{-1}q_0+q^{-1}q_1x_1+\cdots+q^{-1}q_Nx_N,$$

thus $x\in\mathbf{Q}[x_1,\dots,x_N]$. We conclude that E_N has full measure.

Next, we define the probability measure

$$\mu_N=\frac{1}{N}\sum_{j=1}^{N}\delta_{x_j},$$

where δ_{x_j} are Dirac delta masses at the points x_j. For this measure we have

$$\begin{aligned}\left|\sum_{k=-L}^{L}\widehat{\mu_N}(k)e^{2\pi ikx}\right|&=\left|\sum_{k=-L}^{L}\left(\frac{1}{N}\sum_{j=1}^{N}e^{-2\pi ikx_j}\right)e^{2\pi ikx}\right|\\&=\left|\frac{1}{N}\sum_{j=1}^{N}D_L(x-x_j)\right|\\&=\left|\frac{1}{N}\sum_{j=1}^{N}\frac{\sin(2\pi(L+\frac12)(x-x_j))}{\sin(\pi(x-x_j))}\right|\\&=\left|\frac{1}{N}\sum_{j=1}^{N}\frac{\operatorname{Im}\left[e^{2\pi i(L+\frac12)(x-x_j)}\right]\operatorname{sgn}\big(\sin(\pi(x-x_j))\big)}{|\sin(\pi(x-x_j))|}\right|,\end{aligned} \tag{4.2.4}$$

where the signum function is defined as $\operatorname{sgn}a=1$ for $a>0$, -1 for $a<0$, and zero if $a=0$. By Lemma 4.2.2, for all $x\in E_N$ there exists an $L\in\mathbf{Z}^+$ such that

$$\left|e^{2\pi iL(x-x_j)}-ie^{-2\pi i\frac12(x-x_j)}\operatorname{sgn}\big(\sin(\pi(x-x_j))\big)\right|<\frac{1}{2},$$

which can be equivalently written as

$$\left|e^{2\pi i(L+\frac{1}{2})(x-x_j)}\operatorname{sgn}\big(\sin(\pi(x-x_j))\big)-i\right|<\frac{1}{2}. \tag{4.2.5}$$

It follows from (4.2.5) that

$$\operatorname{Im}\left[e^{2\pi i(L+\frac{1}{2})(x-x_j)}\right]\operatorname{sgn}\big(\sin(\pi(x-x_j))\big)>\frac{1}{2}.$$

Combining this with the result of the calculation in (4.2.4), we obtain that

$$\left|\sum_{k=-L}^{L}\widehat{\mu_N}(k)e^{2\pi ikx}\right|>\frac{1}{2N}\sum_{j=1}^{N}\frac{1}{|\sin(\pi(x-x_j))|}\geq\frac{1}{2\pi N}\sum_{j=1}^{N}\frac{1}{|x-x_j|}.$$

But for every $x\in[0,1)$, there exists a j_0 such that $x\in[x_{j_0},x_{j_0+1})$. It follows from (4.2.3) that $|x-x_j|\leq C(|j-j_0|+1)N^{-1}$, and thus

$$\sum_{j=1}^{N}\frac{1}{|x-x_j|}\geq c'N\log N.$$

Thus for every $x\in E_N$ there exists an $L\in\mathbf{Z}^+$ such that

$$\left|D_L*\mu_N(x)\right|=\left|\sum_{k=-L}^{L}\widehat{\mu_N}(k)e^{2\pi ikx}\right|>c\log N,$$

which proves the required conclusion since E_N is a set of measure 1. □

Lemma 4.2.4. *For each $0<M<\infty$ there exists a trigonometric polynomial g_M and a measurable subset A_M of $\mathbf{T}^1$ with measure $|A_M|>1-2^{-M}$ such that $\|g_M\|_{L^1}=1$, and such that*

$$\inf_{x\in A_M}\sup_{L\geq 1}\left|(D_L*g_M)(x)\right|=\inf_{x\in A_M}\sup_{L\geq 1}\left|\sum_{k=-L}^{L}\widehat{g_M}(k)e^{2\pi ikx}\right|>2^M. \tag{4.2.6}$$

Proof. Given an $M\in\mathbf{Z}^+$, we pick an integer $N(M)$ such that $c\log N(M)>2^{M+2}$, where c is as in (4.2.2), and we also pick the measure $\mu_{N(M)}$, which satisfies (4.2.2). By Fatou's lemma we have

$$\begin{aligned}1&=\left|\left\{x\in\mathbf{T}^1:\ \sup_{L\geq 1}|(D_L*\mu_{N(M)})(x)|\geq 2^{M+2}\right\}\right|\\&=\left|\bigcup_{L\geq 1}\left\{x\in\mathbf{T}^1:\ \sup_{1\leq j\leq L}|(D_j*\mu_{N(M)})(x)|\geq 2^{M+2}\right\}\right|\\&=\int_{\mathbf{T}^1}\lim_{L\to\infty}\chi_{\{x\in\mathbf{T}^1:\ \sup_{1\leq j\leq L}|(D_j*\mu_{N(M)})(x)|\geq 2^{M+2}\}}\,dx\\&\leq\liminf_{L\to\infty}\left|\left\{x\in\mathbf{T}^1:\ \sup_{1\leq j\leq L}|(D_j*\mu_{N(M)})(x)|\geq 2^{M+2}\right\}\right|,\end{aligned}$$

and thus we can find a positive integer $L(M)$ such that the set

$$A_M = \left\{x \in \mathbf{T}^1 : \sup_{1 \le L \le L(M)} |(D_L * \mu_{N(M)})(x)| \ge 2^{M+2}\right\}$$

has measure greater than $1 - 2^{-M}$. We pick a positive integer $K(M)$ such that

$$\sup_{1 \le j \le L(M)} \left\|F_{K(M)} * D_j - D_j\right\|_{L^\infty} \le 1,$$

where F_K is the Fejér kernel. This is possible, since the Fejér kernel is an approximate identity and $\{D_j : 1 \le j \le L(M)\}$ is a finite family of continuous functions. Then we define $g_M = \mu_{N(M)} * F_{K(M)}$. Since $\mu_{N(M)}$ is a probability measure, we obtain

$$|(D_j * g_M)(x) - (D_j * \mu_{N(M)})(x)| \le \left\|D_j * F_{K(M)} - D_j\right\|_{L^\infty} \le 1$$

for all $x \in [0,1]$ and $1 \le j \le L(M)$. But given $x \in A_M$ there exists an L in $\{1,\dots,L(M)\}$ such that $|(D_L * \mu_{N(M)})(x)| \ge 2^{M+2}$ and for this L we have

$$|(D_L * g_M)(x)| \ge |(D_L * \mu_{N(M)})(x)| - 1 \ge 2^{M+2} - 1 \ge 2^{M+1} > 2^M.$$

Therefore, (4.2.6) is satisfied for this g_M and A_M. Since μ_N is a nonnegative measure and $F_{K(M)}$ is nonnegative and has L^1 norm 1, we have that

$$\left\|g_M\right\|_{L^1} = \left\|\mu_{N(M)} * F_{K(M)}\right\|_{L^1} = \left\|\mu_{N(M)}\right\|_{\mathcal{M}} \left\|F_{K(M)}\right\|_{L^1} = 1,$$

showing that g_M has L^1 norm equal to one. □

We now have the tools needed to construct an example of a function whose Fourier series diverges almost everywhere. The example is given as a series of functions whose behavior worsens as its index becomes bigger. The function we wish to construct is a sum of the form

$$g = \sum_{j=1}^{\infty} \varepsilon_j g_{M_j}, \tag{4.2.7}$$

for a choice of sequences $\varepsilon_j \to 0$ and $M_j \to \infty$, where g_M are as in Lemma 4.2.4.

Let us be specific. First, we set $d_0 = 1$ and for $N \ge 1$

$$d_N = \max_{1 \le s \le N} \text{degree}(g_{M_s}), \tag{4.2.8}$$

where g_M is the trigonometric polynomial of Lemma 4.2.4. We set $\varepsilon_0 = M_0 = 1$. Assume that we have defined ε_j and M_j for all $1 \le j < N$ for some $N \ge 2$. We set

$$\varepsilon_N = 2^{-N}(3d_{N-1})^{-1} \tag{4.2.9}$$

and then we pick M_N such that

$$2^{M_N} \geq \left(2^N + d_{N-1} + 1\right)\varepsilon_N^{-1}. \tag{4.2.10}$$

This defines ε_N and M_N for a given positive integer N, provided ε_j and M_j are known for all $j < N$. This way we define ε_N and M_N for all natural numbers N.

We observe that the selections of ε_j and M_j force the inequalities $\varepsilon_j \leq 2^{-j}$ and $d_j \leq d_{j+1}$ for all $j \geq 1$. Since each g_{M_j} has L^1 norm 1 and $\varepsilon_j \leq 2^{-j}$, the function g in (4.2.7) is integrable and has L^1 norm at most 1.

For a given $j \geq 1$ and $x \in A_{M_j}$, by Lemma 4.2.4 there exists an $L \geq 1$ such that $|(D_L * g_{M_j})(x)| > 2^{M_j}$. Set

$$k = k(x) = \min(L, d_j).$$

Then we have

$$|(D_k * g)(x)| \geq \varepsilon_j |(D_k * g_{M_j})(x)| - \sum_{1 \leq s < j} \varepsilon_s |(D_k * g_{M_s})(x)| - \sum_{s > j} \varepsilon_s |(D_k * g_{M_s})(x)|.$$

We make the following observations:

(i) $|(D_k * g_{M_j})(x)| = |(D_L * g_{M_j})(x)| > 2^{M_j}$.

(ii) $|(D_k * g_{M_s})(x)| = |(D_{\min(d_s,k)} * g_{M_s})(x)| \leq \left\|D_{\min(d_s,L)}\right\|_{L^\infty} \leq 3d_s$, when $s < j$.

(iii) $|(D_k * g_{M_s})(x)| = |(D_{\min(d_s,k)} * g_{M_s})(x)| \leq \left\|D_{\min(d_j,L)}\right\|_{L^\infty} \leq 3d_j$, when $s > j$.

In these estimates we have used that $k = \min(L, d_j)$, $\left\|D_m\right\|_{L^\infty} \leq 2m + 1 \leq 3m$, and that

$$D_r * g_{M_s} = D_{\min(r,d_s)} * g_{M_s},$$

which follows easily by examining the corresponding Fourier coefficients.

Using the estimates in (i), (ii), and (iii), for a fixed $x \in A_{M_j}$ and $k = k(x)$ we obtain

$$|(D_k * g)(x)| \geq \varepsilon_j 2^{M_j} - 3 \sum_{1 \leq s < j} \varepsilon_s d_s - 3 \sum_{s > j} \varepsilon_s d_j. \tag{4.2.11}$$

Our selection of ε_j and M_j now ensures that (4.2.11) is a large number. In fact, we have

$$3 \sum_{s > j} \varepsilon_s d_j = \sum_{s > j} 2^{-s} d_j (d_{s-1})^{-1} \leq \sum_{s > j} 2^{-s} \leq 1$$

and

$$3 \sum_{1 \leq s < j} \varepsilon_s d_s \leq 3d_{j-1} \sum_{1 \leq s < j} \varepsilon_s \leq d_{j-1} \sum_{1 \leq s < j} 2^{-s} (d_{s-1})^{-1} \leq d_{j-1}.$$

Therefore, the expression in (4.2.11) is at least $\varepsilon_j 2^{M_j} - d_{j-1} - 1 \geq 2^j$. It follows that for every $j \geq 1$ and every $x \in A_{M_j}$ there exists a $k = k(x) \in \mathbf{Z}^+$ such that

$$|(D_k * g)(x)| \geq 2^j .$$

We conclude that for every $r \geq 1$ and $x \in \bigcup_{j=r}^{\infty} A_{M_j}$,

$$\sup_{k \geq 1} |(D_k * g)(x)| \geq 2^j \geq 2^r , \tag{4.2.12}$$

since x belongs to some A_{M_j} with $j \geq r$. For given $r \geq 1$, Lemma 4.2.4 yields that

$$1 \geq \left| \bigcup_{j=r}^{\infty} A_{M_j} \right| \geq \liminf_{j \to \infty} |A_{M_j}| \geq \lim_{j \to \infty} (1 - 2^{-M_j}) = 1.$$

Then the set

$$A = \bigcap_{r=0}^{\infty} \bigcup_{j=r}^{\infty} A_{M_j}$$

has measure 1, since it is a countable intersection of subsets of $\mathbf{T}^1$ of full measure. In view of (4.2.12) we have that for all x in A

$$\sup_{k \geq 1} |(D_k * g)(x)| \geq \sup_{r \geq 1} 2^r = \infty \tag{4.2.13}$$

and thus the required conclusion follows. □

4.2.2 Divergence of Bochner–Riesz Means of Integrable Functions

We now turn to the corresponding n-dimensional problem for spherical summability of Fourier series. The situation here is quite similar at the critical index $\alpha = \frac{n-1}{2}$.

Theorem 4.2.5. *Let $n > 1$. There exists an integrable function f on $\mathbf{T}^n$ such that*

$$\limsup_{R \to \infty} \left| B_R^{\frac{n-1}{2}}(f)(x) \right| = \limsup_{R \to \infty} \left| \sum_{\substack{m \in \mathbf{Z}^n \\ |m| \leq R}} \left(1 - \tfrac{|m|^2}{R^2}\right)^{\frac{n-1}{2}} \widehat{f}(m) e^{2\pi i m \cdot x} \right| = \infty$$

for almost all $x \in \mathbf{T}^n$. Furthermore, such a function can be constructed such that it is supported in an arbitrarily small given neighborhood of the origin.

We will need a couple of lemmas.

Lemma 4.2.6. *Let $n \geq 2$. The complement of the set*

$$S = \left\{ x \in \mathbf{R}^n : \{1\} \cup \{|x - m| : m \in \mathbf{Z}^n\} \text{ is linearly independent over } \mathbf{Q} \right\}$$

has n-dimensional Lebesgue measure zero.

Proof. Recall that a function g defined on an open subset Ω of $\mathbf{R}^n$ is called *real analytic* if for every point x_0 in Ω there is a ball $B(x_0,\varepsilon)$ is contained in Ω and there exist coefficients $c_\beta(x_0)$ such that $g(x)=\sum_\beta c_\beta(x_0)(x-x_0)^\beta$ for all $|x-x_0|<\varepsilon$, where the sum is taken over all multiindices. We will need two facts about real analytic functions. First, the function $x\to|x|$ is real analytic on $\mathbf{R}^n\setminus\{0\}$. Indeed, given $x_0\neq 0$, for $|x-x_0|<|x_0|/3$ we have that

$$\frac{|x-x_0|^2}{|x_0|^2}+2(x-x_0)\cdot\frac{x_0}{|x_0|^2}<1\,.$$

This allows us to write

$$\begin{aligned}|x|&=|x_0|\left(1+\frac{|x-x_0|^2}{|x_0|^2}+2(x-x_0)\cdot\frac{x_0}{|x_0|^2}\right)^{\frac{1}{2}}\\&=|x_0|\sum_{k=0}^{\infty}\binom{1/2}{k}\left(\frac{|x-x_0|^2}{|x_0|^2}+2(x-x_0)\cdot\frac{x_0}{|x_0|^2}\right)^{k},\end{aligned}$$

which is a power series of the form $\sum_\beta c_\beta(x_0)(x-x_0)^\beta$.

Secondly, we need the fact that a real analytic function defined on an open connected subset of $\mathbf{R}^n$ cannot vanish on a set of positive measure, unless it is identically equal to zero; a proof of this in dimension one and an outline of the proof in higher dimensions is contained in [205].

We return to the proof of the lemma which requires us to show that S has full measure in $\mathbf{R}^n$. Indeed, if $x\in\mathbf{R}^n\setminus S$, then there exist $k\in\mathbf{Z}^+$, $m_1,\dots m_k\in\mathbf{Z}^n$, and $q_0,q_1,\dots,q_k$ nonzero rational numbers such that

$$q_0+\sum_{j=1}^{k}q_j|x-m_j|=0\,. \tag{4.2.14}$$

Since the function

$$y\mapsto q_0+\sum_{j=1}^{k}q_j|y-m_j|$$

is nonzero and real analytic on $\mathbf{R}^n\setminus\mathbf{Z}^n$, it must vanish only on a set of Lebesgue measure zero. Therefore, there exists a set $A_{m_1,\dots,m_k,q_0,q_1,\dots,q_k}$ of Lebesgue measure zero such that (4.2.14) holds exactly when x is in this set. Then

$$\mathbf{R}^n\setminus S\subseteq\bigcup_{k=1}^{\infty}\bigcup_{m_1,\dots,m_k\in\mathbf{Z}^n}\bigcup_{q_0,q_1,\dots,q_k\in\mathbf{Q}}A_{m_1,\dots,m_k,q_0,q_1,\dots,q_k}\,,$$

from which it follows that $\mathbf{R}^n\setminus S$ has Lebesgue measure zero. □

Let us denote the Bochner–Riesz kernel by

$$K_R^\alpha(x) = \sum_{|m|\le R} \left(1 - \tfrac{|m|^2}{R^2}\right)^\alpha e^{2\pi i m\cdot x}$$

when $x \in \mathbf{T}^n$. We need the following lemma regarding K_R^α:

Lemma 4.2.7. *Let $n \ge 2$. For almost every $x \in \mathbf{T}^n$ we have*

$$\limsup_{R\to\infty} |K_R^{\frac{n-1}{2}}(x)| = \infty.$$

It is noteworthy to compare the result of this lemma with the analogous one-dimensional statement

$$\limsup_{R\to\infty} |D_R(x)| = \infty$$

for the Dirichlet kernel, which holds exactly when $x = 0$. Thus the uniform ill behavior of the kernel $K_R^{\frac{n-1}{2}}$ reflects in some sense its lack of localization.

Proof. Fix $n \ge 2$ and fix $x_0 \in \big([-1/2,1/2)^n \setminus \{0\}\big) \cap S$, where S is as in Lemma 4.2.6. Using (4.1.18) and the Poisson summation formula (Theorem 3.2.8), for each $\alpha > \frac{n-1}{2}$ we obtain the identity

$$K_R^\alpha(x_0) = \frac{\Gamma(\alpha+1)}{\pi^\alpha} R^n \sum_{m\in\mathbf{Z}^n} \frac{J_{\frac{n}{2}+\alpha}(2\pi R|x_0-m|)}{(R|x_0-m|)^{\frac{n}{2}+\alpha}} \tag{4.2.15}$$

and the sum converges absolutely because of the asymptotics for the Bessel functions in Appendix B.8. The term with $m = 0$ in the sum in (4.2.15) is a finite constant since by Appendix B.6 the function

$$y \mapsto \frac{J_{\frac{n}{2}+\alpha}(2\pi R|y|)}{|y|^{\frac{n}{2}+\alpha}}$$

is smooth and therefore bounded. But for $m \ne 0$ in (4.2.15) we have $|m - x_0| \ge 1/2$. The asymptotics in Appendix B.8 imply that for $R \ge 2$ we have

$$\begin{aligned} J_{\frac{n}{2}+\alpha}(2\pi R|x_0-m|) &= \frac{e^{2\pi i R|x_0-m|}e^{-i\frac{\pi}{2}(\frac{n}{2}+\alpha)-i\frac{\pi}{4}} + e^{-2\pi i R|x_0-m|}e^{i\frac{\pi}{2}(\frac{n}{2}+\alpha)+i\frac{\pi}{4}}}{\pi\sqrt{R|x_0-m|}} \\ &\quad + O\big((R|x_0-m|)^{-\frac{3}{2}}\big) \end{aligned}$$

for all $\alpha > \frac{n-1}{2}$. We insert this expression in (4.2.15), we multiply by $e^{2\pi i\lambda R}$ for some λ real, and then we average in R from 1 to T, for some $T > 10$. We obtain

$$
\begin{aligned}
&\frac{1}{T}\int_1^T K_R^\alpha(x_0)e^{2\pi i\lambda R}\,dR\\
&=\frac{\Gamma(\alpha+1)}{\pi^\alpha}\sum_{m\in\mathbf{Z}^n\setminus\{0\}}\frac{e^{-i\frac{\pi}{2}(\frac{n}{2}+\alpha)-i\frac{\pi}{4}}}{|x_0-m|^{\frac{n+1}{2}+\alpha}}\frac{1}{T}\int_1^T e^{2\pi iR(\lambda+|x_0-m|)}R^{\frac{n-1}{2}-\alpha}\,dR\\
&\quad+\frac{\Gamma(\alpha+1)}{\pi^\alpha}\sum_{m\in\mathbf{Z}^n\setminus\{0\}}\frac{e^{i\frac{\pi}{2}(\frac{n}{2}+\alpha)+i\frac{\pi}{4}}}{|x_0-m|^{\frac{n+1}{2}+\alpha}}\frac{1}{T}\int_1^T e^{2\pi iR(\lambda-|x_0-m|)}R^{\frac{n-1}{2}-\alpha}\,dR \qquad (4.2.16)\\
&\quad+\frac{\Gamma(\alpha+1)}{\pi^\alpha}\sum_{m\in\mathbf{Z}^n\setminus\{0\}}O\Big(\frac{1}{|x_0-m|^{\frac{n+3}{2}+\alpha}}\Big)\frac{1}{T}\int_1^T R^{\frac{n-3}{2}-\alpha}\,dR\\
&\quad+\frac{\Gamma(\alpha+1)}{\pi^\alpha}\frac{1}{T}\int_1^T\frac{J_{\frac{n}{2}+\alpha}(2\pi R|x_0|)}{(R|x_0|)^{\frac{n}{2}+\alpha}}e^{2\pi i\lambda R}R^n\,dR.
\end{aligned}
$$

Assume that we are able to pass the limit as $\alpha\to\frac{n-1}{2}+$ through the sums and integrals in the preceding identity; we justify this step momentarily. Then we obtain

$$
\begin{aligned}
&\frac{1}{T}\int_1^T K_R^{\frac{n-1}{2}}(x_0)e^{2\pi i\lambda R}\,dR\\
&\quad=\frac{\Gamma(\frac{n+1}{2})}{\pi^{\frac{n-1}{2}}}\sum_{m\in\mathbf{Z}^n\setminus\{0\}}\frac{e^{-i\frac{\pi}{2}(\frac{2n-1}{2})-i\frac{\pi}{4}}}{|x_0-m|^n}\frac{1}{T}\int_1^T e^{2\pi iR(\lambda+|x_0-m|)}\,dR\\
&\quad\quad+\frac{\Gamma(\frac{n+1}{2})}{\pi^{\frac{n-1}{2}}}\sum_{m\in\mathbf{Z}^n\setminus\{0\}}\frac{e^{i\frac{\pi}{2}(\frac{2n-1}{2})+i\frac{\pi}{4}}}{|x_0-m|^n}\frac{1}{T}\int_1^T e^{2\pi iR(\lambda-|x_0-m|)}\,dR \qquad (4.2.17)\\
&\quad\quad+\frac{\Gamma(\frac{n+1}{2})}{\pi^{\frac{n-1}{2}}}\sum_{m\in\mathbf{Z}^n\setminus\{0\}}O\Big(\frac{1}{|x_0-m|^{n+1}}\Big)\frac{1}{T}\int_1^T\frac{dR}{R}\\
&\quad\quad+\frac{\Gamma(\frac{n+1}{2})}{\pi^{\frac{n-1}{2}}}\frac{1}{T}\int_1^T\frac{J_{n-\frac{1}{2}}(2\pi R|x_0|)}{(R|x_0|)^{n-\frac{1}{2}}}e^{2\pi i\lambda R}R^n\,dR.
\end{aligned}
$$

We now justify the passage of the limit in α inside the sums and the integrals in (4.2.16) to obtain (4.2.17). First, when $|m|\le R\le T$ and $\alpha>\frac{n-1}{2}$, the mean value theorem gives

$$
\begin{aligned}
&\left|\Big(1-\frac{|m|^2}{R^2}\Big)^{\frac{n-1}{2}}e^{2\pi ix_0\cdot m}-\Big(1-\frac{|m|^2}{R^2}\Big)^{\alpha}e^{2\pi ix_0\cdot m}\right|\\
&\qquad\le\Big(\alpha-\frac{n-1}{2}\Big)\Big(1-\frac{|m|^2}{R^2}\Big)^{\frac{n-1}{2}}\log\frac{1}{1-\frac{|m|^2}{R^2}}\\
&\qquad\le\Big(\alpha-\frac{n-1}{2}\Big)\sup_{0<t\le1}t^{\frac{n-1}{2}}\log\frac{1}{t},
\end{aligned}
$$

thus $K_R^{\alpha}(x_0)$ converges to $K_R^{\frac{n-1}{2}}(x_0)$ uniformly in $R \in [1,T]$ as $\alpha \to \frac{n-1}{2}$ and therefore the integral over $[1,T]$ of the former converges to the integral over $[1,T]$ of the latter. Next, an integration by parts shows that the integral

$$\int_1^T e^{2\pi i R(\lambda+|x_0-m|)} R^{\frac{n-1}{2}-\alpha}\, dR$$

is bounded by a constant multiple of $(\lambda+|x_0-m|)^{-1}$, which makes the first infinite sum in (4.2.16) converge absolutely and uniformly in $\alpha \geq \frac{n-1}{2}$, thus one may pass the limit in α inside the sum. Also, the integral

$$\int_1^T e^{2\pi i R(\lambda-|x_0-m|)} R^{\frac{n-1}{2}-\alpha}\, dR$$

is bounded by a constant multiple of $(\lambda-|x_0-m|)^{-1}$ whenever λ is not in the set

$$\Lambda_{x_0} = \{|x_0-m| : m \in \mathbf{Z}^n\} = \{\lambda_1, \lambda_2, \lambda_3, \dots\},$$

where $0 < \lambda_1 < \lambda_2 < \lambda_3 < \cdots$. Thus for $\lambda \notin \Lambda_{x_0}$, the preceding argument explains the passage of the limit in α inside the second infinite sum in (4.2.16). If λ happens to be in Λ_{x_0}, then there is at most one $m_0 \neq 0$, such that $\lambda = |x_0-m_0|$ and the second sum in (4.2.16) restricted to $m \in \mathbf{Z}^n \setminus \{0, m_0\}$ converges absolutely, while for the single term with $m = m_0$, letting $\alpha \to \frac{n-1}{2}+$ is trivial. Finally, for the term involving the Bessel function $J_{n-\frac{1}{2}}$, the passage of the limit in α inside the integral is straightforward since the function

$$(\alpha, R) \mapsto \frac{J_{\frac{n}{2}+\alpha}(2\pi R|x_0|)}{(R|x_0|)^{\frac{n}{2}+\alpha}}$$

is continuous on the compact set $[\frac{n-1}{2}, \frac{n}{2}] \times [1,T]$. This completes the proof of (4.2.17).

There are four terms to the right of (4.2.17) and we observe that if $\lambda \neq \pm|x_0-m_0|$ for any $m_0 \in \mathbf{Z}^n$, then all these terms converge to zero as $T \to \infty$. This assertion is trivial for the first three of these four terms, while for the last we assume that $T > |x_0|^{-1}$. We split the integral

$$\frac{1}{T}\int_1^T \frac{J_{n-\frac{1}{2}}(2\pi R|x_0|)}{(R|x_0|)^{n-\frac{1}{2}}} e^{2\pi i\lambda R} R^n\, dR \tag{4.2.18}$$

as a sum of the integral over $[1, |x_0|^{-1}]$, which obviously converges to zero as $T \to \infty$ by Appendix B.6, and of the integral over $[|x_0|^{-1}, T]$. For the latter, we use the asymptotics in Appendix B.8 to write

$$J_{n-\frac{1}{2}}(2\pi R|x_0|) = \frac{e^{2\pi i R|x_0|} e^{-i\frac{\pi}{2}(\frac{2n-1}{2})-i\frac{\pi}{4}} + e^{-2\pi i R|x_0|} e^{i\frac{\pi}{2}(\frac{2n-1}{2})+i\frac{\pi}{4}}}{\pi\sqrt{R|x_0|}} + O\big((R|x_0|)^{-\frac{3}{2}}\big).$$

The part of the integral in (4.2.18) over $[|x_0|^{-1}, T]$ corresponding to $O\big((R|x_0|)^{-\frac{3}{2}}\big)$ grows like $\log(T|x_0|)$ which divided by T obviously tends to zero. The part of the integral in (4.2.18) over $[|x_0|^{-1}, T]$ corresponding to the main term is

$$\frac{1}{T}\frac{1}{\pi\sqrt{|x_0|}}\int_{|x_0|^{-1}}^{T} e^{2\pi i R(\lambda+|x_0|)}e^{-i\frac{\pi}{2}(\frac{2n-1}{2})-i\frac{\pi}{4}} + e^{2\pi i R(\lambda-|x_0|)}e^{i\frac{\pi}{2}(\frac{2n-1}{2})+i\frac{\pi}{4}}\,dR$$

which tends to zero as $T\to\infty$ by an integration by parts, since $\lambda \neq |x_0|$ because we are considering the case where $\lambda \neq \pm|x_0 - m|$ for any $m \in \mathbf{Z}^n$.

Now consider the case where $\lambda = \pm|x_0 - m_0|$ for some $m_0 \in \mathbf{Z}^n$. In this case the expression to the right in (4.2.17) converges to

$$\frac{\Gamma(\frac{n+1}{2})}{\pi^{\frac{n-1}{2}}}\frac{e^{\pm i(\frac{\pi}{2}(\frac{2n-1}{2})+\frac{\pi}{4})}}{|x_0-m_0|^n} = \frac{\Gamma(\frac{n+1}{2})}{\pi^{\frac{n-1}{2}}}\frac{e^{\pm i\frac{\pi n}{2}}}{|x_0-m_0|^n}$$

as $T\to\infty$. Next observe that

$$\sum_{j=1}^{\infty}\frac{1}{\lambda_j^n} = \infty. \tag{4.2.19}$$

We have now shown that

$$\lim_{T\to\infty}\frac{1}{T}\int_1^T K_t^{\frac{n-1}{2}}(x_0)e^{2\pi i\lambda t}\,dt = \begin{cases} \frac{\Gamma(\frac{n+1}{2})}{\pi^{\frac{n-1}{2}}}\frac{e^{i\frac{\pi n}{2}}}{\lambda_j^n} & \text{if } \lambda = \lambda_j,\\ 0 & \text{if } \lambda \neq \pm\lambda_j,\\ \frac{\Gamma(\frac{n+1}{2})}{\pi^{\frac{n-1}{2}}}\frac{e^{-i\frac{\pi n}{2}}}{\lambda_j^n} & \text{if } \lambda = -\lambda_j. \end{cases} \tag{4.2.20}$$

Since x_0 lies in S, the set $\{1\}\cup\{\lambda_1,\lambda_2,\lambda_3,\dots\}$ is linearly independent over the rationals and thus no expression of the form $\pm\lambda_{j_1}\pm\cdots\pm\lambda_{j_s}$ is equal to an integer. It follows from this fact and (4.2.20) that

$$\lim_{T\to\infty}\frac{1}{T}\int_1^T K_t^{\frac{n-1}{2}}(x_0)\prod_{j=1}^{N}\left[1+\frac{e^{-i\frac{\pi n}{2}}e^{2\pi i\lambda_j t}+e^{i\frac{\pi n}{2}}e^{-2\pi i\lambda_j t}}{2}\right]dt = \frac{\Gamma(\frac{n+1}{2})}{\pi^{\frac{n-1}{2}}}\sum_{j=1}^{N}\frac{1}{\lambda_j^n}.$$

Suppose we had that

$$\sup_{R\geq 1}|K_R^{\frac{n-1}{2}}(x_0)| \leq A_{x_0} < \infty.$$

Then, setting $c_n = \frac{\Gamma(\frac{n+1}{2})}{\pi^{\frac{n-1}{2}}}$, we would have

$$\begin{aligned} c_n\sum_{j=1}^{N}\frac{1}{\lambda_j^n} &= \lim_{T\to\infty}\frac{1}{T}\int_1^T K_t^{\frac{n-1}{2}}(x_0)\prod_{j=1}^{N}\left[1+\frac{e^{-i\frac{\pi n}{2}}e^{2\pi i\lambda_j t}+e^{i\frac{\pi n}{2}}e^{-2\pi i\lambda_j t}}{2}\right]dt\\ &= \limsup_{T\to\infty}\frac{1}{T}\int_1^T |K_t^{\frac{n-1}{2}}(x_0)|\prod_{j=1}^{N}\left|1+\frac{e^{-i\frac{\pi n}{2}}e^{2\pi i\lambda_j t}+e^{i\frac{\pi n}{2}}e^{-2\pi i\lambda_j t}}{2}\right|dt \end{aligned}$$

$$\leq A_{x_0} \limsup_{T\to\infty} \frac{1}{T} \int_1^T \prod_{j=1}^N \left[1 + \frac{e^{-i\frac{\pi n}{2}} e^{2\pi i \lambda_j t} + e^{i\frac{\pi n}{2}} e^{-2\pi i \lambda_j t}}{2}\right] dt$$

$$= A_{x_0},$$

which contradicts (4.2.19) by letting $N \to \infty$. Here, once again, we used the fact that no expression of the form $\pm\lambda_{j_1} \pm \cdots \pm \lambda_{j_s}$ is equal to an integer and thus the preceding lim sup is a limit and is equal to 1, since the integral of all the exponentials produces another exponential which remains bounded.

We deduce that $\sup_{R\geq 1} |K_R^{\frac{n-1}{2}}(x_0)| = \infty$ for every point $x_0 \in S \cap [-\frac{1}{2}, \frac{1}{2})^n \setminus \{0\}$ and this concludes the proof of Lemma 4.2.7. □

Proof. We now prove Theorem 4.2.5. This part of the proof is similar to the proof of Theorem 4.2.1. Lemma 4.2.7 says that the means $B_R^{\frac{n-1}{2}}(\delta_0)(x)$, where δ_0 is the Dirac mass at 0, do not converge for almost all $x \in \mathbf{T}^n$. Our goal is to replace this Dirac mass by a series of integrable functions on $\mathbf{T}^n$ that have a peak at the origin.

Let us fix a nonnegative $\mathscr{C}^\infty$ radial function $\widehat{\Phi}$ on $\mathbf{R}^n$ that is supported in the unit ball $|\xi| \leq 1$ and has integral equal to 1. We set

$$\varphi_\varepsilon(x) = \sum_{m\in\mathbf{Z}^n} \tfrac{1}{\varepsilon^n} \widehat{\Phi}(\tfrac{x+m}{\varepsilon}) = \sum_{m\in\mathbf{Z}^n} \Phi(\varepsilon m) e^{2\pi i m\cdot x},$$

where the identity is valid because of the Poisson summation formula. It follows that the mth Fourier coefficient of φ_ε is $\Phi(\varepsilon m)$. Therefore, we have the estimate

$$\sup_{x\in\mathbf{T}^n} \sup_{R>0} |B_R^{\frac{n-1}{2}}(\varphi_\varepsilon)(x)| \leq \sum_{m\in\mathbf{Z}^n} |\Phi(\varepsilon m)| \leq \sum_{m\in\mathbf{Z}^n} \frac{C_n'}{(1+\varepsilon|m|)^{n+1}} \leq \frac{C_n}{\varepsilon^n}. \tag{4.2.21}$$

For any $k \geq 1$, we construct measurable subsets E_k of $\mathbf{T}^n$ with $|E_k| \geq 1 - \frac{1}{k}$, a sequence of positive numbers $R_1 < R_2 < \cdots$, with $R_k \uparrow \infty$, and two sequences of positive numbers $\varepsilon_k \downarrow 0$ and $\gamma_k \downarrow 0$ such that $\varepsilon_k \leq \gamma_k$ for all k and

$$\sup_{R\leq R_k} \left| B_R^{\frac{n-1}{2}} \Big(\sum_{s=1}^\infty 2^{-s} (\varphi_{\varepsilon_s} - \varphi_{\gamma_s}) \Big)(x) \right| \geq k \qquad \text{for } x \in E_k. \tag{4.2.22}$$

We pick $E_1 = \emptyset$, $R_1 = 1$, and $\varepsilon_1 = \gamma_1 = 1$. Let $k > 1$ and suppose that we have selected E_j, R_j, γ_j, and ε_j for all $1 \leq j \leq k-1$ such that (4.2.22) is satisfied. We construct E_k, R_k, γ_k, and ε_k such that (4.2.22) is satisfied with $j = k$. We begin by choosing γ_k. Let B be a constant such that

$$|\Phi(x) - \Phi(y)| \leq B|x-y|$$

for all $x, y \in \mathbf{R}^n$. Define γ_k such that

$$B\gamma_k \sum_{|m|\leq R_{k-1}} |m| = 1. \tag{4.2.23}$$

Then define

$$A_k = C_n 2^{-k}\gamma_k^{-n} + C_n \sum_{j=1}^{k-1} 2^{-j}(\varepsilon_j^{-n} + \gamma_j^{-n}),$$

where C_n is the constant in (4.2.21), and observe that in view of (4.2.21) we have

$$\sup_{x\in\mathbf{T}^n}\sup_{R>0}\left|B_R^{\frac{n-1}{2}}\Big(-2^{-k}\varphi_{\gamma_k} + \sum_{j=1}^{k-1} 2^{-j}(\varphi_{\varepsilon_j} - \varphi_{\gamma_j})\Big)(x)\right| \le A_k. \tag{4.2.24}$$

Let δ_0 be the Dirac mass at the origin in $\mathbf{T}^n$. Since by Fatou's lemma and Lemma 4.2.7 we have

$$\liminf_{R'\to\infty}\left|\left\{x\in\mathbf{T}^n:\ \sup_{0<R\le R'}\left|B_R^{\frac{n-1}{2}}(\delta_0)(x)\right| > 2^k(A_k+k+2)\right\}\right| = 1,$$

there exists an $R_k > \max(R_{k-1},k)$ such that the set

$$E_k = \left\{x\in\mathbf{T}^n:\ \sup_{0<R\le R_k}\left|B_R^{\frac{n-1}{2}}(2^{-k}\delta_0)(x)\right| > A_k+k+2\right\}$$

has measure at least $1-\frac{1}{k}$. Note that since R_k is increasing and tends to infinity, (4.2.23) yields that γ_k is decreasing and tends to zero.

We now choose ε_k such that $\varepsilon_k \le \gamma_k$, $\varepsilon_k \le \varepsilon_{k-1}$, and that

$$\sup_{x\in\mathbf{T}^n}\sup_{R\le R_k} 2^{-k}\left|B_R^{\frac{n-1}{2}}(\delta_0)(x) - B_R^{\frac{n-1}{2}}(\varphi_{\varepsilon_k})(x)\right| \le \sum_{|m|\le R_k} 2^{-k}\big(1-\tfrac{|m|^2}{R_k^2}\big)^{\frac{n-1}{2}}\big|1-\widehat{\varphi_{\varepsilon_k}}(m)\big| \le 1.$$

This is possible, since for a fixed R_k, the preceding sum tends to zero as $\varepsilon_k \to 0$. Then for $x\in E_k$ we have

$$\inf_{x\in E_k}\sup_{R\le R_k} 2^{-k}\left|B_R^{\frac{n-1}{2}}(\varphi_{\varepsilon_k})(x)\right| \ge A_k+k+1. \tag{4.2.25}$$

The inductive selection of the parameters can be schematically described as follows:

$$\{\gamma_{k-1},R_{k-1},E_{k-1},\varepsilon_{k-1}\} \Longrightarrow \gamma_k \Longrightarrow A_k \Longrightarrow \{R_k,E_k\} \Longrightarrow \varepsilon_k \Longrightarrow \{\gamma_k,R_k,E_k,\varepsilon_k\}.$$

Observe that the construction of γ_k gives for all $s\ge k+1$ the estimate

$$\begin{aligned}\sup_{x\in\mathbf{T}^n}\sup_{R\le R_k}\left|B_R^{\frac{n-1}{2}}(\varphi_{\varepsilon_s}-\varphi_{\gamma_s})(x)\right| &\le \sum_{|m|\le R_k}|\Phi(\varepsilon_s m)-\Phi(\gamma_s m)|\\ &\le B(\gamma_s-\varepsilon_s)\sum_{|m|\le R_k}|m|\\ &\le B\gamma_s\sum_{|m|\le R_k}|m|\\ &\le B\gamma_{k+1}\sum_{|m|\le R_k}|m| = 1,\end{aligned} \tag{4.2.26}$$

using (4.2.23) and the fact that the sequence γ_k is decreasing.

We now prove (4.2.22). For $x \in \mathbf{T}^n$ write

$$
\begin{aligned}
B_R^{\frac{n-1}{2}}\Big(\sum_{s=1}^{\infty} 2^{-s}(\varphi_{\varepsilon_s}-\varphi_{\gamma_s})\Big)(x) &= B_R^{\frac{n-1}{2}}\Big(-2^{-k}\varphi_{\gamma_k}+\sum_{s=1}^{k-1} 2^{-s}(\varphi_{\varepsilon_s}-\varphi_{\gamma_s})\Big)(x)\\
&\quad + B_R^{\frac{n-1}{2}}(2^{-k}\varphi_{\varepsilon_k})(x)\\
&\quad + B_R^{\frac{n-1}{2}}\Big(\sum_{s=k+1}^{\infty} 2^{-s}(\varphi_{\varepsilon_s}-\varphi_{\gamma_s})\Big)(x),
\end{aligned}
$$

from which it follows that

$$
\begin{aligned}
\sup_{R\le R_k}\Big|B_R^{\frac{n-1}{2}}\Big(\sum_{s=1}^{\infty} 2^{-s}(\varphi_{\varepsilon_s}-\varphi_{\gamma_s})\Big)(x)\Big| &\ge \sup_{R\le R_k}\Big|B_R^{\frac{n-1}{2}}(2^{-k}\varphi_{\varepsilon_k})(x)\Big|\\
&\quad - \sup_{R\le R_k}\Big|B_R^{\frac{n-1}{2}}\Big(-2^{-k}\varphi_{\gamma_k}+\sum_{s=1}^{k-1} 2^{-s}(\varphi_{\varepsilon_s}-\varphi_{\gamma_s})\Big)(x)\Big|\\
&\quad - \sup_{R\le R_k}\Big|B_R^{\frac{n-1}{2}}\Big(\sum_{s=k+1}^{\infty} 2^{-s}(\varphi_{\varepsilon_s}-\varphi_{\gamma_s})\Big)(x)\Big|.
\end{aligned}
$$

In view of (4.2.25), (4.2.24), and (4.2.26) for all $x \in E_k$, we obtain

$$
\sup_{R\le R_k}\Big|B_R^{\frac{n-1}{2}}\Big(\sum_{s=1}^{\infty} 2^{-s}(\varphi_{\varepsilon_s}-\varphi_{\gamma_s})\Big)(x)\Big| \ge (A_k+k+1)-A_k-\sum_{s=k+1}^{\infty} 2^{-s} \ge k,
$$

which clearly implies (4.2.22). Setting

$$
f=\sum_{s=1}^{\infty} 2^{-s}(\varphi_{\varepsilon_s}-\varphi_{\gamma_s}) \in L^1(\mathbf{T}^n),
$$

we deduce that $\sup_{R>0}\big|B_R^{\frac{n-1}{2}}(f)(x)\big| \ge k$ for all x in $\bigcup_{r=k}^{\infty} E_r$, and thus

$$
\sup_{R>0}\big|B_R^{\frac{n-1}{2}}(f)(x)\big| = \infty
$$

for all x in

$$
\bigcap_{k=1}^{\infty}\bigcup_{r=k}^{\infty} E_r .
$$

Since this set has full measure in $\mathbf{T}^n$, the required conclusion follows.

By taking ε_1 arbitrarily small (instead of picking $\varepsilon_1 = 1$), we force f to be supported in an arbitrarily small neighborhood of the origin. □

The previous argument shows that the Bochner–Riesz means B_R^{α} are badly behaved on $L^1(\mathbf{T}^n)$ when $\alpha = \frac{n-1}{2}$. It follows that the "rougher" spherical Dirichlet means $\overset{\circ}{D}{}_N^n * f$ (which correspond to $\alpha = 0$) are also ill behaved on $L^1(\mathbf{T}^n)$. See Exercise 4.2.2.

Exercises

4.2.1. Using Theorem 4.2.1 construct a function F on $\mathbf{T}^n$ such that

$$\limsup_{N\to\infty} |(D_N^n * F)(x_1,\dots,x_n)| = \infty$$

for almost all $(x_1,\dots,x_n) \in \mathbf{T}^n$.

4.2.2. For any $0 \le \alpha < \infty$ and $R > 0$ consider the Bochner–Riesz kernel

$$K_R^\alpha(x) = \sum_{|m|\le R} \left(1 - \tfrac{|m|^2}{R^2}\right)^\alpha e^{2\pi i m\cdot x}.$$

Use Exercise 4.1.8 to obtain that if for some $x_0 \in \mathbf{T}^n$ we have

$$\limsup_{R\to\infty} |K_R^\alpha(x_0)| < \infty,$$

then for all $\beta > \alpha$ we have

$$\sup_{R>0} |K_R^\beta(x_0)| < \infty.$$

Conclude that whenever $0 \le \alpha \le \frac{n-1}{2}$, the Bochner–Riesz means of order α of the function f constructed in the proof of Theorem 4.2.5, in particular the circular (spherical) Dirichlet means of this function, diverge a.e.

4.2.3. (a) Show that for M,N positive integers we have

$$(F_M * D_N)(x) = \begin{cases} F_M(x) & \text{for } M \le N, \\ F_N(x) + \frac{M-N}{(M+1)(N+1)} \sum_{|k|\le N} |k|\, e^{2\pi i k x} & \text{for } M > N. \end{cases}$$

(b) Prove that for some constant $c > 0$ we have

$$\int_{\mathbf{T}^1} \left| \sum_{|k|\le N} |k|\, e^{2\pi i k x} \right| dx \ge c N \log N$$

as $N \to \infty$.
[*Hint:* Part (b): Show that for $x \in \left[-\frac{1}{2}, \frac{1}{2}\right]$ we have

$$\sum_{|k|\le N} |k|\, e^{2\pi i k x} = (N+1)(D_N(x) - F_N(x))$$

and use the result of Exercise 3.1.5.]

4.2.4. Given the integrable functions

$$f_1(x) = \sum_{j=0}^{\infty} 2^{-j} F_{2^{2^{2j}}}(x), \qquad f_2(x) = \sum_{j=1}^{\infty} \frac{1}{j^2} F_{2^{2^j}}(x), \qquad x \in \mathbf{T}^1,$$

show that $\|f_1 * D_N\|_{L^1} \to \infty$ and $\|f_2 * D_N\|_{L^1} \to \infty$ as $N \to \infty$.
[*Hint:* Let $M_j = 2^{2^{2^j}}$ or $M_j = 2^{2^j}$ depending on the situation. For fixed N let j_N be the least integer j such that $M_j > N$. Then for $j \geq j_N + 1$ we have $M_j \geq M_{j_N}^2 > N^2 \geq 2N + 1$, hence $\frac{M_j - N}{M_j + 1} \geq \frac{1}{2}$. Split the summation indices into the sets $j \geq j_N$ and $j < j_N$. Conclude that $\|f_1 * D_N\|_{L^1}$ and $\|f_2 * D_N\|_{L^1}$ tend to infinity as $N \to \infty$ using Exercise 4.2.3.]

4.3 Multipliers, Transference, and Almost Everywhere Convergence

In Chapter 2 we saw that bounded operators from $L^p(\mathbf{R}^n)$ to $L^q(\mathbf{R}^n)$ that commute with translations are given by convolution with tempered distributions on $\mathbf{R}^n$. In particular, when $p = q$, these tempered distributions have bounded Fourier transforms, called Fourier multipliers. Convolution operators that commute with translations can also be defined on the torus. These lead to Fourier multipliers on the torus.

4.3.1 Multipliers on the Torus

In analogy with the nonperiodic case, we could identify convolution operators on $\mathbf{T}^n$ with appropriate distributions on the torus; see Exercise 4.3.2 for an introduction to this topic. However, it is simpler to avoid this point of view and consider the study of multipliers directly, bypassing the discussion of distributions on the torus.

For $h \in \mathbf{T}^n$ we define the *translation operator* τ^h acting on a periodic function f as follows: $\tau^h(f)(x) = f(x - h)$ for $x \in \mathbf{T}^n$. We say that a linear operator T acting on functions on the torus *commutes with translations* if for all $h \in \mathbf{T}^n$ we have $\tau^h(T(f))(x) = T(\tau^h f)(x)$ for almost all $x \in \mathbf{T}^n$.

Theorem 4.3.1. *Suppose that T is a linear operator that commutes with translations and maps $L^p(\mathbf{T}^n)$ to $L^q(\mathbf{T}^n)$ for some $1 \leq p, q \leq \infty$. Then there exists a bounded sequence $\{a_m\}_{m \in \mathbf{Z}^n}$ such that*

$$T(f)(x) = \sum_{m \in \mathbf{Z}^n} a_m \widehat{f}(m) e^{2\pi i m \cdot x} \tag{4.3.1}$$

for all $f \in \mathscr{C}^\infty(\mathbf{T}^n)$. Moreover, we have

$$\big\|\{a_m\}\big\|_{\ell^\infty} \le \big\|T\big\|_{L^p\to L^q}.$$

Proof. Consider the functions $e_m(x) = e^{2\pi i m\cdot x}$ defined on $\mathbf{T}^n$ for m in $\mathbf{Z}^n$. Since T commutes with translations, for every $h \in \mathbf{T}^n$ there is a subset F_h of $\mathbf{T}^n$ of full measure such that

$$T(e_m)(x-h) = T(\tau^h(e_m))(x) = e^{-2\pi i m\cdot h}T(e_m)(x)$$

for every $x \in F_h$. Note that

$$\begin{aligned}\int_{\mathbf{T}^n} |\{h\in\mathbf{T}^n:\, x\in F_h\}|\,dx &= \int_{\mathbf{T}^n}\int_{\mathbf{T}^n} \chi_{\{(h,x)\in\mathbf{T}^n\times\mathbf{T}^n:\, x\in F_h\}}\,dh\,dx\\ &= \int_{\mathbf{T}^n}\int_{\mathbf{T}^n} \chi_{\{(h,x)\in\mathbf{T}^n\times\mathbf{T}^n:\, x\in F_h\}}\,dx\,dh\\ &= \int_{\mathbf{T}^n} |F_h|\,dh = 1.\end{aligned}$$

Therefore there exists an $x_0 \in \mathbf{T}^n$ such that $|\{h\in\mathbf{T}^n:\, x_0\in F_h\}| = 1$. It follows that for almost all $h \in \mathbf{T}^n$ we have $T(e_m)(x_0-h) = e^{-2\pi i m\cdot h}T(e_m)(x_0)$. Replacing $x_0 - h$ by x, we obtain

$$T(e_m)(x) = e^{2\pi i m\cdot x}\big(e^{-2\pi i m\cdot x_0}T(e_m)(x_0)\big) = a_m e_m(x) \tag{4.3.2}$$

for almost all $x \in \mathbf{T}^n$, where we set $a_m = e^{-2\pi i m\cdot x_0}T(e_m)(x_0)$, for $m \in \mathbf{Z}^n$. Taking L^q norms in (4.3.2), we deduce $|a_m| = \|T(e_m)\|_{L^q} \le \|T\|_{L^p\to L^q}$, and thus a_m is bounded. Moreover, since $T(e_m) = a_m e_m$ for all m in $\mathbf{Z}^n$, it follows that (4.3.1) holds for all trigonometric polynomials. By density this extends to all $f \in \mathscr{C}^\infty(\mathbf{T}^n)$ and the theorem is proved. □

Definition 4.3.2. Let $1 \le p,q \le \infty$. We call a bounded sequence $\{a_m\}_{m\in\mathbf{Z}^n}$ an *(L^p,L^q) multiplier* if the corresponding operator given by (4.3.1) maps $L^p(\mathbf{T}^n)$ to $L^q(\mathbf{T}^n)$. If $p=q$, (L^p,L^p) multipliers are simply called L^p multipliers. When $1 \le p < \infty$, the space of all L^p multipliers on $\mathbf{T}^n$ is denoted by $\mathscr{M}_p(\mathbf{Z}^n)$. This notation follows the convention that $\mathscr{M}_p(\widehat{G})$ denote the space of L^p multipliers on $L^p(G)$, where G is a locally compact group and $\widehat{G}$ is its dual group. The norm of an element $\{a_m\}$ in $\mathscr{M}_p(\mathbf{Z}^n)$ is the norm of the operator T given by (4.3.1) from $L^p(\mathbf{T}^n)$ to itself. This norm is denoted by $\big\|\{a_m\}\big\|_{\mathscr{M}_p}$.

We now examine some special cases. We begin with the case $p=q=2$. As expected, it turns out that $\mathscr{M}_2(\mathbf{Z}^n) = \ell^\infty(\mathbf{Z}^n)$.

Theorem 4.3.3. *A linear operator T that commutes with translations maps $L^2(\mathbf{T}^n)$ to itself if and only if there exists a sequence $\{a_m\}_{m\in\mathbf{Z}^n}$ in $\ell^\infty(\mathbf{Z}^n)$ such that*

$$T(f)(x) = \sum_{m\in\mathbf{Z}^n} a_m\widehat{f}(m)e^{2\pi i m\cdot x} \tag{4.3.3}$$

for all $f \in \mathscr{C}^\infty(\mathbf{T}^n)$. Moreover, in this case we have $\big\|T\big\|_{L^2\to L^2} = \big\|\{a_m\}_m\big\|_{\ell^\infty}$.

Proof. The existence of such a sequence is guaranteed by Theorem 4.3.1, which also gives $\|\{a_m\}_m\|_{\ell^\infty} \le \|T\|_{L^2\to L^2}$. Conversely, any operator of the form (4.3.3) satisfies

$$\|T(f)\|_{L^2}^2 = \sum_{m\in\mathbf{Z}^n} |a_m\widehat{f}(m)|^2 \le \|\{a_m\}_m\|_{\ell^\infty}^2 \sum_{m\in\mathbf{Z}^n} |\widehat{f}(m)|^2,$$

and thus $\|T\|_{L^2\to L^2} \le \|\{a_m\}_m\|_{\ell^\infty}$. □

We continue with the case $p = q = 1$. Recall the definition of a finite Borel measure on $\mathbf{T}^n$. Given such a measure μ, its Fourier coefficients are defined by

$$\widehat{\mu}(m) = \int_{\mathbf{T}^n} e^{-2\pi i x\cdot m}\, d\mu(x), \qquad m \in \mathbf{Z}^n.$$

Clearly, all the Fourier coefficients of the measure μ are bounded by the total variation $\|\mu\|$ of μ. See Exercise 4.3.3 for basic properties of Fourier transforms of distributions on the torus.

Theorem 4.3.4. *A linear operator T that commutes with translations maps $L^1(\mathbf{T}^n)$ to itself if and only if there exists a finite Borel measure μ on the torus such that*

$$T(f)(x) = \sum_{m\in\mathbf{Z}^n} \widehat{\mu}(m)\widehat{f}(m)e^{2\pi i m\cdot x} \tag{4.3.4}$$

for all $f \in \mathscr{C}^\infty(\mathbf{T}^n)$. Moreover, in this case we have $\|T\|_{L^1\to L^1} = \|\mu\|$. In other words, $\mathscr{M}_1(\mathbf{Z}^n)$ is the set of all sequences given by Fourier coefficients of finite Borel measures on $\mathbf{T}^n$.

Proof. Fix $f \in L^1(\mathbf{T}^n)$. If (4.3.4) is valid, then $\widehat{T(f)}(m) = \widehat{f}(m)\widehat{\mu}(m)$ for all $m \in \mathbf{Z}^n$. But Exercise 4.3.3 gives that $\widehat{f*\mu}(m) = \widehat{f}(m)\widehat{\mu}(m)$ for all $m \in \mathbf{Z}^n$; therefore, the integrable functions $f*\mu$ and $T(f)$ have the same Fourier coefficients, and they must be equal. Thus $T(f) = f*\mu$, which implies that T is bounded on L^1 and $\|T(f)\|_{L^1} \le \|\mu\|\|f\|_{L^1}$.

To prove the converse direction, we suppose that T commutes with translations and maps $L^1(\mathbf{T}^n)$ to itself. We recall the Poisson kernel P_ε defined on $\mathbf{T}^n$, which can be expressed in the following two ways:

$$P_\varepsilon(x) = \sum_{m\in\mathbf{Z}^n} e^{-2\pi|m|\varepsilon} e^{2\pi i m\cdot x} = \frac{\Gamma(\frac{n+1}{2})}{\pi^{\frac{n+1}{2}}} \sum_{m\in\mathbf{Z}^n} \frac{\varepsilon^{-n}}{(1+|\frac{x+m}{\varepsilon}|^2)^{\frac{n+1}{2}}} \ge 0 \tag{4.3.5}$$

for all $x \in \mathbf{T}^n$, in view of the identity obtained in (3.2.4). The preceding identity says that $P_\varepsilon \ge 0$; hence, $\|P_\varepsilon\|_{L^1} = \int_{\mathbf{T}^n} P_\varepsilon(x)\,dx$. Integrating the first series in (4.3.5) over $\mathbf{T}^n$ we conclude that $\|P_\varepsilon\|_{L^1(\mathbf{T}^n)} = 1$. The boundedness of T now gives

$$\|T(P_\varepsilon)\|_{L^1(\mathbf{T}^n)} \le \|T\|_{L^1\to L^1}$$

for all $\varepsilon > 0$. The Banach–Alaoglu theorem implies that there exist a sequence $\varepsilon_j \downarrow 0$ and a finite Borel measure μ on $\mathbf{T}^n$ such that $T(P_{\varepsilon_j})$ tends to μ weakly as $j \to \infty$. This means that for all continuous functions g on $\mathbf{T}^n$ we have

$$\lim_{j\to\infty} \int_{\mathbf{T}^n} g(x) T(P_{\varepsilon_j})(x)\,dx = \int_{\mathbf{T}^n} g(x)\,d\mu(x). \tag{4.3.6}$$

It follows from (4.3.6) that for all g continuous on $\mathbf{T}^n$ we have

$$\left| \int_{\mathbf{T}^n} g(x)\,d\mu(x) \right| \le \sup_j \big\|T(P_{\varepsilon_j})\big\|_{L^1} \big\|g\big\|_{L^\infty} \le \big\|T\big\|_{L^1\to L^1} \big\|g\big\|_{L^\infty}.$$

Since, by the Riesz representation theorem we have that the norm of the linear functional

$$g \mapsto \int_{\mathbf{T}^n} g(x)\,d\mu(x)$$

on the space of continuous functions $\mathscr{C}(\mathbf{T}^n)$ is $\|\mu\|$, it follows that

$$\|\mu\| \le \big\|T\big\|_{L^1\to L^1}. \tag{4.3.7}$$

It remains to prove that T has the form given in (4.3.4). By Theorem 4.3.1 we have that there exists a bounded sequence $\{a_m\}$ on $\mathbf{Z}^n$ such that (4.3.1) is satisfied. Taking $g(x) = e^{-2\pi i k\cdot x}$ in (4.3.6) and using the representation for T in (4.3.1), we obtain

$$\begin{aligned}
\widehat{\mu}(k) &= \int_{\mathbf{T}^n} e^{-2\pi i k\cdot x}\,d\mu(x) \\
&= \lim_{j\to\infty} \int_{\mathbf{T}^n} e^{-2\pi i k\cdot x} \sum_{m\in\mathbf{Z}^n} a_m e^{-2\pi\varepsilon_j |m|} e^{2\pi i m\cdot x}\,dx \\
&= \lim_{j\to\infty} \sum_{m\in\mathbf{Z}^n} \int_{\mathbf{T}^n} e^{-2\pi i k\cdot x} a_m e^{-2\pi\varepsilon_j |m|} e^{2\pi i m\cdot x}\,dx \\
&= \lim_{j\to\infty} a_k e^{-2\pi\varepsilon_j |k|} = a_k\,.
\end{aligned}$$

This proves assertion (4.3.4). It follows from (4.3.4) that $T(f) = f * \mu$ and thus $\|T\|_{L^1\to L^1} \le \|\mu\|$. This fact combined with (4.3.7) gives $\|T\|_{L^1\to L^1} = \|\mu\|$. □

Remark 4.3.5. It is not hard to see that most basic properties of the space $\mathscr{M}_p(\mathbf{R}^n)$ of L^p Fourier multipliers on $\mathbf{R}^n$ are also valid for $\mathscr{M}_p(\mathbf{Z}^n)$. In particular, $\mathscr{M}_p(\mathbf{Z}^n)$ is a closed subspace of $\ell^\infty(\mathbf{Z}^n)$ and thus a Banach space itself. Moreover, sums, scalar multiples, and products of elements of $\mathscr{M}_p(\mathbf{Z}^n)$ are also in $\mathscr{M}_p(\mathbf{Z}^n)$, which makes this space a Banach algebra. As in the nonperiodic case, we also have $\mathscr{M}_p(\mathbf{Z}^n) = \mathscr{M}_{p'}(\mathbf{Z}^n)$ when $1 < p < \infty$.

4.3.2 Transference of Multipliers

It is clear by now that multipliers on $L^1(\mathbf{T}^n)$ and $L^1(\mathbf{R}^n)$ are very similar, and the same is true for $L^2(\mathbf{T}^n)$ and $L^2(\mathbf{R}^n)$. These similarities became obvious when we characterized L^1 and L^2 multipliers on both $\mathbf{R}^n$ and $\mathbf{T}^n$. So far, it is not known if a nontrivial characterization of $\mathscr{M}_p(\mathbf{R}^n)$ exists, but we might ask whether this space is related to $\mathscr{M}_p(\mathbf{Z}^n)$. There are several connections of this type and there are general ways to produce multipliers on the torus from multipliers on $\mathbf{R}^n$ and vice versa. General methods of this sort are called *transference of multipliers*.

We begin with a useful definition.

Definition 4.3.6. Let $t_0 \in \mathbf{R}^n$. A bounded function b on $\mathbf{R}^n$ is called *regulated at the point t_0* if

$$\lim_{\varepsilon\to 0}\frac{1}{\varepsilon^n}\int_{|t|\le\varepsilon}\big(b(t_0-t)-b(t_0)\big)\,dt = 0\,. \tag{4.3.8}$$

The function b is called *regulated* if it is regulated at every $t_0 \in \mathbf{R}^n$.

Clearly, if t_0 is a Lebesgue point of b, then b is regulated at t_0. In particular, this is the case if b is continuous at t_0. If $b(t_0)=0$, condition (4.3.8) also holds when $b(t_0-t) = -b(t_0+t)$ whenever $|t|\le\varepsilon$ for some $\varepsilon>0$; for instance the function $b(t) = -i\,\mathrm{sgn}\,(t-t_0)$ has this property.

An example of a regulated function is the following modification of the characteristic function of the cube $[-1,1]^n$

$$\widetilde{\chi}_{[-1,1]^n}(x_1,\dots,x_n) = \begin{cases} 1 & \text{when all } |x_j|<1,\\ 2^{k-n} & \text{if } (x_1,\dots,x_n) \text{ belongs to some } k\text{-dimensional}\\ & \text{face of the boundary of } [-1,1]^n,\\ 0 & \text{when some } |x_j|>1, \end{cases}$$

with the understanding that points are zero-dimensional.

The first transference result we discuss is the following.

Theorem 4.3.7. *Suppose that b is a regulated function at every point $m\in\mathbf{Z}^n$ and that b lies in $\mathscr{M}_p(\mathbf{R}^n)$ for some $1<p<\infty$. Then the sequence $\{b(m)\}_{m\in\mathbf{Z}^n}$ is in $\mathscr{M}_p(\mathbf{Z}^n)$ and moreover,*

$$\big\|\{b(m)\}_{m\in\mathbf{Z}^n}\big\|_{\mathscr{M}_p(\mathbf{Z}^n)} \le \|b\|_{\mathscr{M}_p(\mathbf{R}^n)}\,.$$

If b is regulated everywhere, then for all $R>0$ the sequences $\{b(m/R)\}_{m\in\mathbf{Z}^n}$ are in $\mathscr{M}_p(\mathbf{Z}^n)$ and we have

$$\sup_{R>0}\big\|\{b(m/R)\}_{m\in\mathbf{Z}^n}\big\|_{\mathscr{M}_p(\mathbf{Z}^n)} \le \|b\|_{\mathscr{M}_p(\mathbf{R}^n)}\,.$$

The second conclusion of the theorem is a consequence of the first, since for a given $R>0$ the function $b(\xi/R)$ is regulated on $\mathbf{Z}^n$ and has the same $\mathscr{M}_p(\mathbf{R}^n)$ norm as $b(\xi)$. Before we prove this result, we state and prove a couple of lemmas.

Lemma 4.3.8. *Suppose that the function b on $\mathbf{R}^n$ is regulated at the point x_0. Let $K_\varepsilon(x) = \varepsilon^{-n}e^{-\pi|x/\varepsilon|^2}$ for $\varepsilon > 0$. Then we have that $(b*K_\varepsilon)(x_0) \to b(x_0)$ as $\varepsilon \to 0$.*

Proof. For $r > 0$ define the function

$$F_{x_0}(r) = \frac{1}{r^n}\int_{|t|\le r}\big(b(x_0-t)-b(x_0)\big)\,dt = \frac{1}{r^n}\int_0^r\int_{\mathbf{S}^{n-1}}\big(b(x_0-s\theta)-b(x_0)\big)\,d\theta s^{n-1}ds.$$

Let $\eta > 0$. Since b is regulated at x_0 there is a $\delta > 0$ such that for $r \le \delta$ we have $|F_{x_0}(r)| \le \eta$. Fix such a δ and write

$$(b*K_\varepsilon)(x_0) - b(x_0) = \int_{y\in\mathbf{R}^n}\big(b(x_0-y)-b(x_0)\big)K_\varepsilon(y)\,dy = A_1^\varepsilon + A_2^\varepsilon,$$

where

$$A_1^\varepsilon = \int_{|y|\ge\delta}\big(b(x_0-y)-b(x_0)\big)K_\varepsilon(y)\,dy$$

and

$$\begin{aligned} A_2^\varepsilon &= \int_{|y|<\delta}(b(x_0-y)-b(x_0))K_\varepsilon(y))\,dy \\ &= \int_0^\delta \frac{1}{\varepsilon^n}e^{-\pi(r/\varepsilon)^2}\int_{\mathbf{S}^{n-1}}\big(b(x_0-r\theta)-b(x_0)\big)r^n dr \\ &= \int_0^\delta \frac{1}{\varepsilon^n}e^{-\pi(r/\varepsilon)^2}\frac{d}{dr}(r^nF_{x_0}(r))\,dr. \end{aligned}$$

For our given $\eta > 0$ there is an $\varepsilon_0 > 0$ such that for $\varepsilon < \varepsilon_0$ we have

$$|A_1^\varepsilon| \le 2\|b\|_{L^\infty}\int_{|y|\ge\frac{\delta}{\varepsilon}} e^{-\pi|y|^2}dy < \eta\,.$$

Via an integration by parts A_2^ε can be written as

$$\begin{aligned} |A_2^\varepsilon| &= \left|\delta^nF_{x_0}(\delta)\frac{1}{\varepsilon^n}e^{-\pi(\delta/\varepsilon)^2} - 0 + 2\pi\int_0^\delta \frac{r}{\varepsilon^{n+2}}e^{-\pi(r/\varepsilon)^2}r^nF_{x_0}(r)\,dr\right| \\ &= \left|F_{x_0}(\delta)\frac{\delta^n}{\varepsilon^n}e^{-\pi(\delta/\varepsilon)^2} + 2\pi\int_0^{\delta/\varepsilon} r^{n+1}F_{x_0}(\varepsilon r)e^{-\pi r^2}dr\right| \\ &\le |F_{x_0}(\delta)|\frac{\delta^n}{\varepsilon^n}e^{-\pi(\delta/\varepsilon)^2} + \sup_{0<r\le\frac{\delta}{\varepsilon}}|F_{x_0}(\varepsilon r)|2\pi\int_0^{\delta/\varepsilon} r^{n+1}e^{-\pi r^2}dr \\ &\le |F_{x_0}(\delta)|C_n + \sup_{0<r\le\delta}|F_{x_0}(r)|C_n' \\ &\le (C_n + C_n')\eta\,, \end{aligned}$$

where we set $C_n = \sup_{t>0}t^ne^{-\pi t^2}$ and $C_n' = 2\pi\int_0^\infty r^{n+1}e^{-\pi r^2}dr$. Then for $\varepsilon < \varepsilon_0$ we have $|(b*K_\varepsilon)(x_0 - b(x_0)| < (C_n + C_n' + 1)\eta$, thus $(b*K_\varepsilon)(x_0) \to b(x_0)$ as $\varepsilon \to 0$. □

Lemma 4.3.9. *Let T be the operator on $\mathbf{R}^n$ whose multiplier is $b(\xi)$, and let S be the operator on $\mathbf{T}^n$ whose multiplier is the sequence $\{b(m)\}_{m\in\mathbf{Z}^n}$. Assume that $b(\xi)$ is regulated at every point $\xi = m \in \mathbf{Z}^n$. Suppose that P and Q are trigonometric polynomials on $\mathbf{T}^n$ and let $L_\varepsilon(x) = e^{-\pi\varepsilon|x|^2}$ for $x \in \mathbf{R}^n$ and $\varepsilon > 0$. Then the following identity is valid whenever $\alpha, \beta > 0$ and $\alpha + \beta = 1$:*

$$\lim_{\varepsilon\to 0} \varepsilon^{\frac{n}{2}} \int_{\mathbf{R}^n} T(PL_{\varepsilon\alpha})(x)\overline{Q(x)L_{\varepsilon\beta}(x)}\,dx = \int_{\mathbf{T}^n} S(P)(x)\overline{Q(x)}\,dx. \tag{4.3.9}$$

Proof. It suffices to prove the required assertion for $P(x) = e^{2\pi i m\cdot x}$ and $Q(x) = e^{2\pi i k\cdot x}$, $k, m \in \mathbf{Z}^n$, since the general case follows from this case by linearity. In view of Parseval's relation (Proposition 3.2.7 (3)), we have

$$\int_{\mathbf{T}^n} S(P)(x)\overline{Q(x)}\,dx = \sum_{r\in\mathbf{Z}^n} b(r)\widehat{P}(r)\overline{\widehat{Q}(r)} = \begin{cases} b(m) & \text{when } k = m, \\ 0 & \text{when } k \neq m. \end{cases} \tag{4.3.10}$$

On the other hand, using the identity in Theorem 2.2.14 (3), we obtain

$$\begin{aligned}
&\varepsilon^{\frac{n}{2}} \int_{\mathbf{R}^n} T(PL_{\varepsilon\alpha})(x)\overline{Q(x)L_{\varepsilon\beta}(x)}\,dx \\
&= \varepsilon^{\frac{n}{2}} \int_{\mathbf{R}^n} b(\xi)\widehat{PL_{\varepsilon\alpha}}(\xi)\overline{\widehat{QL_{\varepsilon\beta}}(\xi)}\,d\xi \\
&= \varepsilon^{\frac{n}{2}} \int_{\mathbf{R}^n} b(\xi)(\varepsilon\alpha)^{-\frac{n}{2}} e^{-\pi\frac{|\xi-m|^2}{\varepsilon\alpha}} (\varepsilon\beta)^{-\frac{n}{2}} e^{-\pi\frac{|\xi-k|^2}{\varepsilon\beta}}\,d\xi \\
&= (\varepsilon\alpha\beta)^{-\frac{n}{2}} \int_{\mathbf{R}^n} b(\xi) e^{-\pi\frac{|\xi-m|^2}{\varepsilon\alpha}} e^{-\pi\frac{|\xi-k|^2}{\varepsilon\beta}}\,d\xi\,.
\end{aligned} \tag{4.3.11}$$

Now if $m = k$, since $\alpha + \beta = 1$, the expression in (4.3.11) is equal to

$$(\varepsilon\alpha\beta)^{-\frac{n}{2}} \int_{\mathbf{R}^n} b(\xi) e^{-\pi\frac{|\xi-m|^2}{\varepsilon\alpha\beta}}\,d\xi\,, \tag{4.3.12}$$

which tends to $b(m)$ in view of Lemma 4.3.8, since b is regulated at every point $m \in \mathbf{Z}^n$.

We now consider the case $m \neq k$ in (4.3.11). Since $|m-k| \geq 1$, then every ξ in $\mathbf{R}^n$ must satisfy either $|\xi - m| \geq 1/2$ or $|\xi - k| \geq 1/2$. Therefore, the expression in (4.3.11) is controlled by

$$(\varepsilon\alpha\beta)^{-\frac{n}{2}} \left(\int_{|\xi-m|\geq\frac{1}{2}} b(\xi) e^{-\frac{\pi}{4\varepsilon\alpha}} e^{-\pi\frac{|\xi-k|^2}{\varepsilon\beta}}\,d\xi + \int_{|\xi-k|\geq\frac{1}{2}} b(\xi) e^{-\frac{\pi}{4\varepsilon\beta}} e^{-\pi\frac{|\xi-m|^2}{\varepsilon\alpha}}\,d\xi \right),$$

which is in turn controlled by

$$\|b\|_{L^\infty} \big(\alpha^{-\frac{n}{2}} e^{-\frac{\pi}{4\varepsilon\alpha}} + \beta^{-\frac{n}{2}} e^{-\frac{\pi}{4\varepsilon\beta}} \big),$$

which tends to zero as $\varepsilon \to 0$. This proves that the expression in (4.3.10) is equal to the limit of the expression in (4.3.11) as $\varepsilon \to 0$. This completes the proof of Lemma 4.3.9 □

Proof (Theorem 4.3.7). We are assuming that T maps $L^p(\mathbf{R}^n)$ to itself and we need to show that S maps $L^p(\mathbf{T}^n)$ to itself. We prove this using duality. For P and Q trigonometric polynomials, using Lemma 4.3.9, we have

$$\begin{aligned}
&\left|\int_{\mathbf{T}^n} S(P)(x)\overline{Q(x)}\,dx\right| \\
&= \left|\lim_{\varepsilon\to 0} \varepsilon^{\frac{n}{2}} \int_{\mathbf{R}^n} T(PL_{\varepsilon/p})(x)\overline{Q(x)L_{\varepsilon/p'}(x)}\,dx\right| \\
&\le \big\|T\big\|_{L^p\to L^p} \limsup_{\varepsilon\to 0} \varepsilon^{\frac{n}{2}} \big\|PL_{\varepsilon/p}\big\|_{L^p(\mathbf{R}^n)} \big\|QL_{\varepsilon/p'}\big\|_{L^{p'}(\mathbf{R}^n)} \\
&= \big\|T\big\|_{L^p\to L^p} \limsup_{\varepsilon\to 0} \left(\varepsilon^{\frac{n}{2}}\int_{\mathbf{R}^n} |P(x)|^p e^{-\varepsilon\pi|x|^2}\,dx\right)^{\frac{1}{p}} \left(\varepsilon^{\frac{n}{2}}\int_{\mathbf{R}^n} |Q(x)|^{p'} e^{-\varepsilon\pi|x|^2}\,dx\right)^{\frac{1}{p'}} \\
&= \big\|T\big\|_{L^p\to L^p} \left(\int_{\mathbf{T}^n} |P(x)|^p\,dx\right)^{\frac{1}{p}} \left(\int_{\mathbf{T}^n} |Q(x)|^{p'}\,dx\right)^{\frac{1}{p'}},
\end{aligned}$$

provided for all continuous 1-periodic functions g on $\mathbf{R}^n$ we have that

$$\lim_{\varepsilon\to 0} \varepsilon^{\frac{n}{2}} \int_{\mathbf{R}^n} g(x)e^{-\varepsilon\pi|x|^2}\,dx = \int_{\mathbf{T}^n} g(x)\,dx. \tag{4.3.13}$$

Assuming (4.3.13) for the moment, we take the supremum over all trigonometric polynomials Q on $\mathbf{T}^n$ with $L^{p'}$ norm at most 1 to obtain that S maps $L^p(\mathbf{T}^n)$ to itself with norm at most $\|T\|_{L^p\to L^p}$, yielding the required conclusion.

We now prove (4.3.13). Use the Poisson summation formula to write the left-hand side of (4.3.13) as

$$\begin{aligned}
\varepsilon^{\frac{n}{2}} \sum_{k\in\mathbf{Z}^n} \int_{\mathbf{T}^n} g(x-k)e^{-\varepsilon\pi|x-k|^2}\,dx &= \int_{\mathbf{T}^n} g(x)\varepsilon^{\frac{n}{2}} \sum_{k\in\mathbf{Z}^n} e^{-\varepsilon\pi|x-k|^2}\,dx \\
&= \int_{\mathbf{T}^n} g(x) \sum_{k\in\mathbf{Z}^n} e^{-\pi|k|^2/\varepsilon} e^{2\pi i x\cdot k}\,dx \\
&= \int_{\mathbf{T}^n} g(x)\,dx + A_\varepsilon,
\end{aligned}$$

where

$$|A_\varepsilon| \le \big\|g\big\|_{L^\infty} \sum_{|k|\ge 1} e^{-\pi|k|^2/\varepsilon} \to 0$$

as $\varepsilon \to 0$. This completes the proof of Theorem 4.3.7. □

We now obtain a converse of Theorem 4.3.7. If $b(\xi)$ is a bounded function on $\mathbf{R}^n$ and the sequence $\{b(m)\}_{m\in\mathbf{Z}^n}$ is in $\mathscr{M}_p(\mathbf{Z}^n)$, then we cannot necessarily obtain that

b is in $\mathscr{M}_p(\mathbf{R}^n)$, since such a conclusion would depend on the values of b on the integer lattice, which is a set of measure zero. However, a converse can be formulated if we assume that for all $R > 0$, the sequences $\{b(m/R)\}_{m\in\mathbf{Z}^n}$ are in $\mathscr{M}_p(\mathbf{Z}^n)$ uniformly in R. Then we obtain that $b(\xi/R)$ is in $\mathscr{M}_p(\mathbf{R}^n)$ uniformly in $R > 0$, which is equivalent to saying that $b \in \mathscr{M}_p(\mathbf{R}^n)$, since dilations of multipliers on $\mathbf{R}^n$ do not affect their norms (see Proposition 2.5.14). These remarks can be precisely expressed in the following theorem.

Theorem 4.3.10. *Suppose that $b(\xi)$ is a bounded function defined on $\mathbf{R}^n$ which is Riemann integrable over any cube. Suppose that the sequences $\{b(\frac{m}{R})\}_{m\in\mathbf{Z}^n}$ are in $\mathscr{M}_p(\mathbf{Z}^n)$ uniformly in $R > 0$ for some $1 < p < \infty$. Then b is in $\mathscr{M}_p(\mathbf{R}^n)$ and we have*

$$\|b\|_{\mathscr{M}_p(\mathbf{R}^n)} \le \sup_{R>0} \big\|\{b(\tfrac{m}{R})\}_{m\in\mathbf{Z}^n}\big\|_{\mathscr{M}_p(\mathbf{Z}^n)}. \tag{4.3.14}$$

Proof. Suppose that f and g are smooth functions with compact support on $\mathbf{R}^n$. Then there is an $R_0 > 0$ such that for $R \ge R_0$, the functions $x \mapsto f(Rx)$ and $x \mapsto g(Rx)$ are supported in $[-1/2, 1/2]^n$. We define periodic functions

$$F_R(x) = \sum_{k\in\mathbf{Z}^n} f(R(x-k)) \quad \text{and} \quad G_R(x) = \sum_{k\in\mathbf{Z}^n} g(R(x-k))$$

on $\mathbf{T}^n$. Observe that the mth Fourier coefficient of F_R is $\widehat{F_R}(m) = R^{-n}\widehat{f}(m/R)$ and that of G_R is $\widehat{G_R}(m) = R^{-n}\widehat{g}(m/R)$.

Now for $R \ge R_0$ we have

$$\Big|\sum_{m\in\mathbf{Z}^n} b(m/R)\widehat{f}(m/R)\overline{\widehat{g}(m/R)}\,\text{Volume}\left(\tfrac{m}{R} + [0,\tfrac{1}{R}]^n\right)\Big| \tag{4.3.15}$$

$$\begin{aligned}
&= \Big| R^n \sum_{m\in\mathbf{Z}^n} b(m/R)\widehat{F_R}(m)\overline{\widehat{G_R}(m)}\Big| \\
&= \Big| R^n \int_{\mathbf{T}^n} \Big(\sum_{m\in\mathbf{Z}^n} b(m/R)\widehat{F_R}(m)e^{2\pi i m\cdot x}\Big)\overline{G_R(x)}\,dx\Big| \\
&\le R^n \big\|\{b(m/R)\}_m\big\|_{\mathscr{M}_p(\mathbf{Z}^n)} \big\|F_R\big\|_{L^p(\mathbf{T}^n)} \big\|G_R\big\|_{L^{p'}(\mathbf{T}^n)} \\
&\le \sup_{R>0} \big\|\{b(m/R)\}_{m\in\mathbf{Z}^n}\big\|_{\mathscr{M}_p(\mathbf{Z}^n)} R^n \big\|F_R\big\|_{L^p(\mathbf{R}^n)} \big\|G_R\big\|_{L^{p'}(\mathbf{R}^n)} \\
&= \sup_{R>0} \big\|\{b(m/R)\}_{m\in\mathbf{Z}^n}\big\|_{\mathscr{M}_p(\mathbf{Z}^n)} \big\|f\big\|_{L^p(\mathbf{R}^n)} \big\|g\big\|_{L^{p'}(\mathbf{R}^n)}.
\end{aligned} \tag{4.3.16}$$

Since b is bounded and Riemann integrable over any cube in $\mathbf{R}^n$, the function $b(\xi)\widehat{f}(\xi)\overline{\widehat{g}(\xi)}$ is Riemann integrable over $\mathbf{R}^n$. The expressions in (4.3.15) are sums associated with the partition $\{[\frac{m}{R}, \frac{m+1}{R})^n\}_{m\in\mathbf{Z}^n}$ of $\mathbf{R}^n$ which tend to

$$\Big|\int_{\mathbf{R}^n} b(\xi)\widehat{f}(\xi)\overline{\widehat{g}(\xi)}\,d\xi\Big|$$

as $R \to \infty$ by the definition of the Riemann integral. We deduce that the absolute value of

$$\int_{\mathbf{R}^n} b(\xi)\widehat{f}(\xi)\overline{\widehat{g}(\xi)}\,d\xi = \int_{\mathbf{R}^n} (b\widehat{f})^\vee(x)\,\overline{g(x)}\,dx$$

is bounded by the expression in (4.3.16). This proves the theorem via duality. □

4.3.3 Applications of Transference

Having established two main transference theorems, we turn to an application.

Corollary 4.3.11. *Let* $1 < p < \infty$, $f \in L^p(\mathbf{T}^n)$, *and* $\alpha \geq 0$. *Then*

(a) $\big\|D_R^n * f - f\big\|_{L^p(\mathbf{T}^n)} \to 0$ *as* $R \to \infty$ *if and only if* $\chi_{[-1,1]^n} \in \mathscr{M}_p(\mathbf{R}^n)$.
(b) $\big\|\mathring{D}_R^n * f - f\big\|_{L^p(\mathbf{T}^n)} \to 0$ *as* $R \to \infty$ *if and only if* $\chi_{B(0,1)} \in \mathscr{M}_p(\mathbf{R}^n)$.
(c) $\big\|B_R^\alpha(f) - f\big\|_{L^p(\mathbf{T}^n)} \to 0$ *as* $R \to \infty$ *if and only if* $(1-|\xi|^2)_+^\alpha \in \mathscr{M}_p(\mathbf{R}^n)$.

Proof. First observe that in view of Corollary 4.1.3, the assertions on the left in (a), (b), and (c) are equivalent to the statements

$$\sup_{R>0}\big\|D_R^n * f\big\|_{L^p(\mathbf{T}^n)} \leq C_p\big\|f\big\|_{L^p(\mathbf{T}^n)},$$

$$\sup_{R>0}\big\|\mathring{D}_R^n * f\big\|_{L^p(\mathbf{T}^n)} \leq C_p\big\|f\big\|_{L^p(\mathbf{T}^n)},$$

$$\sup_{R>0}\big\|B_R^\alpha(f)\big\|_{L^p(\mathbf{T}^n)} \leq C_p\big\|f\big\|_{L^p(\mathbf{T}^n)},$$

for some constant $0 < C_p < \infty$ and all f in $L^p(\mathbf{T}^n)$. These statements can be rephrased as

$$\sup_{R>0}\big\|\{\chi_{[-1,1]^n}(m/R)\}_{m\in\mathbf{Z}^n}\big\|_{\mathscr{M}_p(\mathbf{Z}^n)} < \infty,$$

$$\sup_{R>0}\big\|\{\chi_{B(0,1)}(m/R)\}_{m\in\mathbf{Z}^n}\big\|_{\mathscr{M}_p(\mathbf{Z}^n)} < \infty,$$

$$\sup_{R>0}\big\|\{(1-|m/R|^2)_+^\alpha\}_{m\in\mathbf{Z}^n}\big\|_{\mathscr{M}_p(\mathbf{Z}^n)} < \infty.$$

If these statements hold, then Theorem 4.3.10 gives that the functions $\chi_{[-1,1]^n}(\xi)$, $\chi_{B(0,1)}(\xi)$, and $(1-|\xi|^2)_+^\alpha$ lie in $\mathscr{M}_p(\mathbf{R}^n)$.

To prove the converse implication, for any given $R' \in \mathbf{R}^+ \setminus \{|m| : m \in \mathbf{Z}^n\}$, the functions $\chi_{[-1,1]^n}(\xi/R')$, $\chi_{B(0,1)}(\xi/R')$ are Riemann integrable over $\mathbf{R}^n$ and are regulated (actually continuous) at every point in $\mathbf{Z}^n$. Moreover, the function $(1-|\xi|^2)_+^\alpha$ is continuous, regulated, and Riemann integrable over $\mathbf{R}^n$. Then the hypotheses of Theorem 4.3.7 are satisfied and its conclusion yields that

$$\big\|\{\chi_{[-1,1]^n}(m/R')\}_{m\in\mathbf{Z}^n}\big\|_{\mathscr{M}_p(\mathbf{Z}^n)} \leq \big\|\chi_{[-1,1]^n}\big\|_{\mathscr{M}_p}, \tag{4.3.17}$$

$$\left\|\{\chi_{B(0,1)}(m/R')\}_{m\in\mathbf{Z}^n}\right\|_{\mathscr{M}_p(\mathbf{Z}^n)} \le \left\|\chi_{B(0,1)}\right\|_{\mathscr{M}_p}, \tag{4.3.18}$$
$$\sup_{R>0}\left\|\{(1-|m/R|^2)_+^\alpha\}_{m\in\mathbf{Z}^n}\right\|_{\mathscr{M}_p(\mathbf{Z}^n)} \le \left\|(1-|\cdot|^2)_+^\alpha\right\|_{\mathscr{M}_p}.$$

Notice that the first and second estimates are uniform in R', so one may insert a supremum over $R'\in\mathbf{R}^+\setminus\{|m|: m\in\mathbf{Z}^n\}$ in (4.3.17) and (4.3.18). To replace R' by a general $R\in\mathbf{Z}^+$ simply notice that for any $R>0$ there is an $R'\in\mathbf{R}^+\setminus\{|m|: m\in\mathbf{Z}^n\}$ such that

$$D_R^n * f = D_{R'}^n * f \qquad \text{and} \qquad \overset{\circ}{D}{}_R^n * f = \overset{\circ}{D}{}_{R'}^n * f$$

for any $f\in L^p(\mathbf{T}^n)$. Then using (4.3.17) we obtain

$$\sup_{R>0}\left\|D_R^n * f\right\|_{L^p} = \sup_{R>0}\left\|D_{R'}^n * f\right\|_{L^p} = \sup_{R'>0}\left\|D_{R'}^n * f\right\|_{L^p} \le \left\|\chi_{[-1,1]^n}\right\|_{\mathscr{M}_p}\left\|f\right\|_{L^p}$$

and likewise for $\overset{\circ}{D}{}_R^n$. □

4.3.4 Transference of Maximal Multipliers

We now prove a theorem concerning maximal multipliers analogous to Theorems 4.3.7 and 4.3.10. This enables us to reduce problems related to almost everywhere convergence of Fourier series on the torus to problems of boundedness of maximal operators on $\mathbf{R}^n$.

Let b be a bounded function defined on all of $\mathbf{R}^n$. For $R>0$, we introduce the multiplier operators

$$S_{b,R}(F)(x) = \sum_{m\in\mathbf{Z}^n} b(m/R)\widehat{F}(m)e^{2\pi i m\cdot x}, \tag{4.3.19}$$
$$T_{b,R}(f)(x) = \int_{\mathbf{R}^n} b(\xi/R)\widehat{f}(\xi)e^{2\pi i\xi\cdot x}\,d\xi, \tag{4.3.20}$$

initially defined for smooth functions with compact support f on $\mathbf{R}^n$ and smooth functions F on $\mathbf{T}^n$.

We introduce the maximal operators

$$M_b(F)(x) = \sup_{R>0}\left|S_{b,R}(F)(x)\right|, \tag{4.3.21}$$
$$N_b(f)(x) = \sup_{R>0}\left|T_{b,R}(f)(x)\right|, \tag{4.3.22}$$

defined for smooth functions F on $\mathbf{T}^n$ and smooth functions with compact support f on $\mathbf{R}^n$. Let $\tau^y(b)(\xi)=b(\xi-y)$ be a translation operator defined for $y\in\mathbf{R}^n$. We have the following result concerning these operators.

Theorem 4.3.12. *Let b be a function defined on $\mathbf{R}^n$. Suppose that b is bounded, regulated, Riemann integrable over any cube, and assume that for all $\xi\in\mathbf{R}^n$ the* *function $t \mapsto b(\xi/t)$ has only countably many discontinuities on $\mathbf{R}^+$. Let $1 < p < \infty$ and $C_p < \infty$, and suppose that b lies in $\mathscr{M}_p(\mathbf{R}^n)$. Let M_b and N_b be as in (4.3.21) and (4.3.22). Then the following assertions are equivalent:*

$$\big\|M_b(F)\big\|_{L^p(\mathbf{T}^n)} \le C_p\big\|b\big\|_{\mathscr{M}_p}\big\|F\big\|_{L^p(\mathbf{T}^n)}, \qquad F \in \mathscr{C}^\infty(\mathbf{T}^n), \tag{4.3.23}$$

$$\big\|N_b(f)\big\|_{L^p(\mathbf{R}^n)} \le C_p\big\|b\big\|_{\mathscr{M}_p}\big\|f\big\|_{L^p(\mathbf{R}^n)}, \qquad f \in \mathscr{C}_0^\infty(\mathbf{R}^n). \tag{4.3.24}$$

Proof. Let $\mathscr{F} = \{t_1, \dots, t_k\}$ be a finite subset of $\mathbf{R}^+$. We prove the claimed equivalences for the maximal operators

$$\begin{aligned} M_b^{\mathscr{F}}(G)(x) &= \sup_{t\in\mathscr{F}} \big|S_{b,t}(G)(x)\big|, \\ N_b^{\mathscr{F}}(g)(x) &= \sup_{t\in\mathscr{F}} \big|T_{b,t}(g)(x)\big|, \end{aligned}$$

with constants that are uniform in the finite set $\mathscr{F}$. Then $M_b^{\mathscr{F}}$ may be viewed as an operator defined on the dense subspace $\mathscr{C}^\infty(\mathbf{T}^n)$ of $L^p(\mathbf{T}^n)$ and taking values in $L^p(\mathbf{T}^n, \ell^\infty(\mathscr{F}))$, which is the dual space of $L^{p'}(\mathbf{T}^n, \ell^1(\mathscr{F}))$. Likewise, $N_b^{\mathscr{F}}$ is defined on the dense subspace $\mathscr{C}_0^\infty(\mathbf{R}^n)$ of $L^p(\mathbf{R}^n)$ and takes values in $L^p(\mathbf{R}^n, \ell^\infty(\mathscr{F}))$, which is the dual space of $L^{p'}(\mathbf{R}^n, \ell^1(\mathscr{F}))$. Using duality, with respect to the complex inner product, estimates (4.3.23) and (4.3.24) are equivalent to the pair of inequalities

$$\bigg|\sum_{m\in\mathbf{Z}^n} \overline{\widehat{G}(m)} \sum_{j=1}^k b\Big(\frac{m}{t_j}\Big)\widehat{F_j}(m)\bigg| \le C_p\big\|b\big\|_{\mathscr{M}_p}\big\|G\big\|_{L^p(\mathbf{T}^n)}\bigg\|\sum_{j=1}^k |F_j|\bigg\|_{L^{p'}(\mathbf{T}^n)}, \tag{4.3.25}$$

$$\bigg|\int_{\mathbf{R}^n} \overline{\widehat{g}(\xi)} \sum_{j=1}^k b\Big(\frac{\xi}{t_j}\Big)\widehat{f_j}(\xi)\,d\xi\bigg| \le C_p\big\|b\big\|_{\mathscr{M}_p}\big\|g\big\|_{L^p(\mathbf{R}^n)}\bigg\|\sum_{j=1}^k |f_j|\bigg\|_{L^{p'}(\mathbf{R}^n)}, \tag{4.3.26}$$

where $g, f_j \in \mathscr{C}_0^\infty(\mathbf{R}^n)$, and $G, F_j \in \mathscr{C}^\infty(\mathbf{T}^n)$. In proving the equivalence of (4.3.25) and (4.3.26), by density, we work with smooth functions with compact support g, f_j and trigonometric polynomials G, F_j.

Suppose that (4.3.25) holds and let $f_1, \dots, f_k, g$ be smooth functions with compact support on $\mathbf{R}^n$. Then there is an $R_0 > 0$ such that for $R \ge R_0$ the functions $F_{j,R}(x) = f_j(Rx)$ and $G_R(x) = g(Rx)$ are supported in $[-1/2, 1/2]^n$ and thus they can be viewed as functions on $\mathbf{T}^n$ once they are periodized. Also, the mth Fourier coefficient of $F_{j,R}$ is $\widehat{F_{j,R}}(m) = R^{-n}\widehat{f_j}(m/R)$ and that of G_R is $\widehat{G_R}(m) = R^{-n}\widehat{g}(m/R)$. Since b lies in $\mathscr{M}_p(\mathbf{R}^n)$ we have

$$\|b\|_{\mathscr{M}_p(\mathbf{R}^n)} = \sup_{R>0} \big\|\{b(m/R)\}_{m\in\mathbf{Z}^n}\big\|_{\mathscr{M}_p(\mathbf{Z}^n)}$$

in view of Theorems 4.3.7 and 4.3.10, which are both applicable in view of the hypotheses of b.

As in the proof of Theorem 4.3.10, for $R \geq R_0$ we have

$$\left| \sum_{m\in\mathbf{Z}^n} \sum_{j=1}^{k} b(m/Rt_j)\widehat{f_j}(m/R)\overline{\widehat{g}(m/R)}\,\text{Volume}\left(\tfrac{m}{R}+[0,\tfrac{1}{R}]^n\right)\right| \tag{4.3.27}$$

$$\begin{aligned}
&= \left| R^n \sum_{m\in\mathbf{Z}^n} \sum_{j=1}^{k} b(m/Rt_j)\widehat{F_{j,R}}(m)\overline{\widehat{G_R}(m)}\right| \\
&\leq C_p \|b\|_{\mathscr{M}_p} R^n \left\| \sum_{j=1}^{k} |F_{j,R}| \right\|_{L^{p'}(\mathbf{T}^n)} \|G_R\|_{L^p(\mathbf{T}^n)} \\
&= C_p \|b\|_{\mathscr{M}_p} \left\| \sum_{j=1}^{k} |f_j| \right\|_{L^{p'}(\mathbf{R}^n)} \|g\|_{L^p(\mathbf{R}^n)},
\end{aligned}$$

where we applied (4.3.25) in the first inequality above for the function $\xi \mapsto b(\xi/R)$, which has the same $\mathscr{M}_p$ norm as b.

Since b is bounded and Riemann integrable over any cube in $\mathbf{R}^n$, the functions $b(\xi/t_j)\widehat{f_j}(\xi)\overline{\widehat{g_j}(\xi)}$ are Riemann integrable over $\mathbf{R}^n$. Realizing the limit of the partial sums in (4.3.27) when $R \to \infty$ as a Riemann integral, we obtain

$$\left| \int_{\mathbf{R}^n} \sum_{j=1}^{k} b(\xi/t_j)\,\widehat{f_j}(\xi)\overline{\widehat{g}(\xi)}\,d\xi \right| \leq C_p \left\|b\right\|_{\mathscr{M}_p} \left\| \sum_{j=1}^{k} |f_j| \right\|_{L^{p'}(\mathbf{R}^n)} \|g\|_{L^p(\mathbf{R}^n)}$$

and thus we showed that (4.3.25) implies (4.3.26).

We now turn to the converse. Assume that (4.3.26) holds. We will prove (4.3.25) for trigonometric polynomials and then by density we extend it to all $\mathscr{C}^\infty$ functions on $\mathbf{T}^n$. Expressing $\widehat{g}$ in terms of g in (4.3.26) and taking the supremum in (4.3.26) over all $\mathscr{C}_0^\infty$ functions g with L^p norm 1 we deduce that

$$\left\| \int_{\mathbf{R}^n} \sum_{j=1}^{k} b\Big(\frac{\xi}{t_j}\Big)\widehat{f_j}(\xi)e^{2\pi i(\cdot)\cdot\xi}\,d\xi \right\|_{L^{p'}} \leq C_p \left\|b\right\|_{\mathscr{M}_p} \left\| \sum_{j=1}^{k} |f_j| \right\|_{L^{p'}(\mathbf{R}^n)}. \tag{4.3.28}$$

Let $P_1,\dots,P_k$ and Q be trigonometric polynomials on $\mathbf{T}^n$. Set $L_\varepsilon(x) = e^{-\pi\varepsilon|x|^2}$. Since b is regulated at every point in $\mathbf{R}^n$, Lemma 4.3.9 gives

$$\begin{aligned}
&\left| \sum_{m\in\mathbf{Z}^n} \sum_{j=1}^{k} \overline{\widehat{Q}(m)} b(m/t_j)\widehat{P_j}(m) \right| \\
&= \left| \int_{\mathbf{T}^n} \Big(\sum_{m\in\mathbf{Z}^n} \sum_{j=1}^{k} \widehat{P_j}(m) b(m/t_j) e^{2\pi i m\cdot x} \Big) \overline{Q(x)}\,dx \right| \\
&= \left| \lim_{\varepsilon\to 0} \varepsilon^{\frac{n}{2}} \int_{\mathbf{R}^n} \Big(\int_{\mathbf{R}^n} \sum_{j=1}^{k} \widehat{P_j L_{\varepsilon/p'}}(\xi) b(\xi/t_j) e^{2\pi i \xi\cdot x} d\xi \Big) \overline{Q(x) L_{\varepsilon/p}(x)}\,dx \right|
\end{aligned}$$

$$\leq C_p \|b\|_{\mathcal{M}_p} \limsup_{\varepsilon\to 0} \left[\varepsilon^{\frac{n}{2}} \Big\| \sum_{j=1}^{k} |P_j L_{\varepsilon/p'}| \Big\|_{L^{p'}(\mathbf{R}^n)} \big\| Q L_{\varepsilon/p} \big\|_{L^p(\mathbf{R}^n)} \right]$$

$$= C_p \|b\|_{\mathcal{M}_p} \limsup_{\varepsilon\to 0} \left[\varepsilon^{\frac{n}{2p'}} \Big\| \Big(\sum_{j=1}^{k} |P_j| \Big) L_{\varepsilon/p'} \Big\|_{L^{p'}(\mathbf{R}^n)} \Big(\varepsilon^{\frac{n}{2}} \int_{\mathbf{R}^n} |Q(x)|^p e^{-\varepsilon\pi|x|^2} dx \Big)^{\frac{1}{p}} \right]$$

$$= C_p \|b\|_{\mathcal{M}_p} \Big\| \sum_{j=1}^{k} |P_j| \Big\|_{L^{p'}(\mathbf{R}^n)} \big\| Q \big\|_{L^p(\mathbf{T}^n)},$$

where we used Hölder's inequality and (4.3.28) in the only inequality above and (4.3.13) in the last equality. Thus we obtain that (4.3.26) implies (4.3.25), and this completes the equivalence of boundedness of $M_b^{\mathscr{F}}$ and $N_b^{\mathscr{F}}$.

We now prove the claimed equivalence for the operators M_b and N_b. We first show that if $M_b^{\mathscr{F}}$ is bounded on $(\mathscr{C}^\infty(\mathbf{T}^n), \|\cdot\|_{L^p})$ with bound independent of the finite set $\mathscr{F}$, then M_b is bounded on $(\mathscr{C}^\infty(\mathbf{T}^n), \|\cdot\|_{L^p})$.

For each $\xi \in \mathbf{R}^n$, let A_ξ be the null subset of $\mathbf{R}^+$ such that $t \mapsto b(\xi/t)$ is continuous on $\mathbf{R}^+ \setminus A_\xi$. We fix a function F in $\mathscr{C}^\infty(\mathbf{T}^n)$, and we note that for each $x \in \mathbf{T}^n$ the function

$$t \mapsto S_{b,t}(F)(x) = \sum_{m\in\mathbf{Z}^n} b(m/t)\widehat{F}(m) e^{2\pi i m\cdot x} \tag{4.3.29}$$

is continuous on the set $\mathbf{R}^+ \setminus \bigcup_{m\in\mathbf{Z}^n} A_m$. We pick a countable dense subset D' of $\mathbf{R}^+ \setminus \bigcup_{m\in\mathbf{Z}^n} A_m$, and we let $D = D' \cup \bigcup_{m\in\mathbf{Z}^n} A_m$. Then D is a countable set and the Lebesgue monotone convergence theorem gives that

$$\Big\| \sup_{t\in D} |S_{b,t}(F)| \Big\|_{L^p(\mathbf{T}^n)} = \lim_{k\to\infty} \big\| M_b^{\mathscr{F}_k}(F) \big\|_{L^p(\mathbf{T}^n)} \leq C_p \|b\|_{\mathcal{M}_p} \big\| F \big\|_{L^p(\mathbf{T}^n)}, \tag{4.3.30}$$

where $\mathscr{F}_k$ is an increasing sequence of finite sets whose union is D. Using that the function in (4.3.29) is continuous on $\mathbf{R}^+ \setminus D$, we conclude that the supremum over $t \in D$ in (4.3.30) can be replaced by the supremum over $t \in \mathbf{Z}^+$ (Exercise 4.3.7).

Assume now that $N_b^{\mathscr{F}}$ is bounded on $(\mathscr{C}_0^\infty(\mathbf{R}^n), \|\cdot\|_{L^p})$ with bound independent of the finite set $\mathscr{F}$. We show that N_b is bounded on $(\mathscr{C}_0^\infty(\mathbf{R}^n), \|\cdot\|_{L^p})$. Let f be in $\mathscr{C}_0^\infty(\mathbf{R}^n)$. We have that the map

$$t \mapsto T_{b,t}(f)(x) = \int_{\mathbf{R}^n} b(\xi/t)\widehat{f}(\xi) e^{2\pi i \xi\cdot x} d\xi = t^n \int_{\mathbf{R}^n} b(\xi)\widehat{f}(t\xi) e^{2\pi i \xi\cdot tx} d\xi \tag{4.3.31}$$

is a continuous function on $\mathbf{R}^+$ since $\widehat{f}$ is continuous. Thus the estimate

$$\Big\| \sup_{t\in D} |T_{b,t}(f)| \Big\|_{L^p(\mathbf{R}^n)} \leq C_p \|b\|_{\mathcal{M}_p} \big\| f \big\|_{L^p(\mathbf{R}^n)} \tag{4.3.32}$$

for a countable dense subset D of $\mathbf{R}^+$ (such as $D = \mathbf{Q}^+$) can be easily extended by replacing the supremum over D by the supremum over $\mathbf{R}^+$. And estimate (4.3.32) for $D = \mathbf{Q}^+$ follows from the corresponding estimate on finite sets via the Lebesgue monotone convergence theorem. □

Remark 4.3.13. Under the hypotheses of Theorem 4.3.12, the following two inequalities are also equivalent:

$$\big\|M_b(G)\big\|_{L^{p,\infty}(\mathbf{T}^n)} \le C_p \|b\|_{\mathscr{M}_p} \big\|G\big\|_{L^p(\mathbf{T}^n)}, \qquad G \in \mathscr{C}^\infty(\mathbf{T}^n), \tag{4.3.33}$$

$$\big\|N_b(g)\big\|_{L^{p,\infty}(\mathbf{R}^n)} \le C'_p \|b\|_{\mathscr{M}_p} \|g\|_{L^p(\mathbf{R}^n)}, \qquad g \in \mathscr{C}_0^\infty(\mathbf{R}^n), \tag{4.3.34}$$

with $C'_p \le C_p \le C(n,p)C'_p$ for some other constant $C(n,p)$. Indeed, Exercise 1.4.12 gives that the pair of inequalities (4.3.33) and (4.3.34) is equivalent to the pair of inequalities

$$\Big|\sum_{m\in\mathbf{Z}^n}\sum_{j=1}^{k} \widehat{F_j}(m) b(m/t_j)\overline{\widehat{G}(m)}\Big| \le C_p \|b\|_{\mathscr{M}_p} \big\|G\big\|_{L^p(\mathbf{T}^n)} \Big\|\sum_{j=1}^{k}|F_j|\Big\|_{L^{p',1}(\mathbf{T}^n)}, \tag{4.3.35}$$

$$\Big|\int_{\mathbf{R}^n}\sum_{j=1}^{k} \widehat{f_j}(\xi) b(\xi/t_j)\overline{\widehat{g}(\xi)}\,d\xi\Big| \le C'_p \|b\|_{\mathscr{M}_p} \|g\|_{L^p(\mathbf{R}^n)} \Big\|\sum_{j=1}^{k}|f_j|\Big\|_{L^{p',1}(\mathbf{R}^n)}, \tag{4.3.36}$$

where $L^{p',1}$ is the Lorentz space and $f_j \in \mathscr{C}_0^\infty(\mathbf{R}^n)$ and $F_j \in \mathscr{C}^\infty(\mathbf{T}^n)$.

Now (4.3.36) follows from (4.3.35) just like (4.3.26) follows from (4.3.25) with the only exception being that Hölder's inequality for L^p and $L^{p'}$ is replaced by Hölder's inequality for $L^{p,\infty}$ and $L^{p',1}$ and we use that $\|g\|_{L^{p,\infty}} \le \|g\|_{L^p}$. Conversely, assuming (4.3.36), in order to prove (4.3.35) it will suffice to know that

$$\sup_{0<\varepsilon<1} \varepsilon^{\frac{n}{2q}} \Big\|\Big(\sum_{j=1}^{k}|P_j|\Big) L_{\varepsilon/q}\Big\|_{L^{q,1}(\mathbf{R}^n)} \le C(n,q) \Big\|\sum_{j=1}^{k}|P_j|\Big\|_{L^{q,1}(\mathbf{T}^n)}. \tag{4.3.37}$$

For this we refer to Exercise 4.3.6.

4.3.5 Applications to Almost Everywhere Convergence

As an application of the preceding results, we relate the almost everywhere convergence of Fourier series of functions on $\mathbf{T}^1$ with the almost everywhere convergence of Fourier integrals of functions on $\mathbf{R}$. In this subsection we show that the following two results are equivalent:

Theorem 4.3.14. *For every $1<p<\infty$ there exists a finite constant C_p such that for all $F \in \mathscr{C}^\infty(\mathbf{T}^1)$ we have*

$$\Big\|\sup_{N\in\mathbf{Z}^+}|F*D_N|\Big\|_{L^p} \le C_p \big\|F\big\|_{L^p}. \tag{4.3.38}$$

Theorem 4.3.15. *For every $1<p<\infty$ there exists a finite constant C_p such that for all $f \in \mathscr{C}_0^\infty(\mathbf{R})$ we have*

$$\big\|\mathscr{C}_{**}(f)\big\|_{L^p(\mathbf{R})} \le C_p \|f\|_{L^p(\mathbf{R})} \tag{4.3.39}$$

where

$$\mathscr{C}_{**}(f)(x) = \sup_{R>0}\left|\int_{|\xi|\le R} \widehat{f}(\xi)e^{2\pi i x\xi}\,d\xi\right|$$

is the Carleson operator.

As a consequence of Theorem 4.3.14, we obtain that for any $F \in L^p(\mathbf{T}^1)$, we have

$$\lim_{N\to\infty} \sum_{|m|\le N} \widehat{F}(m)e^{2\pi i m x} = F(x)$$

for almost every $x \in [0,1]$.

Theorem 4.3.14 can be proved directly, but we do not pursue this here. Instead, we show the equivalence of the two theorems and refer the interested reader to [131], which contains the proof of Theorem 4.3.15.

We observe that both operators $F \mapsto \mathscr{C}_*(F) = \sup_{N>0}|F * D_N|$ and $f \mapsto \mathscr{C}_{**}(f)$ are sublinear and take nonnegative values. Thus they satisfy the inequalities

$$|\mathscr{C}_*(F) - \mathscr{C}_*(G)| \le \mathscr{C}_*(F-G) \qquad |\mathscr{C}_{**}(f) - \mathscr{C}_{**}(g)| \le \mathscr{C}_{**}(f-g)$$

for all F,G in $\mathscr{C}^\infty(\mathbf{T}^1)$ and f,g in $\mathscr{C}_0^\infty(\mathbf{R})$. Then, by density (see the argument in the proof of Theorem 1.4.19 or Exercise 1.4.17), they admit bounded extensions to $L^p(\mathbf{T}^1)$ and $L^p(\mathbf{R})$, respectively, so that (4.3.38) and (4.3.39) hold for all $F \in L^p(\mathbf{T}^1)$ and $f \in L^p(\mathbf{R})$.

Next, we discuss the details of the transference argument that claims the equivalence of Theorems 4.3.14 and 4.3.15.

Consider the following function defined on $\mathbf{R}$:

$$b(x) = \begin{cases} 1 & \text{when } |x| < 1, \\ 1/2 & \text{when } |x| = 1, \\ 0 & \text{when } |x| > 1. \end{cases} \tag{4.3.40}$$

Then b is bounded and Riemann integrable over any interval, and is easily seen to be regulated; also, given any $x \in \mathbf{R}$, the function $t \mapsto b(x/t)$ is discontinuous only for $t \in \{x,-x\}$.

Let $S_{b,R}$ be as in (4.3.19), where b is defined in (4.3.40). We note that inequality (4.3.38) is equivalent to

$$\Big\|\sup_{R>0}|S_{b,R}(F)|\Big\|_{L^p} \le C_p'\big\|F\big\|_{L^p} \tag{4.3.41}$$

for all $F \in \mathscr{C}^\infty(\mathbf{T}^1)$, where $\{D_R\}_{R>0}$ is the family of Dirichlet kernels as defined in (3.1.16), depending on the continuous parameter R. Indeed, we have

$$S_{b,R}(F)(x) = \begin{cases} \sum_{|m|\le[R]} \widehat{F}(m)e^{2\pi imx} & \text{if } R \notin \mathbf{Z}^+, \\ \\ (D_{R-1} * F)(x) + \dfrac{\widehat{F}(R)e^{2\pi ixR} + \widehat{F}(-R)e^{-2\pi ixR}}{2} & \text{if } R \in \mathbf{Z}^+. \end{cases} \tag{4.3.42}$$

Since $\sup_{R>0}|\widehat{F}(\pm R)| \le \|F\|_{L^1} \le \|F\|_{L^p}$, it follows that if (4.3.38) holds, then (4.3.41) also holds with $C'_p = C_p + 1$.

The only hypothesis of Theorem 4.3.12 missing is that b lies in $\mathscr{M}_p(\mathbf{R})$. We obtain this from the fact that $\sup_{R>0}\|b(\cdot/R)\|_{\mathscr{M}_p(\mathbf{Z})} < \infty$ via Theorem 4.3.10, since

$$\sup_{R>0}\big\|F * D_R\big\|_{L^p(\mathbf{T}^1)} = \sup_{N\in\mathbf{Z}^+}\big\|F * D_N\big\|_{L^p(\mathbf{T}^1)} \le C''_p\big\|F\big\|_{L^p(\mathbf{T}^1)}, \tag{4.3.43}$$

where the last estimate follows from Proposition 4.1.6, Theorem 4.1.7, and Corollary 4.1.3. The preceding equality is due to the fact that $D_R = D_{R+\varepsilon}$ whenever $0 < \varepsilon < 1$.

Now all hypotheses of Theorem 4.3.12 are valid. As a consequence we obtain the equivalence of the boundedness of the the maximal operator

$$N_b(f)(x) = \mathscr{C}_{**}(f)(x) = \sup_{R>0}\left|\int_{-R}^{+R} \widehat{f}(\xi)e^{2\pi ix\xi}\,d\xi\right|$$

on $L^p(\mathbf{R})$ and of

$$M_b(F)(x) = \sup_{R>0}\left|\sum_{m\in\mathbf{Z}} \widehat{F}(m)e^{2\pi imx} b\Big(\frac{m}{R}\Big)\right| = \sup_{R>0}|S_{b,R}(F)(x)|,$$

on $L^p(\mathbf{T}^1)$. But in view of (4.3.42) and of the fact that $\sup_{R>0}|\widehat{F}(\pm R)| \le \|F\|_{L^p}$, the L^p boundedness of M_b is equivalent to the L^p boundedness of $\mathscr{C}_*$ on $L^p(\mathbf{T}^1)$. This discussion concludes the equivalence of Theorems 4.3.14 and 4.3.15.

4.3.6 Almost Everywhere Convergence of Square Dirichlet Means

The extension of Theorem 4.3.14 to higher dimensions is a rather straightforward consequence of the one-dimensional result.

Theorem 4.3.16. *For every $1 < p < \infty$, there exists a finite constant $C_{p,n}$ such that for all $f \in L^p(\mathbf{T}^n)$ we have*

$$\Big\|\sup_{N>0}|D_N^n * f|\Big\|_{L^p(\mathbf{T}^n)} \le C_{p,n}\big\|f\big\|_{L^p(\mathbf{T}^n)} \tag{4.3.44}$$

and consequently

$$\lim_{N\to\infty}\sum_{\substack{m\in\mathbf{Z}^n\\ |m_j|\le N}} \widehat{f}(m)e^{2\pi im\cdot x} = f(x)$$

for almost every $x \in \mathbf{T}^n$ and $f \in L^p(\mathbf{T}^n)$.

Proof. We prove Theorem 4.3.16 when $n = 2$. Fix a p with $1 < p < \infty$. Since the Riesz projection P_+ is bounded on $L^p(\mathbf{T}^1)$ (Theorem 4.1.7 and identity (4.1.9)), applying Theorem 4.3.10 with $b(\xi) = \chi_{(0,\infty)}$, we obtain that that the function $\chi_{(0,\infty)}$ is in $\mathscr{M}_p(\mathbf{R})$. It follows that the characteristic function of the half-space $\xi_1 > 0$ in $\mathbf{R}^2$ lies in $\mathscr{M}_p(\mathbf{R}^2)$. Since rotations and translations of multipliers preserve their $\mathscr{M}_p$ norms (Proposition 2.5.14), it follows that the characteristic function of any half space created by a line in $\mathbf{R}^2$ lies in $\mathscr{M}_p(\mathbf{R}^2)$ with a fixed norm. The product of three multipliers is a multiplier (Proposition 2.5.13); thus the characteristic function of the triangle T created by the lines $\xi_2 = \xi_1 - \frac{1}{4}$, $\xi_2 = -\xi_1 - \frac{1}{4}$, $\xi_2 = L + \frac{1}{4}$ lies also in $\mathscr{M}_p(\mathbf{R}^2)$ with norm independent of $L \in \mathbf{Z}^+$. The regulated function

$$\sigma(\xi_1,\xi_2) = \begin{cases} 1 & \text{if } (\xi_1,\xi_2) \in \overline{T} \setminus \partial T \\ 0 & \text{if } (\xi_1,\xi_2) \notin \overline{T} \\ \frac{1}{2} & \text{if } (\xi_1,\xi_2) \in \partial T \setminus \{(0,-\frac{1}{4}),(L+\frac{1}{2},L+\frac{1}{4}),(-L-\frac{1}{2},L+\frac{1}{4})\} \\ \frac{1}{8} & \text{if } (\xi_1,\xi_2) \in \{(L+\frac{1}{2},L+\frac{1}{4}),(-L-\frac{1}{2},L+\frac{1}{4})\} \\ \frac{1}{4} & \text{if } (\xi_1,\xi_2) = (0,-\frac{1}{4}) \end{cases}$$

is a.e. equal to the characteristic function of T. Thus Theorem 4.3.7 gives that the restriction of σ on $\mathbf{Z}^2$, i.e., the sequence $\{a_{m_1,m_2}\}_{m_1,m_2}$ defined by $a_{m_1,m_2} = 1$ when $|m_1| \le |m_2| \le L$ and zero otherwise, lies in $\mathscr{M}_p(\mathbf{Z}^2)$ with norm independent of L in $\mathbf{Z}^+$. This means that for some constant B_p we have the following inequality for all f in $L^p(\mathbf{T}^2)$:

$$\int_{\mathbf{T}^2} \Bigg| \sum_{\substack{m_2\in\mathbf{Z} \\ |m_2|\le L}} \sum_{\substack{m_1\in\mathbf{Z} \\ |m_1|\le|m_2|}} \widehat{f}(m_1,m_2) e^{2\pi i(m_1x_1+m_2x_2)} \Bigg|^p dx_2\,dx_1 \le B_p^p \|f\|_{L^p(\mathbf{T}^2)}^p, \tag{4.3.45}$$

where B_p is independent of $L \in \mathbf{Z}^+$. There is also a a version of (4.3.45), proved similarly, in which $|m_1| \le |m_2|$ is replaced by the strict inequality $|m_1| < |m_2|$.

Now let $1 < p < \infty$, $L \in \mathbf{Z}^+$, and $f \in L^p(\mathbf{T}^2)$. For fixed $x_1 \in \mathbf{T}^1$ define

$$f_{x_1}^L(x_2) = \sum_{\substack{m_2\in\mathbf{Z} \\ |m_2|\le L}} \Bigg[\sum_{\substack{m_1\in\mathbf{Z} \\ |m_1|\le|m_2|}} \widehat{f}(m_1,m_2) e^{2\pi i m_1 x_1} \Bigg] e^{2\pi i m_2 x_2} = \sum_{\substack{m_2\in\mathbf{Z} \\ |m_2|\le L}} \widehat{f_{x_1}^L}(m_2) e^{2\pi i m_2 x_2}$$

and for fixed $x_2 \in \mathbf{T}^1$ define

$$f_L^{x_2}(x_1) = \sum_{\substack{m_1\in\mathbf{Z} \\ |m_1|\le L}} \Bigg[\sum_{\substack{m_2\in\mathbf{Z} \\ |m_2|<|m_1|}} \widehat{f}(m_1,m_2) e^{2\pi i m_2 x_2} \Bigg] e^{2\pi i m_1 x_1} = \sum_{\substack{m_1\in\mathbf{Z} \\ |m_1|\le L}} \widehat{f_L^{x_2}}(m_1) e^{2\pi i m_1 x_1}.$$

We have

$$
\begin{aligned}
&\int_{\mathbf{T}^1}\int_{\mathbf{T}^1}\sup_{0<N\le L}\bigg|\sum_{|m_1|\le N}\sum_{|m_2|\le N}\widehat{f}(m_1,m_2)e^{2\pi i m_1x_1}e^{2\pi i m_2x_2}\bigg|^p dx_2\,dx_1\\
&\le 2^{p-1}\int_{\mathbf{T}^1}\int_{\mathbf{T}^1}\sup_{0<N\le L}\bigg|\sum_{|m_2|\le N}\bigg[\sum_{|m_1|\le|m_2|}\widehat{f}(m_1,m_2)e^{2\pi i m_1x_1}\bigg]e^{2\pi i m_2x_2}\bigg|^p\\
&\qquad\qquad+\sup_{0<N\le L}\bigg|\sum_{|m_1|\le N}\bigg[\sum_{|m_2|<|m_1|}\widehat{f}(m_1,m_2)e^{2\pi i m_2x_2}\bigg]e^{2\pi i m_1x_1}\bigg|^p dx_1\,dx_2\\
&=2^{p-1}\bigg[\int_{\mathbf{T}^1}\int_{\mathbf{T}^1}\sup_{0<N\le L}\big|(D_N*f_{x_1}^L)(x_2)\big|^p dx_2\,dx_1\\
&\qquad\qquad+\int_{\mathbf{T}^1}\int_{\mathbf{T}^1}\sup_{0<N\le L}\big|(D_N*f_L^{x_2})(x_1)\big|^p dx_1\,dx_2\bigg]\\
&\le 2^{p-1}\bigg[\int_{\mathbf{T}^1}\int_{\mathbf{T}^1}\sup_{N\in\mathbf{Z}^+}\big|(D_N*f_{x_1}^L)(x_2)\big|^p dx_2dx_1\\
&\qquad\qquad+\int_{\mathbf{T}^1}\int_{\mathbf{T}^1}\sup_{N\in\mathbf{Z}^+}\big|(D_N*f_L^{x_2})(x_1)\big|^p dx_1\,dx_2\bigg]\\
&\le 2^{p-1}C_p^p\int_{\mathbf{T}^1}\int_{\mathbf{T}^1}|f_{x_1}^L(x_2)|^p dx_2\,dx_1+2^{p-1}C_p^p\int_{\mathbf{T}^1}\int_{\mathbf{T}^1}|f_L^{x_2}(x_1)|^p dx_1\,dx_2\\
&\le 2^pC_p^pB_p^p\big\|f\big\|_{L^p(\mathbf{T}^2)}^p,
\end{aligned}
$$

where we used Theorem 4.3.14 in the penultimate inequality and estimate (4.3.45) in the last inequality. Since the last estimate we obtained is independent of $L\in\mathbf{Z}^+$, letting $L\to\infty$ and applying Fatou's lemma, we obtain the conclusion (4.3.44) for $n=2$. When $n\ge 3$ the idea of the proof is similar, but the notation a bit more cumbersome. □

Exercises

4.3.1. Let $\alpha\ge 0$. Prove that the function $(1-|\xi|^2)_+^\alpha$ is in $\mathscr{M}_p(\mathbf{R}^n)$ if and only if the function $(1-|\xi|)_+^\alpha$ is in $\mathscr{M}_p(\mathbf{R}^n)$.
[*Hint:* Use that smooth functions with compact support lie in $\mathscr{M}_p(\mathbf{R}^n)$.]

4.3.2. The purpose of this exercise is to introduce distributions on the torus. The set of test functions on the torus is $\mathscr{C}^\infty(\mathbf{T}^n)$ equipped with the following topology. Given f_j, f in $\mathscr{C}^\infty(\mathbf{T}^n)$, we say that $f_j\to f$ in $\mathscr{C}^\infty(\mathbf{T}^n)$ if

$$\big\|\partial^\alpha f_j-\partial^\alpha f\big\|_{L^\infty(\mathbf{T}^n)}\to 0\quad\text{as } j\to\infty,\ \forall\,\alpha.$$

Under this notion of convergence, $\mathscr{C}^\infty(\mathbf{T}^n)$ is a topological vector space with topology induced by the family of seminorms $\rho_\alpha(\varphi)=\sup_{x\in\mathbf{T}^n}|(\partial^\alpha f)(x)|$, where α

ranges over all multi-indices. The dual space of $\mathscr{C}^\infty(\mathbf{T}^n)$ under this topology is the set of all distributions on $\mathbf{T}^n$ and is denoted by $\mathscr{D}'(\mathbf{T}^n)$. The definition implies that for u_j and u in $\mathscr{D}'(\mathbf{T}^n)$ we have $u_j \to u$ in $\mathscr{D}'(\mathbf{T}^n)$ if and only if

$$\langle u_j, f\rangle \to \langle u, f\rangle \quad \text{as } j \to \infty \text{ for all } f \in \mathscr{C}^\infty(\mathbf{T}^n).$$

The following operations can be defined on elements of $\mathscr{D}'(\mathbf{T}^n)$: differentiation (as in Definition 2.3.6), translation and reflection (as in Definition 2.3.11), convolution with a $\mathscr{C}^\infty$ function (as in Definition 2.3.13), multiplication by a $\mathscr{C}^\infty$ function (as in Definition 2.3.15), the support of a distribution (as in Definition 2.3.16). Use the same ideas as in $\mathbf{R}^n$ to prove the following:
(a) Prove that if $u \in \mathscr{D}'(\mathbf{T}^n)$ and $f \in \mathscr{C}^\infty(\mathbf{T}^n)$, then $(f*u)(x) = \langle u, \tau^x(\widetilde{f})\rangle$ is a $\mathscr{C}^\infty$ function.
(b) In contrast to $\mathbf{R}^n$, the convolution of two distributions on $\mathbf{T}^n$ can be defined. For $u, v \in \mathscr{D}'(\mathbf{T}^n)$ and $f \in \mathscr{C}^\infty(\mathbf{T}^n)$ define

$$\langle u*v, f\rangle = \langle u, f*\widetilde{v}\rangle .$$

Check that convolution of distributions on $\mathscr{D}'(\mathbf{T}^n)$ is associative, commutative, and distributive.
(c) Prove the analogue of Proposition 2.3.23, i.e., that $\mathscr{C}^\infty(\mathbf{T}^n)$ is dense in $\mathscr{D}'(\mathbf{T}^n)$.

4.3.3. For $u \in \mathscr{D}'(\mathbf{T}^n)$ and $m \in \mathbf{Z}^n$ define the Fourier coefficient $\widehat{u}(m)$ by

$$\widehat{u}(m) = u(e^{-2\pi i m\cdot(\cdot)}) = \langle u, e^{-2\pi i m\cdot(\cdot)}\rangle .$$

Prove properties (1), (2), (4), (5), (6), (8), (9), (11), and (12) of Proposition 2.3.22 regarding the Fourier coefficients of distributions on the circle. Moreover, prove that for any u, v in $\mathscr{D}'(\mathbf{T}^n)$ we have $(u*v)\widehat{\ }(m) = \widehat{u}(m)\widehat{v}(m)$. In particular, this is valid for finite Borel measures.

4.3.4. Let μ be a finite Borel measure on $\mathbf{R}^n$ and let ν be the periodization of μ, that is, ν is a measure on $\mathbf{T}^n$ defined by

$$\nu(A) = \sum_{m\in\mathbf{Z}^n} \mu(A+m)$$

for all measurable subsets A of $\mathbf{T}^n$. Prove that the restriction of the Fourier transform of μ on $\mathbf{Z}^n$ coincides with the sequence of the Fourier coefficients of the measure ν.

4.3.5. Let v_n be the volume of the unit ball in $\mathbf{R}^n$ and $e_1 = (1,0,\dots,0)$. Prove that

$$\lim_{\varepsilon\to 0} \frac{1}{v_n \varepsilon^n} \int_{|x-e_1|\le\varepsilon} \chi_{|x|\le 1}\, dx = \frac{1}{2}.$$

Conclude that the function

$$\widetilde{\chi}_{B(0,1)}(x) = \begin{cases} 1 & \text{when } |x| < 1, \\ 1/2 & \text{when } |x| = 1, \\ 0 & \text{when } |x| > 1 \end{cases}$$

is regulated.

4.3.6. Let $L_\varepsilon(x) = e^{-\pi\varepsilon|x|^2}$ be defined for $\varepsilon > 0$ and $x \in \mathbf{R}^n$ and let $1 < q < \infty$. Prove that there is a constant $C(n,q) < \infty$ such that for any 1-periodic continuous function g on $\mathbf{R}^n$ we have

$$\sup_{0<\varepsilon<1} \varepsilon^{\frac{n}{2q}} \big\|g L_{\varepsilon/q}\big\|_{L^{q,1}(\mathbf{R}^n)} \le C(n,q)\|g\|_{L^{q,1}(\mathbf{T}^n)}.$$

[*Hint:* Reduce matters to the situation where $g = \sum_{k\in\mathbf{Z}^n} \chi_{k+E}$, where E is a measurable subset of $[-1/2,1/2)^n$. Express the $L^{q,1}$ norm of $gL_{\varepsilon/q}$ in terms of its distribution function and for $0 < \lambda < 1$ estimate the measure

$$\Big|\{L_{\varepsilon/q} > \lambda\} \bigcap \bigcup_{k\in\mathbf{Z}^n} (k+E)\Big| = \Big|B\Big(0, \Big(\frac{q}{\pi\varepsilon}\log\frac{1}{\lambda}\Big)^{\frac{1}{2}}\Big) \bigcap \bigcup_{k\in\mathbf{Z}^n} (k+E)\Big|$$

by $C_n\big(\sqrt{n} + \big(\frac{q}{\pi\varepsilon}\log\frac{1}{\lambda}\big)^{1/2}\big)^n |E|$.]

4.3.7. Let $0 < C_0 < \infty$. Suppose that $\{f_t\}_{t\in\mathbf{R}^+}$ is a family of measurable functions on a measure space X that satisfies

$$\big\|\sup_{t\in F}|f_t|\big\|_{L^p} \le C_0$$

for every finite subset F of $\mathbf{R}^+$.
(a) Suppose that for each $x \in X$, the function $t \mapsto f_t(x)$ is continuous. Show that

$$\big\|\sup_{t>0}|f_t|\big\|_{L^p} \le C_0.$$

(b) Prove that for any $t > 0$ there is a measurable function $\widetilde{f}_t$ on X that is a.e. equal to f_t such that

$$\big\|\sup_{t\in\mathbf{R}^+}|\widetilde{f}_t|\big\|_{L^p} \le C_0.$$

[*Hint:* Part (a): Notice that in view of the Lebesgue monotone convergence theorem, we have $\|\sup_{t\in\mathbf{Q}}|f_t|\|_{L^p} \le C_0$. Also, for each $x \in X$ we have $\sup_{t\in\mathbf{Q}}|f_t(x)| = \sup_{t\in\mathbf{R}^+}|f_t(x)|$ by continuity. Part (b): Let $a = \sup_F \|\sup_{t\in F}|f_t|\|_{L^p} \le C_0$, where the supremum is taken over all finite subsets F of $\mathbf{R}$. Pick an increasing sequence of finite sets F_n such that $\|\sup_{t\in F_n}|f_t|\|_{L^p} \to a$ as $n \to \infty$. Let $g = \sup_n \sup_{t\in F_n}|f_t|$ and note that $\|g\|_{L^p} = a$. Then for any $s \in \mathbf{R}$ we have

$$\Big\| \max(|f_s|, \sup_{t\in F_k} |f_t|) \Big\|_{L^p} \le a\,.$$

This implies $\| \max(|f_s|, g)\|_{L^p} \le a = \|g\|_{L^p}$, so that $|f_s| \le g$ a.e. for all $s \in \mathbf{R}$.]

4.3.8. (*E. Prestini*) Show that for $f \in L^2(\mathbf{T}^2)$ we have that

$$\sum_{\substack{|m_1|\le N\\ |m_2|\le N^2}} \widehat{f}(m_1,m_2)e^{2\pi i(m_1x_1+m_2x_2)} \to f(x_1,x_2)$$

for almost all (x_1,x_2) in $\mathbf{T}^2$.
[*Hint:* Use the splitting $\widehat{f}(m_1,m_2) = \widehat{f}(m_1,m_2)\chi_{|m_2|\le |m_1|^2} + \widehat{f}(m_1,m_2)\chi_{|m_2|>|m_1|^2}$ and apply the idea of the proof of Theorem 4.3.16.]

4.4 Applications to Geometry and Partial Differential Equations

In this section we discuss two applications of Fourier series. The first concerns a classical result in planar geometry and the other the heat equation.

4.4.1 The Isoperimetric Inequality

Suppose we are given a closed positively oriented nonself intersecting $\mathscr{C}^1$ curve C in the (x,y) plane of length L that encloses a region R of area A. The curve can be described in terms of its parametric equations $x = x(t)$ and $y = y(t)$, where $t \in [0,1]$. Since the curve is closed, we have $(x(0),y(0)) = (x(1),y(1))$ and the $\mathscr{C}^1$ functions $x(t), y(t)$ can be thought of as 1-periodic functions on the circle. The perimeter L of the curve is given by the equation

$$L = \int_0^1 \sqrt{|x'(t)|^2 + |y'(t)|^2}\, dt$$

while the area of the region R enclosed by the curve is equal to

$$\begin{aligned} A &= \iint_R 1\, dxdy \\ &= \iint_R \frac{\partial}{\partial x}\frac{x}{2} - \frac{\partial}{\partial y}\Big(-\frac{y}{2}\Big)dxdy \\ &= \frac{1}{2}\oint_C x\,dy - y\,dx \\ &= \frac{1}{2}\int_0^1 x(t)y'(t) - x'(t)y(t)\, dt\,, \end{aligned}$$

where we made use of Green's theorem in the third equality above.

A $\mathscr{C}^1$ curve $\gamma(t)$ is regular if $\gamma'(t) \neq 0$ for all t. We have the following result relating the perimeter and the enclosed area of a region enclosed by a closed $\mathscr{C}^1$ curve.

Theorem 4.4.1. *Given a closed, positively oriented, nonself intersecting, regular, $\mathscr{C}^1$ planar curve of length L that encloses a region of area A, we have that*

$$A \leq L^2/4\pi \tag{4.4.1}$$

with equality holding if and only if the curve is a circle.

Proof. Assume that the curve has parametric equations $x = x(t)$, $y = y(t)$, $0 \leq t \leq 1$. We may assume that the curve has constant speed, i.e., it satisfies

$$\sqrt{|x'(t)|^2 + |y'(t)|^2} = L$$

for all $t \in [0,1]$. This is achieved via the reparametrization of the curve in terms of the inverse function $s^{-1}(t) = \gamma(t)$ of the normalized arc length function

$$s(t) = \frac{1}{L}\int_0^t \sqrt{|x'(u)|^2 + |y'(u)|^2}\, du\,.$$

Since $|(x'(t), y'(t))| \neq 0$, $t \mapsto s(t)$ is a one-to-one and onto continuous map from $[0,1]$ to $[0,1]$. Then the curve $t \mapsto (x(\gamma(t)), y(\gamma(t)))$ has constant speed, since

$$|x'(\gamma(t))|^2 |\gamma'(t)|^2 + |y'(\gamma(t))|^2 |\gamma'(t)|^2 = \frac{|x'(s^{-1}(t))|^2 + |y'(s^{-1}(t))|^2}{|s'(s^{-1}(t))|^2} = L^2\,.$$

So we can replace the map $(x(t), y(t))$ by $(x(\gamma(t)), y(\gamma(t)))$ which produces the same curve. Let

$$f(t) = x(t) + iy(t)$$

for $t \in [0,1]$. Then in view of the preceding discussion, we may assume that the function $f(t) = x(t) + iy(t)$ satisfies $|f'(t)| = L$ for all $t \in [0,1]$.

Under the assumption $|f'(t)| = L$ for all $t \in [0,1]$, we now show that (4.4.1) holds, with equality if and only if $f(t) = c_0 e^{2\pi i t} + C_0$ for some $c_0, C_0 \in \mathbf{C}$ with $|c_0| = \frac{L}{2\pi}$. To prove this claim we argue as follows:

$$\begin{aligned} A &= \frac{1}{2}\operatorname{Im} \int_0^1 f'(t)\overline{f(t)}\,dt \\ &= \frac{1}{2}\operatorname{Im} \int_0^1 f'(t)\overline{\big(f(t) - \widehat{f}(0)\big)}\,dt \\ &\leq \frac{1}{2} L \big\| f - \widehat{f}(0) \big\|_{L^2} \\ &\leq \frac{L}{2}\frac{1}{2\pi} \big\| f' \big\|_{L^2} \\ &= \frac{L^2}{4\pi}, \end{aligned}$$

establishing (4.4.1), but we need to explain why the inequality

$$\|f-\widehat{f}(0)\|_{L^2} \le \frac{1}{2\pi}\|f'\|_{L^2} \tag{4.4.2}$$

is valid. Indeed, we have

$$f'(t) = \sum_{m\in\mathbf{Z}} 2\pi i m \widehat{f}(m) e^{2\pi i m t}$$

where the series converges in L^2. Thus we have

$$\big\|f'\big\|_{L^2} = 2\pi\bigg[\sum_{m\in\mathbf{Z}} |m\widehat{f}(m)|^2\bigg]^{\frac{1}{2}} \ge 2\pi\bigg[\sum_{m\in\mathbf{Z}\setminus\{0\}} |\widehat{f}(m)|^2\bigg]^{\frac{1}{2}} = 2\pi\big\|f-\widehat{f}(0)\big\|_{L^2}, \tag{4.4.3}$$

which proves (4.4.2).

Now suppose that equality holds in (4.4.1), then we must have equality in (4.4.2) and thus in (4.4.3), which implies that $\widehat{f}(m)=0$ when $|m|\ge 2$; hence for all $t\in[0,1]$ we must have

$$f(t) = ce^{2\pi i t} + c'e^{-2\pi i t} + \widehat{f}(0) \tag{4.4.4}$$

where c, c' are complex numbers. But since $\|f'\|_{L^2} = L$, it follows that

$$4\pi^2(|c|^2+|c'|^2) = L^2, \tag{4.4.5}$$

and since $|f'(t)| = L$ for all $t\in[0,1]$, it follows that

$$\left(\frac{L}{2\pi}\right)^2 = |c|^2+|c'|^2 - 2\,\mathrm{Re}\left[c\overline{c'}e^{2\pi i 2t}\right] \tag{4.4.6}$$

for all $t\in[0,1]$. Combining (4.4.5) and (4.4.6) we obtain

$$\mathrm{Re}\left[c\overline{c'}e^{2\pi i 2t}\right] = 0. \tag{4.4.7}$$

Inserting $t=0$ and $t=1/8$ in (4.4.7) and using that $\mathrm{Im}\,(iz) = -\mathrm{Re}\,z$, we deduce that $\mathrm{Re}\,c\overline{c'} = \mathrm{Im}\,c\overline{c'} = 0$. This implies that either c or c' is zero. In either case (4.4.4) and (4.4.5) imply that $f(t)$ is a circle of radius $L/2\pi$ centered at the point $\widehat{f}(0)$. □

4.4.2 The Heat Equation with Periodic Boundary Condition

Let $k>0$ be a fixed quantity. Consider the partial differential equation

$$\frac{\partial}{\partial t}F(x,t) = k\sum_{j=1}^{n}\frac{\partial^2}{\partial x_j^2}F(x,t) \qquad t\in(0,\infty), \quad x\in\mathbf{R}^n, \tag{4.4.8}$$

which is called the *heat equation*. Assume that there is an initial condition

$$F(0,x) = f(x) \qquad\qquad x \in \mathbf{R}^n \tag{4.4.9}$$

for a given $\mathscr{C}^\infty$ function f on $\mathbf{R}^n$ which is assumed to be 1-periodic in every variable.

We would like to find a continuous function $F(t,x)$ on $[0,\infty) \times \mathbf{R}^n$ which is $\mathscr{C}^\infty$ on $(0,\infty) \times \mathbf{R}^n$ such that

$$F(t, x+e_j) = F(t,x)$$

for all $t \geq 0$ and all $e_j = (0,\dots,0,1,0,\dots,0)$, so that F solves the equation (4.4.8).

The function $F(t,x)$ represents the temperature of a body at time $t > 0$ at the location $(x_1,\dots,x_n)$. Since the initial temperature f is 1-periodic in each variable, we expect $F(t,\cdot)$ to also be periodic in each variable. For example, $F(t,x)$ is a good model for the temperature of the torus $\{(e^{2\pi i x_1},\dots,e^{2\pi i x_n}) : x_j \in \mathbf{R}\}$ at time $t > 0$, given that its temperature at time $t = 0$ is $f(x)$. When $n = 1$, $F(t,x)$ models the temperature of a infinitesimally thin ring, thought of as the unit circle, at time $t > 0$ at the location $e^{2\pi i x}$.

Let us suppose there is a continuous function $F(t,x)$ on $[0,\infty) \times \mathbf{R}^n$ which is $\mathscr{C}^\infty$ on $(0,\infty) \times \mathbf{R}^n$ that solves the equation (4.4.8) and satisfies $F(t,x+e_j) = F(t,x)$ for all $x \in \mathbf{R}^n$ and $t \geq 0$. Denote by $c_m(t)$ the Fourier coefficient of the function $x \mapsto F(t,x)$ defined by

$$c_m(t) = \int_{\mathbf{T}^n} F(t,x)\, e^{-2\pi i m\cdot x}\, dx.$$

Then $c_m(t)$ is a continuous function on $[0,\infty)$ since F is continuous in the variable t. For the same reason, c_m is a smooth function on $(0,\infty)$ whose jth derivative is given by

$$\frac{d^j}{dt^j} c_m(t) = \int_{\mathbf{T}^n} \frac{\partial^j}{\partial t^j} F(t,x)\, e^{-2\pi i m\cdot x}\, dx$$

for any $j = 1,2,\dots$. Using equation (4.4.8) we obtain that

$$c_m'(t) = \int_{\mathbf{T}^n} \frac{\partial}{\partial t} F(t,x)\, e^{-2\pi i m\cdot x}\, dx = \int_{\mathbf{T}^n} k \frac{\partial^2}{\partial^2 x} F(t,x)\, e^{-2\pi i m\cdot x}\, dx = -4\pi^2 |m|^2 k\, c_m(t),$$

where the last identity is due to an integration by parts in which the boundary terms cancel each other in view of the periodicity of the integrand in x. Also $c_m(0) = \widehat{f}(m)$. The ordinary differential equation $c_m'(t) = -4\pi^2 |m|^2 k\, c_m(t)$ with initial condition $c_m(0) = \widehat{f}(m)$ is easily solved by separating the variables

$$\frac{dc_m(t)}{c_m(t)} = -4\pi^2 |m|^2 k dt\,, \tag{4.4.10}$$

yielding the solution

$$c_m(t) = \widehat{f}(m) e^{-4\pi^2 |m|^2 kt}.$$

We may therefore define the function

$$F(t,x) = \sum_{m\in\mathbf{Z}^n} \widehat{f}(m)e^{-4\pi^2|m|^2kt}e^{2\pi i m\cdot x} \tag{4.4.11}$$

on $[0,\infty)\times\mathbf{R}^n$ and observe the following:

(a) F is continuous on $[0,\infty)\times\mathbf{R}^n$ and $\mathscr{C}^\infty$ on $(0,\infty)\times\mathbf{R}^n$.
(b) F satisfies the heat equation (4.4.8) and the initial condition (4.4.9).
(c) F is 1-periodic in each of the last n variables.

These statements can be easily proved by passing the differentiation inside the sum, in view of the rapid convergence of the series in (4.4.11) due to the fact that the periodic function f is $\mathscr{C}^\infty(\mathbf{R}^n)$. Furthermore, F is unique with properties (a), (b), and (c), since any other function $G(x,t)$ with these properties is derived in the preceding way, and so it has to be equal to $F(x,t)$.

Definition 4.4.2. Define the *heat kernel*

$$H_t(x) = \sum_{m\in\mathbf{Z}^n} e^{-4\pi^2|m|^2kt}e^{2\pi i m\cdot x}$$

for $t>0$. Notice that the series defining H_t is absolutely convergent for any $t>0$. The importance of the heat kernel lies in the fact that one can express the solution $F(x,t)$ of (4.4.8) in terms of the convolution $F(x,t) = (f*H_t)(x)$.

We summarize these facts in the following proposition.

Proposition 4.4.3. *Let $k>0$ be fixed and let f be in $\mathscr{C}^\infty(\mathbf{R}^n)$. Assume that f is 1-periodic function in each variable. Then the heat equation*

$$\frac{\partial}{\partial t}F(x,t) = k\Delta_x F(x,t) \qquad t\in(0,\infty), \quad x\in\mathbf{R}^n \tag{4.4.12}$$

under the initial condition

$$F(0,x) = f(x) \qquad x\in\mathbf{R}^n \tag{4.4.13}$$

has a unique solution which is continuous on $[0,\infty)\times\mathbf{R}^n$ and $\mathscr{C}^\infty$ on $(0,\infty)\times\mathbf{R}^n$ given by

$$F(x,t) = (f*H_t)(x) = \sum_{m\in\mathbf{Z}^n} \widehat{f}(m)e^{-4\pi^2|m|^2kt}e^{2\pi i m\cdot x}. \tag{4.4.14}$$

Proof. Since f is $\mathscr{C}^\infty$, the series in (4.4.14) is rapidly convergent in m and thus it gives a continuous function on $[0,\infty)\times\mathbf{R}^n$. Moreover, the series can be differentiated term by term in the variable $t>0$, and thus it produces a $\mathscr{C}^\infty$ function on $(0,\infty)\times\mathbf{R}^n$. By Fourier inversion (Proposition 3.2.5), F satisfies the initial condition (4.4.13). Finally, to verify (4.4.12), we simply notice that

$$
\begin{aligned}
\frac{\partial}{\partial t}F(x,t) &= \sum_{m\in\mathbf{Z}^n}\frac{\partial}{\partial t}\widehat{f}(m)e^{-4\pi^2|m|^2kt}e^{2\pi i m\cdot x}\\
&= -4\pi^2 k\sum_{m\in\mathbf{Z}^n}\widehat{f}(m)e^{-4\pi^2|m|^2kt}|m|^2e^{2\pi i m\cdot x}\\
&= k\sum_{m\in\mathbf{Z}^n}\widehat{f}(m)e^{-4\pi^2|m|^2kt}\Big(\frac{\partial^2}{\partial^2_{x_1}}+\cdots+\frac{\partial^2}{\partial^2_{x_n}}\Big)e^{2\pi i m\cdot x}\\
&= k\Big(\frac{\partial^2}{\partial^2_{x_1}}+\cdots+\frac{\partial^2}{\partial^2_{x_n}}\Big)\sum_{m\in\mathbf{Z}^n}\widehat{f}(m)e^{-4\pi^2|m|^2kt}e^{2\pi i m\cdot x}\\
&= k\frac{\partial^2}{\partial^2 x}F(x,t),
\end{aligned}
$$

where the rapid convergence of the series in m makes it possible to pass the differentiations in and out of the sum. Finally, to show uniqueness, assume that there is another solution $G(t,x)$, continuous on $[0,\infty)\times\mathbf{R}^n$ and $\mathscr{C}^\infty$ on $(0,\infty)\times\mathbf{R}^n$, that can be expanded in Fourier series as follows:

$$
G(t,x) = \sum_{m\in\mathbf{Z}^n} c_m(t)e^{2\pi i m\cdot x}.
$$

Conditions (4.4.12) and the rapid decay of the coefficients $c_m(t)$ yield the ordinary differential equation (4.4.10) with initial condition $c_m(0)=\widehat{f}(m)$, which has the solution $c_m(t)=\widehat{f}(m)e^{-4\pi^2|m|^2kt}$. Thus $G=F$ on $[0,\infty)\times\mathbf{R}^n$. □

It is important to observe that the family $\{H_t\}_{t>0}$ is an approximate identity on $\mathbf{T}^1$. Indeed, the Poisson summation formula (Theorem 3.2.8) and the fact that the inverse Fourier transform of $e^{-4\pi^2kt|\xi|^2}$ is $e^{-|x|^2/4kt}/(2\sqrt{\pi kt})^n$ [Example 2.2.9 and Proposition 2.2.11 (8)] yield that for all $x\in[0,1]^n$ we have

$$
H_t(x) = \frac{1}{(2\sqrt{\pi kt})^n}\sum_{\ell\in\mathbf{Z}} e^{-\frac{|x+\ell|^2}{4kt}}.
$$

This identity implies that $H_t(x)\geq 0$ for all $t>0$ and that

$$
\int_{\mathbf{T}^n} H_t(x)\,dx = \int_{\mathbf{R}^n}\frac{1}{(2\sqrt{\pi kt})^n}e^{-\frac{|x|^2}{4kt}}\,dx = \int_{\mathbf{R}^n} e^{-\pi|x|^2}\,dx = 1
$$

for all $t>0$ and that

$$
\int_{\delta\leq|u|} H_t(u)\,du \leq \int_{|x|\geq\delta}\frac{1}{(2\sqrt{\pi kt})^n}e^{-\frac{|x|^2}{4kt}}\,dx = \int_{|x|\geq\frac{\delta}{2\sqrt{k\pi t}}} e^{-\pi|x|^2}\,dx
$$

which tends to zero as $t\to 0$ for any $\delta>0$ in view of the Lebesgue differentiation theorem. Thus properties (i), (ii), and (iii) of approximate identities hold.

As a consequence, we have that $\|F(t,\cdot)-f\|_{L^p(\mathbf{T}^n)}\to 0$ as $t\to 0$ for $1\leq p<\infty$ and $F(t,\cdot)$ converges to f uniformly on $\mathbf{T}^n$; see Theorem 1.2.19.

Exercises

4.4.1. Let f,F be as in Proposition 4.4.3. Prove that the total heat on the torus remains constant in time by showing that for all $t \geq 0$ we have

$$\int_{\mathbf{T}^n} F(t,x)\,dx = \int_{\mathbf{T}^n} f(x)\,dx.$$

Moreover, show that the temperature at any fixed point $x \in \mathbf{T}^n$ on the torus tends to the average initial temperature, i.e., it satisfies

$$\lim_{t\to\infty} F(t,x) = \int_{\mathbf{T}^n} f(y)\,dy.$$

4.4.2. Derive the following property of the heat kernel,

$$H_t * H_s = H_{t+s}$$

for all $t,s > 0$.

4.4.3. Consider the heat equation on $[0,\infty)\times\mathbf{R}$

$$\frac{\partial}{\partial t}u(x,t) = \frac{\partial^2}{\partial x^2}u(x,t)$$

without a boundary condition. Show that $u = 2t + x^2$ and $u(t,x) = e^{-q^2 t}e^{iqx}$, as well as constant functions, are solutions of this equation. Prove that the set of solutions is a vector space over the field of complex numbers $\mathbf{C}$.

4.4.4. Suppose that a square-integrable function $g(x)$ on $\mathbf{R}^n$ is supported in a cube $[-A,A]^n$ for some $A > 0$. Then we have the following representation:

$$g(x) = \sum_{m\in\mathbf{Z}^n} \left(\frac{1}{(2A)^n}\int_{\mathbf{R}^n} g(y)e^{-2\pi i\frac{m\cdot y}{2A}}\,dy\right)e^{2\pi i\frac{x\cdot m}{2A}}\chi_{[-A,A]^n},$$

where the series converges in L^2

4.4.5. This exercise provides an application of Fourier series in complex analysis. Let $z \in \mathbf{C}\setminus\mathbf{Z}$. Consider the function $h_z(x) = \cos(2\pi z x)$ defined on $[-\frac{1}{2},\frac{1}{2}]$ extended periodically on the entire line [notice $h_z(-\frac{1}{2}) = h_z(\frac{1}{2})$].
(a) Compute the Fourier coefficients of h_z.
(b) Obtain a Fourier series expansion of h_z noticing that it is a Lipschitz function.
(c) Plug in $x = 1/2$ to prove that

$$\cot(\pi z) = \frac{1}{\pi z} + \frac{1}{\pi}\sum_{m=1}^{\infty}\frac{2z}{z^2 - m^2}.$$

4.5 Applications to Number theory and Ergodic theory

In this section we discuss three applications of Fourier series techniques to number theory and ergodic theory.

4.5.1 Evaluation of the Riemann Zeta Function at even Natural numbers

Definition 4.5.1. We define the *Bernoulli polynomials* $\{B_k\}_{k=0}^{\infty}$ on $[0,1]$ recursively as follows:

$$B_0(x) = 1$$

$$B_k'(x) = kB_{k-1}(x)$$

for $k = 1, 2, \ldots$, and

$$\int_0^1 B_k(x)\,dx = 0.$$

In view of this definition we find the first few polynomials $B_1(x) = x - \frac{1}{2}$, $B_2(x) = x^2 - x + \frac{1}{6}$, $B_3(x) = x^3 - \frac{3}{2}x^2 + \frac{1}{2}x$, etc. Unlike orthogonal polynomials, the Bernoulli polynomials have the remarkable property that their number of zeros in the unit interval does not increase as the degree of the polynomials increases; in fact all Bernoulli polynomials have at most three zeros in $[0,1]$.

Notice that for $k \geq 2$ we have

$$B_k(1) - B_k(0) = \int_0^1 B_k'(x)\,dx = k\int_0^1 B_{k-1}(x)\,dx = 0,$$

thus we may think of these polynomials as functions on the circle $\mathbf{T}^1$. We extend the Bernoulli polynomials to the whole line periodically by setting $B_k(x+l) = B_k(x)$ for $x \in [0,1]$. We now compute the Fourier coefficients of B_k. We have

$$\widehat{B_1}(m) = \int_0^1 (t - \tfrac{1}{2})e^{-2\pi imt}\,dt = \begin{cases} 0 & \text{if } m = 0 \\ -\frac{1}{2\pi im} & \text{if } m \neq 0. \end{cases}$$

Therefore, using Corollary 3.4.10, we can write

$$B_1(x) = \sum_{m \neq 0} -\frac{1}{2\pi im} e^{2\pi imx}$$

where the series converges at every $x \in (0,1)$.

We have the following result concerning the Fourier expansion of the Bernoulli polynomials.

Theorem 4.5.2. *For each $k \geq 2$ we have*

$$B_k(x) = -k! \sum_{m\in\mathbf{Z}\setminus\{0\}} \frac{1}{(2\pi i m)^k} e^{2\pi i m x}, \tag{4.5.1}$$

where the series converges absolutely and uniformly on $[0,1]$. When $k = 1$ we have

$$B_1(x) = -\sum_{m\in\mathbf{Z}\setminus\{0\}} \frac{1}{2\pi i m} e^{2\pi i m x} \tag{4.5.2}$$

for all $x \in (0,1)$ and the series converges conditionally.

Proof. We have already proved (4.5.2) and we focus attention to the case $k \geq 2$. As a consequence of $B_k' = kB_{k-1}$ we obtain

$$B_k(x) = k\int_0^x B_{k-1}(t)\,dt + C_k\,.$$

Using the property that B_k has integral zero over $[0,1]$ we evaluate the constant C_k. We have

$$\begin{aligned} 0 &= \int_0^1 \left[k\int_0^x B_{k-1}(t)\,dt + C_k\right] dx \\ &= k\int_0^1 \left(\int_t^1 dx\right) B_{k-1}(t)\,dt + C_k \\ &= -k\int_0^1 tB_{k-1}(t)\,dt + C_k\,. \end{aligned}$$

Thus

$$C_k = k\int_0^1 tB_{k-1}(t)\,dt\,.$$

The Fourier series of $B_k(x)$ can be obtained by integrating the one for $B_{k-1}(x)$ for all $k \geq 2$ by induction via the identity

$$B_k(x) = k\int_0^x B_{k-1}(t)\,dt + k\int_0^1 tB_{k-1}(t)\,dt\,. \tag{4.5.3}$$

Indeed, assume that (4.5.1) holds for some $k \geq 2$. Then using (4.5.3) we obtain

$$\begin{aligned} B_k(x) &= \int_0^x \lim_{N\to\infty} \sum_{\substack{|m|\leq N\\ m\neq 0}} -k\frac{(k-1)!}{(2\pi i m)^{k-1}} e^{2\pi i m t}\,dt - \int_0^1 t \lim_{N\to\infty} \sum_{\substack{|m|\leq N\\ m\neq 0}} k\frac{(k-1)!}{(2\pi i m)^{k-1}} e^{2\pi i m t}\,dt \\ &= \lim_{N\to\infty} \int_0^x \sum_{\substack{|m|\leq N\\ m\neq 0}} -k\frac{(k-1)!}{(2\pi i m)^{k-1}} e^{2\pi i m t}\,dt - \lim_{N\to\infty} \int_0^1 t \sum_{\substack{|m|\leq N\\ m\neq 0}} k\frac{(k-1)!}{(2\pi i m)^{k-1}} e^{2\pi i m t}\,dt \end{aligned}$$

$$= -\lim_{N\to\infty}\sum_{\substack{|m|\le N\\ m\neq 0}}\frac{k!}{(2\pi i m)^{k-1}}\frac{e^{2\pi i m x}-1}{2\pi i m} - \lim_{N\to\infty}\sum_{\substack{|m|\le N\\ m\neq 0}}\frac{k!}{(2\pi i m)^{k-1}}\left[t\frac{e^{2\pi i m t}}{2\pi i m}\right]_{t=0}^{t=1}$$

$$= -k!\sum_{m\neq 0}\frac{e^{2\pi i m x}}{(2\pi i m)^k}.$$

Passing the limit from inside the integral to outside is allowed due to the uniform convergence of the series when $k \geq 3$. In the case $k = 2$, one may use Exercise 3.5.6 which says that for all $[a,b] \subseteq \mathbf{T}^1$ and g integrable functions over $[a,b]$ one has

$$\lim_{N\to\infty}\int_a^b (g * D_N)(t)\,dt = \int_a^b \lim_{N\to\infty}(g * D_N)(t)\,dt = \int_a^b g(t)\,dt.$$

This argument proves identity (4.5.1) for all $k \geq 2$ by induction and concludes the proof. □

We recall the following definition from number theory.

Definition 4.5.3. For $s > 1$ we define

$$\zeta(s) = \sum_{k=1}^{\infty}\frac{1}{k^s}$$

called the *Riemann zeta function.*

We use the Fourier expansions of the Bernoulli polynomials to obtain the values of the Riemann zeta function for integers. When k is an even integer, identity (4.5.1) can also be written as

$$B_k(x) = 2(-1)^{1+\frac{k}{2}}k!\sum_{n=1}^{\infty}\frac{\cos(2\pi n x)}{(2\pi n)^k}$$

and inserting $x = 0$ yields

$$\zeta(k) = \sum_{n=1}^{\infty}\frac{1}{n^k} = \frac{B_k(0)(2\pi)^k}{2(-1)^{1+\frac{k}{2}}k!}.$$

The polynomial $B_1(x) = x - 1/2$ has rational coefficients and thus so do all the B_k by a straightforward inductive argument that uses the identity (4.5.3). Thus $B_k(0)$ is a rational number for all $k \geq 1$. We conclude that

$$\zeta(2m) = \frac{B_{2m}(0)(2\pi)^{2m}}{2(-1)^{1+m}(2m)!} \tag{4.5.4}$$

which is a rational multiple of $(2\pi)^{2m}$, hence transcendental, since π is a transcendental number. We have therefore obtained the following.

Corollary 4.5.4. ***(Euler)*** *The value of the Riemann zeta function* $\zeta(2m)$, $m = 1,2,\dots$, *is equal to a rational multiple of* $(2\pi)^{2m}$; *hence it is a transcendental number.*

The corresponding statement for odd integers remains unresolved in general, as of this writing.

4.5.2 Equidistributed sequences

Here we discuss Weyl's theorem on equidistributed sequences.

Definition 4.5.5. A sequence $\{a_k\}_{k=0}^{\infty}$ with values in $\mathbf{T}^n$ is called *equidistributed* if for every cube Q in $\mathbf{T}^n$ we have

$$\lim_{N\to\infty} \frac{\#\{k:\ 0\le k\le N-1,\quad a_k\in Q\}}{N} = |Q|\,.$$

Theorem 4.5.6. *The following statements are equivalent:*
(a) The sequence $\{a_k\}_{k=0}^{\infty}$ *is equidistributed.*
(b) For every smooth function f *on* $\mathbf{T}^n$ *we have that*

$$\lim_{N\to\infty} \frac{1}{N}\sum_{k=0}^{N-1} f(a_k) = \int_{\mathbf{T}^n} f(x)\,dx\,.$$

(c) For every $m \in \mathbf{Z}^n \setminus \{0\}$ *we have*

$$\lim_{N\to\infty} \frac{1}{N}\sum_{k=0}^{N-1} e^{2\pi i m\cdot a_k} = 0\,.$$

Proof. We first prove the equivalence of (a) and (b). We begin by observing that (b) is a restatement of (a) if $f = \chi_Q$ and Q is a cube in $\mathbf{T}^n$. Thus, if (a) holds, then (b) holds for all step functions, i.e., finite linear combinations of characteristic functions of cubes. We prove that (a) implies (b) for smooth functions. Given a smooth function f on $\mathbf{T}^n$ and given $\varepsilon > 0$, by the uniform continuity of f, there is a step function $g = \sum_{j=1}^{m} c_j \chi_{Q_j}$ ($c_j \in \mathbf{C}$ and Q_j are cubes in $\mathbf{T}^n$) such that $\|f-g\|_{L^\infty} < \frac{\varepsilon}{3}$. Since g is a finite linear combination of step functions, there is an N_0 such that for $N \ge N_0$ we have

$$\left|\frac{1}{N}\sum_{k=0}^{N-1} g(a_k) - \int_{\mathbf{T}^n} g(x)\,dx\right| < \frac{\varepsilon}{3}\,.$$

Since

$$\left|\int_{\mathbf{T}^n} f(x)\,dx - \int_{\mathbf{T}^n} g(x)\,dx\right| \le \|f-g\|_{L^\infty} < \frac{\varepsilon}{3}$$

and

$$\left|\frac{1}{N}\sum_{k=0}^{N-1} g(a_k) - \frac{1}{N}\sum_{k=0}^{N-1} f(a_k)\right| \le \|f-g\|_{L^\infty} < \frac{\varepsilon}{3},$$

it follows that for $N \ge N_0$ we have

$$\left|\frac{1}{N}\sum_{k=0}^{N-1} f(a_k) - \int_{\mathbf{T}^n} f(x)\,dx\right| < \varepsilon,$$

thus (b) holds.

To prove that (b) implies (a) given a cube Q in $\mathbf{T}^n$ pick two smooth functions g and h such that

$$0 \le h \le \chi_Q \le g$$

and such that g is equal to 1 on Q and vanishes off $(1+\varepsilon)Q$ while h is equal to 1 on $(1-\varepsilon)Q$ and vanishes off Q. Observe that

$$|Q| - c_n\,\varepsilon \le \int_{\mathbf{T}^n} h(x)\,dx \le |Q| \le \int_{\mathbf{T}^n} g(x)\,dx \le |Q| + c_n\,\varepsilon.$$

for some $c_n > 0$. Since

$$\frac{1}{N}\sum_{k=0}^{N-1} h(a_k) \le \frac{1}{N}\sum_{k=0}^{N-1} \chi_Q(a_k) \le \frac{1}{N}\sum_{k=0}^{N-1} g(a_k),$$

the sandwich theorem implies that

$$|Q| - c_n\,\varepsilon \le \liminf_{N\to\infty} \frac{1}{N}\sum_{k=0}^{N-1} \chi_Q(a_k) \le \limsup_{N\to\infty} \frac{1}{N}\sum_{k=0}^{N-1} \chi_Q(a_k) \le |Q| + c_n\,\varepsilon.$$

Since $\varepsilon > 0$ was arbitrary the conclusion follows.

The implication $(b) \implies (c)$ is trivial.

We now prove that $(c) \implies (b)$.

Given a smooth function f on $\mathbf{T}^n$ we write

$$\frac{1}{N}\sum_{k=0}^{N-1} f(a_k) = \frac{1}{N}\sum_{k=0}^{N-1}\sum_{m\in\mathbf{Z}^n} \widehat{f}(m)e^{2\pi i m\cdot a_k} = \widehat{f}(0) + \sum_{m\in\mathbf{Z}^n\setminus\{0\}} \widehat{f}(m)\left(\frac{1}{N}\sum_{k=0}^{N-1} e^{2\pi i m\cdot a_k}\right).$$

Because of the rapid decay of the Fourier coefficients of f we can pass the limit as $N\to\infty$ inside the sum in m. It follows that

$$\lim_{N\to\infty}\frac{1}{N}\sum_{k=0}^{N-1} f(a_k) = \widehat{f}(0) = \int_{\mathbf{T}^n} f(x)\,dx.$$

□

Example 4.5.7. The sequence $\{k\sqrt{2}-[k\sqrt{2}]\}_{k=0}^{\infty}$ is equidistributed on $\mathbf{T}^1$. We check this by verifying condition (c) of Theorem 4.5.6. Indeed if $m \in \mathbf{Z}\setminus\{0\}$ then

$$\lim_{N\to\infty}\frac{1}{N}\sum_{k=0}^{N-1} e^{2\pi i m(k\sqrt{2}-[k\sqrt{2}])} = \lim_{N\to\infty}\frac{1}{N}\,\frac{e^{2\pi i N(m\sqrt{2})}-1}{e^{2\pi i(m\sqrt{2})}-1} = 0,$$

since $m\sqrt{2}$ is never a rational and thus the denominator never vanishes.

Naturally, the same conclusion is valid for any other irrational number in place of $\sqrt{2}$.

Example 4.5.8. We examine the sequence of the first digits of powers of 2. Consider the following sequence of numbers defined for $m = 1,2,\dots$

$$d_m = \text{first digit of } 2^m.$$

For instance we have $d_1 = 2, d_2 = 4, d_3 = 8, d_4 = 1, d_5 = 3, \dots$.

Fix an integer $k \in \{1,2,3,4,5,6,7,8,9\}$. We would like to find the frequency in which k appears as a first digit of 2^m, precisely, we would like to compute

$$\lim_{N\to\infty}\frac{\#\{m\in\{1,2,\dots,N\}:\ d_m = k\}}{N}.$$

The crucial observation is that the first digit of 2^m is equal to k if and only if there is a nonnegative integer s such that

$$k10^s \le 2^m < (k+1)\,10^s.$$

Taking logarithms with base 10 we obtain

$$s+\log_{10}(k) \le m\log_{10}2 < s+\log_{10}(k+1),$$

but since $0 \le \log_{10}(k)$ and $\log_{10}(k+1) \le 1$, taking fractional parts we obtain that

$$s = [m\log_{10}2]$$

and that

$$\log_{10}(k) \le m\log_{10}2 - [m\log_{10}2] < \log_{10}(k+1).$$

Since the number $\log_{10}2$ is irrational, it follows from Example 4.5.7 that the sequence

$$\{m\log_{10}2 - [m\log_{10}2]\}_{m=1}^{\infty}$$

is equidistributed in $[0,1)$. Using Definition 4.5.5 in dimension $n = 1$ with

$$Q = [a,b] = \big[\log_{10}(k),\log_{10}(k+1)\big]$$

we obtain that

$$\lim_{N\to\infty}\frac{\#\{m\in\{1,2,\dots,N\}:\ d_m=k\}}{N}=\log_{10}(k+1)-\log_{10}(k)=\log_{10}(1+\tfrac{1}{k}).$$

This gives the frequency in which k appears as first digit of 2^m. Notice that

$$\sum_{k=1}^{9}\log_{10}(1+\tfrac{1}{k})=\sum_{k=1}^{9}\big(\log_{10}(k+1)-\log_{10}(k)\big)=1,$$

as expected, and that the digit with the highest frequency that appears first in a term of the sequence $\{1,2,4,8,16,32,64,\dots\}$ is 1, while the one with the lowest frequency is 9.

4.5.3 The Number of Lattice Points inside a Ball

Points in $\mathbf{Z}^n$ are called *lattice points*. In this subsection we obtain the number of lattice points $N(R)$ inside a closed ball of radius R in $\mathbf{R}^n$ centered at the origin, precisely, we compute the asymptotic behavior of

$$N(R)=|\overline{B(0,R)}\cap\mathbf{Z}^n|$$

as $R\to\infty$. We denote by v_n the volume of the closed unit ball in $\mathbf{R}^n$. We have the following result.

Theorem 4.5.9. *Let $n\geq 2$. If $N(R)$ is the number of lattice points inside the closed ball of radius R centered at zero in $\mathbf{R}^n$, then we have that*

$$N(R)=v_nR^n+O(R^{n\frac{n-1}{n+1}}),$$

as $R\to\infty$.

Proof. Let B be the closed unit ball in $\mathbf{R}^n$ and χ_B its characteristic function. Using the result in Appendix B.5 we have $\widehat{\chi_B}(\xi)=\frac{J_{n/2}(2\pi|\xi|)}{|\xi|^{n/2}}$. Now in view of the behavior of the Bessel function given in Appendix B.6 for $|\xi|<\frac{1}{2\pi}$ we have $J_{n/2}(2\pi|\xi|)\leq C|\xi|^{\frac{n}{2}}$. Also for $|\xi|\geq\frac{1}{2\pi}$ we have $J_{n/2}(2\pi|\xi|)\leq C|\xi|^{-\frac{1}{2}}$, in view of the result in Appendix B.7. Consequently, there is a constant C_n such that for all $\xi\in\mathbf{R}^n$ we have

$$|\widehat{\chi_B}(\xi)|\leq C_n(1+|\xi|)^{-\frac{n+1}{2}}.$$

Fix a smooth nonnegative radial function ζ supported in $|x|\leq\frac{1}{2}$ with integral equal to 1 and define $\zeta_\varepsilon(x)=\frac{1}{\varepsilon^n}\zeta(\frac{x}{\varepsilon})$ for $\varepsilon>0$. For $0<\varepsilon<\frac{1}{10}$, define functions

$$\Phi^\varepsilon=\chi_{(1-\frac{\varepsilon}{2})B}*\zeta_\varepsilon$$
$$\Psi^\varepsilon=\chi_{(1+\frac{\varepsilon}{2})B}*\zeta_\varepsilon.$$

These functions are even, hence their Fourier transforms are real-valued. We observe that

$$\Phi^\varepsilon(x)=1 \text{ when } |x|\le 1-\varepsilon \quad\text{and}\quad \Phi^\varepsilon(x)=0 \text{ when } |x|\ge 1. \tag{4.5.5}$$

Indeed, we have

$$\begin{aligned}\Phi^\varepsilon(x)&=\int_{\mathbf{R}^n}\chi_{(1-\frac{\varepsilon}{2})B}(y)\zeta_\varepsilon(x-y)dy\\&=\int_{|y|\le 1-\frac{\varepsilon}{2}}\frac{1}{\varepsilon^n}\zeta(\tfrac{x-y}{\varepsilon})dy\\&=\int_{|y|\le\frac{1}{\varepsilon}-\frac12}\zeta(\tfrac{x}{\varepsilon}-y)dy.\end{aligned}$$

For $|x|\le 1-\varepsilon$, $|\frac{x}{\varepsilon}|\le\frac{1}{\varepsilon}-1$, so $|\frac{x}{\varepsilon}-t|\le\frac{1}{\varepsilon}-1+\frac12=\frac{1}{\varepsilon}-\frac12$ for $|t|\le\frac12$, which means

$$\Phi^\varepsilon(x)=\int_{|y|\le\frac{1}{\varepsilon}-\frac12}\zeta(\tfrac{x}{\varepsilon}-y)dy=\int_{|t|\le\frac12}\zeta(t)dt=1.$$

For $|x|\ge 1, |\frac{x}{\varepsilon}-y|\ge\frac{1}{\varepsilon}-\frac{1}{\varepsilon}+\frac12=\frac12$, so

$$\Phi^\varepsilon(x)=\int_{|y|\le\frac{1}{\varepsilon}-\frac12}\zeta(\tfrac{x}{\varepsilon}-y)dy=0,$$

proving (4.5.5). Likewise one can show that

$$\Psi^\varepsilon(x)=1 \text{ when } |x|\le 1 \quad\text{and}\quad \Psi^\varepsilon(x)=0 \text{ when } |x|\ge 1+\varepsilon. \tag{4.5.6}$$

Next we claim that

$$\big|\widehat{\Phi^\varepsilon}(\xi)\big|+\big|\widehat{\Psi^\varepsilon}(\xi)\big|\le C_{n,N}(1+|\xi|)^{-\frac{n+1}{2}}(1+\varepsilon|\xi|)^{-N} \tag{4.5.7}$$

for every $\xi\in\mathbf{R}^n$ and N a large positive number. Indeed to show (4.5.7) for Φ^ε we write

$$\begin{aligned}|\widehat{\Phi^\varepsilon}(\xi)|&=|\widehat{\chi_{(1-\frac{\varepsilon}{2})B}}(\xi)\widehat{\zeta}(\xi\varepsilon)|\\&\le(1-\tfrac{\varepsilon}{2})^nC_n(1+|\xi|(1-\tfrac{\varepsilon}{2}))^{-\frac{n+1}{2}}|\widehat{\zeta}(\xi\varepsilon)|\\&\le C_{n,N}(1+|\xi|)^{-\frac{n+1}{2}}(1+\varepsilon|\xi|)^{-N}\end{aligned} \tag{4.5.8}$$

since $\zeta\in\mathscr{S}(\mathbf{R})$ and $0<\varepsilon<\frac{1}{10}$. The proof for Ψ^ε is completely similar.

We now notice that for $R>0$, $m\in\mathbf{Z}^n\setminus\{0\}$, and $x\in[0,1]^n$ we have

$$1+|m+x|R\le 1+(\sqrt{n}+|m|)R\le 2\sqrt{n}(1+|m|R).$$

This implies that for $\varepsilon < 1/10$ we have

$$\begin{aligned}
&\sum_{m\in\mathbf{Z}^n\setminus\{0\}} R^n(1+R|m|)^{-\frac{n+1}{2}}(1+\varepsilon R|m|)^{-N} \\
&\qquad\le C'\int_{[0,1]^n}\sum_{m\in\mathbf{Z}^n\setminus\{0\}} R^n(1+R|m+x|)^{-\frac{n+1}{2}}(1+\varepsilon R|m+x|)^{-N}dx \\
&\qquad\le C'\int_{\mathbf{R}^n} R^n(1+R|x|)^{-\frac{n+1}{2}}(1+\varepsilon R|x|)^{-N}dx \\
&\qquad\le C''\varepsilon^{-\frac{n-1}{2}}, \qquad (4.5.9)
\end{aligned}$$

where the proof of (4.5.9) is easily deduced by considering the cases (a) $|x| \le R^{-1}$ which yields a constant, (b) $R^{-1} \le |x| \le (R\varepsilon)^{-1}$ which yields a constant multiple of $\varepsilon^{-\frac{n-1}{2}}$, and (c) $(R\varepsilon)^{-1} \le |x|$ which also produces a constant multiple of $\varepsilon^{-\frac{n-1}{2}}$ if we pick $N > \frac{n-1}{2}$.

Using (4.5.5) and the Poisson summation formula we write

$$\begin{aligned}
\sum_{m\in\mathbf{Z}^n}\chi_B(\tfrac{m}{R}) &\ge \sum_{m\in\mathbf{Z}^n}\Phi^\varepsilon(\tfrac{m}{R}) \\
&= R^n\widehat{\Phi^\varepsilon}(0)+\sum_{m\in\mathbf{Z}^n\setminus\{0\}} R^n\widehat{\Phi^\varepsilon}(Rm) \\
&\ge v_nR^n(1-\varepsilon)^n - C_{n,N}\sum_{m\in\mathbf{Z}^n\setminus\{0\}} R^n(1+R|m|)^{-\frac{n+1}{2}}(1+\varepsilon R|m|)^{-N} \\
&\ge v_nR^n - nv_nR^n\varepsilon - C'_{n,N}\varepsilon^{-\frac{n-1}{2}},
\end{aligned}$$

where we used that $(1-\varepsilon)^n \ge 1-n\varepsilon$, (4.5.8), and (4.5.9). Now pick ε such that $\varepsilon R^n = \varepsilon^{-\frac{n-1}{2}}$, or equivalently $\varepsilon = R^{-\frac{2n}{n+1}}$ to deduce the estimate

$$N(R) \ge v_nR^n - O(R^{n\frac{n-1}{n+1}})$$

as $R \to \infty$.

Finally, making use of (4.5.6), and via a similar argument we write

$$\begin{aligned}
\sum_{m\in\mathbf{Z}^n}\chi_B(\tfrac{m}{R}) &\le \sum_{m\in\mathbf{Z}^n}\Psi^\varepsilon(\tfrac{m}{R}) = R^n\widehat{\Psi^\varepsilon}(0)+\sum_{m\in\mathbf{Z}^n\setminus\{0\}} R^n\widehat{\Psi^\varepsilon}(Rm) \\
&\le v_nR^n(1+\varepsilon)^n + C_{n,N}\sum_{m\in\mathbf{Z}^n\setminus\{0\}} R^n(1+R|m|)^{-\frac{n+1}{2}}(1+\varepsilon R|m|)^{-N} \\
&\le v_nR^n + v_n2^nR^n\varepsilon + C_{n,N}\varepsilon^{-\frac{n-1}{2}}.
\end{aligned}$$

The same choice of $\varepsilon = R^{-\frac{2n}{n+1}}$, yields the upper estimate for $N(R)$.

Combining the upper and lower estimates for $N(R)$ we obtain

$$N(R) = v_n R^n + O(R^{n\frac{n-1}{n+1}}),$$

as $R \to \infty$. $\square$

Exercise 4.5.8 contains an application of Theorem 4.5.9.

Exercises

4.5.1. Prove that for all $x \in [0,1]$ we have

$$\sum_{j=1}^{\infty} \frac{\sin(2\pi jx)}{j^{2m+1}} = \frac{(-1)^{m+1}}{2} \frac{(2\pi)^{2m+1}}{(2m+1)!} B_{2m+1}(x).$$

4.5.2. Show that for all $z \in \mathbf{C}$ with $|z| < 1$ we have

$$\pi z \cot(\pi z) = 1 - 2\sum_{k=0}^{\infty} z^{2k+2} \zeta(2k+2).$$

[*Hint:* Use the result of Exercise 4.4.5.]

4.5.3. Suppose that a point $x = (x_1, \dots, x_n) \in [0,1]^n$ has the property that $m \cdot x$ is irrational for all $m \in \mathbf{Z}^n \setminus \{0\}$. Show that the sequence $\{(kx_1 - [kx_1], \dots, kx_n - [kx_n])\}_{k=0}^{\infty}$ is equidistributed in $\mathbf{T}^n$.

4.5.4. ([191]) Let $N(x,R)$ be the number of lattice points inside the closed ball of radius $R > 0$ centered at $x \in \mathbf{R}^n$. Show that

$$\int_{\mathbf{T}^n} \left| N(x,R) - v_n R^n \right|^2 dx = O(R^{n-1})$$

as $R \to \infty$, where v_n is the volume of the unit ball on $\mathbf{R}^n$.

4.5.5. (*Minkowski*) Let S be an open convex symmetric set in $\mathbf{R}^n$ and assume that the Fourier transform of its characteristic function satisfies the decay estimate

$$|\widehat{\chi_S}(\xi)| \le C(1+|\xi|)^{-\frac{n+1}{2}}.$$

(This is the case if the boundary of S has nonzero Gaussian curvature.) Assume that $|S| > 2^n$. Prove that S contains at least one lattice point other than the origin.
[*Hint:* Assume the contrary, set $f = \chi_{\frac{1}{2}S} * \chi_{\frac{1}{2}S}$, and apply the Poisson summation formula to f to prove that $f(0) \ge \widehat{f}(0)$.]

4.5.6. For $t \in [0,\infty)$ let

$$N(t) = \#\{m \in \mathbf{Z}^n : \ |m| \le t\}.$$

Let $0 = r_0 < r_1 < r_2 < \cdots$ be the sequence all of numbers r for which there exist $m \in \mathbf{Z}^n$ such that $|m| = r$.
(a) Observe that N is right continuous and constant on intervals of the form $[r_j, r_{j+1})$.
(b) Show that the distributional derivative of N is the measure

$$\mu(t) = \#\{m \in \mathbf{Z}^n : \ |m| = t\},$$

defined via the identity $\langle \mu, \varphi \rangle = \sum_{j=0}^{\infty} \#\{m \in \mathbf{Z}^n : \ |m| = r_j\}\, \varphi(r_j)$.

4.5.7. Let $f \in \mathscr{C}^1((0,\infty))$, and let $0 < a < b < \infty$. Derive the identity

$$\sum_{\substack{m \in \mathbf{Z}^n \\ a<|m|\le b}} f(|m|) = \int_a^b f(t)\, dN(t) = f(b)N(b) - f(a)N(a) - \int_a^b f'(x)N(x)\, dx,$$

where N is defined in Exercise 4.5.6 and $\int_a^b f(t)\, dN(t)$ is the Riemann-Stieltjes integral of f with respect to N.

4.5.8. Let $n \in \mathbf{Z}^+$ and $0 < \lambda < \infty$.
(a) Prove that for $k \in \mathbf{Z}^+ \cup \{0\}$ we have

$$\sum_{\substack{m \in \mathbf{Z}^n \\ a<|m|\le b}} \frac{e^{i|m|}}{|m|^{\lambda}} = \frac{-i\,\omega_{n-1}\, e^{ib}}{b^{\lambda-(n-1)}} - \frac{-i\,\omega_{n-1}\, e^{ia}}{a^{\lambda-(n-1)}} + O(a^{-\lambda+(n-1)-\frac{n-1}{n+1}})$$

for all $0 < a < b < \infty$ with $b - a \le 1$, where ω_{n-1} is the volume of $\mathbf{S}^{n-1}$.
(b) Show that when $\lambda > n - \frac{n-1}{n+1}$, the limit

$$\lim_{R\to\infty} \sum_{\substack{m \in \mathbf{Z}^n \setminus \{0\} \\ |m|\le R}} \frac{e^{i|m|}}{|m|^{\lambda}}$$

exists.
(c) Prove, however, that when $n - 1 - \frac{n-1}{n+1} < \lambda \le n-1$, the limit in part (b) does not exist.
[*Hint:* Use Exercise 4.5.7 and Theorem 4.5.9. Part (b): For $R > 1$ use the identity

$$\sum_{\substack{m \in \mathbf{Z}^n \setminus \{0\} \\ |m|\le R}} \frac{e^{i|m|}}{|m|^{\lambda}} = \sum_{k=0}^{[R-\frac{1}{2}]-1} \Bigg(\sum_{\substack{m \in \mathbf{Z}^n \\ k+\frac{1}{2}<|m|\le k+\frac{3}{2}}} \frac{e^{i|m|}}{|m|^{\lambda}} \Bigg) + \Bigg\{ \sum_{\substack{m \in \mathbf{Z}^n \\ [R-\frac{1}{2}]+\frac{1}{2}<|m|\le R}} \frac{e^{i|m|}}{|m|^{\lambda}} \Bigg\}.$$

Notice that the main term in the first sum on the right is telescoping. Part (c): Show that $\lim_{R\to\infty} \sum_{\substack{m \in \mathbf{Z}^n \setminus \{0\} \\ R<|m|\le R+1}} \frac{e^{i|m|}}{|m|^{\lambda}}$ does not tend to zero.]

HISTORICAL NOTES

The boundedness of the conjugate function on the circle (Theorem 4.1.7) and, hence, the L^p convergence of one-dimensional Fourier series was announced by Riesz in [292], but its proof appeared a little later in [293]. The proof of Theorem 4.1.7 in the text is attributed to S. Bochner. Luzin's conjecture [235] on almost everywhere convergence of the Fourier series of continuous functions was announced in 1913 and settled by Carleson [54] in 1965 for the more general class of square summable functions (Theorem 4.3.14). Carleson's theorem was later extended by Hunt [165] for the class of L^p functions for all $1 < p < \infty$ (Theorem 4.3.15). Sjölin [325] sharpened this result by showing that the Fourier series of functions f with $|f|(\log^+|f|)(\log^+\log^+|f|)$ integrable over $\mathbf{T}^1$ converge almost everywhere. Antonov [5] improved Sjölin's result by extending it to functions f with $|f|(\log^+|f|)(\log^+\log^+\log^+|f|)$ integrable over $\mathbf{T}^1$. One should also consult the related results of Soria [330] and Arias de Reyna [9]. The book [10] of Arias de Reyna contains a historically motivated comprehensive study of topics related to the Carleson–Hunt theorem. Counterexamples due to Konyagin [200] show that Fourier series of functions f with $|f|(\log^+|f|)^{\frac{1}{2}}(\log^+\log^+|f|)^{-\frac{1}{2}-\varepsilon}$ integrable over $\mathbf{T}^1$ may diverge when $\varepsilon > 0$. Examples of continuous functions whose Fourier series diverge exactly on given sets of measure zero are given in Katznelson [189] and Kahane and Katznelson [183].

The extension of the Carleson–Hunt theorem to higher dimensions for square summability of Fourier series (Theorem 4.3.16) is a rather straightforward consequence of the one-dimensional result and was independently obtained by Fefferman [112], Sjölin [325], and Tevzadze [359]. An example showing that the circular partial sums of a Fourier series may not converge in $L^p(\mathbf{T}^n)$ for $n \geq 2$ and $p \neq 2$ was obtained by Fefferman [113]. This example also shows that there exist L^p functions on $\mathbf{T}^n$ for $n \geq 2$ whose circular partial sums do not converge almost everywhere when $1 \leq p < 2$. Indeed, if the opposite happened, then the maximal operator $f \to \sup_{N\geq 0} |\widetilde{D}(n,N) * f|$ would have to be finite a.e. for all $f \in L^p(\mathbf{T}^n)$, and by Stein's theorem [335] it would have to be of weak type (p,p) for some $1 < p < 2$. But this would contradict Fefferman's counterexample on L^{p_1} for some $p < p_1 < 2$. On the other hand, almost everywhere is valid for the square partial sums of functions f with $|f|(\log^+|f|)^n(\log^+\log^+\log^+|f|)$ integrable over $\mathbf{T}^n$, as shown by Antonov [6]; see also Sjölin and Soria [327].

The development of the complex methods in the study of Fourier series was pioneered by the Russian school, especially Luzin and his students Kolmogorov, Menshov, and Privalov. The existence of an integrable function on $\mathbf{T}^1$ whose Fourier series diverges almost everywhere (Theorem 4.2.1) is due to Kolmogorov [195]. An example of an integrable function whose Fourier series diverges everywhere was also produced by Kolmogorov [198] three years later. Localization of the Bochner–Riesz means at the critical exponent $\alpha = \frac{n-1}{2}$ fails for L^1 functions on $\mathbf{T}^n$ (see Bochner [30]) but holds for functions f such that $|f|\log^+|f|$ is integrable over $\mathbf{T}^n$ (see Stein [333]). The latter article also contains the L^p boundedness of the maximal Bochner–Riesz operator $\sup_{R>0}|B_R^\alpha(f)|$ for $1<p<\infty$ when $\alpha > |\frac{1}{p}-\frac{1}{2}|$. Proposition 4.1.9 is due to Stein [331] and Theorem 4.2.5 is also due to Stein [335]. The technique that involves the points for which the set $\{|x-m| : m \in \mathbf{Z}^n\}$ is linearly independent over the rationals was introduced by Bochner [30].

Transference of regulated multipliers originated in the article of de Leeuw [94]. The methods of transference in Section 4.3 were beautifully placed into the framework of a general theory by Coifman and Weiss [70]. The key Lemma 4.3.8 is attributed to G. Weiss. Transference of maximal multipliers (Theorem 4.3.12) was first obtained by Kenig and Tomas [192] and later elaborated by Asmar, Berkson, and Gillespie [12], [13].

Paraphrasing Pappus of Alexandria, *bees know than a hexagon will hold more honey than a triangle or square of the same length, but people claim a greater share of wisdom knowing that the circle of a given length holds the maximum area among all geometric shapes of equal perimeter.* This reflection captures the isoperimetric inequality, which was first recorded by Pappus in the fourth century A.D. and was credited it to Zenodorus (second century B.C.). Archimedes also studied the problem, but his work on the subject, like the original writings of Zenodorus, has been lost. Rigorous modern-day proofs of this inequality can be traced to J. Steiner, K. Weierstrass, and

F. Edler, whose methods are based in geometry and calculus. The proof in the text is due to A. Hurwitz. On the history of the isoperimetric inequality see [322].

The mean square error for lattice points (Exercise 4.5.4) is due to Kendall [191] while the more delicate pointwise asymptotic formula of Theorem 4.5.9 was obtained by Landau [212]. Using Landau's formula Pinsky, Stanton, and Trapa [284] showed that the spherical partial sums of the Fourier series of the characteristic function of a sufficiently small ball in $\mathbf{T}^n$ converge at the center of the ball if and only if the dimension n is strictly less than three; this property is valid for the characteristic function of any ball as shown in Pinsky [283].

Chapter 5
Singular Integrals of Convolution Type

The topic of singular integrals is motivated by its intimate connection with some of the most important problems in Fourier analysis, such as that of the convergence of Fourier series. As we have seen, the L^p boundedness of the conjugate function on the circle is equivalent to the L^p convergence of Fourier series of L^p functions. And since the Hilbert transform on the real line provides an analogue of the conjugate function on the circle, it is deeply connected with the L^p convergence of Fourier integrals. It also appears in the theory of harmonic functions on the upper half space and has so many remarkable properties that deserve a careful investigation. The Hilbert transform is the prototype of all singular integrals and provides inspiration for subsequent development of the subject.

Historically, the theory of the Hilbert transform depended on techniques of complex analysis. With the development of the Calderón–Zygmund school, and the extension of one-dimensional theory to higher dimensions, real-variable methods slowly replaced complex analysis. The higher-dimensional framework proved to be flexible enough for generalizations and led to the introduction of singular integrals in other areas of mathematics. Singular integrals are nowadays intimately connected with partial differential equations, operator theory, several complex variables, and other fields. In this chapter we study singular integrals given by convolution with tempered distributions. We call such operators singular integrals of convolution type.

5.1 The Hilbert Transform and the Riesz Transforms

We begin the investigation of singular integrals with a careful study of the Hilbert transform which provides inspiration for the subsequent development of the theory.

L. Grafakos, *Classical Fourier Analysis*, Graduate Texts in Mathematics 249,
DOI 10.1007/978-1-4939-1194-3_5, © Springer Science+Business Media New York 2014

5.1.1 Definition and Basic Properties of the Hilbert Transform

There are several equivalent ways to introduce the Hilbert transform; in this exposition we first define it as a convolution operator with a certain principal value distribution, but we later discuss other equivalent definitions.

We begin by defining a distribution W_0 in $\mathscr{S}'(\mathbf{R})$ as follows:

$$\langle W_0, \varphi \rangle = \frac{1}{\pi} \lim_{\varepsilon \to 0} \int_{\varepsilon \le |x| \le 1} \frac{\varphi(x)}{x}\, dx + \frac{1}{\pi} \int_{|x| \ge 1} \frac{\varphi(x)}{x}\, dx, \tag{5.1.1}$$

for φ in $\mathscr{S}(\mathbf{R})$. The function $1/x$ integrated over $[-1,-\varepsilon] \bigcup [\varepsilon, 1]$ has mean value zero, and we may replace $\varphi(x)$ by $\varphi(x) - \varphi(0)$ in the first integral in (5.1.1). Since $(\varphi(x) - \varphi(0))x^{-1}$ is controlled by $\|\varphi'\|_{L^\infty}$, it follows that the limit in (5.1.1) exists. To see that W_0 is indeed in $\mathscr{S}'(\mathbf{R})$, we note that the estimate

$$\big|\langle W_0, \varphi \rangle\big| \le \frac{2}{\pi} \big\|\varphi'\big\|_{L^\infty} + \frac{2}{\pi} \sup_{x \in \mathbf{R}} |x\varphi(x)| \tag{5.1.2}$$

is valid. This says that $W_0 \in \mathscr{S}'(\mathbf{R})$.

Definition 5.1.1. The *truncated Hilbert transform* (at height ε) of a function f in $L^p(\mathbf{R})$, $1 \le p < \infty$, is defined by

$$H^{(\varepsilon)}(f)(x) = \frac{1}{\pi} \int_{|y| \ge \varepsilon} \frac{f(x-y)}{y}\, dy = \frac{1}{\pi} \int_{|x-y| \ge \varepsilon} \frac{f(y)}{x-y}\, dy. \tag{5.1.3}$$

The *Hilbert transform* of $\varphi \in \mathscr{S}(\mathbf{R})$ is defined by

$$H(\varphi)(x) = (W_0 * \varphi)(x) = \lim_{\varepsilon \to 0} H^{(\varepsilon)}(\varphi)(x). \tag{5.1.4}$$

Observe that $H^{(\varepsilon)}(f)$ is well defined for all $f \in L^p$, $1 \le p < \infty$. This follows from Hölder's inequality, since $1/x$ is integrable to the power p' on the set $|x| \ge \varepsilon$.

For Schwartz functions φ, the integral

$$\int_{-\infty}^{+\infty} \frac{\varphi(x-y)}{y}\, dy$$

may not converge absolutely for any real number x, but is defined as a limit of the absolutely convergent integrals

$$\int_{|y| \ge \varepsilon} \frac{\varphi(x-y)}{y}\, dy,$$

as $\varepsilon \to 0$. Such limits are called *principal value integrals* and are denoted by the letters p.v. Using this notation, the Hilbert transform of a Schwartz function φ is

$$H(\varphi)(x) = \frac{1}{\pi} \text{p.v.} \int_{-\infty}^{+\infty} \frac{\varphi(x-y)}{y}\, dy = \frac{1}{\pi} \text{p.v.} \int_{-\infty}^{+\infty} \frac{\varphi(y)}{x-y}\, dy. \tag{5.1.5}$$

Remark 5.1.2. We extend the definition of the Hilbert transform to a bigger class of functions. Suppose that f is an integrable function on $\mathbf{R}$ that satisfies a Hölder condition near every point x; that is, for any $x \in \mathbf{R}$ there are $C_x > 0$ and $\varepsilon_x > 0$ such that

$$|f(x) - f(y)| \le C_x |x-y|^{\varepsilon_x}$$

whenever $|y-x| < \delta_x$. Then we write

$$\begin{aligned} H^{(\varepsilon)}(f)(x) &= \frac{1}{\pi} \int_{\varepsilon<|x-y|<\delta_x} \frac{f(y)}{x-y}\,dy + \frac{1}{\pi} \int_{|x-y|\ge\delta_x} \frac{f(y)}{x-y}\,dy \\ &= \frac{1}{\pi} \int_{\varepsilon<|x-y|<\delta_x} \frac{f(y)-f(x)}{x-y}\,dy + \frac{1}{\pi} \int_{|x-y|\ge\delta_x} \frac{f(y)}{x-y}\,dy. \end{aligned}$$

Both integrals converge absolutely; hence the limit of $H^{(\varepsilon)}(f)(x)$ exists as $\varepsilon \to 0$.

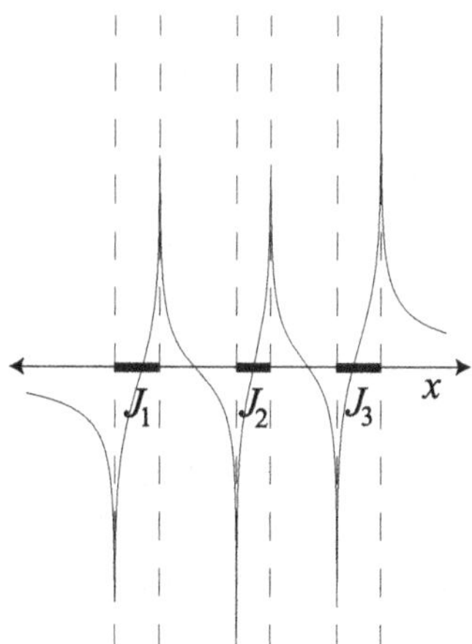

Fig. 5.1 The graph of the function $H(\chi_E)$ when E is a union of three disjoint intervals $J_1 \cup J_2 \cup J_3$.

Example 5.1.3. For the characteristic function $\chi_{[a,b]}$ of an interval $[a,b]$ we show that

$$H(\chi_{[a,b]})(x) = \frac{1}{\pi} \log \frac{|x-a|}{|x-b|}. \tag{5.1.6}$$

Let us verify this identity. Pick $\varepsilon < \min(|x-a|, |x-b|)$. To show (5.1.6) consider the three cases $0 < x-b$, $x-a < 0$, and $x-b < 0 < x-a$. In the first two cases, (5.1.6) follows immediately. In the third case we have

$$H(\chi_{[a,b]})(x) = \frac{1}{\pi} \lim_{\varepsilon\to 0} \left(\log \frac{|x-a|}{\varepsilon} + \log \frac{\varepsilon}{|x-b|} \right), \tag{5.1.7}$$

which yields (5.1.6). Observe that the cancellation of ε in (5.1.7) reflects the fact that $1/x$ has integral zero on symmetric intervals $\varepsilon < |x| < c$. Note that $H(\chi_{[a,b]})(x)$ blows up logarithmically in x near the points a and b and decays like $|x|^{-1}$ as $x \to \infty$. See Figure 5.1.

Example 5.1.4. Let $\log^+ x = \log x$ when $x \geq 1$ and zero otherwise. Observe that the calculation in the previous example actually gives

$$H^{(\varepsilon)}(\chi_{[a,b]})(x) = \begin{cases} \dfrac{1}{\pi}\log^+ \dfrac{|x-a|}{\max(\varepsilon, |x-b|)} & \text{when } x > b, \\[2ex] -\dfrac{1}{\pi}\log^+ \dfrac{|x-b|}{\max(\varepsilon, |x-a|)} & \text{when } x < a, \\[2ex] \dfrac{1}{\pi}\log^+ \dfrac{|x-a|}{\varepsilon} - \dfrac{1}{\pi}\log^+ \dfrac{|x-b|}{\varepsilon} & \text{when } a < x < b. \end{cases}$$

We now give an alternative characterization of the Hilbert transform using the Fourier transform. To achieve this we need to compute the Fourier transform of the distribution W_0 defined in (5.1.1). Fix a Schwartz function φ on $\mathbf{R}$. Then

$$\begin{aligned} \langle \widehat{W_0}, \varphi \rangle &= \langle W_0, \widehat{\varphi} \rangle \qquad (5.1.8) \\ &= \frac{1}{\pi} \lim_{\varepsilon \to 0} \int_{|\xi| \geq \varepsilon} \widehat{\varphi}(\xi) \frac{d\xi}{\xi} \\ &= \frac{1}{\pi} \lim_{\varepsilon \to 0} \int_{\frac{1}{\varepsilon} \geq |\xi| \geq \varepsilon} \int_{\mathbf{R}} \varphi(x) e^{-2\pi i x \xi} \, dx \frac{d\xi}{\xi} \\ &= \lim_{\varepsilon \to 0} \int_{\mathbf{R}} \varphi(x) \left[\frac{1}{\pi} \int_{\frac{1}{\varepsilon} \geq |\xi| \geq \varepsilon} e^{-2\pi i x \xi} \frac{d\xi}{\xi} \right] dx \\ &= \lim_{\varepsilon \to 0} \int_{\mathbf{R}} \varphi(x) \left[\frac{-i}{\pi} \int_{\frac{1}{\varepsilon} \geq |\xi| \geq \varepsilon} \sin(2\pi x \xi) \frac{d\xi}{\xi} \right] dx \\ &= \lim_{\varepsilon \to 0} \int_{\mathbf{R}} \varphi(x) \left[\left(\frac{-i}{\pi} \operatorname{sgn} x \right) \int_{\frac{1}{2\pi\varepsilon} \geq |\xi| \geq \frac{\varepsilon}{2\pi}} \sin(|x|\xi) \frac{d\xi}{\xi} \right] dx. \qquad (5.1.9) \end{aligned}$$

Here we used the signum function

$$\operatorname{sgn} x = \begin{cases} +1 & \text{when } x > 0, \\ 0 & \text{when } x = 0, \\ -1 & \text{when } x < 0. \end{cases} \qquad (5.1.10)$$

Using the results (a) and (b) in Exercise 5.1.1 we obtain that the integrals inside the square brackets in (5.1.9) are uniformly bounded by 8 and converge to $2\frac{\pi}{2} = \pi$ as $\varepsilon \to 0$, whenever $x \neq 0$. These observations make possible the use of the Lebesgue dominated convergence theorem that allows the passage of the limit inside the integral in (5.1.9). We obtain that

$$\langle \widehat{W_0}, \varphi \rangle = \int_{\mathbf{R}} \varphi(x)(-i \operatorname{sgn}(x)) \, dx. \qquad (5.1.11)$$

This implies that

$$\widehat{W_0}(\xi) = -i\,\mathrm{sgn}\,\xi\,. \tag{5.1.12}$$

In particular, identity (5.1.12) says that $\widehat{W_0}$ is a (bounded) function.

We now use identity (5.1.12) to write

$$H(f)(x) = \big(\widehat{f}(\xi)(-i\,\mathrm{sgn}\,\xi)\big)^{\vee}(x)\,. \tag{5.1.13}$$

This formula can be used to give an alternative definition of the Hilbert transform. An immediate consequence of (5.1.13) is that

$$\big\|H(f)\big\|_{L^2} = \big\|f\big\|_{L^2}\,, \tag{5.1.14}$$

that is, H is an isometry on $L^2(\mathbf{R})$. Moreover, H satisfies

$$H^2 = HH = -I\,, \tag{5.1.15}$$

where I is the identity operator. Equation (5.1.15) is a simple consequence of the fact that $(-i\,\mathrm{sgn}\,\xi)^2 = -1$. The adjoint operator H^* of H is uniquely defined via the identity

$$\big\langle f\,|\,H(g)\big\rangle = \int_{\mathbf{R}} f\,\overline{H(g)}\,dx = \int_{\mathbf{R}} H^*(f)\,\overline{g}\,dx = \big\langle H^*(f)\,|\,g\big\rangle\,,$$

and we can easily obtain that H^* has multiplier $\overline{-i\,\mathrm{sgn}\,\xi} = i\,\mathrm{sgn}\,\xi$. We conclude that $H^* = -H$. Likewise, we obtain $H^t = -H$.

5.1.2 Connections with Analytic Functions

We now investigate connections of the Hilbert transform with the Poisson kernel. Recall the definition of the Poisson kernel P_y given in Example 1.2.17. Then for a real-valued function f in $L^p(\mathbf{R})$, $1 \le p < \infty$, we have

$$(P_y * f)(x) = \frac{y}{\pi}\int_{-\infty}^{+\infty} \frac{f(t)}{(x-t)^2+y^2}\,dt\,, \tag{5.1.16}$$

and the integral in (5.1.16) converges absolutely by Hölder's inequality, since the function $t \mapsto ((x-t)^2+y^2)^{-1}$ is in $L^{p'}(\mathbf{R})$ whenever $y > 0$.

Let $\mathrm{Re}\,z$ and $\mathrm{Im}\,z$ denote the real and imaginary parts of a complex number z. Observe that

$$(P_y * f)(x) = \mathrm{Re}\left(\frac{i}{\pi}\int_{-\infty}^{+\infty} \frac{f(t)}{x-t+iy}\,dt\right) = \mathrm{Re}\left(\frac{i}{\pi}\int_{-\infty}^{+\infty} \frac{f(t)}{z-t}\,dt\right),$$

where $z = x + iy$. The function

$$F_f(z) = \frac{i}{\pi} \int_{-\infty}^{+\infty} \frac{f(t)}{z-t}\,dt$$

defined on

$$\mathbf{R}^2_+ = \{z = x + iy : \, y > 0\}$$

is analytic, since its $\partial/\partial\bar{z}$ derivative is zero. The real part of $F_f(x+iy)$ is $(P_y * f)(x)$. The imaginary part of $F_f(x+iy)$ is

$$\operatorname{Im}\left(\frac{i}{\pi} \int_{-\infty}^{+\infty} \frac{f(t)}{x-t+iy}\,dt\right) = \frac{1}{\pi} \int_{-\infty}^{+\infty} \frac{f(t)(x-t)}{(x-t)^2+y^2}\,dt = (f * Q_y)(x),$$

where Q_y is called the *conjugate Poisson kernel* and is given by

$$Q_y(x) = \frac{1}{\pi} \frac{x}{x^2+y^2}. \tag{5.1.17}$$

The function $u_f + iv_f$ is analytic and thus $u_f(x+iy) = (f * P_y)(x)$ and $v_f(x+iy) = (f * Q_y)(x)$ are *conjugate harmonic functions*. Since the family P_y, $y > 0$, is an approximate identity, it follows from Theorem 1.2.19 that $P_y * f \to f$ in $L^p(\mathbf{R})$ as $y \to 0$. The following question therefore arises: What is the limit of $f * Q_y$ as $y \to 0$? The next result addresses this question.

Theorem 5.1.5. *Let $1 \le p < \infty$. For any $f \in L^p(\mathbf{R})$ we have*

$$f * Q_\varepsilon - H^{(\varepsilon)}(f) \to 0 \tag{5.1.18}$$

in L^p and almost everywhere as $\varepsilon \to 0$. Moreover, for φ in $\mathscr{S}(\mathbf{R})$ we have

$$F_\varphi(x+iy) = \frac{i}{\pi} \int_{-\infty}^{+\infty} \frac{\varphi(t)}{x+iy-t}\,dt \to \varphi(x) + iH(\varphi)(x) \tag{5.1.19}$$

as $y \to 0+$ for all $x \in \mathbf{R}$.

Proof. We see that

$$(Q_\varepsilon * f)(x) - \frac{1}{\pi} \int_{|t|\ge\varepsilon} \frac{f(x-t)}{t}\,dt = \frac{1}{\pi}(f * \psi_\varepsilon)(x),$$

where $\psi_\varepsilon(x) = \varepsilon^{-1}\psi(\varepsilon^{-1}x)$ and

$$\psi(t) = \begin{cases} \frac{t}{t^2+1} - \frac{1}{t} & \text{when } |t| \ge 1, \\ \frac{t}{t^2+1} & \text{when } |t| < 1. \end{cases} \tag{5.1.20}$$

Note that ψ is integrable over the line and has integral zero. Furthermore, the integrable function

$$\Psi(t) = \begin{cases} \frac{1}{t^2+1} & \text{when } |t| \geq 1, \\ 1 & \text{when } |t| < 1, \end{cases} \tag{5.1.21}$$

is a radially decreasing majorant of ψ, i.e., it is even, decreasing on $[0,\infty)$, and satisfies $|\psi| \leq \Psi$. It follows from Theorem 1.2.21 (with $a = 0$) that $f * \psi_\varepsilon \to 0$ in L^p. Also Corollary 2.1.19 (with $a = 0$) implies that $f * \psi_\varepsilon \to 0$ almost everywhere as $\varepsilon \to 0$.

Assertion (5.1.19) is a consequence of (5.1.18), the discussion preceding Theorem 5.1.5, and the observation that $H^{(\varepsilon)}(\varphi)$ converges to $H(\varphi)$ pointwise everywhere as $\varepsilon \to 0$. □

Remark 5.1.6. We will show later that for $f \in L^p(\mathbf{R})$, $1 \leq p < \infty$, the expressions $H^{(\varepsilon)}(f)$ converge a.e. (and also in L^p when $p > 1$) to a function $\widetilde{H}(f)$. This will be a consequence of Theorem 5.1.12 (or Corollary 5.3.6 when $p = 1$), combined with Theorem 2.1.14 and the observation that for Schwartz functions φ, $H^{(\varepsilon)}(\varphi)$ converge to $H(\varphi)$ as $\varepsilon \to 0$. The linear operator $\widetilde{H}$ defined in this way extends the Hilbert transform H initially defined on Schwartz functions and will still be denoted by H. Thus for $f \in L^p(\mathbf{R})$, $1 \leq p < \infty$, one has

$$\lim_{\varepsilon \to 0} f * Q_\varepsilon = H(f) \quad \text{a.e.}$$

This convergence is also valid in L^p in view of the preceding observations and Theorem 5.1.5.

5.1.3 L^p Boundedness of the Hilbert Transform

As a consequence of the result in Exercise 5.1.4 and of the fact that

$$x \leq \tfrac{1}{2}(e^x - e^{-x}), \qquad x \geq 0,$$

we obtain that

$$|\{x : |H(\chi_E)(x)| > \alpha\}| \leq \frac{2}{\pi}\frac{|E|}{\alpha}, \qquad \alpha > 0, \tag{5.1.22}$$

for all subsets E of the real line of finite measure. Theorem 1.4.19 with $p_0 = q_0 = 1$ and $p_1 = q_1 = 2$ now implies that H is bounded on L^p for $1 < p < 2$. Duality gives that $H^* = -H$ is bounded on L^p for $2 < p < \infty$ and hence so is H.

We give another proof of the boundedness of the Hilbert transform H on $L^p(\mathbf{R})$, which has the advantage that it gives the best possible constant in the resulting norm inequality when p is a power of 2.

Theorem 5.1.7. *For all $1 < p < \infty$, there exists a positive constant C_p such that*

$$\big\|H(f)\big\|_{L^p} \le C_p \big\|f\big\|_{L^p}$$

for all f in $\mathscr{S}(\mathbf{R})$. Moreover, the constant C_p satisfies $C_p \le 2p$ for $2 \le p < \infty$ and $C_p \le 2p/(p-1)$ for $1 < p \le 2$. Therefore, the Hilbert transform H admits an extension to a bounded operator on $L^p(\mathbf{R})$ when $1 < p < \infty$.

Proof. The proof we give is based on the interesting identity

$$H(f)^2 = f^2 + 2H(fH(f)), \tag{5.1.23}$$

which is valid whenever f is a real-valued Schwartz function. We prove (5.1.23) in two different ways. First we consider the analytic function

$$F_f(z) = \frac{i}{\pi}\int_{\mathbf{R}} \frac{f(t)}{z-t}\,dt$$

defined on the upper half space. We compute its square. Fix $z \in \mathbf{C}$ with $\operatorname{Re} z > 0$ and f a real-valued Schwartz function. Then for $\varepsilon > 0$ we have

$$\begin{aligned}
F_f(z)^2 &= \Big(\frac{i}{\pi}\Big)^2 \int_{\mathbf{R}}\int_{\mathbf{R}} \frac{f(t)f(s)}{(z-t)(z-s)}\,dt\,ds \\
&= \Big(\frac{i}{\pi}\Big)^2 \int_{\mathbf{R}} \int_{|t-s|>\varepsilon} \frac{f(t)f(s)}{t-s}\Big(\frac{1}{z-t} - \frac{1}{z-s}\Big)\,dt\,ds - \frac{1}{\pi^2}\iint_{|t-s|\le\varepsilon} \frac{f(t)f(s)\,dt\,ds}{(z-t)(z-s)} \\
&= \Big(\frac{i}{\pi}\Big)^2 \int_{\mathbf{R}} f(t) \int_{|t-s|>\varepsilon} \frac{f(s)}{t-s}\,ds\,\frac{dt}{z-t} - \Big(\frac{i}{\pi}\Big)^2 \int_{\mathbf{R}} f(s) \int_{|t-s|>\varepsilon} \frac{f(t)}{t-s}\,dt\,\frac{ds}{z-s} \\
&\qquad - \frac{1}{\pi^2}\iint_{|t-s|\le\varepsilon} \frac{f(t)f(s)\,dt\,ds}{(z-t)(z-s)}.
\end{aligned}$$

Letting $\varepsilon \to 0$ and passing the limit inside the integral by the Lebesgue dominated convergence theorem, we deduce

$$F_f(z)^2 = i\frac{i}{\pi}\int_{\mathbf{R}} \frac{2f(t)H(f)(t)}{z-t}\,dt\,. \tag{5.1.24}$$

We now let $\operatorname{Im} z \to 0+$ in (5.1.24) and use (5.1.19) in Theorem 5.1.5. We obtain

$$f^2 - H(f)^2 + i2fH(f) = \big(f + iH(f)\big)^2 = i\Big(2fH(f) + iH\big(2fH(f)\big)\Big),$$

and equating the real parts we deduce (5.1.23).

To give an alternative proof of (5.1.23) we take Fourier transforms. Let

$$m(\xi) = -i\operatorname{sgn}\xi$$

be the symbol of the Hilbert transform. We have

$$
\begin{aligned}
&\widehat{f^2}(\xi)+2[H(fH(f))]\widehat{\ }(\xi)\\
&\quad=(\widehat{f}*\widehat{f})(\xi)+2m(\xi)(\widehat{f}*\widehat{H(f)})(\xi)\\
&\quad=\int_{\mathbf{R}}\widehat{f}(\eta)\widehat{f}(\xi-\eta)\,d\eta+2m(\xi)\int_{\mathbf{R}}\widehat{f}(\eta)\widehat{f}(\xi-\eta)m(\eta)\,d\eta \qquad (5.1.25)\\
&\quad=\int_{\mathbf{R}}\widehat{f}(\eta)\widehat{f}(\xi-\eta)\,d\eta+2m(\xi)\int_{\mathbf{R}}\widehat{f}(\eta)\widehat{f}(\xi-\eta)m(\xi-\eta)\,d\eta. \qquad (5.1.26)
\end{aligned}
$$

Averaging (5.1.25) and (5.1.26) we obtain

$$\widehat{f^2}(\xi)+2[H(fH(f))]\widehat{\ }(\xi)=\int_{\mathbf{R}}\widehat{f}(\eta)\widehat{f}(\xi-\eta)\big[1+m(\xi)\big(m(\eta)+m(\xi-\eta)\big)\big]\,d\eta\,.$$

But the last displayed expression is equal to

$$\int_{\mathbf{R}}\widehat{f}(\eta)\widehat{f}(\xi-\eta)m(\eta)m(\xi-\eta)\,d\eta=(\widehat{H(f)}*\widehat{H(f)})(\xi)$$

in view of the identity

$$m(\eta)m(\xi-\eta)=1+m(\xi)m(\eta)+m(\xi)m(\xi-\eta),$$

which is valid for all $(\xi,\eta)\in\mathbf{R}^2\setminus\{(0,0)\}$ for the function $m(\xi)=-i\operatorname{sgn}\xi$.

Having established (5.1.23), we can easily obtain L^p bounds for H when $p=2^k$ is a power of 2. We already know that H is bounded on L^p with norm one when $p=2^k$ and $k=1$. Suppose that H is bounded on L^p with bound c_p for $p=2^k$ for some $k\in\mathbf{Z}^+$. Then for a nonzero real-valued function f in $\mathscr{C}_0^\infty$ we have

$$
\begin{aligned}
\big\|H(f)\big\|_{L^{2p}}=\big\|H(f)^2\big\|_{L^p}^{\frac12}&\le\big(\big\|f^2\big\|_{L^p}+\big\|2H(fH(f))\big\|_{L^p}\big)^{\frac12}\\
&\le\big(\big\|f\big\|_{L^{2p}}^2+2c_p\big\|fH(f)\big\|_{L^p}\big)^{\frac12}\\
&\le\big(\big\|f\big\|_{L^{2p}}^2+2c_p\big\|f\big\|_{L^{2p}}\big\|H(f)\big\|_{L^{2p}}\big)^{\frac12}.
\end{aligned}
$$

Since $\|H(f)\|_{L^{2p}}<\infty$, we obtain that

$$\left(\frac{\big\|H(f)\big\|_{L^{2p}}}{\big\|f\big\|_{L^{2p}}}\right)^2-2c_p\frac{\big\|H(f)\big\|_{L^{2p}}}{\big\|f\big\|_{L^{2p}}}-1\le 0.$$

If follows that

$$\frac{\big\|H(f)\big\|_{L^{2p}}}{\big\|f\big\|_{L^{2p}}}\le c_p+\sqrt{c_p^2+1}\,,$$

and from this we conclude that H is bounded on L^{2p} with bound

$$c_{2p}\le c_p+\sqrt{c_p^2+1}\,. \qquad (5.1.27)$$

This completes the induction. We have proved that H maps L^p to L^p when $p = 2^k$, $k = 1,2,\dots$. Interpolation now gives that H maps L^p to L^p for all $p \ge 2$. Since $H^* = -H$, duality gives that H is also bounded on L^p for $1 < p \le 2$.

The previous proof of the boundedness of the Hilbert transform provides us with some useful information about the norm of this operator on $L^p(\mathbf{R})$. Let us begin with the identity

$$\cot\frac{x}{2} = \cot x + \sqrt{1+\cot^2 x},$$

valid for $0 < x < \frac{\pi}{2}$. If $c_p \le \cot\frac{\pi}{2p}$, then (5.1.27) gives that

$$c_{2p} \le c_p + \sqrt{c_p^2+1} \le \cot\frac{\pi}{2p} + \sqrt{1+\cot^2\frac{\pi}{2p}} = \cot\frac{\pi}{2\cdot 2p},$$

and since $1 = \cot\frac{\pi}{4} = \cot\frac{\pi}{2\cdot 2}$, we obtain by induction that the numbers $\cot\frac{\pi}{2p}$ are indeed bounds for the norm of H on L^p when $p = 2^k$, $k = 1,2,\dots$. Duality now gives that the numbers $\cot\frac{\pi}{2p'} = \tan\frac{\pi}{2p}$ are bounds for the norm of H on L^p when $p = \frac{2^k}{2^k-1}$, $k = 1,2,\dots$. These bounds allow us to derive good estimates for the norm $\|H\|_{L^p\to L^p}$ as $p \to 1$ and $p \to \infty$. Indeed, since $\cot\frac{\pi}{2p} \le p$ when $p \ge 2$, the Riesz–Thorin interpolation theorem gives that $\|H\|_{L^p\to L^p} \le 2p$ for $2 \le p < \infty$ and by duality $\|H\|_{L^p\to L^p} \le \frac{2p}{p-1}$ for $1 < p \le 2$. This completes the proof which is worth comparing with that of Theorem 4.1.7. □

Remark 5.1.8. The numbers $\cot\frac{\pi}{2p}$ for $2 \le p < \infty$ and $\tan\frac{\pi}{2p}$ for $1 < p \le 2$ are indeed equal to the norms of the Hilbert transform H on $L^p(\mathbf{R})$. This requires a more delicate argument; see Exercise 5.1.12.

Remark 5.1.9. We may wonder what happens when $p = 1$ or $p = \infty$. The Hilbert transform of $\chi_{[a,b]}$ computed in Example 5.1.3 is easily seen to be unbounded and not integrable, since it behaves like $1/|x|$ as $x \to \infty$. This behavior near infinity suggests that the Hilbert transform may map L^1 to $L^{1,\infty}$. This is indeed the case, but this will not be shown until Section 5.3.

We now introduce the maximal Hilbert transform.

Definition 5.1.10. The *maximal Hilbert transform* is the operator

$$H^{(*)}(f)(x) = \sup_{\varepsilon>0}\left|H^{(\varepsilon)}(f)(x)\right| \tag{5.1.28}$$

defined for all f in L^p, $1 \le p < \infty$. For such f, $H^{(\varepsilon)}(f)$ is well defined as a convergent integral by Hölder's inequality. Hence $H^{(*)}(f)$ makes sense for $f \in L^p(\mathbf{R})$, although for some values of x, $H^{(*)}(f)(x)$ may be infinite.

Example 5.1.11. Using the result of Example 5.1.4, we obtain that

$$H^{(*)}(\chi_{[a,b]})(x) = \frac{1}{\pi}\left|\log\frac{|x-a|}{|x-b|}\right|. \tag{5.1.29}$$

We see that in general, $H^{(*)}(f)(x) \neq |H(f)(x)|$ by taking f to be the characteristic function of the union of two disjoint closed intervals.

The definition of H gives that $H^{(\varepsilon)}(f)$ converges pointwise to $H(f)$ whenever f lies in $\mathscr{C}_0^\infty(\mathbf{R})$. If we have the estimate $\|H^{(*)}(f)\|_{L^p} \le C_p\|f\|_{L^p}$ for $f \in L^p(\mathbf{R})$, Theorem 2.1.14 yields that $H^{(\varepsilon)}(f)$ converges to $H(f)$ a.e. as $\varepsilon \to 0$ for any $f \in L^p$. This almost everywhere limit provides a way to describe $H(f)$ for general $f \in L^p(\mathbf{R})$. Note that Theorem 5.1.7 implies only that H has a (unique) bounded extension on L^p, but it does not provide a way to describe $H(f)$ when f is a general L^p function.

The next theorem is a simple consequence of these ideas.

Theorem 5.1.12. *There exists a constant C such that for all $1 < p < \infty$ we have*

$$\left\|H^{(*)}(f)\right\|_{L^p} \le C\max\left(p,(p-1)^{-2}\right)\left\|f\right\|_{L^p}. \tag{5.1.30}$$

Moreover, for all f in $L^p(\mathbf{R})$, $H^{(\varepsilon)}(f)$ converges to $H(f)$ a.e. and in L^p.

Proof. Another proof of this theorem is given in Theorem 4.2.4 in [131] in which the asserted bound is improved.

Recall the kernels P_ε and Q_ε defined in (5.1.16) and (5.1.17). Fix $1 < p < \infty$ and suppose momentarily that

$$f * Q_\varepsilon = H(f) * P_\varepsilon\,, \qquad \varepsilon > 0, \tag{5.1.31}$$

holds whenever f is an L^p function. Then we have

$$H^{(\varepsilon)}(f) = H^{(\varepsilon)}(f) - f * Q_\varepsilon + H(f) * P_\varepsilon\,. \tag{5.1.32}$$

Using the identity

$$H^{(\varepsilon)}(f)(x) - (f * Q_\varepsilon)(x) = -\frac{1}{\pi}\int_{\mathbf{R}} f(x-t)\,\psi_\varepsilon(t)\,dt\,, \tag{5.1.33}$$

where ψ is as in (5.1.20), and applying Corollary 2.1.12, we obtain the estimate

$$\sup_{\varepsilon>0}|H^{(\varepsilon)}(f)(x) - (f * Q_\varepsilon)(x)| \le \frac{1}{\pi}\left\|\Psi\right\|_{L^1}M(f)(x)\,, \tag{5.1.34}$$

where Ψ is as in (5.1.21) and M is the Hardy–Littlewood maximal function. In view of (5.1.32) and (5.1.34), we obtain for $f \in L^p(\mathbf{R}^n)$ that

$$|H^{(*)}(f)(x)| \le \left\|\Psi\right\|_{L^1}M(f)(x) + M(H(f))(x)\,. \tag{5.1.35}$$

It follows immediately from (5.1.35) that $H^{(*)}$ is L^p bounded with norm at most $C\max\left(p,(p-1)^{-2}\right)$.

We now turn to the proof of (5.1.31). It suffices to prove (5.1.31) for Schwartz functions since, given $f \in L^p$ there is a sequence $\phi_j \in \mathscr{S}$ such that $\|f - \phi_j\|_{L^p} \to 0$ as $j \to \infty$ and P_ε, Q_ε lie in $L^{p'}$. Taking Fourier transforms, we see that (5.1.31) is a consequence of the identity

$$\big((-i\,\mathrm{sgn}\,\xi)e^{-2\pi|\xi|}\big)^\vee(x) = \frac{1}{\pi}\frac{x}{x^2+1}. \tag{5.1.36}$$

To prove (5.1.36) we write

$$\begin{aligned}
\big((-i\,\mathrm{sgn}\,\xi)e^{-2\pi|\xi|}\big)^\vee(x) &= \int_{-\infty}^{+\infty} e^{-2\pi|\xi|}(-i\,\mathrm{sgn}\,\xi)e^{2\pi i x\xi}\,d\xi \\
&= 2\int_0^\infty e^{-2\pi\xi}\sin(2\pi x\xi)\,d\xi \\
&= \frac{1}{\pi}\int_0^\infty e^{-\xi}\sin(x\xi)\,d\xi \qquad (5.1.37)\\
&= \frac{1}{\pi}\int_0^\infty (e^{-\xi})''\sin(x\xi)\,d\xi \\
&= -\frac{x}{\pi}\int_0^\infty (e^{-\xi})'\cos(x\xi)\,d\xi \\
&= -\frac{x}{\pi}\left[-1 + x\int_0^\infty e^{-\xi}\sin(x\xi)\,d\xi\right] \qquad (5.1.38)
\end{aligned}$$

and we equate (5.1.38) and (5.1.37).

The statement in the theorem about the almost everywhere convergence of $H^{(\varepsilon)}(f)$ to $H(f)$ is a consequence of (5.1.30), of the fact that the alleged convergence holds for Schwartz functions, and of Theorem 2.1.14. Finally, the L^p convergence follows from the almost everywhere convergence and the Lebesgue dominated convergence theorem in view of the validity of (5.1.35). □

5.1.4 The Riesz Transforms

We now study an n-dimensional analogue of the Hilbert transform. It turns out that there exist n operators in $\mathbf{R}^n$, called the Riesz transforms, with properties analogous to those of the Hilbert transform on $\mathbf{R}$.

To define the Riesz transforms, we first introduce tempered distributions W_j on $\mathbf{R}^n$, for $1 \le j \le n$, as follows. For $\varphi \in \mathscr{S}(\mathbf{R}^n)$, let

$$\langle W_j, \varphi\rangle = \frac{\Gamma(\frac{n+1}{2})}{\pi^{\frac{n+1}{2}}}\lim_{\varepsilon\to 0}\int_{|y|\ge\varepsilon}\frac{y_j}{|y|^{n+1}}\varphi(y)\,dy.$$

One should check that indeed $W_j \in \mathscr{S}'(\mathbf{R}^n)$. Observe that the normalization of W_j is similar to that of the Poisson kernel.

Definition 5.1.13. For $1 \le j \le n$, the jth *Riesz transform* of f is given by convolution with the distribution W_j, that is,

$$R_j(f)(x) = (f * W_j)(x) = \frac{\Gamma(\frac{n+1}{2})}{\pi^{\frac{n+1}{2}}} \,\text{p.v.} \int_{\mathbf{R}^n} \frac{x_j - y_j}{|x-y|^{n+1}} f(y)\,dy, \tag{5.1.39}$$

for all $f \in \mathscr{S}(\mathbf{R}^n)$. Definition 5.1.13 makes sense for any integrable function f that has the property that for all x there exist $C_x > 0$, $\varepsilon_x > 0$, and $\delta_x > 0$ such that for y satisfying $|y-x| < \delta_x$ we have $|f(x) - f(y)| \le C_x |x-y|^{\varepsilon_x}$. The principal value integral in (5.1.39) is as in Definition 5.1.1.

We now give a characterization of R_j using the Fourier transform. For this we need to compute the Fourier transform of W_j.

Proposition 5.1.14. *The jth Riesz transform R_j is given on the Fourier transform side by multiplication by the function $-i\xi_j/|\xi|$. That is, for any f in $\mathscr{S}(\mathbf{R}^n)$ we have*

$$R_j(f)(x) = \Big(-\frac{i\xi_j}{|\xi|}\widehat{f}(\xi)\Big)^{\vee}(x). \tag{5.1.40}$$

Proof. The proof is essentially a reprise of the corresponding proof for the Hilbert transform, but it involves a few technical difficulties. Fix a Schwartz function φ on $\mathbf{R}^n$. Then for $1 \le j \le n$ we have

$$\begin{aligned}
\langle \widehat{W_j}, \varphi \rangle &= \langle W_j, \widehat{\varphi} \rangle \qquad\qquad (5.1.41)\\
&= \frac{\Gamma(\frac{n+1}{2})}{\pi^{\frac{n+1}{2}}} \lim_{\varepsilon\to 0} \int_{|\xi|\ge\varepsilon} \widehat{\varphi}(\xi) \frac{\xi_j}{|\xi|^{n+1}}\,d\xi \\
&= \frac{\Gamma(\frac{n+1}{2})}{\pi^{\frac{n+1}{2}}} \lim_{\varepsilon\to 0} \int_{\frac{1}{\varepsilon}\ge|\xi|\ge\varepsilon} \int_{\mathbf{R}^n} \varphi(x) e^{-2\pi i x\cdot\xi}\,dx \frac{\xi_j}{|\xi|^{n+1}}\,d\xi \\
&= \lim_{\varepsilon\to 0} \int_{\mathbf{R}^n} \varphi(x) \left[\frac{\Gamma(\frac{n+1}{2})}{\pi^{\frac{n+1}{2}}} \int_{\frac{1}{\varepsilon}\ge|\xi|\ge\varepsilon} e^{-2\pi i x\cdot\xi} \frac{\xi_j}{|\xi|^{n+1}}\,d\xi \right] dx \\
&= \lim_{\varepsilon\to 0} \int_{\mathbf{R}^n} \varphi(x) \left[\frac{\Gamma(\frac{n+1}{2})}{\pi^{\frac{n+1}{2}}} \int_{\mathbf{S}^{n-1}} \int_{\varepsilon\le r\le\frac{1}{\varepsilon}} e^{-2\pi i r x\cdot\theta} \frac{r}{r^{n+1}} r^{n-1}\,dr\,\theta_j\,d\theta \right] dx \\
&= \int_{\mathbf{R}^n} \varphi(x) \left[-i \frac{\Gamma(\frac{n+1}{2})}{\pi^{\frac{n+1}{2}}} \int_{\mathbf{S}^{n-1}} \int_0^\infty \sin(2\pi r x\cdot\theta) \frac{dr}{r}\,\theta_j\,d\theta \right] dx \\
&= \int_{\mathbf{R}^n} \varphi(x) \left[-i \frac{\pi}{2} \frac{\Gamma(\frac{n+1}{2})}{\pi^{\frac{n+1}{2}}} \int_{\mathbf{S}^{n-1}} \operatorname{sgn}(x\cdot\theta)\,\theta_j\,d\theta \right] dx \\
&= \int_{\mathbf{R}^n} -i\varphi(x) \frac{x_j}{|x|}\,dx,
\end{aligned}$$

where in the penultimate equality we used the identity $\int_0^\infty \frac{\sin t}{t}\,dt = \frac{\pi}{2}$, for which we refer to Exercise 5.1.1, while in the last equality we used the identity

$$-i\frac{\pi}{2}\frac{\Gamma(\frac{n+1}{2})}{\pi^{\frac{n+1}{2}}}\int_{\mathbf{S}^{n-1}} \operatorname{sgn}(x\cdot\theta)\theta_j\,d\theta = -i\frac{x_j}{|x|}, \tag{5.1.42}$$

which needs to be established. The passage of the limit inside the integral in the previous calculation is a consequence of the Lebesgue dominated convergence theorem, which is justified from the fact that

$$\left|\int_\varepsilon^{1/\varepsilon} \frac{\sin(2\pi r\theta)}{r}\,dr\right| \le 4 \tag{5.1.43}$$

for all $\varepsilon > 0$. For a proof of (5.1.43) we again refer to Exercise 5.1.1. □

It remains to establish (5.1.42). Let us recall that $O(n)$ is the set of all orthogonal $n\times n$ matrices with real entries. An invertible matrix A is called orthogonal if its transpose A^t is equal to its inverse A^{-1}, that is, $AA^t = A^tA = I$.

Lemma 5.1.15. *The following identity is valid for all* $\xi \in \mathbf{R}^n\setminus\{0\}$*:*

$$\int_{\mathbf{S}^{n-1}} \operatorname{sgn}(\xi\cdot\theta)\,\theta_j\,d\theta = \frac{2\pi^{\frac{n-1}{2}}}{\Gamma(\frac{n+1}{2})}\frac{\xi_j}{|\xi|}. \tag{5.1.44}$$

Therefore (5.1.42) *holds.*

Proof. We begin with the identity

$$\int_{\mathbf{S}^{n-1}} \operatorname{sgn}(\theta_k)\,\theta_j\,d\theta = \begin{cases} 0 & \text{if } k\neq j,\\ \displaystyle\int_{\mathbf{S}^{n-1}} |\theta_j|\,d\theta & \text{if } k=j, \end{cases} \tag{5.1.45}$$

which can be proved by noting that for $k\neq j$, $\operatorname{sgn}(\theta_k)$ has a constant sign on the hemispheres $\theta_k>0$ and $\theta_k<0$, on either of which the function $\theta\mapsto\theta_j$ has integral zero.

It suffices to prove (5.1.44) for a unit vector ξ. Given $\xi\in\mathbf{S}^{n-1}$, pick an orthogonal $n\times n$ matrix $A=(a_{kl})_{k,l}$ such that $Ae_j=\xi$. Then the jth column of the matrix A is the vector $(\xi_1,\xi_2,\dots,\xi_n)^t$. We have

$$\begin{aligned}
\int_{\mathbf{S}^{n-1}} \operatorname{sgn}(\xi\cdot\theta)\,\theta_j\,d\theta &= \int_{\mathbf{S}^{n-1}} \operatorname{sgn}(Ae_j\cdot\theta)\,\theta_j\,d\theta\\
&= \int_{\mathbf{S}^{n-1}} \operatorname{sgn}(e_j\cdot A^t\theta)\,(AA^t\theta)_j\,d\theta\\
&= \int_{\mathbf{S}^{n-1}} \operatorname{sgn}(e_j\cdot\theta)\,(A\theta)_j\,d\theta\\
&= \int_{\mathbf{S}^{n-1}} \operatorname{sgn}(\theta_j)\,(a_{j1}\theta_1+\cdots+\xi_j\theta_j+\cdots+a_{jn}\theta_n)\,d\theta
\end{aligned}$$

$$
\begin{aligned}
&= \xi_j \int_{\mathbf{S}^{n-1}} \operatorname{sgn}(\theta_j)\, \theta_j \, d\theta + \sum_{1\le m\ne j\le n} 0 \\
&= \frac{\xi_j}{|\xi|} \int_{\mathbf{S}^{n-1}} |\theta_j| \, d\theta \, .
\end{aligned}
$$

Next, for all $j \in \{1,2,\dots,n\}$, we compute the value of the integral

$$
\int_{\mathbf{S}^{n-1}} |\theta_j| \, d\theta = \int_{\mathbf{S}^{n-1}} |\theta_1| \, d\theta \, ,
$$

which is obviously independent of j by symmetry. In view of the result of Appendix D.2, we write

$$
\begin{aligned}
\int_{\mathbf{S}^{n-1}} |\theta_1| \, d\theta &= \int_{-1}^{1} |s| \int_{\sqrt{1-s^2}\,\mathbf{S}^{n-2}} d\varphi \, \frac{ds}{(1-s^2)^{\frac{1}{2}}} \\
&= \omega_{n-2} \int_{-1}^{1} |s| (1-s^2)^{\frac{n-3}{2}} \, ds \\
&= \omega_{n-2} \int_{0}^{1} u^{\frac{n-3}{2}} \, du \\
&= \frac{2\omega_{n-2}}{n-1} \\
&= \frac{2\pi^{\frac{n-1}{2}}}{\Gamma(\frac{n-1}{2})\frac{n-1}{2}} \\
&= \frac{2\pi^{\frac{n-1}{2}}}{\Gamma(\frac{n+1}{2})} \, ,
\end{aligned}
$$

having used the expression for ω_{n-2} in Appendix A.3. This proves (5.1.44). The proof of the lemma and hence that of Proposition 5.1.14 is complete. □

Proposition 5.1.16. *The Riesz transforms satisfy*

$$
-I = \sum_{j=1}^{n} R_j^2 \, , \qquad \text{on } L^2(\mathbf{R}^n), \tag{5.1.46}
$$

where I is the identity operator.

Proof. Use the Fourier transform and the identity $\sum_{j=1}^{n}(-i\xi_j/|\xi|)^2 = -1$ to obtain that $\sum_{j=1}^{n} R_j^2(f) = -f$ for any f in $L^2(\mathbf{R}^n)$. □

We can express the mixed derivatives of Schwartz function in terms of its Laplacian using the Riesz transforms.

Proposition 5.1.17. *For φ in $\mathscr{S}(\mathbf{R}^n)$ and $1 \le j,k \le n$ we have*

$$\partial_j \partial_k \varphi(x) = -R_j R_k \Delta \varphi(x) \tag{5.1.47}$$

for all $x \in \mathbf{R}^n$.

Proof. We verify the claimed identity by taking Fourier transforms. We have

$$\begin{aligned}(\partial_j \partial_k \varphi)\widehat{\ }(\xi) &= (2\pi i \xi_j)(2\pi i \xi_k)\widehat{\varphi}(\xi) \\ &= -\Big(-\frac{i\xi_j}{|\xi|}\Big)\Big(-\frac{i\xi_k}{|\xi|}\Big)(-4\pi^2|\xi|^2)\widehat{\varphi}(\xi) \\ &= -(R_j R_k \Delta \varphi)\widehat{\ }(\xi)\end{aligned}$$

and taking the inverse Fourier transform, identity (5.1.47) follows. □

Next we discuss a use of the Riesz transforms to partial differential equations.

Example 5.1.18. Suppose that f is a given function in $L^2(\mathbf{R}^n)$ and that u is a tempered distribution on $\mathbf{R}^n$ that solves *Laplace's equation*

$$\Delta u = f\,. \tag{5.1.48}$$

We express all second-order derivatives of u in terms of the Riesz transforms of f.

To solve equation (5.1.48) we first show that the tempered distribution

$$\big(\partial_j \partial_k u + R_j R_k (f)\big)\widehat{\ }$$

is supported at $\{0\}$. In view of Proposition 2.4.1, this implies that

$$\partial_j \partial_k u = -R_j R_k (f) + P$$

where P is a polynomial of n variables (that depends on j and k) and provides a way to express the mixed partials of u in terms of the Riesz transforms of f.

To verify that $\big(\partial_j \partial_k u + R_j R_k (f)\big)\widehat{\ }$ is supported at $\{0\}$, we fix a Schwartz function ψ whose support does not contain the origin. Then ψ vanishes in a neighborhood of zero and we can pick $\mathscr{C}^\infty$ function η which vanishes in a smaller neighborhood of zero and is equal to 1 on the support of ψ. We define

$$\zeta(\xi) = -\eta(\xi)\Big(-\frac{i\xi_j}{|\xi|}\Big)\Big(-\frac{i\xi_k}{|\xi|}\Big)$$

and we notice that ζ is a bounded $\mathscr{C}^\infty$ function and so are all of its derivatives; also

$$\eta(\xi)(2\pi i \xi_j)(2\pi i \xi_k) = \zeta(\xi)(-4\pi^2|\xi|^2)\,.$$

Taking the Fourier transform of both sides of (5.1.48) we obtain

$$(-4\pi^2|\xi|^2)\,\widehat{u}(\xi) = \widehat{f}(\xi)$$

and multiplying by ζ (which is allowed since all derivatives of ζ lie in $L^\infty \cap \mathscr{C}^\infty$)

$$\zeta(\xi)(-4\pi^2|\xi|^2)\,\widehat{u} = \zeta(\xi)\,\widehat{\Delta u} = \zeta(\xi)\widehat{f}(\xi).$$

It follows that for all $1 \le j,k \le n$ we have

$$\begin{aligned}
\langle (\partial_j\partial_k u)^\frown, \psi\rangle &= \langle (2\pi i\xi_j)(2\pi i\xi_k)\widehat{u}, \psi\rangle \\
&= \langle (2\pi i\xi_j)(2\pi i\xi_k)\widehat{u}, \eta\psi\rangle \\
&= \langle \eta(\xi)(2\pi i\xi_j)(2\pi i\xi_k)\widehat{u}, \psi\rangle \\
&= \langle \zeta(\xi)(-4\pi^2|\xi|^2)\widehat{u}, \psi\rangle \\
&= \langle \zeta(\xi)\widehat{f}(\xi), \psi\rangle \\
&= \langle -\eta(\xi)(-\tfrac{i\xi_j}{|\xi|})(-\tfrac{i\xi_k}{|\xi|})\widehat{f}(\xi), \psi\rangle \\
&= \langle -\eta(\xi)(R_jR_k(f))^\frown(\xi), \psi\rangle \\
&= -\langle (R_jR_k(f))^\frown, \eta\psi\rangle \\
&= -\langle (R_jR_k(f))^\frown, \psi\rangle
\end{aligned}$$

and since this holds for all Schwartz functions ψ whose support does not contain the origin, it follows that $\big(\partial_j\partial_k u + R_jR_k(f)\big)^\frown$ is supported at $\{0\}$.

Exercises

5.1.1. (a) Show that for all $0 < a < b < \infty$ we have

$$\left|\int_a^b \frac{\sin x}{x}\,dx\right| \le 4.$$

(b) For $a > 0$ define

$$I(a) = \int_0^\infty \frac{\sin x}{x} e^{-ax}\,dx$$

and show that $I(a)$ is continuous at zero. Differentiate in a and look at the behavior of $I(a)$ as $a \to \infty$ to obtain the identity

$$I(a) = \frac{\pi}{2} - \arctan(a).$$

Deduce that $I(0) = \frac{\pi}{2}$ and also derive the following identity used in (5.1.10):

$$\int_{-\infty}^{+\infty} \frac{\sin(bx)}{x}\,dx = \pi\,\mathrm{sgn}\,(b).$$

(c) Argue as in part (b) to prove for $a \geq 0$ the identity

$$\int_0^\infty \frac{1-\cos x}{x^2} e^{-ax}\, dx = \frac{\pi}{2} - \arctan(a) + a \log \frac{a}{\sqrt{1+a^2}}.$$

[*Hint:* Part (a): Consider the cases $b \leq 1$, $a \leq 1 \leq b$, $1 \leq a$. When $a \geq 1$, integrate by parts.]

5.1.2. (a) Let φ be a compactly supported $\mathscr{C}^{m+1}$ function on $\mathbf{R}$ for some m in $\mathbf{Z}^+ \bigcup \{0\}$. Prove that if $\varphi^{(m)}$ is the mth derivative of φ, then

$$|H(\varphi^{(m)})(x)| \leq C_{m,\varphi}\,(1+|x|)^{-m-1}$$

for some $C_{m,\varphi} > 0$.
(b) Let φ be a compactly supported $\mathscr{C}^{m+1}$ function on $\mathbf{R}^n$ for some $m \in \mathbf{Z}^+$. Show that

$$|R_j(\partial^\alpha \varphi)(x)| \leq C_{n,m,\varphi}\,(1+|x|)^{-n-m}$$

for some $C_{n,m,\varphi} > 0$ and all multi-indices α with $|\alpha| = m$.
(c) Let I be an interval on the line and assume that a function h is equal to 1 on the left half of I, is equal to -1 on the right half of I, and vanishes outside I. Prove that for $x \notin 2I$ we have

$$|H(h)(x)| \leq 4|I|^2 |x - \text{center}(I)|^{-2}.$$

[*Hint:* Use that when $|t| \leq \frac{1}{2}$ we have $\log(1+t) = t + R_1(t)$, where $|R_1(t)| \leq 2|t|^2$.]

5.1.3. (a) Using identity (5.1.13) one may define $H(f)$ as an element of $\mathscr{S}'(\mathbf{R})$ for bounded functions f on the line whose Fourier transform vanishes in a neighborhood of the origin. Using this interpretation, prove that

$$\begin{aligned} H(e^{ix}) &= -ie^{ix}, \\ H(\cos x) &= \sin x, \\ H(\sin x) &= -\cos x, \\ H(\sin(\pi x)/\pi x) &= (1-\cos(\pi x))/\pi x. \end{aligned}$$

(b) Show that the operators given by convolution with the smooth function $\sin(t)/t$ and the distribution p.v. $\cos(t)/t$ are bounded on $L^p(\mathbf{R})$ whenever $1 < p < \infty$.
[*Hint:* Use that the Fourier transform of the distribution e^{ix} is $\delta_{1/2\pi}$.]

5.1.4. ([347]) Show that the distribution function of the Hilbert transform of the characteristic function of a measurable subset E of the real line of finite measure is

$$d_{H(\chi_E)}(\alpha) = \frac{4|E|}{e^{\pi\alpha} - e^{-\pi\alpha}}, \qquad \alpha > 0.$$

[*Hint:* First take $E = \bigcup_{j=1}^N (a_j, b_j)$, where $b_j < a_{j+1}$. Show that the equation $H(\chi_E)(x) = \alpha$ has exactly one root ρ_j in each open interval (a_j, b_j) for $1 \leq j \leq N$ and exactly one root r_j in each interval (b_j, a_{j+1}) for $1 \leq j \leq N$, $(a_{N+1} = \infty)$. Then

$|\{x \in \mathbf{R} : H(\chi_E)(x) > \alpha\}| = \sum_{j=1}^{N} r_j - \sum_{j=1}^{N} \rho_j$, and this can be expressed in terms of $\sum_{j=1}^{N} a_j$ and $\sum_{j=1}^{N} b_j$. Argue similarly for the set $\{x \in \mathbf{R} : H(\chi_E)(x) < -\alpha\}$. For a general measurable set E, find sets E_n such that each E_n is a finite union of intervals and that $\chi_{E_n} \to \chi_E$ in L^2. Then $H(\chi_{E_n}) \to H(\chi_E)$ in measure; thus $H(\chi_{E_{n_k}}) \to H(\chi_E)$ a.e. for some subsequence n_k. The Lebesgue dominated convergence theorem gives $d_{H(\chi_{E_{n_k}})} \to d_{H(\chi_E)}$. See Figure 5.1.]

5.1.5. Let $1 \le p < \infty$ and let T be a linear operator defined on the space of Schwartz functions that commutes with dilations, i.e., $T(\delta^\lambda f) = \delta^\lambda T(f)$ for all $f \in \mathscr{S}(\mathbf{R}^n)$ and all $\lambda > 0$. (Here $\delta^\lambda(f)(x) = f(\lambda x)$.) Suppose that there exists a constant $C > 0$ such that for all $f \in \mathscr{S}(\mathbf{R}^n)$ with L^p norm one we have

$$|\{x : |T(f)(x)| > 1\}| \le C.$$

Prove that T admits a bounded extension from $L^p(\mathbf{R}^n)$ to $L^{p,\infty}(\mathbf{R}^n)$ with norm at most $C^{1/p}$.
[*Hint:* Try functions of the form $\lambda^{-n/p} f(\lambda^{-1}x)/\|f\|_{L^p}$ with $\lambda > 0$.]

5.1.6. Let φ be in $\mathscr{S}(\mathbf{R})$. Prove that

$$\lim_{N\to\infty} \text{p.v.} \int_{\mathbf{R}} \frac{e^{2\pi i N x}}{x} \varphi(x)\,dx = \varphi(0)\pi i,$$

$$\lim_{N\to-\infty} \text{p.v.} \int_{\mathbf{R}} \frac{e^{2\pi i N x}}{x} \varphi(x)\,dx = -\varphi(0)\pi i.$$

5.1.7. Let T_α, $\alpha \in \mathbf{R}$, be the operator given by convolution with the distribution whose Fourier transform is the function

$$u_\alpha(\xi) = e^{-\pi i \alpha \operatorname{sgn} \xi}.$$

(a) Show that the T_α's are isometries on $L^2(\mathbf{R})$ that satisfy

$$(T_\alpha)^{-1} = T_{2-\alpha}.$$

(b) Express T_α in terms of the identity operator and the Hilbert transform.

5.1.8. Let $Q_y^{(j)}$ be the jth *conjugate Poisson kernel* of P_y defined by

$$Q_y^{(j)}(x) = \frac{\Gamma(\frac{n+1}{2})}{\pi^{\frac{n+1}{2}}} \frac{x_j}{(|x|^2 + y^2)^{\frac{n+1}{2}}}.$$

Prove that

$$(Q_y^{(j)})^\wedge(\xi) = -i\frac{\xi_j}{|\xi|} e^{-2\pi y|\xi|}.$$

Conclude that $R_j(P_y) = Q_y^{(j)}$ and that for f in $L^2(\mathbf{R}^n)$ we have $R_j(f) * P_y = f * Q_y^{(j)}$. These results are analogous to the statements $\widehat{Q_y}(\xi) = -i\operatorname{sgn}(\xi)\widehat{P_y}(\xi)$, $H(P_y) = Q_y$, and $H(f) * P_y = f * Q_y$.

5.1.9. Fix $n \geq 2$. Let $f_0, f_1, \dots, f_n$ be in $L^2(\mathbf{R}^n)$ and, for $0 \leq j \leq n$, let $u_j(x,x_0) = (P_{x_0} * f_j)(x)$ be the Poisson integrals of f_j where $x = (x_1, \dots, x_n) \in \mathbf{R}^n$ and $x_0 > 0$. Show that a necessary and sufficient condition for

$$f_j = R_j(f_0), \qquad j = 1, \dots, n,$$

is that the following system of *generalized Cauchy-Riemann equations* holds:

$$\sum_{j=0}^{n} \frac{\partial u_j}{\partial x_j}(x,x_0) = 0,$$

$$\frac{\partial u_j}{\partial x_k}(x,x_0) = \frac{\partial u_k}{\partial x_j}(x,x_0), \quad 0 \leq j \neq k \leq n.$$

5.1.10. Prove the distributional identity

$$\partial_j |x|^{-n+1} = (1-n)\text{p.v.}\frac{x_j}{|x|^{n+1}}.$$

Then take Fourier transforms of both sides and use Theorem 2.4.6 to obtain another proof of Proposition 5.1.14.

5.1.11. (a) Prove that if T is a bounded linear operator on $L^2(\mathbf{R})$ that commutes with translations and dilations and anticommutes with the reflection $f(x) \mapsto \widetilde{f}(x) = f(-x)$, then T is a constant multiple of the Hilbert transform.
(b) Prove that if T is a bounded linear operator on $L^2(\mathbf{R})$ that commutes with translations and dilations and vanishes when applied to functions whose Fourier transform is supported in $[0,\infty)$, then T is a constant multiple of the operator $f \mapsto \big(\widehat{f}\chi_{(-\infty,0]}\big)^{\vee}$.

5.1.12. ([282]) Fix $1 < p \leq 2$.
(a) Show that the function $G(x,y) = \text{Re}\,(|x| + iy)^p$ is subharmonic on $\mathbf{R}^2$.
(b) Let $u(x,y), v(x,y)$ be real-valued functions on $\mathbf{R}^2$ such that $u + iv$ is a holomorphic function of $x + iy$. Prove that $G(u,v)$ is a subharmonic function on $\mathbf{R}^2$.
(c) Prove that there is a constant B_p such that for all a and b reals we have

$$|b|^p \leq \left(\tan\frac{\pi}{2p}\right)^p |a|^p - B_p \text{Re}\,(|a| + ib)^p.$$

(d) Prove that for f in $\mathscr{C}_0^\infty(\mathbf{R})$ we have

$$\int_{\mathbf{R}} \text{Re}\,(|f(x)| + iH(f)(x))^p\,dx \geq 0.$$

(e) Combine the results in parts (d) and (c) with $a = f(x)$, $b = H(f)(x)$ to obtain that

$$\big\|H\big\|_{L^p \to L^p} \leq \tan\frac{\pi}{2p}.$$

(f) To deduce that this constant is sharp, take $\pi/2p' < \gamma < \pi/2p$ and let $f_\gamma(x) = (x+1)^{-1}|x+1|^{2\gamma/\pi}|x-1|^{-2\gamma/\pi}\cos\gamma$. Then

$$H(f_\gamma)(x) = \begin{cases} \frac{1}{x+1}\left|\frac{x+1}{x-1}\right|^{2\gamma/\pi}\sin\gamma & \text{when } |x|>1, \\ -\frac{1}{x+1}\left|\frac{x+1}{x-1}\right|^{2\gamma/\pi}\sin\gamma & \text{when } |x|<1. \end{cases}$$

[*Hint:* Part (d): Let C_R be the circle of radius R centered at $(0,R)$ in $\mathbf{R}^2$. Use that the integral of the subharmonic function $G((P_y * f)(x), Q_y * f)(x))$ over C_R is at least $2\pi R \operatorname{Re}\,(|(P_R * f)(0)| + i(Q_R * f)(0))^p$ and let $R \to \infty$. Part (f): The formula for $H(f_\gamma)$ is best derived by considering the restriction of the analytic function

$$F(z) = (z+1)^{-1}\left(\frac{iz+i}{z-1}\right)^{2\gamma/\pi}$$

on the real line.]

5.2 Homogeneous Singular Integrals and the Method of Rotations

So far we have introduced the Hilbert and the Riesz transforms and we have derived the L^p boundedness of the former. The boundedness properties of the Riesz transforms on L^p spaces are consequences of the results discussed in this section.

5.2.1 Homogeneous Singular and Maximal Singular Integrals

We introduce singular integral operators on $\mathbf{R}^n$ that appropriately generalize the Riesz transforms on $\mathbf{R}^n$. Here is the setup. We fix Ω to be an integrable function of the unit sphere $\mathbf{S}^{n-1}$ with mean value zero. Observe that the kernel

$$K_\Omega(x) = \frac{\Omega(x/|x|)}{|x|^n}, \qquad x \neq 0, \tag{5.2.1}$$

is homogeneous of degree $-n$ just like the functions $x_j/|x|^{n+1}$. Since K_Ω is not in $L^1(\mathbf{R}^n)$, convolution with K_Ω cannot be defined as an operation on Schwartz functions on $\mathbf{R}^n$. For this reason we introduce a distribution W_Ω in $\mathscr{S}'(\mathbf{R}^n)$ by setting

$$\langle W_\Omega, \varphi\rangle = \lim_{\varepsilon\to 0}\int_{|x|\geq\varepsilon} K_\Omega(x)\varphi(x)\,dx = \lim_{\varepsilon\to 0}\int_{\varepsilon\leq|x|\leq\varepsilon^{-1}} K_\Omega(x)\varphi(x)\,dx \tag{5.2.2}$$

for $\varphi \in \mathscr{S}(\mathbf{R}^n)$. Using the fact that Ω has mean value zero, we can easily see that W_Ω is a well defined tempered distribution on $\mathbf{R}^n$. Indeed, since K_Ω has integral zero over all annuli centered at the origin, we have

$$\begin{aligned}
\big|\langle W_\Omega, \varphi\rangle\big| &= \left|\lim_{\varepsilon\to 0}\int_{\varepsilon\le|x|\le 1}\frac{\Omega(x/|x|)}{|x|^n}(\varphi(x)-\varphi(0))\,dx + \int_{|x|\ge 1}\frac{\Omega(x/|x|)}{|x|^n}\varphi(x)\,dx\right| \\
&\le \big\|\nabla\varphi\big\|_{L^\infty}\int_{|x|\le 1}\frac{|\Omega(x/|x|)|}{|x|^{n-1}}\,dx + \sup_{y\in\mathbf{R}^n}|y|\,|\varphi(y)|\int_{|x|\ge 1}\frac{|\Omega(x/|x|)|}{|x|^{n+1}}\,dx \\
&\le C_1\big\|\nabla\varphi\big\|_{L^\infty}\big\|\Omega\big\|_{L^1} + C_2\sum_{|\alpha|\le 1}\big\|\varphi(x)x^\alpha\big\|_{L^\infty}\big\|\Omega\big\|_{L^1},
\end{aligned}$$

for suitable C_1 and C_2, where we used (2.2.2) in the last estimate. Note that the distribution W_Ω coincides with the function K_Ω on $\mathbf{R}^n\setminus\{0\}$.

The Hilbert transform and the Riesz transforms are examples of these general operators T_Ω. For instance, the function $\Omega(\theta) = \frac{\theta}{\pi|\theta|} = \frac{1}{\pi}\operatorname{sgn}\theta$ defined on the unit sphere $\mathbf{S}^0 = \{-1,1\} \subsetneq \mathbf{R}$ gives rise to the Hilbert transform, while the function

$$\Omega(\theta) = \frac{\Gamma(\frac{n+1}{2})}{\pi^{\frac{n+1}{2}}}\frac{\theta_j}{|\theta|}$$

defined on $\mathbf{S}^{n-1} \subsetneq \mathbf{R}^n$ gives rise to the jth Riesz transform.

Definition 5.2.1. Let Ω be integrable on the sphere $\mathbf{S}^{n-1}$ with mean value zero. For $0<\varepsilon<N$ and $f\in\bigcup_{1\le p<\infty}L^p(\mathbf{R}^n)$ we define the *truncated singular integral*

$$T_\Omega^{(\varepsilon,N)}(f)(x) = \int_{\varepsilon\le|y|\le N} f(x-y)\,\frac{\Omega(y/|y|)}{|y|^n}\,dy. \tag{5.2.3}$$

Note that for $f\in L^p(\mathbf{R}^n)$ we have

$$\big\|T_\Omega^{(\varepsilon,N)}(f)\big\|_{L^p} \le \big\|\Omega\big\|_{L^1}\log(N/\varepsilon)\,\big\|f\big\|_{L^p(\mathbf{R}^n)},$$

which implies that (5.2.3) is finite a.e. and therefore well defined. We denote by T_Ω the singular integral operator whose kernel is the distribution W_Ω, that is,

$$T_\Omega(f)(x) = (f*W_\Omega)(x) = \lim_{\substack{\varepsilon\to 0\\ N\to\infty}} T_\Omega^{(\varepsilon,N)}(f)(x),$$

defined for $f\in\mathscr{S}(\mathbf{R}^n)$. The associated *maximal singular integral* is defined by

$$T_\Omega^{(**)}(f) = \sup_{0<N<\infty}\sup_{0<\varepsilon<N}\big|T_\Omega^{(\varepsilon,N)}(f)\big|. \tag{5.2.4}$$

We note that if Ω is bounded, there is no need to use the upper truncations in the definition of $T_\Omega^{(\varepsilon,N)}$ given in (5.2.3). In this case the maximal singular integrals could be defined as

$$T_\Omega^{(*)}(f) = \sup_{\varepsilon>0} \left| T_\Omega^{(\varepsilon)}(f) \right|, \tag{5.2.5}$$

where for $f \in \bigcup_{1\le p<\infty} L^p(\mathbf{R})$, $\varepsilon > 0$, and $x \in \mathbf{R}^n$, $T_\Omega^{(\varepsilon)}(f)(x)$ is defined in terms of the absolutely convergent integral

$$T_\Omega^{(\varepsilon)}(f)(x) = \int_{|y|\ge\varepsilon} f(x-y)\frac{\Omega(y/|y|)}{|y|^n}\,dy.$$

To examine the relationship between $T_\Omega^{(*)}$ and $T_\Omega^{(**)}$ for $\Omega \in L^\infty(\mathbf{S}^{n-1})$, notice that

$$\left| \int_{\varepsilon\le|y|\le N} f(x-y)\frac{\Omega(y/|y|)}{|y|^n}\,dy \right| \le \sup_{0<N<\infty} \left| T_\Omega^{(\varepsilon,N)}(f)(x) \right|. \tag{5.2.6}$$

Then for $f \in L^p(\mathbf{R}^n)$, $1 \le p < \infty$, we let $N \to \infty$ on the left in (5.2.6) and we note that the limit exists in view of the absolute convergence of the integral. Then we take the supremum over $\varepsilon > 0$ to deduce that $T_\Omega^{(*)}$ is pointwise bounded by $T_\Omega^{(**)}$. Since $T_\Omega^{(\varepsilon,N)} = T_\Omega^{(\varepsilon)} - T_\Omega^{(N)}$, it also follows that $T_\Omega^{(**)} \le 2T_\Omega^{(*)}$; thus $T_\Omega^{(*)}$ and $T_\Omega^{(**)}$ are pointwise comparable when Ω lies in $L^\infty(\mathbf{S}^{n-1})$. This is the case with the Hilbert transform, that is, $H^{(**)}$ is comparable to $H^{(*)}$; likewise with the Riesz transforms.

A certain class of multipliers can be realized as singular integral operators of the kind discussed. Recall from Proposition 2.4.7 that if m is homogeneous of degree 0 and infinitely differentiable on the sphere, then $m^\vee$ is given by

$$m^\vee = c\,\delta_0 + W_\Omega,$$

for some complex constant c and some smooth Ω on $\mathbf{S}^{n-1}$ with mean value zero. Therefore, all convolution operators whose multipliers are homogeneous of degree zero smooth functions on $\mathbf{S}^{n-1}$ can be realized as a constant multiple of the identity plus an operator of the form T_Ω.

Example 5.2.2. Let $P(\xi) = \sum_{|\alpha|=k} b_\alpha \xi^\alpha$ be a homogeneous polynomial of degree k in $\mathbf{R}^n$ that vanishes only at the origin. Let α be a multi-index of order k. Then the function

$$m(\xi) = \frac{\xi^\alpha}{P(\xi)} \tag{5.2.7}$$

is infinitely differentiable on the sphere and homogeneous of degree zero. The operator given by multiplication on the Fourier transform by $m(\xi)$ is a constant multiple of the identity plus an operator given by convolution with a distribution of the form W_Ω for some Ω in $\mathscr{C}^\infty(\mathbf{S}^{n-1})$ with mean value zero. In this section we establish the L^p boundedness of such operators when Ω has appropriate smoothness on the sphere. This, in particular, implies that $m(\xi)$ defined by (5.2.7) lies in the space $\mathscr{M}_p(\mathbf{R}^n)$, defined in Section 2.5, for $1 < p < \infty$.

5.2.2 L^2 Boundedness of Homogeneous Singular Integrals

Next we would like to compute the Fourier transform of W_Ω. This provides information as to whether the operator given by convolution with K_Ω is L^2 bounded. We have the following result.

Proposition 5.2.3. *Let $n \ge 2$ and $\Omega \in L^1(\mathbf{S}^{n-1})$ have mean value zero. Then the Fourier transform of W_Ω is a (finite a.e.) function given by the formula*

$$\widehat{W_\Omega}(\xi) = \int_{\mathbf{S}^{n-1}} \Omega(\theta)\left(\log\frac{1}{|\xi\cdot\theta|} - \frac{i\pi}{2}\operatorname{sgn}(\xi\cdot\theta)\right)d\theta. \tag{5.2.8}$$

Remark 5.2.4. We need to show that the function of ξ on the right in (5.2.8) is well defined and finite for almost all ξ in $\mathbf{R}^n$. Write $\xi = |\xi|\xi'$ where $\xi' \in \mathbf{S}^{n-1}$ and notice that

$$\log\frac{1}{|\xi\cdot\theta|} = \log\frac{1}{|\xi|} + \log\frac{1}{|\xi'\cdot\theta|}.$$

Since Ω has mean value zero, the term $\log\frac{1}{|\xi|}$ multiplied by $\Omega(\theta)$ vanishes when integrated over the sphere.

We need to show that

$$\int_{\mathbf{S}^{n-1}} |\Omega(\theta)|\log\frac{1}{|\xi'\cdot\theta|}\,d\theta < \infty \tag{5.2.9}$$

for almost all $\xi' \in \mathbf{S}^{n-1}$. Integrate (5.2.9) over $\xi' \in \mathbf{S}^{n-1}$ and apply Fubini's theorem to obtain

$$\begin{aligned}
\int_{\mathbf{S}^{n-1}} |\Omega(\theta)| \int_{\mathbf{S}^{n-1}} &\log\frac{1}{|\xi'\cdot\theta|}\,d\xi'\,d\theta \\
&= \int_{\mathbf{S}^{n-1}} |\Omega(\theta)| \int_{\mathbf{S}^{n-1}} \log\frac{1}{|\xi_1|}\,d\xi\,d\theta \\
&= \omega_{n-2}\int_{\mathbf{S}^{n-1}} |\Omega(\theta)| \int_{-1}^{+1}\left(\log\frac{1}{|s|}\right)(1-s^2)^{\frac{n-3}{2}}\,ds\,d\theta \\
&= C_n\big\|\Omega\big\|_{L^1(\mathbf{S}^{n-1})} < \infty,
\end{aligned}$$

since we are assuming that $n \ge 2$. (The second-to-last identity follows from the identity in Appendix D.2.) We conclude that (5.2.9) holds for almost all $\xi' \in \mathbf{S}^{n-1}$.

Since the function of ξ on the right in (5.2.8) is homogeneous of degree zero, it follows that it is a locally integrable function on $\mathbf{R}^n$.

Before we return to the proof of Proposition 5.2.3, we discuss the following lemma:

Lemma 5.2.5. *Let a be a nonzero real number. Then for $0 < \varepsilon < N < \infty$ we have*

$$\lim_{\substack{\varepsilon\to 0\\ N\to\infty}} \int_\varepsilon^N \frac{\cos(ra)-\cos(r)}{r}\,dr = \log\frac{1}{|a|}, \tag{5.2.10}$$

$$\left|\int_\varepsilon^N \frac{\cos(ra)-\cos(r)}{r}\,dr\right| \le 2\left|\log\frac{1}{|a|}\right| \qquad \textit{for all } N>\varepsilon>0, \tag{5.2.11}$$

$$\lim_{\substack{\varepsilon\to 0\\ N\to\infty}} \int_\varepsilon^N \frac{e^{-ira}-\cos(r)}{r}\,dr = \log\frac{1}{|a|} - i\frac{\pi}{2}\operatorname{sgn} a, \tag{5.2.12}$$

$$\left|\int_\varepsilon^N \frac{e^{-ira}-\cos(r)}{r}\,dr\right| \le 2\left|\log\frac{1}{|a|}\right| + 4 \qquad \textit{for all } N>\varepsilon>0. \tag{5.2.13}$$

Proof. We first prove (5.2.10) and (5.2.11). By the fundamental theorem of calculus we write

$$\begin{aligned}
\int_\varepsilon^N \frac{\cos(ra)-\cos(r)}{r}\,dr &= \int_\varepsilon^N \frac{\cos(r|a|)-\cos(r)}{r}\,dr \\
&= -\int_\varepsilon^N \int_1^{|a|} \sin(tr)\,dt\,dr \\
&= -\int_1^{|a|} \int_\varepsilon^N \sin(tr)\,dr\,dt \\
&= -\int_1^{|a|} \frac{\cos(\varepsilon t)}{t}\,dt + \int_N^{N|a|} \frac{\cos(t)}{t}\,dt,
\end{aligned}$$

and from this expression, we clearly obtain (5.2.11). But the first integral of the same expression converges to $-\log|a|$ as $\varepsilon\to 0$ while the second integral converges to zero as $N\to\infty$ by an integration by parts. This proves (5.2.10).

To prove (5.2.12) and (5.2.13) we need to know that the expressions

$$\left|\int_\varepsilon^N \frac{\sin(ra)}{r}\,dr\right| = \left|\int_{\varepsilon|a|}^{N|a|} \frac{\sin(r)}{r}\,dr\right| \tag{5.2.14}$$

tend to $\frac{\pi}{2}$ as $\varepsilon\to 0$ and $N\to\infty$ and are bounded by 4. Both statements follow from Exercise 5.1.1. $\square$

Let us now prove Proposition 5.2.3.

Proof. Let us set $\xi' = \xi/|\xi|$. We have the following:

$$\begin{aligned}
\langle \widehat{W_\Omega}, \varphi\rangle &= \langle W_\Omega, \widehat{\varphi}\rangle \\
&= \lim_{\varepsilon\to 0} \int_{|x|\ge\varepsilon} \frac{\Omega(x/|x|)}{|x|^n}\,\widehat{\varphi}(x)\,dx \\
&= \lim_{\substack{\varepsilon\to 0\\ N\to\infty}} \int_{\varepsilon\le|x|\le N} \frac{\Omega(x/|x|)}{|x|^n}\,\widehat{\varphi}(x)\,dx \\
&= \lim_{\substack{\varepsilon\to 0\\ N\to\infty}} \int_{\mathbf{R}^n} \varphi(\xi) \int\limits_{\varepsilon\le|x|\le N} \frac{\Omega(x/|x|)}{|x|^n} e^{-2\pi i x\cdot\xi}\,dx\,d\xi \\
&= \lim_{\substack{\varepsilon\to 0\\ N\to\infty}} \int_{\mathbf{R}^n} \varphi(\xi) \int_{\mathbf{S}^{n-1}} \Omega(\theta) \int\limits_{\varepsilon\le r\le N} e^{-2\pi i r\theta\cdot\xi}\,\frac{dr}{r}\,d\theta\,d\xi
\end{aligned}$$

$$= \lim_{\substack{\varepsilon\to 0\\ N\to\infty}} \int_{\mathbf{R}^n} \varphi(\xi) \int_{\mathbf{S}^{n-1}} \Omega(\theta) \int_{\varepsilon\le r\le N} (e^{-2\pi r|\xi| i\theta\cdot\xi'} - \cos(2\pi r|\xi|)) \frac{dr}{r}\, d\theta\, d\xi$$

$$= \lim_{\substack{\varepsilon\to 0\\ N\to\infty}} \int_{\mathbf{R}^n} \varphi(\xi) \int_{\mathbf{S}^{n-1}} \Omega(\theta) \int_{\frac{\varepsilon}{2\pi|\xi|}\le r\le \frac{N}{2\pi|\xi|}} \frac{e^{-ir\theta\cdot\xi'} - \cos(r)}{r}\, dr\, d\theta\, d\xi$$

$$= \int_{\mathbf{R}^n} \varphi(\xi) \int_{\mathbf{S}^{n-1}} \Omega(\theta) \Big(\log\frac{1}{|\xi'\cdot\theta|} - \frac{i\pi}{2} \operatorname{sgn}(\xi\cdot\theta) \Big)\, d\theta\, d\xi,$$

where we used the Lebesgue dominated convergence theorem to pass the limit inside, Lemma 5.2.5, and Remark 5.2.4. We were able to subtract $\cos(2\pi r|\xi|)$ from the r integral in the previous calculation, since Ω has mean value zero over the sphere. Also, the use of the dominated convergence theorem is justified from the fact that the function

$$(\theta,\xi) \mapsto |\Omega(\theta)|\,|\varphi(\xi)| \Big(\log\frac{1}{|\xi'\cdot\theta|} + 4 \Big)$$

lies in $L^1(\mathbf{S}^{n-1}\times\mathbf{R}^n)$. Moreover, all the interchanges of integrals are well justified by Fubini's theorem. □

Corollary 5.2.6. *Let $\Omega \in L^1(\mathbf{S}^{n-1})$ have mean value zero. Then for almost all ξ' in $\mathbf{S}^{n-1}$ the integral*

$$\int_{\mathbf{S}^{n-1}} \Omega(\theta) \log\frac{1}{|\xi'\cdot\theta|}\, d\theta \tag{5.2.15}$$

converges absolutely. Moreover, the associated operator T_Ω maps $L^2(\mathbf{R}^n)$ to itself if and only if

$$\operatorname*{ess.sup}_{\xi'\in\mathbf{S}^{n-1}} \left| \int_{\mathbf{S}^{n-1}} \Omega(\theta) \log\frac{1}{|\xi'\cdot\theta|}\, d\theta \right| < \infty. \tag{5.2.16}$$

Proof. To obtain the absolute convergence of the integral in (5.2.15) we integrate over $\xi' \in \mathbf{S}^{n-1}$ and we apply Fubini's theorem. The assertion concerning the boundedness of T_Ω on L^2 is an immediate consequence of Proposition 5.2.3 and Theorem 2.5.10. □

There exist functions Ω in $L^1(\mathbf{S}^{n-1})$ with mean value zero such that the expressions in (5.2.16) are equal to infinity; consequently, not all such Ω give rise to bounded operators on $L^2(\mathbf{R}^n)$. Observe, however, that for Ω odd i.e., $\Omega(-\theta) = -\Omega(\theta)$ for all $\theta \in \mathbf{S}^{n-1}$, (5.2.16) trivially holds, since $\log\frac{1}{|\xi\cdot\theta|}$ is even and its product against an odd function must have integral zero over $\mathbf{S}^{n-1}$. We conclude that singular integrals T_Ω with odd Ω are always L^2 bounded.

5.2.3 The Method of Rotations

Having settled the issue of L^2 boundedness for singular integrals of the form T_Ω with Ω odd, we turn our attention to their L^p boundedness. A simple procedure called the method of rotations plays a crucial role in the study of operators T_Ω when Ω is an odd function. This method is based on the use of the directional Hilbert transforms. Fix a unit vector θ in $\mathbf{R}^n$. For a Schwartz function f on $\mathbf{R}^n$ let

$$\mathscr{H}_\theta(f)(x) = \frac{1}{\pi}\,\text{p.v.}\int_{-\infty}^{+\infty} f(x-t\theta)\frac{dt}{t}. \tag{5.2.17}$$

We call $\mathscr{H}_\theta(f)$ the *directional Hilbert transform* of f in the direction θ. For functions $f \in \mathscr{S}(\mathbf{R}^n)$ the integral in (5.2.17) is well defined, since it converges rapidly at infinity and by subtracting the constant $f(x)$, it also converges near zero.

Likewise, we define the *directional maximal Hilbert transforms*. For a function f in $\bigcup_{1\le p<\infty} L^p(\mathbf{R}^n)$ and $0<\varepsilon<N<\infty$ we let

$$\begin{aligned}
\mathscr{H}_\theta^{(\varepsilon,N)}(f)(x) &= \frac{1}{\pi}\int_{\varepsilon\le|t|\le N} f(x-t\theta)\frac{dt}{t},\\
\mathscr{H}_\theta^{(**)}(f)(x) &= \sup_{0<\varepsilon<N<\infty}\left|\mathscr{H}_\theta^{(\varepsilon,N)}(f)(x)\right|.
\end{aligned}$$

We observe that for any fixed $0<\varepsilon<N<\infty$ and $f\in L^p(\mathbf{R}^n)$, $\mathscr{H}_\theta^{(\varepsilon,N)}(f)$ is well defined almost everywhere. Indeed, by Minkowski's integral inequality we obtain

$$\left\|\mathscr{H}_\theta^{(\varepsilon,N)}(f)\right\|_{L^p(\mathbf{R}^n)} \le \frac{2}{\pi}\left\|f\right\|_{L^p(\mathbf{R}^n)}\log\frac{N}{\varepsilon}<\infty,$$

which implies that $\mathscr{H}_\theta^{(\varepsilon,N)}(f)(x)$ is finite for almost all $x\in\mathbf{R}^n$. Thus $\mathscr{H}_\theta^{(**)}(f)$ is well defined for f in $\bigcup_{1\le p<\infty} L^p(\mathbf{R}^n)$.

Theorem 5.2.7. *If Ω is odd and integrable over $\mathbf{S}^{n-1}$, then T_Ω and $T_\Omega^{(**)}$ are L^p bounded for all $1<p<\infty$. More precisely, T_Ω initially defined on Schwartz functions has a bounded extension on $L^p(\mathbf{R}^n)$ (which is also denoted by T_Ω).*

Proof. Let e_j be the usual unit vectors in $\mathbf{S}^{n-1}$. The operator $\mathscr{H}_{e_1}$ is obtained by applying the Hilbert transform in the first variable followed by the identity operator in the remaining variables. Clearly, $\mathscr{H}_{e_1}$ is bounded on $L^p(\mathbf{R}^n)$ with norm equal to that of the Hilbert transform on $L^p(\mathbf{R})$. Next observe that the following identity is valid for all matrices $A\in O(n)$:

$$\mathscr{H}_{A(e_1)}(f)(x) = \mathscr{H}_{e_1}(f\circ A)(A^{-1}x). \tag{5.2.18}$$

This implies that the L^p boundedness of $\mathscr{H}_\theta$ can be reduced to that of $\mathscr{H}_{e_1}$. We conclude that $\mathscr{H}_\theta$ is L^p bounded for $1<p<\infty$ with norm bounded by the norm of the Hilbert transform on $L^p(\mathbf{R})$ for every $\theta\in\mathbf{S}^{n-1}$.

Identity (5.2.18) is also valid for $\mathscr{H}_\theta^{(\varepsilon,N)}$ and $\mathscr{H}_\theta^{(**)}$. Consequently, $\mathscr{H}_\theta^{(**)}$ is bounded on $L^p(\mathbf{R}^n)$ for $1<p<\infty$ with norm at most that of $H^{(**)}$ on $L^p(\mathbf{R})$.

Next we realize a general singular integral T_Ω with Ω odd as an average of the directional Hilbert transforms $\mathscr{H}_\theta$. We start with f in $\bigcup_{1\le p<\infty} L^p(\mathbf{R}^n)$ and the following identities:

$$\begin{aligned}\int_{\varepsilon\le|y|\le N}\frac{\Omega(y/|y|)}{|y|^n}f(x-y)\,dy &= +\int_{\mathbf{S}^{n-1}}\Omega(\theta)\int_{r=\varepsilon}^{N}f(x-r\theta)\frac{dr}{r}\,d\theta\\ &= -\int_{\mathbf{S}^{n-1}}\Omega(\theta)\int_{r=\varepsilon}^{N}f(x+r\theta)\frac{dr}{r}\,d\theta,\end{aligned}$$

where the first follows by switching to polar coordinates and the second one is a consequence of the first one and the fact that Ω is odd via the change variables $\theta\mapsto-\theta$. Averaging the two identities, we obtain

$$\begin{aligned}&\int_{\varepsilon\le|y|\le N}\frac{\Omega(y/|y|)}{|y|^n}f(x-y)\,dy\\ &\quad=\frac{1}{2}\int_{\mathbf{S}^{n-1}}\Omega(\theta)\int_{r=\varepsilon}^{N}\frac{f(x-r\theta)-f(x+r\theta)}{r}\,dr\,d\theta \qquad (5.2.19)\\ &\quad=\frac{\pi}{2}\int_{\mathbf{S}^{n-1}}\Omega(\theta)\,\mathscr{H}_\theta^{(\varepsilon,N)}(f)(x)\,d\theta.\end{aligned}$$

It follows from the identity in (5.2.19) that

$$\int_{\varepsilon\le|y|\le N}\frac{\Omega(y/|y|)}{|y|^n}f(x-y)\,dy=\frac{\pi}{2}\int_{\mathbf{S}^{n-1}}\Omega(\theta)\,\mathscr{H}_\theta^{(\varepsilon,N)}(f)(x)\,d\theta\,, \qquad (5.2.20)$$

from which we conclude that

$$T_\Omega^{(**)}(f)(x)\le\frac{\pi}{2}\int_{\mathbf{S}^{n-1}}|\Omega(\theta)|\,\mathscr{H}_\theta^{(**)}(f)(x)\,d\theta\,. \qquad (5.2.21)$$

Using the Lebesgue dominated convergence theorem, we see that for f in $\mathscr{S}(\mathbf{R}^n)$, we can pass the limits as $\varepsilon\to0$ and $N\to\infty$ inside the integral in (5.2.20), concluding that

$$T_\Omega(f)(x)=\frac{\pi}{2}\int_{\mathbf{S}^{n-1}}\Omega(\theta)\,\mathscr{H}_\theta(f)(x)\,d\theta\,, \qquad (5.2.22)$$

for $f\in\mathscr{S}(\mathbf{R}^n)$. The L^p boundedness of T_Ω and $T_\Omega^{(**)}$ for Ω odd are then trivial consequences of (5.2.22) and (5.2.21) via Minkowski's integral inequality. □

Corollary 5.2.8. *The Riesz transforms R_j and the maximal Riesz transforms $R_j^{(*)}$ are bounded on $L^p(\mathbf{R}^n)$ for $1<p<\infty$.*

Proof. The assertion follows from the fact that the Riesz transforms have odd kernels. Since the kernel of R_j decays like $|x|^{-n}$ near infinity, it follows that $R_j^{(*)}(f)$ is well defined for $f\in L^p(\mathbf{R}^n)$. Since $R_j^{(*)}$ is pointwise bounded by $2R_j^{(**)}$, the conclusion follows from Theorem 5.2.7. □

Remark 5.2.9. It follows from the proof of Theorem 5.2.7 and from Theorems 5.1.7 and 5.1.12 that whenever Ω is an odd function on $\mathbf{S}^{n-1}$, we have

$$\big\|T_\Omega\big\|_{L^p\to L^p} \le \big\|\Omega\big\|_{L^1}\begin{cases} a\,p & \text{when } p\ge 2,\\ a\,(p-1)^{-1} & \text{when } 1<p\le 2,\end{cases}$$

$$\big\|T_\Omega^{(**)}\big\|_{L^p\to L^p} \le \big\|\Omega\big\|_{L^1}\begin{cases} a\,p & \text{when } p\ge 2,\\ a\,(p-1)^{-1} & \text{when } 1<p\le 2,\end{cases}$$

for some $a>0$ independent of p and the dimension.

5.2.4 Singular Integrals with Even Kernels

Since a general integrable function Ω on $\mathbf{S}^{n-1}$ with mean value zero can be written as a sum of an odd and an even function, it suffices to study singular integral operators T_Ω with even kernels. For the rest of this section, fix an integrable even function Ω on $\mathbf{S}^{n-1}$ with mean value zero. The following idea is fundamental in the study of such singular integrals. Proposition 5.1.16 implies that

$$T_\Omega = -\sum_{j=1}^{n} R_jR_jT_\Omega\,. \tag{5.2.23}$$

If R_jT_Ω were another singular integral operator of the form T_{Ω_j} for some odd Ω_j, then the boundedness of T_Ω would follow from that of T_{Ω_j} via the identity (5.2.23) and Theorem 5.2.7. It turns out that R_jT_Ω does have an odd kernel, but it may not be integrable on $\mathbf{S}^{n-1}$ unless Ω itself possesses an additional amount of integrability. The amount of extra integrability needed is logarithmic, more precisely of this sort:

$$c_\Omega = \int_{\mathbf{S}^{n-1}} |\Omega(\theta)|\log^+|\Omega(\theta)|\,d\theta<\infty\,. \tag{5.2.24}$$

Observe that

$$\big\|\Omega\big\|_{L^1}\le c_\Omega + e\,\omega_{n-1}\le C_n\,(c_\Omega+1)\,,$$

which says that the norm $\|\Omega\|_{L^1}$ is always controlled by a dimensional constant multiple of $c_\Omega+1$. The following theorem is the main result of this section.

Theorem 5.2.10. *Let $n\ge 2$ and let Ω be an even integrable function on $\mathbf{S}^{n-1}$ with mean value zero that satisfies (5.2.24). Then the corresponding singular integral T_Ω is bounded on $L^p(\mathbf{R}^n)$, $1<p<\infty$, with norm at most a dimensional constant multiple of the quantity* $\max\big((p-1)^{-2},p^2\big)(c_\Omega+1)$.

If the operator T_Ω in Theorem 5.2.10 is weak type $(1,1)$, then the estimate on the L^p operator norm of T_Ω can be improved to $\|T_\Omega\|_{L^p\to L^p}\le C_n(p-1)^{-1}$ as $p\to 1$. This is indeed the case; see the historical comments at the end of this chapter.

Proof. Let W_Ω be the distributional kernel of T_Ω. We have that W_Ω coincides with the function $\Omega(x/|x|)|x|^{-n}$ on $\mathbf{R}^n \setminus \{0\}$. Using Proposition 5.2.3 and the fact that Ω is an even function, we obtain the formula

$$\widehat{W_\Omega}(\xi) = \int_{\mathbf{S}^{n-1}} \Omega(\theta) \log \frac{1}{|\xi \cdot \theta|} \, d\theta \,, \tag{5.2.25}$$

which implies that $\widehat{W_\Omega}$ is itself an even function. Now, using Exercise 5.2.3 and condition (5.2.24), we conclude that $\widehat{W_\Omega}$ is a bounded function. Therefore, T_Ω is L^2 bounded. To obtain the L^p boundedness of T_Ω, we use the idea mentioned earlier involving the Riesz transforms. In view of (5.1.46), we have that

$$T_\Omega = -\sum_{j=1}^{n} R_j T_j, \tag{5.2.26}$$

where $T_j = R_j T_\Omega$. Equality (5.2.26) makes sense as an operator identity on $L^2(\mathbf{R}^n)$, since T_Ω and each R_j are well defined and bounded on $L^2(\mathbf{R}^n)$.

The kernel of the operator T_j is the inverse Fourier transform of the distribution $-i\frac{\xi_j}{|\xi|}\widehat{W_\Omega}(\xi)$, which we denote by K_j. At this point we know only that K_j is a tempered distribution whose Fourier transform is the function $-i\frac{\xi_j}{|\xi|}\widehat{W_\Omega}(\xi)$. Our first goal is to show that K_j coincides with an integrable function on an annulus. To prove this assertion we write

$$W_\Omega = W_\Omega^0 + W_\Omega^1 + W_\Omega^\infty \,,$$

where W_Ω^0 is a distribution and $W_\Omega^1, W_\Omega^\infty$ are functions defined by

$$\begin{aligned}
\langle W_\Omega^0, \varphi \rangle &= \lim_{\varepsilon \to 0} \int_{\varepsilon < |x| \le \frac{1}{2}} \frac{\Omega(x/|x|)}{|x|^n} \varphi(x) \, dx \,, \\
W_\Omega^1(x) &= \frac{\Omega(x/|x|)}{|x|^n} \chi_{\frac{1}{2} \le |x| \le 2} \,, \\
W_\Omega^\infty(x) &= \frac{\Omega(x/|x|)}{|x|^n} \chi_{2 < |x|} \,.
\end{aligned}$$

We now fix a $j \in \{1, 2, \dots, n\}$ and we write

$$K_j = K_j^0 + K_j^1 + K_j^\infty \,,$$

where

$$\begin{aligned}
K_j^0 &= \big(-i\tfrac{\xi_j}{|\xi|}\widehat{W_\Omega^0}(\xi)\big)^\vee \,, \\
K_j^1 &= \big(-i\tfrac{\xi_j}{|\xi|}\widehat{W_\Omega^1}(\xi)\big)^\vee \,, \\
K_j^\infty &= \big(-i\tfrac{\xi_j}{|\xi|}\widehat{W_\Omega^\infty}(\xi)\big)^\vee \,.
\end{aligned}$$

Notice that K_j^0 is well defined via Theorem 2.3.21.

Define the annulus

$$A = \{x \in \mathbf{R}^n : \, 2/3 < |x| < 3/2\}.$$

For a smooth function ϕ supported in the annulus $2/3 < |x| < 3/2$ we have

$$\begin{aligned}
\langle K_j^0, \phi\rangle &= \langle (-i\tfrac{\xi_j}{|\xi|}\widehat{W_\Omega^0}(\xi))^\vee, \phi\rangle \\
&= \langle -i\tfrac{\xi_j}{|\xi|}\widehat{W_\Omega^0}(\xi), \phi^\vee(\xi)\rangle \\
&= \langle \widehat{W_\Omega^0}(\xi), -i\tfrac{\xi_j}{|\xi|}\phi^\vee(\xi)\rangle \\
&= \langle W_\Omega^0, (-i\tfrac{\xi_j}{|\xi|}\phi^\vee(\xi))^\wedge\rangle \\
&= -\langle W_\Omega^0, \widetilde{R_j(\phi)}\rangle \\
&= -\lim_{\varepsilon\to 0}\int_{\varepsilon<|y|<1/2} \frac{\Omega(y/|y|)}{|y|^n} R_j(\phi)(-y)\,dy \qquad (\Omega \text{ is even}) \\
&= -\frac{\Gamma(\frac{n+1}{2})}{\pi^{\frac{n+1}{2}}}\lim_{\varepsilon\to 0}\int_{\varepsilon<|y|<1/2} \frac{\Omega(y/|y|)}{|y|^n}\int_{R^n}\frac{y_j-x_j}{|y-x|^{n+1}}\phi(x)\,dx\,dy,
\end{aligned}$$

where the action of the distribution W_Ω^0 on $R_j(\phi)$ is justified by fact that $R_j(\phi)(y)$ is smooth on the support of W_Ω^0; note $|x-y| \geq 1/6$. Moreover, $\langle \widehat{W_\Omega^0}(\xi), -i\frac{\xi_j}{|\xi|}\phi^\vee(\xi)\rangle$ should be interpreted as a convergent integral.

It follows that for $x \in A$, the absolute value of the convolution of W_Ω^0 with the kernel of the Riesz transform R_j is

$$\left|\frac{\Gamma(\frac{n+1}{2})}{\pi^{\frac{n+1}{2}}}\lim_{\varepsilon\to 0}\int_{\varepsilon<|y|<\frac{1}{2}} \frac{x_j-y_j}{|x-y|^{n+1}}\frac{\Omega(y/|y|)}{|y|^n}\,dy\right| \tag{5.2.27}$$

$$= \left|\frac{\Gamma(\frac{n+1}{2})}{\pi^{\frac{n+1}{2}}}\int_{|y|<\frac{1}{2}}\left(\frac{x_j-y_j}{|x-y|^{n+1}} - \frac{x_j}{|x|^{n+1}}\right)\frac{\Omega(y/|y|)}{|y|^n}\,dy\right| \tag{5.2.28}$$

$$\begin{aligned}
&\leq \int_{|y|\leq\frac{1}{2}} C_n|y|\,\frac{|\Omega(y/|y|)|}{|y|^n}\,dy \\
&= C_n'\,\|\Omega\|_{L^1},
\end{aligned}$$

where we used the fact that $\Omega(y/|y|)|y|^{-n}$ has integral zero over annuli of the form $\varepsilon < |y| < \frac{1}{2}$, the mean value theorem applied to the function $x_j|x|^{-(n+1)}$, and the fact that $|x-y| \geq 1/6$ for x in the annulus A. We conclude that on A, K_j^0 coincides with the bounded function inside the absolute value in (5.2.27).

Likewise, for $x \in A$ we have

$$\begin{aligned}
&\frac{\Gamma(\frac{n+1}{2})}{\pi^{\frac{n+1}{2}}}\left|\int_{|y|>2} \frac{x_j-y_j}{|x-y|^{n+1}} \frac{\Omega(y/|y|)}{|y|^n}\,dy\right| \qquad (5.2.29)\\
&\qquad \le \frac{\Gamma(\frac{n+1}{2})}{\pi^{\frac{n+1}{2}}}\int_{|y|>2} \frac{1}{|x-y|^{n}} \frac{|\Omega(y/|y|)|}{|y|^n}\,dy\\
&\qquad \le \frac{\Gamma(\frac{n+1}{2})}{\pi^{\frac{n+1}{2}}}\int_{|y|>2} \frac{4^n}{|y|^{2n}} |\Omega(y/|y|)|\,dy\\
&\qquad = C\big\|\Omega\big\|_{L^1},
\end{aligned}$$

from which it follows that on the annulus A, K_j^∞ coincides with the bounded function inside the absolute value in (5.2.29) or in (5.2.28).

Now observe that condition (5.2.24) gives that the function W_Ω^1 satisfies

$$\begin{aligned}
&\int_{|x|\le 2} |W_\Omega^1(x)|\,\log^+|W_\Omega^1(x)|\,dx\\
&\qquad \le \int_{1/2}^{2}\int_{\mathbf{S}^{n-1}} \frac{|\Omega(\theta)|}{r^n}\log^+[2^n|\Omega(\theta)|]\,d\theta\, r^{n-1}\frac{dr}{r}\\
&\qquad \le (\log 4)\,\big[n(\log 2)\big\|\Omega\big\|_{L^1}+c_\Omega\big] < \infty .
\end{aligned}$$

Since the Riesz transform R_j is countably subadditive and maps L^p to L^p with norm at most $4(p-1)^{-1}$ for $1<p<2$, it follows from Exercise 1.3.7 that $K_j^1 = R_j(W_\Omega^1)$ is integrable over the ball $|x|\le 3/2$ and moreover, it satisfies

$$\int_A |K_j^1(x)|\,dx \le C_n\left[\int_{|x|\le 2} |W_\Omega^1(x)|\,\log^+|W_\Omega^1(x)|\,dx+1\right] \le C_n'(c_\Omega+1).$$

Furthermore, since $\widehat{K_j}$ is homogeneous of degree zero, K_j is a homogeneous distribution of degree $-n$ (Exercise 2.3.9). This means that for all test functions φ and all $\lambda>0$ we have

$$\big\langle K_j, \delta^\lambda(\varphi)\big\rangle = \big\langle K_j,\varphi\big\rangle, \qquad (5.2.30)$$

where $\delta^\lambda(\varphi)(x)=\varphi(\lambda x)$. But for $\varphi\in\mathscr{C}_0^\infty$ supported in the annulus $3/4<|x|<4/3$ and for λ in $(8/9,9/8)$ we have that $\delta^{\lambda^{-1}}(\varphi)$ is supported in A and thus we can express (5.2.30) as convergent integrals as follows:

$$\int_{\mathbf{R}^n} K_j(x)\varphi(x)\,dx = \int_{\mathbf{R}^n} K_j(x)\varphi(\lambda^{-1}x)\,dx = \int_{\mathbf{R}^n} \lambda^n K_j(\lambda x)\varphi(x)\,dx . \qquad (5.2.31)$$

From this it would be ideal to be able to directly obtain that $K_j(x)=\lambda^n K_j(\lambda x)$ for all $8/9<|x|<9/8$ and $8/9<\lambda<9/8$, in particular when $\lambda=|x|^{-1}$. But unfortunately, we can only deduce that for every $\lambda\in(8/9,9/8)$, $K_j(x)=\lambda^n K_j(\lambda x)$ holds for all x in the annulus except a set of measure zero that depends on λ. To be able to define the restriction of K_j on $\mathbf{S}^{n-1}$, we employ a more delicate argument.

For any J subinterval of $[8/9,9/8]$ we obtain from (5.2.31) that

$$\int_{\mathbf{R}^n} K_j(x)\varphi(x)\,dx = \int_{\mathbf{R}^n} \fint_J \lambda^n K_j(\lambda x)\,d\lambda\,\varphi(x)\,dx,$$

where integral with the slashed integral denotes the average of a function over the set J. Since φ was an arbitrary $\mathscr{C}_0^\infty$ function supported in the annulus $3/4 < |x| < 4/3$, it follows that for every J subinterval of $[8/9,9/8]$, there is a null subset E_J of the annulus $A' = \{x:\ 27/32 < |x| < 32/27\}$ such that

$$K_j(x) = \fint_J \lambda^n K_j(\lambda x)\,d\lambda \tag{5.2.32}$$

for all $x \in A' \setminus E_J$.

Let $J_0 = [\sqrt{8/9}, \sqrt{9/8}]$. We claim that there is a set of null subset E of A' such that for all $x \in A' \setminus E$ we have

$$\fint_{J_0} \lambda^n K_j(\lambda x)\,d\lambda = \fint_{rJ_0} \lambda^n K_j(\lambda x)\,d\lambda \tag{5.2.33}$$

for every r in J_0. Indeed, let E be the union of E_{rJ_0} over all r in $J_0 \cap \mathbf{Q}$. Then in view of (5.2.32), identity (5.2.33) holds for $x \in A' \setminus E$ and $J_0 \cap \mathbf{Q}$. But for a fixed x in $A' \setminus E$, the function of r on the right hand side of (5.2.33) is constant on the rationals and is also continuous (in r), hence it must be constant for all $r \in J_0$. Thus the claim follows since both sides of (5.2.33) are equal to (5.2.32).

Writing $x = \delta\theta$, where $27/32 < \delta < 32/27$ and $\theta \in \mathbf{S}^{n-1}$, it follows by Fubini's theorem that there is a $\delta \in (27/32, 32/27)$ (in fact almost all δ have this property) such that

$$\fint_{J_0} \lambda^n K_j(\lambda\delta\theta)\,d\lambda = \fint_{rJ_0} \lambda^n K_j(\lambda\delta\theta)\,d\lambda \tag{5.2.34}$$

for almost all $\theta \in \mathbf{S}^{n-1}$ and all $r \in J_0$. We fix such a δ, which we denote δ_0.

We now define a function Ω_j on $\mathbf{S}^{n-1}$ by setting

$$\Omega_j(\theta) = \fint_{J_0} \delta_0^n \lambda^n K_j(\lambda\delta_0\theta)\,d\lambda = \fint_{rJ_0} \delta_0^n \lambda^n K_j(\lambda\delta_0\theta)\,d\lambda$$

for all $r \in J_0$. The function Ω_j is defined almost everywhere and is integrable over $\mathbf{S}^{n-1}$, since K_j is integrable over the annulus A.

Let $e_1 = (1,0,\dots,0)$. Let Ψ be a $\mathscr{C}_0^\infty(\mathbf{R}^n)$ nonzero, nonnegative, radial, and supported in the annulus $32/(27\sqrt{2}) < |x| < 27\sqrt{2}/32$ around $\mathbf{S}^{n-1}$. We start with

$$\Omega_j(\theta) = \fint_{r^{-1}J_0} \delta_0^n \lambda^n K_j(\lambda\delta_0\theta)\,d\lambda = \fint_{J_0} \delta_0^n r^n \lambda^n K_j(r\lambda\delta_0\theta)\,d\lambda,$$

which holds for all $r \in J_0$, we multiply by $\Psi(re_1)$, and we integrate over $\mathbf{S}^{n-1}$ and over $(0,\infty)$ with respect to the measure dr/r. We obtain

$$\begin{aligned}\int_0^\infty \Psi(re_1)\frac{dr}{r}\int_{\mathbf{S}^{n-1}}\Omega_j(\theta)\,d\theta &= \fint_{J_0}\int_0^\infty\int_{\mathbf{S}^{n-1}}\delta_0^n\lambda^n K_j(\lambda\delta_0 r\theta)\Psi(re_1)r^n d\theta\frac{dr}{r}d\lambda\\ &= \fint_{J_0}\int_{\mathbf{R}^n}\delta_0^n\lambda^n K_j(\lambda\delta_0 x)\,\Psi(x)dxd\lambda\\ &= \fint_{J_0}\int_{\mathbf{R}^n}K_j(x)\Psi((\lambda\delta_0)^{-1}x)\,dxd\lambda\\ &= \fint_{J_0}\langle K_j,\Psi\rangle\,d\lambda\,,\\ &= \langle K_j,\Psi\rangle\end{aligned}$$

in view of the homogeneity of K_j. But also, for some constant c'_Ψ we have

$$\langle K_j,\Psi\rangle = \langle\widehat{K_j},\widehat{\Psi}\rangle = \int_{\mathbf{R}^n}\frac{-i\xi_j}{|\xi|}\widehat{W_\Omega}(\xi)\widehat{\Psi}(\xi)\,d\xi = c'_\Psi\int_{\mathbf{S}^{n-1}}\frac{-i\theta_j}{|\theta|}\widehat{W_\Omega}(\theta)\,d\theta = 0,$$

since by (5.2.25), $\frac{-i\xi_j}{|\xi|}\widehat{W_\Omega}(\xi)$ is an odd function. We conclude that Ω_j has mean value zero over $\mathbf{S}^{n-1}$.

Thus $\Omega\in L^1(\mathbf{S}^{n-1})$ has mean value zero and the distribution W_{Ω_j} is well defined. We claim that

$$K_j = W_{\Omega_j}\,. \tag{5.2.35}$$

To establish (5.2.35), we show first that $\langle K_j,\varphi\rangle = \langle W_{\Omega_j},\varphi\rangle$ whenever φ is supported in the annulus $8/9<|x|<9/8$. Using (5.2.32) we have

$$\begin{aligned}\int_{\mathbf{R}^n}K_j(x)\varphi(x)\,dx &= \int_{\mathbf{R}^n}\fint_{J_0}K_j(\delta_0\lambda x)\delta_0^n\lambda^n d\lambda\,\varphi(x)\,dx\\ &= \int_0^\infty\int_{\mathbf{S}^{n-1}}\fint_{J_0}K_j(\delta_0\lambda r\theta)\delta_0^n\lambda^n r^n\,d\lambda\,\varphi(r\theta)\,d\theta\frac{dr}{r}\\ &= \int_0^\infty\int_{\mathbf{S}^{n-1}}\fint_{rJ_0}K_j(\delta_0\lambda'\theta)\delta_0^n(\lambda')^n d\lambda'\,\varphi(r\theta)\,d\theta\frac{dr}{r}\\ &= \int_0^\infty\int_{\mathbf{S}^{n-1}}\Omega_j(\theta)\varphi(r\theta)\,d\theta\frac{dr}{r}\\ &= \langle W_{\Omega_j},\varphi\rangle,\end{aligned}$$

having used (5.2.34) in the second to last equality.

Given a general $\mathscr{C}_0^\infty$ function φ whose support is contained in an annulus of the form $M^{-1}<|x|<M$, for some $M>0$, via a smooth partition of unity, we write φ as a finite sum of smooth functions φ_k whose supports are contained in annuli of the form $8s/9<|x|<9s/8$ for some $s>0$. These annuli can be brought inside the annulus $8/9<|x|<9/8$ by a dilation. Since both K_j and W_{Ω_j} are homogeneous distributions of degree $-n$ and agree on the annulus $8/9<|x|<9/8$ they must agree on annuli $8s/9<|x|<9s/8$. Consequently, $\langle K_j,\varphi\rangle = \langle W_{\Omega_j},\varphi\rangle$ for all $\varphi\in\mathscr{C}_0^\infty(\mathbf{R}^n\setminus\{0\})$. Therefore, $K_j - W_{\Omega_j}$ is supported at the origin, and since it is homogeneous of degree $-n$, it must be equal to $b\delta_0$, a constant multiple of the Dirac mass. But $\widehat{K_j}$ is an odd function and hence K_j is also odd. It follows that W_{Ω_j} is an odd function on $\mathbf{R}^n \setminus \{0\}$, which implies that Ω_j is an odd function. We say that $u \in \mathscr{S}'(\mathbf{R}^n)$ is odd if $\widetilde{u} = -u$, where $\widetilde{u}$ is defined by $\langle \widetilde{u}, \psi \rangle = \langle u, \widetilde{\psi} \rangle$ for all $\psi \in \mathscr{S}(\mathbf{R}^n)$ and $\widetilde{\psi}(x) = \psi(-x)$. We have that $K_j - W_{\Omega_j}$ is an odd distribution, and thus $b\delta_0$ must be an odd distribution. But if $b\delta_0$ is odd, then $b = 0$. We conclude that for each j there exists an odd integrable function Ω_j on $\mathbf{S}^{n-1}$ with $\|\Omega_j\|_{L^1}$ controlled by a constant multiple of $c_\Omega + 1$ such that (5.2.35) holds.

Then we use (5.2.26) and (5.2.35) to write

$$T_\Omega = -\sum_{j=1}^{n} R_j T_{\Omega_j},$$

and appealing to the boundedness of each T_{Ω_j} (Theorem 5.2.7) and to that of the Riesz transforms, we obtain the required L^p boundedness for T_Ω. □

We note that Theorem 5.2.10 holds for all $\Omega \in L^1(\mathbf{S}^{n-1})$ that satisfy (5.2.24), not necessarily even Ω. Simply write $\Omega = \Omega_e + \Omega_o$, where Ω_e is even and Ω_o is odd, and check that condition (5.2.24) holds for Ω_e.

5.2.5 Maximal Singular Integrals with Even Kernels

We have the corresponding theorem for maximal singular integrals.

Theorem 5.2.11. *Let Ω be an even integrable function on $\mathbf{S}^{n-1}$ with mean value zero that satisfies (5.2.24). Then the corresponding maximal singular integral $T_\Omega^{(**)}$, defined in (5.2.4), is bounded on $L^p(\mathbf{R}^n)$ for $1 < p < \infty$ with norm at most a dimensional constant multiple of* $\max(p^2, (p-1)^{-2})(c_\Omega + 1)$.

Proof. For $f \in L^1_{\mathrm{loc}}(\mathbf{R}^n)$, x define the maximal function of f in the direction θ by setting

$$M_\theta(f)(x) = \sup_{a>0} \frac{1}{2a} \int_{|r|\le a} |f(x - r\theta)|\, dr. \tag{5.2.36}$$

In view of Exercise 5.2.5 we have that M_θ is bounded on $L^p(\mathbf{R}^n)$ with norm at most $3\,p(p-1)^{-1}$.

Fix Φ a smooth radial function such that $\Phi(x) = 0$ for $|x| \le 1/4$, $\Phi(x) = 1$ for $|x| \ge 3/4$, and $0 \le \Phi(x) \le 1$ for all x in $\mathbf{R}^n$. For $f \in L^p(\mathbf{R}^n)$ and $0 < \varepsilon < N < \infty$ we introduce the smoothly truncated singular integral

$$\widetilde{T}_\Omega^{(\varepsilon,N)}(f)(x) = \int_{\mathbf{R}^n} \frac{\Omega\big(\frac{y}{|y|}\big)}{|y|^n} \Big(\Phi\big(\tfrac{y}{\varepsilon}\big) - \Phi\big(\tfrac{y}{N}\big) \Big) f(x-y)\, dy$$

and the corresponding maximal singular integral operator

$$\widetilde{T}_\Omega^{(**)}(f) = \sup_{0<N<\infty} \sup_{0<\varepsilon<N} |\widetilde{T}_\Omega^{(\varepsilon,N)}(f)|. \tag{5.2.37}$$

Computing the supremum in (5.2.37), we first consider the case where $N > 4\varepsilon$.

For f in $L^p(\mathbf{R}^n)$ (for some $1 < p < \infty$), we have

$$\begin{aligned}
&\big|\widetilde{T}_\Omega^{(\varepsilon,N)}(f)(x) - T_\Omega^{(\varepsilon,N)}(f)(x)\big| \\
&= \Bigg| \int\limits_{\frac{\varepsilon}{4}\le|y|\le\varepsilon} \frac{\Omega\big(\frac{y}{|y|}\big)}{|y|^n}\Phi\big(\tfrac{y}{\varepsilon}\big) f(x-y)\,dy - \int\limits_{\frac{N}{4}\le|y|\le N} \frac{\Omega\big(\frac{y}{|y|}\big)}{|y|^n}\Phi\big(\tfrac{y}{N}\big) f(x-y)\,dy \Bigg| \\
&\le \Bigg[\int\limits_{\frac{\varepsilon}{4}\le|y|\le\varepsilon} \frac{|\Omega\big(\frac{y}{|y|}\big)|}{|y|^n}|f(x-y)|\,dy + \int\limits_{\frac{N}{4}\le|y|\le N} \frac{|\Omega\big(\frac{y}{|y|}\big)|}{|y|^n}|f(x-y)|\,dy \Bigg] \\
&\le \int\limits_{\mathbf{S}^{n-1}} |\Omega(\theta)| \left[\frac{4}{\varepsilon}\int_{\frac{\varepsilon}{4}}^{\varepsilon} |f(x-r\theta)|\,dr + \frac{4}{N}\int_{\frac{N}{4}}^{N} |f(x-r\theta)|\,dr \right] d\theta \\
&\le 16 \int_{\mathbf{S}^{n-1}} |\Omega(\theta)| M_\theta(f)(x)\,d\theta\,.
\end{aligned}$$

Now if $N \le 4\varepsilon$, then the function $\Phi\big(\frac{y}{\varepsilon}\big) - \Phi\big(\frac{y}{N}\big) - \chi_{\varepsilon\le|y|\le N}$ is bounded by 3 and is supported in the annulus $\frac{\varepsilon}{4} \le |y| \le 4\varepsilon$. In this case we obtain

$$\begin{aligned}
\big|\widetilde{T}_\Omega^{(\varepsilon,N)}(f)(x) - T_\Omega^{(\varepsilon,N)}(f)(x)\big| &\le 3 \int\limits_{\mathbf{S}^{n-1}} |\Omega(\theta)| \int_{\frac{\varepsilon}{4}}^{4\varepsilon} |f(x-r\theta)| \frac{dr}{r}\,d\theta \\
&\le 96 \int_{\mathbf{S}^{n-1}} |\Omega(\theta)| M_\theta(f)(x)\,d\theta\,.
\end{aligned}$$

We deduce from these estimates that

$$\sup_{0<\varepsilon<N<\infty} \big|\widetilde{T}_\Omega^{(\varepsilon,N)}(f)(x) - T_\Omega^{(\varepsilon,N)}(f)(x)\big| \le 96 \int_{\mathbf{S}^{n-1}} |\Omega(\theta)| M_\theta(f)(x)\,d\theta\,.$$

Using the result of Exercise 5.2.5 we conclude that

$$\big\|\widetilde{T}_\Omega^{(**)}(f) - T_\Omega^{(**)}(f)\big\|_{L^p} \le 600 \big\|\Omega\big\|_{L^1} \max(p, (p-1)^{-1}) \big\|f\big\|_{L^p}\,.$$

This implies that it suffices to obtain the required L^p bound for the smoothly truncated maximal singular integral operator $\widetilde{T}_\Omega^{(**)}$.

Let K_j, Ω_j, and T_j be as in the previous theorem, and let F_j be the Riesz transform of the function $\Omega(x/|x|)\Phi(x)|x|^{-n}$. Let $f \in L^p(\mathbf{R}^n)$. A calculation yields the identity

$$\begin{aligned}
\widetilde{T}_\Omega^{(\varepsilon,N)}(f)(x) &= \int_{\mathbf{R}^n} \left[\frac{1}{\varepsilon^n}\frac{\Omega(\frac{y}{\varepsilon}/|\frac{y}{\varepsilon}|)}{|\frac{y}{\varepsilon}|^n}\Phi(\tfrac{y}{\varepsilon}) - \frac{1}{N^n}\frac{\Omega(\frac{y}{N}/|\frac{y}{N}|)}{|\frac{y}{N}|^n}\Phi(\tfrac{y}{N}) \right] f(x-y)\,dy \\
&= -\Big(\sum_{j=1}^n \Big[\tfrac{1}{\varepsilon^n}F_j\big(\tfrac{\cdot}{\varepsilon}\big) - \tfrac{1}{N^n}F_j\big(\tfrac{\cdot}{N}\big) \Big] * R_j(f) \Big)(x)\,,
\end{aligned}$$

where in the last step we used Proposition 5.1.16. Therefore we may write

$$\begin{aligned}
-\widetilde{T}_{\Omega}^{(\varepsilon,N)}(f)(x) &= \sum_{j=1}^{n}\int_{\mathbf{R}^n}\Big[\tfrac{1}{\varepsilon^n}F_j\big(\tfrac{x-y}{\varepsilon}\big)-\tfrac{1}{N^n}F_j\big(\tfrac{x-y}{N}\big)\Big]R_j(f)(y)\,dy \\
&= A_1^{(\varepsilon,N)}(f)(x)+A_2^{(\varepsilon,N)}(f)(x)+A_3^{(\varepsilon,N)}(f)(x),
\end{aligned} \tag{5.2.38}$$

where

$$\begin{aligned}
A_1^{(\varepsilon,N)}(f)(x) &= \sum_{j=1}^{n}\frac{1}{\varepsilon^n}\int_{|x-y|\le\varepsilon}F_j\big(\tfrac{x-y}{\varepsilon}\big)R_j(f)(y)\,dy \\
&\qquad\qquad -\sum_{j=1}^{n}\frac{1}{N^n}\int_{|x-y|\le N}F_j\big(\tfrac{x-y}{N}\big)R_j(f)(y)\,dy, \\
A_2^{(\varepsilon,N)}(f)(x) &= \sum_{j=1}^{n}\int_{\mathbf{R}^n}\Big[\tfrac{1}{\varepsilon^n}\chi_{|x-y|>\varepsilon}\big\{F_j\big(\tfrac{x-y}{\varepsilon}\big)-K_j\big(\tfrac{x-y}{\varepsilon}\big)\big\} \\
&\qquad\qquad -\tfrac{1}{N^n}\chi_{|x-y|>N}\big\{F_j\big(\tfrac{x-y}{N}\big)-K_j\big(\tfrac{x-y}{N}\big)\big\}\Big]R_j(f)(y)\,dy, \\
A_3^{(\varepsilon,N)}(f)(x) &= \sum_{j=1}^{n}\int_{\mathbf{R}^n}\Big[\tfrac{1}{\varepsilon^n}\chi_{|x-y|>\varepsilon}K_j\big(\tfrac{x-y}{\varepsilon}\big)-\tfrac{1}{N^n}\chi_{|x-y|>N}K_j\big(\tfrac{x-y}{N}\big)\Big]R_j(f)(y)\,dy.
\end{aligned}$$

It follows from the definitions of F_j and K_j that

$$\begin{aligned}
F_j(z)-K_j(z) &= \frac{\Gamma(\frac{n+1}{2})}{\pi^{\frac{n+1}{2}}}\lim_{\varepsilon\to0}\int_{\varepsilon\le|y|}\frac{\Omega(y/|y|)}{|y|^n}(\Phi(y)-1)\frac{z_j-y_j}{|z-y|^{n+1}}\,dy \\
&= \frac{\Gamma(\frac{n+1}{2})}{\pi^{\frac{n+1}{2}}}\int_{|y|\le\frac34}\frac{\Omega(y/|y|)}{|y|^n}(\Phi(y)-1)\left\{\frac{z_j-y_j}{|z-y|^{n+1}}-\frac{z_j}{|z|^{n+1}}\right\}dy
\end{aligned}$$

whenever $|z|\ge1$. But using the mean value theorem, the last expression is easily seen to be bounded by

$$C_n\int_{|y|\le\frac34}\frac{\Omega(y/|y|)}{|y|^n}\frac{|y|}{|z|^{n+1}}\,dy = C_n'\big\|\Omega\big\|_{L^1}|z|^{-(n+1)},$$

whenever $|z|\ge1$. Using this estimate, we obtain that the jth term in $A_2^{(\varepsilon,N)}(f)(x)$ is bounded by

$$C_n\frac{\|\Omega\|_{L^1}}{\varepsilon^n}\int\limits_{|x-y|>\varepsilon}\frac{|R_j(f)(y)|\,dy}{(|x-y|/\varepsilon)^{n+1}}\le C_n\frac{2\|\Omega\|_{L^1}}{2^{-n}\varepsilon^n}\int\limits_{\mathbf{R}^n}\frac{|R_j(f)(y)|\,dy}{\big(1+\frac{|x-y|}{\varepsilon}\big)^{n+1}}.$$

It follows that for functions f in L^p we have

$$\sup_{0<\varepsilon<N<\infty}|A_2^{(\varepsilon,N)}(f)|\le C_n\big\|\Omega\big\|_{L^1}M(R_j(f)),$$

in view of Theorem 2.1.10. (M here is the Hardy–Littlewood maximal operator.) By Theorem 2.1.6, M maps $L^p(\mathbf{R}^n)$ to itself with norm bounded by a dimensional constant multiple of $\max(1,(p-1)^{-1})$. Since by Remark 5.2.9 the norm $\|R_j\|_{L^p\to L^p}$ is controlled by a dimensional constant multiple of $\max(p,(p-1)^{-1})$, it follows that

$$\Big\| \sup_{0<\varepsilon<N<\infty} |A_2^{(\varepsilon,N)}(f)|\Big\|_{L^p} \le C_n \|\Omega\|_{L^1} \max(p,(p-1)^{-1})\|f\|_{L^p}. \tag{5.2.39}$$

Next, recall that in the proof of Theorem 5.2.10 we showed that

$$K_j(x)=\frac{\Omega_j(x/|x|)}{|x|^n},$$

where Ω_j are integrable functions on $\mathbf{S}^{n-1}$ that satisfy

$$\|\Omega_j\|_{L^1}\le C_n(c_\Omega+1). \tag{5.2.40}$$

Consequently, for functions f in $L^p(\mathbf{R}^n)$ we have

$$\sup_{0<\varepsilon<N<\infty} |A_3^{(\varepsilon,N)}(f)| \le 2\sum_{j=1}^n T_{\Omega_j}^{(**)}(R_j(f)),$$

and by Remark 5.2.9 this last expression has L^p norm at most a dimensional constant multiple of $\|\Omega_j\|_{L^1}\max(p,(p-1)^{-1})\|R_j(f)\|_{L^p}$. It follows that

$$\Big\| \sup_{0<\varepsilon<N<\infty} |A_3^{(\varepsilon,N)}(f)|\Big\|_{L^p} \le C_n \max(p^2,(p-1)^{-2})(c_\Omega+1)\,\|f\|_{L^p}. \tag{5.2.41}$$

Finally, we turn our attention to the term $A_1^{(\varepsilon,N)}(f)$. To prove the required estimate, we first show that there exist nonnegative homogeneous of degree zero functions G_j on $\mathbf{R}^n$ that satisfy

$$|F_j(x)|\le G_j(x) \qquad \text{when } |x|\le 1 \tag{5.2.42}$$

and

$$\int_{\mathbf{S}^{n-1}} |G_j(\theta)|\,d\theta \le C_n(c_\Omega+1). \tag{5.2.43}$$

To prove (5.2.42), first note that if $|x|\le 1/8$, then

$$\begin{aligned} |F_j(x)| &= \frac{\Gamma(\frac{n+1}{2})}{\pi^{\frac{n+1}{2}}}\left|\int_{\mathbf{R}^n}\frac{\Omega(y/|y|)}{|y|^n}\Phi(y)\frac{x_j-y_j}{|x-y|^{n+1}}dy\right| \\ &\le C_n\int_{|y|\ge\frac14}\frac{|\Omega(y/|y|)|}{|y|^{2n}}dy \\ &\le C_n'\|\Omega\|_{L^1}. \end{aligned}$$

We now fix an x satisfying $1/8 \le |x| \le 1$ and we write

$$\begin{aligned}
|F_j(x)| &\le \Phi(x)|K_j(x)| + |F_j(x) - \Phi(x)K_j(x)| \\
&\le |K_j(x)| + \frac{\Gamma(\frac{n+1}{2})}{\pi^{\frac{n+1}{2}}} \left| \lim_{\varepsilon \to 0} \int_{|y|>\varepsilon} \frac{x_j - y_j}{|x-y|^{n+1}} \big(\Phi(y) - \Phi(x)\big) \frac{\Omega(y/|y|)}{|y|^n} dy \right| \\
&= |K_j(x)| + \frac{\Gamma(\frac{n+1}{2})}{\pi^{\frac{n+1}{2}}} \big(P_1(x) + P_2(x) + P_3(x)\big),
\end{aligned}$$

where

$$\begin{aligned}
P_1(x) &= \left| \int_{|y| \le \frac{1}{16}} \left(\frac{x_j - y_j}{|x-y|^{n+1}} - \frac{x_j}{|x|^{n+1}} \right) \big(\Phi(y) - \Phi(x)\big) \frac{\Omega(y/|y|)}{|y|^n} dy \right|, \\
P_2(x) &= \left| \int_{\frac{1}{16} \le |y| \le 2} \frac{x_j - y_j}{|x-y|^{n+1}} \big(\Phi(y) - \Phi(x)\big) \frac{\Omega(y/|y|)}{|y|^n} dy \right|, \\
P_3(x) &= \left| \int_{|y| \ge 2} \frac{x_j - y_j}{|x-y|^{n+1}} \big(\Phi(y) - \Phi(x)\big) \frac{\Omega(y/|y|)}{|y|^n} dy \right|.
\end{aligned}$$

But since $1/8 \le |x| \le 1$, we see that

$$P_1(x) \le C_n \int_{|y| \le \frac{1}{16}} \frac{|y|}{|x|^{n+1}} \frac{|\Omega(y/|y|)|}{|y|^n} dy \le C_n' \big\|\Omega\big\|_{L^1}$$

and that

$$P_3(x) \le C_n \int_{|y| \ge 2} \frac{|\Omega(y/|y|)|}{|y|^{2n}} dy \le C_n' \big\|\Omega\big\|_{L^1}.$$

For $P_2(x)$ we use the estimate $|\Phi(y) - \Phi(x)| \le C|x-y|$ to obtain

$$\begin{aligned}
P_2(x) &\le \int_{\frac{1}{16} \le |y| \le 2} \frac{C}{|x-y|^{n-1}} \frac{|\Omega(y/|y|)|}{|y|^n} dy \\
&\le 4C \int_{\frac{1}{16} \le |y| \le 2} \frac{|\Omega(y/|y|)|}{|x-y|^{n-1} |y|^{n-\frac{1}{2}}} dy \\
&\le 4C \int_{\mathbf{R}^n} \frac{|\Omega(y/|y|)|}{|x-y|^{n-1} |y|^{n-\frac{1}{2}}} dy.
\end{aligned}$$

Recall that $K_j(x) = \Omega_j(x/|x|)|x|^{-n}$. We now set

$$G_j(x) = C_n \left(\big\|\Omega\big\|_{L^1} + \left| \Omega_j\Big(\frac{x}{|x|}\Big) \right| + |x|^{n-\frac{3}{2}} \int_{\mathbf{R}^n} \frac{|\Omega(y/|y|)|\, dy}{|x-y|^{n-1} |y|^{n-\frac{1}{2}}} \right) \tag{5.2.44}$$

and we observe that G_j is a homogeneous of degree zero function, it satisfies (5.2.42), and it is integrable over the annulus $\frac{1}{2} \le |x| \le 2$. To verify the last assertion, we split up the double integral

$$I = \int_{\frac{1}{2}\le|x|\le 2}\int_{\mathbf{R}^n} \frac{|\Omega(y/|y|)|\,dy}{|x-y|^{n-1}|y|^{n-\frac{1}{2}}}\,dx$$

into the pieces $1/4 \le |y| \le 4$, $|y| > 4$, and $|y| < 1/4$. The part of I where $1/4 \le |y| \le 4$ is pointwise bounded by a constant multiple of

$$\int\limits_{\frac{1}{4}\le|y|\le 4} \Big|\Omega\Big(\frac{y}{|y|}\Big)\Big| \int\limits_{\frac{1}{2}\le|x|\le 2} \frac{dx}{|y-x|^{n-1}}\,dy \le \int\limits_{\frac{1}{4}\le|y|\le 4} \Big|\Omega\Big(\frac{y}{|y|}\Big)\Big| \int\limits_{|x-y|\le 6} \frac{dx}{|y-x|^{n-1}}\,dy,$$

which is pointwise controlled by a constant multiple of $\|\Omega\|_{L^1}$. In the part of I where $|y| > 4$ we use that $|x-y|^{-n+1} \le (|y|/2)^{-n+1}$ to obtain rapid decay in y and hence a bound by a constant multiple of $\|\Omega\|_{L^1}$. Finally, in the part of I where $|y| < 1/4$ we use that $|x-y|^{-n+1} \le (1/4)^{-n+1}$, and then we also obtain a similar bound. It follows from (5.2.44) and (5.2.40) that

$$\int_{\frac{1}{2}\le|x|\le 2} |G_j(x)|\,dx \le C_n\big(\|\Omega\|_{L^1} + \|\Omega_j\|_{L^1} + \|\Omega\|_{L^1}\big) \le C_n(c_\Omega + 1).$$

Since G_j is homogeneous of degree zero, we deduce (5.2.43).

To complete the proof, we argue as follows:

$$\begin{aligned}
&\sup_{0<\varepsilon<N<\infty} |A_1^{(\varepsilon,N)}(f)(x)| \\
&\qquad \le 2\sup_{\varepsilon>0}\sum_{j=1}^n \frac{1}{\varepsilon^n}\int_{|z|\le\varepsilon} |F_j(z)|\,|R_j(f)(x-z)|\,dz \\
&\qquad \le 2\sup_{\varepsilon>0}\sum_{j=1}^n \frac{1}{\varepsilon^n}\int_{r=0}^{\varepsilon}\int_{\mathbf{S}^{n-1}} |F_j(r\theta)|\,|R_j(f)(x-r\theta)|\,r^{n-1}\,d\theta\,dr \\
&\qquad \le 2\sum_{j=1}^n \int_{\mathbf{S}^{n-1}} |G_j(\theta)| \Big\{ \sup_{\varepsilon>0} \frac{1}{\varepsilon^n}\int_{r=0}^{\varepsilon} |R_j(f)(x-r\theta)|\,r^{n-1}\,dr \Big\}\,d\theta \\
&\qquad \le 4\sum_{j=1}^n \int_{\mathbf{S}^{n-1}} |G_j(\theta)| M_\theta(R_j(f))(x)\,d\theta\,.
\end{aligned}$$

Using (5.2.43) together with the L^p boundedness of the Riesz transforms and of M_θ we obtain

$$\Big\|\sup_{0<\varepsilon<N<\infty} |A_1^{(\varepsilon,N)}(f)|\Big\|_{L^p} \le C_n \max(p,(p-1)^{-2})(c_\Omega+1)\big\|f\big\|_{L^p}\,. \tag{5.2.45}$$

Combining (5.2.45), (5.2.39), and (5.2.41), we obtain the required conclusion. □

The following corollary is a consequence of Theorem 5.2.11.

Corollary 5.2.12. *Let Ω be as in Theorem 5.2.11. Then for $1 < p < \infty$ and f in $L^p(\mathbf{R}^n)$ the functions $T_\Omega^{(\varepsilon,N)}(f)$ converge to $T_\Omega(f)$ in L^p and almost everywhere as $\varepsilon \to 0$ and $N \to \infty$.*

Proof. The a.e. convergence is a consequence of Theorem 2.1.14. The L^p convergence is a consequence of the Lebesgue dominated convergence theorem since for $f \in L^p(\mathbf{R}^n)$ we have that $|T_\Omega^{(\varepsilon,N)}(f)| \le T_\Omega^{(**)}(f)$ and $T_\Omega^{(**)}(f)$ is in $L^p(\mathbf{R}^n)$. $\square$

Exercises

5.2.1. Show that the directional Hilbert transform $\mathscr{H}_\theta$ is given by convolution with the distribution w_θ in $\mathscr{S}'(\mathbf{R}^n)$ defined by

$$\langle w_\theta, \varphi \rangle = \frac{1}{\pi} \text{p.v.} \int_{-\infty}^{+\infty} \frac{\varphi(t\theta)}{t}\, dt.$$

Compute the Fourier transform of w_θ and prove that $\mathscr{H}_\theta$ maps $L^1(\mathbf{R}^n)$ to $L^{1,\infty}(\mathbf{R}^n)$.

5.2.2. Extend the definitions of W_Ω and T_Ω to $\Omega = d\mu$ a finite signed Borel measure on $\mathbf{S}^{n-1}$ with mean value zero. Compute the Fourier transform of $W_{d\mu}$ and find a necessary and sufficient condition on measures $d\mu$ so that $T_{d\mu}$ is L^2 bounded. Notice that the directional Hilbert transform $\mathscr{H}_\theta$ is a special case of such an operator $T_{d\mu}$.

5.2.3. Use the inequality $AB \le A \log A + e^B$ for $A \ge 1$ and $B > 0$ to prove that if Ω satisfies (5.2.24) then it must satisfy (5.2.16). Conclude that if $|\Omega| \log^+ |\Omega|$ is in $L^1(\mathbf{S}^{n-1})$, then T_Ω is L^2 bounded.
[*Hint:* Use that $\int_{\mathbf{S}^{n-1}} |\xi \cdot \theta|^{-\alpha}\, d\theta$ converges when $\alpha < 1$. See Appendix D.3.]

5.2.4. Let Ω be a nonzero integrable function on $\mathbf{S}^{n-1}$ with mean value zero. Let $f \ge 0$ be nonzero and integrable over $\mathbf{R}^n$. Prove that $T_\Omega(f)$ is not in $L^1(\mathbf{R}^n)$.
[*Hint:* Show that $\widehat{T_\Omega(f)}$ cannot be continuous at zero.]

5.2.5. Let $\theta \in \mathbf{S}^{n-1}$. Use an identity similar to (5.2.18) to show that the maximal operators

$$\sup_{a>0} \frac{1}{a} \int_0^a |f(x - r\theta)|\, dr, \qquad \sup_{a>0} \frac{1}{2a} \int_{-a}^{+a} |f(x - r\theta)|\, dr$$

are $L^p(\mathbf{R}^n)$ bounded for $1 < p < \infty$ with norm at most $3p(p-1)^{-1}$.

5.2.6. For $\Omega \in L^1(\mathbf{S}^{n-1})$ and f locally integrable on $\mathbf{R}^n$, define

$$M_\Omega(f)(x) = \sup_{R>0} \frac{1}{v_n R^n} \int_{|y| \le R} |\Omega(y/|y|)|\, |f(x-y)|\, dy.$$

Apply the method of rotations to prove that M_Ω maps $L^p(\mathbf{R}^n)$ to itself for $1 < p < \infty$.

5.2.7. Let $\Omega(x, \theta)$ be a function on $\mathbf{R}^n \times \mathbf{S}^{n-1}$ satisfying
(a) $\Omega(x, -\theta) = -\Omega(x, \theta)$ for all x and θ.
(b) $\sup_x |\Omega(x, \theta)|$ is in $L^1(\mathbf{S}^{n-1})$.

Use the method of rotations to prove that

$$T_\Omega(f)(x) = \text{p.v.} \int_{\mathbf{R}^n} \frac{\Omega(x, y/|y|)}{|y|^n} f(x-y)\,dy$$

is bounded on $L^p(\mathbf{R}^n)$ for $1 < p < \infty$.

5.2.8. Let $\Omega \in L^1(\mathbf{S}^{n-1})$ have mean value zero. Prove that if T_Ω maps $L^p(\mathbf{R}^n)$ to $L^q(\mathbf{R}^n)$, then $p = q$.
[*Hint:* Use dilations.]

5.2.9. Prove that for all $1 < p < \infty$ there exists a constant $A_p > 0$ such that for every complex-valued $\mathscr{C}^2(\mathbf{R}^2)$ function f with compact support we have the bound

$$\big\|\partial_{x_1} f\big\|_{L^p} + \big\|\partial_{x_2} f\big\|_{L^p} \le A_p \big\|\partial_{x_1} f + i\partial_{x_2} f\big\|_{L^p}.$$

5.2.10. (a) Let $\Delta = \sum_{j=1}^n \partial_{x_j}^2$ be the usual Laplacian on $\mathbf{R}^n$. Prove that for all $1 < p < \infty$ there exists a constant $A_p > 0$ such that for all $\mathscr{C}^2$ functions f with compact support we have the bound

$$\big\|\partial_{x_j}\partial_{x_k} f\big\|_{L^p} \le A_p \big\|\Delta f\big\|_{L^p}.$$

(b) Let $\Delta^m = \overbrace{\Delta \circ \cdots \circ \Delta}^{m \text{ times}}$. Show that for any $1 < p < \infty$ there exists a $C_p > 0$ such that for all f of class $\mathscr{C}^{2m}$ with compact support and all differential monomials ∂_x^α of order $|\alpha| = 2m$ we have

$$\big\|\partial_x^\alpha f\big\|_{L^p} \le C_p \big\|\Delta^m f\big\|_{L^p}.$$

5.2.11. Use the same idea as in Lemma 5.2.5 to show that if f is continuous on $[0, \infty)$, differentiable in $(0, \infty)$, and satisfies

$$\lim_{N\to\infty} \int_N^{Na} \frac{f(u)}{u}\,du = 0$$

for all $a > 0$, then

$$\lim_{\substack{\varepsilon\to 0\\ N\to\infty}} \int_\varepsilon^N \frac{f(at) - f(t)}{t}\,dt = f(0)\log\frac{1}{a}.$$

5.2.12. Let Ω_o be an odd integrable function on $\mathbf{S}^{n-1}$ and Ω_e an even function on $\mathbf{S}^{n-1}$ that satisfies (5.2.24). Let f be a function supported in a ball B in $\mathbf{R}^n$. Prove that
(a) If $|f|\log^+|f|$ is integrable over a ball B, then $T_{\Omega_o}(f)$ and $T_{\Omega_o}^{(**)}(f)$ are integrable over B.
(b) If $|f|(\log^+|f|)^2$ is integrable over a ball B, then $T_{\Omega_e}(f)$ and $T_{\Omega_e}^{(**)}(f)$ are integrable over B.
[*Hint:* Use Exercise 1.3.7.]

5.2.13. ([324]) Let Ω be integrable on $\mathbf{S}^{n-1}$ with mean value zero. Use Jensen's inequality to show that for some $C > 0$ and every radial function $f \in L^2(\mathbf{R}^n)$ we have

$$\big\|T_\Omega(f)\big\|_{L^2} \le C\big\|f\big\|_{L^2}\,.$$

This inequality subsumes that T_Ω is well defined on radial $L^2(\mathbf{R}^n)$ functions.

5.3 The Calderón–Zygmund Decomposition and Singular Integrals

The behavior of singular integral operators on $L^1(\mathbf{R}^n)$ is a more subtle issue than that on L^p for $1 < p < \infty$. It turns out that singular integrals are not bounded from L^1 to L^1. See Example 5.1.3 and also Exercise 5.2.4. In this section we see that singular integrals map L^1 into the larger space $L^{1,\infty}$. This result strengthens their L^p boundedness.

5.3.1 The Calderón–Zygmund Decomposition

To make some advances in the theory of singular integrals, we need to introduce the Calderón–Zygmund decomposition. This is a powerful stopping-time construction that has many other interesting applications. We have already encountered an example of a stopping-time argument in Section 2.1.

Recall that a dyadic cube in $\mathbf{R}^n$ is the set

$$[2^k m_1, 2^k(m_1+1)) \times \cdots \times [2^k m_n, 2^k(m_n+1))\,,$$

where $k, m_1, \dots, m_n \in \mathbf{Z}$. Two dyadic cubes are either disjoint or related by inclusion.

Theorem 5.3.1. *Let $f \in L^1(\mathbf{R}^n)$ and $\alpha > 0$. Then there exist functions g and b on $\mathbf{R}^n$ such that*

(1) $f = g + b$.

(2) $\|g\|_{L^1} \le \|f\|_{L^1}$ *and* $\|g\|_{L^\infty} \le 2^n\alpha$.

(3) $b = \sum_j b_j$, *where each b_j is supported in a dyadic cube Q_j. Furthermore, the cubes Q_k and Q_j are disjoint when $j \neq k$.*

(4) $\displaystyle\int_{Q_j} b_j(x)\,dx = 0$.

(5) $\|b_j\|_{L^1} \le 2^{n+1}\alpha|Q_j|$.

(6) $\sum_j |Q_j| \le \alpha^{-1}\|f\|_{L^1}$.

Remark 5.3.2. This decomposition is called the *Calderón–Zygmund decomposition* of f at height α. The function g is called the *good function* of the decomposition, since it is both integrable and bounded; hence the letter g. The function b is called the *bad function*, since it contains the singular part of f (hence the letter b), but it is carefully chosen to have mean value zero. It follows from (1) and (2) that the bad function b is integrable and satisfies

$$\big\|b\big\|_{L^1} \le \big\|f\big\|_{L^1} + \big\|g\big\|_{L^1} \le 2\big\|f\big\|_{L^1}.$$

By (2) the good function is integrable and bounded; hence it lies in all the L^p spaces for $1 \le p \le \infty$. More specifically, we have the following estimate:

$$\big\|g\big\|_{L^p} \le \big\|g\big\|_{L^1}^{\frac{1}{p}} \big\|g\big\|_{L^\infty}^{1-\frac{1}{p}} \le \big\|f\big\|_{L^1}^{\frac{1}{p}} (2^n\alpha)^{1-\frac{1}{p}} = 2^{\frac{n}{p'}} \alpha^{\frac{1}{p'}} \big\|f\big\|_{L^1}^{\frac{1}{p}}. \tag{5.3.1}$$

Proof. Decompose $\mathbf{R}^n$ into a mesh of disjoint dyadic cubes of the same size such that

$$|Q| \ge \frac{1}{\alpha}\big\|f\big\|_{L^1}$$

for every cube Q in the mesh. Call these cubes of zero generation. Subdivide each cube of zero generation into 2^n congruent cubes by bisecting each of its sides. We now have a new mesh of dyadic cubes, which we call of generation one. Select a cube Q of generation one if

$$\frac{1}{|Q|}\int_Q |f(x)|\,dx > \alpha. \tag{5.3.2}$$

Let $S^{(1)}$ be the set of all selected cubes of generation one. Now subdivide each nonselected cube of generation one into 2^n congruent subcubes by bisecting each side and call these cubes of generation two. Then select all cubes Q of generation two if (5.3.2) holds. Let $S^{(2)}$ be the set of all selected cubes of generation two. Repeat this procedure indefinitely.

The set of all selected cubes $\bigcup_{m=1}^{\infty} S^{(m)}$ is countable and is exactly the set of the cubes Q_j proclaimed in the proposition. Note that in some instances this set may be empty, in which case $b = 0$ and $g = f$. Let us observe that the selected cubes are disjoint, for otherwise some Q_k would be a proper subset of some Q_j, which is impossible since the selected cube Q_j was never subdivided. Now define

$$b_j = \left(f - \frac{1}{|Q_j|}\int_{Q_j} f\,dx\right)\chi_{Q_j},$$

$b = \sum_j b_j$, and $g = f - b$.

For a selected cube Q_j there exists a unique nonselected cube Q' with twice its side length that contains Q_j. Let us call this cube the parent of Q_j. Since the parent Q' of Q_j was not selected, we have $|Q'|^{-1}\int_{Q'}|f|\,dx \le \alpha$. Then

$$\frac{1}{|Q_j|}\int_{Q_j}|f(x)|\,dx \le \frac{1}{|Q_j|}\int_{Q'}|f(x)|\,dx = \frac{2^n}{|Q'|}\int_{Q'}|f(x)|\,dx \le 2^n\alpha.$$

Consequently,

$$\int_{Q_j}|b_j|\,dx \le \int_{Q_j}|f|\,dx + |Q_j|\left|\frac{1}{|Q_j|}\int_{Q_j} f\,dx\right| \le 2\int_{Q_j}|f|\,dx \le 2^{n+1}\alpha|Q_j|,$$

which proves (5). To prove (6), simply observe that

$$\sum_j |Q_j| \le \frac{1}{\alpha}\sum_j \int_{Q_j}|f|\,dx = \frac{1}{\alpha}\int_{\bigcup_j Q_j}|f|\,dx \le \frac{1}{\alpha}\|f\|_{L^1}.$$

Next we need to obtain the estimates concerning g. We obviously have

$$g = \begin{cases} f & \text{on } \mathbf{R}^n \setminus \bigcup_j Q_j, \\ \frac{1}{|Q_j|}\int_{Q_j} f\,dx & \text{on } Q_j. \end{cases} \tag{5.3.3}$$

On the cube Q_j, g is equal to the constant $|Q_j|^{-1}\int_{Q_j} f\,dx$, and this is bounded by $2^n\alpha$. It suffices to show that g is bounded outside the union of the Q_j's. Indeed, for each $x \in \mathbf{R}^n \setminus \bigcup_j Q_j$ and for each $k = 0, 1, 2, \dots$ there exists a unique nonselected dyadic cube $Q_x^{(k)}$ of generation k that contains x. Then for each $k \ge 0$, we have

$$\left|\frac{1}{|Q_x^{(k)}|}\int_{Q_x^{(k)}} f(y)\,dy\right| \le \frac{1}{|Q_x^{(k)}|}\int_{Q_x^{(k)}}|f(y)|\,dy \le \alpha.$$

The intersection of the closures of the cubes $Q_x^{(k)}$ is the singleton $\{x\}$. Using Corollary 2.1.16, we deduce that for almost all $x \in \mathbf{R}^n \setminus \bigcup_j Q_j$ we have

$$f(x) = \lim_{k\to\infty}\frac{1}{|Q_x^{(k)}|}\int_{Q_x^{(k)}} f(y)\,dy.$$

Since these averages are at most α, we conclude that $|f| \le \alpha$ a.e. on $\mathbf{R}^n \setminus \bigcup_j Q_j$, hence $|g| \le \alpha$ a.e. on this set. Finally, it follows from (5.3.3) that $\|g\|_{L^1} \le \|f\|_{L^1}$. This finishes the proof of the theorem. □

We now apply the Calderón–Zygmund decomposition to obtain weak type $(1,1)$ bounds for a wide class of singular integral operators that includes the operators T_Ω we studied in the previous section.

5.3.2 General Singular Integrals

The kernels of the general singular integrals we will study are tempered distributions that coincide with functions away from the origin. The setup as follows. Let K be a measurable function defined on $\mathbf{R}^n \setminus \{0\}$ that is integrable on compact subsets of $\mathbf{R}^n \setminus \{0\}$ and satisfies the size condition

$$\sup_{R>0} \int_{R\le|x|\le 2R} |K(x)|\,dx = A_1 < \infty. \tag{5.3.4}$$

This condition is less restrictive than the standard size estimate

$$\sup_{x\in\mathbf{R}^n} |x|^n |K(x)| < \infty, \tag{5.3.5}$$

but it is strong enough to capture size properties of kernels $K(x) = \Omega(x/|x|)/|x|^n$, where $\Omega \in L^1(\mathbf{S}^{n-1})$. We also note that condition (5.3.4) is equivalent to

$$\sup_{R>0} \frac{1}{R} \int_{|x|\le R} |K(x)|\,|x|\,dx < \infty. \tag{5.3.6}$$

See Exercise 5.3.1.

The size condition (5.3.4) is sufficient to make the restriction of $K(x)$ on $|x| > \delta$ a tempered distribution (for any $\delta > 0$). Indeed, for $\varphi \in \mathscr{S}(\mathbf{R}^n)$ we have

$$\begin{aligned}\int_{|x|\ge 1} |K(x)\varphi(x)|\,dx &\le \sum_{m=0}^{\infty} \int_{2^{m+1}\ge|x|\ge 2^m} \frac{|K(x)|(1+|x|)^N|\varphi(x)|}{(1+2^m)^N}\,dx \\ &\le \sum_{m=0}^{\infty} \frac{A_1}{(1+2^m)^N} \sup_{x\in\mathbf{R}^n} (1+|x|)^N|\varphi(x)|,\end{aligned}$$

and this expression is bounded by a constant times a finite sum of Schwartz seminorms of φ.

We are interested in tempered distributions W on $\mathbf{R}^n$ that extend the function K defined on $\mathbf{R}^n \setminus \{0\}$ and have the form

$$\langle W, \varphi\rangle = \lim_{j\to\infty} \int_{|x|\ge\delta_j} K(x)\varphi(x)\,dx, \qquad \varphi \in \mathscr{S}(\mathbf{R}^n), \tag{5.3.7}$$

for some sequence $\delta_j \downarrow 0$ as $j \to \infty$. It is not hard to see that there exists a tempered distribution W satisfying (5.3.7) for all $\varphi \in \mathscr{S}(\mathbf{R}^n)$ if and only if

$$\lim_{j\to\infty} \int_{1\ge|x|\ge\delta_j} K(x)\,dx = L \tag{5.3.8}$$

exists. See Exercise 5.3.2. If such a distribution W exists it may not be unique, since it depends on the choice of the sequence δ_j. Two different sequences tending to zero may give two different tempered distributions W of the form (5.3.7), both coinciding with the function K on $\mathbf{R}^n \setminus \{0\}$. See Example 5.4.2 and Remark 5.4.3.

If condition (5.3.8) is satisfied, we can define

$$\langle W, \varphi \rangle = \lim_{j\to\infty} \int_{j\geq |x| \geq \delta_j} K(x)\varphi(x)\,dx \tag{5.3.9}$$

and the limit exists as $j \to \infty$ for all $\varphi \in \mathscr{S}(\mathbf{R}^n)$ and is equal to

$$\langle W, \varphi \rangle = \int_{|x|\leq 1} K(x)(\varphi(x) - \varphi(0))\,dx + \varphi(0)L + \int_{|x|\geq 1} K(x)\varphi(x)\,dx.$$

Moreover, the previous calculations show that W is an element of $\mathscr{S}'(\mathbf{R}^n)$.

Next we assume that the given function K on $\mathbf{R}^n \setminus \{0\}$ satisfies a certain smoothness condition. There are three kinds of smoothness conditions that we encounter: first, the *gradient condition*

$$|\nabla K(x)| \leq A_2 |x|^{-n-1}, \qquad x \neq 0; \tag{5.3.10}$$

next, the weaker *Lipschitz condition*,

$$|K(x-y) - K(x)| \leq A_2 \frac{|y|^\delta}{|x|^{n+\delta}}, \qquad \text{whenever } |x| \geq 2|y|; \tag{5.3.11}$$

and finally the even weaker smoothness condition

$$\sup_{y\neq 0} \int_{|x|\geq 2|y|} |K(x-y) - K(x)|\,dx = A_2, \tag{5.3.12}$$

for some $A_2 < \infty$. One should verify that (5.3.12) is a weaker condition than (5.3.11), which in turn is weaker than (5.3.10). Condition (5.3.12) is often referred to as *Hörmander's condition.*

5.3.3 L^r Boundedness Implies Weak Type $(1,1)$ Boundedness

This next theorem provides the most classical application of the Calderón–Zygmund decomposition.

Theorem 5.3.3. *Let K be a function on $\mathbf{R}^n \setminus \{0\}$ that satisfies* (5.3.4)[1] *and (5.3.12) for some $A_1, A_2 < \infty$. Let W be an element of $\mathscr{S}'(\mathbf{R}^n)$ related to K as in* (5.3.7)*. Suppose that the operator T given by convolution with W has a bounded extension that maps $L^r(\mathbf{R}^n)$ to itself with norm B for some $1 < r \leq \infty$. Then T has an extension that maps $L^1(\mathbf{R}^n)$ to $L^{1,\infty}(\mathbf{R}^n)$ with norm*

[1] this condition could be replaced by the assumption that K is integrable over any compact set that does not contain the origin.

$$\big\|T\big\|_{L^1\to L^{1,\infty}} \le C_n(A_2+B), \tag{5.3.13}$$

and T also extends to a bounded operator from $L^p(\mathbf{R}^n)$ to itself for $1<p<\infty$ with norm

$$\big\|T\big\|_{L^p\to L^p} \le C_n' \max\big(p,(p-1)^{-1}\big)\,(A_2+B), \tag{5.3.14}$$

where C_n, C_n' are constants that depend on the dimension but not on r or p.

Proof. We discuss the case $r<\infty$ and we refer to Exercise 5.3.7 for the case $r=\infty$. Let $\alpha>0$ be given. We fix a step function f given as a finite linear combination of characteristic functions of disjoint dyadic intervals. The class of such functions is dense in all the L^p spaces. Once (5.3.13) is obtained for such functions, a density argument gives that T admits an extension on L^1 that also satisfies (5.3.13). Therefore it suffices to prove (5.3.13) for such a function f.

Apply the Calderón–Zygmund decomposition to f at height $\gamma\alpha$, where γ is a positive constant to be chosen later. That is, write the function f as the sum

$$f = g + b = g + \sum_j b_j,$$

where conditions (1)–(6) of Theorem 5.3.1 are satisfied with the constant α replaced by $\gamma\alpha$. Since f is a finite linear combination of characteristic functions of disjoint dyadic cubes, there are only finitely many cubes Q_j that appear in the Calderón–Zygmund decomposition to f. Each b_j is supported in a dyadic cube Q_j with center y_j and the Q_j's are pairwise disjoint. We denote by $\ell(Q)$ the side length of a cube Q. Let Q_j^* be the unique cube with sides parallel to the axes having the same center as Q_j and having side length $\ell(Q_j^*) = 2\sqrt{n}\,\ell(Q_j)$. Because of the form of f, each b_j is a bounded function supported in $\overline{Q_j}$, hence it is in L^r, thus each $T(b_j)$ is a well-defined L^r function. We observe that for all j and all $x\notin Q_j^*$ we have

$$T(b_j)(x) = \lim_{k\to\infty}\int_{k\ge|x-y|\ge\delta_k} K(x-y)b_j(y)\,dy = \int_{Q_j} K(x-y)b_j(y)\,dy,$$

where the last integral converges absolutely. This is a consequence of the Lebesgue dominated convergence theorem, based on the facts that b_j is bounded, that K is integrable over any compact annulus that does not contain the origin (cf. (5.3.4)), and that $x-Q_j$ is contained in such a compact annulus, since $x\notin Q_j^*$.

Next we use the cancellation of b_j in the following way:

$$\begin{aligned}
&\int_{(\cup_i Q_i^*)^c} \sum_j \big|T(b_j)(x)\big|\,dx \\
&= \int_{(\cup_i Q_i^*)^c} \sum_j \left|\int_{Q_j} b_j(y)\big(K(x-y)-K(x-y_j)\big)\,dy\right| dx \\
&\le \sum_j \int_{(Q_j^*)^c}\int_{Q_j} |b_j(y)||K(x-y)-K(x-y_j)|\,dy\,dx
\end{aligned}$$

$$
\begin{aligned}
&= \sum_j \int_{Q_j} |b_j(y)| \int_{(Q_j^*)^c} |K(x-y) - K(x-y_j)|\,dx\,dy \\
&= \sum_j \int_{Q_j} |b_j(y)| \int_{-y_j+(Q_j^*)^c} |K(x-(y-y_j)) - K(x)|\,dx\,dy \\
&\le \sum_j \int_{Q_j} |b_j(y)| \int_{|x|\ge 2|y-y_j|} |K(x-(y-y_j)) - K(x)|\,dx\,dy \\
&\le A_2 \sum_j \big\|b_j\big\|_{L^1} \\
&\le A_2 2^{n+1} \big\|f\big\|_{L^1} < \infty,
\end{aligned}
$$

where we used (5.3.12) since if $x \in -y_j + (Q_j^*)^c$ then $|x| \ge \frac{1}{2}\ell(Q_j^*) = \sqrt{n}\,\ell(Q_j)$ and since $y - y_j \in -y_j + Q_j$ we have $|y-y_j| \le \frac{\sqrt{n}}{2}\,\ell(Q_j)$, thus $|x| \ge 2|y-y_j|$. See Figure 5.2.

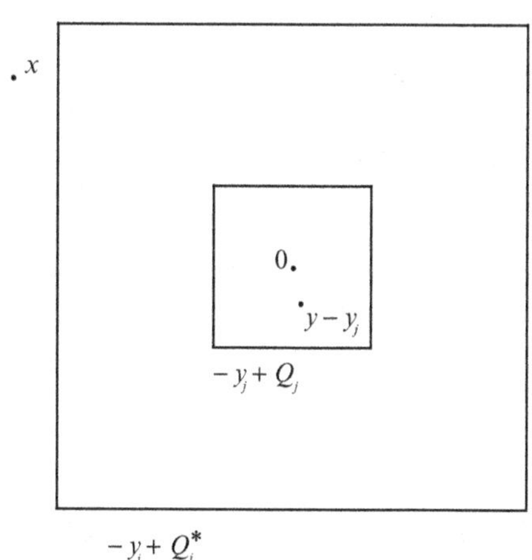

Fig. 5.2 The cubes $-y_j + Q_j$ and $-y_j + Q_j^*$.

Thus we proved that

$$
\int_{(\bigcup_i Q_i^*)^c} \sum_j |T(b_j)(x)|\,dx \le 2^{n+1} A_2 \big\|f\big\|_{L^1},
$$

an inequality we use below. We have

$$
\begin{aligned}
&\big|\big\{x \in \mathbf{R}^n : |T(f)(x)| > \alpha\big\}\big| \\
&\quad\le \Big|\Big\{x \in \mathbf{R}^n :\ |T(g)(x)| > \frac{\alpha}{2}\Big\}\Big| + \Big|\Big\{x \in \mathbf{R}^n :\ |T(b)(x)| > \frac{\alpha}{2}\Big\}\Big| \\
&\quad\le \frac{2^r}{\alpha^r}\big\|T(g)\big\|_{L^r}^r + \Big|\bigcup_i Q_i^*\Big| + \Big|\Big\{x \notin \bigcup_i Q_i^* :\ |T\big(\sum_j b_j\big)(x)| > \frac{\alpha}{2}\Big\}\Big|
\end{aligned}
$$

$$
\begin{aligned}
&= \frac{2^r}{\alpha^r}\|T(g)\|_{L^r}^r + \Big|\bigcup_i Q_i^*\Big| + \Big|\Big\{x \notin \bigcup_i Q_i^* : |\sum_j T(b_j)(x)| > \frac{\alpha}{2}\Big\}\Big| \\
&\leq \frac{2^r}{\alpha^r} B^r \|g\|_{L^r}^r + \sum_i |Q_i^*| + \frac{2}{\alpha}\int_{(\bigcup_i Q_i^*)^c} \sum_j |T(b_j)(x)|\,dx \\
&\leq \frac{2^r}{\alpha^r} 2^{\frac{nr}{r'}} B^r (\gamma\alpha)^{\frac{r}{r'}} \|f\|_{L^1} + (2\sqrt{n})^n \frac{\|f\|_{L^1}}{\gamma\alpha} + \frac{2}{\alpha} 2^{n+1} A_2 \|f\|_{L^1} \\
&\leq \left(\frac{(2^{n+1}B\gamma)^r}{2^n\gamma} + \frac{(2\sqrt{n})^n}{\gamma} + 2^{n+2}A_2 \right) \frac{\|f\|_{L^1}}{\alpha}.
\end{aligned}
$$

Choosing $\gamma = 2^{-(n+1)}B^{-1}$, we deduce the weak type $(1,1)$ estimate (5.3.13) for T with $C_n = 2 + 2^{n+1}(2\sqrt{n})^n + 2^{n+2}$.

It follows from Exercise 1.3.2 that T has an extension that maps L^p to L^p with bound at most $C_n'(A_2+B)(p-1)^{-1/p}$ when $1 < p < r$. The adjoint operator T^* of T, defined by

$$\langle T(f)\,|\,g\rangle = \langle f\,|\,T^*(g)\rangle,$$

has a kernel that coincides with the function $K^*(x) = \overline{K(-x)}$ on $\mathbf{R}^n \setminus \{0\}$. We notice that since K satisfies (5.3.12), then so does K^* and with the same bound. Therefore, T^*, which maps $L^{r'}$ to $L^{r'}$, has a kernel that satisfies (5.3.12). By the preceding argument, T^* maps $L^{p'}$ to $L^{p'}$ with bound at most $C_n'(A_2+B)(p-1)^{-1/p}$ whenever $1 < p' < r'$. By duality this yields that T maps $L^p(\mathbf{R}^n)$ to $L^p(\mathbf{R}^n)$ with bound at most $C_n'(A_2+B)(p-1)^{1-1/p}$ whenever $r' < p < \infty$. Using interpolation we obtain that T maps L^p to itself with norm at most $C_n'(A_2+B)\max((p-1)^{-1/p},(p-1)^{1-1/p})$ for p in the interval (r,r'), which is nonempty only if $r < 2$. Then (5.3.14) holds since $\max((p-1)^{-1/p},(p-1)^{1-1/p}) \leq \max((p-1)^{-1},p)$. □

5.3.4 Discussion on Maximal Singular Integrals

In this subsection we introduce maximal singular integrals and we derive their boundedness under certain smoothness conditions on the kernels, assuming boundedness of the associated linear operator.

Suppose that K is a kernel on $\mathbf{R}^n \setminus \{0\}$ that satisfies the size condition

$$|K(x)| \leq A_1 |x|^{-n} \tag{5.3.15}$$

for $x \neq 0$. Then for any $\varepsilon > 0$ the function $K^{(\varepsilon)}(x) = K(x)\chi_{|x|\geq\varepsilon}$ lies in $L^{p'}(\mathbf{R}^n)$ (with norm bounded by $c_{p,n}A_1\varepsilon^{-n/p}$) for all $1 \leq p < \infty$. Consequently, by Hölder's inequality, the integral

$$(f * K^{(\varepsilon)})(x) = \int_{|y|\geq\varepsilon} f(x-y)K(y)\,dy$$

converges absolutely for all $x \in \mathbf{R}^n$ and all $f \in L^p(\mathbf{R}^n)$, when $1 \leq p < \infty$.

Let $f \in \bigcup_{1\le p<\infty} L^p(\mathbf{R}^n)$. We define the truncated singular integrals $T^{(\varepsilon)}(f)$ associated with the kernel K by setting

$$T^{(\varepsilon)}(f) = f * K^{(\varepsilon)};$$

we also define the *maximal truncated singular integral operator* associated with K by setting

$$T^{(*)}(f) = \sup_{\varepsilon>0} |(f * K^{(\varepsilon)})| = \sup_{\varepsilon>0} |T^{(\varepsilon)}(f)|.$$

This operator is well defined, but possibly infinite, for certain points in $\mathbf{R}^n$.

We now consider the situation in which the kernel K satisfies an integrability condition over concentric annuli centered at the origin, a condition that is certainly a weaker condition than (5.3.15). Precisely, suppose that K is a measurable function on $\mathbf{R}^n \setminus \{0\}$, that is integrable on compact subsets of $\mathbf{R}^n \setminus \{0\}$, for which there is a constant $A_1 < \infty$ such that

$$\sup_{R>0} \int_{R\le|x|\le 2R} |K(x)|\,dx \le A_1 < \infty. \tag{5.3.16}$$

Such kernels may not be integrable to the power $p' > 1$ over the region $|x| \ge \varepsilon$. For this reason, it is not possible to define $T^{(\varepsilon)}$ as an absolutely convergent integral. To overcome this difficulty, we consider double truncations. We define the doubly truncated kernel $K^{(\varepsilon,N)}$ by setting

$$K^{(\varepsilon,N)}(x) = K(x)\chi_{\varepsilon\le|x|\le N}(x). \tag{5.3.17}$$

A repeated application of (5.3.16) yields that

$$\int |K^{(\varepsilon,N)}(x)|\,dx \le A_1\Big(\Big[\log_2 \frac{N}{\varepsilon}\Big] + 1\Big),$$

which implies that $K^{(\varepsilon,N)}$ is integrable over concentric annuli centered at the origin. Next, we define the doubly truncated singular integrals $T^{(\varepsilon,N)}$ by setting

$$T^{(\varepsilon,N)}(f) = f * K^{(\varepsilon,N)},$$

and we observe that these operators are well defined when f in L^p, for $1 \le p \le \infty$. Indeed, Theorem 1.2.10 yields that

$$\big\|T^{(\varepsilon,N)}(f)\big\|_{L^p} \le \big\|f\big\|_{L^p} \int |K^{(\varepsilon,N)}(x)|\,dx < \infty$$

for functions f in L^p, $1 \le p \le \infty$. Consequently, for almost every $x \in \mathbf{R}^n$ we have

$$|T^{(\varepsilon,N)}(f)(x)| < \infty.$$

For functions in $\bigcup_{1\le p\le\infty} L^p(\mathbf{R}^n)$ we define the *doubly truncated maximal singular integral operator* $T^{(**)}$ associated with K by setting

$$T^{(**)}(f) = \sup_{0<\varepsilon<N<\infty} |T^{(\varepsilon,N)}(f)|. \tag{5.3.18}$$

For such functions and for almost all $x \in \mathbf{R}^n$, $T^{(**)}(f)(x)$ is well defined, but potentially infinite.

One observation is that under condition (5.3.16), one can also define $T^{(*)}(g)$ for general integrable functions g with compact support. In this case, say that the ball $B(0,R)$ contains the support of g. Let $x \in B(0,M)$ and $N = M + R$. Then $|T^{(\varepsilon)}(g)(x)| \le |g| * |K^{(\varepsilon,N)}|(x)$, which is finite a.e. as the convolution of two L^1 functions; consequently, the integral defining $T^{(\varepsilon)}(g)(x)$ converges absolutely for all $x \in B(0,R)$. Since $R > 0$ is arbitrary, $T^{(\varepsilon)}(g)(x)$ is defined and finite for almost all $x \in \mathbf{R}^n$.

Obviously $T^{(*)}$ and $T^{(**)}$ are related. If K satisfies condition (5.3.15), then

$$\left| \int_{\varepsilon \le |y|} f(x-y)K(y)\, dy \right| \le \sup_{N>0} \left| \int_{\varepsilon \le |y| \le N} f(x-y)K(y)\, dy \right|,$$

which implies that

$$T^{(*)}(f) \le T^{(**)}(f)$$

for all $f \in \bigcup_{1 \le p < \infty} L^p$. Also, $T^{(\varepsilon,N)}(f) = T^{(\varepsilon)}(f) - T^{(N)}(f)$; hence

$$T^{(**)}(f) \le 2T^{(*)}(f).$$

Therefore, for kernels satisfying (5.3.15), $T^{(**)}$ and $T^{(*)}$ are comparable and the boudnedness properties of $T^{(**)}$ and $T^{(*)}$ are equivalent

Theorem 5.3.4. ***(Cotlar's inequality)*** *Let $0 < A_1, A_2, A_3 < \infty$ and suppose that K is defined on $\mathbf{R}^n \setminus \{0\}$ and satisfies the size condition,*

$$|K(x)| \le A_1 |x|^{-n}, \qquad x \ne 0, \tag{5.3.19}$$

the smoothness condition

$$|K(x-y) - K(x)| \le A_2 |y|^{\delta}\, |x|^{-n-\delta}, \tag{5.3.20}$$

whenever $|x| \ge 2|y| > 0$, and the cancellation condition

$$\sup_{0<r<R<\infty} \left| \int_{r<|x|<R} K(x)\, dx \right| \le A_3. \tag{5.3.21}$$

Let W in $\mathscr{S}'(\mathbf{R}^n)$ be related to K via (5.3.7) and let T be the operator given by convolution with W. Then there is a constant $C_{n,\delta}$ such that the following inequality is valid:

$$T^{(*)}(f) \le M(T(f)) + C_{n,\delta}\,(A_1 + A_2 + A_3)\, M(f), \tag{5.3.22}$$

for all $f \in \mathscr{S}(\mathbf{R}^n)$*, where* M *is the Hardy–Littlewood maximal operator. Consequently,* $T^{(*)}$ *is bounded on* $L^p(\mathbf{R}^n)$ *when* $1 < p < \infty$*.*

Proof. Let φ be a radially decreasing smooth function with integral 1 supported in the ball $B(0,1/2)$. For a function g and $\varepsilon > 0$ we use the notation $g_\varepsilon(x) = \varepsilon^{-n} g(\varepsilon^{-1}x)$. For a distribution W we define W_ε analogously, i.e. as the unique distribution with the property $\langle W_\varepsilon, \psi\rangle = \varepsilon^{-n}\langle W, \psi_{\varepsilon^{-1}}\rangle$. We begin by observing that $K_{\varepsilon^{-1}}(x) = \varepsilon^n K(\varepsilon x)$ satisfies (5.3.19), (5.3.20), and (5.3.21) uniformly in $\varepsilon > 0$.

Set, as before, $K^{(\varepsilon)}(x) = K(x)\chi_{|x|\ge\varepsilon}$. Fix $f \in \mathscr{S}(\mathbf{R}^n)$ for some $1 < p < \infty$. Obviously we have

$$f * K^{(\varepsilon)} = f * \big((K_{\varepsilon^{-1}})^{(1)}\big)_\varepsilon = f * W * \varphi_\varepsilon + f * \big((K_{\varepsilon^{-1}})^{(1)} - W_{\varepsilon^{-1}} * \varphi\big)_\varepsilon . \tag{5.3.23}$$

Next we prove the following estimate for all $\varepsilon > 0$:

$$\big|\big((K_{\varepsilon^{-1}})^{(1)} - W_{\varepsilon^{-1}} * \varphi\big)(x)\big| \le C(A_1 + A_2 + A_3)(1 + |x|)^{-n-\delta} \tag{5.3.24}$$

for all $x \in \mathbf{R}^n$. Indeed, for $|x| \ge 1$ we express the left-hand side in (5.3.24) as

$$\left| \int_{\mathbf{R}^n} \big(K_{\varepsilon^{-1}}(x) - K_{\varepsilon^{-1}}(x-y)\big)\varphi(y)\,dy \right| .$$

Since φ is supported in $|y| \le 1/2$, we have $|x| \ge 2|y|$, and condition (5.3.20) yields that the expression on the left-hand side of (5.3.24) is bounded by

$$\frac{A_2}{|x|^{n+\delta}} \int_{\mathbf{R}^n} |y|^\delta |\varphi(y)|\,dy \le c\,\frac{A_2}{(1+|x|)^{n+\delta}} ,$$

which proves (5.3.24) in the case $|x| \ge 1$. When $|x| < 1$, the left-hand side of (5.3.24) equals

$$(W_{\varepsilon^{-1}} * \varphi)(x) = \lim_{\delta_j \to 0} \int_{|x-y|\ge\delta_j} K_{\varepsilon^{-1}}(x-y)\varphi(y)\,dy \tag{5.3.25}$$

for some sequence $\delta_j \downarrow 0$; see the discussion in Section 5.3.2. The expression in (5.3.25) is equal to

$$I_1 + I_2 + I_3 ,$$

where

$$\begin{aligned} I_1 &= \int_{|x-y|>\frac18} K_{\varepsilon^{-1}}(x-y)\varphi(y)\,dy , \\ I_2 &= \int_{|x-y|\le\frac18} K_{\varepsilon^{-1}}(x-y)\big(\varphi(y) - \varphi(x)\big)\,dy , \\ I_3 &= \varphi(x) \lim_{\delta_j\to 0} \int_{\frac18 \ge |x-y| \ge \delta_j} K_{\varepsilon^{-1}}(x-y)\,dy . \end{aligned}$$

In I_1 we have $1/8 \le |x-y| \le 1 + 1/2 = 3/2$; hence I_1 is bounded by a multiple of A_1. Since $|\varphi(x) - \varphi(y)| \le c|x-y|$, the same is valid for I_2. Finally, I_3 is bounded by a multiple of A_3. Combining these facts yields the proof of (5.3.24) in the case $|x| < 1$ as well.

Use Corollary 2.1.12 to deduce that

$$\sup_{\varepsilon>0}\left|f*\left((K_{\varepsilon^{-1}})^{(1)}-K_{\varepsilon^{-1}}*\varphi\right)_\varepsilon\right|\le c\,(A_1+A_2+A_3)M(f)\,.$$

Finally, take the supremum over $\varepsilon>0$ in (5.3.23) and use (5.3.24) and Corollary 2.1.12 one more time to deduce the estimate

$$T^{(*)}(f)\le M(f*W)+C\,(A_1+A_2+A_3)M(f)\,,$$

where C depends on n and δ; this concludes the proof of (5.3.22) for all functions $f\in\mathscr{S}(\mathbf{R}^n)$. Thus $T^{(*)}$ is bounded on L^p, $1<p<\infty$, when restricted to Schwartz functions.

Now given a general function g in $L^p(\mathbf{R}^n)$ we find a sequence h_j in $\mathscr{S}(\mathbf{R}^n)$ such that $\|h_j-g\|_{L^p}\to 0$ as $j\to\infty$. Then we have the pointwise estimate

$$|T^{(\varepsilon)}(g)|\le|T^{(\varepsilon)}(g-h_j)|+|T^{(\varepsilon)}(h_j)|\le c_{p,n}A_1\varepsilon_0^{-\frac{n}{p}}\|g-h_j\|_{L^p}+|T^{(*)}(h_j)|$$

for all $\varepsilon\ge\varepsilon_0$. Taking the supremum over $\varepsilon\ge\varepsilon_0$ and then L^p norm over the ball $B(0,R)$, we obtain

$$\Big\|\sup_{\varepsilon\ge\varepsilon_0}|T^{(\varepsilon)}(g)|\Big\|_{L^p(B(0,R))}\le c'_{p,n}A_1\varepsilon_0^{-\frac{n}{p}}R^{\frac{n}{p}}\|g-h_j\|_{L^p}+C'\,(A_1+A_2+A_3)\|h_j\|_{L^p}\,.$$

Now we let $j\to\infty$ first, and then $R\to\infty$ and $\varepsilon_0\to 0$ to deduce the boundedness of $T^{(*)}$ on $L^p(\mathbf{R}^n)$ via the Lebesgue monotone convergence theorem. □

5.3.5 *Boundedness for Maximal Singular Integrals Implies Weak Type* $(1,1)$ *Boundedness*

We now state and prove a result analogous to that in Theorem 5.3.3 for maximal singular integrals.

Theorem 5.3.5. *Let $K(x)$ be function on $\mathbf{R}^n\setminus\{0\}$ satisfying (5.3.4) with constant $A_1<\infty$ and Hörmander's condition (5.3.12) with constant $A_2<\infty$. Suppose that the operator $T^{(**)}$ as defined in (5.3.18) maps $L^2(\mathbf{R}^n)$ to itself with norm B. Then $T^{(**)}$ maps $L^1(\mathbf{R}^n)$ to $L^{1,\infty}(\mathbf{R}^n)$ with norm*

$$\|T^{(**)}\|_{L^1\to L^{1,\infty}}\le C_n(A_1+A_2+B),$$

where C_n is some dimensional constant.

Proof. The proof of this theorem is only a little more involved than the proof of Theorem 5.3.3. We fix an $L^1(\mathbf{R}^n)$ function f. We apply the Calderón–Zygmund decomposition of f at height $\gamma\alpha$ for some $\gamma,\alpha>0$. We then write $f=g+b$, where

$b = \sum_j b_j$ and each b_j is supported in some cube Q_j. We define Q_j^* as the cube with the same center as Q_j and with sides parallel to the sides of Q_j having length $\ell(Q_j^*) = 5\sqrt{n}\,\ell(Q_j)$. This is only a minor change compared with the definition of Q_j in Theorem 5.3.3. The main change in the proof is in the treatment of the term

$$\left|\left\{x \in \Big(\bigcup_j Q_j^*\Big)^c : \, |T^{(**)}(b)(x)| > \frac{\alpha}{2}\right\}\right|. \tag{5.3.26}$$

We show that for all $\gamma \le (2^{n+5}A_1)^{-1}$ we have

$$\left|\left\{x \in \Big(\bigcup_j Q_j^*\Big)^c : \, |T^{(**)}(b)(x)| > \frac{\alpha}{2}\right\}\right| \le 2^{n+8} A_2 \frac{\|f\|_{L^1}}{\alpha}. \tag{5.3.27}$$

Let us conclude the proof of the theorem assuming for the moment the validity of (5.3.27). As in the proof of Theorem 5.3.3, we can show that

$$\left|\left\{x \in \mathbf{R}^n : \, |T^{(**)}(g)(x)| > \frac{\alpha}{2}\right\}\right| + \Big|\bigcup_j Q_j^*\Big| \le \Big(2^{n+2}B^2\gamma + \frac{(5\sqrt{n})^n}{\gamma}\Big)\frac{\|f\|_{L^1}}{\alpha}.$$

Combining this estimate with (5.3.27) and choosing

$$\gamma = (2^{n+5}(A_1 + A_2 + B))^{-1},$$

we obtain the required estimate

$$\left|\left\{x \in \mathbf{R}^n : \, |T^{(**)}(f)(x)| > \alpha\right\}\right| \le C_n (A_1 + A_2 + B)\frac{\|f\|_{L^1}}{\alpha}$$

with $C_n = 2^{-3} + (5\sqrt{n})^n 2^{n+5} + 2^{n+8}$.

It remains to prove (5.3.27). This estimate will be a consequence of the fact that for $x \in \big(\bigcup_j Q_j^*\big)^c$ we have the key inequality

$$T^{(**)}(b)(x) \le 4E_1(x) + 2^{n+2}\alpha\gamma E_2(x) + 2^{n+3}\alpha\gamma A_1, \tag{5.3.28}$$

where

$$\begin{aligned} E_1(x) &= \sum_j \int_{Q_j} |K(x-y) - K(x-y_j)|\,|b_j(y)|\,dy, \\ E_2(x) &= \sum_j \int_{Q_j} |K(x-y) - K(x-y_j)|\,dy, \end{aligned}$$

and y_j is the center of Q_j.

If we had (5.3.28), then we could easily derive (5.3.27). Indeed, fix a γ satisfying $\gamma \le (2^{n+5}A_1)^{-1}$. Then we have $2^{n+3}\alpha\gamma A_1 < \frac{\alpha}{3}$, and using (5.3.28), we obtain

$$
\begin{aligned}
&\left|\left\{x\in\Big(\bigcup_j Q_j^*\Big)^c : |T^{(**)}(b)(x)|>\frac{\alpha}{2}\right\}\right| \\
&\qquad\le\left|\left\{x\in\Big(\bigcup_j Q_j^*\Big)^c : 4E_1(x)>\frac{\alpha}{12}\right\}\right| \\
&\qquad\qquad\qquad+\left|\left\{x\in\Big(\bigcup_j Q_j^*\Big)^c : 2^{n+2}\alpha\gamma E_2(x)>\frac{\alpha}{12}\right\}\right| \\
&\qquad\le\frac{48}{\alpha}\int_{(\bigcup_j Q_j^*)^c}E_1(x)\,dx+2^{n+6}\gamma\int_{(\bigcup_j Q_j^*)^c}E_2(x)\,dx,
\end{aligned}
\tag{5.3.29}
$$

since $\frac{\alpha}{2}=\frac{\alpha}{3}+\frac{\alpha}{12}+\frac{\alpha}{12}$. We have

$$
\begin{aligned}
&\int_{(\bigcup_j Q_j^*)^c}E_1(x)\,dx \\
&\qquad\le\sum_j\int_{Q_j}|b_j(y)|\int_{(Q_j^*)^c}|K(x-y)-K(x-y_j)|\,dx\,dy \\
&\qquad\le\sum_j\int_{Q_j}|b_j(y)|\int_{|x-y_j|\ge 2|y-y_j|}|K(x-y)-K(x-y_j)|\,dx\,dy \\
&\qquad\le A_2\sum_j\int_{Q_j}|b_j(y)|\,dy=A_2\sum_j\big\|b_j\big\|_{L^1}\le A_2 2^{n+1}\big\|f\big\|_{L^1},
\end{aligned}
\tag{5.3.30}
$$

where we used the fact that if $x\in(Q_j^*)^c$, then $|x-y_j|\ge\frac{1}{2}\ell(Q_j^*)=\frac{5}{2}\sqrt{n}\,\ell(Q_j)$. But since $|y-y_j|\le\frac{\sqrt{n}}{2}\ell(Q_j)$, this implies that $|x-y_j|\ge 2|y-y_j|$. Here we used the fact that the diameter of a cube is equal to $\sqrt{n}$ times its side length. Likewise, we obtain that

$$
\int_{(\bigcup_j Q_j^*)^c}E_2(x)\,dx\le A_2\sum_j|Q_j|\le A_2\frac{\big\|f\big\|_{L^1}}{\alpha\gamma}.
\tag{5.3.31}
$$

Combining (5.3.30) and (5.3.31) with (5.3.29) yields (5.3.27).

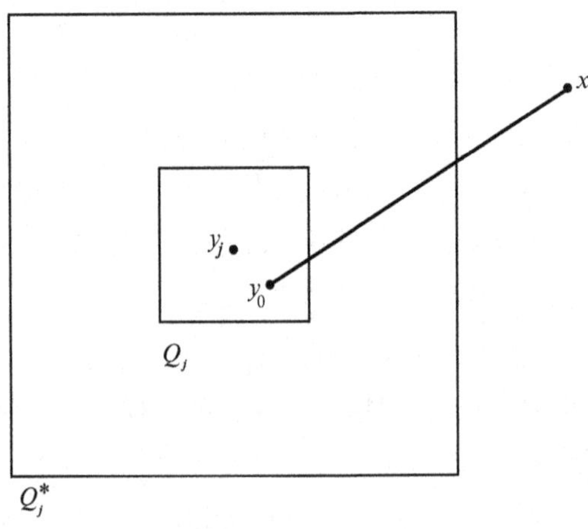

Fig. 5.3 The cubes Q_j and Q_j^*.

Therefore, the main task of the proof is to prove (5.3.28). Since $b = \sum_j b_j$, to estimate $T^{(**)}(b)$, it suffices to estimate each $|T^{(\varepsilon,N)}(b_j)|$ uniformly in ε and N. To achieve this we use the estimate

$$|T^{(\varepsilon,N)}(b_j)| \le |T^{(\varepsilon)}(b_j)| + |T^{(N)}(b_j)|, \tag{5.3.32}$$

noting that the truncated singular integrals $T^{(\varepsilon)}(b_j)$ are well defined. Indeed, say x lies in a compact set K_0. Pick M such that $K_0 - Q_j$ is contained in a ball $B(0,M)$. Then

$$|T^{(\varepsilon)}(b_j)(x)| \le |b_j| * |K^{(\varepsilon,M)}|(x),$$

which is finite a.e. as the convolution of two L^1 functions; thus the integral defining $T^{(\varepsilon)}(b_j)(x)$ converges absolutely and the expression $T^{(\varepsilon)}(b_j)(x)$ is well defined for almost all x.

We work with $T^{(\varepsilon)}$ and we note that $T^{(N)}$ can be treated similarly. Fix $x \notin \bigcup_j Q_j^*$ and $\varepsilon > 0$ and define

$$\begin{aligned} J_1(x,\varepsilon) &= \{j : \forall y \in Q_j \text{ we have } |x-y| < \varepsilon\}, \\ J_2(x,\varepsilon) &= \{j : \forall y \in Q_j \text{ we have } |x-y| > \varepsilon\}, \\ J_3(x,\varepsilon) &= \{j : \exists y \in Q_j \text{ we have } |x-y| = \varepsilon\}. \end{aligned}$$

Note that

$$T^{(\varepsilon)}(b_j)(x) = 0$$

whenever $x \notin \bigcup_j Q_j^*$ and $j \in J_1(x,\varepsilon)$. Also note that

$$K^{(\varepsilon)}(x-y) = K(x-y)$$

whenever $x \notin \bigcup_j Q_j^*$, $j \in J_2(x,\varepsilon)$ and $y \in Q_j$. Therefore,

$$\sup_{\varepsilon>0} |T^{(\varepsilon)}(b)(x)| \le \sup_{\varepsilon>0} \Big| \sum_{j \in J_2(x,\varepsilon)} T(b_j)(x) \Big| + \sup_{\varepsilon>0} \Big| \sum_{j \in J_3(x,\varepsilon)} T(b_j \chi_{|x-\cdot| \ge \varepsilon})(x) \Big|,$$

but since

$$\sup_{\varepsilon>0} \Big| \sum_{j \in J_2(x,\varepsilon)} T(b_j)(x) \Big| \le \sum_j |T(b_j)(x)| \le E_1(x), \tag{5.3.33}$$

it suffices to estimate the term

$$\sup_{\varepsilon>0} \Big| \sum_{j \in J_3(x,\varepsilon)} T(b_j \chi_{|x-\cdot| \ge \varepsilon})(x) \Big|.$$

We now make some geometric observations; see Figure 5.3. Fix $\varepsilon > 0$ and a cube Q_j with $j \in J_3(x,\varepsilon)$; recall that x lies in $(\bigcup_j Q_j^*)^c$. Then we have

$$\varepsilon \ge \frac{1}{2}\big(\ell(Q_j^*) - \ell(Q_j)\big) = \frac{1}{2}(5\sqrt{n}-1)\ell(Q_j) \ge 2\sqrt{n}\,\ell(Q_j). \tag{5.3.34}$$

Since $j \in J_3(x,\varepsilon)$, there exists a $y_0 \in Q_j$ with

$$|x-y_0| = \varepsilon .$$

Using (5.3.34), we obtain that for any $y \in Q_j$ we have

$$\frac{\varepsilon}{2} \le \varepsilon - \sqrt{n}\,\ell(Q_j) \le |x-y_0| - |y-y_0| \le |x-y| ,$$
$$|x-y| \le |x-y_0| + |y-y_0| \le \varepsilon + \sqrt{n}\,\ell(Q_j) \le \frac{3\varepsilon}{2} .$$

Therefore, we have proved that

$$\bigcup_{j\in J_3(x,\varepsilon)} Q_j \subseteqq B(x,\tfrac{3\varepsilon}{2}) \setminus B(x,\tfrac{\varepsilon}{2}) .$$

Letting

$$c_j(\varepsilon) = \frac{1}{|Q_j|}\int_{Q_j} b_j(y)\chi_{|x-y|\ge\varepsilon}(y)\,dy ,$$

we note that in view of property (5) of the Calderón–Zygmund decomposition (Theorem 5.3.1), the estimate

$$|c_j(\varepsilon)| \le 2^{n+1}\alpha\gamma$$

holds. Then

$$\begin{aligned}
&\sup_{\varepsilon>0}\bigg|\sum_{j\in J_3(x,\varepsilon)}\int_{Q_j} K(x-y)b_j(y)\chi_{|x-y|\ge\varepsilon}(y)\,dy\bigg| \\
&\quad\le \sup_{\varepsilon>0}\bigg|\sum_{j\in J_3(x,\varepsilon)}\int_{Q_j} K(x-y)\big(b_j(y)\chi_{|x-y|\ge\varepsilon}(y)-c_j(\varepsilon)\big)\,dy\bigg| \\
&\qquad\qquad + \sup_{\varepsilon>0}\bigg|\sum_{j\in J_3(x,\varepsilon)} c_j(\varepsilon)\int_{Q_j} K(x-y)\,dy\bigg| \\
&\quad\le \sup_{\varepsilon>0}\bigg|\sum_{j\in J_3(x,\varepsilon)}\int_{Q_j} \big(K(x-y)-K(x-y_j)\big)\big(b_j(y)\chi_{|x-y|\ge\varepsilon}(y)-c_j(\varepsilon)\big)\,dy\bigg| \\
&\qquad\qquad + 2^{n+1}\alpha\gamma\sup_{\varepsilon>0}\int_{B(x,\frac{3\varepsilon}{2})\setminus B(x,\frac{\varepsilon}{2})} |K(x-y)|\,dy \\
&\quad\le \sum_j \int_{Q_j} |K(x-y)-K(x-y_j)|\big(|b_j(y)|+2^{n+1}\alpha\gamma\big)\,dy \\
&\qquad\qquad + 2^{n+1}\alpha\gamma\sup_{\varepsilon>0}\int_{\frac{\varepsilon}{2}\le|x-y|\le\frac{3\varepsilon}{2}} |K(x-y)|\,dy \\
&\quad\le E_1(x) + 2^{n+1}\alpha\gamma E_2(x) + 2^{n+1}\alpha\gamma(2A_1) .
\end{aligned}$$

The last estimate, together with (5.3.33), with (5.3.32), and with the analogous estimate for $\sup_{N>0}|T^{(N)}(b_j)(x)|$ (which is similarly obtained), yields (5.3.28). □

The value of the previous theorem lies in the following: Since we know that for some sequences $\varepsilon_j \downarrow 0$, $N_j \uparrow \infty$ the pointwise limit $T^{(\varepsilon_j,N_j)}(f)$ exists a.e. for all f in a dense subclass of L^1, then Theorem 5.3.5 allows us to deduce that $T^{(\varepsilon_j,N_j)}(f)$ exists a.e. for all f in $L^1(\mathbf{R}^n)$.

If the singular integrals have kernels of the form $\Omega(x/|x|)|x|^{-n}$ with Ω in L^∞, such as the Hilbert transform and the Riesz transforms, then the upper truncations are not needed for K in (5.3.17). In this case

$$T_\Omega^{(\varepsilon)}(f)(x) = \int_{|y|\geq\varepsilon} f(x-y)\frac{\Omega(y/|y|)}{|y|^n}\,dy$$

is well defined for $f \in \bigcup_{1\leq p<\infty} L^p(\mathbf{R}^n)$ by Hölder's inequality and is equal to

$$\lim_{N\to\infty}\int_{\varepsilon\leq|y|\leq N} f(x-y)\frac{\Omega(y/|y|)}{|y|^n}\,dy\,.$$

Corollary 5.3.6. *The maximal Hilbert transform $H^{(*)}$ and the maximal Riesz transforms $R_j^{(*)}$ are weak type $(1,1)$. Secondly, $\lim_{\varepsilon\to 0} H^{(\varepsilon)}(f)$ and $\lim_{\varepsilon\to 0} R_j^{(\varepsilon)}(g)$ exist a.e. for all $f \in L^1(\mathbf{R})$ and $g \in L^1(\mathbf{R}^n)$, as $\varepsilon \to 0$.*

Proof. Since the kernels $1/x$ on $\mathbf{R}$ and $x_j/|x|^n$ on $\mathbf{R}^n$ satisfy (5.3.10), the first statement in the corollary is an immediate consequence of Theorem 5.3.5. The second statement follows from Theorem 2.1.14 and Corollary 5.2.8, since these limits exist for Schwartz functions. □

Corollary 5.3.7. *Under the hypotheses of Theorem 5.3.5, $T^{(**)}$ maps $L^p(\mathbf{R}^n)$ to itself for $1 < p < 2$ with norm*

$$\big\|T^{(**)}\big\|_{L^p\to L^p} \leq \frac{C_n(A_1+A_2+B)}{p-1}\,,$$

where C_n is some dimensional constant.

Exercises

5.3.1. Let A_1 be defined in (5.3.4). Prove that

$$\frac{1}{2}A_1 \leq \sup_{R>0}\frac{1}{R}\int_{|x|\leq R}|K(x)|\,|x|\,dx \leq 2A_1\,;$$

thus the expressions in (5.3.6) and (5.3.4) are equivalent.

5.3.2. Suppose that K is a locally integrable function on $\mathbf{R}^n\setminus\{0\}$ that satisfies (5.3.4). Suppose that $\delta_j \downarrow 0$. Prove that the principal value operation

$$\langle W,\varphi\rangle = \lim_{j\to\infty}\int_{\delta_j\leq|x|\leq 1} K(x)\varphi(x)\,dx$$

defines a distribution in $\mathscr{S}'(\mathbf{R}^n)$ if and only if the following limit exists:

$$\lim_{j\to\infty}\int_{\delta_j\le|x|\le 1} K(x)\,dx.$$

5.3.3. Suppose that a function K on $\mathbf{R}^n\setminus\{0\}$ satisfies condition (5.3.4) with constant A_1 and condition (5.3.12) with constant A_2.
(a) Show that the functions $K(x)\chi_{|x|\ge\varepsilon}$ also satisfy condition (5.3.12) uniformly in $\varepsilon>0$ with constant A_1+A_2.
(b) Obtain the same conclusion for the upper truncations $K(x)\chi_{|x|\le N}$.
(c) Deduce a similar conclusion for the double truncations $K^{(\varepsilon,N)}(x)=K(x)\chi_{\varepsilon\le|x|\le N}$.

5.3.4. Modify the proof of Theorem 5.3.5 to prove that if $T^{(**)}$ maps L^r to $L^{r,\infty}$ for some $1<r<\infty$, and K satisfies condition (5.3.12), then $T^{(**)}$ maps L^1 to $L^{1,\infty}$.

5.3.5. Assume that T is a linear operator acting on measurable functions on $\mathbf{R}^n$ such that whenever a function f is supported in a cube Q, then $T(f)$ is supported in a fixed multiple of Q.
(a) Suppose that T maps L^p to itself for some $1<p<\infty$ with norm B. Prove that T extends to a bounded operator from L^1 to $L^{1,\infty}$ with norm a constant multiple of B.
(b) Suppose that T maps L^p to L^q for some $1<q<p<\infty$ with norm B. Prove that T extends to a bounded operator from L^1 to $L^{s,\infty}$ with norm a multiple of B, where

$$\frac{1}{p'}+\frac{1}{q}=\frac{1}{s}.$$

5.3.6. (a) Let $1<q<\infty$. Show that the good function g in Theorem 5.3.1 lies in the Lorentz space $L^{q,1}$ and that $\|g\|_{L^{q,1}}\le C_{n,q}\alpha^{1/q'}\|f\|_{L^1}^{1/q}$ for some constant $C_{n,q}$.
(b) Let $1<r<\infty$. Obtain a generalization of Theorem 5.3.3 in which the assumption that T maps L^r to L^r is replaced by that T maps $L^{r,1}$ to $L^{r,\infty}$ with norm B.
(c) Let $1<r<\infty$. Obtain a further generalization of Theorem 5.3.3 in which the assumption that T maps L^r to L^r is replaced by that it is of restricted weak type (r,r), i.e., it satisfies

$$|\{x:\ |T(\chi_E)(x)|>\alpha\}|\le B^r\,\frac{|E|}{\alpha^r}$$

for all subsets E of $\mathbf{R}^n$ with finite measure.

5.3.7. Let K satisfy (5.3.12) for some $A_2>0$, let $W\in\mathscr{S}'(\mathbf{R}^n)$ be an extension of K on $\mathbf{R}^n$ as in (5.3.7), and let T be the operator given by convolution with W. Obtain the case $r=\infty$ in Theorem 5.3.3. Precisely, prove that if T maps $L^\infty(\mathbf{R}^n)$ to itself with constant B, then T has an extension on L^1+L^∞ that satisfies

$$\big\|T\big\|_{L^1\to L^{1,\infty}}\le C_n'\,(A_2+B),$$

and for $1 < p < \infty$ it satisfies

$$\|T\|_{L^p \to L^p} \le C_n \frac{1}{(p-1)^{1/p}} (A_2 + B),$$

where C_n, C_n' are constants that depend only on the dimension.
[*Hint:* Apply the Calderón–Zygmund decomposition $f = g + b$ at height $\alpha\gamma$, where $\gamma = (2^{n+1}B)^{-1}$. Since $|g| \le 2^n \alpha\gamma$, observe that

$$|\{x : |T(f)(x)| > \alpha\}| \le |\{x : |T(b)(x)| > \alpha/2\}|.$$

For the interpolation use the result of Exercise 1.3.2.]

5.3.8. (*Calderón–Zygmund decomposition on* L^q) Fix a function $f \in L^q(\mathbf{R}^n)$ for some $1 \le q < \infty$ and let $\alpha > 0$. Then there exist functions g and b on $\mathbf{R}^n$ such that

(1) $f = g + b$.

(2) $\|g\|_{L^q} \le \|f\|_{L^q}$ and $\|g\|_{L^\infty} \le 2^{\frac{n}{q}} \alpha$.

(3) $b = \sum_j b_j$, where each b_j is supported in a cube Q_j. Furthermore, the cubes Q_k and Q_j have disjoint interiors when $j \ne k$.

(4) $\|b_j\|_{L^q}^q \le 2^{n+q} \alpha^q |Q_j|$.

(5) $\int_{Q_j} b_j(x)\, dx = 0$.

(6) $\sum_j |Q_j| \le \alpha^{-q} \|f\|_{L^q}^q$.

(7) $\|b\|_{L^q} \le 2^{\frac{n+q}{q}} \|f\|_{L^q}$ and $\|b\|_{L^1} \le 2\alpha^{1-q} \|f\|_{L^q}^q$.

[*Hint:* Imitate the basic idea of the proof of Theorem 5.3.1, but select a cube Q if $\left(\frac{1}{|Q|} \int_Q |f(x)|^q\, dx\right)^{1/q} > \alpha$. Define g and b as in the proof of Theorem 5.3.1.]

5.3.9. Let $f \in L^1(\mathbf{R}^n)$. Then for any $\alpha > 0$, prove that there exist disjoint cubes Q_j in $\mathbf{R}^n$ such that the set $E_\alpha = \{x \in \mathbf{R}^n : M_c(f)(x) > \alpha\}$ is contained in $\bigcup_j 3Q_j$ and $\frac{\alpha}{4^n} < \frac{1}{|Q_j|} \int_{Q_j} |f(t)|\, dt \le \frac{\alpha}{2^n}$.
[*Hint:* For given $\alpha > 0$, select all maximal dyadic cubes $Q_j(\alpha)$ such that the average of f over them is bigger than α. Given $x \in E_\alpha$, pick a cube R that contains x such that the average of $|f|$ over R is bigger than α and find a dyadic cube Q such that $2^{-n}|Q| < |R| \le |Q|$ and that $\int_{R \cap Q} |f|\, dx > 2^{-n}\alpha|R|$. Conclude that Q is contained in some $Q_k(4^{-n}\alpha)$ and thus R is contained in $3Q_k(4^{-n}\alpha)$. The collection of all $Q_j = Q_j(4^{-n}\alpha)$ is the required one.]

5.3.10. Let $K(x)$ be a function on $\mathbf{R}^n \setminus \{0\}$ that satisfies $|K(x)| \le A|x|^{-n}$. Let $\eta(x)$ be a smooth function equal to 1 when $|x| \ge 2$ and vanishing when $|x| \le 1$. For $f \in L^p$, $1 \le p < \infty$, define truncated singular integral operators

$$T^{(\varepsilon)}(f)(x) = \int_{|y|\geq\varepsilon} K(y)f(x-y)\,dy,$$
$$T_\eta^{(\varepsilon)}(f)(x) = \int_{\mathbf{R}^n} \eta(y/\varepsilon)K(y)f(x-y)\,dy.$$

Show that the truncated maximal singular integral $T^{(*)}(f) = \sup_{\varepsilon>0}|T^{(\varepsilon)}(f)|$ is L^p bounded for $1 < p < \infty$ if and only if the smoothly truncated maximal singular integral $T_\eta^{(*)}(f) = \sup_{\varepsilon>0}|T_\eta^{(\varepsilon)}(f)|$ is L^p bounded. Formulate an analogous statement for $p = 1$.

5.3.11. (M. Mastyło) Let $1 \leq p < \infty$. Suppose that T_ε are linear operators defined on $L^p(\mathbf{R}^n)$ such that for all $f \in L^p(\mathbf{R}^n)$ we have $|T_\varepsilon(f)| \leq A\,\varepsilon^{-a}\|f\|_{L^p}$ for some $0 < a, A < \infty$. Also suppose that there is a constant $C < \infty$ such that the maximal operator $T_*(f) = \sup_{\varepsilon>0}|T_\varepsilon(f)|$ satisfies $\big\|T_*(h)\big\|_{L^p} \leq C\|h\|_{L^p}$ for all $h \in \mathscr{S}(\mathbf{R}^n)$. Prove that the same inequality is valid for all $f \in L^p(\mathbf{R}^n)$.
$\big[$*Hint:* For a fixed $\delta > 0$ define $S_\delta(f) = \sup_{\varepsilon>\delta}|T_\varepsilon(f)|$, which is a subadditive functional on $L^p(\mathbf{R}^n)$. For a fixed $f_0 \in L^p(\mathbf{R}^n)$ define a linear space $X_0 = \{\lambda f_0 : \lambda \in \mathbf{C}\}$ and a linear functional T_0 on X_0 by setting $T_0(\lambda f_0) = \lambda S_\delta(f_0)$. By the Hahn–Banach theorem there is an extension $\widetilde{T_0}$ of T_0 that satisfies $|\widetilde{T_0}(f)| \leq S_\delta(f)$ for all $f \in L^p(\mathbf{R}^n)$. Since S_δ is L^p is bounded on Schwartz functions with norm at most C, then so is $\widetilde{T_0}$. But $\widetilde{T_0}$ is linear and by density it is bounded on $L^p(\mathbf{R}^n)$ with norm at most C; consequently, $\|S_\delta(f_0)\|_{L^p} = \|T_0(f_0)\|_{L^p} = \|\widetilde{T_0}(f_0)\|_{L^p} \leq C\|f_0\|_{L^p}$. The required conclusion for T_* follows by Fatou's lemma.$\big]$

5.4 Sufficient Conditions for L^p Boundedness

We have used the Calderón–Zygmund decomposition to prove weak type $(1,1)$ boundedness for singular integral and maximal singular integral operators, assuming that these operators are already L^2 bounded. It is therefore natural to ask for sufficient conditions that imply L^2 boundedness for such operators. Precisely, what are sufficient conditions on functions K on $\mathbf{R}^n \setminus \{0\}$ so that the corresponding singular and maximal singular integral operators associated with K are L^2 bounded? We saw in Section 5.2 that if K has the special form $K(x) = \Omega(x/|x|)/|x|^n$ for some $\Omega \in L^1(\mathbf{S}^{n-1})$ with mean value zero, then condition (5.2.16) is necessary and sufficient for the L^2 boundedness of T, while the L^2 boundedness of $T^{(*)}$ requires the stronger smoothness condition (5.2.24).

For the general K considered in this section (for which the corresponding operator does not necessarily commute with dilations), we only give some sufficient conditions for L^2 boundedness of T and $T^{(**)}$.

Throughout this section K denotes a locally integrable function on $\mathbf{R}^n \setminus \{0\}$ that satisfies the "size" condition

$$\sup_{R>0}\int_{R\leq|x|\leq 2R} |K(x)|\,dx = A_1 < \infty, \tag{5.4.1}$$

the "smoothness" condition

$$\sup_{y\neq 0}\int_{|x|\geq 2|y|}|K(x-y)-K(x)|\,dx = A_2 < \infty, \tag{5.4.2}$$

and the "cancellation" condition

$$\sup_{0<R_1<R_2<\infty}\left|\int_{R_1<|x|<R_2} K(x)\,dx\right| = A_3 < \infty, \tag{5.4.3}$$

for some $A_1,A_2,A_3>0$. As mentioned earlier, condition (5.4.2) is often referred to as Hörmander's condition. In this section we show that these three conditions give rise to convolution operators that are bounded on L^p.

5.4.1 Sufficient Conditions for L^p Boundedness of Singular Integrals

We first note that under conditions (5.4.1), (5.4.2), and (5.4.3), there exists a tempered distribution W of the form (5.3.7) that coincides with K on $\mathbf{R}^n\setminus\{0\}$. Indeed, condition (5.4.3) implies that there exists a sequence $\delta_j\downarrow 0$ such that

$$\lim_{j\to\infty}\int_{\delta_j<|x|\leq 1} K(x)\,dx = L$$

exists. Using (5.3.8), we conclude that there exists such a tempered distribution W. Note that we must have $|L|\leq A_3$.

We observe that the difference of two distributions W and W' that coincide with K on $\mathbf{R}^n\setminus\{0\}$ must be supported at the origin.

Theorem 5.4.1. *Assume that K satisfies (5.4.1), (5.4.2), and (5.4.3), and let W be a tempered distribution of the form* (5.3.7) *that coincides with K on $\mathbf{R}^n\setminus\{0\}$. Then we have*

$$\sup_{0<\varepsilon<N<\infty}\sup_{\xi\in\mathbf{R}^n}|(K\chi_{\varepsilon<|\cdot|<N})^\wedge(\xi)| \leq 15(A_1+A_2+A_3) \tag{5.4.4}$$

and consequently

$$\sup_{\xi\in\mathbf{R}^n}|\widehat{W}(\xi)| \leq 15(A_1+A_2+A_3)\,. \tag{5.4.5}$$

Thus the operator given by convolution with W maps $L^2(\mathbf{R}^n)$ to itself with norm at most $15(A_1+A_2+A_3)$. Consequently, it also maps $L^1(\mathbf{R}^n)$ to $L^{1,\infty}(\mathbf{R}^n)$ with bound at most a dimensional constant multiple of $A_1+A_2+A_3$ and $L^p(\mathbf{R}^n)$ to itself with bound at most $C_n\max(p,(p-1)^{-1})(A_1+A_2+A_3)$, when $1<p<\infty$, where C_n is a dimensional constant.

Proof. Let us set $K^{(\varepsilon,N)}(x) = K(x)\chi_{\varepsilon<|x|<N}$. Estimate (5.4.4) implies that for all f in $\mathscr{S}(\mathbf{R}^n)$ we have

$$\big\|f * K^{(\delta_j,j)}\big\|_{L^2} \le 15(A_1+A_2+A_3)\big\|f\big\|_{L^2}$$

uniformly in j. Using this, (5.3.9), and Fatou's lemma, we obtain that

$$\big\|f * W\big\|_{L^2} \le 15(A_1+A_2+A_3)\big\|f\big\|_{L^2},$$

for all $f \in \mathscr{S}(\mathbf{R}^n)$, and this is equivalent to (5.4.5).
Case 1: Suppose that $\varepsilon < |\xi|^{-1} < N$. Then we write

$$\widehat{K^{(\varepsilon,N)}}(\xi) = I_1(\xi) + I_2(\xi),$$

where

$$\begin{aligned} I_1(\xi) &= \int_{\varepsilon<|x|<|\xi|^{-1}} K(x)e^{-2\pi i x\cdot\xi}\,dx, \\ I_2(\xi) &= \int_{|\xi|^{-1}<|x|<N} K(x)e^{-2\pi i x\cdot\xi}\,dx. \end{aligned}$$

We now have

$$I_1(\xi) = \int_{\varepsilon<|x|<|\xi|^{-1}} K(x)\,dx + \int_{\varepsilon<|x|<|\xi|^{-1}} K(x)\,(e^{-2\pi i x\cdot\xi} - 1)\,dx. \tag{5.4.6}$$

It follows that

$$|I_1(\xi)| \le A_3 + 2\pi|\xi| \int_{|x|<|\xi|^{-1}} |x|\,|K(x)|\,dx \le A_3 + 2\pi(2A_1)$$

uniformly in ε. Let us now examine $I_2(\xi)$. Let $z = \frac{\xi}{2|\xi|^2}$ so that $e^{2\pi i z\cdot\xi} = -1$ and $2|z| = |\xi|^{-1}$. By changing variables $x = x' - z$, rewrite I_2 as

$$I_2(\xi) = -\int_{|\xi|^{-1}<|x'-z|<N} K(x'-z)\,e^{-2\pi i x'\cdot\xi}\,dx';$$

hence averaging gives

$$I_2(\xi) = \frac{1}{2}\int_{|\xi|^{-1}<|x|<N} K(x)e^{-2\pi i x\cdot\xi}dx - \frac{1}{2}\int_{|\xi|^{-1}<|x-z|<N} K(x-z)\,e^{-2\pi i x\cdot\xi}\,dx.$$

Now use that

$$\int_A F\,dx - \int_B G\,dx = \int_B (F-G)\,dx + \int_{A\setminus B} F\,dx - \int_{B\setminus A} F\,dx \tag{5.4.7}$$

to write $I_2(\xi) = J_1(\xi) + J_2(\xi) + J_3(\xi) + J_4(\xi) + J_5(\xi)$, where

$$J_1(\xi) = +\frac{1}{2} \int_{|\xi|^{-1}<|x-z|<N} \big(K(x) - K(x-z)\big) e^{-2\pi i x\cdot\xi}\, dx, \tag{5.4.8}$$

$$J_2(\xi) = +\frac{1}{2} \int_{\substack{|\xi|^{-1}<|x|<N \\ |x-z|\le|\xi|^{-1}}} K(x)\, e^{-2\pi i x\cdot\xi}\, dx, \tag{5.4.9}$$

$$J_3(\xi) = +\frac{1}{2} \int_{\substack{|\xi|^{-1}<|x|<N \\ |x-z|\ge N}} K(x)\, e^{-2\pi i x\cdot\xi}\, dx, \tag{5.4.10}$$

$$J_4(\xi) = -\frac{1}{2} \int_{\substack{|\xi|^{-1}<|x-z|<N \\ |x|\le|\xi|^{-1}}} K(x)\, e^{-2\pi i x\cdot\xi}\, dx, \tag{5.4.11}$$

$$J_5(\xi) = -\frac{1}{2} \int_{\substack{|\xi|^{-1}<|x-z|<N \\ |x|\ge N}} K(x)\, e^{-2\pi i x\cdot\xi}\, dx. \tag{5.4.12}$$

Since $2|z| = |\xi|^{-1}$, $J_1(\xi)$ is bounded in absolute value by $\frac{1}{2}A_2$, in view of (5.4.2).

Next observe that $|\xi|^{-1} \le |x| \le \frac{3}{2}|\xi|^{-1}$ in (5.4.9), while $\frac{1}{2}|\xi|^{-1} \le |x| \le |\xi|^{-1}$ in (5.4.11); hence both J_2 and J_4 are bounded by $\frac{1}{2}A_1$. Finally, we have $\frac{1}{2}N < |x| < N$ in (5.4.10) (since $|x| > N - \frac{1}{2}|\xi|^{-1}$), and similarly we have $N \le |x| < \frac{3}{2}N$ in (5.4.12). Thus both J_3 and J_5 are bounded above by $\frac{1}{2}A_1$.
Case 2: If $\varepsilon < N \le |\xi|^{-1}$, then we write

$$\int_{\varepsilon<|x|<N} K(x) e^{-2\pi i x\cdot\xi}\, dx = \int_{\varepsilon<|x|<N} K(x)\, dx + \int_{\varepsilon<|x|<N} K(x)(e^{-2\pi i x\cdot\xi} - 1)\, dx$$

which is bounded in absolute value by

$$A_3 + 2\pi|\xi| \int_{|x|\le|\xi|^{-1}} |K(x)|\,|x|\, dx \le A_3 + 4\pi A_1.$$

Notice that if $\xi = 0$, only the first term appears which is bounded by A_3.
Case 2: If $|\xi|^{-1} \le \varepsilon < N$, we write

$$\int_{\varepsilon<|x|<N} K(x) e^{-2\pi i x\cdot\xi} dx = \int_{|\xi|^{-1}<|x|<N} K(x) e^{-2\pi i x\cdot\xi} dx - \int_{|\xi|^{-1}<|x|<\varepsilon} K(x) e^{-2\pi i x\cdot\xi} dx,$$

and both of the terms on the right are similar to $I_2(\xi)$ which was shown to be bounded by $2A_1 + \frac{1}{2}A_2$.

In all cases (5.4.4) holds. □

5.4.2 An Example

We now give an example of a distribution that satisfies conditions (5.4.1), (5.4.2), and (5.4.3).

Example 5.4.2. Let τ be a nonzero real number and let $K(x) = \frac{1}{|x|^{n+i\tau}}$ defined for $x \in \mathbf{R}^n \setminus \{0\}$. For a sequence $\delta_k \downarrow 0$ and φ a Schwartz function on $\mathbf{R}^n$, define

$$\langle W, \varphi \rangle = \lim_{k\to\infty} \int_{\delta_k \le |x|} \varphi(x) \frac{dx}{|x|^{n+i\tau}}, \tag{5.4.13}$$

whenever the limit exists. We claim that for some choices of sequences δ_k, W is a well defined tempered distribution on $\mathbf{R}^n$. Take, for example, $\delta_k = e^{-2\pi k/\tau}$. For this sequence δ_k, observe that

$$\int_{\delta_k \le |x| \le 1} \frac{1}{|x|^{n+i\tau}}\, dx = \omega_{n-1} \frac{1 - (e^{-2\pi k/\tau})^{-i\tau}}{-i\tau} = 0,$$

and thus

$$\langle W, \varphi \rangle = \int_{|x|\le 1} (\varphi(x) - \varphi(0)) \frac{dx}{|x|^{n+i\tau}} + \int_{|x|\ge 1} \varphi(x) \frac{dx}{|x|^{n+i\tau}}, \tag{5.4.14}$$

which implies that $W \in \mathscr{S}'(\mathbf{R}^n)$, since

$$|\langle W, \varphi \rangle| \le C\big[\|\nabla\varphi\|_{L^\infty} + \big\| |x|\, \varphi(x) \big\|_{L^\infty} \big].$$

If φ is supported in $\mathbf{R}^n \setminus \{0\}$, then

$$\langle W, \varphi \rangle = \int_{\mathbf{R}^n} K(x)\varphi(x)\, dx.$$

Therefore W coincides with the function K away from the origin. Moreover, (5.4.1) and (5.4.2) are clearly satisfied for K, while (5.4.3) is also satisfied, since

$$\left| \int_{R_1 < |x| < R_2} \frac{1}{|x|^{n+i\tau}}\, dx \right| = \omega_{n-1} \left| \frac{R_1^{-i\tau} - R_2^{-i\tau}}{-i\tau} \right| \le \frac{2\omega_{n-1}}{|\tau|}.$$

Remark 5.4.3. It is important to emphasize that the limit in (5.4.13) *may not* exist for all sequences $\delta_k \to 0$. For example, the limit in (5.4.13) does not exist if $\delta_k = e^{-\pi k/\tau}$. Moreover, for a different choice of a sequence δ_k for which the limit in (5.4.13) exists (for example, $\delta_k = e^{-\pi(2k+1)/\tau}$), we obtain a different distribution W_1 that coincides with the function $K(x) = |x|^{-n-i\tau}$ on $\mathbf{R}^n \setminus \{0\}$.

We discuss a point of caution. We can directly check that the distributions W defined by (5.4.13) are not homogeneous distributions of degree $-n-i\tau$. In fact, the only homogeneous distribution of degree $-n-i\tau$ that coincides with the function $|x|^{-n-i\tau}$ away from zero is a multiple of the distribution $u_{-n-i\tau}$, where u_z is defined in (2.4.7). Let us investigate the relationship between $u_{-n-i\tau}$ and W defined in (5.4.14). Recall that (2.4.8) gives

$$\begin{aligned}\langle u_{-n-i\tau}, \varphi\rangle = & \int_{|x|\geq 1} \varphi(x) \frac{\pi^{-i\frac{\tau}{2}}}{\Gamma(-i\frac{\tau}{2})} |x|^{-n-i\tau} \, dx \\ & + \int_{|x|\leq 1} (\varphi(x) - \varphi(0)) \frac{\pi^{-i\frac{\tau}{2}}}{\Gamma(-i\frac{\tau}{2})} |x|^{-n-i\tau} \, dx + \frac{\omega_{n-1} \pi^{-i\frac{\tau}{2}}}{-i\tau \Gamma(-i\frac{\tau}{2})} \varphi(0) \, .\end{aligned}$$

Using (5.4.14), we conclude that $u_{-n-i\tau} - c_1 W = c_2 \delta_0$ for suitable nonzero constants c_1 and c_2. Since the Dirac mass at the origin is not a homogeneous distribution of degree $-n-i\tau$, it follows that neither is W.

Since

$$\widehat{u_{-n-i\tau}} = u_{i\tau} = c_3 |\xi|^{i\tau} \, ,$$

the identity $u_{-n-i\tau} - c_1 W = c_2 \delta_0$ can be used to obtain a formula for the Fourier transform of W and thus produce a different proof that convolution with W is a bounded operator on $L^2(\mathbf{R}^n)$.

5.4.3 Necessity of the Cancellation Condition

Although conditions (5.4.1), (5.4.2), and (5.4.3) are sufficient for L^2 boundedness, they might not necessary. However, we show that (5.4.3) is a necessary condition, given (5.4.1).

Proposition 5.4.4. *Suppose that K is a function on $\mathbf{R}^n \setminus \{0\}$ that satisfies (5.4.1). Let W be a tempered distribution on $\mathbf{R}^n$ related to K as in (5.3.7). If the operator T given by convolution with W maps $L^2(\mathbf{R}^n)$ to itself (equivalently if $\widehat{W}$ is an L^∞ function), then the function K must satisfy (5.4.3).*

Proof. Pick a radial $\mathscr{C}^\infty$ function φ supported in the ball $|x| \leq 2$ with $0 \leq \varphi \leq 1$, and $\varphi(x) = 1$ when $|x| \leq 1$. For $R > 0$ let $\varphi^R(x) = \varphi(x/R)$. Fourier inversion for distributions gives the second equality,

$$(W * \varphi^R)(0) = \langle W, \varphi^R\rangle = \langle \widehat{W}, \widehat{\varphi^R}\rangle = \int_{\mathbf{R}^n} \widehat{W}(\xi) R^n \widehat{\varphi}(R\xi) \, d\xi \, ,$$

and the preceding identity implies that

$$|(W * \varphi^R)(0)| \leq \big\|\widehat{W}\big\|_{L^\infty} \big\|\widehat{\varphi}\big\|_{L^1} = \big\|T\big\|_{L^2 \to L^2} \big\|\widehat{\varphi}\big\|_{L^1}$$

uniformly in $R > 0$. Fix $0 < R_1 < R_2 < \infty$. If $R_2 \leq 2R_1$, we have

$$\left| \int_{R_1 < |x| < R_2} K(x) \, dx \right| \leq \int_{R_1 < |x| < 2R_1} |K(x)| \, dx \leq A_1 \, ,$$

which implies the required conclusion. We may therefore assume that $2R_1 < R_2$. Since the part of the integral in (5.4.3) over the set $R_1 < |x| < 2R_1$ is controlled by A_1, it suffices to control the integral of $K(x)$ over the set $2R_1 < |x| < R_2$. Since the function $\varphi^{R_2} - \varphi^{R_1}$ is supported away from the origin, the action of the distribution W on it can be written as integration against the function K. We have

$$\begin{aligned}
&\int_{\mathbf{R}^n} K(x)(\varphi^{R_2}(x) - \varphi^{R_1}(x))\,dx \\
&= \int_{2R_1<|x|<R_2} K(x)\,dx + \int_{R_1<|x|<2R_1} K(x)(1-\varphi^{R_1}(x))\,dx + \int_{R_2<|x|<2R_2} K(x)\varphi^{R_2}(x)\,dx.
\end{aligned}$$

The sum of the last two integrals is bounded by $3A_1$ (since $0 \le \varphi \le 1$), while the first integral is equal to

$$(W * \varphi^{R_2})(0) - (W * \varphi^{R_1})(0)$$

and is therefore bounded by $2\|T\|_{L^2\to L^2}\|\widehat{\varphi}\|_{L^1}$. We conclude that the function K must satisfy (5.4.3) with constant

$$A_3 \le 3A_1 + 2\|\widehat{\varphi}\|_{L^1}\|T\|_{L^2\to L^2} \le c\big(A_1 + \|T\|_{L^2\to L^2}\big).$$

This establishes the assertion. □

5.4.4 Sufficient Conditions for L^p Boundedness of Maximal Singular Integrals

We now discuss the analogous result to Theorem 5.4.1 for the maximal singular integral operator $T^{(**)}$.

Theorem 5.4.5. *Suppose that K satisfies (5.4.1), (5.4.2), and (5.4.3) and let $T^{(**)}$ be as in* (5.3.18). *Then $T^{(**)}$ is bounded on $L^p(\mathbf{R}^n)$, $1 < p < \infty$, with norm*

$$\big\|T^{(**)}\big\|_{L^p\to L^p} \le C_n \max(p,(p-1)^{-1})(A_1 + A_2 + A_3),$$

where C_n is a dimensional constant.

Proof. We first define an operator T associated with K that satisfies (5.4.1), (5.4.2), and (5.4.3). Because of condition (5.4.3), there exists a sequence $\delta_j \downarrow 0$ such that

$$\lim_{j\to\infty} \int_{\delta_j<|x|\le 1} K(x)\,dx$$

exists. Therefore, for $\varphi \in \mathscr{S}(\mathbf{R}^n)$ we can define a tempered distribution

$$\langle W, \varphi\rangle = \lim_{j\to\infty} \int_{\delta_j\le|x|\le j} K(x)\varphi(x)\,dx$$

and an operator T given by $T(f) = f * W$ for $f \in \mathscr{S}(\mathbf{R}^n)$. In view of Theorems 5.4.1 and 5.3.3, T admits an L^p bounded extension $(1 < p < \infty)$ with

$$\|T\|_{L^p \to L^p} \leq c_n \max(p, (p-1)^{-1})(A_1 + A_2 + A_3) \tag{5.4.15}$$

and is weak type $(1,1)$. This extension is still denoted by T.

Fix $1 < p < \infty$ and $f \in L^p(\mathbf{R}^n) \cap L^\infty(\mathbf{R}^n)$ with compact support. We have

$$\begin{aligned}
&T^{(\varepsilon,N)}(f)(x)\\
&= \int_{\varepsilon \leq |x-y| < N} K(x-y) f(y)\, dy = T^{(\varepsilon)}(f)(x) - T^{(N)}(f)(x)\\
&= \int_{\varepsilon \leq |x-y|} K(x-y) f(y)\, dy - \int_{N \leq |x-y|} K(x-y) f(y)\, dy\\
&= \int_{\varepsilon \leq |x-y|} (K(x-y) - K(z_1 - y)) f(y)\, dy + \int_{\varepsilon \leq |x-y|} K(z_1 - y) f(y)\, dy\\
&\quad - \int_{N \leq |x-y|} (K(x-y) - K(z_2 - y)) f(y)\, dy - \int_{N \leq |x-y|} K(z_2 - y) f(y)\, dy\\
&= \int_{\varepsilon \leq |x-y|} (K(x-y) - K(z_1 - y)) f(y)\, dy + T(f)(z_1) - T(f\chi_{|x-\cdot|<\varepsilon})(z_1)\\
&\quad - \int_{N \leq |x-y|} (K(x-y) - K(z_2 - y)) f(y)\, dy - T(f)(z_2) + T(f\chi_{|x-\cdot|<N})(z_2),
\end{aligned}$$

where z_1 and z_2 are arbitrary points in $\mathbf{R}^n$ that satisfy $|z_1 - x| \leq \frac{\varepsilon}{2}$ and $|z_2 - x| \leq \frac{N}{2}$. We used that f has compact support in order to be able to write $T^{(\varepsilon)}(f)(x)$ and $T^{(N)}(f)(x)$ as convergent integrals for almost every x.

At this point we take absolute values, average over $|z_1 - x| \leq \frac{\varepsilon}{2}$ and $|z_2 - x| \leq \frac{N}{2}$, and we apply Hölder's inequality in two terms. We obtain the estimate

$$\begin{aligned}
&|T^{(\varepsilon,N)}(f)(x)|\\
&\leq \frac{1}{v_n}\Big(\frac{2}{\varepsilon}\Big)^n \int_{|z_1-x|\leq\frac{\varepsilon}{2}} \int_{|x-y|\geq\varepsilon} |K(x-y) - K(z_1-y)|\, |f(y)|\, dy\, dz_1\\
&\quad + \frac{1}{v_n}\Big(\frac{2}{\varepsilon}\Big)^n \int_{|z_1-x|\leq\frac{\varepsilon}{2}} |T(f)(z_1)|\, dz_1\\
&\quad + \bigg(\frac{1}{v_n}\Big(\frac{2}{\varepsilon}\Big)^n \int_{|z_1-x|\leq\frac{\varepsilon}{2}} |T(f\chi_{|x-\cdot|<\varepsilon})(z_1)|^p\, dz_1\bigg)^{\frac{1}{p}}\\
&\quad + \frac{1}{v_n}\Big(\frac{2}{N}\Big)^n \int_{|z_2-x|\leq\frac{N}{2}} \int_{|x-y|\geq N} |K(x-y) - K(z_2-y)|\, |f(y)|\, dy\, dz_2\\
&\quad + \frac{1}{v_n}\Big(\frac{2}{N}\Big)^n \int_{|z_2-x|\leq\frac{N}{2}} |T(f)(z_2)|\, dz_2\\
&\quad + \bigg(\frac{1}{v_n}\Big(\frac{2}{N}\Big)^n \int_{|z_2-x|\leq\frac{N}{2}} |T(f\chi_{|x-\cdot|<N})(z_2)|^p\, dz_2\bigg)^{\frac{1}{p}},
\end{aligned}$$

where v_n is the volume of the unit ball in $\mathbf{R}^n$. Applying condition (5.4.2) and estimate (5.4.15), we obtain for f in $L^p(\mathbf{R}^n)\cap L^\infty(\mathbf{R}^n)$ with compact support that

$$
\begin{aligned}
&|T^{(\varepsilon,N)}(f)(x)| \\
&\quad\le \frac{1}{v_n}\Big(\frac{2}{\varepsilon}\Big)^n\int_{|z_1-x|\le\frac{\varepsilon}{2}}|T(f)(z_1)|\,dz_1+\frac{1}{v_n}\Big(\frac{2}{N}\Big)^n\int_{|z_2-x|\le\frac{N}{2}}|T(f)(z_2)|\,dz_2\\
&\qquad+c_n\Big(\sum_{j=1}^3 A_j\Big)\max(p,(p-1)^{-1})\Big(\frac{1}{v_n}\Big(\frac{2}{\varepsilon}\Big)^n\int_{|z_1-x|\le\varepsilon}|f(z_1)|^p\,dz_1\Big)^{\frac{1}{p}}\\
&\qquad+c_n\Big(\sum_{j=1}^3 A_j\Big)\max(p,(p-1)^{-1})\Big(\frac{1}{v_n}\Big(\frac{2}{N}\Big)^n\int_{|z_2-x|\le N}|f(z_2)|^p\,dz_2\Big)^{\frac{1}{p}}\\
&\qquad+2A_2\big\|f\big\|_{L^\infty}.
\end{aligned}
$$

We now use density to remove the compact support condition on f and obtain the last displayed estimate for all functions f in $L^p(\mathbf{R}^n)\cap L^\infty(\mathbf{R}^n)$. Taking the supremum over all $0<\varepsilon<N$ and over all $N>0$, we deduce that for all f in $L^p(\mathbf{R}^n)\cap L^\infty(\mathbf{R}^n)$ we have the estimate

$$
T^{(**)}(f)(x)\le 2A_2\big\|f\big\|_{L^\infty}+S_p(f)(x), \tag{5.4.16}
$$

where S_p is the sublinear operator defined by

$$
S_p(f)(x)=2M(T(f))(x)+3^{n+1}c_n\Big(\sum_{j=1}^3 A_j\Big)\max(p,(p-1)^{-1})(M(|f|^p)(x))^{\frac{1}{p}},
$$

and M is the Hardy–Littlewood maximal operator.

Recalling that M maps L^1 to $L^{1,\infty}$ with bound at most 3^n and also L^p to $L^{p,\infty}$ with bound at most $2\cdot 3^{n/p}$ for $1<p<\infty$ (Exercise 2.1.4), we conclude that S_p maps $L^p(\mathbf{R}^n)$ to $L^{p,\infty}(\mathbf{R}^n)$ with norm at most

$$
\big\|S_p\big\|_{L^p\to L^{p,\infty}}\le \widetilde{c}_n(A_1+A_2+A_3)\max(p,(p-1)^{-1}), \tag{5.4.17}
$$

where $\widetilde{c}_n$ is another dimensional constant.

Now write $f=f_\alpha+f^\alpha$, where

$$
f_\alpha=f\chi_{|f|\le\alpha/(16A_2)} \qquad\text{and}\qquad f^\alpha=f\chi_{|f|>\alpha/(16A_2)}.
$$

The function f_α is in $L^\infty\cap L^p$ and f^α is in $L^1\cap L^p$. Moreover, we see that

$$
\big\|f^\alpha\big\|_{L^1}\le(16A_2/\alpha)^{p-1}\big\|f\big\|_{L^p}^p. \tag{5.4.18}
$$

Apply the Calderón–Zygmund decomposition (Theorem 5.3.1) to the function f^α at height $\alpha\gamma$ to write $f^\alpha = g^\alpha + b^\alpha$, where g^α is the good function and b^α is the bad function of this decomposition. Using (5.3.1), we obtain

$$\big\|g^\alpha\big\|_{L^p} \le 2^{n/p'}(\alpha\gamma)^{1/p'}\big\|f^\alpha\big\|_{L^1}^{1/p} \le 2^{(n+4)/p'}(A_2\gamma)^{1/p'}\big\|f\big\|_{L^p}. \tag{5.4.19}$$

We now use (5.4.16) to get

$$\big|\{x\in\mathbf{R}^n:\ T^{(**)}(f)(x)>\alpha\}\big| \le b_1+b_2+b_3, \tag{5.4.20}$$

where

$$\begin{aligned} b_1 &= \big|\{x\in\mathbf{R}^n:\ 2A_2\big\|f_\alpha\big\|_{L^\infty}+S_p(f_\alpha)(x)>\alpha/4\}\big|,\\ b_2 &= \big|\{x\in\mathbf{R}^n:\ 2A_2\big\|g^\alpha\big\|_{L^\infty}+S_p(g^\alpha)(x)>\alpha/4\}\big|,\\ b_3 &= \big|\{x\in\mathbf{R}^n:\ T^{(**)}(b^\alpha)(x)>\alpha/2\}\big|. \end{aligned}$$

Observe that $2A_2\|f_\alpha\|_{L^\infty}\le\alpha/8$. Selecting $\gamma = 2^{-n-5}(A_1+A_2)^{-1}$ and using property (2) in Theorem 5.3.1, we obtain

$$2A_2\big\|g^\alpha\big\|_{L^\infty} \le A_2 2^{n+1}\alpha\gamma \le \alpha 2^{-4} < \frac{\alpha}{8}$$

and therefore

$$\begin{aligned} b_1 &\le \big|\{x\in\mathbf{R}^n:\ S_p(f_\alpha)(x)>\alpha/8\}\big|,\\ b_2 &\le \big|\{x\in\mathbf{R}^n:\ S_p(g^\alpha)(x)>\alpha/8\}\big|. \end{aligned} \tag{5.4.21}$$

Since $\gamma \le (2^{n+5}A_1)^{-1}$, it follows from (5.3.27) that

$$b_3 \le \Big|\bigcup_j Q_j^*\Big| + 2^{n+8}A_2\frac{\big\|f^\alpha\big\|_{L^1}}{\alpha} \le \left(\frac{(5\sqrt{n})^n}{\gamma}+2^{n+8}A_2\right)\frac{\big\|f^\alpha\big\|_{L^1}}{\alpha},$$

and using (5.4.18), we obtain

$$b_3 \le C_n(A_1+A_2)^p\alpha^{-p}\big\|f\big\|_{L^p}^p.$$

Using Chebyshev's inequality in (5.4.21) and (5.4.17), we finally obtain that

$$b_1+b_2 \le (8/\alpha)^p\,(\widetilde{c}_n)^p(A_1+A_2+A_3)^p\max(p,(p-1)^{-1})^p\big(\big\|f\big\|_{L^p}^p+\big\|g^\alpha\big\|_{L^p}^p\big).$$

Combining the estimates for b_1, b_2, and b_3 and using (5.4.19), we deduce

$$\big\|T^{(**)}(f)\big\|_{L^{p,\infty}} \le C_n(A_1+A_2+A_3)\max(p,(p-1)^{-1})\big\|f\big\|_{L^p(\mathbf{R}^n)}. \tag{5.4.22}$$

Finally, we need to obtain a similar estimate to (5.4.22), in which the weak L^p norm on the left is replaced by the L^p norm. This is a consequence of Theorem 1.3.2 via interpolation between the estimates $L^{\frac{p+1}{2}} \to L^{\frac{p+1}{2},\infty}$ and $L^{2p} \to L^{2p,\infty}$ for $2 < p < \infty$ and between the estimates $L^{2p} \to L^{2p,\infty}$ and $L^1 \to L^{1,\infty}$ for $1 < p < 2$. The latter estimate follows from Theorem 5.3.5. See also Corollary 5.3.7. □

Exercises

5.4.1. Let T be a convolution operator that is L^2 bounded. Suppose that a given function $f_0 \in L^1(\mathbf{R}^n) \cap L^2(\mathbf{R}^n)$ has vanishing integral and that $T(f_0)$ is integrable. Prove that $T(f_0)$ also has vanishing integral.

5.4.2. Let K satisfy (5.4.1), (5.4.2), and (5.4.3) and let $W \in \mathscr{S}'$ be an extension of K on $\mathbf{R}^n$. Let f be a compactly supported $\mathscr{C}^1$ function on $\mathbf{R}^n$ with mean value zero. Prove that the function $f * W$ is in $L^1(\mathbf{R}^n)$.

5.4.3. Suppose K is a function on $\mathbf{R}^n \setminus \{0\}$ that satisfies (5.4.1), (5.4.2), and (5.4.3). Let $K^{(\varepsilon,N)}(x) = K(x)\chi_{\varepsilon<|x|<N}$ for $0 < \varepsilon < N < \infty$ and let $T^{(\varepsilon,N)}$ be the operator given by convolution with $K^{(\varepsilon,N)}$. Let $1 < p < \infty$ and $f \in L^p(\mathbf{R}^n)$. Prove that for some sequence $\varepsilon_j \downarrow 0$, $T^{(\varepsilon_j,N)}(f)$ converges almost everywhere as $j \to \infty$ and $N \to \infty$.
[*Hint:* Use Theorems 5.4.5 and 2.1.14.]

5.4.4. (a) Prove that for all $x, y \in \mathbf{R}^n$ that satisfy $0 \neq x \neq y$ we have

$$\left| \frac{x-y}{|x-y|} - \frac{x}{|x|} \right| \le 2\frac{|y|}{|x|}.$$

(b) Let Ω be an integrable function with mean value zero on the sphere $\mathbf{S}^{n-1}$. Suppose that Ω satisfies a *Lipschitz (Hölder) condition* of order $0 < \alpha < 1$ on $\mathbf{S}^{n-1}$. This means that

$$|\Omega(\theta_1) - \Omega(\theta_2)| \le B_0|\theta_1 - \theta_2|^\alpha$$

for all $\theta_1, \theta_2 \in \mathbf{S}^{n-1}$. Prove that $K(x) = \Omega(x/|x|)/|x|^n$ satisfies Hörmander's condition (5.4.3) with constant at most a multiple of $B_0 + \|\Omega\|_{L^\infty}$.

5.4.5. Let Ω be an L^1 function on $\mathbf{S}^{n-1}$ with mean value zero.
(a) Let $\omega_\infty(t) = \sup\{|\Omega(\theta_1) - \Omega(\theta_2)| : \theta_1, \theta_2 \in \mathbf{S}^{n-1}, |\theta_1 - \theta_2| \le t\}$ and suppose that the following *Dini condition* holds:

$$\int_0^1 \omega_\infty(t)\frac{dt}{t} < \infty.$$

Prove that the function $K(x) = \Omega(x/|x|)|x|^{-n}$ satisfies Hörmander's condition.
(b) (*A. Calderón and A. Zygmund*) For $A \in O(n)$, let

$$\|A\| = \sup\{|\theta - A(\theta)| : \theta \in \mathbf{S}^{n-1}\}.$$

Suppose that Ω satisfies the more general Dini-type condition

$$\int_0^1 \omega_1(t)\frac{dt}{t} < \infty,$$

where

$$\omega_1(t) = \sup_{\substack{A\in O(n)\\ \|A\|\le t}} \int_{\mathbf{S}^{n-1}} |\Omega(A(\theta)) - \Omega(\theta)|\,d\theta\,.$$

Prove the same conclusion as in part (a).
[*Hint:* Part (b): Use the result in part (a) of Exercise 5.4.4 and switch to polar coordinates.]

5.5 Vector-Valued Inequalities

Certain nonlinear expressions that appear in Fourier analysis, such as maximal functions and square functions, can be viewed as linear quantities taking values in some Banach space. This point of view provides the motivation for a systematic study of Banach-valued operators. Let us illustrate this line of thinking via an example. Let T be a linear operator acting on L^p of some measure space (X,μ) and taking values in the set of measurable functions of another measure space (Y,ν). The seemingly nonlinear inequality

$$\Big\|\Big(\sum_j |T(f_j)|^2\Big)^{\frac{1}{2}}\Big\|_{L^p} \le C_p\Big\|\Big(\sum_j |f_j|^2\Big)^{\frac{1}{2}}\Big\|_{L^p} \tag{5.5.1}$$

can be transformed to a linear one with only a slight change of view. Let us denote by $L^p(X,\ell^2)$ the Banach space of all sequences $\{f_j\}_j$ of measurable functions on X that satisfy

$$\big\|\{f_j\}_j\big\|_{L^p(X,\ell^2)} = \Big(\int_X \Big(\sum_j |f_j|^2\Big)^{\frac{p}{2}} d\mu\Big)^{\frac{1}{p}} < \infty. \tag{5.5.2}$$

Define a linear operator acting on such sequences by setting

$$\vec{T}(\{f_j\}_j) = \{T(f_j)\}_j. \tag{5.5.3}$$

Then (5.5.1) is equivalent to the inequality

$$\big\|\vec{T}(\{f_j\}_j)\big\|_{L^p(Y,\ell^2)} \le C_p\big\|\{f_j\}_j\big\|_{L^p(X,\ell^2)}, \tag{5.5.4}$$

in which $\vec{T}$ is thought of as a linear operator acting on the L^p space of ℓ^2-valued functions on X. This is the basic idea of vector-valued inequalities. A nonlinear inequality such as (5.5.1) can be viewed as a linear norm estimate for an operator acting and taking values in suitable Banach spaces.

5.5.1 ℓ^2-Valued Extensions of Linear Operators

The following result is classical and fundamental in the subject of vector-valued inequalities.

Theorem 5.5.1. *Let $0 < p, q < \infty$ and let (X, μ) and (Y, ν) be two σ-finite measure spaces. The following are valid:*
(a) Suppose that T is a bounded linear operator from $L^p(X)$ to $L^q(Y)$ with norm N. Then T has an ℓ^2-valued extension, that is, for all complex-valued functions f_j in $L^p(X)$ we have

$$\Big\|\Big(\sum_j |T(f_j)|^2\Big)^{\frac{1}{2}}\Big\|_{L^q(Y)} \le C_{p,q} N \Big\|\Big(\sum_j |f_j|^2\Big)^{\frac{1}{2}}\Big\|_{L^p(X)} \tag{5.5.5}$$

for some constant $C_{p,q}$ that satisfies $C_{p,q} = 1$ if $p \le q$. Moreover, if T maps real-valued L^p functions to real-valued L^q functions with norm N_{real}, then (5.5.5) *holds for real-valued functions f_j with N_{real} in place of N.*
(b) Suppose that T is a bounded linear operator from $L^p(X)$ to $L^{q,\infty}(Y)$ with norm M. Then T has an ℓ^2-valued extension, that is,

$$\Big\|\Big(\sum_j |T(f_j)|^2\Big)^{\frac{1}{2}}\Big\|_{L^{q,\infty}(Y)} \le D_{p,q} M \Big\|\Big(\sum_j |f_j|^2\Big)^{\frac{1}{2}}\Big\|_{L^p(X)} \tag{5.5.6}$$

for some constant $D_{p,q}$ that depends only on p and q. Moreover, if T maps real-valued L^p functions to real-valued $L^{q,\infty}$ functions with norm M_{real}, then (5.5.5) *holds for real-valued functions f_j with M_{real} in place of M.*

To prove this theorem, we need the following identities.

Lemma 5.5.2. *For any $0 < r < \infty$, define constants*

$$A_r = \left(\frac{\Gamma(\frac{r+1}{2})}{\pi^{\frac{r+1}{2}}}\right)^{\frac{1}{r}} \qquad \text{and} \qquad B_r = \left(\frac{\Gamma(\frac{r}{2}+1)}{\pi^{\frac{r}{2}}}\right)^{\frac{1}{r}}. \tag{5.5.7}$$

Then for any $\lambda_1, \lambda_2, \dots, \lambda_n \in \mathbf{R}$ we have

$$\left(\int_{\mathbf{R}^n} |\lambda_1 x_1 + \cdots + \lambda_n x_n|^r e^{-\pi|x|^2} dx\right)^{\frac{1}{r}} = A_r (\lambda_1^2 + \cdots + \lambda_n^2)^{\frac{1}{2}}, \tag{5.5.8}$$

where $dx = dx_1 \cdots dx_n$. Also for all $w_1, w_2, \dots, w_n \in \mathbf{C}$ we have

$$\left(\int_{\mathbf{C}^n} |w_1 z_1 + \cdots + w_n z_n|^r e^{-\pi|z|^2} dz\right)^{\frac{1}{r}} = B_r (|w_1|^2 + \cdots + |w_n|^2)^{\frac{1}{2}}, \tag{5.5.9}$$

where $dz = dz_1 \cdots dz_n = dx_1 dy_1 \cdots dx_n dy_n$ if $z_j = x_j + iy_j$.

Proof. Dividing both sides of (5.5.8) by $(\lambda_1^2+\cdots+\lambda_n^2)^{\frac{1}{2}}$, we reduce things to the situation in which $\lambda_1^2+\cdots+\lambda_n^2=1$. Let $e_1=(1,0,\dots,0)^t$ be the standard basis column unit vector on $\mathbf{R}^n$ and find an orthogonal $n\times n$ matrix $A\in O(n)$ (orthogonal means a real matrix satisfying $A^t=A^{-1}$) such that $A^{-1}e_1=(\lambda_1,\dots,\lambda_n)^t$. Then the first coordinate of Ax is

$$(Ax)_1=Ax\cdot e_1=x\cdot A^te_1=x\cdot A^{-1}e_1=\lambda_1x_1+\cdots+\lambda_nx_n\,.$$

Now change variables $y=Ax$ in the integral in (5.5.8) and use the fact that $|Ax|=|x|$ to obtain

$$\begin{aligned}\left(\int_{\mathbf{R}^n}|\lambda_1x_1+\cdots+\lambda_nx_n|^re^{-\pi|x|^2}dx\right)^{\frac{1}{r}}&=\left(\int_{\mathbf{R}^n}|y_1|^re^{-\pi|y|^2}dy\right)^{\frac{1}{r}}\\&=\left(2\int_0^\infty t^re^{-\pi t^2}dt\right)^{\frac{1}{r}}\\&=\left(\int_0^\infty s^{\frac{r-1}{2}}e^{-\pi s}ds\right)^{\frac{1}{r}}\\&=\left(\frac{\Gamma(\frac{r+1}{2})}{\pi^{\frac{r+1}{2}}}\right)^{\frac{1}{r}}\\&=A_r\,,\end{aligned}$$

which proves (5.5.8). We used the fact that $\int_{\mathbf{R}}e^{-\pi|x|^2}dx=1$.

The proof of (5.5.9) is almost identical. We normalize by assuming that

$$|w_1|^2+\cdots+|w_n|^2=1\,,$$

and we let ε_1 be the column vector of $\mathbf{C}^n$ having 1 in the first entry and zero elsewhere. We find a unitary $n\times n$ matrix $\mathscr{A}$ such that $\mathscr{A}^{-1}\varepsilon_1=(\overline{w_1},\dots,\overline{w_n})^t$. Unitary means $\mathscr{A}^{-1}=\mathscr{A}^*$, where $\mathscr{A}^*$ is the conjugate transpose matrix of $\mathscr{A}$, i.e., the matrix whose entries are the complex conjugates of $\mathscr{A}^t$ and that satisfies $u\cdot\overline{\mathscr{A}v}=\mathscr{A}^*u\cdot\overline{v}$ for all $u,v\in\mathbf{C}^n$. Then $(\mathscr{A}z)_1=w_1z_1+\cdots+w_nz_n$ and also $|\mathscr{A}z|=|z|$; therefore, changing variables $\zeta=\mathscr{A}z$ in the integral in (5.5.9), we can rewrite that integral as

$$\begin{aligned}\left(\int_{\mathbf{C}^n}|\zeta_1|^re^{-\pi|\zeta|^2}d\zeta\right)^{\frac{1}{r}}&=\left(\int_{\mathbf{C}}|\zeta_1|^re^{-\pi|\zeta_1|^2}d\zeta_1\right)^{\frac{1}{r}}\\&=\left(2\pi\int_0^\infty t^re^{-\pi t^2}t\,dt\right)^{\frac{1}{r}}\\&=\left(\pi\int_0^\infty s^{\frac{r}{2}}e^{-\pi s}ds\right)^{\frac{1}{r}}\end{aligned}$$

$$
\begin{aligned}
&= \left(\frac{\Gamma(\frac{r}{2}+1)}{\pi^{\frac{r}{2}}} \right)^{\frac{1}{r}} \\
&= B_r ,
\end{aligned}
$$

where the second equality follows by polar coordinates and the third by the change of variables $s = t^2$. □

We now continue with the proof of Theorem 5.5.1.

Proof. The proof is based on conclusion (5.5.9) of Lemma 5.5.2.

Part (a): Assume first that $q < p$ and let B_r be as in (5.5.7). We may assume that the sequence $\{f_j\}_j$ is indexed by $\mathbf{Z}^+$. Use successively identity (5.5.9), the boundedness of T, Hölder's inequality with exponents p/q and $(p/q)'$ with respect to the measure $e^{-\pi|z|^2}dz$, and identity (5.5.9) again to deduce for $n \in \mathbf{Z}^+$

$$
\begin{aligned}
\Big\| \Big(\sum_{j=1}^n |T(f_j)|^2 \Big)^{\frac{1}{2}} \Big\|_{L^q(Y)}^q &= (B_q)^{-q} \int_Y \int_{\mathbf{C}^n} |z_1 T(f_1) + \cdots + z_n T(f_n)|^q e^{-\pi|z|^2} dz\, d\nu \\
&= (B_q)^{-q} \int_{\mathbf{C}^n} \int_Y |T(z_1 f_1 + \cdots + z_n f_n)|^q \, d\nu \, e^{-\pi|z|^2} dz \\
&\le (B_q)^{-q} N^q \int_{\mathbf{C}^n} \Big(\int_X |z_1 f_1 + \cdots + z_n f_n|^p \, d\mu \Big)^{\frac{q}{p}} e^{-\pi|z|^2} dz \\
&\le (B_q)^{-q} N^q \Big(\int_{\mathbf{C}^n} \int_X |z_1 f_1 + \cdots + z_n f_n|^p \, d\mu \, e^{-\pi|z|^2} dz \Big)^{\frac{q}{p}} \\
&= (B_q)^{-q} N^q \Big(B_p^p \int_X \Big(\sum_{j=1}^n |f_j|^2 \Big)^{\frac{p}{2}} d\mu \Big)^{\frac{q}{p}} \\
&= (B_p B_q^{-1})^q N^q \Big\| \Big(\sum_{j=1}^n |f_j|^2 \Big)^{\frac{1}{2}} \Big\|_{L^p(X)}^q .
\end{aligned}
$$

Now, letting $n \to \infty$ in the previous inequality, we obtain the required conclusion with $C_{p,q} = B_p B_q^{-1}$ when $q < p$.

We now turn to the case $q \ge p$. Using similar reasoning, we obtain

$$
\begin{aligned}
\Big\| \Big(\sum_{j=1}^n |T(f_j)|^2 \Big)^{\frac{1}{2}} \Big\|_{L^q(Y)}^q &= (B_q)^{-q} \int_Y \int_{\mathbf{C}^n} |z_1 T(f_1) + \cdots + z_n T(f_n)|^q e^{-\pi|z|^2} dz\, d\nu \\
&= (B_q)^{-q} \int_{\mathbf{C}^n} \int_Y |T(z_1 f_1 + \cdots + z_n f_n)|^q \, d\nu \, e^{-\pi|z|^2} dz \\
&\le (N B_q^{-1})^q \int_{\mathbf{C}^n} \Big(\int_X |z_1 f_1 + \cdots + z_n f_n|^p \, d\mu \Big)^{\frac{q}{p}} e^{-\pi|z|^2} dz \\
&= (N B_q^{-1})^q \Big\| \int_X |z_1 f_1 + \cdots + z_n f_n|^p \, d\mu \Big\|_{L^{\frac{q}{p}}(\mathbf{C}^n, e^{-\pi|z|^2} dz)}^{q/p} \\
&\le (N B_q^{-1})^q \Big\{ \int_X \Big\| |z_1 f_1 + \cdots + z_n f_n|^p \Big\|_{L^{\frac{q}{p}}(\mathbf{C}^n, e^{-\pi|z|^2} dz)} d\mu \Big\}^{\frac{q}{p}}
\end{aligned}
$$

$$= (NB_q^{-1})^q \Big\{ \int_X \Big(\int_{\mathbf{C}^n} |z_1 f_1 + \cdots + z_n f_n|^q e^{-\pi|z|^2} dz \Big)^{\frac{p}{q}} d\mu \Big\}^{\frac{q}{p}}$$

$$= (NB_q^{-1})^q \Big\{ \int_X (B_q)^p \Big(\sum_{j=1}^n |f_j|^2 \Big)^{\frac{p}{2}} d\mu \Big\}^{\frac{q}{p}}$$

$$= N^q \Big\| \Big(\sum_{j=1}^n |f_j|^2 \Big)^{\frac{1}{2}} \Big\|_{L^p(X)}^q .$$

Note that we made use of Minkowski's integral inequality (Exercise 1.1.6) in the last inequality. Letting $n \to \infty$ proves the required conclusion with $C_{p,q} = 1$ if $q \ge p$.

If T happens to map real-valued functions to real-valued functions, then we adapt the preceding argument by taking f_j to be real-valued functions, we replace B_p by A_p, B_q by A_q, N by N_{real} and we use identity (5.5.8) instead of (5.5.9).

Part (b): Inequality (5.5.6) will be a consequence of (5.5.5) and of the following result of Exercise 1.1.12 which holds when Y is σ-finite:

$$\|g\|_{L^{q,\infty}} \le \sup_{0<\nu(E)<\infty} \nu(E)^{\frac{1}{q}-\frac{1}{r}} \Big(\int_E |g|^r d\nu \Big)^{\frac{1}{r}} \le \Big(\frac{q}{q-r} \Big)^{\frac{1}{r}} \|g\|_{L^{q,\infty}}, \tag{5.5.10}$$

where $0 < r < q$ and the supremum is taken over all subsets E of Y of finite measure. Using (5.5.10), we obtain

$$\Big\| \Big(\sum_j |T(f_j)|^2 \Big)^{\frac{1}{2}} \Big\|_{L^{q,\infty}(Y)}$$

$$\le \sup_{0<\nu(E)<\infty} \nu(E)^{\frac{1}{q}-\frac{1}{r}} \Big(\int_E \Big(\sum_j |T(f_j)|^2 \Big)^{\frac{r}{2}} d\nu \Big)^{\frac{1}{r}}$$

$$= \sup_{0<\nu(E)<\infty} \nu(E)^{\frac{1}{q}-\frac{1}{r}} \Big(\int_Y \Big(\sum_j |\chi_E T(f_j)|^2 \Big)^{\frac{r}{2}} d\nu \Big)^{\frac{1}{r}}$$

$$\le \sup_{0<\nu(E)<\infty} \nu(E)^{\frac{1}{q}-\frac{1}{r}} \big\| T_E \big\|_{L^p \to L^r} C_{p,r} \Big(\int_X \Big(\sum_j |f_j|^2 \Big)^{\frac{p}{2}} d\mu \Big)^{\frac{1}{p}}, \tag{5.5.11}$$

where T_E is defined by $T_E(f) = \chi_E T(f)$. Since for any function f in $L^p(X)$ we have

$$\nu(E)^{\frac{1}{q}-\frac{1}{r}} \big\| T_E(f) \big\|_{L^r} \le \Big(\frac{q}{q-r} \Big)^{\frac{1}{r}} \big\| T(f) \big\|_{L^{q,\infty}} \le \Big(\frac{q}{q-r} \Big)^{\frac{1}{r}} M \big\| f \big\|_{L^p},$$

it follows that for any measurable set E of finite measure the estimate

$$\nu(E)^{\frac{1}{q}-\frac{1}{r}} \big\| T_E \big\|_{L^p \to L^r} \le \Big(\frac{q}{q-r} \Big)^{\frac{1}{r}} M \tag{5.5.12}$$

is valid. Inserting (5.5.12) in (5.5.11), we obtain (5.5.6) with $D_{p,q} = C_{p,r} \big(\frac{q}{q-r} \big)^{\frac{1}{r}}$ and $0 < r < q$. Recall that $C_{p,r} = 1$ if $r \ge p$ and $C_{p,r} = B_p B_r^{-1}$ if $r < p$.

If T happens to map real-valued functions to real-valued functions, then we take f_j to be real-valued and we replace M by M_{real}, B_r by A_r, and B_q by A_q in the preceding argument. □

5.5.2 Applications and ℓ^r-Valued Extensions of Linear Operators

Here is an application of Theorem 5.5.1:

Example 5.5.3. On the real line consider the intervals $I_j = [b_j, \infty)$ for $j \in \mathbf{Z}$. Let T_j be the operator given by multiplication on the Fourier transform by the characteristic function of I_j. Then we have the following two inequalities:

$$\Big\|\Big(\sum_{j\in\mathbf{Z}}|T_j(f_j)|^2\Big)^{\frac{1}{2}}\Big\|_{L^p(\mathbf{R})} \le C_p\Big\|\Big(\sum_{j\in\mathbf{Z}}|f_j|^2\Big)^{\frac{1}{2}}\Big\|_{L^p(\mathbf{R})}, \tag{5.5.13}$$

$$\Big\|\Big(\sum_{j\in\mathbf{Z}}|T_j(f_j)|^2\Big)^{\frac{1}{2}}\Big\|_{L^{1,\infty}(\mathbf{R})} \le C\Big\|\Big(\sum_{j\in\mathbf{Z}}|f_j|^2\Big)^{\frac{1}{2}}\Big\|_{L^1(\mathbf{R})}, \tag{5.5.14}$$

for $1 < p < \infty$. To prove these, first observe that the operator $T = \frac{1}{2}(I + iH)$ is given on the Fourier transform by multiplication by the characteristic function of the half-axis $[0, \infty)$ [precisely, the Fourier multiplier of T is equal to 1 on the set $(0, \infty)$ and $1/2$ at the origin; this function is almost everywhere equal to the characteristic function of the half-axis $[0, \infty)$]. Moreover, each T_j is given by

$$T_j(f)(x) = e^{2\pi i b_j x} T(e^{-2\pi i b_j (\cdot)} f)(x)$$

and thus with $g_j(x) = e^{-2\pi i b_j x} f(x)$, (5.5.13) and (5.5.14) can be written respectively as

$$\Big\|\Big(\sum_{j\in\mathbf{Z}}|T(g_j)|^2\Big)^{\frac{1}{2}}\Big\|_{L^p(\mathbf{R})} \le C_p\Big\|\Big(\sum_{j\in\mathbf{Z}}|g_j|^2\Big)^{\frac{1}{2}}\Big\|_{L^p(\mathbf{R})},$$

$$\Big\|\Big(\sum_{j\in\mathbf{Z}}|T(g_j)|^2\Big)^{\frac{1}{2}}\Big\|_{L^{1,\infty}(\mathbf{R})} \le C\Big\|\Big(\sum_{j\in\mathbf{Z}}|g_j|^2\Big)^{\frac{1}{2}}\Big\|_{L^1(\mathbf{R})}.$$

Theorem 5.5.1 gives that both of the previous estimates are valid by in view of the boundedness of $T = \frac{1}{2}(I + iH)$ from L^p to L^p and from $L^1 \to L^{1,\infty}$. For a slight generalization and an extension to higher dimensions, see Exercise 5.6.1.

We have now seen that bounded operators from L^p to L^q (or to $L^{q,\infty}$) always admit ℓ^2-valued extensions. It is natural to ask whether they also admit ℓ^r-valued extensions for some $r \neq 2$. For some values of r we may answer this question. Here is a straightforward corollary of Theorem 5.5.1.

Corollary 5.5.4. *Let (X,μ) and (Y,ν) be σ-finite measure spaces. Suppose that T is a linear bounded operator from $L^p(X)$ to $L^p(Y)$ with norm A for some $1<p<\infty$. Let r be a number between p and 2. Then we have*

$$\Big\|\Big(\sum_j |T(f_j)|^r\Big)^{\frac{1}{r}}\Big\|_{L^p(Y)} \le A\Big\|\Big(\sum_j |f_j|^r\Big)^{\frac{1}{r}}\Big\|_{L^p(X)}. \tag{5.5.15}$$

Proof. Using Exercise 5.5.2 we interpolate between the trivial bound $L^p(X,\ell^p)\to L^p(Y,\ell^p)$ and the bound $L^p(X,\ell^2)\to L^p(Y,\ell^2)$, which follows from Theorem 5.5.1. We obtain the bound $L^p(X,\ell^r)\to L^p(Y,\ell^r)$ since r lies between p and 2. □

Example 5.5.5. The result of Corollary 5.5.4 may fail if r does not lie in the interval with endpoints p and 2. Let us take, for example, $1<p<2$ and consider an $r<p$. Take $X=Y=\mathbf{R}$ and define a linear operator T by setting

$$T(f)(x)=\widehat{f}(x)\chi_{[0,1]}(x).$$

Then T is L^p bounded, since $\|T(f)\|_{L^p}\le\|T(f)\|_{L^{p'}}\le\|f\|_{L^p}$. Now take $f_j=\chi_{[j-1,j]}$ for $j=1,\dots,N$. A simple calculation gives

$$\Big(\sum_{j=1}^N |T(f_j)(x)|^r\Big)^{\frac{1}{r}} = N^{\frac{1}{r}}\left|\frac{e^{-2\pi ix}-1}{-2\pi ix}\chi_{[0,1]}(x)\right|,$$

while

$$\Big(\sum_{j=1}^N |f_j|^r\Big)^{\frac{1}{r}} = \chi_{[0,N]}.$$

It follows that $N^{1/r}\le CN^{1/p}$ for all $N>1$, and hence (5.5.15) cannot hold if $p>r$.

We have now seen that ℓ^r-valued extensions for $r\neq 2$ may fail in general. But do they fail for some specific operators of interest in Fourier analysis? For instance, is the inequality

$$\Big\|\Big(\sum_{j\in\mathbf{Z}} |H(f_j)|^r\Big)^{\frac{1}{r}}\Big\|_{L^p} \le C_{p,r}\Big\|\Big(\sum_{j\in\mathbf{Z}} |f_j|^r\Big)^{\frac{1}{r}}\Big\|_{L^p} \tag{5.5.16}$$

true for the Hilbert transform H whenever $1<p,r<\infty$? The answer to this question is affirmative. Inequality (5.5.16) is indeed valid and was first proved using complex function theory. In the next section we plan to study inequalities such as (5.5.16) for general singular integrals using the Calderón–Zygmund theory of the previous section applied to the context of Banach-valued functions.

5.5.3 General Banach-Valued Extensions

Let $\mathscr{B}$ be a Banach space over the field of complex numbers with norm $\|\ \|_{\mathscr{B}}$, and let $\mathscr{B}^*$ be its dual (with norm $\|\ \|_{\mathscr{B}^*}$). A function F defined on a measure space (X,μ)

and taking values in $\mathscr{B}$ is called $\mathscr{B}$-measurable if there exists a measurable subset X_0 of X such that $\mu(X \setminus X_0) = 0$, $F[X_0]$ is contained in some separable subspace $\mathscr{B}_0$ of $\mathscr{B}$, and for every $u^* \in \mathscr{B}^*$ the complex-valued map

$$x \mapsto \langle u^*, F(x) \rangle$$

is measurable. A consequence of this definition is that the positive function $x \mapsto \|F(x)\|_{\mathscr{B}}$ on X is measurable; to see this, use the relevant result in [382, p. 131].

For $0 < p \le \infty$ we denote by $L^p(X)$ the space $L^p(X, \mathbf{C})$. Let $L^p(X) \otimes \mathscr{B}$ be the set of all finite linear combinations of elements of $\mathscr{B}$ with coefficients in $L^p(X)$, that is, elements of the form

$$F = f_1 u_1 + \cdots + f_m u_m, \tag{5.5.17}$$

where $f_j \in L^p(X)$, $u_j \in \mathscr{B}$, and $m \in \mathbf{Z}^+$. We define $L^p(X, \mathscr{B})$ to be the space of all $\mathscr{B}$-measurable functions F on X satisfying

$$\left(\int_X \|F(x)\|_{\mathscr{B}}^p \, d\mu(x) \right)^{\frac{1}{p}} < \infty, \tag{5.5.18}$$

with the obvious modification when $p = \infty$. Similarly define $L^{p,\infty}(X, \mathscr{B})$ as the space of all $\mathscr{B}$-measurable functions F on X satisfying

$$\Big\| \|F(\cdot)\|_{\mathscr{B}} \Big\|_{L^{p,\infty}(X)} < \infty. \tag{5.5.19}$$

Then $L^p(X, \mathscr{B})$ (respectively, $L^{p,\infty}(X, \mathscr{B})$) is called the L^p (respectively, $L^{p,\infty}$) space of functions on X with values in $\mathscr{B}$. The quantity in (5.5.18) (respectively, in (5.5.19)) is the *norm* of F in $L^p(X, \mathscr{B})$ (respectively, in $L^{p,\infty}(X, \mathscr{B})$).

Proposition 5.5.6. *Let $\mathscr{B}$ be a Banach space and (X, μ) a σ-finite measure space.*
(a) The set $\{\sum_{j=1}^m \chi_{E_j} u_j : u_j \in \mathscr{B}, E_j \subseteq X$ are pairwise disjoint and $\mu(E_j) < \infty\}$ is dense in $L^p(X, \mathscr{B})$ whenever $0 < p < \infty$.
(b) The set $\{\sum_{j=0}^\infty \chi_{E_j} u_j : u_j \in \mathscr{B}, E_j \subseteq X$ are pairwise disjoint and $X = \bigcup_{j=0}^\infty E_j\}$ is dense in $L^\infty(X, \mathscr{B})$.
(c) The space $\mathscr{C}_0^\infty \otimes \mathscr{B}$ of functions of the form $\sum_{j=1}^m \varphi_j u_j$, where $u_j \in \mathscr{B}$, φ_j are in $\mathscr{C}_0^\infty(\mathbf{R}^n)$, is dense in $L^p(\mathbf{R}^n, \mathscr{B})$ for $1 \le p < \infty$.

Proof. If $F \in L^p(X, \mathscr{B})$ for $0 < p \le \infty$, then F is $\mathscr{B}$-measurable; thus there exists $X_0 \subseteq X$ satisfying $\mu(X \setminus X_0) = 0$ and $F[X_0] \subseteq \mathscr{B}_0$, where $\mathscr{B}_0$ is some separable subspace of $\mathscr{B}$. Choose a countable dense sequence $\{u_j\}_{j=1}^\infty$ of $\mathscr{B}_0$.

(a) First assume that $p < \infty$. Since X is σ-finite, for any $\varepsilon > 0$, there exists a measurable subset X_1 of X_0 with $\mu(X_1) < \infty$ such that

$$\int_{X \setminus X_1} \|F(x)\|_{\mathscr{B}}^p \, d\mu < \frac{\varepsilon^p}{3}.$$

Setting

$$\widetilde{B}(u_j, \varepsilon) = \left\{ u \in \mathscr{B}_0 : \|u - u_j\|_{\mathscr{B}} < \varepsilon (3\mu(X_1))^{-\frac{1}{p}} \right\},$$

we have $\mathscr{B}_0 \subseteqq \bigcup_{j=1}^{\infty} \widetilde{B}(u_j,\varepsilon)$. Let $A_1 = \widetilde{B}(u_1,\varepsilon)$ and $A_j = \widetilde{B}(u_j,\varepsilon) \setminus (\bigcup_{i=1}^{j-1} \widetilde{B}(u_i,\varepsilon))$ for $j \geq 2$. It is easily seen that $\{A_j\}_{j=1}^{\infty}$ are pairwise disjoint and $\bigcup_{j=1}^{\infty} A_j = \bigcup_{j=1}^{\infty} \widetilde{B}(u_j,\varepsilon)$. Set $E_j = F^{-1}[A_j] \cap X_1$. Then $X_1 = \bigcup_{j=1}^{\infty} E_j$ and $\{E_j\}_{j=1}^{\infty}$ are pairwise disjoint. Since $\mu(X_1) = \sum_{j=1}^{\infty} \mu(E_j) < \infty$, it follows that $\mu(E_j) < \infty$ and also that for some $m \in \mathbf{Z}^+$,

$$\int_{\bigcup_{j=m+1}^{\infty} E_j} \|F(x)\|_{\mathscr{B}}^p \, d\mu < \frac{\varepsilon^p}{3}. \tag{5.5.20}$$

Moreover, one can easily verify that $\sum_{j=1}^{m} \chi_{E_j} u_j$ is $\mathscr{B}$-measurable. Notice that $\|F(x) - u_j\|_{\mathscr{B}} < \varepsilon(3\mu(X_1))^{-1/p}$ for any $x \in E_j$ and $j \in \{1,\dots,m\}$. This fact combined with (5.5.20) and the mutual disjointness of $\{E_j\}_{j=1}^{m}$ yields that

$$\begin{aligned}\int_X \bigg\| F(x) - \sum_{j=1}^{m} \chi_{E_j}(x) u_j \bigg\|_{\mathscr{B}}^p d\mu &= \int_{X\setminus X_1} \|F(x)\|_{\mathscr{B}}^p \, d\mu + \int_{\bigcup_{j=m+1}^{\infty} E_j} \|F(x)\|_{\mathscr{B}}^p \, d\mu \\ &\quad + \int_{\bigcup_{j=1}^{m} E_j} \bigg\| \sum_{j=1}^{m} \chi_{E_j}(x)[F(x) - u_j] \bigg\|_{\mathscr{B}}^p d\mu \\ &< \frac{\varepsilon^p}{3} + \frac{\varepsilon^p}{3} + \frac{\varepsilon^p}{3} = \varepsilon^p .\end{aligned}$$

(b) Now consider the case $p = \infty$. Obviously we have $\mathscr{B}_0 \subseteqq \bigcup_{j=1}^{\infty} B(u_j,\varepsilon)$, where $B(u_j,\varepsilon) = \{u \in \mathscr{B}_0 : \|u - u_j\|_{\mathscr{B}} < \varepsilon\}$. Let $A_1 = B(u_1,\varepsilon)$ and for $j \geq 2$ define sets $A_j = B(u_j,\varepsilon) \setminus (\bigcup_{i=1}^{j-1} B(u_i,\varepsilon))$. Let $E_j = F^{-1}[A_j]$ for $j \geq 1$ and $E_0 = X \setminus (\bigcup_{j=1}^{\infty} E_j)$. Then $\mu(E_0) = 0$. As in the proof of the case $p < \infty$, we have that $\{E_j\}_{j=0}^{\infty}$ are pairwise disjoint and $X_0 \subseteqq \bigcup_{j=0}^{\infty} E_j$. Pick $u_0 = 0$. Notice that $\sum_{j=0}^{\infty} \chi_{E_j} u_j$ is $\mathscr{B}$-measurable. Since $\|F(x) - u_j\|_{\mathscr{B}} < \varepsilon$ for any $x \in E_j$ and $j \geq 0$, we have

$$\bigg\| F - \sum_{j=0}^{\infty} \chi_{E_j} u_j \bigg\|_{L^\infty(X,\mathscr{B})} = \bigg\| \sum_{j=0}^{\infty} \chi_{E_j} (F - u_j) \bigg\|_{L^\infty(X,\mathscr{B})} < \varepsilon,$$

which completes the proof in the case $p = \infty$.

(c) For the last assertion, we fix a smooth function with supported in the unit ball of $\mathbf{R}^n$ with integral one. Let $\varphi_\delta(x) = \delta^{-n}\varphi(x/\delta)$ for $x \in \mathbf{R}^n$ and $\delta > 0$. Given a function $\sum_{j=1}^{m} \chi_{E_j} u_j$ as in part (a) approximating a given f in $L^p \otimes \mathscr{B}$, we consider the function $\sum_{j=1}^{m} (\chi_{E_j} * \varphi_\delta) u_j$ which lies in $\mathscr{C}_0^\infty \otimes \mathscr{B}$. Since $\|\chi_{E_j} * \varphi_\delta - \chi_{E_j}\|_{L^p} \to 0$ as $\delta \to 0$ when $1 \leq p < \infty$, $\sum_{j=1}^{m} (\chi_{E_j} * \varphi_\delta) u_j$ tends to $\sum_{j=1}^{m} \chi_{E_j} u_j$ in $L^p \otimes \mathscr{B}$ as $\delta \to 0$, and the conclusion follows. □

Let (X,μ) be a measure space. If F is an element of $L^1(X) \otimes \mathscr{B}$ given as in (5.5.17), we define its integral (which is an element of $\mathscr{B}$) by setting

$$\int_X F(x)\, d\mu(x) = \sum_{j=1}^{m} \left(\int_X f_j(x)\, d\mu(x) \right) u_j.$$

Observe that for every $F \in L^1(X) \otimes \mathscr{B}$ we have

$$\begin{aligned}\Big\| \int_X F(x)\, d\mu(x) \Big\|_{\mathscr{B}} &= \sup_{\|u^*\|_{\mathscr{B}^*} \le 1} \Big| \Big\langle u^*, \sum_{j=1}^m \Big(\int_X f_j\, d\mu \Big) u_j \Big\rangle \Big| \\ &= \sup_{\|u^*\|_{\mathscr{B}^*} \le 1} \Big| \int_X \langle u^*, \sum_{j=1}^m f_j u_j \rangle\, d\mu \Big| \\ &\le \int_X \sup_{\|u^*\|_{\mathscr{B}^*} \le 1} \big| \langle u^*, \sum_{j=1}^m f_j u_j \rangle \big|\, d\mu \\ &= \big\| F \big\|_{L^1(X,\mathscr{B})}.\end{aligned}$$

Thus the linear operator

$$F \mapsto I_F = \int_X F(x)\, d\mu(x)$$

is bounded from $L^1(X) \otimes \mathscr{B}$ into $\mathscr{B}$. Since every element of $L^1(X,\mathscr{B})$ is a (norm) limit (Proposition 5.5.6 (c)) of a sequence of elements in $L^1(X) \otimes \mathscr{B}$, by continuity, the operator $F \mapsto I_F$ has a unique extension on $L^1(X,\mathscr{B})$ that we call the *Bochner integral* of F and denote by

$$\int_X F(x)\, d\mu(x).$$

$L^1(X,\mathscr{B})$ is called the space of all Bochner integrable functions from X to $\mathscr{B}$. Since the Bochner integral is an extension of I_F, for each $F \in L^1(X,\mathscr{B})$ we have

$$\Big\| \int_X F(x)\, dx \Big\|_{\mathscr{B}} \le \int_X \big\| F(x) \big\|_{\mathscr{B}}\, dx.$$

Consequently, measurable functions F with $\int_X \|F(x)\|_{\mathscr{B}}\, dx < \infty$ are Bochner integrable over X. It is not difficult to show that the Bochner integral of F is the only element of $\mathscr{B}$ that satisfies

$$\Big\langle u^*, \int_X F(x)\, d\mu(x) \Big\rangle = \int_X \langle u^*, F(x) \rangle\, d\mu(x) \tag{5.5.21}$$

for all $u^* \in \mathscr{B}^*$. The next result concerns duality in this context when $X = \mathbf{R}^n$.

Proposition 5.5.7. *Let $\mathscr{B}$ be a Banach space and $1 \le p \le \infty$.*
(a) For any $F \in L^p(\mathbf{R}^n,\mathscr{B})$ we have

$$\|F\|_{L^p(\mathbf{R}^n,\mathscr{B})} = \sup_{\|G\|_{L^{p'}(\mathbf{R}^n,\mathscr{B}^*)} \le 1} \Big| \int_{\mathbf{R}^n} \langle G(x), F(x) \rangle\, dx \Big|.$$

Consequently, $L^p(\mathbf{R}^n,\mathscr{B})$ isometrically embeds in $(L^{p'}(\mathbf{R}^n,\mathscr{B}^*))^*$.
(b) for any $G \in L^{p'}(\mathbf{R}^n,\mathscr{B}^*)$ one has

$$\|G\|_{L^{p'}(\mathbf{R}^n,\mathscr{B}^*)} = \sup_{\|F\|_{L^p(\mathbf{R}^n,\mathscr{B})} \le 1} \Big| \int_{\mathbf{R}^n} \langle G(x), F(x) \rangle\, dx \Big|$$

and thus $L^{p'}(\mathbf{R}^n,\mathscr{B}^*)$ isometrically embeds in $(L^p(\mathbf{R}^n,\mathscr{B}))^*$.

Proof. Hölder's inequality yields that the right-hand side of (a) is controlled by its left-hand side. It remains to establish the reverse inequality.

For $F \in L^p(\mathbf{R}^n, \mathscr{B})$ and $\varepsilon > 0$, by Proposition 5.5.6, there is $F_\varepsilon(x) = \sum_{j=1}^m \chi_{E_j}(x) u_j$ with $m \in \mathbf{Z}^+$ or $m = \infty$ (when $p = \infty$) such that $\|F_\varepsilon - F\|_{L^p(\mathbf{R}^n, \mathscr{B})} < \varepsilon/2$, where $\{E_j\}_{j=1}^m$ are pairwise disjoint subsets of $\mathbf{R}^n$ and $u_j \in \mathscr{B}$. Since $F_\varepsilon \in L^p(\mathbf{R}^n, \mathscr{B})$, we choose a nonnegative function h satisfying $\|h\|_{L^{p'}(\mathbf{R}^n)} \le 1$ such that

$$\|F_\varepsilon\|_{L^p(\mathbf{R}^n, \mathscr{B})} = \left(\int_{\mathbf{R}^n} \|F_\varepsilon(x)\|_{\mathscr{B}}^p \, dx \right)^{\frac{1}{p}} < \int_{\mathbf{R}^n} h(x) \|F_\varepsilon(x)\|_{\mathscr{B}} \, dx + \frac{\varepsilon}{4}. \tag{5.5.22}$$

When $1 \le p < \infty$, we can further choose $h \in L^{p'}(\mathbf{R}^n)$ to be a function with bounded support, which ensures that it is integrable. For given $u_j \in \mathscr{B}$, there exists $u_j^* \in \mathscr{B}^*$ satisfying $\|u_j^*\|_{\mathscr{B}^*} = 1$ and

$$\|u_j\|_{\mathscr{B}} < \langle u_j^*, u_j \rangle + \frac{\varepsilon}{4(\|h\|_{L^1(\mathbf{R}^n)} + 1)}. \tag{5.5.23}$$

Set $G(x) = \sum_{j=1}^m h(x) \chi_{E_j}(x) u_j^*$. Clearly G is $\mathscr{B}^*$-measurable and $\|G\|_{L^{p'}(\mathbf{R}^n, \mathscr{B}^*)} \le 1$. It follows from (5.5.22) and (5.5.23) that

$$\begin{aligned}
\int_{\mathbf{R}^n} \langle G(x), F_\varepsilon(x) \rangle \, dx &= \int_{\mathbf{R}^n} h(x) \sum_{j=1}^m \chi_{E_j}(x) \langle u_j^*, u_j \rangle \, dx \\
&\ge \int_{\mathbf{R}^n} h(x) \sum_{j=1}^m \left(\|u_j\|_{\mathscr{B}} - \frac{\varepsilon}{4(\|h\|_{L^1(\mathbf{R}^n)} + 1)} \right) \chi_{E_j}(x) \, dx \\
&\ge \|F_\varepsilon\|_{L^p(\mathbf{R}^n, \mathscr{B})} - \frac{\varepsilon}{4} - \frac{\varepsilon}{4}.
\end{aligned}$$

Using Hölder's inequality we have

$$\left| \int_{\mathbf{R}^n} \langle G(x), F_\varepsilon(x) - F(x) \rangle \, dx \right| \le \|G\|_{L^{p'}(\mathbf{R}^n, \mathscr{B}^*)} \|F - F_\varepsilon\|_{L^p(\mathbf{R}^n, \mathscr{B})} < \frac{\varepsilon}{2},$$

hence we obtain

$$\|F_\varepsilon\|_{L^p(\mathbf{R}^n, \mathscr{B})} \le \sup_{\|G\|_{L^{p'}(\mathbf{R}^n, \mathscr{B}^*)} \le 1} \left| \int_{\mathbf{R}^n} \langle G(x), F(x) \rangle \, dx \right| + \varepsilon.$$

Letting $\varepsilon \to 0$ yields the desired inequality in part (a).

(b) The duality statement $|\langle G(x), F(x) \rangle| \le \|G(x)\|_{\mathscr{B}^*} \|F(x)\|_{\mathscr{B}}$ together with Hölder's inequality imply the $\ge$ inequality in part (b). We prove the $\le$ inequality via an argument symmetric with that in case (a). For completeness we include the details. For a given G in $L^{p'}(\mathbf{R}^n, \mathscr{B})$ and $\varepsilon > 0$, by Proposition 5.5.6, there is $G_\varepsilon(x) = \sum_{j=1}^m \chi_{E_j}(x) u_j^*$ with $m \in \mathbf{Z}^+$ or $m = \infty$ (when $p = 1$) such that $\|G_\varepsilon - G\|_{L^{p'}(\mathbf{R}^n, \mathscr{B}^*)} < \varepsilon/2$, where $\{E_j\}_{j=1}^m$ are pairwise disjoint subsets of $\mathbf{R}^n$ and

$u_j^* \in \mathscr{B}^*$. Since G_ε lies in $L^{p'}(\mathbf{R}^n, \mathscr{B}^*)$, we choose a nonnegative function h satisfying $\|h\|_{L^p(\mathbf{R}^n)} \le 1$ such that

$$\big\|G_\varepsilon\big\|_{L^{p'}(\mathbf{R}^n,\mathscr{B}^*)} = \left(\int_{\mathbf{R}^n} \|G_\varepsilon(x)\|_{\mathscr{B}^*}^{p'}\,dx\right)^{\frac{1}{p'}} < \int_{\mathbf{R}^n} h(x)\big\|G_\varepsilon(x)\big\|_{\mathscr{B}^*}\,dx + \frac{\varepsilon}{4}. \tag{5.5.24}$$

When $1 < p \le \infty$, we can further choose $h \in L^p(\mathbf{R}^n)$ to be a function with bounded support, which ensures that it is integrable. For each $u_j^* \in \mathscr{B}$, there exists $u_j \in \mathscr{B}$ satisfying $\|u_j\|_{\mathscr{B}} = 1$ and

$$\|u_j^*\|_{\mathscr{B}^*} < \langle u_j^*, u_j\rangle + \frac{\varepsilon}{4(\|h\|_{L^1(\mathbf{R}^n)}+1)}. \tag{5.5.25}$$

Set $F(x) = \sum_{j=1}^m h(x)\chi_{E_j}(x)u_j$. Clearly F is $\mathscr{B}$-measurable and $\|F\|_{L^p(\mathbf{R}^n,\mathscr{B})} \le 1$. It follows from (5.5.24) and (5.5.25) that

$$\begin{aligned}
\int_{\mathbf{R}^n} \langle G_\varepsilon(x), F(x)\rangle\,dx &= \int_{\mathbf{R}^n} h(x)\sum_{j=1}^m \chi_{E_j}(x)\langle u_j^*, u_j\rangle\,dx \\
&\ge \int_{\mathbf{R}^n} h(x)\sum_{j=1}^m \left(\|u_j^*\|_{\mathscr{B}^*} - \frac{\varepsilon}{4(\|h\|_{L^1(\mathbf{R}^n)}+1)}\right)\chi_{E_j}(x)\,dx \\
&\ge \big\|G_\varepsilon\big\|_{L^{p'}(\mathbf{R}^n,\mathscr{B}^*)} - \frac{\varepsilon}{2}.
\end{aligned}$$

Hence, for any $\varepsilon > 0$, we have

$$\big\|G_\varepsilon\big\|_{L^{p'}(\mathbf{R}^n,\mathscr{B}^*)} \le \sup_{\|F\|_{L^p(\mathbf{R}^n,\mathscr{B})}\le 1} \left|\int_{\mathbf{R}^n} \langle G(x), F(x)\rangle\,dx\right| + \varepsilon$$

which implies the reverse inequality in part (b) by letting $\varepsilon \to 0$. □

Definition 5.5.8. Let X, Y be measure spaces. Let T be a linear operator that maps $L^p(X)$ to $L^q(Y)$ (respectively, $L^p(X)$ to $L^{q,\infty}(Y)$) for some $0 < p, q \le \infty$. We define another operator $\vec{T}$ acting on $L^p(X) \otimes \mathscr{B}$ by setting

$$\vec{T}\Big(\sum_{j=1}^m f_j u_j\Big) = \sum_{j=1}^m T(f_j)\,u_j.$$

If $\vec{T}$ happens to have a bounded extension from $L^p(X,\mathscr{B})$ to $L^q(Y,\mathscr{B})$ (respectively from $L^p(X,\mathscr{B})$ to $L^{q,\infty}(Y,\mathscr{B})$), then we say that T has a bounded *$\mathscr{B}$-valued extension*. In this case we also denote by $\vec{T}$ the $\mathscr{B}$-valued extension of T.

Example 5.5.9. Let $\mathscr{B} = \ell^r$ for some $1 \le r < \infty$. Then a measurable function $F : X \to \mathscr{B}$ is just a sequence $\{f_j\}_j$ of measurable functions $f_j : X \to \mathbf{C}$. The space $L^p(X, \ell^r)$ consists of all measurable complex-valued sequences $\{f_j\}_j$ on X that satisfy

$$\big\|\{f_j\}_j\big\|_{L^p(X,\ell^r)} = \Big\|\Big(\sum_j |f_j|^r\Big)^{\frac{1}{r}}\Big\|_{L^p(X)} < \infty.$$

The space $L^p(X)\otimes\ell^r$ is the set of all finite sums

$$\sum_{j=1}^{m}(a_{j1},a_{j2},a_{j3},\dots)\,g_j\,,$$

where $g_j\in L^p(X)$ and $(a_{j1},a_{j2},a_{j3},\dots)\in\ell^r$, $j=1,\dots,m$. This is certainly a subspace of $L^p(X,\ell^r)$. Now given $(f_1,f_2,\dots)\in L^p(X,\ell^r)$, let $F_m=e_1f_1+\dots+e_mf_m$, where e_j is the infinite sequence with zeros everywhere except at the jth entry, where it has 1. Then $F_m\in L^p(X)\otimes\ell^r$ and approximates f in the norm of $L^p(X,\ell^r)$. This shows the density of $L^p(X)\otimes\ell^r$ in $L^p(X,\ell^r)$.

If T is a linear operator bounded from $L^p(X)$ to $L^q(Y)$, then $\vec{T}$ is defined by

$$\vec{T}(\{f_j\}_j)=\{T(f_j)\}_j.$$

According to Definition 5.5.8, T has a bounded ℓ^r-extension if and only if the inequality

$$\Big\|\Big(\sum_j|T(f_j)|^r\Big)^{\frac{1}{r}}\Big\|_{L^q}\le C\Big\|\Big(\sum_j|f_j|^r\Big)^{\frac{1}{r}}\Big\|_{L^p}$$

is valid.

A linear operator T acting on measurable functions is called *positive* if it satisfies $f\ge 0\implies T(f)\ge 0$. It is straightforward to verify that positive operators satisfy

$$\begin{aligned}&f\le g\implies T(f)\le T(g)\,,\\&|T(f)|\le T(|f|)\,,\\&\sup_j|T(f_j)|\le T\big(\sup_j|f_j|\big)\,,\end{aligned}\tag{5.5.26}$$

for all f,g,f_j measurable functions. We have the following result regarding vector-valued extensions of positive operators:

Proposition 5.5.10. *Let $0<p,q\le\infty$ and (X,μ), (Y,ν) be two measure spaces. Let T be a positive linear operator mapping $L^p(X)$ to $L^q(Y)$ (respectively, to $L^{q,\infty}(Y)$) with norm A. Let $\mathscr{B}$ be a Banach space. Then T has a $\mathscr{B}$-valued extension $\vec{T}$ that maps $L^p(X,\mathscr{B})$ to $L^q(Y,\mathscr{B})$ (respectively, to $L^{q,\infty}(Y,\mathscr{B})$) with the same norm.*

Proof. Let us first understand this theorem when $\mathscr{B}=\ell^r$ for $1\le r\le\infty$. The two endpoint cases $r=1$ and $r=\infty$ can be checked easily using the properties in (5.5.26). For instance, for $r=1$ we have

$$\Big\|\sum_j|T(f_j)|\Big\|_{L^q}\le\Big\|\sum_j T(|f_j|)\Big\|_{L^q}=\Big\|T\Big(\sum_j|f_j|\Big)\Big\|_{L^q}\le A\Big\|\sum_j|f_j|\Big\|_{L^p},$$

while for $r=\infty$ we have

$$\Big\|\sup_j|T(f_j)|\Big\|_{L^q}\le\Big\|T(\sup_j|f_j|)\Big\|_{L^q}\le A\,\big\|\sup_j|f_j|\big\|_{L^p}.$$

The required inequality for $1 < r < \infty$,

$$\Big\|\Big(\sum_j |T(f_j)|^r\Big)^{\frac{1}{r}}\Big\|_{L^q} \le A \Big\|\Big(\sum_j |f_j|^r\Big)^{\frac{1}{r}}\Big\|_{L^p},$$

follows from the Riesz–Thorin interpolation theorem (see Exercise 5.5.2).

The result for a general Banach space $\mathscr{B}$ can be proved using the following inequality:

$$\big\|\vec{T}(F)(x)\big\|_{\mathscr{B}} \le T\big(\|F\|_{\mathscr{B}}\big)(x), \qquad x \in X, \tag{5.5.27}$$

by simply taking L^q norms. To prove (5.5.27), let us take $F = \sum_{j=1}^n f_j u_j$. Then

$$\begin{aligned}
\big\|\vec{T}(F)(x)\big\|_{\mathscr{B}} = \Big\|\sum_{j=1}^n T(f_j)(x)u_j\Big\|_{\mathscr{B}} &= \sup_{\|u^*\|_{\mathscr{B}^*}\le 1} \Big|\Big\langle u^*, \sum_{j=1}^n T(f_j)(x)u_j\Big\rangle\Big| \\
&= \sup_{\|u^*\|_{\mathscr{B}^*}\le 1} \Big|T\Big(\sum_{j=1}^n f_j \langle u^*, u_j\rangle\Big)(x)\Big| \\
&\le T\Big(\sup_{\|u^*\|_{\mathscr{B}^*}\le 1} \Big|\Big\langle u^*, \sum_{j=1}^n f_j u_j\Big\rangle\Big|\Big)(x) \\
&= T\Big(\Big\|\sum_{j=1}^n f_j u_j\Big\|_{\mathscr{B}}\Big)(x) \\
&= T\big(\|F\|_{\mathscr{B}}\big)(x),
\end{aligned}$$

where the inequality makes use of the fact that T is a positive operator. □

Exercises

5.5.1. ([207]) Let (X,μ) and (Y,ν) be σ-finite measure spaces and suppose that $0 < p_1, p_1 \le \infty$, $1 \le q_0, q_1 \le \infty$, and that $p_0 > p_1$. For $0 < \theta < 1$ define p,q by

$$\frac{1}{p} = \frac{1-\theta}{p_0} + \frac{\theta}{p_1}, \qquad \frac{1}{q} = \frac{1-\theta}{q_0} + \frac{\theta}{q_1}.$$

Let $\mathscr{B}_1$ and $\mathscr{B}_2$ be Banach spaces and let $\vec{T}$ be a linear operator that maps $L^{p_0}(X,\mathscr{B}_1)$ to $L^{q_0}(Y,\mathscr{B}_2)$ with norm A_0 and $L^{p_1}(X,\mathscr{B}_1)$ to $L^{q_1}(Y,\mathscr{B}_2)$ with norm A_1. Show that $\vec{T}$ has an extension that maps $L^p(X,\mathscr{B}_1)$ to $L^q(Y,\mathscr{B}_2)$ with norm at most $9A_0^{1-\theta}A_1^{\theta}$, by following the steps below:

(a) Let $i \in \{0,1\}$. If F_j, $j = 1,\dots,m$ are in $L^{p_i}(\mathscr{B}_1)$ with disjoint supports, using $\max_{j\in\{1,\dots,m\}} \|\vec{T}(F_j)(y)\|_{\mathscr{B}_2} \le \frac{1}{2^m}\sum_{\varepsilon_k=\pm 1}\big\|\sum_{k=1}^m \varepsilon_k \vec{T}(F_k)(y)\big\|_{\mathscr{B}_2}$ for $y \in Y$ show that

$$\Big\|\max_{j=1,\dots,m} \|\vec{T}(F_j)\|_{\mathscr{B}_2}\Big\|_{L^{q_i}} \le A_i \Big\|\sum_{j=1}^m F_j\Big\|_{L^{p_i}(X,\mathscr{B}_1)}.$$

(b) Assume that F lies in a dense subspace of $\mathscr{B}_1$, it satisfies $\| \|F\|_{\mathscr{B}_1} \|_{L^p} = 1$, and it takes only finitely many values. For a $\lambda > 1$ pick a large integer N such that $\lambda^{-N} < \|F(x)\|_{\mathscr{B}_1} \le \lambda^N$ for all $x \in X$ such that $\|F(x)\|_{\mathscr{B}_1} \neq 0$ and define $F_j = F\chi_{\Omega_j}$, where $\Omega_j = \{x:\ 2^j < \|F\|_{\mathscr{B}_1} \le 2^{j+1}\}$. Let $a = \frac{p}{p_1} - \frac{p}{p_0}$. Prove the inequalities

$$\Big\| \sum_j \lambda^{-ja\theta} F_j \Big\|_{L^{p_0}(X,\mathscr{B}_1)}^{p_0} \le \lambda^{a\theta p_0} \qquad \text{and} \qquad \Big\| \sum_j \lambda^{ja\theta} F_j \Big\|_{L^{p_1}(X,\mathscr{B}_1)}^{p_1} \le 1 .$$

(c) Define $g_0(y) = \max_j \lambda^{-ja\theta} \|\vec{T}(F_j)(y)\|_{\mathscr{B}_2}$, $g_1(y) = \max_j \lambda^{ja(1-\theta)} \|\vec{T}(F_j)(y)\|_{\mathscr{B}_2}$ for $y \in Y$ and show that

$$\|g_0\|_{L^{q_0}(Y)} \le A_0 \lambda^{a\theta} \qquad \text{and} \qquad \|g_1\|_{L^{q_1}(Y)} \le A_1 .$$

(d) Prove that for all $y \in Y$ we have

$$\|\vec{T}(F)(y)\|_{\mathscr{B}_2} \le \sum_j \|\vec{T}(F_j)(y)\|_{\mathscr{B}_2} \le g_0(y)^{1-\theta} g_1(y)^{\theta} \Big(2 + \frac{1}{\lambda^{a\theta}-1} + \frac{1}{\lambda^{a(1-\theta)}-1} \Big)$$

and conclude that $\|\vec{T}(F)\|_{L^p(Y,\mathscr{B}_2)} \le 9 A_0^{1-\theta} A_1^{\theta}$ by picking $\lambda = (1+\sqrt{2})^{a\theta}$.
[*Hint:* Part (d): Split the sum according to whether $\lambda^{ja} > \frac{g_1(y)}{g_0(y)}$ and $\lambda^{ja} \le \frac{g_1(y)}{g_0(y)}$.]

5.5.2. Prove the following version of the Riesz–Thorin interpolation theorem. Let (X,μ) and (Y,ν) be σ-finite measure spaces. Let $1 < p_0, q_0, , p_1, q_1, r_0, s_0, r_1, s_1 < \infty$ and $0 < \theta < 1$ satisfy

$$\frac{1-\theta}{p_0} + \frac{\theta}{p_1} = \frac{1}{p}, \qquad \frac{1-\theta}{q_0} + \frac{\theta}{q_1} = \frac{1}{q},$$
$$\frac{1-\theta}{r_0} + \frac{\theta}{r_1} = \frac{1}{r}, \qquad \frac{1-\theta}{s_0} + \frac{\theta}{s_1} = \frac{1}{s}.$$

Suppose that T is a linear operator that maps $L^{p_0}(X)$ to $L^{q_0}(Y)$ and $L^{p_1}(X)$ to $L^{q_1}(Y)$. Define a vector-valued operator $\vec{T}$ by setting $\vec{T}(\{f_j\}_j) = \{T(f_j)\}_j$ acting on sequences of complex-valued functions defined on X. Suppose that $\vec{T}$ maps $L^{p_0}(X,\ell^{r_0}(\mathbf{C}))$ to $L^{q_0}(Y,\ell^{s_0}(\mathbf{C}))$ with norm M_0 and $L^{p_1}(X,\ell^{r_1}(\mathbf{C}))$ to $L^{q_1}(Y,\ell^{s_1}(\mathbf{C}))$ with norm M_1. Prove that $\vec{T}$ maps $L^p(X,\ell^r(\mathbf{C}))$ to $L^q(Y,\ell^s(\mathbf{C}))$ with norm at most $M_0^{1-\theta} M_1^{\theta}$.
[*Hint:* Use the idea of the proof of Theorem 1.3.4. Apply Lemma 1.3.5 to the function

$$F(z) = \sum_{k=1}^{m} \sum_{j=1}^{n} \sum_{l \in \mathbf{Z}} \frac{a_{k,l}^{P(z)} e^{i\alpha_{k,l}}}{\|\{a_{k,l}\}_l\|_{\ell^r}^{R(z)-P(z)}} \frac{b_{j,l}^{Q(z)} e^{i\beta_{j,l}}}{\|\{b_{k,l}\}_l\|_{\ell^{s'}}^{S(z)-Q(z)}} \int_Y T(\chi_{A_k})(y)\, \chi_{B_j}(y)\, d\nu(y)$$

where $P(z) = \frac{p}{p_0}(1-z) + \frac{p}{p_1}z$, $Q(z) = \frac{q'}{q'_0}(1-z) + \frac{q'}{q'_1}z$, $R(z) = \frac{r}{r_0}(1-z) + \frac{r}{r_1}z$, and $S(z) = \frac{s'}{s'_0}(1-z) + \frac{s'}{s'_1}z$, $\{a_{k,l}\}_{l\in\mathbf{Z}}, \{b_{j,l}\}_{l\in\mathbf{Z}}$ are finitely supported sequences of positive reals, $\alpha_{k,l}, \beta_{k,l} \in \mathbf{R}$, A_k are pairwise disjoint subsets of X with finite measure, and B_j are pairwise disjoint subsets of Y with finite measure.]

5.5.3. Prove the following version of the Marcinkiewicz interpolation theorem. Let (X,μ) and (Y,ν) be σ-finite measure spaces and let $0 < p_0 < p < p_1 \le \infty$ and $0 < \theta < 1$ satisfy

$$\frac{1-\theta}{p_0} + \frac{\theta}{p_1} = \frac{1}{p}.$$

Let $\mathscr{B}_1, \mathscr{B}_2$ be Banach spaces and $\vec{T}$ be defined on $L^{p_0}(X,\mathscr{B}_1) + L^{p_1}(X,\mathscr{B}_1)$ such that for every $y \in Y$ and for all $F,G \in L^{p_0}(X,\mathscr{B}_1) + L^{p_1}(X,\mathscr{B}_1)$ we have

$$\big\|\vec{T}(F+G)(y)\big\|_{\mathscr{B}_2} \le \big\|\vec{T}(F)(y)\big\|_{\mathscr{B}_2} + \big\|\vec{T}(G)(y)\big\|_{\mathscr{B}_2}.$$

(a) Suppose that $\vec{T}$ maps $L^{p_0}(X,\mathscr{B}_1)$ to $L^{p_0,\infty}(Y,\mathscr{B}_2)$ with norm A_0 and $L^{p_1}(X,\mathscr{B}_1)$ to $L^{p_1,\infty}(Y,\mathscr{B}_2)$ with norm A_1. Show that $\vec{T}$ maps $L^p(X,\mathscr{B}_1)$ to $L^p(Y,\mathscr{B}_2)$ with norm at most $2\big(\frac{p}{p-p_0} + \frac{p}{p_1-p}\big)^{\frac{1}{p}} A_0^{1-\theta} A_1^{\theta}$.
(b) Let $p_0 = 1$. If $\vec{T}$ is linear and maps $L^1(X,\mathscr{B}_1)$ to $L^{1,\infty}(Y,\mathscr{B}_2)$ with norm A_0 and $L^{p_1}(X,\mathscr{B}_1)$ to $L^{p_1}(Y,\mathscr{B}_2)$ with norm A_1, show that $\vec{T}$ maps $L^p(X,\mathscr{B}_1)$ to $L^p(Y,\mathscr{B}_2)$ with norm at most $72(p-1)^{-1/p} A_0^{1-\theta} A_1^{\theta}$.
[*Hint:* Part (a): Copy the proof of Theorem 1.3.2. Part (b): Use Exercise 5.5.1.]

5.5.4. For all $x \in \mathbf{R}^n$ let $\vec{K}(x)$ be a bounded linear operator from $\mathscr{B}_1$ to $\mathscr{B}_2$ and let $\mathscr{Q} \otimes \mathscr{B}_1$ the space of all finite linear combinations of elements of the form $F = \sum_{i=1}^m \chi_{E_i} u_i$, where E_i are disjoint measurable subsets of $\mathbf{R}^n$ of finite measure, u_i in $\mathscr{B}_1$, and $m \in \mathbf{Z}^+$.
(a) Suppose that $\vec{K}$ satisfies

$$\int_{\mathbf{R}^n} \big\|\vec{K}(x)\big\|_{\mathscr{B}_1\to\mathscr{B}_2}\, dx = C_1 < \infty.$$

Prove that the operator

$$\vec{T}(F)(x) = \int_{\mathbf{R}^n} \vec{K}(x-y)(F(y))\, dy,$$

initially defined on $\mathscr{Q} \otimes \mathscr{B}_1$, has an extension that maps $L^p(\mathbf{R}^n,\mathscr{B}_1)$ to $L^p(\mathbf{R}^n,\mathscr{B}_2)$ with norm at most C_1 for $1 \le p < \infty$.
(b) (*Young's inequality*) Let $1 \le p,q,s \le \infty$ be such that $p < \infty$, $s > 1$, $1/q + 1 = 1/s + 1/p$. Suppose that $\vec{K}$ satisfies

$$\Big\| \big\|\vec{K}(\cdot)\big\|_{\mathscr{B}_1\to\mathscr{B}_2} \Big\|_{L^s(\mathbf{R}^n)} = C_2 < \infty.$$

Prove that the $\vec{T}$ defined in part (a) has an extension that maps $L^p(\mathbf{R}^n, \mathscr{B}_1)$ to $L^q(\mathbf{R}^n, \mathscr{B}_2)$ with norm at most C_2.
(c) (*Young's inequality for weak type spaces*) Suppose that $1 < p, s < \infty$, $1/q + 1 = 1/s + 1/p$, and that $\vec{K}$ satisfies

$$\Big\| \big\| \vec{K}(\cdot) \big\|_{\mathscr{B}_1 \to \mathscr{B}_2} \Big\|_{L^{s,\infty}(\mathbf{R}^n)} = C_3 < \infty .$$

Prove that the $\vec{T}$ defined in part (a) has an extension that maps $L^p(\mathbf{R}^n, \mathscr{B}_1)$ to $L^q(\mathbf{R}^n, \mathscr{B}_2)$.
(d) Prove the following (slight) generalization of the assertion in part (a). Suppose that $\vec{K}$ satisfies

$$\int_{\mathbf{R}^n} \big\| \vec{K}(x)(u) \big\|_{\mathscr{B}_2} \, dx \le C_1 \|u\|_{\mathscr{B}_1}$$

for all $u \in \mathscr{B}_1$. Then $\vec{T}$ has an extension that maps $L^1(\mathbf{R}^n, \mathscr{B}_1)$ to $L^1(\mathbf{R}^n, \mathscr{B}_2)$ with norm at most C_1.

5.5.5. Use the inequality for the Rademacher functions in Appendix C.2 instead of Lemma 5.5.2 to prove part (a) of Theorem 5.5.1 in the special case $p = q$. Notice that this approach does not yield a sharp constant.

5.5.6. Let $0 < p \ne 2 \le \infty$ and suppose that T_j are uniformly bounded linear operators from $L^p(\mathbf{R})$ to $L^p(\mathbf{R})$. Show that the inequality

$$\Big\| \Big(\sum_{j \in \mathbf{Z}} |T_j(f_j)|^2 \Big)^{\frac{1}{2}} \Big\|_{L^p} \le C_p \Big\| \Big(\sum_{j \in \mathbf{Z}} |f_j|^2 \Big)^{\frac{1}{2}} \Big\|_{L^p}$$

may fail.
[*Hint:* Let $T_j(g)(x) = g(x - j)$. When $p > 2$ take $f_j(x) = \chi_{[-j, 1-j]}$ for $j = 1, 2, \dots, N$. When $p < 2$ take $f_j = \chi_{[0,1]}$ for $j = 1, 2, \dots, N$.]

5.5.7. Suppose that T is a linear operator that takes real-valued functions to real-valued functions. Use Theorem 5.5.1(a) with $p = q$ to prove that

$$\sup_{\substack{f \text{ real-valued} \\ f \ne 0}} \frac{\big\| T(f) \big\|_{L^p}}{\big\| f \big\|_{L^p}} = \sup_{\substack{f \text{ complex-valued} \\ f \ne 0}} \frac{\big\| T(f) \big\|_{L^p}}{\big\| f \big\|_{L^p}} .$$

5.6 Vector-Valued Singular Integrals

We now discuss some results about vector-valued singular integrals. By this we mean singular integral operators acting on functions defined on $\mathbf{R}^n$ and taking values in Banach spaces.

5.6.1 Banach-Valued Singular Integral Operators

Suppose that $\mathscr{B}_1$ and $\mathscr{B}_2$ are Banach spaces. We denote by $L(\mathscr{B}_1,\mathscr{B}_2)$ the space of all bounded linear operators from $\mathscr{B}_1$ to $\mathscr{B}_2$. We consider a kernel $\vec{K}$ defined on $\mathbf{R}^n\setminus\{0\}$ that takes values in $L(\mathscr{B}_1,\mathscr{B}_2)$. In other words, for all $x\in\mathbf{R}^n\setminus\{0\}$, $\vec{K}(x)$ is a bounded linear operator from $\mathscr{B}_1$ to $\mathscr{B}_2$ with norm $\|\vec{K}(x)\|_{\mathscr{B}_1\to\mathscr{B}_2}$. Thus for any $v\in\mathscr{B}_1$ and any $x\in\mathbf{R}^n\setminus\{0\}$ we have

$$\big\|\vec{K}(x)(v)\big\|_{\mathscr{B}_2}\le\|\vec{K}(x)\|_{\mathscr{B}_1\to\mathscr{B}_2}\|v\|_{\mathscr{B}_1}\,.$$

We assume that there is a constant $A<\infty$ such that the size condition holds

$$\big\|\vec{K}(x)\big\|_{\mathscr{B}_1\to\mathscr{B}_2}\le A\,|x|^{-n}, \tag{5.6.1}$$

and also the regularity condition

$$\sup_{y\in\mathbf{R}^n\setminus\{0\}}\int_{|x|\ge 2|y|}\big\|\vec{K}(x-y)-\vec{K}(x)\big\|_{\mathscr{B}_1\to\mathscr{B}_2}\,dx\le A<\infty. \tag{5.6.2}$$

Moreover, we assume that there is a sequence $\varepsilon_k\downarrow 0$ as $k\to\infty$ and an element $\vec{K}_0$ of $L(\mathscr{B}_1,\mathscr{B}_2)$ such that

$$\lim_{k\to\infty}\bigg\|\int_{\varepsilon_k\le|y|\le 1}\vec{K}(y)\,dy-\vec{K}_0\bigg\|_{\mathscr{B}_1\to\mathscr{B}_2}=0. \tag{5.6.3}$$

Given these assumptions, we define an operator $\vec{T}$ on $\mathscr{C}_0^\infty\otimes\mathscr{B}_1$ as follows:

For functions $f_i\in\mathscr{C}_0^\infty(\mathbf{R}^n)$ and $u_i\in\mathscr{B}_1$ we define

$$\begin{aligned}\vec{T}\Big(\sum_{i=1}^m f_iu_i\Big)(x)&=\lim_{k\to\infty}\int_{\varepsilon_k\le|y|}\vec{K}(y)\Big(\sum_{i=1}^m f_i(x-y)u_i\Big)\,dy &(5.6.4)\\ &=\sum_{i=1}^m\int_{|y|\le 1}(f_i(x-y)-f_i(x))\vec{K}(y)(u_i)\,dy+\sum_{i=1}^m f_i(x)\vec{K}_0(u_i)\\ &\quad+\int_{|y|>1}\sum_{i=1}^m f_i(x-y)\vec{K}(y)(u_i)\,dy. &(5.6.5)\end{aligned}$$

Notice that for each $i\in\{1,\dots,m\}$ we have

$$\int_{|y|\le 1}|f_i(x-y)-f_i(x)|\,\|\vec{K}(y)(u_i)\|_{\mathscr{B}_2}dy\le\|\nabla f_i\|_{L^\infty}\|u_i\|_{\mathscr{B}_1}\int_{|y|\le 1}|y|\,\|\vec{K}(y)\|_{\mathscr{B}_1\to\mathscr{B}_2}dy$$

and this is a finite integral in view of (5.6.1). Thus the function

$$(f_i(x-y)-f_i(x))\,\vec{K}(y)(u_i)$$

is $\mathscr{B}_2$-integrable and the expression

$$\int_{|y|\le 1} (f_i(x-y) - f_i(x))\, \|\vec{K}(y)(u_i)\|_{\mathscr{B}_2}\, dy$$

is a well-defined element of $\mathscr{B}_2$. Also the integral in (5.6.5) is over the compact set $1 \le |y| \le |x| + M$, where the ball $B(0,M)$ contains the supports of all f_i, and thus it also converges in $\mathscr{B}_2$, using (5.6.1).

The following vector-valued extension of Theorem 5.3.3 is the main result of this section.

Theorem 5.6.1. *Let $\mathscr{B}_1$ and $\mathscr{B}_2$ be Banach spaces. Suppose that $\vec{K}(x)$ satisfies (5.6.1), (5.6.2), and (5.6.3) for some $A > 0$ and $\vec{K_0} \in L(\mathscr{B}_1, \mathscr{B}_2)$. Let $\vec{T}$ be the operator associated with $\vec{K}$ as in (5.6.4). Assume that $\vec{T}$ is a bounded linear operator from $L^r(\mathbf{R}^n, \mathscr{B}_1)$ to $L^r(\mathbf{R}^n, \mathscr{B}_2)$ with norm $B_\star$ for some $1 < r \le \infty$. Then $\vec{T}$ has well-defined extensions on $L^p(\mathbf{R}^n, \mathscr{B}_1)$ for all $1 \le p < \infty$. Moreover, there exist dimensional constants C_n and C_n' such that*

$$\big\|\vec{T}(F)\big\|_{L^{1,\infty}(\mathbf{R}^n,\mathscr{B}_2)} \le C_n'(A + B_\star)\big\|F\big\|_{L^1(\mathbf{R}^n,\mathscr{B}_1)} \tag{5.6.6}$$

for all F in $L^1(\mathbf{R}^n, \mathscr{B}_1)$ and

$$\big\|\vec{T}(F)\big\|_{L^p(\mathbf{R}^n,\mathscr{B}_2)} \le C_n \max\big(p, (p-1)^{-1}\big)(A + B_\star)\big\|F\big\|_{L^p(\mathbf{R}^n,\mathscr{B}_1)} \tag{5.6.7}$$

whenever $1 < p < \infty$ and F is in $L^p(\mathbf{R}^n, \mathscr{B}_1)$.

Proof. Although $\vec{T}$ is defined on the entire $L^1(\mathbf{R}^n, \mathscr{B}_1) \cap L^r(\mathbf{R}^n, \mathscr{B}_1)$, it will be convenient to work with its restriction to a smaller dense subspace of $L^1(\mathbf{R}^n, \mathscr{B}_1)$. We make the observation that the space $\mathscr{D} \otimes \mathscr{B}_1$ of all functions of the form $\sum_{i=1}^m \chi_{R_i} u_i$, where R_i are disjoint dyadic cubes and $u_i \in \mathscr{B}_1$, is dense in $L^1(\mathbf{R}^n, \mathscr{B}_1)$. Indeed, by Proposition 5.5.6 (b) it suffices to approximate a $\mathscr{C}_0^\infty \otimes \mathscr{B}_1$-valued function with a $\mathscr{D} \otimes \mathscr{B}_1$-valued function. But this is immediate since any function in $\mathscr{C}_0^\infty(\mathbf{R}^n)$ can be approximated in $L^1(\mathbf{R}^n)$ by finite linear combinations of characteristic functions of disjoint dyadic cubes.

Case 1: $r = \infty$. We fix $F = \sum_{i=1}^m \chi_{R_i} u_i$ in $\mathscr{D} \otimes \mathscr{B}_1$ and we notice that for each $x \in \mathbf{R}^n$ we have $\|F(x)\|_{\mathscr{B}_1} = \sum_{i=1}^m \chi_{R_i}(x)\|u_i\|_{\mathscr{B}_1}$, which is also a finite linear combination of characteristic functions of dyadic cubes. Apply the Calderón-Zygmund decomposition to $\|F\|_{\mathscr{B}_1}$ at height $\gamma\alpha$, where $\gamma = 2^{-n-1}B_\star^{-1}$ as in the proof of Theorem 6.3.1. We extract a finite collection of closed dyadic cubes $\{Q_j\}_j$ satisfying $\sum_j |Q_j| \le (\gamma\alpha)^{-1}\|F\|_{L^1(\mathbf{R}^n,\mathscr{B}_1)}$ and we define the good function of the decomposition

$$G(x) = \begin{cases} F(x) & \text{for } x \notin \cup_j Q_j \\ |Q_j|^{-1} \int_{Q_j} F(x)\, dx & \text{for } x \in Q_j. \end{cases}$$

Also define the bad function $B(x) = F(x) - G(x)$. Then $B(x) = \sum_j B_j(x)$, where each B_j is supported in Q_j and has mean value zero over Q_j. Moreover,

$$\|G\|_{L^1(\mathbf{R}^n,\mathscr{B}_1)} \le \|F\|_{L^1(\mathbf{R}^n,\mathscr{B}_1)} \tag{5.6.8}$$

$$\|G\|_{L^\infty(\mathbf{R}^n,\mathscr{B}_1)} \le 2^n\gamma\alpha \tag{5.6.9}$$

and $\|B_j\|_{L^1(\mathbf{R}^n,\mathscr{B}_1)} \le 2^{n+1}\gamma\alpha|Q_j|$, by an argument similar to that given in the proof of Theorem 5.3.1. We only verify (5.6.9). On the cube Q_j, G is equal to the constant $|Q_j|^{-1}\int_{Q_j} F(x)\,dx$, and this is bounded by $2^n\gamma\alpha$. For each $x \in \mathbf{R}^n \setminus \bigcup_j Q_j$ and for each $k = 0,1,2,\dots$ there exists a unique nonselected dyadic cube $Q_x^{(k)}$ of generation k that contains x. Then for each $k \ge 0$, we have

$$\left\|\frac{1}{|Q_x^{(k)}|}\int_{Q_x^{(k)}} F(y)\,dy\right\|_{\mathscr{B}_1} \le \frac{1}{|Q_x^{(k)}|}\int_{Q_x^{(k)}} \|F(y)\|_{\mathscr{B}_1}\,dy \le \gamma\alpha.$$

The intersection of the closures of the cubes $Q_x^{(k)}$ is the singleton $\{x\}$. Using Corollary 2.1.16, we deduce that for almost all $x \in \mathbf{R}^n \setminus \bigcup_j Q_j$ we have

$$F(x) = \sum_{i=1}^m \chi_{R_i}(x)u_i = \sum_{i=1}^m \lim_{k\to\infty}\left(\frac{1}{|Q_x^{(k)}|}\int_{Q_x^{(k)}} \chi_{R_i}(y)\,dy\right)u_i = \lim_{k\to\infty}\frac{1}{|Q_x^{(k)}|}\int_{Q_x^{(k)}} F(y)\,dy.$$

Since these averages are at most $\gamma\alpha$, we conclude that $\|F\|_{\mathscr{B}_1} \le \gamma\alpha$ almost everywhere on $\mathbf{R}^n \setminus \bigcup_j Q_j$, hence $\|G\|_{\mathscr{B}_1} \le \gamma\alpha$ a.e. on this set. This proves (5.6.9).

By assumption we have

$$\|\vec{T}(G)\|_{L^\infty(\mathbf{R}^n,\mathscr{B})} \le B_\star\|G\|_{L^\infty(\mathbf{R}^n,\mathscr{B})} \le 2^n\gamma\alpha B_\star = \alpha/2.$$

Then the set $\{x \in \mathbf{R}^n : \|\vec{T}(G)(x)\|_{\mathscr{B}_2} > \alpha/2\}$ is null and we have

$$\left|\{x \in \mathbf{R}^n : \|\vec{T}(F)(x)\|_{\mathscr{B}_2} > \alpha\}\right| \le \left|\{x \in \mathbf{R}^n : \|\vec{T}(B)(x)\|_{\mathscr{B}_2} > \alpha/2\}\right|.$$

Let $Q_j^* = 2\sqrt{n}\,Q_j$. We have

$$\begin{aligned}
&\left|\{x \in \mathbf{R}^n : \|T(B)(x)\|_{\mathscr{B}_2} > \alpha/2\}\right| \\
&\le \Big|\bigcup_j Q_j^*\Big| + \Big|\{x \notin \bigcup_j Q_j^* : \|\vec{T}(B)(x)\|_{\mathscr{B}_2} > \alpha/2\}\Big| \\
&\le \frac{(2\sqrt{n})^n}{\gamma}\frac{\|F\|_{L^1(\mathbf{R}^n,\mathscr{B}_1)}}{\alpha} + \frac{2}{\alpha}\int_{(\bigcup_j Q_j^*)^c} \|\vec{T}(B)(x)\|_{\mathscr{B}_2}\,dx \\
&\le \frac{(2\sqrt{n})^n}{\gamma}\frac{\|F\|_{L^1(\mathbf{R}^n,\mathscr{B}_1)}}{\alpha} + \frac{2}{\alpha}\sum_j\int_{(Q_j^*)^c} \|T(B_j)(x)\|_{\mathscr{B}_2}\,dx,
\end{aligned}$$

since $B = \sum_j B_j$. It suffices to estimate the last sum. Denoting by y_j is the center of the cube Q_j and using the fact that B_j has mean value zero over Q_j, we write

$$
\begin{aligned}
&\sum_j \int_{(Q_j^*)^c} \big\|\vec{T}(B_j)(x)\big\|_{\mathscr{B}_2}\, dx \\
&\quad = \sum_j \int_{(Q_j^*)^c} \bigg\| \int_{Q_j} \big(\vec{K}(x-y)-\vec{K}(x-y_j)\big)(B_j(y))\, dy \bigg\|_{\mathscr{B}_2} dx \\
&\quad \le \sum_j \int_{Q_j} \big\|B_j(y)\big\|_{\mathscr{B}_1} \int_{(Q_j^*)^c} \big\|\vec{K}(x-y)-\vec{K}(x-y_j)\big\|_{\mathscr{B}_1\to\mathscr{B}_2}\, dx dy \\
&\quad \le \sum_j \int_{Q_j} \big\|B_j(y)\big\|_{\mathscr{B}_1} \int_{|x-y_j|\ge 2|y-y_j|} \big\|\vec{K}(x-y)-\vec{K}(x-y_j)\big\|_{\mathscr{B}_1\to\mathscr{B}_2}\, dx dy \\
&\quad \le A \sum_j \|B_j\|_{L^1(Q_j,\mathscr{B}_1)} \\
&\quad \le 2^{n+1} A \|F\|_{L^1(\mathbf{R}^n,\mathscr{B}_1)},
\end{aligned}
$$

where we used the fact that $|x-y_j| \ge 2|y-y_j|$ for all $x \notin Q_j^*$ and $y \in Q_j$ and (5.6.2). Consequently,

$$
\begin{aligned}
\big|\{x\in\mathbf{R}^n :\, \|\vec{T}(f)(x)\|_{\mathscr{B}_2} > \alpha\}\big| &\le \frac{(2\sqrt{n})^n}{\gamma}\frac{\|F\|_{L^1(\mathbf{R}^n,\mathscr{B}_1)}}{\alpha} + \frac{2}{\alpha}2^{n+1}A\|F\|_{L^1(\mathbf{R}^n,\mathscr{B}_1)} \\
&= \big((2\sqrt{n})^n 2^{n+1} B_\star + 2^{n+1}A\big)\frac{\|F\|_{L^1(\mathbf{R}^n,\mathscr{B}_1)}}{\alpha} \\
&\le C_n'(A+B_\star)\frac{\|F\|_{L^1(\mathbf{R}^n,\mathscr{B}_1)}}{\alpha},
\end{aligned}
$$

where $C_n' = (2\sqrt{n})^n 2^{n+1} + 2^{n+1}$. Thus $\vec{T}$ has an extension that maps $L^1(\mathbf{R}^n,\mathscr{B}_1)$ to $L^{1,\infty}(\mathbf{R}^n,\mathscr{B}_2)$ with constant $C_n(A+B_\star)$. By interpolation (Exercise 5.5.3 (b)) it has an extension that satisfies (5.6.7).

Case 2: $1<r<\infty$. We fix $F=\sum_{i=1}^m \chi_{R_i} u_i$ in $\mathscr{Q}\otimes\mathscr{B}_1$ and we notice that for each $x\in\mathbf{R}^n$ we have $\|F(x)\|_{\mathscr{B}_1} = \sum_{i=1}^m \chi_{R_i}(x)\|u_i\|_{\mathscr{B}_1}$. Thus the function $x\mapsto \|F(x)\|_{\mathscr{B}_1}$ is a finite linear combination of characteristic functions of disjoint dyadic cubes. We prove the weak type estimate (5.6.6) by applying the Calderón–Zygmund decomposition to the function $x\mapsto \|F(x)\|_{\mathscr{B}_1}$ defined on $\mathbf{R}^n$. Then we decompose $F=G+B$, where G and B satisfy properties analogous to the case $r=\infty$. The new ingredient in this case is that the set $\{x\in\mathbf{R}^n :\, \|\vec{T}(G)(x)\|_{\mathscr{B}_2} > \alpha/2\}$ is not null but its measure can be estimated as follows:

$$
\big|\{x\in\mathbf{R}^n :\, \|\vec{T}(G)(x)\|_{\mathscr{B}_2} > \alpha/2\}\big| \le \Big(\frac{2B_\star}{\alpha}\Big)^r \|G\|^r_{L^2(\mathbf{R}^n,\mathscr{B}_1)} \le \frac{2}{\alpha}\|F\|_{L^1(\mathbf{R}^n,\mathscr{B}_1)},
$$

where the first inequality is a consequence of the boundedness of $\vec{T}$ on L^r and the second is obtained by combining (5.6.8) and (5.6.9). Combining this estimate for the good function with the one for the bad function obtained in the preceding case, it follows that $\vec{T}$ has an extension that satisfies (5.6.6), i.e., it maps $\vec{T}:\, L^1(\mathbf{R}^n,\mathscr{B}_1)$ to $L^{1,\infty}(\mathbf{R}^n,\mathscr{B}_2)$ with constant $C_n'(A+B_\star)$, where $C_n' = 2+(2\sqrt{n})^n 2^{n+1} + 2^{n+1}$.

Next we interpolate between the estimates $\vec{T} : L^1(\mathbf{R}^n, \mathscr{B}_1) \to L^{1,\infty}(\mathbf{R}^n, \mathscr{B}_2)$ and $\vec{T} : L^r(\mathbf{R}^n, \mathscr{B}_1) \to L^r(\mathbf{R}^n, \mathscr{B}_2)$. Using Exercise 5.5.3 (b) and the fact that $(p-1)^{-1/p} \le (p-1)^{-1}$ when $1 < p < 2$, we obtain

$$\big\|\vec{T}(F)\big\|_{L^p(\mathbf{R}^n,\mathscr{B}_2)} \le C_n (p-1)^{-1}(A+B_\star)\|F\|_{L^p(\mathbf{R}^n,\mathscr{B}_1)}, \tag{5.6.10}$$

when $1 < p < \min(r,2)$, where C_n is independent of r, p, $\mathscr{B}_1$, and $\mathscr{B}_2$.

We prove (5.6.7) for $p > r$ via duality. Since $\vec{K}(x)$ is an operator from $\mathscr{B}_1$ to $\mathscr{B}_2$, its adjoint $\vec{K}^*(x)$ is an operator from $\mathscr{B}_2^*$ to $\mathscr{B}_1^*$. Obviously $\vec{K}^*(x)$ and $\vec{K}(x)$ have the same norm, so (5.6.1) holds. For the same reason, condition (5.6.2) also holds for $\vec{K}^*$, Finally condition (5.6.3) also holds since for any $\varepsilon_k \downarrow$ as $k \to \infty$ we have

$$\left\| \int_{\varepsilon_k \le |y| \le 1} \vec{K}^*(y)\,dy - \vec{K_0}^* \right\|_{\mathscr{B}_2^* \to \mathscr{B}_1^*} = \left\| \int_{\varepsilon_k \le |y| \le 1} \vec{K}(y)\,dy - \vec{K_0} \right\|_{\mathscr{B}_1 \to \mathscr{B}_2} \to 0.$$

Let $\vec{T}'$ be the Banach-valued operator with kernel $\vec{K}^*(-x)$. Clearly $\vec{T}'$ is well defined on $\mathscr{C}_0^\infty \otimes \mathscr{B}_2^*$. For $F(y) = \sum_{i=1}^m f_i(y) w_i^*$ in $\mathscr{C}_0^\infty \otimes \mathscr{B}_2^*$ and $G(z) = \sum_{j=1}^l g_j(z) v_j$ in $\mathscr{C}_0^\infty \otimes \mathscr{B}_1$ we prove the following duality relation

$$\int_{\mathbf{R}^n} \big\langle \vec{T}'(F)(x), G(x) \big\rangle\,dx = \int_{\mathbf{R}^n} \big\langle F(z), \vec{T}(G)(z) \big\rangle\,dz. \tag{5.6.11}$$

Indeed, for each index $i \in \{1,\dots,m\}$ and $j \in \{1,\dots,l\}$ we have

$$\begin{aligned}
&\int_{\mathbf{R}^n} \Big\langle \lim_{k\to\infty} \int_{|y|\ge\varepsilon_k} \vec{K}^*(-y)(f_i(x-y)w_i^*)\,dy, g_j(x)v_j \Big\rangle\,dx \\
&= \lim_{k\to\infty} \int_{|y|\ge\varepsilon_k} \int_{\mathbf{R}^n} \Big\langle \vec{K}^*(-y)(f_i(x-y)w_i^*), g_j(x)v_j \Big\rangle\,dx\,dy \\
&= \lim_{k\to\infty} \int_{|y|\ge\varepsilon_k} \int_{\mathbf{R}^n} \Big\langle \vec{K}^*(-y)(f_i(z)w_i^*), g_j(z+y)v_j \Big\rangle\,dz\,dy \\
&= \lim_{k\to\infty} \int_{|y|\ge\varepsilon_k} \int_{\mathbf{R}^n} \Big\langle f_i(z)w_i^*, \vec{K}(-y)(g_j(z+y)v_j) \Big\rangle\,dz\,dy \\
&= \int_{\mathbf{R}^n} \Big\langle f_i(z)w_i^*, \lim_{k\to\infty} \int_{|y|\ge\varepsilon_k} \vec{K}(y)(g_j(z-y)v_j)\,dy \Big\rangle\,dz,
\end{aligned}$$

proving (5.6.11), provided we can justify the interchange the x (or z)-integral with the y-integral paired with the limit. These justifications can be given using the definition in (5.6.4). For the part of the y-integral where $|y| \ge 1$ the interchange is easily justified in view of the absolute convergence of the double integral and (5.5.21). For the part of the y-integral where $|y| \le 1$ we introduce the operators $\vec{K_0}^*$ and $\vec{K_0}$ and we use the facts that $|g_j(z-y) - g_j(z)| \le \|\nabla g_j\|_{L^\infty}|y|$ and $|f_i(x-y) - f_i(x)| \le \|\nabla f_i\|_{L^\infty}|y|$ together with the assumption $\|\vec{K}(y)\|_{\mathscr{B}_1\to\mathscr{B}_2} \le A|y|^{-n}$ and (5.5.21) to obtain the absolute convergence of the double integral, and thus justify the interchange.

We claim that $\vec{T}'$ is bounded from $L^{r'}(\mathbf{R}^n,\mathscr{B}_2^*)$ to $L^{r'}(\mathbf{R}^n,\mathscr{B}_1^*)$. Indeed, to verify this assertion, we fix F in $\mathscr{C}_0^\infty\otimes\mathscr{B}_2^*$ and use Proposition 5.5.7 (b). Using (5.6.11), for each $G\in\mathscr{C}_0^\infty\otimes\mathscr{B}_1$, we write

$$\begin{aligned}\left|\int_{\mathbf{R}^n}\langle\vec{T}'(F)(x),G(x)\rangle\,dx\right| &= \left|\int_{\mathbf{R}^n}\langle F(x),\vec{T}(G)(x)\rangle\,dx\right| \\ &\le \int_{\mathbf{R}^n}\big\|F(x)\big\|_{\mathscr{B}_2^*}\big\|\vec{T}(G)(x)\big\|_{\mathscr{B}_2}\,dx \\ &\le \big\|F\big\|_{L^{r'}(\mathbf{R}^n,\mathscr{B}_2^*)}\big\|\vec{T}(G)\big\|_{L^r(\mathbf{R}^n,\mathscr{B}_2)} \\ &\le \big\|F\big\|_{L^{r'}(\mathbf{R}^n,\mathscr{B}_2^*)}B_\star\big\|G\big\|_{L^r(\mathbf{R}^n,\mathscr{B}_1)},\end{aligned}$$

so taking the supremum over all $G\in\mathscr{C}_0^\infty\otimes\mathscr{B}_1$ with $\|G\|_{L^r(\mathbf{R}^n,\mathscr{B}_1)}\le 1$ we deduce that

$$\big\|\vec{T}'(F)\big\|_{L^{r'}(\mathbf{R}^n,\mathscr{B}_1^*)}\le B_\star\big\|F\big\|_{L^{r'}(\mathbf{R}^n,\mathscr{B}_2^*)}.$$

Collecting these facts, we have that $\vec{T}'$ is associated with a kernel $\vec{K}^*(-x)$ which satisfies (5.6.1), (5.6.2), and (5.6.3) (with $\vec{K}_0^*$ in place of $\vec{K}_0$), and moreover it has a bounded extension that maps $L^{r'}(\mathbf{R}^n,\mathscr{B}_2^*)$ to $L^{r'}(\mathbf{R}^n,\mathscr{B}_1^*)$. The Calderón-Zygmund decomposition in the vector-valued setting (discussed in the first paragraph of the proof) yields that $\vec{T}'$ has an extension that satisfies

$$\big\|\vec{T}'(F)\big\|_{L^{1,\infty}(\mathbf{R}^n,\mathscr{B}_1^*)}\le C_n'(A+B_\star)\big\|F\big\|_{L^1(\mathbf{R}^n,\mathscr{B}_2^*)}.$$

Using interpolation (Exercise 5.5.3 (b)) and the fact that $(p'-1)^{1/p'}\le p$, we obtain that for $1<p'<r'$, $\vec{T}'$ has an extension on $L^{p'}(\mathbf{R}^n,\mathscr{B}_2^*)$ that satisfies

$$\big\|\vec{T}'(F)\big\|_{L^{p'}(\mathbf{R}^n,\mathscr{B}_1^*)}\le C_n\,p\,(A+B_\star)\big\|F\big\|_{L^{p'}(\mathbf{R}^n,\mathscr{B}_2^*)}. \tag{5.6.12}$$

Let $F=\sum_{i=1}^m\varphi_i u_i$ be in the dense subspace $\mathscr{C}_0^\infty\otimes\mathscr{B}_1$ of $L^p(\mathbf{R}^n,\mathscr{B}_1)$. We observe that $\|\vec{T}(F)\|_{L^p(\mathbf{R}^n,\mathscr{B}_2)}<\infty$. Indeed, all φ_i are supported in $|x|\le R$, then for $|x|\ge 2R$ we have

$$\left\|\int_{|y|\le R}\vec{K}(x-y)\Big(\sum_{i=1}^m\varphi_i u_i\Big)\,dy\right\|_{\mathscr{B}_2}\le A\Big(\frac{|x|}{2}\Big)^{-n}\sum_{i=1}^m\|\varphi_i\|_{L^1}\|u_i\|_{\mathscr{B}_1} \tag{5.6.13}$$

which is integrable to the power $p>1$ in the region $|x|\ge 2R$. Also using the definition in (5.6.4) we see that the expression on the left in (5.6.13) is bounded, hence integrable to the power p in the region $|x|\le 2R$. For a fixed $r<p<\infty$, we are now able to apply Proposition 5.5.7 (a) to write

$$\begin{aligned}\big\|\vec{T}(F)\big\|_{L^p(\mathbf{R}^n,\mathscr{B}_2)} &\le \sup_{\|G\|_{L^{p'}(\mathbf{R}^n,\mathscr{B}_2^*)}\le 1}\left|\int_{\mathbf{R}^n}\langle G(x),\vec{T}(F)(x)\rangle\,dx\right| \\ &= \sup_{\|G\|_{L^{p'}(\mathbf{R}^n,\mathscr{B}_2^*)}\le 1}\left|\int_{\mathbf{R}^n}\langle\vec{T}'(G)(x),F(x)\rangle\,dx\right|\end{aligned}$$

$$\leq \sup_{\|G\|_{L^{p'}(\mathbf{R}^n,\mathscr{B}_2^*)}\leq 1} \|\vec{T}'(G)\|_{L^{p'}(\mathbf{R}^n,\mathscr{B}_1^*)} \|F\|_{L^p(\mathbf{R}^n,\mathscr{B}_1)}$$

$$\leq C_n p(A+B_\star)\|F\|_{L^p(\mathbf{R}^n,\mathscr{B}_1)},$$

where we used (5.6.12). This combined with (5.6.10) implies the required conclusion whenever $r<\infty$ and $p\in\big(1,\min(r,2)\big)\cup(r,\infty)$. The remaining p's follow by interpolation (Exercise 5.5.3 (a)). □

5.6.2 Applications

We proceed with some applications. An important consequence of Theorem 5.6.1 is the following:

Corollary 5.6.2. *Let $A,B>0$ and let W_j be a sequence of tempered distributions on $\mathbf{R}^n$ whose Fourier transforms are uniformly bounded functions (i.e., $|\widehat{W_j}|\leq B$). Suppose that for each j, W_j coincides with a function K_j on $\mathbf{R}^n\setminus\{0\}$ that satisfies*

$$|K_j(x)|\leq A\,|x|^{-n}, \qquad x\neq 0, \tag{5.6.14}$$

$$\lim_{\varepsilon_k\to 0}\int_{|x|\geq\varepsilon_k} K_j(x)\,dx = L_j\,, \tag{5.6.15}$$

for some complex constant L_j, and

$$\sup_{y\in\mathbf{R}^n\setminus\{0\}}\int_{|x|\geq 2|y|}\sup_j|K_j(x-y)-K_j(x)|\,dx\leq A\,. \tag{5.6.16}$$

Then there are constants $C_n, C_n'>0$ such that for all $1<p,r<\infty$ we have

$$\Big\|\Big(\sum_j|W_j*f_j|^r\Big)^{\frac{1}{r}}\Big\|_{L^{1,\infty}}\leq C_n'\max(r,(r-1)^{-1})(A+B)\Big\|\Big(\sum_j|f_j|^r\Big)^{\frac{1}{r}}\Big\|_{L^1},$$

$$\Big\|\Big(\sum_j|W_j*f_j|^r\Big)^{\frac{1}{r}}\Big\|_{L^p}\leq C_n c(p,r)(A+B)\Big\|\Big(\sum_j|f_j|^r\Big)^{\frac{1}{r}}\Big\|_{L^p},$$

where $c(p,r)=\max(p,(p-1)^{-1})\max(r,(r-1)^{-1})$.

Proof. Let T_j be the operator given by convolution with the distribution W_j. Clearly T_j is L^2 bounded with norm at most B. It follows from Theorem 5.3.3 that the T_j's are of weak type $(1,1)$ and also bounded on L^r with bounds at most a dimensional constant multiple of $\max(r,(r-1)^{-1})(A+B)$, uniformly in j. We set $\mathscr{B}_1=\mathscr{B}_2=\ell^r$ and define

$$\vec{T}(\{f_j\}_j)=\{W_j*f_j\}_j$$

for $\{f_j\}_j \in L^r(\mathbf{R}^n, \ell^r)$. It is immediate to verify that $\vec{T}$ maps $L^r(\mathbf{R}^n, \ell^r)$ to itself with norm at most a dimensional constant multiple of $\max(r, (r-1)^{-1})(A+B)$. The kernel of $\vec{T}$ is $\vec{K}$ in $L(\ell^r, \ell^r)$ defined by

$$\vec{K}(x)(\{t_j\}_j) = \{K_j(x)t_j\}_j, \qquad \{t_j\}_j \in \ell^r.$$

Obviously, we have

$$\big\|\vec{K}(x-y) - \vec{K}(x)\big\|_{\ell^r \to \ell^r} \le \sup_j |K_j(x-y) - K_j(x)|,$$

and therefore condition (5.6.3) holds for $\vec{K}$ as a consequence of (5.6.16). Moreover, (5.6.1) and (5.6.2) with $\vec{K}_0 = \{L_j\}_j$ are also valid for this $\vec{K}$, in view of assumptions (5.6.14) and (5.6.15). The desired conclusion follows from Theorem 5.6.1. □

If all the W_j's are equal, we obtain the following corollary, which contains in particular the inequality (5.5.16) mentioned earlier.

Corollary 5.6.3. *Let W be an element of $\mathscr{S}'(\mathbf{R}^n)$ whose Fourier transform is a function bounded in absolute value by some $B > 0$. Suppose that W coincides with some locally integrable function K on $\mathbf{R}^n \setminus \{0\}$ that satisfies*

$$|K(x)| \le A\,|x|^{-n}, \qquad x \neq 0,$$

$$\lim_{\varepsilon_k \to 0} \int_{\varepsilon_k \le |x| \le 1} K(x)\,dx = L,$$

and

$$\sup_{y \in \mathbf{R}^n \setminus \{0\}} \int_{|x| \ge 2|y|} |K(x-y) - K(x)|\,dx \le A. \tag{5.6.17}$$

Let T be the operator given by convolution with W. Then there exist constants $C_n, C_n' > 0$ such that for all $1 < p, r < \infty$ we have that

$$\Big\|\Big(\sum_j |T(f_j)|^r\Big)^{\frac{1}{r}}\Big\|_{L^{1,\infty}} \le C_n' \max(r,(r-1)^{-1})(A+B)\Big\|\Big(\sum_j |f_j|^r\Big)^{\frac{1}{r}}\Big\|_{L^1},$$

$$\Big\|\Big(\sum_j |T(f_j)|^r\Big)^{\frac{1}{r}}\Big\|_{L^p} \le C_n c(p,r)\,(A+B)\Big\|\Big(\sum_j |f_j|^r\Big)^{\frac{1}{r}}\Big\|_{L^p},$$

where $c(p,r) = \max(p,(p-1)^{-1})\max(r,(r-1)^{-1})$. In particular, these inequalities are valid for the Hilbert transform and the Riesz transforms.

Interestingly enough, we can use the very statement of Theorem 5.6.1 to obtain its corresponding vector-valued version.

Proposition 5.6.4. *Let let $1 < p, r < \infty$ and let $\mathscr{B}_1$ and $\mathscr{B}_2$ be two Banach spaces. Suppose that $\vec{T}$ given by (5.6.4) is a bounded linear operator from $L^r(\mathbf{R}^n, \mathscr{B}_1)$ to $L^r(\mathbf{R}^n, \mathscr{B}_2)$ with norm $B = B(r)$. Also assume that for all $x \in \mathbf{R}^n \setminus \{0\}$, $\vec{K}(x)$ is a* *bounded linear operator from $\mathscr{B}_1$ to $\mathscr{B}_2$ that satisfies conditions (5.6.1) , (5.6.2), (5.6.3) for some $A > 0$ and $\vec{K}_0 \in L(\mathscr{B}_1, \mathscr{B}_2)$. Then there exist positive constants C_n, C_n' such that for all $\mathscr{B}_1$-valued functions F_j we have*

$$\Big\|\Big(\sum_j \|\vec{T}(F_j)\|_{\mathscr{B}_2}^r\Big)^{\frac{1}{r}}\Big\|_{L^{1,\infty}(\mathbf{R}^n)} \le C_n'(A+B)\Big\|\Big(\sum_j \|F_j\|_{\mathscr{B}_1}^r\Big)^{\frac{1}{r}}\Big\|_{L^1(\mathbf{R}^n)},$$

$$\Big\|\Big(\sum_j \|\vec{T}(F_j)\|_{\mathscr{B}_2}^r\Big)^{\frac{1}{r}}\Big\|_{L^p(\mathbf{R}^n)} \le C_n(A+B)c(p)\Big\|\Big(\sum_j \|F_j\|_{\mathscr{B}_1}^r\Big)^{\frac{1}{r}}\Big\|_{L^p(\mathbf{R}^n)},$$

where $c(p) = \max(p, (p-1)^{-1})$.

Proof. Let us denote by $\ell^r(\mathscr{B}_1)$ the Banach space of all $\mathscr{B}_1$-valued sequences $\{u_j\}_j$ that satisfy

$$\big\|\{u_j\}_j\big\|_{\ell^r(\mathscr{B}_1)} = \Big(\sum_j \|u_j\|_{\mathscr{B}_1}^r\Big)^{\frac{1}{r}} < \infty.$$

Now consider the operator $\vec{S}$ defined on $L^r(\mathbf{R}^n, \ell^r(\mathscr{B}_1))$ by

$$\vec{S}(\{F_j\}_j) = \{\vec{T}(F_j)\}_j.$$

It is obvious that $\vec{S}$ maps $L^r(\mathbf{R}^n, \ell^r(\mathscr{B}_1))$ to $L^r(\mathbf{R}^n, \ell^r(\mathscr{B}_2))$ with norm at most B. Moreover, $\vec{S}$ has kernel $\widetilde{K}(x) \in L(\ell^r(\mathscr{B}_1), \ell^r(\mathscr{B}_2))$ given by

$$\widetilde{K}(x)(\{u_j\}_j) = \{\vec{K}(x)(u_j)\}_j,$$

where $\vec{K}$ is the kernel of $\vec{T}$. It is not hard to verify that for $x \in \mathbf{R}^n \setminus \{0\}$ we have

$$\big\|\widetilde{K}(x)\big\|_{\ell^r(\mathscr{B}_1)\to\ell^r(\mathscr{B}_2)} = \big\|\vec{K}(x)\big\|_{\mathscr{B}_1\to\mathscr{B}_2},$$

hence for $x \neq y \in \mathbf{R}^n$ we also have

$$\big\|\widetilde{K}(x-y) - \widetilde{K}(x)\big\|_{\ell^r(\mathscr{B}_1)\to\ell^r(\mathscr{B}_2)} = \big\|\vec{K}(x-y) - \vec{K}(x)\big\|_{\mathscr{B}_1\to\mathscr{B}_2}.$$

Moreover, if we define $\widetilde{K_0} \in L\big(\ell^r(\mathscr{B}_1), \ell^r(\mathscr{B}_2)\big)$ by

$$\widetilde{K_0}(\{u_j\}_j) = \big\{\vec{K}_0(y)(u_j)\big\}_j.$$

for $\{u_j\}_j \in \ell^r(\mathscr{B}_1)$, then we have

$$\lim_{k\to\infty} \int_{\varepsilon_k \le |y| \le 1} \widetilde{K}(y)\, dy = \widetilde{K_0},$$

in $L\big(\ell^r(\mathscr{B}_1), \ell^r(\mathscr{B}_2)\big)$.

We conclude that $\widetilde{K}$ satisfies conditions (5.6.1) , (5.6.2), (5.6.3). Hence the operator $\vec{S}$ associate with $\widetilde{K}$ satisfies the conclusion of Theorem 5.6.1, that is, the desired inequalities for $\vec{T}$. □

5.6.3 Vector-Valued Estimates for Maximal Functions

Next, we discuss applications of vector-valued inequalities to some nonlinear operators. We fix an integrable function Φ on $\mathbf{R}^n$ and for $t>0$ define $\Phi_t(x)=t^{-n}\Phi(t^{-1}x)$. We suppose that Φ satisfies the following *regularity* condition:

$$\sup_{y\in\mathbf{R}^n\setminus\{0\}}\int_{|x|\geq 2|y|}\sup_{t>0}|\Phi_t(x-y)-\Phi_t(x)|\,dx=A_\Phi<\infty. \tag{5.6.18}$$

We consider the maximal operator

$$M_\Phi(f)(x)=\sup_{t>0}|(f*\Phi_t)(x)|$$

defined for f in L^1+L^∞. We are interested in obtaining L^p estimates for M_Φ. We observe that the trivial estimate

$$\big\|M_\Phi(f)\big\|_{L^\infty}\leq\|\Phi\|_{L^1}\|f\|_{L^\infty} \tag{5.6.19}$$

holds when $p=\infty$. It is natural to set

$$\mathscr{B}_1=\mathbf{C}\qquad\text{and}\qquad\mathscr{B}_2=L^\infty(\mathbf{R}^+)$$

and view M_Φ as the linear operator $f\mapsto\{f*\Phi_\delta\}_{\delta>0}$ that maps $\mathscr{B}_1$-valued functions to $\mathscr{B}_2$-valued functions.

To do this precisely, for each $x\in\mathbf{R}^n$ we define a bounded linear operator $\vec{K}_\Phi(x)$ from $\mathscr{B}_1$ to $\mathscr{B}_2$ by setting for $c\in\mathbf{C}$

$$\vec{K}_\Phi(x)(c)=\{c\,\Phi_\delta(x)\}_{\delta\in\mathbf{R}^+}.$$

Clearly we have

$$\big\|\vec{K}_\Phi(x)\big\|_{\mathbf{C}\to L^\infty(\mathbf{R}^+)}=\sup_{\delta>0}|\Phi_\delta(x)|.$$

Now (5.6.18) implies condition (5.6.2) for the kernel $\vec{K}_\Phi$. Also condition (5.6.1) holds (for some $A<$ depending on n) since

$$\sup_{\delta>0}|\Phi_\delta(x)|\leq A\,|x|^{-n}$$

and also condition (5.6.3) holds since for every $\delta>0$ we have

$$\lim_{\varepsilon\to 0}\int_{\varepsilon\leq|y|\leq 1}\Phi_\delta(y)\,dy=\int_{|y|\leq 1}\Phi_\delta(y)\,dy.$$

We also define a $\mathscr{B}_2$-valued linear operator acting on complex-valued functions on $\mathbf{R}^n$ by

$$\vec{M}_\Phi(f)=f*\vec{K}_\Phi=\{f*\Phi_\delta\}_{\delta\in\mathbf{R}^+}.$$

Obvisouly $\vec{M}_\Phi$ maps $L^\infty(\mathbf{R}^n,\mathscr{B}_1)$ to $L^\infty(\mathbf{R}^n,\mathscr{B}_2)$ with norm at most $\|\Phi\|_{L^1}$.

Applying Theorem 5.6.1 with $r = \infty$ we obtain for $1 < p < \infty$,

$$\big\|\vec{M}_\Phi(f)\big\|_{L^p(\mathbf{R}^n,\mathscr{B}_2)} \le C_n \max(p,(p-1)^{-1})\big(A_\Phi + \|\Phi\|_{L^1}\big)\big\|f\big\|_{L^p(\mathbf{R}^n)}, \tag{5.6.20}$$

which can be immediately improved to

$$\big\|\vec{M}_\Phi(f)\big\|_{L^r(\mathbf{R}^n,\mathscr{B}_2)} \le C_n \max(1,(r-1)^{-1})\big(A_\Phi + \|\Phi\|_{L^1}\big)\big\|f\big\|_{L^r(\mathbf{R}^n)} \tag{5.6.21}$$

via interpolation with estimate (5.6.19) for all $1 < r < \infty$.

Next we use estimate (5.6.21) to obtain vector-valued estimates for the sublinear operator M_Φ.

Corollary 5.6.5. *Let Φ be an integrable function on $\mathbf{R}^n$ that satisfies (5.6.18). Then there exist dimensional constants C_n and C_n' such that for all $1 < p,r < \infty$ the following vector-valued inequalities are valid:*

$$\Big\|\Big(\sum_j |M_\Phi(f_j)|^r\Big)^{\frac{1}{r}}\Big\|_{L^{1,\infty}} \le C_n' c(r)\big(A_\Phi + \|\Phi\|_{L^1}\big)\Big\|\Big(\sum_j |f_j|^r\Big)^{\frac{1}{r}}\Big\|_{L^1}, \tag{5.6.22}$$

where $c(r) = 1+(r-1)^{-1}$, and

$$\Big\|\Big(\sum_j |M_\Phi(f_j)|^r\Big)^{\frac{1}{r}}\Big\|_{L^p} \le C_n c(p,r)\big(A_\Phi + \|\Phi\|_{L^1}\big)\Big\|\Big(\sum_j |f_j|^r\Big)^{\frac{1}{r}}\Big\|_{L^p}, \tag{5.6.23}$$

where $c(p,r) = \big(1+(r-1)^{-1}\big)\big(p+(p-1)^{-1}\big)$.

Proof. We set $\mathscr{B}_1 = \mathbf{C}$ and $\mathscr{B}_2 = L^\infty(\mathbf{R}^+)$. We use estimate (5.6.21) as a starting point in Proposition 5.6.4, which immediately yields the required conclusions (5.6.22) and (5.6.23). □

Similar estimates hold for the Hardy–Littlewood maximal operator.

Theorem 5.6.6. *For $1 < p,r < \infty$ the Hardy–Littlewood maximal function M satisfies the vector-valued inequalities*

$$\Big\|\Big(\sum_j |M(f_j)|^r\Big)^{\frac{1}{r}}\Big\|_{L^{1,\infty}} \le C_n'\big(1+(r-1)^{-1}\big)\Big\|\Big(\sum_j |f_j|^r\Big)^{\frac{1}{r}}\Big\|_{L^1}, \tag{5.6.24}$$

$$\Big\|\Big(\sum_j |M(f_j)|^r\Big)^{\frac{1}{r}}\Big\|_{L^p} \le C_n c(p,r)\Big\|\Big(\sum_j |f_j|^r\Big)^{\frac{1}{r}}\Big\|_{L^p}, \tag{5.6.25}$$

where $c(p,r) = \big(1+(r-1)^{-1}\big)\big(p+(p-1)^{-1}\big)$.

Proof. Let us fix a positive radial symmetrically decreasing Schwartz function Φ on $\mathbf{R}^n$ that satisfies $\Phi(x) \ge 1$ when $|x| \le 1$. Then the Hardy–Littlewood maximal function $M(f)$ is pointwise controlled by a constant multiple of the function $M_\Phi(|f|)$.

In view of Corollary 5.6.5, it suffices to check that for such a Φ, (5.6.18) holds. First observe that in view of the decreasing character of Φ, we have

$$\sup_j |f| * \Phi_{2^j} \le M_\Phi(|f|) \le 2^n \sup_j |f| * \Phi_{2^j},$$

and for this reason we choose to work with the easier dyadic maximal operator

$$M_\Phi^d(f) = \sup_j |f * \Phi_{2^j}|.$$

We observe the validity of the simple inequalties

$$2^{-n} M(f) \le \mathcal{M}(f) \le M_\Phi(|f|) \le 2^n M_\Phi^d(|f|). \tag{5.6.26}$$

If we can show that

$$\sup_{y \in \mathbf{R}^n \setminus \{0\}} \int_{|x| \ge 2|y|} \sup_{j \in \mathbf{Z}} |\Phi_{2^j}(x-y) - \Phi_{2^j}(x)|\, dx = C_n < \infty, \tag{5.6.27}$$

then (5.6.22) and (5.6.23) are satisfied with M_Φ^d replacing M_Φ. We therefore turn our attention to (5.6.27). We have

$$\begin{aligned}
&\int_{|x| \ge 2|y|} \sup_{j \in \mathbf{Z}} |\Phi_{2^j}(x-y) - \Phi_{2^j}(x)|\, dx \\
&\quad \le \sum_{j \in \mathbf{Z}} \int_{|x| \ge 2|y|} |\Phi_{2^j}(x-y) - \Phi_{2^j}(x)|\, dx \\
&\quad \le \sum_{2^j > |y|} \int_{|x| \ge 2|y|} \frac{|y|\, |\nabla \Phi(\frac{x-\theta y}{2^j})|}{2^{(n+1)j}}\, dx + \sum_{2^j \le |y|} \int_{|x| \ge 2|y|} (|\Phi_{2^j}(x-y)| + |\Phi_{2^j}(x)|)\, dx \\
&\quad \le \sum_{2^j > |y|} \int_{|x| \ge 2|y|} \frac{|y|}{2^{(n+1)j}} \frac{C_N\, dx}{(1+|2^{-j}(x-\theta y)|)^N} + 2 \sum_{2^j \le |y|} \int_{|x| \ge |y|} |\Phi_{2^j}(x)|\, dx \\
&\quad \le \sum_{2^j > |y|} \int_{|x| \ge 2|y|} \frac{|y|}{2^{(n+1)j}} \frac{C_N}{(1+|2^{-j-1}x|)^N}\, dx + 2 \sum_{2^j \le |y|} \int_{|x| \ge 2^{-j}|y|} |\Phi(x)|\, dx \\
&\quad \le \sum_{2^j > |y|} \int_{|x| \ge 2^{-j}|y|} \frac{|y|}{2^j} \frac{C_N}{(1+|x|)^N}\, dx + 2 \sum_{2^j \le |y|} C_N (2^{-j}|y|)^{-N} \\
&\quad \le C_N \sum_{2^j > |y|} \frac{|y|}{2^j} + C_N \\
&\quad \le 3C_N,
\end{aligned}$$

where $C_N > 0$ depends on $N > n$, $\theta \in [0,1]$, and $|x - \theta y| \ge |x|/2$ when $|x| \ge 2|y|$.

Now apply (5.6.22) and (5.6.23) to M_Φ^d and use (5.6.26) to obtain the desired vector-valued inequalities. $\square$

Remark 5.6.7. Observe that (5.6.24) and (5.6.25) also hold for $r=\infty$. These endpoint estimates can be proved directly by observing that

$$\sup_j M(f_j) \le M(\sup_j |f_j|)\,.$$

The same is true for estimates (5.6.22) and (5.6.23). Finally, estimates (5.6.25) and (5.6.23) also hold for $p=\infty$.

Exercises

5.6.1. (a) For all $j \in \mathbf{Z}$, let I_j be an interval in $\mathbf{R}$ and let T_j be the operator given on the Fourier transform by multiplication by the characteristic function of I_j. Prove that there exists a constant $C>0$ such that for all $1<p,r<\infty$ and for all square integrable functions f_j on $\mathbf{R}$ we have

$$\Big\|\Big(\sum_j |T_j(f_j)|^r\Big)^{\frac{1}{r}}\Big\|_{L^p(\mathbf{R})} \le C\max\Big(r,\frac{1}{r-1}\Big)\max\Big(p,\frac{1}{p-1}\Big)\Big\|\Big(\sum_j |f_j|^r\Big)^{\frac{1}{r}}\Big\|_{L^p(\mathbf{R})},$$

$$\Big\|\Big(\sum_j |T_j(f_j)|^r\Big)^{\frac{1}{r}}\Big\|_{L^{1,\infty}(\mathbf{R})} \le C\max\Big(r,\frac{1}{r-1}\Big)\Big\|\Big(\sum_j |f_j|^r\Big)^{\frac{1}{r}}\Big\|_{L^1(\mathbf{R})}.$$

(b) Let R_j be arbitrary rectangles on $\mathbf{R}^n$ with sides parallel to the axes and let S_j be the operators given on the Fourier transform by multiplication by the characteristic functions of R_j. Prove that there exists a dimensional constant $C_n<\infty$ such that for all indices $1<p,r<\infty$ and for all square integrable functions f_j in $L^p(\mathbf{R}^n)$ we have

$$\Big\|\Big(\sum_j |S_j(f_j)|^r\Big)^{\frac{1}{r}}\Big\|_{L^p(\mathbf{R}^n)} \le C_n\max\Big(r,\frac{1}{r-1}\Big)^n\max\Big(p,\frac{1}{p-1}\Big)^n\Big\|\Big(\sum_j |f_j|^r\Big)^{\frac{1}{r}}\Big\|_{L^p(\mathbf{R}^n)}.$$

[*Hint:* Part (a): Use Theorem 5.5.1 and the identity $T_j = \frac{i}{2}\big(M^aHM^{-a} - M^bHM^{-b}\big)$, if I_j is $\chi_{(a,b)}$, where $M^a(f)(x) = f(x)e^{2\pi i a x}$ and H is the Hilbert transform. Part (b): Apply the result in part (a) in each variable.]

5.6.2. Let $(T,d\mu)$ be a σ-finite measure space. For every $t\in T$, let $R(t)$ be a rectangle in $\mathbf{R}^n$ with sides parallel to the axes such that the map $t\mapsto R(t)$ is measurable. Then there is a constant $C_n>0$ such that for all $1<p<\infty$ and for all families of square integrable functions $\{f_t\}_{t\in T}$ on $\mathbf{R}^n$ such that $t\mapsto f_t(x)$ is measurable for all $x\in\mathbf{R}^n$ we have

$$\Big\|\Big(\int_T |(\widehat{f_t}\chi_{R(t)})^\vee|^2\,d\mu(t)\Big)^{\frac{1}{2}}\Big\|_{L^p} \le C_n\max(p,(p-1)^{-1})^n\Big\|\Big(\int_T |f_t|^2\,d\mu(t)\Big)^{\frac{1}{2}}\Big\|_{L^p},$$

[*Hint:* When $n=1$ reduce matters to an $L^p(L^2(T,d\mu),L^2(T,d\mu))$ inequality for the Hilbert transform, via the hint in the preceding exercise. Verify the inequality $p=2$ and then use Theorem 5.6.1 for the other p's. Obtain the n-dimensional inequality by iterating the one-dimensional.]

5.6.3. Let Φ be a function on $\mathbf{R}^n$ that satisfies $\sup_{x\in\mathbf{R}^n}|x|^n|\Phi(x)|\le A$ and

$$\int_{\mathbf{R}^n}|\Phi(x-y)-\Phi(x)|\,dx\le\eta(y),\qquad \int_{|x|\ge R}|\Phi(x)|\,dx\le\eta(R^{-1}),$$

for all $R\ge 1$, where η is a continuous increasing function on $[0,2]$ that satisfies $\eta(0)=0$ and $\int_0^2\frac{\eta(t)}{t}\,dt<\infty$.
(a) Prove that (5.6.27) holds.
(b) Show that if Φ lies in $L^1(\mathbf{R}^n)$, then the maximal function $f\mapsto\sup_{j\in\mathbf{Z}}|f*\Phi_{2^j}|$ maps $L^p(\mathbf{R}^n)$ to itself for $1<p\le\infty$.
[*Hint:* Part (a): Modify the calculation in the proof of Theorem 5.6.6. Part (b): Use Theorem 5.6.1 with $r=\infty$.]

5.6.4. (a) On $\mathbf{R}$, take $f_j=\chi_{[2^{j-1},2^j]}$ to prove that inequality (5.6.25) fails when $p=\infty$ and $1<r<\infty$.
(b) Again on $\mathbf{R}$, take $N>2$ and $f_j=\chi_{[\frac{j-1}{N},\frac{j}{N}]}$ for $j=1,2,\dots,N$ to prove that (5.6.25) fails when $1<p<\infty$ and $r=1$.

5.6.5. Let K be an integrable function on the real line and assume that the operator $f\mapsto f*K$ is bounded on $L^p(\mathbf{R})$ for some $1<p<\infty$. Prove that the vector-valued inequality

$$\Big\|\Big(\sum_j|K*f_j|^q\Big)^{\frac{1}{q}}\Big\|_{L^p}\le C_{p,q}\Big\|\Big(\sum_j|f_j|^q\Big)^{\frac{1}{q}}\Big\|_{L^p}$$

may fail in general when $q<1$.
[*Hint:* Take $K=\chi_{[-1,1]}$ and $f_j=\chi_{[\frac{j-1}{N},\frac{j}{N}]}$ for $1\le j\le N$.]

5.6.6. Let $\{Q_j\}_j$ be a countable collection of cubes in $\mathbf{R}^n$ with disjoint interiors. Let c_j be the center of the cube Q_j and d_j its diameter. For $\varepsilon>0$, define the *Marcinkiewicz function* associated with the family $\{Q_j\}_j$ as follows:

$$M_\varepsilon(x)=\sum_j\frac{d_j^{n+\varepsilon}}{|x-c_j|^{n+\varepsilon}+d_j^{n+\varepsilon}}\,.$$

Prove that for some constants $C_{n,\varepsilon,p}$ and $C_{n,\varepsilon}$ one has

$$\big\|M_\varepsilon\big\|_{L^p}\le C_{n,\varepsilon,p}\Big(\sum_j|Q_j|\Big)^{\frac{1}{p}},\qquad p>\frac{n}{n+\varepsilon},$$

$$\big\|M_\varepsilon\big\|_{L^{\frac{n}{n+\varepsilon},\infty}}\le C_{n,\varepsilon}\Big(\sum_j|Q_j|\Big)^{\frac{n+\varepsilon}{n}},$$

and consequently $\int_{\mathbf{R}^n}M_\varepsilon(x)\,dx\le C_{n,\varepsilon}\sum_j|Q_j|$.

[*Hint:* Verify that

$$\frac{d_j^{n+\varepsilon}}{|x-c_j|^{n+\varepsilon}+d_j^{n+\varepsilon}} \le CM(\chi_{Q_j})(x)^{\frac{n+\varepsilon}{n}}$$

and use Theorem 5.6.6.]

HISTORICAL NOTES

The L^p boundedness of the conjugate function on the circle was announced in 1924 by Riesz [292], but its first proof appeared three years later in [294]. In view of the identification of the Hilbert transform with the conjugate function, the L^p boundedness of the Hilbert transform is also attributed to M. Riesz. Riesz's proof was first given for $p = 2k$, $k \in \mathbf{Z}^+$, via an argument similar to that in the proof of Theorem 4.1.7. For $p \neq 2k$ this proof relied on interpolation and was completed with the simultaneous publication of Riesz's article on interpolation of bilinear forms [293]. The weak type $(1,1)$ property of the Hilbert transform is due to Kolmogorov [197]. Additional proofs of the boundedness of the Hilbert transform have been obtained by Stein [350], Loomis [230], and Calderón [41]. The proof of Theorem 5.1.7, based on identity (5.1.23), is a refinement of a proof given by Cotlar [75].

The norm of the conjugate function on $L^p(\mathbf{T}^1)$, and consequently that of the Hilbert transform on $L^p(\mathbf{R})$, was shown by Gohberg and Krupnik [129] to be $\cot(\pi/2p)$ when p is a power of 2. Duality gives that this norm is $\tan(\pi/2p)$ for $1 < p \le 2$ whenever p' is a power of 2. Pichorides [282] extended this result to all $1 < p < \infty$ by refining Calderón's proof of Riesz's theorem. This result was also independently obtained by B. Cole (unpublished). The direct and simplified proof for the Hilbert transform given in Exercise 5.1.12 is in Grafakos [130]. The norm of the operators $\frac{1}{2}(I \pm iH)$ for real-valued functions was found to be $\frac{1}{2}\big[\min(\cos(\pi/2p), \sin(\pi/2p))\big]^{-1}$ by Verbitsky [366] and later independently by Essén [108]. The norm of the same operators for complex-valued functions was shown to be equal to $[\sin(\pi/p)]^{-1}$ by Hollenbeck and Verbitsky [156]. Exact formulas for the L^p norm, $1 \le p < \infty$ of the Hilbert transform acting on a characteristic function were obtained by Laeng [211]. The best constant in the weak type $(1,1)$ estimate for the Hilbert transform is equal to $(1+\frac{1}{3^2}+\frac{1}{5^2}+\cdots)(1-\frac{1}{3^2}+\frac{1}{5^2}-\cdots)^{-1}$ as shown by Davis [91] using Brownian motion; an alternative proof was later obtained by Baernstein [17]. Iwaniec and Martin [175] showed that the norms of the Riesz transforms on $L^p(\mathbf{R}^n)$ coincide with that of the Hilbert transform on $L^p(\mathbf{R})$ for $1 < p < \infty$.

Operators of the kind T_Ω as well as the stopping-time decomposition of Theorem 5.3.1 were introduced by Calderón and Zygmund [46]. In the same article, Calderón and Zygmund used this decomposition to prove Theorem 5.3.3 for operators of the form T_Ω when Ω satisfies a certain weak smoothness condition. The more general condition (5.3.12) first appeared in Hörmander's article [159]. A more flexible condition sufficient to yield weak type $(1,1)$ bounds is contained in the article of Duong and McIntosh [104]. Theorems 5.2.10 and 5.2.11 are also due to Calderón and Zygmund [48]. The latter article contains the method of rotations. Algebras of operators of the form T_Ω were studied in [49]. For more information on algebras of singular integrals see the article of Calderón [44]. Theorem 5.4.1 is due to Benedek, Calderón, and Panzone [22], while Example 5.4.2 is taken from Muckenhoupt [259]. Theorem 5.4.5 is due to Riviere [296]. A weaker version of this theorem, applicable for smoother singular integrals such as the maximal Hilbert transform, was obtained by Cotlar [75] (Theorem 5.3.4). Improvements of the main inequality in Theorem 5.3.4 for homogeneous singular integrals were obtained by Mateu and Verdera [245] and Mateu, Orobitg, and Verdera [244]. For a general overview of singular integrals and their applications, one may consult the expository article of Calderón [43].

Part (a) of Theorem 5.5.1 is due to Marcinkiewicz and Zygmund [242], although the case $p = q$ was proved earlier by Paley [273] with a larger constant. The values of r for which a general linear operator of weak or strong type (p,q) admits bounded ℓ^r extensions are described in Rubio de Francia and Torrea [304]. The L^p and weak L^p spaces in Theorem 5.5.1 can be replaced by general Banach lattices, as shown by Krivine [206] using Grothendieck's inequality. Hilbert-space-valued estimates for singular integrals were obtained by Benedek, Calderón, and Panzone [22]. Other operator-valued singular integral operators were studied by Rubio de Francia, Ruiz, and Torrea [303]. Banach-valued singular integrals are studied in great detail in the book of García-Cuerva and Rubio de Francia [122], which provides an excellent presentation of the subject. The ℓ^r-valued estimates (5.5.16) for the Hilbert transform were first obtained by Boas and Bochner [28]. The corresponding vector-valued estimates for the Hardy–Littlewood maximal function in Theorem 5.6.6 are due to Fefferman and Stein [115]. Conditions of the form (5.6.18) have been applied to several situations and can be traced in Zo [386].

The sharpness of the logarithmic condition (5.2.24) was indicated by Weiss and Zygmund [372], who constructed an example of an integrable function Ω with vanishing integral on $\mathbf{S}^1$ satisfying $\int_{\mathbf{S}^{n-1}} |\Omega(\theta)| \log^+ |\Omega(\theta)| \big(\log(2+\log(2+|\Omega(\theta)|))\big)^{-\delta} d\theta = \infty$ for all $\delta > 0$ and of a continuous function in $L^p(\mathbf{R}^2)$ for all $1 < p < \infty$ such that $\limsup_{\varepsilon\to 0} |T_\Omega^{(\varepsilon)}(f)(x)| = \infty$ for almost all $x \in \mathbf{R}^2$. The proofs of Theorems 5.2.10 and 5.2.11 can be modified to give that if Ω is in the Hardy space H^1 of $\mathbf{S}^{n-1}$, then T_Ω and $T_\Omega^{(*)}$ map L^p to L^p for $1 < p < \infty$. For T_Ω this fact was proved by Connett [72] and independently by Ricci and Weiss [289]; for $T_\Omega^{(*)}$ this was proved by Fan and Pan [110] and independently by Grafakos and Stefanov [139]. The latter authors [138] also obtained that the logarithmic condition $\operatorname{ess.sup}_{|\xi|=1} \int_{\mathbf{S}^{n-1}} |\Omega(\theta)|(\log \frac{1}{|\xi\cdot\theta|}|)^{1+\alpha} d\theta < \infty$, $\alpha > 0$, implies L^p boundedness for T_Ω and $T_\Omega^{(*)}$ for some $p \neq 2$. See also Fan, Guo, and Pan [109] as well as Ryabogin and Rubin [308] for extensions. Examples of functions Ω for which T_Ω maps L^p to L^p for a certain range of p's but not for other ranges of p's is given in Grafakos, Honzík, and Ryabogin [132]. A different example of this sort was provided later by Honzík [158]; the range of p's for which boundedness holds are different for these examples. Honzík [157] also constructed a delicate example of an integrable function Ω with mean value zero over $\mathbf{S}^1$ such that T_Ω is bounded on $L^2(\mathbf{R})$ but $T_\Omega^{(*)}$ is not.

The relatively weak condition $|\Omega| \log^+ |\Omega| \in L^1(\mathbf{S}^{n-1})$ also implies weak type $(1,1)$ boundedness for operators T_Ω. This was obtained by Seeger [317] and later extended by Tao [355] to situations in which there is no Fourier transform structure. Earlier partial results are in Christ and Rubio de Francia [63] and in the simultaneous work of Hofmann [155], both inspired by the work of Christ [60]. Soria and Sjögren [324] showed that for arbitrary Ω in $L^1(\mathbf{S}^{n-1})$, T_Ω is weak type $(1,1)$ when restricted to radial functions. Examples due to Christ (published in [139]) indicate that even for bounded functions Ω on $\mathbf{S}^{n-1}$, T_Ω may not map the endpoint Hardy space $H^1(\mathbf{R}^n)$ to $L^1(\mathbf{R}^n)$. However, Seeger and Tao [318] have showed that T_Ω always maps the Hardy space $H^1(\mathbf{R}^n)$ to the Lorentz space $L^{1,2}(\mathbf{R}^n)$ when $|\Omega|(\log^+|\Omega|)^2$ is integrable over $\mathbf{S}^{n-1}$. This result is sharp in the sense that for such Ω, T_Ω may not map $H^1(\mathbf{R}^n)$ to $L^{1,q}(\mathbf{R}^n)$ when $q < 2$ in general. If T_Ω maps $H^1(\mathbf{R}^n)$ to itself, Daly and Phillips [87] (in dimension $n = 2$) and Daly [86] (in dimensions $n \geq 3$) showed that Ω must lie in the Hardy space $H^1(\mathbf{S}^{n-1})$. There are also results concerning the singular maximal operator $M_\Omega(f)(x) = \sup_{r>0} \frac{1}{v_n r^n} \int_{|y|\leq r} |f(x-y)|\,|\Omega(y)|\, dy$, where Ω is an integrable function on $\mathbf{S}^{n-1}$ of not necessarily vanishing integral. Such operators were studied by Fefferman [116], Christ [60], and Hudson [162]. An excellent treatment of several kinds of singular integral operators with rough kernels is contained in the book of Lu, Ding, and Yan [234].

Chapter 6
Littlewood–Paley Theory and Multipliers

In this chapter we are concerned with orthogonality properties of the Fourier transform. This orthogonality is easily understood on L^2, but at this point it is not clear how it manifests itself on other spaces. Square functions introduce a way to express and quantify orthogonality of the Fourier transform on L^p and other function spaces. The introduction of square functions in this setting was pioneered by Littlewood and Paley, and the theory that subsequently developed is named after them. The extent to which Littlewood–Paley theory characterizes function spaces is remarkable.

Historically, Littlewood–Paley theory first appeared in the context of one-dimensional Fourier series and depended on complex function theory. With the development of real-variable methods, the whole theory became independent of complex methods and was extended to $\mathbf{R}^n$. This is the approach that we follow in this chapter. It turns out that the Littlewood–Paley theory is intimately related to the Calderón–Zygmund theory introduced in the previous chapter. This connection is deep and far-reaching, and its central feature is that one is able to derive the main results of one theory from the other.

The thrust and power of the Littlewood–Paley theory become apparent in some of the applications we discuss in this chapter. Such applications include the derivation of certain multiplier theorems, that is, theorems that yield sufficient conditions for bounded functions to be L^p multipliers. As a consequence of Littlewood–Paley theory we also prove that the lacunary partial Fourier integrals $\int_{|\xi|\le 2^N} \widehat{f}(\xi) e^{2\pi i x\cdot\xi}\, d\xi$ converge almost everywhere to an L^p function f on $\mathbf{R}^n$.

6.1 Littlewood–Paley Theory

We begin by examining more closely what we mean by orthogonality of the Fourier transform. If the functions f_j defined on $\mathbf{R}^n$ have Fourier transforms $\widehat{f_j}$ supported in disjoint sets, then they are *orthogonal* in the sense that

L. Grafakos, *Classical Fourier Analysis*, Graduate Texts in Mathematics 249,
DOI 10.1007/978-1-4939-1194-3_6, © Springer Science+Business Media New York 2014

$$\Big\|\sum_j f_j\Big\|_{L^2}^2 = \sum_j \big\|f_j\big\|_{L^2}^2. \tag{6.1.1}$$

Unfortunately, when 2 is replaced by some $p \neq 2$ in (6.1.1), the previous quantities may not even be comparable, as we show in Examples 6.1.8 and 6.1.9. The Littlewood–Paley theorem provides a substitute inequality to (6.1.1) expressing the fact that certain orthogonality considerations are also valid in $L^p(\mathbf{R}^n)$.

6.1.1 The Littlewood–Paley Theorem

The orthogonality we are searching for is best seen in the context of one-dimensional Fourier series (which was the setting in which Littlewood and Paley formulated their result). The primary observation is that the exponential $e^{2\pi i 2^k x}$ oscillates half as much as $e^{2\pi i 2^{k+1} x}$ and is therefore nearly constant in each period of the latter. This observation was instrumental in the proof of Theorem 3.6.4, which implied in particular that for all $1 < p < \infty$ we have

$$\Big\|\sum_{k=1}^N a_k e^{2\pi i 2^k x}\Big\|_{L^p[0,1]} \approx \Big(\sum_{k=1}^N |a_k|^2\Big)^{\frac{1}{2}}. \tag{6.1.2}$$

In other words, we can calculate the L^p norm of $\sum_{k=1}^N a_k e^{2\pi i 2^k x}$ in almost a precise fashion to obtain (modulo multiplicative constants) the same answer as in the L^2 case. Similar calculations are valid for more general blocks of exponentials in the dyadic range $\{2^k+1,\dots,2^{k+1}-1\}$, since the exponentials in each such block behave independently from those in each previous block. In particular, the L^p integrability of a function on $\mathbf{T}^1$ is not affected by the randomization of the sign of its Fourier coefficients in the previous dyadic blocks. This is the intuition behind the Littlewood–Paley theorem.

Motivated by this discussion, we introduce the Littlewood–Paley operators in the continuous setting.

Definition 6.1.1. Let Ψ be an integrable function on $\mathbf{R}^n$ and $j \in \mathbf{Z}$. We define the *Littlewood–Paley operator* Δ_j associated with Ψ by

$$\Delta_j(f) = f * \Psi_{2^{-j}},$$

where $\Psi_{2^{-j}}(x) = 2^{jn}\Psi(2^j x)$ for all x in $\mathbf{R}^n$. Thus we have $\widehat{\Psi_{2^{-j}}}(\xi) = \widehat{\Psi}(2^{-j}\xi)$ for all ξ in $\mathbf{R}^n$. We note that whenever Ψ is a Schwartz function and f is a tempered distribution, the quantity $\Delta_j(f)$ is a well defined function.

These operators depend on the choice of the function Ψ; in most applications we choose Ψ to be a smooth function with compactly supported Fourier transform. Observe that if $\widehat{\Psi}$ is supported in some annulus $0 < c_1 < |\xi| < c_2 < \infty$, then the Fourier transform of Δ_j is supported in the annulus $c_1 2^j < |\xi| < c_2 2^j$; in other words, it is localized near the frequency $|\xi| \approx 2^j$. Thus the purpose of Δ_j is to isolate the part of frequency of a function concentrated near $|\xi| \approx 2^j$.

The *square function* associated with the Littlewood–Paley operators Δ_j is defined by

$$f \mapsto \Big(\sum_{j\in\mathbf{Z}} |\Delta_j(f)|^2\Big)^{\frac{1}{2}}.$$

This quadratic expression captures the intrinsic orthogonality of the function f.

Theorem 6.1.2. *(Littlewood–Paley theorem) Suppose that Ψ is an integrable $\mathscr{C}^1$ function on $\mathbf{R}^n$ with mean value zero that satisfies*

$$|\Psi(x)| + |\nabla\Psi(x)| \le B(1+|x|)^{-n-1}. \tag{6.1.3}$$

Then there exists a constant $C_n < \infty$ such that for all $1 < p < \infty$ and all f in $L^p(\mathbf{R}^n)$ we have

$$\Big\|\Big(\sum_{j\in\mathbf{Z}} |\Delta_j(f)|^2\Big)^{\frac{1}{2}}\Big\|_{L^p(\mathbf{R}^n)} \le C_n B \max\big(p,(p-1)^{-1}\big)\big\|f\big\|_{L^p(\mathbf{R}^n)}. \tag{6.1.4}$$

There also exists a $C_n' < \infty$ such that for all f in $L^1(\mathbf{R}^n)$ we have

$$\Big\|\Big(\sum_{j\in\mathbf{Z}} |\Delta_j(f)|^2\Big)^{\frac{1}{2}}\Big\|_{L^{1,\infty}(\mathbf{R}^n)} \le C_n' B\big\|f\big\|_{L^1(\mathbf{R}^n)}. \tag{6.1.5}$$

Conversely, let Ψ be a Schwartz function such that either $\widehat{\Psi}(0) = 0$ and

$$\sum_{j\in\mathbf{Z}} |\widehat{\Psi}(2^{-j}\xi)|^2 = 1, \qquad \text{for all } \xi \in \mathbf{R}^n \setminus \{0\}, \tag{6.1.6}$$

or $\widehat{\Psi}$ is compactly supported away from the origin and

$$\sum_{j\in\mathbf{Z}} \widehat{\Psi}(2^{-j}\xi) = 1, \qquad \text{for all } \xi \in \mathbf{R}^n \setminus \{0\}. \tag{6.1.7}$$

Then there is a constant $C_{n,\Psi}$, such that for any $f \in \mathscr{S}'(\mathbf{R}^n)$ with $\big(\sum_{j\in\mathbf{Z}} |\Delta_j(f)|^2\big)^{\frac{1}{2}}$ in $L^p(\mathbf{R}^n)$ for some $1 < p < \infty$, there exists a unique polynomial Q such that the tempered distribution $f - Q$ coincides with an L^p function, and we have

$$\big\|f - Q\big\|_{L^p(\mathbf{R}^n)} \le C_{n,\Psi} B \max\big(p,(p-1)^{-1}\big)\Big\|\Big(\sum_{j\in\mathbf{Z}} |\Delta_j(f)|^2\Big)^{\frac{1}{2}}\Big\|_{L^p(\mathbf{R}^n)}. \tag{6.1.8}$$

Consequently, if g lies in $L^p(\mathbf{R}^n)$ for some $1 < p < \infty$, then

$$\big\|g\big\|_{L^p(\mathbf{R}^n)} \approx \Big\|\Big(\sum_{j\in\mathbf{Z}} |\Delta_j(g)|^2\Big)^{\frac{1}{2}}\Big\|_{L^p(\mathbf{R}^n)}.$$

Proof. We first prove (6.1.4) when $p = 2$. Using Plancherel's theorem, we see that (6.1.4) is a consequence of the inequality

$$\sum_j |\widehat{\Psi}(2^{-j}\xi)|^2 \le C_n B^2 \tag{6.1.9}$$

for some $C_n < \infty$. Because of (6.1.3), Fourier inversion holds for Ψ. Furthermore, Ψ has mean value zero and we may write

$$\widehat{\Psi}(\xi) = \int_{\mathbf{R}^n} e^{-2\pi i x\cdot\xi}\Psi(x)\,dx = \int_{\mathbf{R}^n} (e^{-2\pi i x\cdot\xi} - 1)\Psi(x)\,dx, \tag{6.1.10}$$

from which we obtain the estimate

$$|\widehat{\Psi}(\xi)| \le \sqrt{4\pi|\xi|}\int_{\mathbf{R}^n} |x|^{\frac{1}{2}}|\Psi(x)|\,dx \le C_n B|\xi|^{\frac{1}{2}}. \tag{6.1.11}$$

For $\xi = (\xi_1, \dots, \xi_n) \neq 0$, let j be such that $|\xi_j| \ge |\xi_k|$ for all $k \in \{1, \dots, n\}$. Integrate by parts with respect to ∂_j in (6.1.10) to obtain

$$\widehat{\Psi}(\xi) = -\int_{\mathbf{R}^n} (-2\pi i\xi_j)^{-1} e^{-2\pi i x\cdot\xi}(\partial_j\Psi)(x)\,dx,$$

from which we deduce the estimate

$$|\widehat{\Psi}(\xi)| \le \sqrt{n}|\xi|^{-1}\int_{\mathbf{R}^n} |\nabla\Psi(x)|\,dx \le C_n B|\xi|^{-1}. \tag{6.1.12}$$

We now break the sum in (6.1.9) into the parts where $2^{-j}|\xi| \le 1$ and $2^{-j}|\xi| \ge 1$ and use (6.1.11) and (6.1.12), respectively, to obtain (6.1.9). (See also Exercise 6.1.2.) This proves (6.1.4) when $p = 2$.

We now turn our attention to the case $p \neq 2$ in (6.1.4). We view (6.1.4) and (6.1.5) as vector-valued inequalities in the spirit of Section 5.5. Define an operator $\vec{T}$ acting on functions on $\mathbf{R}^n$ as follows:

$$\vec{T}(f)(x) = \{\Delta_j(f)(x)\}_j.$$

The inequalities (6.1.4) and (6.1.5) we wish to prove say simply that $\vec{T}$ is a bounded operator from $L^p(\mathbf{R}^n, \mathbf{C})$ to $L^p(\mathbf{R}^n, \ell^2)$ and from $L^1(\mathbf{R}^n, \mathbf{C})$ to $L^{1,\infty}(\mathbf{R}^n, \ell^2)$. We just proved that this statement is true when $p = 2$, and therefore the first hypothesis of Theorem 5.6.1 is satisfied. We observe that the operator $\vec{T}$ can be written in the form

$$\vec{T}(f)(x) = \left\{\int_{\mathbf{R}^n} \Psi_{2^{-j}}(x-y)f(y)\,dy\right\}_j = \int_{\mathbf{R}^n} \vec{K}(x-y)(f(y))\,dy,$$

where for each $x \in \mathbf{R}^n$, $\vec{K}(x)$ is a bounded linear operator from $\mathbf{C}$ to ℓ^2 given by

$$\vec{K}(x)(a) = \{\Psi_{2^{-j}}(x)a\}_j. \tag{6.1.13}$$

We clearly have that $\|\vec{K}(x)\|_{\mathbf{C}\to\ell^2} = \big(\sum_j |\Psi_{2^{-j}}(x)|^2\big)^{\frac{1}{2}}$, and to be able to apply Theorem 5.6.1 we need to know that for some constant C_n we have

$$\big\|\vec{K}(x)\big\|_{\mathbf{C}\to\ell^2} \le C_n B\,|x|^{-n}, \tag{6.1.14}$$

$$\lim_{\varepsilon\downarrow 0}\int_{\varepsilon\le|y|\le 1}\vec{K}(y)\,dy = \left\{\int_0^1 \Psi_{2^j}(y)\,dy\right\}_{j\in\mathbf{Z}}, \tag{6.1.15}$$

$$\sup_{y\neq 0}\int_{|x|\ge 2|y|}\big\|\vec{K}(x-y)-\vec{K}(x)\big\|_{\mathbf{C}\to\ell^2}\,dx \le C_n B. \tag{6.1.16}$$

Of these, (6.1.14) is easily obtained using (6.1.3), (6.1.15) i.e. trivial, and so we focus on (6.1.16). Since Ψ is a $\mathscr{C}^1$ function, for $|x|\ge 2|y|$ we have

$$\begin{aligned}
&|\Psi_{2^{-j}}(x-y)-\Psi_{2^{-j}}(x)| \\
&\le 2^{(n+1)j}|\nabla\Psi(2^j(x-\theta y))|\,|y| && \text{for some } \theta\in[0,1], \\
&\le B2^{(n+1)j}\big(1+2^j|x-\theta y|\big)^{-(n+1)}|y| \\
&\le B2^{nj}\big(1+2^{j-1}|x|\big)^{-(n+1)}2^j|y| && \text{since } |x-\theta y|\ge \tfrac{1}{2}|x|.
\end{aligned} \tag{6.1.17}$$

We also have that

$$\begin{aligned}
&|\Psi_{2^{-j}}(x-y)-\Psi_{2^{-j}}(x)| \\
&\qquad\le 2^{nj}|\Psi(2^j(x-y))|+2^{jn}|\Psi(2^jx)| \\
&\qquad\le B2^{nj}\big(1+2^j|x|\big)^{-(n+1)}+B2^{jn}\big(1+2^{j-1}|x|\big)^{-(n+1)} \\
&\qquad\le 2B2^{nj}\big(1+2^{j-1}|x|\big)^{-(n+1)}.
\end{aligned} \tag{6.1.18}$$

Taking the geometric mean of (6.1.17) and (6.1.18), we obtain for any $\gamma\in[0,1]$

$$|\Psi_{2^{-j}}(x-y)-\Psi_{2^{-j}}(x)| \le 2^{1-\gamma}B2^{nj}(2^j|y|)^{\gamma}\big(1+2^{j-1}|x|\big)^{-(n+1)}. \tag{6.1.19}$$

Using this estimate, when $|x|\ge 2|y|$, we obtain

$$\begin{aligned}
\big\|\vec{K}(x-y)-\vec{K}(x)\big\|_{\mathbf{C}\to\ell^2} &= \left(\sum_{j\in\mathbf{Z}}\big|\Psi_{2^{-j}}(x-y)-\Psi_{2^{-j}}(x)\big|^2\right)^{1/2} \\
&\le \sum_{j\in\mathbf{Z}}\big|\Psi_{2^{-j}}(x-y)-\Psi_{2^{-j}}(x)\big| \\
&\le 2B\Big(|y|\sum_{2^j<\frac{2}{|x|}}2^{(n+1)j}+|y|^{\frac{1}{2}}\sum_{2^j\ge\frac{2}{|x|}}2^{(n+\frac{1}{2})j}(2^{j-1}|x|)^{-(n+1)}\Big) \\
&\le C_nB\big(|y||x|^{-n-1}+|y|^{\frac{1}{2}}|x|^{-n-\frac{1}{2}}\big),
\end{aligned}$$

where we used (6.1.19) with $\gamma=1$ in the first sum and (6.1.19) with $\gamma=1/2$ in the second sum. Using this bound, we easily deduce (6.1.16) by integrating over the region $|x| \geq 2|y|$. Finally, using Theorem 5.6.1 we conclude the proofs of (6.1.4) and (6.1.5), which establishes one direction of the theorem.

We now turn to the converse direction. Let Δ_j^* be the adjoint operator of Δ_j given by $\widehat{\Delta_j^* f} = \widehat{f}\,\overline{\widehat{\Psi_{2^{-j}}}}$. Let f be in $\mathscr{S}'(\mathbf{R}^n)$. Then the series $\sum_{j\in\mathbf{Z}} \Delta_j^*\Delta_j(f)$ converges in $\mathscr{S}'(\mathbf{R}^n)$. To see this, it suffices to show that the sequence of partial sums $u_N = \sum_{|j|<N} \Delta_j^*\Delta_j(f)$ converges in $\mathscr{S}'$. This means that if we test this sequence against a Schwartz function g, then it is a Cauchy sequence and hence it converges as $N \to \infty$. But an easy argument using duality and the Cauchy–Schwarz and Hölder's inequalities shows that for $M > N$ we have

$$|\langle u_N, g\rangle - \langle u_M, g\rangle| \leq \Big\| \Big(\sum_j |\Delta_j(f)|^2\Big)^{\frac{1}{2}} \Big\|_{L^p} \Big\| \Big(\sum_{N\leq |j|\leq M} |\Delta_j(g)|^2\Big)^{\frac{1}{2}} \Big\|_{L^{p'}},$$

and this can be made small by picking $M > N \geq N_0(g)$. Since the sequence $\langle u_N, g\rangle$ is Cauchy, it converges to some $\Lambda(g)$. Now it remains to show that the map $g \mapsto \Lambda(g)$ is a tempered distribution. Obviously $\Lambda(g)$ is a linear functional. Also,

$$\begin{aligned} |\Lambda(g)| &\leq \Big\| \Big(\sum_j |\Delta_j(f)|^2\Big)^{\frac{1}{2}} \Big\|_{L^p} \Big\| \Big(\sum_j |\Delta_j(g)|^2\Big)^{\frac{1}{2}} \Big\|_{L^{p'}} \\ &\leq C_{p'} \Big\| \Big(\sum_j |\Delta_j(f)|^2\Big)^{\frac{1}{2}} \Big\|_{L^p} \big\|g\big\|_{L^{p'}}, \end{aligned}$$

and since $\|g\|_{L^{p'}}$ is controlled by a finite number of Schwartz seminorms of g, it follows that Λ is in $\mathscr{S}'$. The distribution Λ is the limit of the series $\sum_j \Delta_j^*\Delta_j$.

Under hypothesis (6.1.6), the Fourier transform of the tempered distribution $f - \sum_{j\in\mathbf{Z}} \Delta_j^*\Delta_j(f)$ is supported at the origin. This implies that there exists a polynomial Q such that $f - Q = \sum_{j\in\mathbf{Z}} \Delta_j^*\Delta_j(f)$. Now let g be a Schwartz function. We have

$$\begin{aligned} \big|\langle f - Q, \overline{g}\rangle\big| &= \Big|\Big\langle \sum_{j\in\mathbf{Z}} \Delta_j^*\Delta_j(f), \overline{g}\Big\rangle\Big| \\ &= \Big|\sum_{j\in\mathbf{Z}} \big\langle \Delta_j^*\Delta_j(f), \overline{g}\big\rangle\Big| \\ &= \Big|\sum_{j\in\mathbf{Z}} \big\langle \Delta_j(f), \overline{\Delta_j(g)}\big\rangle\Big| \\ &= \left|\int_{\mathbf{R}^n} \sum_{j\in\mathbf{Z}} \Delta_j(f)\, \overline{\Delta_j(g)}\, dx\right| \\ &\leq \int_{\mathbf{R}^n} \Big(\sum_{j\in\mathbf{Z}} |\Delta_j(f)|^2\Big)^{\frac{1}{2}} \Big(\sum_{j\in\mathbf{Z}} |\Delta_j(g)|^2\Big)^{\frac{1}{2}} dx \\ &\leq \Big\| \Big(\sum_{j\in\mathbf{Z}} |\Delta_j(f)|^2\Big)^{\frac{1}{2}} \Big\|_{L^p} \Big\| \Big(\sum_{j\in\mathbf{Z}} |\Delta_j(g)|^2\Big)^{\frac{1}{2}} \Big\|_{L^{p'}} \end{aligned}$$

$$\leq \Big\| \Big(\sum_{j\in\mathbf{Z}} |\Delta_j(f)|^2 \Big)^{\frac{1}{2}} \Big\|_{L^p} C_n B \max\big(p',(p'-1)^{-1}\big) \big\|g\big\|_{L^{p'}}, \quad (6.1.20)$$

having used the definition of the adjoint (Section 2.5.2), the Cauchy–Schwarz inequality, Hölder's inequality, and (6.1.4). Taking the supremum over all g in $L^{p'}$ with norm at most one, we obtain that the tempered distribution $f-Q$ is a bounded linear functional on $L^{p'}$. By the Riesz representation theorem, $f-Q$ coincides with an L^p function whose norm satisfies the estimate

$$\big\|f-Q\big\|_{L^p} \leq C_n B \max\big(p,(p-1)^{-1}\big) \Big\| \Big(\sum_{j\in\mathbf{Z}} |\Delta_j(f)|^2 \Big)^{\frac{1}{2}} \Big\|_{L^p}.$$

We now show uniqueness. If Q_1 is another polynomial, with $f-Q_1 \in L^p$, then $Q-Q_1$ must be an L^p function; but the only polynomial that lies in L^p is the zero polynomial. This completes the proof of the converse of the theorem under hypothesis (6.1.6).

To obtain the same conclusion under the hypothesis (6.1.7) we argue in a similar way but we leave the details as an exercise. (One may adapt the argument in the proof of Corollary 6.1.7 to this setting.) □

Remark 6.1.3. We make some observations. If $\widehat{\Psi}$ is real-valued, then the operators Δ_j are self-adjoint. Indeed,

$$\int_{\mathbf{R}^n} \Delta_j(f)\, \overline{g}\, dx = \int_{\mathbf{R}^n} \widehat{f}\, \widehat{\Psi_{2^{-j}}}\, \overline{\widehat{g}}\, d\xi = \int_{\mathbf{R}^n} \widehat{f}\, \overline{\widehat{\Psi_{2^{-j}}}\, \widehat{g}}\, d\xi = \int_{\mathbf{R}^n} f\, \overline{\Delta_j(g)}\, dx.$$

Moreover, if Ψ is a radial function, we see that the operators Δ_j are self-transpose, that is, they satisfy

$$\int_{\mathbf{R}^n} \Delta_j(f)\, g\, dx = \int_{\mathbf{R}^n} f\, \Delta_j(g)\, dx.$$

Assume now that Ψ is both radial and has a real-valued Fourier transform. Suppose also that Ψ satisfies (6.1.3) and that it has mean value zero. Then the inequality

$$\Big\| \sum_{j\in\mathbf{Z}} \Delta_j(f_j) \Big\|_{L^p} \leq C_n B \max\big(p,(p-1)^{-1}\big) \Big\| \Big(\sum_{j\in\mathbf{Z}} |f_j|^2 \Big)^{\frac{1}{2}} \Big\|_{L^p} \quad (6.1.21)$$

is true for sequences of functions $\{f_j\}_j$. To see this we use duality. Let

$$\vec{T}(f) = \{\Delta_j(f)\}_j.$$

Then

$$\vec{T}^*(\{g_j\}_j) = \sum_j \Delta_j(g_j).$$

Inequality (6.1.4) says that the operator $\vec{T}$ maps $L^p(\mathbf{R}^n,\mathbf{C})$ to $L^p(\mathbf{R}^n,\ell^2)$, and its dual statement is that $\vec{T}^*$ maps $L^{p'}(\mathbf{R}^n,\ell^2)$ to $L^{p'}(\mathbf{R}^n,\mathbf{C})$. This is exactly the statement in (6.1.21) if p is replaced by p'. Since p is any number in $(1,\infty)$, (6.1.21) is proved.

6.1.2 Vector-Valued Analogues

We now obtain a vector-valued extension of Theorem 6.1.2. We have the following.

Proposition 6.1.4. *Let Ψ be an integrable $\mathscr{C}^1$ function on $\mathbf{R}^n$ with mean value zero that satisfies (6.1.3) and let Δ_j be the Littlewood–Paley operator associated with Ψ. Then there exists a constant $C_n < \infty$ such that for all $1 < p, r < \infty$ and all sequences of L^p functions f_j we have*

$$\Big\|\Big(\sum_{j\in\mathbf{Z}}\Big(\sum_{k\in\mathbf{Z}}|\Delta_k(f_j)|^2\Big)^{\frac{r}{2}}\Big)^{\frac{1}{r}}\Big\|_{L^p(\mathbf{R}^n)} \le C_n B\widetilde{C}_{p,r}\Big\|\Big(\sum_{j\in\mathbf{Z}}|f_j|^r\Big)^{\frac{1}{r}}\Big\|_{L^p(\mathbf{R}^n)},$$

where $\widetilde{C}_{p,r} = \max(p,(p-1)^{-1})\max(r,(r-1)^{-1})$. Moreover, for some $C_n' > 0$ and all sequences of L^1 functions f_j we have

$$\Big\|\Big(\sum_{j\in\mathbf{Z}}\Big(\sum_{k\in\mathbf{Z}}|\Delta_k(f_j)|^2\Big)^{\frac{r}{2}}\Big)^{\frac{1}{r}}\Big\|_{L^{1,\infty}(\mathbf{R}^n)} \le C_n' B\max(r,(r-1)^{-1})\Big\|\Big(\sum_{j\in\mathbf{Z}}|f_j|^r\Big)^{\frac{1}{r}}\Big\|_{L^1(\mathbf{R}^n)}.$$

In particular,

$$\Big\|\Big(\sum_{j\in\mathbf{Z}}|\Delta_j(f_j)|^r\Big)^{\frac{1}{r}}\Big\|_{L^p(\mathbf{R}^n)} \le C_n B\widetilde{C}_{p,r}\Big\|\Big(\sum_{j\in\mathbf{Z}}|f_j|^r\Big)^{\frac{1}{r}}\Big\|_{L^p(\mathbf{R}^n)}. \tag{6.1.22}$$

Proof. We introduce Banach spaces $\mathscr{B}_1 = \mathbf{C}$ and $\mathscr{B}_2 = \ell^2$ and for $f \in L^p(\mathbf{R}^n)$ define an operator

$$\vec{T}(f) = \{\Delta_k(f)\}_{k\in\mathbf{Z}}.$$

In the proof of Theorem 6.1.2 we showed that $\vec{T}$ has a kernel $\vec{K}$ that satisfies condition (6.1.16). Furthermore, $\vec{T}$ obviously maps $L^r(\mathbf{R}^n,\mathbf{C})$ to $L^r(\mathbf{R}^n,\ell^r)$. Applying Proposition 5.6.4, we obtain the first two statements of the proposition. Restricting to $k = j$ yields (6.1.22). □

6.1.3 L^p Estimates for Square Functions Associated with Dyadic Sums

Let us pick a Schwartz function Ψ whose Fourier transform is compactly supported in the annulus $2^{-1} \le |\xi| \le 2^2$ such that (6.1.6) is satisfied. (Clearly (6.1.6) has no chance of being satisfied if $\widehat{\Psi}$ is supported only in the annulus $1 \le |\xi| \le 2$.) The Littlewood–Paley operation $f \mapsto \Delta_j(f)$ represents the smoothly truncated frequency localization of a function f near the dyadic annulus $|\xi| \approx 2^j$. Theorem 6.1.2 says that the square function formed by these localizations has L^p norm comparable to that of the original function. In other words, this square function characterizes the L^p norm of a function. This is the main feature of Littlewood–Paley theory.

One may ask whether Theorem 6.1.2 still holds if the Littlewood–Paley operators Δ_j are replaced by their nonsmooth versions

$$f \mapsto \big(\chi_{2^j \le |\xi| < 2^{j+1}} \widehat{f}(\xi)\big)^\vee(x). \tag{6.1.23}$$

This question has a surprising answer that already signals that there may be some fundamental differences between one-dimensional and higher-dimensional Fourier analysis. The square function formed by the operators in (6.1.23) can be used to characterize $L^p(\mathbf{R})$ in the same way Δ_j did, but not $L^p(\mathbf{R}^n)$ when $n > 1$ and $p \neq 2$. The problem lies in the fact that the characteristic function of the unit disk is not an L^p multiplier on $\mathbf{R}^n$ when $n \ge 2$ unless $p = 2$; see Section 5.1 in [131]. The one-dimensional result we alluded to earlier is the following.

For $j \in \mathbf{Z}$ we introduce the one-dimensional operator

$$\Delta_j^\#(f)(x) = (\widehat{f}\chi_{I_j})^\vee(x), \tag{6.1.24}$$

where

$$I_j = [2^j, 2^{j+1}) \cup (-2^{j+1}, -2^j],$$

and $\Delta_j^\#$ is a version of the operator Δ_j in which the characteristic function of the set $2^j \le |\xi| < 2^{j+1}$ replaces the function $\widehat{\Psi}(2^{-j}\xi)$.

Theorem 6.1.5. *There exists a constant C_1 such that for all $1 < p < \infty$ and all f in $L^p(\mathbf{R})$ we have*

$$\frac{\|f\|_{L^p(\mathbf{R}^n)}}{C_1(p+\frac{1}{p-1})^2} \le \Big\| \Big(\sum_{j\in\mathbf{Z}} |\Delta_j^\#(f)|^2 \Big)^{\frac{1}{2}} \Big\|_{L^p(\mathbf{R}^n)} \le C_1(p+\tfrac{1}{p-1})^2 \|f\|_{L^p(\mathbf{R}^n)}. \tag{6.1.25}$$

Proof. Pick a Schwartz function ψ on the line whose Fourier transform is supported in the set $2^{-1} \le |\xi| \le 2^2$ and is equal to 1 on the set $1 \le |\xi| \le 2$. Let Δ_j be the Littlewood–Paley operator associated with ψ. Observe that $\Delta_j \Delta_j^\# = \Delta_j^\# \Delta_j = \Delta_j^\#$, since $\widehat{\psi}$ is equal to one on the support of $\Delta_j^\#(f)\widehat{\ }$. We now use Exercise 5.6.1(a) to obtain

$$\begin{aligned}
\Big\| \Big(\sum_{j\in\mathbf{Z}} |\Delta_j^\#(f)|^2 \Big)^{\frac{1}{2}} \Big\|_{L^p} &= \Big\| \Big(\sum_{j\in\mathbf{Z}} |\Delta_j^\#\Delta_j(f)|^2 \Big)^{\frac{1}{2}} \Big\|_{L^p} \\
&\le C\max(p,(p-1)^{-1}) \Big\| \Big(\sum_{j\in\mathbf{Z}} |\Delta_j(f)|^2 \Big)^{\frac{1}{2}} \Big\|_{L^p} \\
&\le CB\max(p,(p-1)^{-1})^2 \|f\|_{L^p},
\end{aligned}$$

where the last inequality follows from Theorem 6.1.2. The reverse inequality for $1 < p < \infty$ follows just like the reverse inequality (6.1.8) of Theorem 6.1.2 by simply replacing the Δ_j's by the $\Delta_j^\#$'s and setting the polynomial Q equal to zero. (There is no need to use the Riesz representation theorem here, just the fact that the L^p norm of f can be realized as the supremum of expressions $|\langle f,g\rangle|$ where g has $L^{p'}$ norm at most 1.) $\square$

There is a higher-dimensional version of Theorem 6.1.5 with dyadic rectangles replacing the dyadic intervals. As has already been pointed out, the higher-dimensional version with dyadic annuli replacing the dyadic intervals is false.

Let us introduce some notation. For $j \in \mathbf{Z}$, we denote by I_j the dyadic set $[2^j,2^{j+1})\bigcup(-2^{j+1},-2^j]$ as in the statement of Theorem 6.1.5. For $j_1,\dots,j_n \in \mathbf{Z}$ define a dyadic rectangle

$$R_{j_1,\dots,j_n} = I_{j_1} \times \cdots \times I_{j_n}$$

in $\mathbf{R}^n$. Actually $R_{j_1,\dots,j_n}$ is not a rectangle but a union of 2^n rectangles; with some abuse of language we still call it a rectangle. For notational convenience we write

$$R_{\mathbf{j}} = R_{j_1,\dots,j_n}, \qquad \text{where } \mathbf{j} = (j_1,\dots,j_n) \in \mathbf{Z}^n.$$

Observe that for different $\mathbf{j},\mathbf{j}' \in \mathbf{Z}^n$ the rectangles $R_{\mathbf{j}}$ and $R_{\mathbf{j}'}$ have disjoint interiors and that the union of all the $R_{\mathbf{j}}$'s is equal to $\mathbf{R}^n \setminus \{0\}$. In other words, the family of $R_{\mathbf{j}}$'s, where $\mathbf{j} \in \mathbf{Z}^n$, forms a *tiling* of $\mathbf{R}^n$, which we call the *dyadic decomposition* of $\mathbf{R}^n$. We now introduce operators

$$\Delta_{\mathbf{j}}^{\#}(f)(x) = (\widehat{f}\chi_{R_{\mathbf{j}}})^{\vee}(x), \tag{6.1.26}$$

and we have the following n-dimensional extension of Theorem 6.1.5.

Theorem 6.1.6. *For a Schwartz function ψ on the line with integral zero we define the operator*

$$\Delta_{\mathbf{j}}(f)(x) = \big(\widehat{\psi}(2^{-j_1}\xi_1)\cdots\widehat{\psi}(2^{-j_n}\xi_n)\widehat{f}(\xi)\big)^{\vee}(x), \tag{6.1.27}$$

where $\mathbf{j} = (j_1,\dots,j_n) \in \mathbf{Z}^n$. Then there is a dimensional constant C_n such that

$$\Big\|\Big(\sum_{\mathbf{j}\in\mathbf{Z}^n} |\Delta_{\mathbf{j}}(f)|^2\Big)^{\frac{1}{2}}\Big\|_{L^p(\mathbf{R}^n)} \le C_n(p+(p-1)^{-1})^n\big\|f\big\|_{L^p(\mathbf{R}^n)}. \tag{6.1.28}$$

Let $\Delta_{\mathbf{j}}^{\#}$ be the operators defined in (6.1.26). Then there exists a positive constant C_n such that for all $1<p<\infty$ and all $f \in L^p(\mathbf{R}^n)$ we have

$$\frac{\big\|f\big\|_{L^p(\mathbf{R}^n)}}{C_n(p+\frac{1}{p-1})^{2n}} \le \Big\|\Big(\sum_{\mathbf{j}\in\mathbf{Z}^n} |\Delta_{\mathbf{j}}^{\#}(f)|^2\Big)^{\frac{1}{2}}\Big\|_{L^p(\mathbf{R}^n)} \le C_n(p+\tfrac{1}{p-1})^{2n}\big\|f\big\|_{L^p(\mathbf{R}^n)}. \tag{6.1.29}$$

Proof. We first prove (6.1.28). Note that if $\mathbf{j} = (j_1,\dots,j_n) \in \mathbf{Z}^n$, then the operator $\Delta_{\mathbf{j}}$ is equal to

$$\Delta_{\mathbf{j}}(f) = \Delta_{j_1}^{(j_1)}\cdots\Delta_{j_n}^{(j_n)}(f),$$

where the $\Delta_{j_r}^{(j_r)}$ are one-dimensional operators given on the Fourier transform by multiplication by $\widehat{\psi}(2^{-j_r}\xi_r)$, with the remaining variables fixed. Inequality in (6.1.28) is a consequence of the one-dimensional case. For instance, we discuss the case $n=2$. Using Proposition 6.1.4, we obtain

$$\begin{aligned}
&\Big\|\Big(\sum_{\mathbf{j}\in\mathbf{Z}^2}|\Delta_{\mathbf{j}}(f)|^2\Big)^{\frac12}\Big\|_{L^p(\mathbf{R}^2)}^p\\
&=\int_{\mathbf{R}}\bigg[\int_{\mathbf{R}}\Big(\sum_{j_1\in\mathbf{Z}}\sum_{j_2\in\mathbf{Z}}|\Delta_{j_1}^{(1)}\Delta_{j_2}^{(2)}(f)(x_1,x_2)|^2\Big)^{\frac{p}{2}}dx_1\bigg]dx_2\\
&\le C^p\max(p,(p-1)^{-1})^p\int_{\mathbf{R}}\bigg[\int_{\mathbf{R}}\Big(\sum_{j_2\in\mathbf{Z}}|\Delta_{j_2}^{(2)}(f)(x_1,x_2)|^2\Big)^{\frac{p}{2}}dx_1\bigg]dx_2\\
&=C^p\max(p,(p-1)^{-1})^p\int_{\mathbf{R}}\bigg[\int_{\mathbf{R}}\Big(\sum_{j_2\in\mathbf{Z}}|\Delta_{j_2}^{(2)}(f)(x_1,x_2)|^2\Big)^{\frac{p}{2}}dx_2\bigg]dx_1\\
&\le C^{2p}\max(p,(p-1)^{-1})^{2p}\int_{\mathbf{R}}\bigg[\int_{\mathbf{R}}|f(x_1,x_2)|^p\,dx_2\bigg]dx_1\\
&=C^{2p}\max(p,(p-1)^{-1})^{2p}\big\|f\big\|_{L^p(\mathbf{R}^2)}^p,
\end{aligned}$$

where we also used Theorem 6.1.2 in the calculation. Higher-dimensional versions of this estimate may easily be obtained by induction.

We now turn to the upper inequality in (6.1.29). We pick a Schwartz function ψ whose Fourier transform is supported in the union $[-4,-1/2]\bigcup[1/2,4]$ and is equal to 1 on $[-2,-1]\bigcup[1,2]$. Then we clearly have

$$\Delta_{\mathbf{j}}^{\#}=\Delta_{\mathbf{j}}^{\#}\Delta_{\mathbf{j}},$$

since $\widehat{\psi}(2^{-j_1}\xi_1)\cdots\widehat{\psi}(2^{-j_n}\xi_n)$ is equal to 1 on the rectangle $R_{\mathbf{j}}$. We now use Exercise 5.6.1(b) and estimate (6.1.28) to obtain

$$\begin{aligned}
\Big\|\Big(\sum_{\mathbf{j}\in\mathbf{Z}^n}|\Delta_{\mathbf{j}}^{\#}(f)|^2\Big)^{\frac12}\Big\|_{L^p}&=\Big\|\Big(\sum_{\mathbf{j}\in\mathbf{Z}^n}|\Delta_{\mathbf{j}}^{\#}\Delta_{\mathbf{j}}(f)|^2\Big)^{\frac12}\Big\|_{L^p}\\
&\le C\max(p,(p-1)^{-1})^n\Big\|\Big(\sum_{\mathbf{j}\in\mathbf{Z}^n}|\Delta_{\mathbf{j}}(f)|^2\Big)^{\frac12}\Big\|_{L^p}\\
&\le CB\max(p,(p-1)^{-1})^{2n}\big\|f\big\|_{L^p}.
\end{aligned}$$

The lower inequality in (6.1.29) for $1<p<\infty$ is proved like inequality (6.1.8) in Theorem 6.1.2. The fundamental ingredient in the proof is that $f=\sum_{\mathbf{j}\in\mathbf{Z}^n}\Delta_{\mathbf{j}}^{\#}\Delta_{\mathbf{j}}^{\#}(f)$ for all Schwartz functions f, where the sum is interpreted as the L^2-limit of the sequence of partial sums. Thus the series converges in $\mathscr{S}'$, and pairing with a Schwartz function $\overline{g}$, we obtain the lower inequality in (6.1.29) for Schwartz functions, by applying the steps that prove (6.1.20) (with $Q=0$). To prove the lower inequality in (6.1.29) for a general function $f \in L^p(\mathbf{R}^n)$ we approximate an L^p function by a sequence of Schwartz functions in the L^p norm. Then both sides of the lower inequality in (6.1.29) for the approximating sequence converge to the corresponding sides of the lower inequality in (6.1.29) for f; the convergence of the sequence of L^p norms of the square functions requires the upper inequality in (6.1.29) that was previously established. This concludes the proof of the theorem. □

Next we observe that if the Schwartz function ψ is suitably chosen, then the reverse inequality in estimate (6.1.28) also holds. More precisely, suppose $\widehat{\psi}(\xi)$ is an even smooth real-valued function supported in the set $\frac{9}{10} \le |\xi| \le \frac{21}{10}$ in $\mathbf{R}$ that satisfies

$$\sum_{j\in\mathbf{Z}} \widehat{\psi}(2^{-j}\xi) = 1, \qquad \xi \in \mathbf{R}\setminus\{0\}; \tag{6.1.30}$$

then we have the following.

Corollary 6.1.7. *Suppose that ψ satisfies (6.1.30) and let $\Delta_{\mathbf{j}}$ be as in (6.1.27). Let f be an L^p function on $\mathbf{R}^n$ such that the function $\big(\sum_{\mathbf{j}\in\mathbf{Z}^n} |\Delta_{\mathbf{j}}(f)|^2\big)^{\frac{1}{2}}$ is in $L^p(\mathbf{R}^n)$. Then there is a constant C_n that depends only on the dimension and ψ such that the lower estimate*

$$\frac{\|f\|_{L^p}}{C_n(p+\frac{1}{p-1})^n} \le \Big\| \Big(\sum_{\mathbf{j}\in\mathbf{Z}^n} |\Delta_{\mathbf{j}}(f)|^2 \Big)^{\frac{1}{2}} \Big\|_{L^p} \tag{6.1.31}$$

holds.

Proof. If we had $\sum_{j\in\mathbf{Z}} |\widehat{\psi}(2^{-j}\xi)|^2 = 1$ instead of (6.1.30), then we could apply the method used in the lower estimate of Theorem 6.1.2 to obtain the required conclusion. In this case we provide another argument that is very similar in spirit.

We first prove (6.1.31) for Schwartz functions f. Then the series $\sum_{j\in\mathbf{Z}^n} \Delta_{\mathbf{j}}(f)$ converges in L^2 (and hence in $\mathscr{S}'$) to f. Now let g be another Schwartz function. We express the inner product $\langle f, \overline{g}\rangle$ as the action of the distribution $\sum_{\mathbf{j}\in\mathbf{Z}^n} \Delta_{\mathbf{j}}(f)$ on the test function $\overline{g}$:

$$\begin{aligned}
|\langle f, \overline{g}\rangle| &= \Big| \Big\langle \sum_{\mathbf{j}\in\mathbf{Z}^n} \Delta_{\mathbf{j}}(f), \overline{g} \Big\rangle \Big| \\
&= \Big| \sum_{\mathbf{j}\in\mathbf{Z}^n} \big\langle \Delta_{\mathbf{j}}(f), \overline{g} \big\rangle \Big| \\
&= \Big| \sum_{\mathbf{j}\in\mathbf{Z}^n} \sum_{\substack{\mathbf{k}=(k_1,\dots,k_n)\in\mathbf{Z}^n \\ \exists r\ |k_r - j_r|\le 1\}}} \big\langle \Delta_{\mathbf{j}}(f), \overline{\Delta_{\mathbf{k}}(g)} \big\rangle \Big| \\
&\le \int_{\mathbf{R}^n} \sum_{\mathbf{j}\in\mathbf{Z}^n} \sum_{\substack{\mathbf{k}=(k_1,\dots,k_n)\in\mathbf{Z}^n \\ \exists r\ |k_r - j_r|\le 1\}}} |\Delta_{\mathbf{j}}(f)|\,|\Delta_{\mathbf{k}}(g)|\,dx \\
&\le 3^n \int_{\mathbf{R}^n} \Big(\sum_{\mathbf{j}\in\mathbf{Z}^n} |\Delta_{\mathbf{j}}(f)|^2 \Big)^{\frac{1}{2}} \Big(\sum_{\mathbf{k}\in\mathbf{Z}^n} |\Delta_{\mathbf{k}}(g)|^2 \Big)^{\frac{1}{2}} dx
\end{aligned}$$

$$\leq 3^n \Big\| \Big(\sum_{\mathbf{j}\in\mathbf{Z}^n} |\Delta_{\mathbf{j}}(f)|^2 \Big)^{\frac{1}{2}} \Big\|_{L^p} \Big\| \Big(\sum_{\mathbf{k}\in\mathbf{Z}^n} |\Delta_{\mathbf{k}}(g)|^2 \Big)^{\frac{1}{2}} \Big\|_{L^{p'}}$$

$$\leq C_n^{-1} \max\big(p', (p'-1)^{-1}\big)^n \|g\|_{L^{p'}} \Big\| \Big(\sum_{\mathbf{j}\in\mathbf{Z}^n} |\Delta_{\mathbf{j}}(f)|^2 \Big)^{\frac{1}{2}} \Big\|_{L^p},$$

where we used the fact that $\Delta_{\mathbf{j}}(f)$ and $\Delta_{\mathbf{k}}(g)$ are orthogonal operators unless every coordinate of $\mathbf{k}$ is within 1 unit of the corresponding coordinate of $\mathbf{j}$; this is an easy consequence of the support properties of $\widehat{\psi}$. We now take the supremum over all g in $L^{p'}$ with norm at most 1, to obtain (6.1.31) for Schwartz functions f.

To extend this estimate to general L^p functions f, we use the density argument described in the last paragraph in the proof of Theorem 6.1.6. □

6.1.4 Lack of Orthogonality on L^p

We discuss two examples indicating why (6.1.1) cannot hold if the exponent 2 is replaced by some other exponent $q \neq 2$. More precisely, we show that if the functions f_j have Fourier transforms supported in disjoint sets, then the inequality

$$\Big\| \sum_j f_j \Big\|_{L^p}^p \leq C_p \sum_j \|f_j\|_{L^p}^p \tag{6.1.32}$$

cannot hold if $p > 2$, and similarly, the inequality

$$\sum_j \|f_j\|_{L^p}^p \leq C_p \Big\| \sum_j f_j \Big\|_{L^p}^p \tag{6.1.33}$$

cannot hold if $p < 2$. In both (6.1.32) and (6.1.33) the constants C_p are supposed to be independent of the functions f_j.

Example 6.1.8. Pick a Schwartz function ζ whose Fourier transform is positive and supported in the interval $|\xi| \leq 1/4$. Let N be a large integer and let

$$f_j(x) = e^{2\pi i j x} \zeta(x).$$

Then

$$\widehat{f_j}(\xi) = \widehat{\zeta}(\xi - j)$$

and the $\widehat{f_j}$'s have disjoint Fourier transforms. We obviously have

$$\sum_{j=0}^{N} \|f_j\|_{L^p}^p = (N+1) \|\zeta\|_{L^p}^p.$$

On the other hand, we have the estimate

$$\begin{aligned}\Big\| \sum_{j=0}^{N} f_j \Big\|_{L^p}^p &= \int_{\mathbf{R}} \Big| \frac{e^{2\pi i (N+1)x} - 1}{e^{2\pi i x} - 1} \Big|^p |\zeta(x)|^p \, dx \\ &\geq c \int_{|x| < \frac{1}{10}(N+1)^{-1}} \frac{(N+1)^p |x|^p}{|x|^p} |\zeta(x)|^p \, dx \\ &= C_\zeta (N+1)^{p-1},\end{aligned}$$

since ζ does not vanish in a neighborhood of zero. We conclude that (6.1.32) cannot hold for this choice of f_j's for $p > 2$.

Example 6.1.9. We now indicate why (6.1.33) cannot hold for $p < 2$. We pick a smooth function Ψ on the line whose Fourier transform $\widehat{\Psi}$ is supported in $\big[\frac{7}{8}, \frac{17}{8}\big]$, is nonnegative, is equal to 1 on $\big[\frac{9}{8}, \frac{15}{8}\big]$, and satisfies

$$\sum_{j \in \mathbf{Z}} \widehat{\Psi}(2^{-j}\xi)^2 = 1, \qquad \xi > 0.$$

Extend $\widehat{\Psi}$ to be an even function on the whole line and let Δ_j be the Littlewood–Paley operator associated with Ψ. Also pick a nonzero Schwartz function φ on the real line whose Fourier transform is nonnegative and supported in the set $\big[\frac{11}{8}, \frac{13}{8}\big]$. Fix N a large positive integer and let

$$f_j(x) = e^{2\pi i \frac{12}{8} 2^j x} \varphi(x), \tag{6.1.34}$$

for $j = 1, 2, \dots, N$. Then the function $\widehat{f_j}(\xi) = \widehat{\varphi}(\xi - \frac{12}{8} 2^j)$ is supported in the set $\big[\frac{11}{8} + \frac{12}{8} 2^j, \frac{13}{8} + \frac{12}{8} 2^j\big]$, which is contained in $\big[\frac{9}{8} 2^j, \frac{15}{8} 2^j\big]$ for $j \geq 3$. In other words, $\widehat{\Psi}(2^{-j}\xi)$ is equal to 1 on the support of $\widehat{f_j}$. This implies that

$$\Delta_j(f_j) = f_j \qquad \text{for} \qquad j \geq 3.$$

This observation combined with (6.1.21) gives for $N \geq 3$,

$$\Big\| \sum_{j=3}^{N} f_j \Big\|_{L^p} = \Big\| \sum_{j=3}^{N} \Delta_j(f_j) \Big\|_{L^p} \leq C_p \Big\| \Big(\sum_{j=3}^{N} |f_j|^2 \Big)^{\frac{1}{2}} \Big\|_{L^p} = C_p \|\varphi\|_{L^p} (N-2)^{\frac{1}{2}},$$

where $1 < p < \infty$. On the other hand, (6.1.34) trivially yields that

$$\Big(\sum_{j=3}^{N} \|f_j\|_{L^p}^p \Big)^{\frac{1}{p}} = \|\varphi\|_{L^p} (N-2)^{\frac{1}{p}}.$$

Letting $N \to \infty$ we see that (6.1.33) cannot hold for $p < 2$ even when the f_j's have Fourier transforms supported in disjoint sets.

Example 6.1.10. A similar idea illustrates the necessity of the ℓ^2 norm in (6.1.4). To see this, let Ψ and Δ_j be as in Example 6.1.9. Let us fix $1 < p < \infty$ and $q < 2$. We show that the inequality

$$\Big\|\Big(\sum_{j\in\mathbf{Z}}|\Delta_j(f)|^q\Big)^{\frac{1}{q}}\Big\|_{L^p} \le C_{p,q}\big\|f\big\|_{L^p} \tag{6.1.35}$$

cannot hold. Take $f = \sum_{j=3}^{N} f_j$, where the f_j are as in (6.1.34) and $N \ge 3$. Then the left-hand side of (6.1.35) is bounded from below by $\|\varphi\|_{L^p}(N-2)^{1/q}$, while the right-hand side is bounded above by $\|\varphi\|_{L^p}(N-2)^{1/2}$. Letting $N \to \infty$, we deduce that (6.1.35) is impossible when $q < 2$.

Example 6.1.11. For $1 < p < \infty$ and $2 < q < \infty$, the inequality

$$\big\|g\big\|_{L^p} \le C_{p,q}\Big\|\Big(\sum_{j\in\mathbf{Z}}|\Delta_j(g)|^q\Big)^{\frac{1}{q}}\Big\|_{L^p} \tag{6.1.36}$$

cannot hold even under assumption (6.1.6) on Ψ. Let Δ_j be as in Example 6.1.9. Let us suppose that (6.1.36) did hold for some $q > 2$ for these Δ_j's. Then the self-adjointness of the Δ_j's and duality would give

$$\begin{aligned}
&\Big\|\Big(\sum_{k\in\mathbf{Z}}|\Delta_k(g)|^{q'}\Big)^{\frac{1}{q'}}\Big\|_{L^{p'}}\\
&= \sup_{\big\|\|\{h_k\}_k\|_{\ell^q}\big\|_{L^p}\le 1}\Big|\int_{\mathbf{R}}\sum_{k\in\mathbf{Z}}\Delta_k(g)\,\overline{h_k}\,dx\Big|\\
&\le \big\|g\big\|_{L^{p'}}\sup_{\big\|\|\{h_k\}_k\|_{\ell^q}\big\|_{L^p}\le 1}\Big\|\sum_{k\in\mathbf{Z}}\overline{\Delta_k(h_k)}\Big\|_{L^p}\\
&\le C\big\|g\big\|_{L^{p'}}\sup_{\big\|\|\{h_k\}_k\|_{\ell^q}\big\|_{L^p}\le 1}\Big\|\Big(\sum_{j\in\mathbf{Z}}\Big|\Delta_j\Big(\sum_{k\in\mathbf{Z}}\Delta_k(h_k)\Big)\Big|^q\Big)^{\frac{1}{q}}\Big\|_{L^p} \qquad \text{by (6.1.36)}\\
&\le C'\big\|g\big\|_{L^{p'}}\sup_{\big\|\|\{h_k\}_k\|_{\ell^q}\big\|_{L^p}\le 1}\Big\{\sum_{l=-1}^{1}\Big\|\Big(\sum_{j\in\mathbf{Z}}|\Delta_j\Delta_{j+l}(h_j)|^q\Big)^{\frac{1}{q}}\Big\|_{L^p}\Big\}\\
&\le C''\big\|g\big\|_{L^{p'}}\sup_{\big\|\|\{h_k\}_k\|_{\ell^q}\big\|_{L^p}\le 1}\Big\|\Big(\sum_{j\in\mathbf{Z}}|h_j|^q\Big)^{\frac{1}{q}}\Big\|_{L^p} = C''\big\|g\big\|_{L^{p'}},
\end{aligned}$$

where the next-to-last inequality follows from (6.1.22) applied twice, while the one before that follows from support considerations. But since $q' < 2$, this exactly proves (6.1.35), previously shown to be false, a contradiction.

We conclude that if both assertions (6.1.4) and (6.1.8) of Theorem 6.1.2 were to hold, then the ℓ^2 norm inside the L^p norm could not be replaced by an ℓ^q norm for some $q \ne 2$. Exercise 6.1.6 indicates the crucial use of the fact that ℓ^2 is a Hilbert space in the converse inequality (6.1.8) of Theorem 6.1.2.

Exercises

6.1.1. Construct a Schwartz function Ψ that satisfies $\sum_{j\in\mathbf{Z}} |\widehat{\Psi}(2^{-j}\xi)|^2 = 1$ for all $\xi \in \mathbf{R}^n \setminus \{0\}$ and whose Fourier transform is supported in the annulus $\frac{6}{7} \le |\xi| \le 2$ and is equal to 1 on the annulus $1 \le |\xi| \le \frac{14}{7}$.
[*Hint:* Set $\widehat{\Psi}(\xi) = \eta(\xi)\big(\sum_{k\in\mathbf{Z}} |\eta(2^{-k}\xi)|^2\big)^{-1/2}$ for a suitable $\eta \in \mathscr{C}_0^\infty(\mathbf{R}^n)$.]

6.1.2. Suppose that Ψ is an integrable function on $\mathbf{R}^n$ that satisfies $|\widehat{\Psi}(\xi)| \le B\min(|\xi|^{\varepsilon}, |\xi|^{-\varepsilon'})$ for some $\varepsilon', \varepsilon > 0$. Show that for some constant $C_{\varepsilon,\varepsilon'} < \infty$ we have

$$\sup_{\xi\in\mathbf{R}^n}\bigg(\int_0^\infty |\widehat{\Psi}(t\xi)|^2 \frac{dt}{t}\bigg)^{\frac{1}{2}} + \sup_{\xi\in\mathbf{R}^n}\bigg(\sum_{j\in\mathbf{Z}} |\widehat{\Psi}(2^{-j}\xi)|^2\bigg)^{\frac{1}{2}} \le C_{\varepsilon,\varepsilon'} B.$$

6.1.3. Let Ψ be an integrable function on $\mathbf{R}^n$ with mean value zero that satisfies

$$|\Psi(x)| \le B(1+|x|)^{-n-\varepsilon}, \qquad \int_{\mathbf{R}^n} |\Psi(x-y) - \Psi(x)|\, dx \le B|y|^{\varepsilon'},$$

for some $B, \varepsilon', \varepsilon > 0$ and for all $y \neq 0$.
(a) Prove that $|\widehat{\Psi}(\xi)| \le c_{n,\varepsilon,\varepsilon'} B \min(|\xi|^{\min(\frac{\varepsilon}{2},1)}, |\xi|^{-\varepsilon})$ for some constant $c_{n,\varepsilon,\varepsilon'}$ and conclude that (6.1.4) holds for $p = 2$.
(b) Deduce the validity of (6.1.4) and (6.1.5).
(c) If $\varepsilon < 1$ and the assumption $|\Psi(x)| \le B(1+|x|)^{-n-\varepsilon}$ is weakened to $|\Psi(x)| \le B|x|^{-n-\varepsilon}$ for all $x \in \mathbf{R}^n$, then show that $|\widehat{\Psi}(\xi)| \le c_{n,\varepsilon,\varepsilon'} B \min(|\xi|^{\frac{\varepsilon}{2}}, |\xi|^{-\varepsilon})$ and thus (6.1.4) and (6.1.5) are valid.
[*Hint:* Part (a): Make use of the identity

$$\widehat{\Psi}(\xi) = \int_{\mathbf{R}^n} e^{-2\pi i x\cdot\xi}\Psi(x)\, dx = -\int_{\mathbf{R}^n} e^{-2\pi i x\cdot\xi}\Psi(x-y)\, dx,$$

where $y = \frac{1}{2}\frac{\xi}{|\xi|^2}$ when $|\xi| \ge 1$. For $|\xi| \le 1$ use the mean value property of Ψ to write $\widehat{\Psi}(\xi) = \int_{|x|\le 1} \Psi(x)(e^{-2\pi i x\cdot\xi} - 1)\, dx$ and split the integral in the regions $|x| \le 1$ and $|x| \ge 1$. Part (b): If $\vec{K}$ is defined by (6.1.13), then control the $\ell^2(\mathbf{Z})$ norm by the $\ell^1(\mathbf{Z})$ norm to prove (6.1.16). Then split the sum $\sum_{j\in\mathbf{Z}} \int_{|x|\ge 2|y|} \big|\Psi_{2^{-j}}(x-y) - \Psi_{2^{-j}}(x)\big|\, dx$ into the parts $\sum_{2^j \le |y|^{-1}}$ and $\sum_{2^j > |y|^{-1}}$. Part (c): Notice that when $\varepsilon < 1$, we have $|\int_{|x|\le 1} \Psi(x)(e^{-2\pi i x\cdot\xi} - 1)\, dx| \le C_n B|\xi|^{\frac{\varepsilon}{2}}$.]

6.1.4. Let Ψ be an integrable function on $\mathbf{R}^n$ with mean value zero that satisfies

$$|\Psi(x)| \le B(1+|x|)^{-n-\varepsilon}, \qquad \int_{\mathbf{R}^n} |\Psi(x-y) - \Psi(x)|\, dx \le B|y|^{\varepsilon'},$$

for some $B, \varepsilon', \varepsilon > 0$ and for all $y \neq 0$. Let $\Psi_t(x) = t^{-n}\Psi(x/t)$. (a) Prove that there are constants c_n, c_n' such that

$$\left(\int_0^\infty |\Psi_t(x)|^2 \frac{dt}{t} dx \right)^{\frac{1}{2}} \leq c_n B |x|^{-n},$$

$$\sup_{y \in \mathbf{R}^n \setminus \{0\}} \int_{|x| \geq 2|y|} \left(\int_0^\infty |\Psi_t(x-y) - \Psi_t(x)|^2 \frac{dt}{t} \right)^{\frac{1}{2}} dx \leq c_n' B.$$

(b) Show that there exist constants C_n, C_n' such that for all $1 < p < \infty$ and for all $f \in L^p(\mathbf{R}^n)$ we have

$$\left\| \left(\int_0^\infty |f * \Psi_t|^2 \frac{dt}{t} \right)^{\frac{1}{2}} \right\|_{L^p(\mathbf{R}^n)} \leq C_n B \max(p, (p-1)^{-1}) \|f\|_{L^p(\mathbf{R}^n)}$$

and also for all $f \in L^1(\mathbf{R}^n)$ we have

$$\left\| \left(\int_0^\infty |f * \Psi_t|^2 \frac{dt}{t} \right)^{\frac{1}{2}} \right\|_{L^{1,\infty}(\mathbf{R}^n)} \leq C_n' B \|f\|_{L^1(\mathbf{R}^n)}.$$

(c) Under the additional hypothesis that $0 < \int_0^\infty |\widehat{\Psi}(t\xi)|^2 \frac{dt}{t} = c_0$ for all $\xi \in \mathbf{R}^n \setminus \{0\}$, prove that for all $f \in L^p(\mathbf{R}^n)$ we have

$$\|f\|_{L^p(\mathbf{R}^n)} \leq C_n'' B \max(p, (p-1)^{-1}) \left\| \left(\int_0^\infty |f * \Psi_t|^2 \frac{dt}{t} \right)^{\frac{1}{2}} \right\|_{L^p(\mathbf{R}^n)}$$

[*Hint:* Part (a): Use the Cauchy-Schwarz inequality to obtain

$$\int_{|x| \geq 2|y|} \left(\int_0^\infty |\Psi_t(x-y) - \Psi_t(x)|^2 \frac{dt}{t} \right)^{\frac{1}{2}} dx$$
$$\leq c_n |y|^{-\frac{\varepsilon}{2}} \left(\int_{|x| \geq 2|y|} |x|^{n+\varepsilon} \int_0^\infty |\Psi_t(x-y) - \Psi_t(x)|^2 \frac{dt}{t} dx \right)^{\frac{1}{2}},$$

and split the integral on the right into the regions $t \leq |y|$ and $t > |y|$. In the second region use that Ψ is bounded to replace the square by the first power. Part (b): Use Exercise 6.1.2 and part (a) of Exercise 6.1.3 and to deduce the inequality when $p = 2$. Then apply Theorem 5.6.1. Part (c): Prove the inequality first for $f \in \mathscr{S}(\mathbf{R}^n)$ using duality.]

6.1.5. Prove the following generalization of Theorem 6.1.2. Let $A > 0$. Suppose that $\{K_j\}_{j \in \mathbf{Z}}$ is a sequence of locally integrable functions on $\mathbf{R}^n \setminus \{0\}$ that satisfies

$$\sup_{x \neq 0} |x|^n \Big(\sum_{j \in \mathbf{Z}} |K_j(x)|^2 \Big)^{\frac{1}{2}} \leq A,$$

$$\sup_{y \in \mathbf{R}^n \setminus \{0\}} \int_{|x| \geq 2|y|} \Big(\sum_{j \in \mathbf{Z}} |K_j(x-y) - K_j(x)|^2 \Big)^{\frac{1}{2}} dx \leq A < \infty,$$

and for each $j \in \mathbf{Z}$ there is a number L_j such that

$$\lim_{\varepsilon_k \downarrow 0} \int_{\varepsilon_k \le |y| \le 1} K_j(y)\,dy = L_j\,.$$

If the K_j coincide with tempered distributions W_j that satisfy

$$\sum_{j\in\mathbf{Z}} |\widehat{W_j}(\xi)|^2 \le B^2\,,$$

then the operator

$$f \to \Big(\sum_{j\in\mathbf{Z}} |K_j * f|^2\Big)^{\frac{1}{2}}$$

maps $L^p(\mathbf{R}^n)$ to itself and is weak type $(1,1)$ norms at most multiples of $A+B$.

6.1.6. Suppose that $\mathscr{H}$ is a Hilbert space with inner product $\langle\cdot\,,\cdot\rangle_{\mathscr{H}}$. Let $A>0$ and $1<p<\infty$. Suppose that an operator T from $L^2(\mathbf{R}^n) \to L^2(\mathbf{R}^n,\mathscr{H})$ is a multiple of an isometry, that is,

$$\big\|T(g)\big\|_{L^2(\mathbf{R}^n,\mathscr{H})} = A\big\|g\big\|_{L^2(\mathbf{R}^n)}$$

for all $g \in L^2(\mathbf{R}^n,\mathscr{H})$. Then the inequality $\|T(f)\|_{L^p(\mathbf{R}^n,\mathscr{H})} \le C_p\|f\|_{L^p(\mathbf{R}^n)}$ for all $f \in \mathscr{S}(\mathbf{R}^n)$ implies

$$\big\|f\big\|_{L^{p'}(\mathbf{R}^n)} \le C_{p'}A^{-2}\big\|T(f)\big\|_{L^{p'}(\mathbf{R}^n,\mathscr{H})}$$

for all in $f \in \mathscr{S}(\mathbf{R}^n)$.
[*Hint:* Use the inner product structure and polarization to obtain

$$A^2\left|\int_{\mathbf{R}^n} f(x)\overline{g(x)}\,dx\right| = \left|\int_{\mathbf{R}^n} \Big\langle T(f)(x), T(g)(x)\Big\rangle_{\mathscr{H}}\,dx\right|$$

and then argue as in the proof of inequality (6.1.8).]

6.1.7. Suppose that $\{m_j\}_{j\in\mathbf{Z}}$ is a sequence of bounded functions supported in the intervals $[2^j,2^{j+1}]$. Let $T_j(f) = (\widehat{f}m_j)^\vee$ be the corresponding multiplier operators. Assume that for all sequences of functions $\{f_j\}_j$ the vector-valued inequality

$$\Big\|\Big(\sum_j |T_j(f_j)|^2\Big)^{\frac{1}{2}}\Big\|_{L^p} \le A_p\Big\|\Big(\sum_j |f_j|^2\Big)^{\frac{1}{2}}\Big\|_{L^p}$$

is valid for some $1<p<\infty$. Prove there is a $C_p>0$ such that for all finite subsets S of $\mathbf{Z}$ we have

$$\Big\|\sum_{j\in S} m_j\Big\|_{\mathscr{M}_p} \le C_pA_p.$$

[*Hint:* Use that $\big\langle \sum_{j\in S} T_j(f), g\big\rangle = \sum_{j\in S}\langle \Delta_j^\# T_j(f), \Delta_j^\#(g)\rangle$.]

6.1.8. Let m be a bounded function on $\mathbf{R}^n$ that is supported in the annulus $1 \le |\xi| \le 2$ and define $T_j(f) = \big(\widehat{f}(\xi)m(2^{-j}\xi)\big)^\vee$. Suppose that the square function $f \mapsto \big(\sum_{j\in\mathbf{Z}} |T_j(f)|^2\big)^{1/2}$ is bounded on $L^p(\mathbf{R}^n)$ for some $1 < p < \infty$. Show that for every finite subset S of the integers we have

$$\Big\|\sum_{j\in S} T_j(f)\Big\|_{L^p(\mathbf{R}^n)} \le C_{p,n}\big\|f\big\|_{L^p(\mathbf{R}^n)}$$

for some constant $C_{p,n}$ independent of S.

6.1.9. Fix a nonzero Schwartz function h on the line whose Fourier transform is supported in the interval $\big[-\frac{1}{8},\frac{1}{8}\big]$. For $\{a_j\}$ a sequence of numbers, set

$$f(x) = \sum_{j=1}^{\infty} a_j e^{2\pi i 2^j x} h(x)\,.$$

Prove that for all $1 < p < \infty$ there exists a constant C_p such that

$$\|f\|_{L^p(\mathbf{R})} \le C_p\Big(\sum_j |a_j|^2\Big)^{\frac{1}{2}}\|h\|_{L^p}\,.$$

$\big[$*Hint:* Write $f = \sum_{j=1}^{\infty} \Delta_j(a_j e^{2\pi i 2^j(\cdot)}h)$, where Δ_j is given by convolution with $\varphi_{2^{-j}}$ for some φ whose Fourier transform is supported in the interval $\big[\frac{6}{8},\frac{10}{8}\big]$ and is equal to 1 on $\big[\frac{7}{8},\frac{9}{8}\big]$. Then use (6.1.21).$\big]$

6.1.10. Let Ψ be a Schwartz function whose Fourier transform is supported in the annulus $\frac{1}{2} \le |\xi| \le 2$ and that satisfies (6.1.7). Define a Schwartz function Φ by setting

$$\widehat{\Phi}(\xi) = \begin{cases} \sum_{j\le 0} \widehat{\Psi}(2^{-j}\xi) & \text{when } \xi \neq 0,\\ 1 & \text{when } \xi = 0.\end{cases}$$

Let S_0 be the operator given by convolution with Φ. Let $1 < p < \infty$ and $f \in L^p(\mathbf{R}^n)$. Show that

$$\big\|f\big\|_{L^p} \approx \big\|S_0(f)\big\|_{L^p} + \Big\|\Big(\sum_{j=1}^{\infty} |\Delta_j(f)|^2\Big)^{\frac{1}{2}}\Big\|_{L^p}.$$

$\big[$*Hint:* Use Theorem 6.1.2 together with the identity $S_0 + \sum_{j=1}^{\infty}\Delta_j = I$.$\big]$

6.2 Two Multiplier Theorems

We now return to the spaces $\mathscr{M}_p$ introduced in Section 2.5. We seek sufficient conditions on L^∞ functions defined on $\mathbf{R}^n$ to be elements of $\mathscr{M}_p$. In this section we are concerned with two fundamental theorems that provide such sufficient conditions.

These are the Marcinkiewicz and the Hörmander–Mihlin multiplier theorems. Both multiplier theorems are consequences of the Littlewood–Paley theory discussed in the previous section.

Using the dyadic decomposition of $\mathbf{R}^n$, we can write any L^∞ function m as the sum

$$m = \sum_{\mathbf{j}\in\mathbf{Z}^n} m\chi_{R_{\mathbf{j}}} \qquad \text{a.e.,}$$

where $\mathbf{j} = (j_1,\dots,j_n)$, $R_{\mathbf{j}} = I_{j_1}\times\cdots\times I_{j_n}$, and $I_k = [2^k,2^{k+1})\bigcup(-2^{k+1},-2^k]$. For $\mathbf{j}$ in $\mathbf{Z}^n$ we set $m_{\mathbf{j}} = m\chi_{R_{\mathbf{j}}}$. A consequence of the ideas developed so far is the following characterization of $\mathscr{M}_p(\mathbf{R}^n)$ in terms of a vector-valued inequality.

Proposition 6.2.1. *Let $m\in L^\infty(\mathbf{R}^n)$ and let $m_{\mathbf{j}} = m\chi_{R_{\mathbf{j}}}$. Then m lies in $\mathscr{M}_p(\mathbf{R}^n)$, that is, for some c_p we have*

$$\big\|(\widehat{f}m)^\vee\big\|_{L^p} \le c_p\big\|f\big\|_{L^p}, \qquad f\in L^p(\mathbf{R}^n),$$

if and only if for some $C_p>0$ we have

$$\Big\|\Big(\sum_{\mathbf{j}\in\mathbf{Z}^n}|(\widehat{f_{\mathbf{j}}}m_{\mathbf{j}})^\vee|^2\Big)^{\frac12}\Big\|_{L^p} \le C_p\Big\|\Big(\sum_{\mathbf{j}\in\mathbf{Z}^n}|f_{\mathbf{j}}|^2\Big)^{\frac12}\Big\|_{L^p} \tag{6.2.1}$$

for all sequences of functions $f_{\mathbf{j}}$ in $L^p(\mathbf{R}^n)$.

Proof. Suppose that $m\in\mathscr{M}_p(\mathbf{R}^n)$. Exercise 5.6.1 gives the first inequality below

$$\Big\|\Big(\sum_{\mathbf{j}\in\mathbf{Z}^n}|(\chi_{R_{\mathbf{j}}}m\widehat{f_{\mathbf{j}}})^\vee|^2\Big)^{\frac12}\Big\|_{L^p} \le C_p\Big\|\Big(\sum_{\mathbf{j}\in\mathbf{Z}^n}|(m\widehat{f_{\mathbf{j}}})^\vee|^2\Big)^{\frac12}\Big\|_{L^p} \le C_p\Big\|\Big(\sum_{\mathbf{j}\in\mathbf{Z}^n}|f_{\mathbf{j}}|^2\Big)^{\frac12}\Big\|_{L^p},$$

while the second inequality follows from Theorem 5.5.1. (Observe that when $p=q$ in Theorem 5.5.1, then $C_{p,q}=1$.) Conversely, suppose that (6.2.1) holds for all sequences of functions $f_{\mathbf{j}}$. Fix a function f and apply (6.2.1) to the sequence $(\widehat{f}\chi_{R_{\mathbf{j}}})^\vee$, where $R_{\mathbf{j}}$ is the dyadic rectangle indexed by $\mathbf{j}=(j_1,\dots,j_n)\in\mathbf{Z}^n$. We obtain

$$\Big\|\Big(\sum_{\mathbf{j}\in\mathbf{Z}^n}|(\widehat{f}m\chi_{R_{\mathbf{j}}})^\vee|^2\Big)^{\frac12}\Big\|_{L^p} \le C_p\Big\|\Big(\sum_{\mathbf{j}\in\mathbf{Z}^n}|(\widehat{f}\chi_{R_{\mathbf{j}}})^\vee|^2\Big)^{\frac12}\Big\|_{L^p}.$$

Using Theorem 6.1.6, we obtain that the previous inequality is equivalent to the inequality

$$\big\|(\widehat{f}m)^\vee\big\|_{L^p} \le c_p\big\|f\big\|_{L^p},$$

which implies that $m\in\mathscr{M}_p(\mathbf{R}^n)$. □

6.2.1 The Marcinkiewicz Multiplier Theorem on R

Proposition 6.2.1 suggests that the behavior of m on each dyadic rectangle R_j should play a crucial role in determining whether m is an L^p multiplier. The Marcinkiewicz multiplier theorem provides such sufficient conditions on m restricted to any dyadic rectangle R_j. Before stating this theorem, we illustrate its main idea via the following example. Suppose that m is a bounded function that vanishes near $-\infty$, that is differentiable at every point, and whose derivative is integrable. Then we may write

$$m(\xi) = \int_{-\infty}^{\xi} m'(t)\,dt = \int_{-\infty}^{+\infty} \chi_{[t,\infty)}(\xi) m'(t)\,dt\,,$$

from which it follows that for a Schwartz function f we have

$$(\widehat{f}m)^{\vee} = \int_{\mathbf{R}} (\widehat{f}\chi_{[t,\infty)})^{\vee} m'(t)\,dt.$$

Since the operators $f \mapsto (\widehat{f}\chi_{[t,\infty)})^{\vee}$ map $L^p(\mathbf{R})$ to itself independently of t, it follows that

$$\big\|(\widehat{f}m)^{\vee}\big\|_{L^p} \le C_p \big\|m'\big\|_{L^1} \big\|f\big\|_{L^p}\,,$$

thus yielding that m is in $\mathscr{M}_p(\mathbf{R})$. The next multiplier theorem is an improvement of this result and is based on the Littlewood–Paley theorem. We begin with the one-dimensional case, which already captures the main ideas.

Theorem 6.2.2. *(Marcinkiewicz multiplier theorem) Let $m:\ \mathbf{R} \to \mathbf{R}$ be a bounded function that is $\mathscr{C}^1$ in every dyadic set $(2^j, 2^{j+1}) \bigcup (-2^{j+1}, -2^j)$ for $j \in \mathbf{Z}$. Assume that the derivative m' of m satisfies*

$$\sup_j \left[\int_{-2^{j+1}}^{-2^j} |m'(\xi)|\,d\xi + \int_{2^j}^{2^{j+1}} |m'(\xi)|\,d\xi \right] \le A < \infty. \tag{6.2.2}$$

Then for all $1 < p < \infty$ we have that $m \in \mathscr{M}_p(\mathbf{R})$ and for some $C > 0$ we have

$$\big\|m\big\|_{\mathscr{M}_p(\mathbf{R})} \le C \max\big(p, (p-1)^{-1}\big)^6 \big(\big\|m\big\|_{L^\infty} + A\big). \tag{6.2.3}$$

Proof. Since the function m has an integrable derivative on $(2^j, 2^{j+1})$, it has bounded variation in this interval and hence it is a difference of two increasing functions. Therefore, m has left and right limits at the points 2^j and 2^{j+1}, and by redefining m at these points we may assume that m is right continuous at the points 2^j and left continuous at the points -2^j.

Set $I_j = [2^j, 2^{j+1}) \cup (-2^{j+1}, -2^j]$ and $I_j^+ = [2^j, 2^{j+1})$ whenever $j \in \mathbf{Z}$. Given an interval I in $\mathbf{R}$, we introduce an operator Δ_I defined by $\Delta_I(f) = (\widehat{f}\chi_I)^{\vee}$. With this notation $\Delta_{I_j^+}(f)$ is "half" of the operator $\Delta_j^\#$ introduced in the previous section. Given m as in the statement of the theorem, we write $m(\xi) = m_+(\xi) + m_-(\xi)$, where $m_+(\xi) = m(\xi)\chi_{\xi \ge 0}$ and $m_-(\xi) = m(\xi)\chi_{\xi < 0}$. We show that both m_+ and m_-

are L^p multipliers. Since m' is integrable over all intervals of the form $[2^j,\xi]$ when $2^j \le \xi < 2^{j+1}$, the fundamental theorem of calculus gives

$$m(\xi) = m(2^j) + \int_{2^j}^{\xi} m'(t)\,dt, \qquad \text{for } 2^j \le \xi < 2^{j+1},$$

from which it follows that for a Schwartz function f on the real line we have

$$m(\xi)\widehat{f}(\xi)\chi_{I_j^+}(\xi) = m(2^j)\widehat{f}(\xi)\chi_{I_j^+}(\xi) + \int_{2^j}^{2^{j+1}} \widehat{f}(\xi)\chi_{[t,\infty)}(\xi)\chi_{I_j^+}(\xi)\,m'(t)\,dt\,.$$

We therefore obtain the identity

$$(\widehat{f}\chi_{I_j}m_+)^\vee = (\widehat{f}m\chi_{I_j^+})^\vee = m(2^j)\Delta_{I_j^+}(f) + \int_{2^j}^{2^{j+1}} \Delta_{[t,\infty)}\Delta_{I_j^+}(f)\,m'(t)\,dt\,,$$

which implies that

$$|(\widehat{f}\chi_{I_j}m_+)^\vee| \le \|m\|_{L^\infty}|\Delta_{I_j^+}(f)| + A^{\frac12}\bigg(\int_{2^j}^{2^{j+1}} \big|\Delta_{[t,\infty)}\Delta_{I_j^+}(f)\big|^2\,|m'(t)|\,dt\bigg)^{\frac12},$$

using the hypothesis (6.2.2). Taking $\ell^2(\mathbf{Z})$ norms we obtain

$$\begin{aligned}\Big(\sum_{j\in\mathbf{Z}}|(\widehat{f}\chi_{I_j}m_+)^\vee|^2\Big)^{\frac12} \le \|m\|_{L^\infty}\Big(\sum_{j\in\mathbf{Z}}|\Delta_{I_j^+}(f)|^2\Big)^{\frac12}& \\ +A^{\frac12}\bigg(\int_0^\infty \big|\Delta_{[t,\infty)}\Delta^{\#}_{[\log_2 t]}(f)\big|^2\,|m'(t)|\,dt\bigg)^{\frac12}&.\end{aligned}$$

Exercise 5.6.2 gives

$$\begin{aligned}&A^{\frac12}\bigg\|\bigg(\int_0^\infty \big|\Delta_{[t,\infty)}\Delta^{\#}_{[\log_2 t]}(f)\big|^2|m'(t)|\,dt\bigg)^{\frac12}\bigg\|_{L^p} \\ &\qquad\le C\max(p,(p-1)^{-1})A^{\frac12}\bigg\|\bigg(\int_0^\infty \big|\Delta^{\#}_{[\log_2 t]}(f)\big|^2|m'(t)|\,dt\bigg)^{\frac12}\bigg\|_{L^p},\end{aligned}$$

while the hypothesis on m' implies the inequality

$$\bigg\|\Big(\sum_{j\in\mathbf{Z}}|\Delta_{I_j^+}(f)|^2\int_{I_j^+}|m'(t)|\,dt\Big)^{\frac12}\bigg\|_{L^p} \le A^{\frac12}\bigg\|\Big(\sum_j|\Delta_{I_j^+}(f)|^2\Big)^{\frac12}\bigg\|_{L^p}.$$

Using Theorem 6.1.5 we obtain that

$$\bigg\|\Big(\sum_j|\Delta_{I_j^+}(f)|^2\Big)^{\frac12}\bigg\|_{L^p} \le C'\max(p,(p-1)^{-1})^2\big\|(\widehat{f}\chi_{(0,\infty)})^\vee\big\|_{L^p},$$

and the latter is at most a constant multiple of $\max(p,(p-1)^{-1})^3\|f\|_{L^p}$. Putting things together we deduce that

$$\Big\|\Big(\sum_j |(\widehat{f}\chi_{I_j}m_+)^\vee|^2\Big)^{\frac{1}{2}}\Big\|_{L^p} \le C'' \max(p,(p-1)^{-1})^4 (A+\|m\|_{L^\infty})\|f\|_{L^p}, \quad (6.2.4)$$

from which we obtain the estimate

$$\|(\widehat{f}m_+)^\vee\|_{L^p} \le C\max(p,(p-1)^{-1})^6 (A+\|m\|_{L^\infty})\|f\|_{L^p},$$

using the lower estimate of Theorem 6.1.5. This proves (6.2.3) for m_+. A similar argument also works for m_-, and this concludes the proof by summing the corresponding estimates for m_+ and m_-. □

We remark that the same proof applies under the more general assumption that m is a function of bounded variation on every interval $[2^j,2^{j+1}]$ and $[-2^{j+1},-2^j]$. In this case the measure $|m'(t)|\,dt$ should be replaced by the total variation $|dm(t)|$ of the Lebesgue–Stieltjes measure $dm(t)$.

Example 6.2.3. Any bounded function that is constant on dyadic intervals is an L^p multiplier. Also, the function

$$m(\xi) = |\xi| 2^{-[\log_2|\xi|]}$$

is an L^p multiplier on $\mathbf{R}$ for $1<p<\infty$.

6.2.2 The Marcinkiewicz Multiplier Theorem on $\mathbf{R}^n$

We now extend this theorem on $\mathbf{R}^n$. As usual we denote the coordinates of a point $\xi\in\mathbf{R}^n$ by $(\xi_1,\dots,\xi_n)$. We recall the notation $I_j = (-2^{j+1},-2^j]\bigcup[2^j,2^{j+1})$ and $R_{\mathbf{j}} = I_{j_1}\times\cdots\times I_{j_n}$ whenever $\mathbf{j}=(j_1,\dots,j_n)\in\mathbf{Z}^n$.

Theorem 6.2.4. *Let m be a bounded function on $\mathbf{R}^n$ such that for all $\alpha=(\alpha_1,\dots,\alpha_n)$ with $|\alpha_1|,\dots,|\alpha_n|\le 1$ the derivatives $\partial^\alpha m$ are continuous up to the boundary of $R_{\mathbf{j}}$ for all $\mathbf{j}\in\mathbf{Z}^n$. Assume that there is a constant $A<\infty$ such that for all partitions $\{s_1,\dots,s_k\}\bigcup\{r_1,\dots,r_\ell\}=\{1,2,\dots,n\}$ with $n=k+\ell$ and all $\xi\in R_{\mathbf{j}}$ we have*

$$\sup_{\xi_{r_1}\in I_{j_{r_1}}}\cdots\sup_{\xi_{r_\ell}\in I_{j_{r_\ell}}}\int_{I_{j_{s_1}}}\cdots\int_{I_{j_{s_k}}}\big|(\partial_{s_1}\cdots\partial_{s_k}m)(\xi_1,\dots,\xi_n)\big|\,d\xi_{s_k}\cdots d\xi_{s_1}\le A \quad (6.2.5)$$

for all $\mathbf{j}=(j_1,\dots,j_n)\in\mathbf{Z}^n$. Then m is in $\mathscr{M}_p(\mathbf{R}^n)$ whenever $1<p<\infty$ and there is a constant $C_n<\infty$ such that

$$\|m\|_{\mathscr{M}_p(\mathbf{R}^n)} \le C_n(A+\|m\|_{L^\infty})\max\big(p,(p-1)^{-1}\big)^{6n}. \quad (6.2.6)$$

Proof. We prove this theorem only in dimension $n = 2$, since the general case presents no substantial differences but only some notational inconvenience. We decompose the given function m as

$$m(\xi) = m_{++}(\xi) + m_{-+}(\xi) + m_{+-}(\xi) + m_{--}(\xi),$$

where each of the last four terms is supported in one of the four quadrants. For instance, the function $m_{+-}(\xi_1,\xi_2)$ is supported in the quadrant $\xi_1 \ge 0$ and $\xi_2 < 0$. As in the one-dimensional case, we work with each of these pieces separately. By symmetry we choose to work with m_{++} in the following argument.

Using the fundamental theorem of calculus, we obtain the following simple identity, valid for $2^{j_1} \le \xi_1 < 2^{j_1+1}$ and $2^{j_2} \le \xi_2 < 2^{j_2+1}$:

$$\begin{aligned} m(\xi_1,\xi_2) &= m(2^{j_1},2^{j_2}) + \int_{2^{j_1}}^{\xi_1} (\partial_1 m)(t_1,2^{j_2})\,dt_1 \\ &\quad + \int_{2^{j_2}}^{\xi_2} (\partial_2 m)(2^{j_1},t_2)\,dt_2 \\ &\quad + \int_{2^{j_1}}^{\xi_1}\int_{2^{j_2}}^{\xi_2} (\partial_1\partial_2 m)(t_1,t_2)\,dt_2\,dt_1 . \end{aligned} \tag{6.2.7}$$

We introduce operators $\Delta_I^{(r)}$, $r \in \{1,2\}$, acting in the rth variable (with the other variable remaining fixed) given by multiplication on the Fourier transform side by the characteristic function of the interval I. Likewise, we introduce operators $\Delta_j^{\#(r)}$, $r \in \{1,2\}$ (also acting in the rth variable), given by multiplication on the Fourier transform side by the characteristic function of the set $(-2^{j+1},-2^j] \bigcup [2^j,2^{j+1})$. For notational convenience, for a given Schwartz function f we write

$$f_{++} = \big(\widehat{f}\chi_{(0,\infty)^2}\big)^\vee,$$

and likewise we define f_{+-}, f_{-+}, and f_{--}.

Multiplying both sides of (6.2.7) by the function $\widehat{f}\chi_{R_{\mathrm{j}}}\chi_{(0,\infty)^2}$ and taking inverse Fourier transforms yields

$$\begin{aligned} (\widehat{f}\chi_{R_{\mathrm{j}}} m_{++})^\vee &= m(2^{j_1},2^{j_2})\Delta_{j_1}^{\#(1)}\Delta_{j_2}^{\#(2)}(f_{++}) \\ &+ \int_{2^{j_1}}^{2^{j_1+1}} \Delta_{j_2}^{\#(2)}\Delta_{[t_1,\infty)}^{(1)}\Delta_{j_1}^{\#(1)}(f_{++})\,(\partial_1 m)(t_1,2^{j_2})\,dt_1 \\ &+ \int_{2^{j_2}}^{2^{j_2+1}} \Delta_{j_1}^{\#(1)}\Delta_{[t_2,\infty)}^{(2)}\Delta_{j_2}^{\#(2)}(f_{++})\,(\partial_2 m)(2^{j_1},t_2)\,dt_2 \\ &+ \int_{2^{j_1}}^{2^{j_1+1}}\int_{2^{j_2}}^{2^{j_2+1}} \Delta_{[t_1,\infty)}^{(1)}\Delta_{j_1}^{\#(1)}\Delta_{[t_2,\infty)}^{(2)}\Delta_{j_2}^{\#(2)}(f_{++})\,(\partial_1\partial_2 m)(t_1,t_2)\,dt_2\,dt_1 . \end{aligned} \tag{6.2.8}$$

We apply the Cauchy–Schwarz inequality in the last three terms of (6.2.8) with respect to the measures $|(\partial_1 m)(t_1,2^{j_2})|dt_1$, $|(\partial_2 m)(2^{j_1},t_2)|dt_2$, $|(\partial_1\partial_2 m)(t_1,t_2)|\,dt_2dt_1$ and we use hypothesis (6.2.5) to deduce

$$
\begin{aligned}
\big|(\widehat{f}\chi_{R_{\mathbf{j}}}m_{++})^\vee\big| &\le \|m\|_{L^\infty}\big|\Delta_{j_1}^{\#(1)}\Delta_{j_2}^{\#(2)}(f_{++})\big| \\
&+ A^{\frac{1}{2}}\bigg(\int_{2^{j_1}}^{2^{j_1+1}} \big|\Delta_{j_2}^{\#(2)}\Delta_{[t_1,\infty)}^{(1)}\Delta_{j_1}^{\#(1)}(f_{++})\big|^2\,|(\partial_1 m)(t_1,2^{j_2})|\,dt_1\bigg)^{\frac{1}{2}} \\
&+ A^{\frac{1}{2}}\bigg(\int_{2^{j_2}}^{2^{j_2+1}} \big|\Delta_{j_1}^{\#(1)}\Delta_{[t_2,\infty)}^{(2)}\Delta_{j_2}^{\#(2)}(f_{++})\big|^2\,|(\partial_2 m)(2^{j_1},t_2)|\,dt_2\bigg)^{\frac{1}{2}} \\
&+ A^{\frac{1}{2}}\bigg(\int_{2^{j_1}}^{2^{j_1+1}}\int_{2^{j_2}}^{2^{j_2+1}} \big|\Delta_{[t_1,\infty)}^{(1)}\Delta_{j_1}^{\#(1)}\Delta_{[t_2,\infty)}^{(2)}\Delta_{j_2}^{\#(2)}(f_{++})\big|^2\,|(\partial_1\partial_2 m)(t_1,t_2)|\,dt_2\,dt_1\bigg)^{\frac{1}{2}}.
\end{aligned}
$$

Both sides of the preceding inequality are sequences indexed by $\mathbf{j}\in\mathbf{Z}^2$. We apply $\ell^2(\mathbf{Z}^2)$ norms and use Minkowski's inequality to deduce the pointwise estimate

$$
\begin{aligned}
\Big(\sum_{\mathbf{j}\in\mathbf{Z}^2}\big|(\widehat{f}\chi_{R_{\mathbf{j}}}m_{++})^\vee\big|^2\Big)^{\frac{1}{2}} &\le \|m\|_{L^\infty}\Big(\sum_{\mathbf{j}\in\mathbf{Z}^2}\big|\Delta_{\mathbf{j}}^{\#}(f_{++})\big|^2\Big)^{\frac{1}{2}} \\
&+A^{\frac{1}{2}}\bigg(\int_0^\infty\!\!\int_0^\infty \big|\Delta_{[t_1,\infty)}^{(1)}\Delta_{[\log_2 t_2]}^{\#(2)}\Delta_{[\log_2 t_1]}^{\#(1)}(f_{++})\big|^2\,\big|(\partial_1 m)(t_1,2^{[\log_2 t_2]})\big|\,dt_1 d\nu(t_2)\bigg)^{\frac{1}{2}} \\
&+A^{\frac{1}{2}}\bigg(\int_0^\infty\!\!\int_0^\infty \big|\Delta_{[t_2,\infty)}^{(2)}\Delta_{[\log_2 t_1]}^{\#(1)}\Delta_{[\log_2 t_2]}^{\#(2)}(f_{++})\big|^2\,\big|(\partial_2 m)(2^{[\log_2 t_1]},t_2)\big|\,d\nu(t_1)dt_2\bigg)^{\frac{1}{2}} \\
&+A^{\frac{1}{2}}\bigg(\int_0^\infty\!\!\int_0^\infty \big|\Delta_{[t_1,\infty)}^{(1)}\Delta_{[t_2,\infty)}^{(2)}\Delta_{[\log_2 t_1]}^{\#(1)}\Delta_{[\log_2 t_2]}^{\#(2)}(f_{++})\big|^2\big|(\partial_1\partial_2 m)(t_1,t_2)\big|\,dt_1dt_2\bigg)^{\frac{1}{2}},
\end{aligned}
$$

where ν is the counting measure $\sum_{j\in\mathbf{Z}}\delta_{2^j}$ defined by $\nu(A)=\#\{j\in\mathbf{Z}:\,2^j\in A\}$ for subsets A of $(0,\infty)$. We now take $L^p(\mathbf{R}^2)$ norms and we estimate separately the contribution of each of the four terms on the right side. Using Exercise 5.6.2 we obtain

$$
\begin{aligned}
\bigg\|\Big(\sum_{\mathbf{j}\in\mathbf{Z}^2}\big|(\widehat{f}\chi_{R_{\mathbf{j}}}m_{++})^\vee\big|^2\Big)^{\frac{1}{2}}\bigg\|_{L^p} &\le \|m\|_{L^\infty}\bigg\|\Big(\sum_{\mathbf{j}\in\mathbf{Z}^2}\big|\Delta_{\mathbf{j}}^{\#}(f_{++})\big|^2\Big)^{\frac{1}{2}}\bigg\|_{L^p} \\
&+ C_2A^{\frac{1}{2}}\max\big(p,(p-1)^{-1}\big)^2 \\
&\times\Bigg\{\bigg\|\bigg(\int_0^\infty\!\!\int_0^\infty \big|\Delta_{[\log_2 t_2]}^{\#(2)}\Delta_{[\log_2 t_1]}^{\#(1)}(f_{++})\big|^2\,\big|(\partial_1 m)(t_1,2^{[\log_2 t_2]})\big|\,dt_1\,d\nu(t_2)\bigg)^{\frac{1}{2}}\bigg\|_{L^p} \\
&+\bigg\|\bigg(\int_0^\infty\!\!\int_0^\infty \big|\Delta_{[\log_2 t_1]}^{\#(1)}\Delta_{[\log_2 t_2]}^{\#(2)}(f_{++})\big|^2\,\big|(\partial_2 m)(2^{[\log_2 t_1]},t_2)\big|\,d\nu(t_1)\,dt_2\bigg)^{\frac{1}{2}}\bigg\|_{L^p} \\
&+\bigg\|\bigg(\int_0^\infty\!\!\int_0^\infty \big|\Delta_{[\log_2 t_1]}^{\#(1)}\Delta_{[\log_2 t_2]}^{\#(2)}(f_{++})\big|^2\big|(\partial_1\partial_2 m)(t_1,t_2)\big|\,dt_1dt_2\bigg)^{\frac{1}{2}}\bigg\|_{L^p}\Bigg\}.
\end{aligned}
$$

But the functions $(t_1,t_2) \mapsto \Delta^{\#(1)}_{[\log_2 t_1]}\Delta^{\#(2)}_{[\log_2 t_2]}(f_{++})$ are constant on products of intervals of the form $[2^{j_1},2^{j_1+1}) \times [2^{j_2},2^{j_2+1})$; hence using hypothesis (6.2.5) again we deduce the estimate

$$\begin{aligned}
&\Big\|\Big(\sum_{\mathbf{j}\in\mathbf{Z}^2} |(\widehat{f}\chi_{R_{\mathbf{j}}} m_{++})^\vee|^2\Big)^{\frac12}\Big\|_{L^p(\mathbf{R}^2)} \\
&\le C_2\big(\|m\|_{L^\infty}+A\big)\max\big(p,(p-1)^{-1}\big)^2\Big\|\Big(\sum_{\mathbf{j}\in\mathbf{Z}^2} |\Delta^\#_{\mathbf{j}}(f_{++})|^2\Big)^{\frac12}\Big\|_{L^p(\mathbf{R}^2)} \\
&\le C_2\big(\|m\|_{L^\infty}+A\big)\max\big(p,(p-1)^{-1}\big)^6\big\|(\widehat{f}\chi_{(0,\infty)^2})^\vee\big\|_{L^p(\mathbf{R}^2)} \\
&\le C_2\big(\|m\|_{L^\infty}+A\big)\max\big(p,(p-1)^{-1}\big)^8\|f\|_{L^p(\mathbf{R}^2)},
\end{aligned}$$

where the penultimate estimate follows from Theorem 6.1.6 and the last estimate by the boundedness of the Hilbert transform (Theorem 5.1.7). We now appeal to inequality (6.1.29) which yields the required estimate for the $L^p(\mathbf{R}^2)$ norm of $(\widehat{f} m_{++})^\vee$. A similar argument also works for the remaining parts of m_{+-}, m_{-+}, m_{--}, and summing concludes the proof of (6.2.6).

The analogous estimate on $\mathbf{R}^n$ is

$$\Big\|\Big(\sum_{\mathbf{j}\in\mathbf{Z}^n} |(\widehat{f}\chi_{R_{\mathbf{j}}} m_{+\cdots+})^\vee|^2\Big)^{\frac12}\Big\|_{L^p(\mathbf{R}^n)} \le C_n\big(\|m\|_{L^\infty}+A\big)\max\big(p,(p-1)^{-1}\big)^{4n}\|f\|_{L^p(\mathbf{R}^n)}$$

which is obtained in a similar fashion. Using (6.1.29), this implies that

$$\|(\widehat{f} m_{+\cdots+})^\vee\|_{L^p(\mathbf{R}^n)} \le C_n\big(\|m\|_{L^\infty}+A\big)\max\big(p,(p-1)^{-1}\big)^{6n}\|f\|_{L^p(\mathbf{R}^n)}.$$

A similar inequality holds when some (or all) $+$'s are replaced by $-$'s. □

We now give a condition that implies (6.2.5) and is well suited for a variety of applications.

Corollary 6.2.5. *Let m be a bounded $\mathscr{C}^n$ function defined away from the coordinate axes on $\mathbf{R}^n$. Assume that for all $k \in \{1,\dots,n\}$, all distinct $j_1,\dots,j_k \in \{1,2,\dots,n\}$, and all $\xi_r \in \mathbf{R}\setminus\{0\}$ for $r \notin \{j_1,\dots,j_k\}$ we have*

$$\big|(\partial_{j_1}\cdots\partial_{j_k} m)(\xi_1,\dots,\xi_n)\big| \le A\,|\xi_{j_1}|^{-1}\cdots|\xi_{j_k}|^{-1}. \tag{6.2.9}$$

Then m satisfies (6.2.6).

Proof. Simply observe that condition (6.2.9) implies (6.2.5). □

Example 6.2.6. The following are examples of functions that satisfy the hypotheses of Corollary 6.2.5:

$$m_1(\xi) = \frac{\xi_1}{\xi_1 + i(\xi_2^2 + \cdots + \xi_n^2)},$$

$$m_2(\xi) = \frac{|\xi_1|^{\alpha_1} \cdots |\xi_n|^{\alpha_n}}{(\xi_1^2 + \xi_2^2 + \cdots + \xi_n^2)^{\alpha/2}},$$

where $\alpha_1 + \alpha_2 + \cdots + \alpha_n = \alpha$, $\alpha_j > 0$,

$$m_3(\xi) = \frac{\xi_2 \xi_3^2}{i\xi_1 + \xi_2^2 + \xi_3^4}.$$

The functions m_1 and m_2 are defined on $\mathbf{R}^n \setminus \{0\}$ and m_3 on $\mathbf{R}^3 \setminus \{0\}$.

The previous examples and many other examples that satisfy the hypothesis (6.2.9) of Corollary 6.2.5 are invariant under a set of dilations in the following sense: suppose that there exist $k_1, \dots, k_n \in \mathbf{R}^+$ and $s \in \mathbf{R}$ such that the smooth function m on $\mathbf{R}^n \setminus \{0\}$ satisfies

$$m(\lambda^{k_1}\xi_1, \dots, \lambda^{k_n}\xi_n) = \lambda^{is} m(\xi_1, \dots, \xi_n)$$

for all $\xi_1, \dots, \xi_n \in \mathbf{R}$ and $\lambda > 0$. Then m satisfies condition (6.2.9). Indeed, differentiation gives

$$\lambda^{\alpha_1 k_1 + \cdots + \alpha_n k_n} \partial^\alpha m(\lambda^{k_1}\xi_1, \dots, \lambda^{k_n}\xi_n) = \lambda^{is} \partial^\alpha m(\xi_1, \dots, \xi_n)$$

for every multi-index $\alpha = (\alpha_1, \dots, \alpha_n)$. Now for every $\xi \in \mathbf{R}^n \setminus \{0\}$ pick the unique $\lambda_\xi > 0$ such that $(\lambda_\xi^{k_1}\xi_1, \dots, \lambda_\xi^{k_n}\xi_n) \in \mathbf{S}^{n-1}$. Then $\lambda_\xi^{k_j \alpha_j} \le |\xi_j|^{-\alpha_j}$, and it follows that

$$|\partial^\alpha m(\xi_1, \dots, \xi_n)| \le \left[\sup_{\mathbf{S}^{n-1}} |\partial^\alpha m| \right] \lambda_\xi^{\alpha_1 k_1 + \cdots + \alpha_n k_n} \le C_\alpha |\xi_1|^{-\alpha_1} \cdots |\xi_n|^{-\alpha_n}.$$

6.2.3 The Mihlin–Hörmander Multiplier Theorem on $\mathbf{R}^n$

We now discuss another multiplier theorem that also requires decay of derivatives. We will consider the situation where each differentiation produces uniform decay in all variables, quantitatively expressed via the condition

$$|\partial_\xi^\alpha m(\xi)| \le C_\alpha |\xi|^{-|\alpha|} \tag{6.2.10}$$

for each multi-index α. The decay can also be expressed in terms of a square integrable estimate that has the form

$$\left(\int_{R<|\xi|<2R}|\partial_\xi^\alpha m(\xi)|^2\,d\xi\right)^{\frac{1}{2}}\le C'_\alpha R^{\frac{n}{2}-|\alpha|}<\infty \tag{6.2.11}$$

for all multi-indices α and all $R>0$. Obviously (6.2.10) implies (6.2.11)

Theorem 6.2.7. *Let $m(\xi)$ be a complex-valued bounded function on $\mathbf{R}^n\setminus\{0\}$ that satisfies for some $A<\infty$*

$$\left(\int_{R<|\xi|<2R}|\partial_\xi^\alpha m(\xi)|^2\,d\xi\right)^{\frac{1}{2}}\le A R^{\frac{n}{2}-|\alpha|}<\infty \tag{6.2.12}$$

for all multi-indices $|\alpha|\le[n/2]+1$ and all $R>0$.
Then for all $1<p<\infty$, m lies in $\mathscr{M}_p(\mathbf{R}^n)$ and the following estimate is valid:

$$\|m\|_{\mathscr{M}_p}\le C_n\max(p,(p-1)^{-1})\big(A+\|m\|_{L^\infty}\big). \tag{6.2.13}$$

Moreover, the operator $f\mapsto(\widehat{f}m)^\vee$ maps $L^1(\mathbf{R}^n)$ to $L^{1,\infty}(\mathbf{R}^n)$ with norm at most a dimensional constant multiple of $A+\|m\|_{L^\infty}$.

We remark that in most applications, condition (6.2.12) appears in the form

$$|\partial_\xi^\alpha m(\xi)|\le C_\alpha|\xi|^{-|\alpha|}, \tag{6.2.14}$$

which should be, in principle, easier to verify.

Proof. Since m is a bounded function, the operator given by convolution with $W=m^\vee$ is bounded on $L^2(\mathbf{R}^n)$. To prove that this operator maps $L^1(\mathbf{R}^n)$ to $L^{1,\infty}(\mathbf{R}^n)$, it suffices to prove that the distribution W coincides with a function K on $\mathbf{R}^n\setminus\{0\}$ that satisfies Hörmander's condition.

Let $\widehat{\zeta}$ be a smooth function supported in the annulus $\frac{1}{2}\le|\xi|\le 2$ such that

$$\sum_{j\in\mathbf{Z}}\widehat{\zeta}(2^{-j}\xi)=1,\qquad\text{when }\xi\neq 0.$$

Set $m_j(\xi)=m(\xi)\widehat{\zeta}(2^{-j}\xi)$ for $j\in\mathbf{Z}$ and $K_j=m_j^\vee$. We begin by observing that $\sum_{-N}^{N}K_j$ converges to W in $\mathscr{S}'(\mathbf{R}^n)$. Indeed, for all $\varphi\in\mathscr{S}(\mathbf{R}^n)$ we have

$$\Big\langle\sum_{j=-N}^{N}K_j,\varphi\Big\rangle=\Big\langle\sum_{j=-N}^{N}m_j,\varphi^\vee\Big\rangle\to\langle m,\varphi^\vee\rangle=\langle W,\varphi\rangle.$$

We set $n_0=[\frac{n}{2}]+1$. We claim that there is a constant $\widetilde{C}_n$ such that

$$\sup_{j\in\mathbf{Z}}\int_{\mathbf{R}^n}|K_j(x)|\,(1+2^j|x|)^{\frac{1}{4}}\,dx\le\widetilde{C}_nA, \tag{6.2.15}$$

$$\sup_{j\in\mathbf{Z}}2^{-j}\int_{\mathbf{R}^n}|\nabla K_j(x)|\,(1+2^j|x|)^{\frac{1}{4}}\,dx\le\widetilde{C}_nA. \tag{6.2.16}$$

To prove (6.2.15) we multiply and divide the integrand in (6.2.15) by the expression $(1+2^j|x|)^{n_0}$. Applying the Cauchy–Schwarz inequality to $|K_j(x)|\,(1+2^j|x|)^{n_0}$ and $(1+2^j|x|)^{-n_0+\frac{1}{4}}$, we control the integral in (6.2.15) by the product

$$\left(\int_{\mathbf{R}^n} |K_j(x)|^2(1+2^j|x|)^{2n_0}\,dx\right)^{\frac{1}{2}}\left(\int_{\mathbf{R}^n}(1+2^j|x|)^{-2n_0+\frac{1}{2}}\,dx\right)^{\frac{1}{2}}. \tag{6.2.17}$$

We now note that $-2n_0+\frac{1}{2}<-n$, and hence the second factor in (6.2.17) is equal to a constant multiple of $2^{-jn/2}$. To estimate the first integral in (6.2.17) we use the simple fact that

$$(1+2^j|x|)^{n_0}\le C(n)\sum_{|\gamma|\le n_0}|(2^jx)^\gamma|.$$

We now have that the expression inside the supremum in (6.2.15) is controlled by

$$C'(n)2^{-jn/2}\sum_{|\gamma|\le n_0}\left(\int_{\mathbf{R}^n}|K_j(x)|^2 2^{2j|\gamma|}|x^\gamma|^2\,dx\right)^{\frac{1}{2}}, \tag{6.2.18}$$

which, by Plancherel's theorem, is equal to

$$2^{-jn/2}\sum_{|\gamma|\le n_0}C_\gamma 2^{j|\gamma|}\left(\int_{\mathbf{R}^n}|(\partial^\gamma m_j)(\xi)|^2\,d\xi\right)^{\frac{1}{2}} \tag{6.2.19}$$

for some constants C_γ.

For multi-indices $\delta=(\delta_1,\dots,\delta_n)$ and $\gamma=(\gamma_1,\dots,\gamma_n)$ we introduce the notation $\delta\le\gamma$ to mean $\delta_j\le\gamma_j$ for all $j=1,\dots,n$. For any $|\gamma|\le n_0$ we use Leibniz's rule to obtain for some constants $C_{\delta,\gamma}$

$$\begin{aligned}
\left(\int_{\mathbf{R}^n}|(\partial^\gamma m_j)(\xi)|^2\,d\xi\right)^{\frac{1}{2}} &\le \sum_{\delta\le\gamma}C_{\delta,\gamma}\left(\int_{\mathbf{R}^n}\left|2^{-j|\gamma-\delta|}(\partial_\xi^{\gamma-\delta}\widehat{\zeta})(2^{-j}\xi)(\partial_\xi^\delta m)(\xi)\right|^2 d\xi\right)^{\frac{1}{2}}\\
&\le \sum_{\delta\le\gamma}C_{\delta,\gamma}2^{-j|\gamma|}2^{j|\delta|}\left(\int_{2^{j-1}\le|\xi|\le 2^{j+1}}\left|(\partial_\xi^\delta m)(\xi)\right|^2 d\xi\right)^{\frac{1}{2}}\\
&\le \sum_{\delta\le\gamma}C_{\delta,\gamma}2^{-j|\gamma|}2^{j|\delta|}\,2A\,2^{jn/2}2^{-j|\delta|}\\
&=\widetilde{C}_n A\,2^{jn/2}2^{-j|\gamma|},
\end{aligned}$$

which inserted in (6.2.19) and combined with (6.2.18) yields (6.2.15). To obtain (6.2.16) we repeat the same argument for every derivative $\partial_r K_j$. Since the Fourier transform of $(\partial_r K_j)(x)x^\gamma$ is equal to a constant multiple of $\partial^\gamma(\xi_r m(\xi)\widehat{\zeta}(2^{-j}\xi))$, we observe that the extra factor 2^{-j} in (6.2.16) can be combined with ξ_r to write $2^{-j}\partial^\gamma(\xi_r m(\xi)\widehat{\zeta}(2^{-j}\xi))$ as $\partial^\gamma(m(\xi)\widehat{\zeta_r}(2^{-j}\xi))$, where $\widehat{\zeta_r}(\xi)=\xi_r\widehat{\zeta}(\xi)$. The previous calculation with $\widehat{\zeta_r}$ replacing $\widehat{\zeta}$ can then be used to complete the proof of (6.2.16).

We now show that for all $x \neq 0$, the series $\sum_{j\in\mathbf{Z}} K_j(x)$ converges to a function, which we denote by $K(x)$. Indeed, as a consequence of (6.2.15) we have that

$$(1+2^j\delta)^{\frac{1}{4}} \int_{|x|\geq\delta} |K_j(x)|\,dx \leq \widetilde{C}_n A\,,$$

for any $\delta > 0$, which implies that the function $\sum_{j>0}|K_j(x)|$ is integrable away from the origin and satisfies $\int_{\delta\leq|x|\leq 2\delta}\sum_{j>0}|K_j(x)|\,dx < \infty$. Now note that (6.2.15) also holds with $-\frac{1}{4}$ in place of $\frac{1}{4}$. Using this observation we obtain

$$(1+2^j 2\delta)^{-\frac{1}{4}} \int_{|x|\leq 2\delta} |K_j(x)|\,dx \leq \int_{|x|\leq 2\delta} |K_j(x)|(1+2^j|x|)^{-\frac{1}{4}}\,dx \leq \widetilde{C}_n A\,,$$

and from this it follows that $\int_{\delta\leq|x|\leq 2\delta}\sum_{j\leq 0}|K_j(x)|\,dx < \infty$.

We conclude that the series $\sum_{j\in\mathbf{Z}} K_j(x)$ converges a.e. on $\mathbf{R}^n\setminus\{0\}$ to a function $K(x)$ that coincides with the distribution $W = m^\vee$ on $\mathbf{R}^n\setminus\{0\}$ and satisfies

$$\sup_{\delta>0}\int_{\delta\leq|x|\leq 2\delta} |K(x)|\,dx < \infty\,.$$

We now prove that the function $K = \sum_{j\in\mathbf{Z}} K_j$ (defined on $\mathbf{R}^n\setminus\{0\}$) satisfies Hörmander's condition. It suffices to prove that for all $y\neq 0$ we have

$$\sum_{j\in\mathbf{Z}}\int_{|x|\geq 2|y|} |K_j(x-y)-K_j(x)|\,dx \leq 2C_n' A\,. \tag{6.2.20}$$

Fix a $y\in\mathbf{R}^n\setminus\{0\}$ and pick a $k\in\mathbf{Z}$ such that $2^{-k}\leq|y|\leq 2^{-k+1}$. The part of the sum in (6.2.20) where $j>k$ is bounded by

$$\begin{aligned}
\sum_{j>k}\int_{|x|\geq 2|y|} |K_j(x-y)|+|K_j(x)|\,dx &\leq 2\sum_{j>k}\int_{|x|\geq|y|} |K_j(x)|\,dx \\
&\leq 2\sum_{j>k}\int_{|x|\geq|y|} |K_j(x)|\frac{(1+2^j|x|)^{\frac{1}{4}}}{(1+2^j|x|)^{\frac{1}{4}}}\,dx \\
&\leq \sum_{j>k}\frac{2\widetilde{C}_n A}{(1+2^j|y|)^{\frac{1}{4}}} \\
&\leq \sum_{j>k}\frac{2\widetilde{C}_n A}{(1+2^j 2^{-k})^{\frac{1}{4}}} = C_n' A\,,
\end{aligned}$$

where we used (6.2.15). The part of the sum in (6.2.20) where $j\leq k$ is bounded by

$$\begin{aligned}
&\sum_{j\leq k}\int_{|x|\geq 2|y|} |K_j(x-y)-K_j(x)|\,dx \\
&\qquad\leq \sum_{j\leq k}\int_{|x|\geq 2|y|}\int_0^1 |-y\cdot\nabla K_j(x-\theta y)|\,d\theta\,dx
\end{aligned}$$

$$\leq \int_0^1 \sum_{j\leq k} 2^{-k+1} \int_{\mathbf{R}^n} |\nabla K_j(x-\theta y)|(1+2^j|x-\theta y|)^{\frac{1}{4}}\,dx\,d\theta$$

$$\leq \int_0^1 \sum_{j\leq k} 2^{-k+1}\widetilde{C}_n A 2^j\,d\theta \leq C_n' A,$$

using (6.2.16). Hörmander's condition is satisfied for K, and we appeal to Theorem 5.3.3 to complete the proof of (6.2.13). □

Example 6.2.8. Let m be a smooth function away from the origin that is homogeneous of imaginary order, i.e., for some fixed τ real and all $\lambda > 0$ we have

$$m(\lambda\xi) = \lambda^{i\tau} m(\xi). \tag{6.2.21}$$

Then m is an L^p Fourier multiplier for $1 < p < \infty$. Indeed, differentiating both sides of (6.2.21) with respect to ∂_ξ^α we obtain

$$\lambda^{|\alpha|}\partial_\xi^\alpha m(\lambda\xi) = \lambda^{i\tau}\partial_\xi^\alpha m(\xi)$$

and taking $\lambda = |\xi|^{-1}$, we deduce condition (6.2.14) with $C_\alpha = \sup_{|\theta|=1} |\partial^\alpha m(\theta)|$. An explicit example of such a function is $m(\xi) = |\xi|^{i\tau}$. Another example is

$$m_0(\xi_1,\xi_2,\xi_3) = \frac{\xi_1^2+\xi_2^2}{\xi_1^2 + i(\xi_2^2+\xi_3^2)}$$

which is homogeneous of degree zero and also smooth on $\mathbf{R}^n \setminus \{0\}$.

Example 6.2.9. Let z be a complex numbers with $\mathrm{Re}\, z \geq 0$. Then the functions

$$m_1(\xi) = \left(\frac{|\xi|^2}{1+|\xi|^2}\right)^z, \qquad m_2(\xi) = \left(\frac{1}{1+|\xi|^2}\right)^z$$

defined on $\mathbf{R}^n$ are L^p Fourier multipliers for $1 < p < \infty$. To prove this assertion for m_1, we verify condition (6.2.14). To achieve this, introduce the function on $\mathbf{R}^{n+1}$

$$M_1(\xi_1,\dots,\xi_n,t) = \left(\frac{|\xi_1|^2+\cdots+|\xi_n|^2}{t^2+|\xi_1|^2+\cdots+|\xi_n|^2}\right)^z = \left(\frac{|\xi|^2}{t^2+|\xi|^2}\right)^z,$$

where $\xi = (\xi_1,\dots,\xi_n)$. Then M is homogeneous of degree 0 and smooth on $\mathbf{R}^{n+1}\setminus\{0\}$. The derivatives $\partial^\beta M_1$ are homogeneous of degree $-|\beta|$ and by the calculation in the preceding example they satisfy $|\partial^\beta M_1(\xi,t)| \leq C_\beta |(\xi,t)|^{-|\beta|}$, with $C_\beta = \sup_{|\theta|=1} |\partial^\beta M_1(\theta)|$, whenever $(\xi,t) \neq 0$ and β is a multi index of $n+1$ variables. In particular, taking $\beta = (\alpha,0)$, we obtain

$$\left|\partial_{\xi_1}^{\alpha_1}\cdots\partial_{\xi_n}^{\alpha_n} M_1(\xi_1,\dots,\xi_n,t)\right| \leq \frac{C_\alpha}{(t^2+|\xi|^2)^{|\alpha|/2}},$$

and setting $t=1$ we deduce that $|\partial^\alpha m_1(\xi)| \leq C_\alpha (1+|\xi|^2)^{-|\alpha|/2} \leq C_\alpha |\xi|^{-|\alpha|}$.

For m_2 we introduce the function

$$M_2(\xi_1,\dots,\xi_n,t) = \left(\frac{1}{t^2+|\xi_1|^2+\cdots+|\xi_n|^2}\right)^z$$

on $\mathbf{R}^{n+1}$, which is homogeneous of degree $-2z$. Then the derivative $\partial^\beta M_2$ is homogeneous of degree $-|\beta|-2z$, hence it satisfies $|\partial^\beta M_2(\xi,t)| \le C_\beta |(\xi,t)|^{-|\beta|-2\mathrm{Re}\,z}$ for all multi-indices β of $n+1$ variables. In particular, taking $\beta=(\alpha,0)$, we obtain

$$\left|\partial_{\xi_1}^{\alpha_1}\cdots\partial_{\xi_n}^{\alpha_n} M_2(\xi_1,\dots,\xi_n,t)\right| \le \frac{C_\alpha}{(t^2+|\xi|^2)^{\frac{|\alpha|}{2}+\mathrm{Re}\,z}},$$

and setting $t=1$, we deduce $|\partial^\alpha m_2(\xi)| \le C_\alpha(1+|\xi|^2)^{-|\alpha|/2} \le C_\alpha |\xi|^{-|\alpha|}$, where in the first inequality we used that $\mathrm{Re}\,z \ge 0$.

We end this section by comparing Theorems 6.2.2 and 6.2.4 with Theorem 6.2.7. It is obvious that in dimension $n=1$, Theorem 6.2.2 is stronger than Theorem 6.2.7 in view of the inequality

$$\int_{2^j<|\xi|<2^{j+1}} |m'(\xi)|\,d\xi \le 2^{j/2}\left(\int_{2^j<|\xi|<2^{j+1}} |m'(\xi)|^2\,d\xi\right)^{\frac{1}{2}},$$

which implies that (6.2.2) is weaker than (6.2.12). Note also that in Theorem 6.2.2 the multiplier m is not required to be differentiable at the points $\pm 2^j$. But in higher dimensions neither theorem includes the other. In Theorem 6.2.4 the multiplier is allowed to be singular on a set of measure zero but is required to be differentiable in every variable, i.e., to be at least $\mathscr{C}^n$ in the complement of this null set. In Theorem 6.2.7, the multiplier is only allowed to be singular only at the origin, but it is assumed to be $\mathscr{C}^{[n/2]+1}$, requiring almost half the differentiability called for by condition (6.2.9). It should be noted that both theorems have their shortcomings. In particular, they are not L^p sensitive, i.e., delicate enough to detect whether m is a bounded Fourier multiplier on some L^p but not on some other L^q.

Exercises

6.2.1. Let $\psi(\xi)$ be a smooth function supported in $[3/4,2]\cup[-2,-3/4]$ and equal to 1 on $[1,3/2]\cup[-3/2,-1]$ that satisfies $\sum_{j\in\mathbf{Z}}\psi(2^{-j}\xi)=1$ for all $\xi\neq 0$. Let $1\le k\le n$. Prove that $m\in\mathscr{M}_p(\mathbf{R}^n)$ if and only if (6.2.1) is satisfied with $m_{\mathbf{j}}(\xi)$ replaced by the function $m(\xi)\psi(2^{-j_1}\xi_1)\cdots\psi(2^{-j_k}\xi_k)$.
[*Hint:* To prove one direction, partition $\mathbf{Z}^k$ in 2^k sets such that for every $\mathbf{j}=(j_1,\dots,j_k)$ in each of these sets, j_i has a fixed remainder modulo 2. For the other direction, use Theorem 6.1.6 in the variables $x_1,\dots,x_k$. Also use the inequality $\|f\|_{L^p(\mathbf{R}^n)} \le C_p \|(\sum_{\mathbf{j}\in\mathbf{Z}^k}|(\widehat{f}\chi_{R_{\mathbf{j}}})^\vee|^2)^{1/2}\|_{L^p(\mathbf{R}^n)}$, $R_{\mathbf{j}}=([-2,-\frac{1}{2}]\cup[\frac{1}{2},2])^k\times\mathbf{R}^{n-k}$, which can be derived by duality from the identity $\sum_{\mathbf{j}\in\mathbf{Z}^k}\chi_{R_{\mathbf{j}}}=2^k$.]

6.2.2. Let φ be a smooth function on the real line supported in the interval $[-1,1]$. Let $\psi(t)$ be a smooth function on the real line that is equal to 1 when $|t| \geq 10$ and vanishes when $|t| \leq 9$. Show that for the function $m(\xi_1,\xi_2) = e^{i\xi_2^2/\xi_1}\varphi(\xi_2)\psi(\xi_1)$ lies in $\mathscr{M}_p(\mathbf{R}^2)$, $1 < p < \infty$, using Theorem 6.2.4. Also show that Theorem 6.2.7 does not apply.

6.2.3. Consider the differential operators

$$\begin{aligned} L_1 &= \partial_1 - \partial_2^2 + \partial_3^4, \\ L_2 &= \partial_1 + \partial_2^2 + \partial_3^2. \end{aligned}$$

Prove that for every $1 < p < \infty$ there exists a constant $C_p < \infty$ such that for all Schwartz functions f on $\mathbf{R}^3$ we have

$$\begin{aligned} \big\|\partial_2\partial_3^2 f\big\|_{L^p} &\leq C_p \big\|L_1(f)\big\|_{L^p}, \\ \big\|\partial_1 f\big\|_{L^p} &\leq C_p \big\|L_2(f)\big\|_{L^p}. \end{aligned}$$

[*Hint:* Use Corollary 6.2.5 and the idea of Example 6.2.6.]

6.2.4. Suppose that $m(\xi)$ is a real-valued function that satisfies either (6.2.9) or $|\partial^\alpha m(\xi)| \leq C_\alpha |\xi|^{-|\alpha|}$ for all multi-indices α with $|\alpha| \leq [\frac{n}{2}]+1$ and all $\xi \in \mathbf{R}^n \setminus \{0\}$. Show that $e^{im(\xi)}$ lies in $\mathscr{M}_p(\mathbf{R}^n)$ for any $1 < p < \infty$.
[*Hint:* Prove by induction and use that

$$\partial^\alpha\left(e^{im(\xi)}\right) = e^{im(\xi)} \sum_{\substack{l_j \geq 0, \beta^j \leq \alpha \\ l_1\beta^1+\cdots+l_k\beta^k=\alpha}} c_{\beta^1,\dots,\beta^k} (\partial^{\beta^1} m(\xi))^{l_1} \cdots (\partial^{\beta^k} m(\xi))^{l_k},$$

where the sum is taken over all partitions of the multi-index α as a linear combination of multi-indices β^j with coefficients $l_j \in \mathbf{Z}^+ \cup \{0\}$.]

6.2.5. Suppose that $\varphi(\xi)$ is a smooth function on $\mathbf{R}^n$ that vanishes in a neighborhood of the origin and is equal to 1 in a neighborhood of infinity. Prove that the function $e^{i\xi_j|\xi|^{-1}}\varphi(\xi)$ is in $\mathscr{M}_p(\mathbf{R}^n)$ for $1 < p < \infty$.

6.2.6. Let $\tau, \tau_1, \dots, \tau_n$ be real numbers and $\rho_1, \dots, \rho_n$ be even natural numbers. Prove that the following functions are L^p multipliers on $\mathbf{R}^n$ for $1 < p < \infty$:

$$\begin{aligned} &|\xi_1|^{i\tau_1} \cdots |\xi_n|^{i\tau_n}, \\ &(|\xi_1|^{\rho_1} + \cdots + |\xi_n|^{\rho_n})^{i\tau}, \\ &(|\xi_1|^{-\rho_1} + |\xi_2|^{-\rho_2})^{i\tau}. \end{aligned}$$

6.2.7. Let $\widehat{\zeta}(\xi)$ be a smooth function on $\mathbf{R}^n$ is supported in a compact set that does not contain the origin and let a_j be a bounded sequence of complex numbers. Prove that the function

$$m(\xi) = \sum_{j\in\mathbf{Z}} a_j \widehat{\zeta}(2^{-j}\xi)$$

is in $\mathscr{M}_p(\mathbf{R}^n)$ for all $1 < p < \infty$.

6.2.8. Let $\widehat{\zeta}(\xi)$ be a smooth function on $\mathbf{R}^n$ supported in a compact set that does not contain the origin and let $\Delta_j^\zeta(f) = \big(\widehat{f}(\xi)\widehat{\zeta}(2^{-j}\xi)\big)^\vee$. Show that the operator

$$f \to \sup_{N\in\mathbf{Z}} \Big| \sum_{j<N} \Delta_j^\zeta(f) \Big|$$

is bounded on $L^p(\mathbf{R})$ when $1 < p < \infty$.
[*Hint:* Pick a Schwartz function φ satisfying $\sum_{j\in\mathbf{Z}} \widehat{\varphi}(2^{-j}\xi) = 1$ on $\mathbf{R}^n \setminus \{0\}$ with $\widehat{\varphi}(\xi)$ supported in $\frac{6}{7} \le |\xi| \le 2$. Then $\Delta_k^\varphi \Delta_j^\zeta = 0$ if $|j-k| < c_0$ and we have

$$\sum_{j<N} \Delta_j^\zeta = \sum_{k<N+c_0} \Delta_k^\varphi \sum_{j<N} \Delta_j^\zeta = \sum_{k<N+c_0} \Delta_k^\varphi \sum_{j} \Delta_j^\zeta - \sum_{k<N+c_0} \Delta_k^\varphi \sum_{j\ge N} \Delta_j^\zeta ,$$

which is a finite sum plus a term controlled by a multiple of the operator

$$f \mapsto M\Big(\sum_{j\in\mathbf{Z}} \Delta_j^\zeta(f)\Big),$$

where M is the Hardy–Littlewood maximal function.]

6.2.9. Let Ψ be a Schwartz function whose Fourier transform is real-valued, supported in a compact set that does not contain the origin, and satisfies

$$\sum_{j\in\mathbf{Z}} \widehat{\Psi}(2^{-j}\xi) = 1 \qquad \text{when } \xi \neq 0.$$

Let Δ_j be the Littlewood–Paley operator associated with Ψ. Prove that

$$\Big\| \sum_{|j|<N} \Delta_j(g) - g \Big\|_{L^p} \to 0$$

as $N \to \infty$ for all functions $g \in \mathscr{S}(\mathbf{R}^n)$. Deduce that Schwartz functions whose Fourier transforms have compact supports that do not contain the origin are dense in $L^p(\mathbf{R}^n)$ for $1 < p < \infty$.
[*Hint:* Use the result of Exercise 6.2.8 and the Lebesgue dominated convergence theorem.]

6.3 Applications of Littlewood–Paley Theory

We now turn our attention to some important applications of Littlewood–Paley theory. We are interested in obtaining bounds for singular and maximal operators. These bounds are obtained by controlling the corresponding operators by quadratic expressions.

6.3.1 Estimates for Maximal Operators

One way to control the maximal operator $\sup_k |T_k(f)|$ is by introducing a good averaging function φ and using the majorization

$$\begin{aligned} \sup_k |T_k(f)| &\le \sup_k |T_k(f) - f * \varphi_{2^{-k}}| + \sup_k |f * \varphi_{2^{-k}}| \\ &\le \Big(\sum_k |T_k(f) - f * \varphi_{2^{-k}}|^2\Big)^{\frac{1}{2}} + C_\varphi M(f) \end{aligned} \tag{6.3.1}$$

for some constant C_φ depending on φ. We apply this idea to prove the following theorem.

Theorem 6.3.1. *Let m be a bounded function on $\mathbf{R}^n$ that is $\mathscr{C}^1$ in a neighborhood of the origin and satisfies $m(0) = 1$ and $|m(\xi)| \le C|\xi|^{-\varepsilon}$ for some $C, \varepsilon > 0$ and all $\xi \neq 0$. For each $k \in \mathbf{Z}$ define $T_k(f)(x) = (\widehat{f}(\xi)m(2^{-k}\xi))^\vee(x)$. Then there is a constant C_n such that for all L^2 functions f on $\mathbf{R}^n$ we have*

$$\Big\| \sup_{k\in\mathbf{Z}} |T_k(f)| \Big\|_{L^2} \le C_n \big\| f \big\|_{L^2}. \tag{6.3.2}$$

Proof. Select a Schwartz function φ such that $\widehat{\varphi}(0) = 1$. Then there are positive constants C_1 and C_2 such that $|m(\xi) - \widehat{\varphi}(\xi)| \le C_1|\xi|^{-\varepsilon}$ for $|\xi|$ away from zero and $|m(\xi) - \widehat{\varphi}(\xi)| \le C_2|\xi|$ for $|\xi|$ near zero. These two inequalities imply that

$$\sum_k |m(2^{-k}\xi) - \widehat{\varphi}(2^{-k}\xi)|^2 \le C_3 < \infty,$$

from which the L^2 boundedness of the operator

$$f \mapsto \Big(\sum_k |T_k(f) - f * \varphi_{2^{-k}}|^2\Big)^{1/2}$$

follows easily. Using estimate (6.3.1) and the well-known L^2 estimate for the Hardy–Littlewood maximal function, we obtain (6.3.2). □

If $m(\xi)$ is the characteristic function of a rectangle with sides parallel to the axes, this result can be extended to L^p.

Theorem 6.3.2. *Let $1 < p < \infty$ and let U be the n-fold product of open intervals that contain zero. For each $k \in \mathbf{Z}$ define $T_k(f)(x) = (\widehat{f}(\xi)\chi_U(2^{-k}\xi))^\vee(x)$. Then there is a constant $C_{p,n}$ such that for all L^p functions f on $\mathbf{R}^n$ we have*

$$\big\| \sup_{k\in\mathbf{Z}} |T_k(f)| \big\|_{L^p(\mathbf{R}^n)} \le C_{p,n} \|f\|_{L^p(\mathbf{R}^n)}.$$

Proof. Let us fix an open annulus A whose interior contains the boundary of U and take a smooth function with compact support $\widehat{\psi}$ that vanishes in a neighborhood of zero and a neighborhood of infinity and is equal to 1 on the annulus A. Then the function $\widehat{\varphi} = (1-\widehat{\psi})\chi_U$ is Schwartz. Since $\chi_U = \chi_U\widehat{\psi} + \widehat{\varphi}$, it follows that for all $f \in L^p(\mathbf{R}^n)$ we have

$$T_k(f) = T_k(f) - f*\varphi_{2^{-k}} + f*\varphi_{2^{-k}} = T_k(f*\psi_{2^{-k}}) + f*\varphi_{2^{-k}}.$$

Taking the supremum over k and using Corollary 2.1.12 we obtain

$$\sup_{k\in\mathbf{Z}} |T_k(f)| \le \Big(\sum_k |T_k(f) - f*\varphi_{2^{-k}}|^2\Big)^{1/2} + C_\varphi M(f). \tag{6.3.3}$$

The operator $T_k(f) - f*\varphi_{2^{-k}}$ is given by multiplication on the Fourier transform side by the multiplier

$$\chi_U(2^{-k}\xi) - \widehat{\varphi}(2^{-k}\xi) = \chi_U(2^{-k}\xi)\widehat{\psi}(2^{-k}\xi) = \chi_{2^kU}(\xi)\widehat{\psi}(2^{-k}\xi).$$

Since $\{2^kU\}_{k\in\mathbf{Z}}$ is a measurable family of rectangles with sides parallel to the axes, Exercise 5.6.1(b) yields the following inequality:

$$\Big\| \Big(\sum_{k\in\mathbf{Z}} |T_k(f) - f*\varphi_{2^{-k}}|^2\Big)^{\frac{1}{2}} \Big\|_{L^p} \le C_{p,n} \Big\| \Big(\sum_{k\in\mathbf{Z}} |f*\psi_{2^{-k}}|^2\Big)^{\frac{1}{2}} \Big\|_{L^p}. \tag{6.3.4}$$

Since $f*\psi_{2^{-k}} = \Delta_j^\psi(f)$, estimate (6.1.4) of Theorem 6.1.2 yields that the expression on the right in (6.3.4) is controlled by a multiple of $\|f\|_{L^p}$. Taking L^p norms in (6.3.3) and using the L^p estimate for the square function yields the required conclusion. □

The following lacunary version of the Carleson–Hunt theorem is yet another indication of the powerful techniques of Littlewood–Paley theory.

Corollary 6.3.3. *(a) Let f be in $L^2(\mathbf{R}^n)$ and let Ω be an open set that contains the origin in $\mathbf{R}^n$. Then*

$$\lim_{k\to\infty} \int_{2^k\Omega} \widehat{f}(\xi) e^{2\pi i x\cdot\xi}\, d\xi = f(x)$$

for almost all $x \in \mathbf{R}^n$.
(b) Let f be in $L^p(\mathbf{R}^n)$ for some $1 < p < \infty$. Then

$$\lim_{k\to\infty} \int_{\substack{|\xi_1|<2^k \\ \cdots \\ |\xi_n|<2^k}} \widehat{f}(\xi) e^{2\pi i x\cdot\xi}\, d\xi = f(x)$$

for almost all $x \in \mathbf{R}^n$.

Proof. Both limits exist everywhere for functions f in the Schwartz class. To obtain almost everywhere convergence for general f in L^p we appeal to Theorem 2.1.14. The required control of the corresponding maximal operator is a consequence of Theorem 6.3.1 with $m = \chi_\Omega$ in case (a) and Theorem 6.3.2 in case (b). □

6.3.2 Estimates for Singular Integrals with Rough Kernels

We now turn to another application of the Littlewood–Paley theory involving singular integrals.

Theorem 6.3.4. *Suppose that μ is a finite Borel measure on $\mathbf{R}^n$ with compact support that satisfies $|\widehat{\mu}(\xi)| \le B\min\big(|\xi|^{-b}, |\xi|^{b}\big)$ for some $b>0$ and all $\xi \neq 0$. Define measures μ_j by setting $\widehat{\mu_j}(\xi) = \widehat{\mu}(2^{-j}\xi)$. Then the operator*

$$T_\mu(f)(x) = \sum_{j\in\mathbf{Z}} (f*\mu_j)(x)$$

is bounded on $L^p(\mathbf{R}^n)$ for all $1<p<\infty$.

Proof. It is natural to begin with the L^2 boundedness of T_μ. The estimate on $\widehat{\mu}$ implies that

$$\sum_{j\in\mathbf{Z}} |\widehat{\mu}(2^{-j}\xi)| \le \sum_{j\in\mathbf{Z}} B\min\big(|2^{-j}\xi|^{b}, |2^{-j}\xi|^{-b}\big) \le C_b B < \infty. \tag{6.3.5}$$

The L^2 boundedness of T_μ is an immediate consequence of (6.3.5).

We now turn to the L^p boundedness of T_μ for $1<p<\infty$. We fix a radial Schwartz function ψ whose Fourier transform is supported in the annulus $\frac{1}{2} < |\xi| < 2$ that satisfies

$$\sum_{j\in\mathbf{Z}} \widehat{\psi}(2^{-j}\xi) = 1 \tag{6.3.6}$$

whenever $\xi \neq 0$. We let $\psi_{2^{-k}}(x) = 2^{kn}\psi(2^k x)$, so that $\widehat{\psi_{2^{-k}}}(\xi) = \widehat{\psi}(2^{-k}\xi)$, and we observe that the identity

$$\mu_j = \sum_{k\in\mathbf{Z}} \mu_j * \psi_{2^{-j-k}}$$

is valid by taking Fourier transforms and using (6.3.6). We now define operators S_k by setting

$$S_k(f) \;=\; \sum_{j\in\mathbf{Z}} \mu_j * \psi_{2^{-j-k}} * f = \sum_{j\in\mathbf{Z}} (\mu * \psi_{2^{-k}})_{2^{-j}} * f\,.$$

Then for f in $\mathscr{S}$ we have that

$$T_\mu(f) = \sum_{j\in\mathbf{Z}} \mu_j * f = \sum_{j\in\mathbf{Z}}\sum_{k\in\mathbf{Z}} \mu_j * \psi_{2^{-j-k}} * f = \sum_{k\in\mathbf{Z}} S_k(f)\,.$$

It suffices therefore to obtain L^p boundedness for the sum of the S_k's. We begin by investigating the L^2 boundedness of each S_k. Since the product $\widehat{\psi_{2^{-j-k}}}\widehat{\psi_{2^{-j'-k}}}$ is nonzero only when $j' \in \{j-1, j, j+1\}$, it follows that

$$\begin{aligned}
\big\|S_k(f)\big\|_{L^2}^2 &\le \sum_{j\in\mathbf{Z}}\sum_{j'\in\mathbf{Z}}\int_{\mathbf{R}^n} |\widehat{\mu_j}(\xi)\overline{\widehat{\mu_{j'}}}(\xi)\widehat{\psi}(2^{-j-k}\xi)\overline{\widehat{\psi}}(2^{-j'-k}\xi)|\,|\widehat{f}(\xi)|^2\,d\xi \\
&\le C_1\sum_{j\in\mathbf{Z}}\sum_{j'=j-1}^{j+1}\int_{|\xi|\approx 2^{j+k}} |\widehat{\mu_j}(\xi)\overline{\widehat{\mu_{j'}}}(\xi)|\,|\widehat{f}(\xi)|^2\,d\xi \\
&\le C_2\sum_{j\in\mathbf{Z}}\int_{|\xi|\approx 2^{j+k}} B^2\min(|2^{-j}\xi|^{b}, |2^{-j}\xi|^{-b})^2|\widehat{f}(\xi)|^2\,d\xi \\
&\le C_3^2B^2 2^{-2|k|b}\sum_{j\in\mathbf{Z}}\int_{|\xi|\approx 2^{j+k}} |\widehat{f}(\xi)|^2\,d\xi \\
&= C_3^2B^2\, 2^{-2|k|b}\big\|f\big\|_{L^2}^2 .
\end{aligned}$$

We have therefore obtained that for all $k\in\mathbf{Z}$ and $f\in\mathscr{S}(\mathbf{R}^n)$ we have

$$\big\|S_k(f)\big\|_{L^2} \le C_3\,B\,2^{-b|k|}\big\|f\big\|_{L^2} . \tag{6.3.7}$$

We notice that for any $R>0$ we have

$$\begin{aligned}
\int_{R\le|x|\le 2R}\sum_{j\in\mathbf{Z}}\big|(\mu*\psi_{2^{-k}})_{2^{-j}}(x)\big|\,dx &= \sum_{j\in\mathbf{Z}}\int_{2^jR\le|x|\le 2^{j+1}R}\big|(\mu*\psi_{2^{-k}})(x)\big|\,dx \\
&= \int_{\mathbf{R}^n}\big|(\mu*\psi_{2^{-k}})(x)\big|\,dx \\
&\le \big\|\mu\big\|\,\|\psi\|_{L^1} ,
\end{aligned}$$

thus condition (5.3.4) of Theorem 5.3.3 is satisfied.

Next we verify that the kernel of each S_k satisfies Hörmander's condition with constant at most a multiple of $(1+|k|)$. Fix $y\neq 0$. Then

$$\begin{aligned}
&\int_{|x|\ge 2|y|}\bigg|\sum_{j\in\mathbf{Z}}\Big((\mu*\psi_{2^{-k}})_{2^{-j}}(x-y)-(\mu*\psi_{2^{-k}})_{2^{-j}}(x)\Big)\bigg|\,dx \\
&\qquad\le \sum_{j\in\mathbf{Z}}\int_{|x|\ge 2|y|} 2^{jn}\big|(\mu*\psi_{2^{-k}})(2^jx-2^jy)-(\mu*\psi_{2^{-k}})(2^jx)\big|\,dx \\
&\qquad= \sum_{j\in\mathbf{Z}} I_{j,k}(y) ,
\end{aligned}$$

where

$$I_{j,k}(y) = \int_{|x|\ge 2^{j+1}|y|}\big|(\mu*\psi_{2^{-k}})(x-2^jy)-(\mu*\psi_{2^{-k}})(x)\big|\,dx .$$

We observe that $I_{j,k}(y) \le C_4 \|\mu\|_{\mathcal{M}}$. Let $|\mu|$ be the total variation of μ. To obtain a more delicate estimate for $I_{j,k}(y)$ we argue as follows:

$$
\begin{aligned}
I_{j,k}(y) &\le \int_{|x|\ge 2^{j+1}|y|} \int_{\mathbf{R}^n} \big|\psi_{2^{-k}}(x-2^j y - z) - \psi_{2^{-k}}(x-z)\big|\, d|\mu|(z)\, dx \\
&= \int_{\mathbf{R}^n} 2^{kn} \int_{|x|\ge 2^{j+1}|y|} \big|\psi(2^k x - 2^k z - 2^{j+k} y) - \psi(2^k x - 2^k z)\big|\, dx\, d|\mu|(z) \\
&\le C_5 \int_{|x|\ge 2^{j+1}|y|} \int_{\mathbf{R}^n} 2^{kn} 2^{j+k}|y| \big|\nabla\psi(2^k x - 2^k z - \theta)\big|\, d|\mu|(z)\, dx \\
&\le C_6 2^{j+k} \int_{\mathbf{R}^n} \int_{|x|\ge 2^{j+1}|y|} 2^{kn}|y| \big(1+|2^k x - 2^k z - \theta|\big)^{-n-2} dx\, d|\mu|(z) \\
&= C_6 2^{j+k}|y| \int_{\mathbf{R}^n} \int_{|x|\ge 2^{j+k+1}|y|} \big(1+|x - 2^k z - \theta|\big)^{-n-2} dx\, d|\mu|(z),
\end{aligned}
$$

where $|\theta| \le 2^{j+k}|y|$. Note that θ depends on j,k, and y. From this and from $I_{j,k}(y) \le C_4\|\mu\|_{\mathcal{M}}$ we obtain

$$
I_{j,k}(y) \le C_7 \left\|\mu\right\|_{\mathcal{M}} \min\big(1, 2^{j+k}|y|\big), \tag{6.3.8}
$$

which is valid for all j,k, and $y \ne 0$. To estimate the last double integral even more delicately, we consider the following two cases: $|x| \ge 2^{k+2}|z|$ and $|x| < 2^{k+2}|z|$. In the first case we have $|x - 2^k z - \theta| \ge \frac{1}{4}|x|$, given the fact that $|x| \ge 2^{j+k+1}|y|$. In the second case we have that $|x| \le 2^{k+2}R$, where $B(0,R)$ contains the support of μ. Applying these observations in the last double integral, we obtain the following estimate:

$$
\begin{aligned}
I_{j,k}(y) &\le C_8 2^{j+k}|y| \int_{\mathbf{R}^n} \Bigg[\int_{\substack{|x|\ge 2^{j+k+1}|y| \\ |x|\ge 2^{k+2}|z|}} \frac{dx}{\big(1+\frac{1}{4}|x|\big)^{n+2}} + \int_{\substack{|x|\ge 2^{j+k+1}|y| \\ |x|< 2^{k+2}R}} dx \Bigg] d|\mu|(z) \\
&\le C_9 2^{j+k}|y| \|\mu\|_{\mathcal{M}} \left[\frac{1}{(2^{j+k}|y|)^2} + 0 \right] \\
&= C_9 (2^{j+k}|y|)^{-1} \left\|\mu\right\|_{\mathcal{M}},
\end{aligned}
$$

provided $2^j|y| \ge 2R$. Combining this estimate with (6.3.8), we obtain

$$
I_{j,k}(y) \le C_{10} \|\mu\|_{\mathcal{M}} \begin{cases} \min\big(1, 2^{j+k}|y|\big) & \text{for all } j,k \text{ and } y, \\ (2^{j+k}|y|)^{-1} & \text{when } 2^j|y| \ge 2R. \end{cases} \tag{6.3.9}
$$

We now estimate $\sum_j I_{j,k}(y)$. When $2^k \geq (2R)^{-1}$ we use (6.3.9) to obtain

$$\sum_j I_{j,k}(y) \leq C_{10}\|\mu\|_{\mathscr{M}}\Big[\sum_{2^j\leq\frac{1}{2^k|y|}} 2^{j+k}|y| + \sum_{\frac{1}{2^k|y|}\leq 2^j\leq\frac{2R}{|y|}} 1 + \sum_{2^j\geq\frac{2R}{|y|}} (2^{j+k}|y|)^{-1}\Big]$$
$$\leq C_{11}\|\mu\|_{\mathscr{M}}(|\log R| + |k|)\,.$$

Also when $2^k < (2R)^{-1}$ we again use (6.3.9) to obtain

$$\sum_j I_{j,k}(y) \leq C_{10}\|\mu\|_{\mathscr{M}}\Big[\sum_{2^j\leq\frac{1}{2^k|y|}} 2^{j+k}|y| + \sum_{2^j\geq\frac{1}{2^k|y|}} (2^{j+k}|y|)^{-1}\Big] \leq C_{12}\|\mu\|_{\mathscr{M}},$$

since in the second sum we have $2^j|y| \geq 2^{-k} > 2R$, which justifies use of the corresponding estimate in (6.3.9). This gives

$$\sum_j I_{j,k}(y) \leq C_{13}\|\mu\|_{\mathscr{M}}(1+|k|)\,, \tag{6.3.10}$$

where the constant C_{13} depends on the dimension and on R. We now use estimates (6.3.7) and (6.3.10) and Theorem 5.3.3 to obtain that each S_k maps $L^1(\mathbf{R}^n)$ to $L^{1,\infty}(\mathbf{R}^n)$ with constant at most

$$C_n(2^{-b|k|} + 1 + |k|)\|\mu\|_{\mathscr{M}} \leq C_n(2+|k|)\|\mu\|_{\mathscr{M}}\,.$$

It follows from the Marcinkiewicz interpolation theorem (Theorem 1.3.2) that S_k maps $L^p(\mathbf{R}^n)$ to itself for $1<p<2$ with bound at most $C_{p,n}2^{-b|k|\theta_p}(1+|k|)^{1-\theta_p}$, when $\frac{1}{p} = \frac{\theta_p}{2} + 1 - \theta_p$. Summing over all $k\in\mathbf{Z}$, we obtain that T_μ maps $L^p(\mathbf{R}^n)$ to itself for $1<p<2$. The boundedness of T_μ for $p>2$ follows by duality. □

An immediate consequence of the previous result is the following.

Corollary 6.3.5. *Suppose that μ is a finite Borel measure on $\mathbf{R}^n$ with compact support that satisfies $|\widehat{\mu}(\xi)| \leq B\min\big(|\xi|^{-b}, |\xi|^{b}\big)$ for some $b>0$ and all $\xi\neq 0$. Define measures μ_j by setting $\widehat{\mu_j}(\xi) = \widehat{\mu}(2^{-j}\xi)$. Then the square function*

$$G(f) = \Big(\sum_{j\in\mathbf{Z}} |\mu_j * f|^2\Big)^{\frac{1}{2}} \tag{6.3.11}$$

maps $L^p(\mathbf{R}^n)$ to itself whenever $1<p<\infty$.

Proof. To obtain the boundedness of the square function in (6.3.11) we use the Rademacher functions $r_j(t)$, introduced in Appendix C.1, reindexed so that their index set is the set of all integers (not the set of nonnegative integers). For each t we introduce the operators

$$T_\mu^t(f) = \sum_{j\in\mathbf{Z}} r_j(t)(f*\mu_j)\,.$$

Next we observe that for each t in $[0,1]$ the operators T_μ^t map $L^p(\mathbf{R}^n)$ to itself with the same constant as the operator T_μ, which is in particular independent of t. Using that the square function in (6.3.11) raised to the power p is controlled by a multiple of the quantity

$$\int_0^1 \Big|\sum_{j\in\mathbf{Z}} r_j(t)(f*\mu_j)\Big|^p dt,$$

a fact stated in Appendix C.2, we obtain the required conclusion by integrating over $\mathbf{R}^n$. □

6.3.3 An Almost Orthogonality Principle on L^p

Suppose that T_j are multiplier operators given by $T_j(f) = (\widehat{f}m_j)^\vee$, for some multipliers m_j. If the functions m_j have disjoint supports and they are bounded uniformly in j, then the operator

$$T = \sum_j T_j$$

is bounded on L^2. The following theorem gives an L^p analogue of this result.

Theorem 6.3.6. *Suppose that $1 < p \le 2 \le q < \infty$. Let m_j be Schwartz functions supported in the annuli $2^{j-1} \le |\xi| \le 2^{j+1}$ and let $T_j(f) = (\widehat{f}m_j)^\vee$. Suppose that the T_j's are uniformly bounded operators from $L^p(\mathbf{R}^n)$ to $L^q(\mathbf{R}^n)$, i.e.,*

$$\sup_j \big\|T_j\big\|_{L^p\to L^q} = A < \infty .$$

Then for each $f \in L^p(\mathbf{R}^n)$, the series

$$T(f) = \sum_j T_j(f)$$

converges in the L^q norm and there exists a constant $C_{p,q,n} < \infty$ such that

$$\big\|T\big\|_{L^p\to L^q} \le C_{p,q,n}A. \tag{6.3.12}$$

Proof. Fix a radial Schwartz function φ whose Fourier transform $\widehat{\varphi}$ is real, equal to one on the annulus $\frac{1}{2} \le |\xi| \le 2$, and vanishes outside the annulus $\frac{1}{4} \le |\xi| \le 4$. We set $\varphi_{2^{-j}}(x) = 2^{jn}\varphi(2^jx)$, so that $\widehat{\varphi_{2^{-j}}}$ is equal to 1 on the support of each m_j. Setting $\Delta_j(f) = f*\varphi_{2^{-j}}$, we observe that

$$T_j = \Delta_j T_j \Delta_j$$

for all $j \in \mathbf{Z}$. For a positive integer N we set

$$T^N = \sum_{|j|\le N} \Delta_j T_j \Delta_j .$$

Fix $f \in L^p(\mathbf{R}^n)$. Clearly for every N, $T^N(f)$ is in $L^q(\mathbf{R}^n)$. Using (6.1.21) we obtain

$$\begin{aligned}
\big\|T^N(f)\big\|_{L^q} &= \Big\| \sum_{|j|\le N} \Delta_j T_j \Delta_j(f)\Big\|_{L^q} \\
&\le C_q' \Big\| \Big(\sum_{j\in\mathbf{Z}} |T_j\Delta_j(f)|^2\Big)^{\frac12}\Big\|_{L^q} \\
&= C_q' \Big\| \sum_{j\in\mathbf{Z}} |T_j\Delta_j(f)|^2\Big\|_{L^{q/2}}^{\frac12} \\
&\le C_q' \Big(\sum_{j\in\mathbf{Z}} \Big\| |T_j\Delta_j(f)|^2\Big\|_{L^{q/2}}\Big)^{\frac12} \\
&= C_q' \Big(\sum_{j\in\mathbf{Z}} \big\| T_j\Delta_j(f)\big\|_{L^q}^2\Big)^{\frac12},
\end{aligned}$$

where we used Minkowski's inequality, since $q/2 \ge 1$. Using the uniform boundedness of the T_j's from L^p to L^q, we deduce that

$$\begin{aligned}
C_q' \Big(\sum_{j\in\mathbf{Z}} \big\| T_j\Delta_j(f)\big\|_{L^q}^2\Big)^{\frac12} &\le C_q' A \Big(\sum_{j\in\mathbf{Z}} \big\| \Delta_j(f)\big\|_{L^p}^2\Big)^{\frac12} \\
&= C_q' A \Big(\sum_{j\in\mathbf{Z}} \big\| |\Delta_j(f)|^2\big\|_{L^{p/2}}\Big)^{\frac12} \\
&\le C_q' A \Big(\Big\| \sum_{j\in\mathbf{Z}} |\Delta_j(f)|^2\Big\|_{L^{p/2}}\Big)^{\frac12} \\
&= C_q' A \Big\| \Big(\sum_{j\in\mathbf{Z}} |\Delta_j(f)|^2\Big)^{\frac12}\Big\|_{L^p} \\
&\le C_q' C_p A \big\| f\big\|_{L^p(\mathbf{R}^n)},
\end{aligned}$$

where we used the result of Exercise 1.1.5(b), since $p \le 2$, and Theorem 6.1.2. We conclude that the operators T^N are uniformly bounded from $L^p(\mathbf{R}^n)$ to $L^q(\mathbf{R}^n)$.

If $\widehat{h}$ is compactly supported in a subset of $\mathbf{R}^n \setminus \{0\}$, then the sequence $T^N(h)$ becomes independent of N for N large enough and hence it is Cauchy in L^q. But in view of Exercise 6.2.9, the set of all such h is dense in $L^p(\mathbf{R}^n)$. Combining these two results with the uniform boundedness of the T^N's from L^p to L^q, a simple $\frac{\varepsilon}{3}$ argument gives that for all $f \in L^p$ the sequence $T^N(f)$ is Cauchy in L^q. Therefore, for all $f \in L^p$ the sequence $\{T^N(f)\}_N$ converges in L^q to some $T(f)$. Fatou's lemma gives

$$\big\|T(f)\big\|_{L^q} \le C_q' C_p A \big\|f\big\|_{L^p},$$

which proves (6.3.12). □

Exercises

6.3.1. (*The g-function*) Let $P_t(x) = \Gamma(\frac{n+1}{2})\pi^{-\frac{n+1}{2}} t(t^2+|x|^2)^{-\frac{n+1}{2}}$ be the Poisson kernel on $\mathbf{R}^n$.
(a) Use Exercise 6.1.4 with $\Psi(x) = \frac{\partial}{\partial t}P_t(x)\big|_{t=1}$ to obtain that the operator

$$f \to \left(\int_0^\infty t\left|\frac{\partial}{\partial t}(P_t * f)(x)\right|^2 dt\right)^{1/2}$$

is bounded from $L^p(\mathbf{R}^n)$ to $L^p(\mathbf{R}^n)$ for $1<p<\infty$.
(b) Use Exercise 6.1.4 with $\Psi(x) = \partial_k P_1(x)$ to obtain that the operator

$$f \to \left(\int_0^\infty t|\partial_k(P_t * f)(x)|^2 dt\right)^{1/2}$$

is bounded from $L^p(\mathbf{R}^n)$ to $L^p(\mathbf{R}^n)$ for $1<p<\infty$.
(c) Conclude that the g-function

$$g(f)(x) = \left(\int_0^\infty t|\nabla_{x,t}(P_t * f)(x)|^2 dt\right)^{1/2}$$

is bounded from $L^p(\mathbf{R}^n)$ to $L^p(\mathbf{R}^n)$ for $1<p<\infty$.

6.3.2. Suppose that μ is a finite Borel measure on $\mathbf{R}^n$ with compact support that satisfies $\widehat{\mu}(0)=0$ and $|\widehat{\mu}(\xi)| \le C|\xi|^{-a}$ for some $a>0$ and all $\xi \neq 0$. Define measures μ_j by setting $\widehat{\mu_j}(\xi) = \widehat{\mu}(2^{-j}\xi)$. Show that the operator

$$T_\mu(f)(x) = \sum_{j\in\mathbf{Z}} (f*\mu_j)(x)$$

is bounded from $L^p(\mathbf{R}^n)$ to $L^p(\mathbf{R}^n)$ for all $1<p<\infty$.
[*Hint:* Use Theorem 6.3.4]

6.3.3. ([50], [71]) (a) Suppose that μ is a finite Borel measure on $\mathbf{R}^n$ with compact support that satisfies $|\widehat{\mu}(\xi)| \le C|\xi|^{-a}$ for some $a>0$ and all $\xi \neq 0$. Show that the maximal function

$$\mathscr{M}_\mu(f)(x) = \sup_{j\in\mathbf{Z}}\left|\int_{\mathbf{R}^n} f(x-2^j y)\,d\mu(y)\right|$$

is bounded from $L^p(\mathbf{R}^n)$ to $L^p(\mathbf{R}^n)$ for all $1<p<\infty$.
(b) Let μ be the surface measure on the sphere $\mathbf{S}^{n-1}$ when $n\ge 2$. Conclude that the *dyadic spherical maximal function* $\mathscr{M}_\mu$ is bounded on $L^p(\mathbf{R}^n)$ for all $1<p<\infty$.
[*Hint:* Part (a): Pick a $\mathscr{C}_0^\infty$ function φ on $\mathbf{R}^n$ with $\widehat{\varphi}(0)=1$. Then the measure $\sigma = \mu - \widehat{\mu}(0)\varphi$ satisfies the hypotheses of Corollary 6.3.5. Since,

$$\mathscr{M}_\mu(f)(x) \le \Big(\sum_j |(\sigma_j * f)(x)|^2\Big)^{1/2} + |\widehat{\mu}(0)|M(f)(x),$$

it follows that $\mathscr{M}_\mu$ is bounded on $L^p(\mathbf{R}^n)$ whenever $1 < p < \infty$. Part (b): If $\mu = d\sigma$ is surface measure on $\mathbf{S}^{n-1}$, then $|\widehat{d\sigma}(\xi)| \le C|\xi|^{-\frac{n-1}{2}}$ (Appendices B.4 and B.7).]

6.3.4. Let Ω be in $L^q(\mathbf{S}^{n-1})$ for some $1 < q < \infty$ and define the absolutely continuous measure

$$d\mu(x) = \frac{\Omega(x/|x|)}{|x|^n}\chi_{1<|x|\le 2}\,dx\,.$$

Show that for all $a < 1/q'$ we have that $|\widehat{\mu}(\xi)| \le C|\xi|^{-a}$. Under the additional hypothesis that Ω has mean value zero, conclude that the singular integral operator

$$T_\Omega(f)(x) = \text{p.v.} \int_{\mathbf{R}^n} \frac{\Omega(y/|y|)}{|y|^n} f(x-y)\,dy = \sum_j f * \mu_j$$

is L^p bounded for all $1 < p < \infty$. This provides an alternative proof of Theorem 5.2.10 under the hypothesis that $\Omega \in L^q(\mathbf{S}^{n-1})$.

6.3.5. For a continuous function F on $\mathbf{R}$ define

$$u(F)(x) = \left(\int_0^\infty |F(x+t) + F(x-t) - 2F(x)|^2 \frac{dt}{t^3} \right)^{\frac{1}{2}}.$$

Given $f \in L^1_{\mathrm{loc}}(\mathbf{R})$ we denote by F_f the indefinite integral of f, that is,

$$F_f(x) = \int_0^x f(t)\,dt\,.$$

Prove that for all $1 < p < \infty$ there exist constants c_p and C_p such that for all functions $f \in L^p(\mathbf{R})$ we have

$$c_p \big\|f\big\|_{L^p} \le \big\|u(F_f)\big\|_{L^p} \le C_p \big\|f\big\|_{L^p}\,.$$

[*Hint:* Let $\varphi = \chi_{[-1,0]} - \chi_{[0,1]}$. Then

$$(\varphi_t * f)(x) = \frac{1}{t}\big(F_f(x+t) + F_f(x-t) - 2F_f(x)\big)$$

and the double inequality follows from parts (b) and (c) of Exercise 6.1.4.]

6.3.6. Let m be a bounded function on $\mathbf{R}^n$ that is $\mathscr{C}^1$ in a neighborhood of zero, it satisfies $m(0) = 1$ and $|m(\xi)| \le B|\xi|^{-\varepsilon}$ for all $\xi \ne 0$, for some $B, \varepsilon > 0$. Define an operator T_t by setting $T_t(f)\widehat{\ }(\xi) = \widehat{f}(\xi)m(t\xi)$. Show that the maximal operator

$$\sup_{N>0} \left(\frac{1}{N} \int_0^N \big|T_t(f)(x)\big|^2 dt \right)^{\frac{1}{2}}$$

maps $L^2(\mathbf{R}^n)$ to itself.

[*Hint:* Majorize this maximal operator by a constant multiple of the sum

$$M(f)(x)+\Big(\int_0^\infty |T_t(f)(x)-(f*\varphi_t)(x)|^2\frac{dt}{t}\Big)^{\frac12},$$

where φ is a $\mathscr{C}_0^\infty$ function such that $\widehat{\varphi}(0)=1$.]

6.3.7. ([150]) Let $0<\beta<1$ and $p_0=(1-\beta/2)^{-1}$. Suppose that $\{f_j\}_{j\in\mathbf{Z}}$ are L^2 functions on the real line with norm at most 1. Assume that each f_j is supported in interval of length 1 and that the orthogonality relation $|\langle f_j\,|\,f_k\rangle|\le(1+|j-k|)^{-\beta}$ holds for all $j,k\in\mathbf{Z}$.
(a) Let $I\subseteqq\mathbf{Z}$ be such that for all $j\in I$ the functions f_j are supported in a fixed interval of length 3. Show that for all p satisfying $0<p\le 2$ there is $C_{p,\beta}<\infty$ such that

$$\Big\|\sum_{j\in I}\varepsilon_j f_j\Big\|_{L^p}\le C_{p,\beta}\,|I|^{1-\frac{\beta}{2}}$$

whenever ε_j are complex numbers with $|\varepsilon_j|\le 1$.
(b) Under the same hypothesis as in part (a), prove that for all $0<p<p_0$ there is a constant $C'_{p,\beta}<\infty$ such that

$$\Big\|\sum_{j\in I}c_j f_j\Big\|_{L^p}\le C'_{p,\beta}\Big(\sum_{j\in\mathbf{Z}}|c_j|^p\Big)^{\frac1p}$$

for all complex-valued sequences $\{c_j\}_j$ in ℓ^p.
(c) Derive the conclusion of part (b) without the assumption that the f_j are supported in a fixed interval of length 3.
[*Hint:* Part (a): Pass from L^p to L^2 and use the hypothesis. Part (b): Assume $\sum_{j\in\mathbf{Z}}|c_j|^p=1$. For each $k=0,1,\dots$, set $I_k=\{j\in\mathbf{Z}:\,2^{-k-1}<|c_j|\le 2^{-k}\}$. Write $\big\|\sum_{j\in\mathbf{Z}}c_jf_j\big\|_{L^p}\le\sum_{k=0}^\infty 2^{-k}\big\|\sum_{j\in I_k}(c_j2^k)f_j\big\|_{L^p}$, use part (b), Hölder's inequality, and the fact that $\sum_{k=0}^\infty 2^{-kp}|I_k|\le 2^p$. Part (c): Write $\sum_{j\in\mathbf{Z}}c_jf_j=\sum_{m\in\mathbf{Z}}F_m$, where F_m is the sum of c_jf_j over all j such that the support of f_j meets the interval $[m,m+1]$. These F_m's are supported in $[m-1,m+2]$ and are almost orthogonal.]

6.4 The Haar System, Conditional Expectation, and Martingales

There is a very strong connection between the Littlewood–Paley operators and certain notions from probability, such as conditional expectation and martingale difference operators. The conditional expectation we are concerned with is with respect to the increasing σ-algebra of all dyadic cubes on $\mathbf{R}^n$.

6.4.1 Conditional Expectation and Dyadic Martingale Differences

We recall the definition of dyadic cubes.

Definition 6.4.1. A *dyadic interval* in $\mathbf{R}$ is an interval of the form

$$\big[m2^{-k},(m+1)2^{-k}\big)$$

where m,k are integers. A *dyadic cube* in $\mathbf{R}^n$ is a product of dyadic intervals of the same length. That is, a dyadic cube is a set of the form

$$\prod_{j=1}^{n}\big[m_j2^{-k},(m_j+1)2^{-k}\big)$$

for some integers $m_1,\dots,m_n,k$.

We defined dyadic intervals to be closed on the left and open on the right, so that different dyadic intervals of the same length are always disjoint sets.

Given a cube Q in $\mathbf{R}^n$ we denote by $|Q|$ its Lebesgue measure and by $\ell(Q)$ its side length. We clearly have $|Q|=\ell(Q)^n$. We introduce some more notation.

Definition 6.4.2. For $k\in\mathbf{Z}$ we denote by $\mathscr{D}_k$ the set of all dyadic cubes in $\mathbf{R}^n$ whose side length is 2^{-k}. We also denote by $\mathscr{D}$ the set of all dyadic cubes in $\mathbf{R}^n$. Then we have

$$\mathscr{D}=\bigcup_{k\in\mathbf{Z}}\mathscr{D}_k,$$

and moreover, the σ-algebra $\sigma(\mathscr{D}_k)$ of measurable subsets of $\mathbf{R}^n$ formed by countable unions and complements of elements of $\mathscr{D}_k$ is increasing as k increases.

We observe the fundamental property of dyadic cubes, which clearly justifies their usefulness. Any two dyadic intervals of the same side length either are disjoint or coincide. Moreover, either two given dyadic intervals are disjoint, or one contains the other. Similarly, either two dyadic cubes are disjoint, or one contains the other.

Definition 6.4.3. Given a locally integrable function f on $\mathbf{R}^n$, we denote by

$$\operatorname*{Avg}_{Q} f=\frac{1}{|Q|}\int_Q f(t)\,dt$$

the average of f over a cube Q.

The *conditional expectation* of a locally integrable function f on $\mathbf{R}^n$ with respect to the increasing family of σ-algebras $\sigma(\mathscr{D}_k)$ generated by $\mathscr{D}_k$ is defined as

$$E_k(f)(x)=\sum_{Q\in\mathscr{D}_k}\big(\operatorname*{Avg}_{Q} f\big)\,\chi_Q(x),$$

for all $k \in \mathbf{Z}$. We also define the *dyadic martingale difference operator* D_k as follows:

$$D_k(f) = E_k(f) - E_{k-1}(f),$$

also for $k \in \mathbf{Z}$.

Next we introduce the family of Haar functions.

Definition 6.4.4. For a dyadic interval $I = [m2^{-k}, (m+1)2^{-k})$ we define $I_L = [m2^{-k}, (m+\frac{1}{2})2^{-k})$ and $I_R = [(m+\frac{1}{2})2^{-k}, (m+1)2^{-k})$ to be the left and right parts of I, respectively. The function

$$h_I(x) = |I|^{-\frac{1}{2}}\chi_{I_L} - |I|^{-\frac{1}{2}}\chi_{I_R}$$

is called the *Haar function associated with the interval I.*

We remark that Haar functions are constructed in such a way that they have L^2 norm equal to 1. Moreover, the Haar functions have the following fundamental orthogonality property:

$$\int_{\mathbf{R}} h_I(x)h_{I'}(x)\,dx = \begin{cases} 0 & \text{when } I \neq I', \\ 1 & \text{when } I = I'. \end{cases} \tag{6.4.1}$$

To see this, observe that the Haar functions have L^2 norm equal to 1 by construction. Moreover, if $I \neq I'$, then I and I' must have different lengths, say we have $|I'| < |I|$. If I and I' are not disjoint, then I' is contained either in the left or in the right half of I, on either of which h_I is constant. Thus (6.4.1) follows.

We recall the notation

$$\langle f, g \rangle = \int_{\mathbf{R}} f(x)g(x)\,dx$$

valid for square integrable functions. Under this notation, (6.4.1) can be rewritten as $\langle h_I, h_{I'} \rangle = \delta_{I,I'}$, where the latter is 1 when $I = I'$ and zero otherwise.

6.4.2 *Relation Between Dyadic Martingale Differences and Haar Functions*

We have the following result relating the Haar functions to the dyadic martingale difference operators in dimension one.

Proposition 6.4.5. *For every locally integrable function f on $\mathbf{R}$ and for all $k \in \mathbf{Z}$ we have the identity*

$$D_k(f) = \sum_{I \in \mathscr{D}_{k-1}} \langle f, h_I \rangle h_I \tag{6.4.2}$$

and also

$$\big\|D_k(f)\big\|_{L^2}^2 = \sum_{I\in\mathscr{D}_{k-1}} \big|\langle f,h_I\rangle\big|^2. \tag{6.4.3}$$

Proof. We observe that every interval J in $\mathscr{D}_k$ is either an I_L or an I_R for some unique $I \in \mathscr{D}_{k-1}$. Thus we can write

$$\begin{aligned} E_k(f) &= \sum_{J\in\mathscr{D}_k} (\operatorname*{Avg}_J f)\,\chi_J \\ &= \sum_{I\in\mathscr{D}_{k-1}} \bigg[\bigg(\frac{2}{|I|}\int_{I_L} f(t)\,dt\bigg)\chi_{I_L} + \bigg(\frac{2}{|I|}\int_{I_R} f(t)\,dt\bigg)\chi_{I_R}\bigg]. \end{aligned} \tag{6.4.4}$$

But we also have

$$\begin{aligned} E_{k-1}(f) &= \sum_{I\in\mathscr{D}_{k-1}} (\operatorname*{Avg}_I f)\,\chi_I \\ &= \sum_{I\in\mathscr{D}_{k-1}} \bigg(\frac{1}{|I|}\int_{I_L} f(t)\,dt + \frac{1}{|I|}\int_{I_R} f(t)\,dt\bigg)(\chi_{I_L}+\chi_{I_R}). \end{aligned} \tag{6.4.5}$$

Now taking the difference between (6.4.4) and (6.4.5) we obtain

$$\begin{aligned} D_k(f) = \sum_{I\in\mathscr{D}_{k-1}} \bigg[&\bigg(\frac{1}{|I|}\int_{I_L} f(t)\,dt\bigg)\chi_{I_L} - \bigg(\frac{1}{|I|}\int_{I_R} f(t)\,dt\bigg)\chi_{I_L} \\ &+ \bigg(\frac{1}{|I|}\int_{I_R} f(t)\,dt\bigg)\chi_{I_R} - \bigg(\frac{1}{|I|}\int_{I_L} f(t)\,dt\bigg)\chi_{I_R}\bigg], \end{aligned}$$

which is easily checked to be equal to

$$\sum_{I\in\mathscr{D}_{k-1}} \bigg(\int_I f(t)h_I(t)\,dt\bigg)h_I = \sum_{I\in\mathscr{D}_{k-1}} \langle f,h_I\rangle\, h_I.$$

Finally, (6.4.3) is a consequence of (6.4.1). □

Theorem 6.4.6. *Every function $f \in L^2(\mathbf{R}^n)$ can be written as*

$$f = \sum_{k\in\mathbf{Z}} D_k(f), \tag{6.4.6}$$

where the series converges almost everywhere and in L^2. We also have

$$\big\|f\big\|_{L^2(\mathbf{R}^n)}^2 = \sum_{k\in\mathbf{Z}} \big\|D_k(f)\big\|_{L^2(\mathbf{R}^n)}^2. \tag{6.4.7}$$

Moreover, when $n=1$ we have the representation

$$f = \sum_{I\in\mathscr{D}} \langle f,h_I\rangle h_I, \tag{6.4.8}$$

where the sum converges a.e. and in L^2 and also

$$\|f\|_{L^2(\mathbf{R})}^2 = \sum_{I\in\mathscr{D}} |\langle f, h_I\rangle|^2 . \tag{6.4.9}$$

Proof. In view of the Lebesgue differentiation theorem, Corollary 2.1.16, given a function $f \in L^2(\mathbf{R}^n)$ there is a set N_f of measure zero on $\mathbf{R}^n$ such that for all $x \in \mathbf{R}^n \setminus N_f$ we have that

$$\operatorname*{Avg}_{Q_j} f \to f(x)$$

whenever Q_j is a sequence of decreasing cubes such that $\bigcap_j \overline{Q_j} = \{x\}$. Given x in $\mathbf{R}^n \setminus N_f$ there exists a unique sequence of dyadic cubes $Q_j(x) \in \mathscr{D}_j$ such that $\bigcap_{j=0}^{\infty} \overline{Q_j(x)} = \{x\}$. Then for all $x \in \mathbf{R}^n \setminus N_f$ we have

$$\lim_{j\to\infty} E_j(f)(x) = \lim_{j\to\infty} \sum_{Q\in\mathscr{D}_j} (\operatorname*{Avg}_{Q} f)\, \chi_Q(x) = \lim_{j\to\infty} \operatorname*{Avg}_{Q_j(x)} f = f(x) .$$

From this we conclude that $E_j(f) \to f$ a.e. as $j \to \infty$. We also observe that since $|E_j(f)| \le M_c(f)$, where M_c denotes the uncentered maximal function with respect to cubes, we have that $|E_j(f) - f| \le 2M_c(f)$, which allows us to obtain from the Lebesgue dominated convergence theorem that $E_j(f) \to f$ in L^2 as $j \to \infty$.

Next we study convergence of $E_j(f)$ as $j \to -\infty$. For a given $x \in \mathbf{R}^n$ and $Q_j(x)$ as before we have that

$$|E_j(f)(x)| = \Big| \operatorname*{Avg}_{Q_j(x)} f \Big| \le \left(\frac{1}{|Q_j(x)|} \int_{Q_j(x)} |f(t)|^2\, dt \right)^{\frac{1}{2}} \le 2^{\frac{jn}{2}} \|f\|_{L^2} ,$$

which tends to zero as $j \to -\infty$, since the side length of each $Q_j(x)$ is 2^{-j}. Since $|E_j(f)| \le M_c(f)$, the Lebesgue dominated convergence theorem allows us to conclude that $E_j(f) \to 0$ in L^2 as $j \to -\infty$. To obtain the conclusion asserted in (6.4.6) we simply observe that

$$\sum_{k=M}^{N} D_k(f) = E_N(f) - E_{M-1}(f) \to f$$

in L^2 and almost everywhere as $N \to \infty$ and $M \to -\infty$.

To prove (6.4.7) we first observe that we can rewrite $D_k(f)$ as

$$\begin{aligned} D_k(f) &= \sum_{Q\in\mathscr{D}_k} (\operatorname*{Avg}_{Q} f)\, \chi_Q - \sum_{R\in\mathscr{D}_{k-1}} (\operatorname*{Avg}_{R} f)\, \chi_R \\ &= \sum_{R\in\mathscr{D}_{k-1}} \Bigg[\sum_{\substack{Q\in\mathscr{D}_k \\ Q\subseteq R}} (\operatorname*{Avg}_{Q} f)\, \chi_Q - (\operatorname*{Avg}_{R} f)\, \chi_R \Bigg] \end{aligned}$$

$$= \sum_{R\in\mathcal{D}_{k-1}} \Bigg[\sum_{\substack{Q\in\mathcal{D}_k \\ Q\subseteq R}} (\operatorname{Avg}_Q f)\,\chi_Q - \frac{1}{2^n} \sum_{\substack{Q\in\mathcal{D}_k \\ Q\subseteq R}} (\operatorname{Avg}_Q f)\,\chi_R \Bigg]$$

$$= \sum_{R\in\mathcal{D}_{k-1}} \sum_{\substack{Q\in\mathcal{D}_k \\ Q\subseteq R}} (\operatorname{Avg}_Q f)\,\big(\chi_Q - 2^{-n}\chi_R\big)\,. \tag{6.4.10}$$

Using this identity we obtain that for given integers $k' > k$ we have

$$\int_{\mathbf{R}^n} D_k(f)(x)\,D_{k'}(f)(x)\,dx$$

$$= \sum_{R\in\mathcal{D}_{k-1}} \sum_{\substack{Q\in\mathcal{D}_k \\ Q\subseteq R}} (\operatorname{Avg}_Q f) \sum_{R'\in\mathcal{D}_{k'-1}} \sum_{\substack{Q'\in\mathcal{D}_{k'} \\ Q'\subseteq R'}} (\operatorname{Avg}_{Q'} f) \int \big(\chi_Q - 2^{-n}\chi_R\big)\big(\chi_{Q'} - 2^{-n}\chi_{R'}\big)\,dx\,.$$

Since $k' > k$, the last integral may be nonzero only when $R' \subsetneqq R$. If this is the case, then $R' \subseteq Q_{R'}$ for some dyadic cube $Q_{R'} \in \mathcal{D}_k$ with $Q_{R'} \subsetneqq R$. See Figure 6.1.

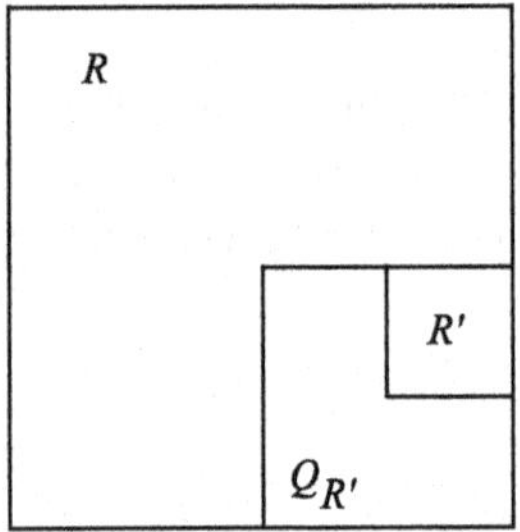

Fig. 6.1 Picture of the cubes R, R', and $Q_{R'}$.

Then the function $\chi_{Q'} - 2^{-n}\chi_{R'}$ is supported in the cube $Q_{R'}$ and the function $\chi_Q - 2^{-n}\chi_R$ is constant on any dyadic subcube Q of R (of half its side length) and in particular is constant on $Q_{R'}$. Then

$$\sum_{\substack{Q'\in\mathcal{D}_{k'} \\ Q'\subseteq R'}} \big(\operatorname{Avg}_{Q'} f\big) \int_{Q_{R'}} \chi_{Q'} - 2^{-n}\chi_{R'}\,dx = \sum_{\substack{Q'\in\mathcal{D}_{k'} \\ Q'\subseteq R'}} \big(\operatorname{Avg}_{Q'} f\big)\,\big(|Q'| - 2^{-n}|R'|\big) = 0\,,$$

since $|R'| = 2^n|Q'|$. We conclude that $\big\langle D_k(f), D_{k'}(f)\big\rangle = 0$ whenever $k \neq k'$, from which we easily derive (6.4.7).

Now observe that (6.4.8) is a direct consequence of (6.4.2), and (6.4.9) is a direct consequence of (6.4.3). □

6.4.3 The Dyadic Martingale Square Function

As a consequence of identity (6.4.7), proved in the previous subsection, we obtain that

$$\Big\| \Big(\sum_{k\in\mathbf{Z}} |D_k(f)|^2 \Big)^{\frac{1}{2}} \Big\|_{L^2(\mathbf{R}^n)} = \big\| f \big\|_{L^2(\mathbf{R}^n)}, \tag{6.4.11}$$

which says that the *dyadic martingale square function*

$$S(f) = \Big(\sum_{k\in\mathbf{Z}} |D_k(f)|^2 \Big)^{\frac{1}{2}}$$

is L^2 bounded. It is natural to ask whether there exist L^p analogues of this result, and this is the purpose of the following theorem.

Theorem 6.4.7. *For any $1 < p < \infty$ there exists a constant $c_{p,n}$ such that for every function f in $L^p(\mathbf{R}^n)$ we have*

$$\frac{1}{c_{p',n}} \big\| f \big\|_{L^p(\mathbf{R}^n)} \le \big\| S(f) \big\|_{L^p(\mathbf{R}^n)} \le c_{p,n} \big\| f \big\|_{L^p(\mathbf{R}^n)}. \tag{6.4.12}$$

The lower inequality subsumes the fact that if $\big\| S(f) \big\|_{L^p(\mathbf{R}^n)} < \infty$, then f must be an L^p function.

Proof. Let $\{r_j\}_j$ be the Rademacher functions (see Appendix C.1) enumerated in such a way that their index set is the set of integers. We rewrite the upper estimate in (6.4.12) as

$$\int_0^1 \int_{\mathbf{R}^n} \Big| \sum_{k\in\mathbf{Z}} r_k(\omega) D_k(f)(x) \Big|^p \, dx \, d\omega \le C_p^p \big\| f \big\|_{L^p}^p. \tag{6.4.13}$$

We prove a stronger estimate than (6.4.13), namely that for all $\omega \in [0,1]$ we have

$$\int_{\mathbf{R}^n} \Big| T_\omega(f)(x) \Big|^p \, dx \le C_p^p \big\| f \big\|_{L^p}^p, \tag{6.4.14}$$

where

$$T_\omega(f)(x) = \sum_{k\in\mathbf{Z}} r_k(\omega) D_k(f)(x).$$

In view of the L^2 estimate (6.4.11), we have that the operator T_ω is L^2 bounded with norm 1. We show that T_ω is weak type $(1,1)$.

To show that T_ω is of weak type $(1,1)$ we fix a function $f \in L^1$ and $\alpha > 0$. We apply the Calderón–Zygmund decomposition (Theorem 5.3.1) to f at height α to write

$$f = g + b, \qquad b = \sum_j \big(f - \operatorname*{Avg}_{Q_j} f\big) \chi_{Q_j},$$

where Q_j are dyadic cubes that satisfy $\sum_j |Q_j| \le \frac{1}{\alpha}\|f\|_{L^1}$ and g has L^2 norm at most $(2^n\alpha\|f\|_{L^1})^{\frac{1}{2}}$; see (5.3.1). To achieve this decomposition, we apply the proof of Theorem 5.3.1 starting with a dyadic mesh of large cubes such that $|Q| \ge \frac{1}{\alpha}\|f\|_{L^1}$ for all Q in the mesh. Then we subdivide each Q in the mesh by halving each side, and we select those cubes for which the average of f over them is bigger than α (and thus at most $2^n\alpha$). Since the original mesh consists of dyadic cubes, the stopping-time argument of Theorem 5.3.1 ensures that each selected cube is dyadic.

We observe (and this is the key observation) that $T_\omega(b)$ is supported in $\bigcup_j Q_j$. To see this, we use identity (6.4.10) to write $T_\omega(b)$ as

$$\sum_j \left[\sum_k r_k(\omega) \sum_{R\in\mathscr{D}_{k-1}} \sum_{\substack{Q\in\mathscr{D}_k\\ Q\subseteq R}} \operatorname*{Avg}_Q[(f-\operatorname*{Avg}_{Q_j} f)\chi_{Q_j}]\,(\chi_Q - 2^{-n}\chi_R)\right]. \tag{6.4.15}$$

We consider the following three cases for the cubes Q that appear in the inner sum in (6.4.15): (i) $Q_j \subseteq Q$, (ii) $Q_j \cap Q = \emptyset$, and (iii) $Q \subsetneq Q_j$. It is simple to see that in cases (i) and (ii) we have $\operatorname{Avg}_Q[(f-\operatorname{Avg}_{Q_j} f)\chi_{Q_j}] = 0$. Therefore the inner sum in (6.4.15) is taken over all Q that satisfy $Q \subsetneq Q_j$. But then we must have that the unique dyadic parent R of Q is also contained in Q_j. It follows that the expression inside the square brackets in (6.4.15) is supported in R and therefore in Q_j. We conclude that $T_\omega(b)$ is supported in $\bigcup_j Q_j$. Using Exercise 5.3.5(a) we obtain that T_ω is weak type $(1,1)$ with norm at most

$$\frac{\alpha\big|\{|T_\omega(g)| > \frac{\alpha}{2}\}\big| + \alpha\big|\bigcup_j Q_j\big|}{\|f\|_{L^1}} \le \frac{\alpha 4\alpha^{-2}\|g\|_{L^2}^2 + \|f\|_{L^1}}{\|f\|_{L^1}} \le 2^{n+2}+1.$$

We have now established that T_ω is weak type $(1,1)$. Since T_ω is L^2 bounded with norm 1, it follows by interpolation that T_ω is L^p bounded for all $1 < p < 2$. The L^p boundedness of T_ω for the remaining $p > 2$ follows by duality. (Note that the operators D_k and E_k are self-transpose.) We conclude the validity of (6.4.14), which implies that of (6.4.13). As observed, this is equivalent to the upper estimate in (6.4.12).

Finally, we notice that the lower estimate in (6.4.12) is a consequence of the upper estimate as in the case of the Littlewood–Paley operators Δ_j. Indeed, we need to observe that in view of (6.4.6) we have

$$\begin{aligned}
|\langle f,g\rangle| &= \Big|\Big\langle \sum_k D_k(f), \sum_{k'} D_{k'}(g)\Big\rangle\Big| \\
&= \Big|\sum_k\sum_{k'}\big\langle D_k(f), D_{k'}(g)\big\rangle\Big| \\
&= \Big|\sum_k \big\langle D_k(f), D_k(g)\big\rangle\Big| && \text{[Exercise 6.4.6(a)]} \\
&\le \int_{\mathbf{R}^n}\sum_k |D_k(f)(x)|\,|D_k(g)(x)|\,dx
\end{aligned}$$

$$
\begin{aligned}
&\leq \int_{\mathbf{R}^n} S(f)(x)\,S(g)(x)\,dx && \text{(Cauchy–Schwarz inequality)}\\
&\leq \big\|S(f)\big\|_{L^p}\big\|S(g)\big\|_{L^{p'}} && \text{(Hölder's inequality)}\\
&\leq \big\|S(f)\big\|_{L^p}\, c_{p',n}\big\|g\big\|_{L^{p'}}\,.
\end{aligned}
$$

Taking the supremum over all functions g on $\mathbf{R}^n$ with $L^{p'}$ norm at most 1, we obtain that f gives rise to a bounded linear functional on $L^{p'}$. It follows by the Riesz representation theorem that f must be an L^p function that satisfies the lower estimate in (6.4.12). □

6.4.4 Almost Orthogonality Between the Littlewood–Paley Operators and the Dyadic Martingale Difference Operators

Next, we discuss connections between the Littlewood–Paley operators Δ_j and the dyadic martingale difference operators D_k. It turns out that these operators are almost orthogonal in the sense that the L^2 operator norm of the composition $D_k\Delta_j$ decays exponentially as the indices j and k get farther away from each other.

For the purposes of the next theorem we define the Littlewood–Paley operators Δ_j as convolution operators with the function $\Psi_{2^{-j}}$, where

$$
\widehat{\Psi}(\xi) = \widehat{\Phi}(\xi) - \widehat{\Phi}(2\xi)
$$

and Φ is a fixed radial Schwartz function whose Fourier transform $\widehat{\Phi}$ is real-valued, supported in the ball $|\xi| < 2$, and equal to 1 on the ball $|\xi| < 1$. In this case we clearly have the identity

$$
\sum_{j\in\mathbf{Z}} \widehat{\Psi}(2^{-j}\xi) = 1, \qquad\qquad \xi \neq 0.
$$

Then we have the following theorem.

Theorem 6.4.8. *There exists a constant C such that for every k, j in $\mathbf{Z}$ the following estimate on the operator norm of $D_k\Delta_j : L^2(\mathbf{R}^n) \to L^2(\mathbf{R}^n)$ is valid:*

$$
\big\|D_k\Delta_j\big\|_{L^2(\mathbf{R}^n)\to L^2(\mathbf{R}^n)} = \big\|\Delta_j D_k\big\|_{L^2(\mathbf{R}^n)\to L^2(\mathbf{R}^n)} \leq C\,2^{-\frac{1}{2}|j-k|}. \tag{6.4.16}
$$

Proof. Since Ψ is a radial function, it follows that Δ_j is equal to its transpose operator on L^2. Moreover, the operator D_k is also equal to its transpose. Thus

$$
(D_k\Delta_j)^t = \Delta_j D_k
$$

and it therefore suffices to prove only that

$$
\big\|D_k\Delta_j\big\|_{L^2\to L^2} \leq C2^{-\frac{1}{2}|j-k|}. \tag{6.4.17}
$$

By a simple dilation argument it suffices to prove (6.4.17) when $k = 0$. In this case we have the estimate

$$\begin{aligned}\big\|D_0\Delta_j\big\|_{L^2\to L^2} &= \big\|E_0\Delta_j - E_{-1}\Delta_j\big\|_{L^2\to L^2}\\ &\le \big\|E_0\Delta_j - \Delta_j\big\|_{L^2\to L^2} + \big\|E_{-1}\Delta_j - \Delta_j\big\|_{L^2\to L^2},\end{aligned}$$

and since the D_k's and Δ_j's are self-transposes, we have

$$\begin{aligned}\big\|D_0\Delta_j\big\|_{L^2\to L^2} = \big\|\Delta_j D_0\big\|_{L^2\to L^2} &= \big\|\Delta_j E_0 - \Delta_j E_{-1}\big\|_{L^2\to L^2}\\ &\le \big\|\Delta_j E_0 - E_0\big\|_{L^2\to L^2} + \big\|\Delta_j E_{-1} - E_0\big\|_{L^2\to L^2}.\end{aligned}$$

Estimate (6.4.17) when $k = 0$ will be a consequence of the pair of inequalities

$$\big\|E_0\Delta_j - \Delta_j\big\|_{L^2\to L^2} + \big\|E_{-1}\Delta_j - \Delta_j\big\|_{L^2\to L^2} \le C2^{\frac{j}{2}} \quad \text{for } j \le 0, \tag{6.4.18}$$

$$\big\|\Delta_j E_0 - E_0\big\|_{L^2\to L^2} + \big\|\Delta_j E_{-1} - E_0\big\|_{L^2\to L^2} \le C2^{-\frac{1}{2}j} \quad \text{for } j \ge 0. \tag{6.4.19}$$

We start by proving (6.4.18). We consider only the term $E_0\Delta_j - \Delta_j$, since the term $E_{-1}\Delta_j - \Delta_j$ is similar. Let $f \in L^2(\mathbf{R}^n)$. Then

$$\begin{aligned}&\big\|E_0\Delta_j(f) - \Delta_j(f)\big\|_{L^2}^2\\ &= \sum_{Q\in\mathscr{D}_0} \Big\|f*\Psi_{2^{-j}} - \operatorname*{Avg}_Q(f*\Psi_{2^{-j}})\Big\|_{L^2(Q)}^2\\ &\le \sum_{Q\in\mathscr{D}_0}\int_Q\int_Q |(f*\Psi_{2^{-j}})(x) - (f*\Psi_{2^{-j}})(t)|^2\,dt\,dx\\ &\le 3\sum_{Q\in\mathscr{D}_0}\int_Q\int_Q\bigg(\int_{5\sqrt{n}Q}|f(y)||\Psi_{2^{-j}}(x-y)|\,dy\bigg)^2 dt\,dx\\ &\qquad + 3\sum_{Q\in\mathscr{D}_0}\int_Q\int_Q\bigg(\int_{5\sqrt{n}Q}|f(y)||\Psi_{2^{-j}}(t-y)|\,dy\bigg)^2 dt\,dx\\ &\qquad + 3\sum_{Q\in\mathscr{D}_0}\int_Q\int_Q\bigg(\int_{(5\sqrt{n}Q)^c}|f(y)|2^{jn+j}|\nabla\Psi(2^j(\xi_{x,t}-y))|\,dy\bigg)^2 dt\,dx,\end{aligned}$$

where $\xi_{x,t}$ lies on the line segment joining x and t. Applying the Cauchy-Schwarz inequality to the first two terms, we see that the last expression is bounded by

$$C2^{jn}\sum_{Q\in\mathscr{D}_0}\int_{5\sqrt{n}Q}|f(y)|^2\,dy + C_M 2^{2j}\sum_{Q\in\mathscr{D}_0}\int_Q\bigg(\int_{\mathbf{R}^n}\frac{2^{jn}|f(y)|\,dy}{(1+2^j|x-y|)^M}\bigg)^2 dx,$$

which is clearly controlled by $C(2^{jn}+2^{2j})\|f\|_{L^2}^2 \le 2C2^j\|f\|_{L^2}^2$. This proves (6.4.18).

We now turn to the proof of (6.4.19). We set $S_j = \sum_{k\le j}\Delta_k$. Since Δ_j is the difference of two S_j's, it suffices to prove (6.4.19), where Δ_j is replaced by S_j. We work only with the term $S_jE_0 - E_0$, since the other term can be treated similarly. We have

$$
\begin{aligned}
\big\|S_jE_0(f)-E_0(f)\big\|_{L^2}^2 &= \Big\|\sum_{Q\in\mathscr{D}_0}(\operatorname*{Avg}_Q f)\,(\Phi_{2^{-j}}*\chi_Q-\chi_Q)\Big\|_{L^2}^2\\
&\le 2\Big\|\sum_{Q\in\mathscr{D}_0}(\operatorname*{Avg}_Q f)\,(\Phi_{2^{-j}}*\chi_Q-\chi_Q)\chi_{5\sqrt{n}Q}\Big\|_{L^2}^2\\
&\qquad +2\Big\|\sum_{Q\in\mathscr{D}_0}(\operatorname*{Avg}_Q f)\,(\Phi_{2^{-j}}*\chi_Q)\chi_{(5\sqrt{n}Q)^c}\Big\|_{L^2}^2.
\end{aligned}
$$

Since the functions appearing inside the sum in the first term have supports with bounded overlap, we obtain

$$
\Big\|\sum_{Q\in\mathscr{D}_0}(\operatorname*{Avg}_Q f)(\Phi_{2^{-j}}*\chi_Q-\chi_Q)\chi_{5\sqrt{n}Q}\Big\|_{L^2}^2\le C\sum_{Q\in\mathscr{D}_0}(\operatorname*{Avg}_Q |f|)^2\big\|\Phi_{2^{-j}}*\chi_Q-\chi_Q\big\|_{L^2}^2,
$$

and the crucial observation is that

$$
\big\|\Phi_{2^{-j}}*\chi_Q-\chi_Q\big\|_{L^2}^2\le C2^{-j},
$$

a consequence of Plancherel's identity and the fact that $|1-\widehat{\Phi}(2^{-j}\xi)|\le\chi_{|\xi|\ge 2^j}$. Putting these observations together, we deduce

$$
\Big\|\sum_{Q\in\mathscr{D}_0}(\operatorname*{Avg}_Q f)\,(\Phi_{2^{-j}}*\chi_Q-\chi_Q)\chi_{3Q}\Big\|_{L^2}^2\le C\sum_{Q\in\mathscr{D}_0}(\operatorname*{Avg}_Q |f|)^2 2^{-j}\le C2^{-j}\|f\|_{L^2}^2,
$$

and the required conclusion will be proved if we can show that

$$
\Big\|\sum_{Q\in\mathscr{D}_0}(\operatorname*{Avg}_Q f)\,(\Phi_{2^{-j}}*\chi_Q)\chi_{(3Q)^c}\Big\|_{L^2}^2\le C2^{-j}\|f\|_{L^2}^2. \tag{6.4.20}
$$

We prove (6.4.20) by using an estimate based purely on size. Let c_Q be the center of the dyadic cube Q. For $x\notin 3Q$ we have the estimate

$$
|(\Phi_{2^{-j}}*\chi_Q)(x)|\le\frac{C_M2^{jn}}{(1+2^j|x-c_Q|)^M}\le\frac{C_M2^{jn}}{(1+2^j)^{M/2}}\frac{1}{(1+|x-c_Q|)^{M/2}},
$$

since both $2^j\ge1$, and $|x-c_Q|\ge1$. We now control the left-hand side of (6.4.20) by

$$
\begin{aligned}
&2^{j(2n-M)}\sum_{Q\in\mathscr{D}_0}\sum_{Q'\in\mathscr{D}_0}(\operatorname*{Avg}_Q|f|)(\operatorname*{Avg}_{Q'}|f|)\int_{\mathbf{R}^n}\frac{C_M\,dx}{(1+|x-c_Q|)^{\frac{M}{2}}(1+|x-c_{Q'}|)^{\frac{M}{2}}}\\
&\le 2^{j(2n-M)}\sum_{Q\in\mathscr{D}_0}\sum_{Q'\in\mathscr{D}_0}\frac{(\operatorname*{Avg}_Q|f|)(\operatorname*{Avg}_{Q'}|f|)}{(1+|c_Q-c_{Q'}|)^{\frac{M}{4}}}\int_{\mathbf{R}^n}\frac{C_M\,dx}{(1+|x-c_Q|)^{\frac{M}{4}}(1+|x-c_{Q'}|)^{\frac{M}{4}}}\\
&\le 2^{j(2n-M)}\sum_{Q\in\mathscr{D}_0}\sum_{Q'\in\mathscr{D}_0}\frac{C_M}{(1+|c_Q-c_{Q'}|)^{\frac{M}{4}}}\Big(\int_Q|f(y)|^2\,dy+\int_{Q'}|f(y)|^2\,dy\Big)
\end{aligned}
$$

$$\leq C_M 2^{j(2n-M)} \sum_{Q\in\mathscr{D}_0} \int_Q |f(y)|^2\,dy$$
$$= C_M 2^{j(2n-M)} \|f\|_{L^2}^2 .$$

By taking M large enough, we obtain (6.4.20) and thus (6.4.19). □

Exercises

6.4.1. (a) Prove that no dyadic cube in $\mathbf{R}^n$ contains the point 0 in its interior.
(b) Prove that every interval $[a,b]$ is contained in the union of three dyadic intervals of length less than $b-a$.
(c) Prove that every cube of length l in $\mathbf{R}^n$ is contained in the union of 3^n dyadic cubes, each having length less than l.

6.4.2. Let $k \in \mathbf{Z}$. Show that the set $[m2^{-k},(m+s)2^{-k})$ is a dyadic interval if and only if $s = 2^p$ for some $p \in \mathbf{Z}$ and m is an integer multiple of s.

6.4.3. Given a cube Q in $\mathbf{R}^n$ of side length $\ell(Q) \leq 2^{k-1}$ for some integer k, prove that there is a dyadic cube D_Q of side length 2^k such that $Q \subseteq \sigma + D_Q$ for some $\sigma = (\sigma_1,\dots,\sigma_n)$, where $\sigma_j \in \{0,1/3,-1/3\}$.

6.4.4. Show that the martingale maximal function $f \mapsto \sup_{k\in\mathbf{Z}} |E_k(f)|$ is weak type $(1,1)$ with constant at most 1.
[*Hint:* Use Exercise 2.1.12.]

6.4.5. (a) Show that $E_N(f) \to f$ a.e. as $N \to \infty$ for all $f \in L^1_{\mathrm{loc}}(\mathbf{R}^n)$.
(b) Prove that $E_N(f) \to f$ in L^p as $N \to \infty$ for all $f \in L^p(\mathbf{R}^n)$ whenever $1 < p < \infty$.

6.4.6. (a) Let $k,k' \in \mathbf{Z}$ be such that $k \neq k'$. Show that for functions f and g in $L^2(\mathbf{R}^n)$ we have
$$\big\langle D_k(f), D_{k'}(g)\big\rangle = 0 .$$
(b) Conclude that for functions f_j in $L^2(\mathbf{R}^n)$ we have
$$\Big\| \sum_{j\in\mathbf{Z}} D_j(f_j) \Big\|_{L^2(\mathbf{R}^n)} = \Big(\sum_{j\in\mathbf{Z}} \|D_j(f_j)\|_{L^2(\mathbf{R}^n)}^2 \Big)^{\frac{1}{2}} .$$
(c) Let Δ_j and C be as in the statement of Theorem 6.4.8. Show that for any $r \in \mathbf{Z}$ we have
$$\Big\| \sum_{j\in\mathbf{Z}} D_j \Delta_{j+r} D_j \Big\|_{L^2(\mathbf{R}^n)\to L^2(\mathbf{R}^n)} \leq C 2^{-\frac{1}{2}|r|} .$$

6.4.7. ([133]) Let D_j, Δ_j be as in Theorem 6.4.8.
(a) Prove that the operator
$$V_r = \sum_{j\in\mathbf{Z}} D_j \Delta_{j+r}$$
is bounded from $L^2(\mathbf{R}^n)$ to itself with norm at most a multiple of $2^{-\frac{1}{2}|r|}$.

(b) Show that V_r is $L^p(\mathbf{R}^n)$ bounded for all $1 < p < \infty$ with a constant depending only on p and n.
(c) Conclude that for each $1 < p < \infty$ there is a constant $c_p > 0$ such that V_r is bounded on $L^p(\mathbf{R}^n)$ with norm at most a multiple of $2^{-c_p|r|}$.
[*Hint:* Part (a): Write $\Delta_j = \Delta_j \widetilde{\Delta}_j$, where $\widetilde{\Delta}_j$ is another family of Littlewood–Paley operators and use Exercise 6.4.6 (b). Part (b): Use duality and (6.1.21).]

6.5 The Spherical Maximal Function

In this section we discuss yet another consequence of the Littlewood–Paley theory, the boundedness of the spherical maximal operator.

6.5.1 Introduction of the Spherical Maximal Function

We denote throughout this section by $d\sigma$ the normalized Lebesgue measure on the sphere $\mathbf{S}^{n-1}$. For f in $L^p(\mathbf{R}^n)$, $1 \le p \le \infty$, we define the maximal operator

$$\mathscr{M}(f)(x) = \sup_{t>0} \left| \int_{\mathbf{S}^{n-1}} f(x - t\theta)\, d\sigma(\theta) \right| \tag{6.5.1}$$

and we observe that by Minkowski's integral inequality each expression inside the supremum in (6.5.1) is well defined for $f \in L^p$ for almost all $x \in \mathbf{R}^n$. The operator $\mathscr{M}$ is called the *spherical maximal function*. It is unclear at this point for which functions f we have $\mathscr{M}(f) < \infty$ a.e. and for which values of $p < \infty$ the maximal inequality

$$\big\|\mathscr{M}(f)\big\|_{L^p(\mathbf{R}^n)} \le C_p \big\|f\big\|_{L^p(\mathbf{R}^n)} \tag{6.5.2}$$

holds for all functions $f \in L^p(\mathbf{R}^n)$.

Spherical averages often make their appearance as solutions of partial differential equations. For instance, the spherical average

$$u(x,t) = \frac{1}{4\pi} \int_{\mathbf{S}^2} t\, f(x - ty)\, d\sigma(y) \tag{6.5.3}$$

is a solution of the *wave equation*

$$\begin{aligned} \Delta_x(u)(x,t) &= \frac{\partial^2 u}{\partial t^2}(x,t), \\ u(x,0) &= 0, \\ \frac{\partial u}{\partial t}(x,0) &= f(x), \end{aligned}$$

in $\mathbf{R}^3$. The introduction of the spherical maximal function is motivated by the fact that the related spherical average

$$u(x,t) = \frac{1}{4\pi}\int_{\mathbf{S}^2} f(x-ty)\,d\sigma(y) \tag{6.5.4}$$

solves *Darboux's equation*

$$\begin{aligned} \Delta_x(u)(x,t) &= \frac{\partial^2 u}{\partial t^2}(x,t) + \frac{2}{t}\frac{\partial u}{\partial t}(x,t),\\ u(x,0) &= f(x),\\ \frac{\partial u}{\partial t}(x,0) &= 0, \end{aligned}$$

in $\mathbf{R}^3$. It is rather remarkable that the Fourier transform can be used to study almost everywhere convergence for several kinds of maximal averaging operators such as the spherical averages in (6.5.4). This is achieved via the boundedness of the corresponding maximal operator; the maximal operator controlling the averages over $\mathbf{S}^{n-1}$ is given in (6.5.1).

Before we begin the analysis of the spherical maximal function, we recall that

$$\widehat{d\sigma}(\xi) = \frac{2\pi}{|\xi|^{\frac{n-2}{2}}} J_{\frac{n-2}{2}}(2\pi|\xi|),$$

as shown in Appendix B.4. Using the estimates in Appendices B.6 and B.7 and the identity

$$\frac{d}{dt}J_\nu(t) = \frac{1}{2}(J_{\nu-1}(t) - J_{\nu+1}(t))$$

derived in Appendix B.2, we deduce the crucial estimate

$$|\widehat{d\sigma}(\xi)| + |\nabla\widehat{d\sigma}(\xi)| \le \frac{C_n}{(1+|\xi|)^{\frac{n-1}{2}}}. \tag{6.5.5}$$

Theorem 6.5.1. *Let $n \ge 3$. For each $\frac{n}{n-1} < p \le \infty$, there is a constant C_p such that*

$$\big\|\mathscr{M}(f)\big\|_{L^p(\mathbf{R}^n)} \le C_p \big\|f\big\|_{L^p(\mathbf{R}^n)} \tag{6.5.6}$$

holds for all f in $L^p(\mathbf{R}^n)$. Consequently, for all $\frac{n}{n-1} < p \le \infty$ and $f \in L^p(\mathbf{R}^n)$ we have

$$\lim_{t\to 0}\frac{1}{\omega_{n-1}}\int_{\mathbf{S}^{n-1}} f(x-t\theta)\,d\sigma(\theta) = f(x) \tag{6.5.7}$$

for almost all $x \in \mathbf{R}^n$. Here we set $\omega_{n-1} = |\mathbf{S}^{n-1}|$.

The proof of this theorem is given in the rest of this section. Before we present the proof we explain the validity of (6.5.7). Clearly this assertion is valid for functions $f \in \mathscr{S}(\mathbf{R}^n)$. Using inequality (6.5.6) and Theorem 2.1.14 we obtain that (6.5.7) holds for all functions in $f \in L^p(\mathbf{R}^n)$.

We now focus on (6.5.6). Define $m(\xi) = \widehat{d\sigma}(\xi)$ and notice that $m(\xi)$ is a $\mathscr{C}^\infty$ function. To study the maximal multiplier operator

$$\sup_{t>0} \left| \left(\widehat{f}(\xi)\, m(t\xi) \right)^\vee \right|$$

we decompose the multiplier $m(\xi)$ into radial pieces as follows: We fix a radial $\mathscr{C}^\infty$ function φ_0 in $\mathbf{R}^n$ such that $\varphi_0(\xi) = 1$ when $|\xi| \le 1$ and $\varphi_0(\xi) = 0$ when $|\xi| \ge 2$. For $j \ge 1$ we let

$$\varphi_j(\xi) = \varphi_0(2^{-j}\xi) - \varphi_0(2^{1-j}\xi) \tag{6.5.8}$$

and we observe that $\varphi_j(\xi)$ is localized near $|\xi| \approx 2^j$. Then we have

$$\sum_{j=0}^{\infty} \varphi_j = 1\,.$$

Set $m_j = \varphi_j m$ for all $j \ge 0$. The m_j's are $\mathscr{C}_0^\infty$ functions that satisfy

$$m = \sum_{j=0}^{\infty} m_j\,.$$

Also, the following estimate is valid:

$$\mathscr{M}(f) \le \sum_{j=0}^{\infty} \mathscr{M}_j(f),$$

where

$$\mathscr{M}_j(f)(x) = \sup_{t>0} \left| \left(\widehat{f}(\xi)\, m_j(t\xi) \right)^\vee (x) \right|.$$

Since the function m_0 is $\mathscr{C}_0^\infty$, we have that $\mathscr{M}_0$ maps L^p to itself for all $1 < p \le \infty$. (See Exercise 6.5.1.)

We define *g-functions* associated with m_j as follows:

$$G_j(f)(x) = \left(\int_0^\infty |A_{j,t}(f)(x)|^2 \frac{dt}{t} \right)^{\frac{1}{2}},$$

where $A_{j,t}(f)(x) = \left(\widehat{f}(\xi)\, m_j(t\xi) \right)^\vee (x)$.

6.5.2 The First Key Lemma

We have the following lemma:

Lemma 6.5.2. *There is a constant $C = C(n) < \infty$ such that for any $j \geq 1$ we have the estimate*

$$\big\|\mathscr{M}_j(f)\big\|_{L^2} \leq C2^{(\frac{1}{2}-\frac{n-1}{2})j}\big\|f\big\|_{L^2}$$

for all functions f in $L^2(\mathbf{R}^n)$.

Proof. We define a function

$$\widetilde{m}_j(\xi) = \xi \cdot \nabla m_j(\xi),$$

we let $\widetilde{A}_{j,t}(f)(x) = \big(\widehat{f}(\xi)\,\widetilde{m}_j(t\xi)\big)^\vee(x)$, and we let

$$\widetilde{G}_j(f)(x) = \bigg(\int_0^\infty |\widetilde{A}_{j,t}(f)(x)|^2\,\frac{dt}{t}\bigg)^{\frac{1}{2}}$$

be the associated g-function. For $f \in L^2(\mathbf{R}^n)$, the identity

$$s\,\frac{dA_{j,s}}{ds}(f) = \widetilde{A}_{j,s}(f)$$

is clearly valid for all j and s. Since $A_{j,s}(f) = f * (m_j^\vee)_s$ and $m_j^\vee$ has integral zero for $j \geq 1$ (here $(m_j^\vee)_s(x) = s^{-n}m_j^\vee(s^{-1}x)$), it follows from Corollary 2.1.19 that

$$\lim_{s\to 0} A_{j,s}(f)(x) = 0$$

for all $x \in \mathbf{R}^n \setminus E_f$, where E_f is some set of Lebesgue measure zero. By the fundamental theorem of calculus for $x \in \mathbf{R}^n \setminus E_f$ we deduce that

$$\begin{aligned}(A_{j,t}(f)(x))^2 &= \int_0^t \frac{d}{ds}(A_{j,s}(f)(x))^2\,ds \\ &= 2\int_0^t A_{j,s}(f)(x)\,s\frac{dA_{j,s}}{ds}(f)(x)\,\frac{ds}{s} \\ &= 2\int_0^t A_{j,s}(f)(x)\widetilde{A}_{j,s}(f)(x)\,\frac{ds}{s},\end{aligned}$$

from which we obtain the estimate

$$\big|A_{j,t}(f)(x)\big|^2 \leq 2\int_0^\infty \big|A_{j,s}(f)(x)\big|\,\big|\widetilde{A}_{j,s}(f)(x)\big|\,\frac{ds}{s}. \tag{6.5.9}$$

Taking the supremum over all $t > 0$ on the left-hand side in (6.5.9) and integrating over $\mathbf{R}^n$, we obtain the estimate

$$\begin{aligned}\big\|\mathcal{M}_j(f)\big\|_{L^2}^2 &\le 2\int_{\mathbf{R}^n}\int_0^\infty \big|A_{j,s}(f)(x)\big|\,\big|\widetilde{A}_{j,s}(f)(x)\big|\,\frac{ds}{s}dx\\ &\le 2\int_{\mathbf{R}^n} G_j(f)(x)\widetilde{G}_j(f)(x)\,dx\\ &\le 2\big\|G_j(f)\big\|_{L^2}\big\|\widetilde{G}_j(f)\big\|_{L^2},\end{aligned}$$

by applying the Cauchy–Schwarz inequality twice. Next we claim that as a consequence of (6.5.5) we have for some $c, \widetilde{c} < \infty$,

$$\big\|m_j\big\|_{L^\infty} \le c\,2^{-j\frac{n-1}{2}} \quad \text{and} \quad \big\|\widetilde{m}_j\big\|_{L^\infty} \le \widetilde{c}\,2^{j(1-\frac{n-1}{2})}.$$

Using these facts together with the facts that the functions m_j and $\widetilde{m}_j$ are supported in the annuli $2^{j-1} \le |\xi| \le 2^{j+1}$, we obtain that the g-functions G_j and $\widetilde{G}_j$ are L^2 bounded with norms at most a constant multiple of the quantities $2^{-j\frac{n-1}{2}}$ and $2^{j(1-\frac{n-1}{2})}$, respectively; see Exercise 6.5.2. Note that since $n \ge 3$, both exponents are negative. We conclude that

$$\big\|\mathcal{M}_j(f)\big\|_{L^2} \le C2^{j(\frac{1}{2}-\frac{n-1}{2})}\big\|f\big\|_{L^2},$$

which is what we needed to prove. □

6.5.3 The Second Key Lemma

Next we need the following lemma.

Lemma 6.5.3. *There exists a constant $C = C(n) < \infty$ such that for all $j \ge 1$ and for all f in $L^1(\mathbf{R}^n)$ we have*

$$\big\|\mathcal{M}_j(f)\big\|_{L^{1,\infty}} \le C2^j\big\|f\big\|_{L^1}.$$

Proof. Let $K^{(j)} = (\varphi_j)^\vee * d\sigma = \Phi_{2^{-j}} * d\sigma$, where Φ is a Schwartz function. Setting

$$(K^{(j)})_t(x) = t^{-n}K^{(j)}(t^{-1}x)$$

we have that

$$\mathcal{M}_j(f) = \sup_{t>0}\big|(K^{(j)})_t * f\big|. \tag{6.5.10}$$

The proof of the lemma is based on the estimate:

$$\mathcal{M}_j(f) \le C2^j\mathcal{M}(f) \tag{6.5.11}$$

and the weak type $(1,1)$ boundedness of the Hardy–Littlewood maximal operator $\mathcal{M}$ (Theorem 2.1.6). To establish (6.5.11), it suffices to show that for any $M > n$ there is a constant $C_M < \infty$ such that

$$|K^{(j)}(x)| = |(\Phi_{2^{-j}} * d\sigma)(x)| \le \frac{C_M 2^j}{(1+|x|)^M}. \tag{6.5.12}$$

Then Theorem 2.1.10 yields (6.5.11) and hence the required conclusion.

Using the fact that Φ is a Schwartz function, we have for every $N > 0$,

$$|(\Phi_{2^{-j}} * d\sigma)(x)| \le C_N \int_{\mathbf{S}^{n-1}} \frac{2^{nj}\, d\sigma(y)}{(1+2^j|x-y|)^N}.$$

We pick an N to depend on M (6.5.12); in fact, any $N > M$ suffices for our purposes. We split the last integral into the regions

$$S_{-1}(x) = \mathbf{S}^{n-1} \cap \{y \in \mathbf{R}^n :\, 2^j|x-y| \le 1\}$$

and for $r \ge 0$,

$$S_r(x) = \mathbf{S}^{n-1} \cap \{y \in \mathbf{R}^n :\, 2^r < 2^j|x-y| \le 2^{r+1}\}.$$

The key observation is that whenever $B(y,R)$ is a ball of radius R in $\mathbf{R}^n$ centered at $y \in \mathbf{S}^{n-1}$, then the spherical measure of the set $\mathbf{S}^{n-1} \cap B(y,R)$ is at most a dimensional constant multiple of R^{n-1}. This implies that the spherical measure of each $S_r(x)$ is at most $c_n 2^{(r+1-j)(n-1)}$, an estimate that is useful only when $r \le j$. Using this observation, together with the fact that for $y \in S_r(x)$ we have $|x| \le 2^{r+1-j}+1$, we obtain the following estimate for the expression $|(\Phi_{2^{-j}} * d\sigma)(x)|$:

$$\begin{aligned}
&\sum_{r=-1}^{j} \int_{S_r(x)} \frac{C_N 2^{nj}\, d\sigma(y)}{(1+2^j|x-y|)^N} + \sum_{r=j+1}^{\infty} \int_{S_r(x)} \frac{C_N 2^{nj}\, d\sigma(y)}{(1+2^j|x-y|)^N} \\
&\le C_N' 2^{nj} \left[\sum_{r=-1}^{j} \frac{d\sigma(S_r(x))\chi_{B(0,3)}(x)}{2^{rN}} + \sum_{r=j+1}^{\infty} \frac{d\sigma(S_r(x))\,\chi_{B(0,2^{r+1-j}+1)}(x)}{2^{rN}} \right] \\
&\le C_N' 2^{nj} \left[\sum_{r=-1}^{j} \frac{c_n 2^{(r+1-j)(n-1)}\chi_{B(0,3)}(x)}{2^{rN}} + \sum_{r=j+1}^{\infty} \frac{\omega_{n-1}\,\chi_{B(0,2^{r+2-j})}(x)}{2^{rN}} \right] \\
&\le C_{N,n} \left[2^j \chi_{B(0,3)}(x) + 2^{nj} \sum_{r=j+1}^{\infty} \frac{1}{2^{rN}} \frac{(1+2^{r+2-j})^M}{(1+|x|)^M} \right] \\
&\le C_{M,n}' \frac{2^j}{(1+|x|)^M} \left[1 + \sum_{r=j+1}^{\infty} \frac{2^{(r-j)(M-N)}}{2^{j(N+1-n)}} \right] \\
&\le \frac{C_{M,n}'' 2^j}{(1+|x|)^M},
\end{aligned}$$

where we used that $N > M > n$. This establishes (6.5.12). □

6.5.4 Completion of the Proof

It remains to combine the previous ingredients to complete the proof of the theorem. Interpolating between the $L^2 \to L^2$ and $L^1 \to L^{1,\infty}$ estimates obtained in Lemmas 6.5.2 and 6.5.3, we obtain

$$\big\|\mathscr{M}_j(f)\big\|_{L^p(\mathbf{R}^n)} \le C_p 2^{(\frac{n}{p}-(n-1))j}\big\|f\big\|_{L^p(\mathbf{R}^n)}$$

for all $1 < p \le 2$. When $p > \frac{n}{n-1}$ the series $\sum_{j=1}^{\infty} 2^{(\frac{n}{p}-(n-1))j}$ converges and we conclude that $\mathscr{M}$ is L^p bounded for these p's. The boundedness of $\mathscr{M}$ on L^p for $p > 2$ follows by interpolation between L^q for $q < 2$ and the estimate $\mathscr{M} : L^\infty \to L^\infty$.

Exercises

6.5.1. Let m be in $L^1(\mathbf{R}^n) \cap L^\infty(\mathbf{R}^n)$ that satisfies $|m^\vee(x)| \le C(1+|x|)^{-n-\delta}$ for some $\delta > 0$. Show that the maximal multiplier

$$\mathscr{M}_m(f)(x) = \sup_{t>0} \big|\big(\widehat{f}(\xi)\, m(t\xi)\big)^\vee(x)\big|$$

is L^p bounded for all $1 < p < \infty$.

6.5.2. Suppose that the function m is supported in the annulus $R \le |\xi| \le 2R$ and is bounded by A. Show that the g-function

$$G(f)(x) = \left(\int_0^\infty |(m(t\xi)\widehat{f}(\xi))^\vee(x)|^2 \frac{dt}{t}\right)^{\frac{1}{2}}$$

maps $L^2(\mathbf{R}^n)$ to $L^2(\mathbf{R}^n)$ with bound at most $A\sqrt{\log 2}$.

6.5.3. ([302]) Let $A, a, b > 0$ with $a + b > 1$. Use the idea of Lemma 6.5.2 to show that if $m(\xi)$ satisfies $|m(\xi)| \le A(1+|\xi|)^{-a}$ and $|\nabla m(\xi)| \le A(1+|\xi|)^{-b}$ for all $\xi \in \mathbf{R}^n$, then the maximal operator

$$\mathscr{M}_m(f)(x) = \sup_{t>0} \big|\big(\widehat{f}(\xi)\, m(t\xi)\big)^\vee(x)\big|$$

is bounded from $L^2(\mathbf{R}^n)$ to itself.
[*Hint:* Use that

$$\mathscr{M}_m \le \sum_{j=0}^{\infty} \mathscr{M}_{m,j}\,,$$

where $\mathscr{M}_{m,j}$ corresponds to the multiplier $\varphi_j m$; here φ_j is as in (6.5.8). Show that

$$\big\|\mathscr{M}_{m,j}(f)\big\|_{L^2} \le C\big\|\varphi_j m\big\|_{L^\infty}^{\frac{1}{2}}\big\|\varphi_j \widetilde{m}\big\|_{L^\infty}^{\frac{1}{2}}\big\|f\big\|_{L^2} \le C 2^{j\frac{1-(a+b)}{2}}\big\|f\big\|_{L^2},$$

where $\widetilde{m}(\xi) = \xi \cdot \nabla m(\xi)$.]

6.5.4. Let $A,c>0$, $a>1/2$, $0<b<n$. Follow the idea of the proof of Theorem 6.5.1 to obtain the following more general result: If $d\mu$ is a finite Borel measure supported in the closed unit ball that satisfies $|\widehat{d\mu}(\xi)|\le A(1+|\xi|)^{-a}$ for all $\xi\in\mathbf{R}^n$ and $d\mu(B(y,R))\le cR^b$ for all $R>0$, then the maximal operator

$$f\mapsto \sup_{t>0}\left|\int_{\mathbf{R}^n} f(x-ty)\,d\mu(y)\right|$$

maps $L^p(\mathbf{R}^n)$ to itself when $p>\frac{2n-2b+2a-1}{n-b+2a-1}$.
[*Hint:* Using the notation of the preceding exercise, show that $\|\mathcal{M}_{m,j}(f)\|_{L^2}\le C2^{j(\frac{1}{2}-a)}\|f\|_{L^2}$ and that $\|\mathcal{M}_{m,j}(f)\|_{L^{1,\infty}}\le C2^{j(n-b)}\|f\|_{L^1}$ for all $j\in\mathbf{Z}^+$, where C is a constant depending on the given parameters.]

6.5.5. Show that Theorem 6.5.1 is false when $n=1$, that is, show that the maximal operator

$$\mathcal{M}_1(f)(x)=\sup_{t>0}\frac{|f(x+t)+f(x-t)|}{2}$$

is unbounded on $L^p(\mathbf{R})$ for all $p<\infty$.

6.5.6. Show that when $n\ge 2$ and $p\le\frac{n}{n-1}$ there exists an $L^p(\mathbf{R}^n)$ function f such that $\mathcal{M}(f)(x)=\infty$ for all $x\in\mathbf{R}^n$. Hence Theorem 6.5.1 is false is this case.
[*Hint:* Choose a compactly supported and radial function equal to $|y|^{1-n}(-\log|y|)^{-1}$ when $|y|\le 1/2$.]

6.6 Wavelets and Sampling

In this section we construct orthonormal bases of $L^2(\mathbf{R})$ generated by translations and dilations of a single function. An example of such base is given by the Haar functions we encountered in Section 6.4. The Haar functions are generated by integer translations and dyadic dilations of the single function $\chi_{[0,\frac{1}{2})}-\chi_{[\frac{1}{2},1)}$. This function is not smooth, and the main question addressed in this section is whether there exist smooth analogues of the Haar functions.

Definition 6.6.1. A square integrable function φ on $\mathbf{R}^n$ is called a *wavelet* if the family of functions

$$\varphi_{\nu,k}(x)=2^{\frac{\nu n}{2}}\varphi(2^\nu x-k),$$

where ν ranges over $\mathbf{Z}$ and k over $\mathbf{Z}^n$, is an orthonormal basis of $L^2(\mathbf{R}^n)$. This means that the functions $\varphi_{\nu,k}$ are mutually orthogonal and span $L^2(\mathbf{R}^n)$, and φ is normalized to have L^2 norm equal to 1. Note that the Fourier transform of $\varphi_{\nu,k}$ is given by

$$\widehat{\varphi_{\nu,k}}(\xi)=2^{-\frac{\nu n}{2}}\widehat{\varphi}(2^{-\nu}\xi)e^{-2\pi i2^{-\nu}\xi\cdot k}. \tag{6.6.1}$$

Rephrasing the question posed earlier, the main issue addressed in this section is whether smooth wavelets actually exist. Before we embark on this topic, we recall that we have already encountered examples of nonsmooth wavelets.

Example 6.6.2. (The Haar wavelet) Recall the family of functions

$$h_I(x) = |I|^{-\frac{1}{2}}(\chi_{I_L} - \chi_{I_R}),$$

where I ranges over $\mathscr{D}$ (the set of all dyadic intervals) and I_L is the left part of I and I_R is the right part of I. Note that if $I = [2^{-\nu}k, 2^{-\nu}(k+1))$, then

$$h_I(x) = 2^{\frac{\nu}{2}}\varphi(2^\nu x - k),$$

where

$$\varphi(x) = \chi_{[0,\frac{1}{2})} - \chi_{[\frac{1}{2},1)}. \tag{6.6.2}$$

The single function φ in (6.6.2) therefore generates the Haar basis by taking translations and dilations. Moreover, we observed in Section 6.4 that the family $\{h_I\}_I$ is orthonormal. Moreover, in Theorem 6.4.6 we obtained the representation

$$f = \sum_{I\in\mathscr{D}} \langle f, h_I\rangle h_I \qquad \text{in } L^2,$$

which proves the completeness of the system $\{h_I\}_{I\in\mathscr{D}}$ in $L^2(\mathbf{R})$.

6.6.1 Some Preliminary Facts

Before we look at more examples, we make some observations. We begin with the following useful fact.

Proposition 6.6.3. *Let $g \in L^1(\mathbf{R}^n)$. Then*

$$\widehat{g}(m) = 0 \qquad \textit{for all } m \in \mathbf{Z}^n \setminus \{0\}$$

if and only if

$$\sum_{k\in\mathbf{Z}^n} g(x+k) = \int_{\mathbf{R}^n} g(t)\,dt$$

for almost all $x \in \mathbf{T}^n$.

Proof. We define the periodic function

$$G(x) = \sum_{k\in\mathbf{Z}^n} g(x+k),$$

which is easily shown to be in $L^1(\mathbf{T}^n)$. Moreover, we have

$$\widehat{G}(m) = \widehat{g}(m)$$

for all $m \in \mathbf{Z}^n$, where $\widehat{G}(m)$ denotes the mth Fourier coefficient of G and $\widehat{g}(m)$ denotes the Fourier transform of g at $\xi = m$. If $\widehat{g}(m) = 0$ for all $m \in \mathbf{Z}^n \setminus \{0\}$, then all the Fourier coefficients of G (except for $m = 0$) vanish, which means that the sequence $\{\widehat{G}(m)\}_{m\in\mathbf{Z}^n}$ lies in $\ell^1(\mathbf{Z}^n)$ and hence Fourier inversion applies. We conclude that for almost all $x \in \mathbf{T}^n$ we have

$$G(x) = \sum_{m\in\mathbf{Z}^n} \widehat{G}(m)e^{2\pi i m\cdot x} = \widehat{G}(0) = \widehat{g}(0) = \int_{\mathbf{R}^n} g(t)\,dt\,.$$

Conversely, if G is a constant, then $\widehat{G}(m) = 0$ for all $m \in \mathbf{Z}^n \setminus \{0\}$, and so the same holds for g. □

A consequence of the preceding proposition is the following.

Proposition 6.6.4. *Let $\varphi \in L^2(\mathbf{R}^n)$. Then the sequence*

$$\{\varphi(x-k)\}_{k\in\mathbf{Z}^n} \tag{6.6.3}$$

forms an orthonormal set in $L^2(\mathbf{R}^n)$ if and only if

$$\sum_{k\in\mathbf{Z}^n} |\widehat{\varphi}(\xi+k)|^2 = 1 \tag{6.6.4}$$

for almost all $\xi \in \mathbf{R}^n$.

Proof. Observe that either (6.6.4) or the hypothesis that the sequence in (6.6.3) is orthonormal implies that $\|\varphi\|_{L^2} = 1$. Also the orthonormality condition

$$\int_{\mathbf{R}^n} \varphi(x-j)\overline{\varphi(x-k)}\,dx = \begin{cases} 1 & \text{when } j = k, \\ 0 & \text{when } j \neq k, \end{cases}$$

is equivalent to

$$\int_{\mathbf{R}^n} e^{-2\pi i k\cdot\xi}\widehat{\varphi}(\xi)\overline{e^{-2\pi i j\cdot\xi}\widehat{\varphi}(\xi)}\,d\xi = (|\widehat{\varphi}|^2)^{\widehat{}}(k-j) = \begin{cases} 1 & \text{when } j = k, \\ 0 & \text{when } j \neq k, \end{cases}$$

in view of Parseval's identity. Proposition 6.6.3 with $g(\xi) = |\widehat{\varphi}(\xi)|^2$ gives that the latter is equivalent to

$$\sum_{k\in\mathbf{Z}^n} |\widehat{\varphi}(\xi+k)|^2 = \int_{\mathbf{R}^n} |\widehat{\varphi}(t)|^2\,dt = 1$$

for almost all $\xi \in \mathbf{R}^n$. □

Corollary 6.6.5. *Let $\varphi \in L^2(\mathbf{R}^n)$ and suppose that the sequence*

$$\{\varphi(x-k)\}_{k\in\mathbf{Z}^n} \tag{6.6.5}$$

forms an orthonormal set in $L^2(\mathbf{R}^n)$. Then the measure of the support of $\widehat{\varphi}$ is at least 1, that is,

$$|\mathrm{supp}\,\widehat{\varphi}| \geq 1\,. \tag{6.6.6}$$

Moreover, if $|\mathrm{supp}\,\widehat{\varphi}| = 1$, then $|\widehat{\varphi}(\xi)| = 1$ for almost all $\xi \in \mathrm{supp}\,\widehat{\varphi}$.

Proof. It follows from (6.6.4) that $|\widehat{\varphi}| \leq 1$ for almost all $\xi \in \mathbf{R}^n$ and thus

$$|\mathrm{supp}\,\widehat{\varphi}| \geq \int_{\mathbf{R}^n} |\widehat{\varphi}(\xi)|^2\,d\xi = \int_{[0,1)^n} \sum_{k\in\mathbf{Z}^n} |\widehat{\varphi}(\xi+k)|^2\,d\xi = \int_{[0,1)^n} 1\,d\xi = 1\,.$$

If equality holds in (6.6.6), then equality holds in the preceding inequality, and since $|\widehat{\varphi}| \leq 1$ a.e., it follows that $|\widehat{\varphi}(\xi)| = 1$ for almost all ξ in $\mathrm{supp}\,\widehat{\varphi}$. □

6.6.2 Construction of a Nonsmooth Wavelet

Having established these preliminary facts, we now start searching for examples of wavelets. It follows from Corollary 6.6.5 that the support of the Fourier transform of a wavelet must have measure at least 1. It is reasonable to ask whether this support can have measure exactly 1. Example 6.6.6 indicates that this can indeed happen. As dictated by the same corollary, the Fourier transform of such a wavelet must satisfy $|\widehat{\varphi}(\xi)| = 1$ for almost all $\xi \in \mathrm{supp}\,\widehat{\varphi}$, so it is natural to look for a wavelet φ such that $\widehat{\varphi} = \chi_A$ for some set A. We can start by asking whether the function

$$\widehat{\varphi} = \chi_{[-\frac{1}{2},\frac{1}{2}]}$$

on $\mathbf{R}$ is an appropriate Fourier transform of a wavelet, but a moment's thought shows that the functions $\varphi_{\mu,0}$ and $\varphi_{\nu,0}$ cannot be orthogonal to each other when $\mu \neq 0$. The problem here is that the Fourier transforms of the functions $\varphi_{\nu,k}$ cluster near the origin and do not allow for the needed orthogonality. We can fix this problem by considering a function whose Fourier transform vanishes near the origin. Among such functions, a natural candidate is

$$\chi_{[-1,-\frac{1}{2})} + \chi_{[\frac{1}{2},1)}\,, \tag{6.6.7}$$

which is indeed the Fourier transform of a wavelet.

Example 6.6.6. Let $A = [-1,-\frac{1}{2}) \bigcup [\frac{1}{2},1)$ and define a function φ on $\mathbf{R}$ by setting

$$\widehat{\varphi} = \chi_A\,.$$

Then we assert that the family of functions

$$\{\varphi_{\nu,k}(x)\}_{k\in\mathbf{Z},\nu\in\mathbf{Z}} = \{2^{\nu/2}\varphi(2^\nu x-k)\}_{k\in\mathbf{Z},\nu\in\mathbf{Z}}$$

is an orthonormal basis of $L^2(\mathbf{R})$ (i.e., the function φ is a wavelet). This is an example of a wavelet with *minimally supported frequency*.

To verify this assertion, first note that $\{\varphi_{0,k}\}_{k\in\mathbf{Z}}$ is an orthonormal set, since (6.6.4) is easily seen to hold. Dilating by 2^ν, it follows that $\{\varphi_{\nu,k}\}_{k\in\mathbf{Z}}$ is also an orthonormal set for every fixed $\nu\in\mathbf{Z}$. Next, observe that if $\mu\neq\nu$, then

$$\operatorname{supp}\widehat{\varphi_{\nu,k}}\cap\operatorname{supp}\widehat{\varphi_{\mu,l}}=\emptyset. \tag{6.6.8}$$

This implies that the family $\{2^{\nu/2}\varphi(2^\nu x-k)\}_{k\in\mathbf{Z},\nu\in\mathbf{Z}}$ is also orthonormal.

Next, we observe that the completeness of $\{\varphi_{\nu,k}\}_{\nu,k\in\mathbf{Z}}$ is equivalent to that of $\{\widehat{\varphi_{\nu,k}}(\xi)\}_{\nu,k\in\mathbf{Z}} = \{2^{-\nu/2}e^{-2\pi ik\xi 2^{-\nu}}\chi_{2^\nu A}(\xi)\}_{\nu,k\in\mathbf{Z}}$. Let $f\in L^2(\mathbf{R})$, fix any $\nu\in\mathbf{Z}$, and define

$$h(\xi)=2^{\nu/2}f(2^\nu\xi).$$

Suppose that for all $k\in\mathbf{Z}$,

$$\begin{aligned} 0=\langle f,\widehat{\varphi_{\nu,k}}\rangle &= \int_{2^\nu A} f(\xi)2^{-\nu/2}e^{-2\pi ik\xi 2^{-\nu}}\,d\xi \\ &= \int_A 2^{\nu/2}f(2^\nu\xi)e^{-2\pi ik\xi}\,d\xi \\ &= \langle \chi_A h, e^{-2\pi ik\xi}\rangle. \end{aligned}$$

Exercise 6.6.1(a) shows $\{e^{-2\pi ik\xi}\}_{k\in\mathbf{Z}}$ is an orthonormal basis of $L^2(A)$, and therefore $\chi_A h=0$ almost everywhere. From the definition of h it follows that $\chi_{2^\nu A}f=0$ almost everywhere. Now suppose for all $\nu,k\in\mathbf{Z}$

$$0=\langle f,\widehat{\varphi_{\nu,k}}\rangle.$$

Then $\chi_{2^\nu A}f=0$ almost everywhere for all $\nu\in\mathbf{Z}$. Since $\cup_{\nu\in\mathbf{Z}}2^\nu A=\mathbf{R}\setminus\{0\}$, it follows that $f=0$ almost everywhere. We conclude $\{\widehat{\varphi_{\nu,k}}\}_{\nu,k\in\mathbf{Z}}$ is complete.

6.6.3 Construction of a Smooth Wavelet

The wavelet basis of $L^2(\mathbf{R}^n)$ constructed in Example 6.6.6 is forced to have slow decay at infinity, since the Fourier transforms of the elements of the basis are nonsmooth. Smoothing out the function $\widehat{\varphi}$ but still expecting φ to be wavelet is a bit tricky, since property (6.6.8) may be violated when $\mu\neq\nu$, and moreover, (6.6.4) may be destroyed. These two obstacles are overcome by the careful construction of the next theorem.

Theorem 6.6.7. *There exists a Schwartz function* φ *on the real line that is a wavelet, that is, the collection of functions* $\{\varphi_{\nu,k}\}_{k,\nu\in\mathbf{Z}}$ *with* $\varphi_{\nu,k}(x) = 2^{\frac{\nu}{2}}\varphi(2^\nu x-k)$ *is an orthonormal basis of* $L^2(\mathbf{R})$*. Moreover, the function* φ *can be constructed so that its Fourier transform satisfies*

$$\operatorname{supp}\widehat{\varphi} \subseteq \left[-\tfrac{4}{3},-\tfrac{1}{3}\right]\cup\left[\tfrac{1}{3},\tfrac{4}{3}\right]. \tag{6.6.9}$$

Note that in view of condition (6.6.9), the function φ must have vanishing moments of all orders.

Proof. We start with an odd smooth real-valued function Θ on the real line such that $\Theta(t) = \frac{\pi}{4}$ for $t \ge \frac{1}{6}$ and such that Θ is strictly increasing on the interval $\left[-\frac{1}{6},\frac{1}{6}\right]$. We set

$$\alpha(t) = \sin(\Theta(t)+\tfrac{\pi}{4}), \qquad \beta(t) = \cos(\Theta(t)+\tfrac{\pi}{4}),$$

and we observe that

$$\alpha(t)^2+\beta(t)^2 = 1$$

and that

$$\alpha(-t) = \beta(t)$$

for all real t. Next we introduce the smooth function ω defined via

$$\omega(t) = \begin{cases} \beta(-\frac{t}{2}-\frac{1}{2}) = \alpha(\frac{t}{2}+\frac{1}{2}) & \text{when } t\in\left[-\frac{4}{3},-\frac{2}{3}\right],\\ \alpha(-t-\frac{1}{2}) = \beta(t+\frac{1}{2}) & \text{when } t\in\left[-\frac{2}{3},-\frac{1}{3}\right],\\ \alpha(t-\frac{1}{2}) & \text{when } t\in\left[\frac{1}{3},\frac{2}{3}\right],\\ \beta(\frac{t}{2}-\frac{1}{2}) & \text{when } t\in\left[\frac{2}{3},\frac{4}{3}\right], \end{cases}$$

on the interval $\left[-\frac{4}{3},-\frac{1}{3}\right]\bigcup\left[\frac{1}{3},\frac{4}{3}\right]$. Note that ω is an even function. Finally we define the function φ by letting

$$\widehat{\varphi}(\xi) = e^{-\pi i\xi}\omega(\xi),$$

and we note that

$$\varphi(x) = \int_{\mathbf{R}}\omega(\xi)e^{2\pi i\xi(x-\frac{1}{2})}d\xi = 2\int_0^\infty \omega(\xi)\cos\left(2\pi(x-\tfrac{1}{2})\xi\right)d\xi.$$

It follows that the function φ is symmetric about the number $\frac{1}{2}$, that is, we have

$$\varphi(x) = \varphi(1-x)$$

for all $x\in\mathbf{R}$. Note that φ is a Schwartz function whose Fourier transform is supported in the set $\left[-\frac{4}{3},-\frac{1}{3}\right]\bigcup\left[\frac{1}{3},\frac{4}{3}\right]$.

Having defined φ, we proceed by showing that it is a wavelet. In view of identity (6.6.1) we have that $\widehat{\varphi_{\nu,k}}$ is supported in the set $\frac{1}{3}2^\nu \le |\xi| \le \frac{4}{3}2^\nu$, while $\widehat{\varphi_{\mu,j}}$ is supported in the set $\frac{1}{3}2^\mu \le |\xi| \le \frac{4}{3}2^\mu$. The intersection of these sets has measure zero when $|\mu-\nu|\ge 2$, which implies that such wavelets are orthogonal to each other. Therefore, it suffices to verify orthogonality between adjacent scales (i.e., when $\nu=\mu$ and $\nu=\mu+1$).

We begin with the case $\nu = \mu$, which, by a simple dilation, is reduced to the case $\nu = \mu = 0$. Thus to obtain the orthogonality of the functions $\varphi_{0,k}(x) = \varphi(x-k)$ and $\varphi_{0,j}(x) = \varphi(x-j)$, in view of Proposition 6.6.4, it suffices to show that

$$\sum_{k\in\mathbf{Z}} |\widehat{\varphi}(\xi+k)|^2 = 1\,. \tag{6.6.10}$$

Since the sum in (6.6.10) is 1-periodic, we check that is equal to 1 only for ξ in $\left[\frac{1}{3},\frac{4}{3}\right]$. First for $\xi \in \left[\frac{1}{3},\frac{2}{3}\right]$, the sum in (6.6.10) is equal to

$$\begin{aligned}
|\widehat{\varphi}(\xi)|^2 + |\widehat{\varphi}(\xi-1)|^2 &= \omega(\xi)^2 + \omega(\xi-1)^2 \\
&= \alpha(\xi - \tfrac{1}{2})^2 + \beta((\xi-1)+\tfrac{1}{2})^2 \\
&= 1
\end{aligned}$$

from the definition of ω. A similar argument also holds for $\xi \in \left[\frac{2}{3},\frac{4}{3}\right]$, and this completes the proof of (6.6.10). As a consequence of this identity we also obtain that the functions $\varphi_{0,k}$ have L^2 norm equal to 1, and thus so have the functions $\varphi_{\nu,k}$, via a change of variables.

Next we prove the orthogonality of the functions $\varphi_{\nu,k}$ and $\varphi_{\nu+1,j}$ for general $\nu,k,j \in \mathbf{Z}$. We begin by observing the validity of the following identity:

$$\widehat{\varphi}(\xi)\overline{\widehat{\varphi}(\tfrac{\xi}{2})} = \begin{cases} e^{-\pi i\xi/2}\beta(\frac{\xi}{2}-\frac{1}{2})\alpha(\frac{\xi}{2}-\frac{1}{2}) & \text{when } \frac{2}{3} \le \xi \le \frac{4}{3}, \\ e^{-\pi i\xi/2}\alpha(\frac{\xi}{2}+\frac{1}{2})\beta(\frac{\xi}{2}+\frac{1}{2}) & \text{when } -\frac{4}{3} \le \xi \le -\frac{2}{3}. \end{cases} \tag{6.6.11}$$

Indeed, from the definition of φ, it follows that

$$\widehat{\varphi}(\xi)\overline{\widehat{\varphi}(\tfrac{\xi}{2})} = e^{-\pi i\xi/2}\omega(\xi)\omega(\tfrac{\xi}{2})\,.$$

This function is supported in

$$\{\xi \in \mathbf{R} : \tfrac{1}{3} \le |\xi| \le \tfrac{4}{3}\} \cap \{\xi \in \mathbf{R} : \tfrac{2}{3} \le |\xi| \le \tfrac{8}{3}\} = \{\xi \in \mathbf{R} : \tfrac{2}{3} \le |\xi| \le \tfrac{4}{3}\},$$

and on this set it is equal to

$$e^{-\pi i\xi/2}\begin{cases} \beta(\frac{\xi}{2}-\frac{1}{2})\alpha(\frac{\xi}{2}-\frac{1}{2}) & \text{when } \frac{2}{3} \le \xi \le \frac{4}{3}, \\ \alpha(\frac{\xi}{2}+\frac{1}{2})\beta(\frac{\xi}{2}+\frac{1}{2}) & \text{when } -\frac{4}{3} \le \xi \le -\frac{2}{3}, \end{cases}$$

by the definition of ω. This establishes (6.6.11).

We now turn to the orthogonality of the functions $\varphi_{\nu,k}$ and $\varphi_{\nu+1,j}$ for general $\nu,k,j \in \mathbf{Z}$. Using (6.6.1) and (6.6.11) we have

$$\begin{aligned}
\langle \varphi_{\nu,k} \,|\, \varphi_{\nu+1,j} \rangle &= \langle \widehat{\varphi_{\nu,k}} \,|\, \widehat{\varphi_{\nu+1,j}} \rangle \\
&= \int_{\mathbf{R}} 2^{-\frac{\nu}{2}}\widehat{\varphi}(2^{-\nu}\xi)e^{-2\pi i\frac{\xi k}{2^\nu}}\, \overline{2^{-\frac{\nu+1}{2}}\widehat{\varphi}(2^{-(\nu+1)}\xi)e^{-2\pi i\frac{\xi j}{2^{\nu+1}}}}\, d\xi
\end{aligned}$$

$$
\begin{aligned}
&= \frac{1}{\sqrt{2}} \int_{\mathbf{R}} \widehat{\varphi}(\xi)\overline{\widehat{\varphi}(\tfrac{\xi}{2})} e^{-2\pi i \xi (k-\frac{j}{2})}\, d\xi \\
&= \frac{1}{\sqrt{2}} \int_{-\frac{4}{3}}^{-\frac{2}{3}} \alpha(\tfrac{\xi}{2}+\tfrac{1}{2})\beta(\tfrac{\xi}{2}+\tfrac{1}{2}) e^{-2\pi i \xi (k-\frac{j}{2}+\frac{1}{4})}\, d\xi \\
&\qquad + \frac{1}{\sqrt{2}} \int_{\frac{2}{3}}^{\frac{4}{3}} \alpha(\tfrac{\xi}{2}-\tfrac{1}{2})\beta(\tfrac{\xi}{2}-\tfrac{1}{2}) e^{-2\pi i \xi (k-\frac{j}{2}+\frac{1}{4})}\, d\xi \\
&= 0,
\end{aligned}
$$

where the last identity follows from the change of variables $\xi = \xi' - 2$ in the second-to-last integral, which transforms its range of integration to $\left[\frac{2}{3}, \frac{4}{3}\right]$ and its integrand to the negative of that of the last displayed integral.

Our final task is to show that the orthonormal system $\{\varphi_{\nu,k}\}_{\nu,k\in\mathbf{Z}}$ is complete. We show this by proving that whenever a square-integrable function f satisfies

$$
\left\langle f \,|\, \varphi_{\nu,k} \right\rangle = 0 \tag{6.6.12}
$$

for all $\nu, k \in \mathbf{Z}$, then f must be zero. Suppose that (6.6.12) holds. Plancherel's identity yields

$$
\int_{\mathbf{R}} \widehat{f}(\xi) 2^{-\frac{\nu}{2}} \overline{\widehat{\varphi}(2^{-\nu}\xi) e^{-2\pi i 2^{-\nu}\xi k}}\, d\xi = 0
$$

for all ν, k and thus

$$
\int_{\mathbf{R}} \widehat{f}(2^{\nu}\xi)\overline{\widehat{\varphi}(\xi)} e^{2\pi i \xi k}\, d\xi = \big(\widehat{f}(2^{\nu}(\cdot))\overline{\widehat{\varphi}}\big)\widehat{\ }(-k) = 0 \tag{6.6.13}
$$

for all $\nu, k \in \mathbf{Z}$. It follows from Proposition 6.6.3 and (6.6.13) (with $k = 0$) that

$$
\sum_{k\in\mathbf{Z}} \widehat{f}(2^{\nu}(\xi+k))\overline{\widehat{\varphi}(\xi+k)} = \int_{\mathbf{R}} \widehat{f}(2^{\nu}\xi)\,\overline{\widehat{\varphi}(\xi)}\, d\xi = \big(\widehat{f}(2^{\nu}(\cdot))\overline{\widehat{\varphi}}\big)\widehat{\ }(0) = 0
$$

for all $\nu \in \mathbf{Z}$.

Next, we show that the identity

$$
\sum_{k\in\mathbf{Z}} \widehat{f}(2^{\nu}(\xi+k))\overline{\widehat{\varphi}(\xi+k)} = 0 \tag{6.6.14}
$$

for all $\nu \in \mathbf{Z}$ implies that $\widehat{f}$ is identically equal to zero. Suppose that $\frac{1}{3} \le \xi \le \frac{2}{3}$. In this case the support properties of $\widehat{\varphi}$ imply that the only terms in the sum in (6.6.14) that do not vanish are $k = 0$ and $k = -1$. Thus for $\frac{1}{3} \le \xi \le \frac{2}{3}$ the identity in (6.6.14) reduces to

$$
\begin{aligned}
0 &= \widehat{f}(2^{\nu}(\xi-1))\overline{\widehat{\varphi}(\xi-1)} + \widehat{f}(2^{\nu}\xi)\overline{\widehat{\varphi}(\xi)} \\
&= \widehat{f}(2^{\nu}(\xi-1)) e^{\pi i(\xi-1)} \beta((\xi-1)+\tfrac{1}{2}) + \widehat{f}(2^{\nu}\xi) e^{\pi i \xi} \alpha(\xi-\tfrac{1}{2});
\end{aligned}
$$

hence

$$-\widehat{f}(2^{\nu}(\xi-1))\beta(\xi-\tfrac{1}{2})+\widehat{f}(2^{\nu}\xi)\alpha(\xi-\tfrac{1}{2})=0, \quad \tfrac{1}{3}\le\xi\le\tfrac{2}{3}. \tag{6.6.15}$$

Next we observe that when $\frac{2}{3}\le\xi\le\frac{4}{3}$, only the terms with $k=0$ and $k=-2$ survive in the identity in (6.6.14). This is because when $k=-1$, $\xi+k=\xi-1\in\left[-\frac{1}{3},\frac{1}{3}\right]$ and this interval has null intersection with the support of $\widehat{\varphi}$. Therefore, (6.6.14) reduces to

$$\begin{aligned}0 &=\widehat{f}(2^{\nu}(\xi-2))\overline{\widehat{\varphi}(\xi-2)}+\widehat{f}(2^{\nu}\xi)\overline{\widehat{\varphi}(\xi)}\\ &=\widehat{f}(2^{\nu}(\xi-2))e^{\pi i(\xi-2)}\alpha(\tfrac{\xi-2}{2}+\tfrac{1}{2})+\widehat{f}(2^{\nu}\xi)e^{\pi i\xi}\beta(\tfrac{\xi}{2}-\tfrac{1}{2});\end{aligned}$$

hence

$$\widehat{f}(2^{\nu}(\xi-2))\alpha(\tfrac{\xi}{2}-\tfrac{1}{2})+\widehat{f}(2^{\nu}\xi)\beta(\tfrac{\xi}{2}-\tfrac{1}{2})=0, \quad \tfrac{2}{3}\le\xi\le\tfrac{4}{3}. \tag{6.6.16}$$

Replacing first ν by $\nu-1$ and then $\frac{\xi}{2}$ by ξ in (6.6.16), we obtain

$$\widehat{f}(2^{\nu}(\xi-1))\alpha(\xi-\tfrac{1}{2})+\widehat{f}(2^{\nu}\xi)\beta(\xi-\tfrac{1}{2})=0, \quad \tfrac{1}{3}\le\xi\le\tfrac{2}{3}. \tag{6.6.17}$$

Now consider the 2×2 system of equations given by (6.6.15) and (6.6.17) with unknown $\widehat{f}(2^{\nu}(\xi-1))$ and $\widehat{f}(2^{\nu}\xi)$. The determinant of the system is

$$\det\begin{pmatrix}-\beta(\xi-1/2) & \alpha(\xi-1/2)\\ \alpha(\xi-1/2) & \beta(\xi-1/2)\end{pmatrix}=-1\neq 0.$$

Therefore, the system has the unique solution

$$\widehat{f}(2^{\nu}(\xi-1))=\widehat{f}(2^{\nu}\xi)=0,$$

which is valid for all $\nu\in\mathbf{Z}$ and all $\xi\in[\frac{1}{3},\frac{2}{3}]$. We conclude that $\widehat{f}(\xi)=0$ for all $\xi\in\mathbf{R}$ and thus $f=0$. This proves the completeness of the system $\{\varphi_{\nu,k}\}$. We conclude that the function φ is a wavelet. □

6.6.4 Sampling

Next we discuss how one can recover a band-limited function by its values at a countable number of points.

Definition 6.6.8. An integrable function on $\mathbf{R}^n$ is called *band limited* if its Fourier transform has compact support.

For every band-limited function there is a $B>0$ such that its Fourier transform is supported in the cube $[-B,B]^n$. In such a case we say that the function is band limited on the cube $[-B,B]^n$.

It is an interesting observation that such functions are completely determined by their values at the points $x = k/2B$, where $k \in \mathbf{Z}^n$. We have the following result.

Theorem 6.6.9. *(a) Let f in $L^1(\mathbf{R}^n)$ be band limited on the cube $[-B,B]^n$. Then f can be sampled by its values at the points $x = k/2B$, where $k \in \mathbf{Z}^n$. In particular, we have*

$$f(x_1,\dots,x_n) = \sum_{k\in\mathbf{Z}^n} f\Big(\frac{k}{2B}\Big)\prod_{j=1}^{n}\frac{\sin(2\pi Bx_j - \pi k_j)}{2\pi Bx_j - \pi k_j} \tag{6.6.18}$$

for almost all $x \in \mathbf{R}^n$.
(b) Suppose that f is band-limited on the cube $[-B',B']^n$ where $0 < B' < B$. Then f can be sampled by its values at the points $x = k/2B$, $k \in \mathbf{Z}^n$ as follows

$$f(x_1,\dots,x_n) = \sum_{k\in\mathbf{Z}^n} f\Big(\frac{k}{2B}\Big)\Phi(x-k), \tag{6.6.19}$$

for some Schwartz function Φ that depends on B, B'.

Proof. Since the function $\widehat{f}$ is supported in $[-B,B]^n$, we use Exercise 6.6.2 to obtain

$$\begin{aligned}\widehat{f}(\xi) &= \frac{1}{(2B)^n}\sum_{k\in\mathbf{Z}^n}\widehat{\widehat{f}}\Big(\frac{k}{2B}\Big)e^{2\pi i\frac{k}{2B}\cdot\xi}\\ &= \frac{1}{(2B)^n}\sum_{k\in\mathbf{Z}^n} f\Big(-\frac{k}{2B}\Big)e^{2\pi i\frac{k}{2B}\cdot\xi}.\end{aligned}$$

Inserting this identity in the inversion formula

$$f(x) = \int_{[-B,B]^n}\widehat{f}(\xi)e^{2\pi i x\cdot\xi}\,d\xi,$$

which holds for almost all $x \in \mathbf{R}^n$ since $\widehat{f}$ is continuous and therefore integrable over $[-B,B]^n$, we obtain

$$\begin{aligned}f(x) &= \int_{[-B,B]^n}\frac{1}{(2B)^n}\sum_{k\in\mathbf{Z}^n} f\Big(-\frac{k}{2B}\Big)e^{2\pi i\frac{k}{2B}\cdot\xi}e^{2\pi i x\cdot\xi}\,d\xi\\ &= \sum_{k\in\mathbf{Z}^n} f\Big(-\frac{k}{2B}\Big)\frac{1}{(2B)^n}\int_{[-B,B]^n}e^{2\pi i(\frac{k}{2B}+x)\cdot\xi}\,d\xi \qquad (6.6.20)\\ &= \sum_{k\in\mathbf{Z}^n} f\Big(-\frac{k}{2B}\Big)\prod_{j=1}^{n}\frac{\sin(2\pi Bx_j+\pi k_j)}{2\pi Bx_j+\pi k_j}. \qquad (6.6.21)\end{aligned}$$

This is exactly (6.6.18) when we change k to $-k$ and thus part (a) is proved. For part (b) we argue similarly, except that we replace $\chi_{[-B,B]^n}$ by $\widehat{\Phi}$, where $\widehat{\Phi}$ is smooth, equal to 1 on $[-B',B']^n$ and vanishes outside $[-B,B]^n$. Then we can insert the function $\widehat{\Phi}(\xi)$ in (6.6.20) and instead of (6.6.21) we obtain the expression on the right in (6.6.19). □

Remark 6.6.10. Identity (6.6.18) holds for any $B'' > B$. In particular, we have

$$\sum_{k\in\mathbf{Z}^n} f\Big(\frac{k}{2B}\Big)\prod_{j=1}^{n}\frac{\sin(2\pi Bx_j-\pi k_j)}{2\pi Bx_j-\pi k_j}=\sum_{k\in\mathbf{Z}^n} f\Big(\frac{k}{2B''}\Big)\prod_{j=1}^{n}\frac{\sin(2\pi B''x_j-\pi k_j)}{2\pi B''x_j-\pi k_j}$$

for all $x\in\mathbf{R}^n$ whenever f is band-limited in $[-B,B]^n$. In particular, band-limited functions in $[-B,B]^n$ can be sampled by their values at the points $k/2B''$ for any $B''\geq B$.

However, band-limited functions in $[-B,B]^n$ cannot be sampled by the points $k/2B'$ for any $B'<B$, as the following example indicates.

Example 6.6.11. For $0<B'<B$, let $f(x)=g(x)\sin(2\pi B'x)$, where $\widehat{g}$ is supported in the interval $[-(B-B'),B-B']$. Then f is band limited in $[-B,B]$, but it cannot be sampled by its values at the points $k/2B'$, since it vanishes at these points and f is not identically zero if g is not the zero function.

Next, we give a couple of results that relate the L^p norm of a given function with the ℓ^p norm (or quasi-norm) of its sampled values.

Theorem 6.6.12. *Let f be a tempered[1] function whose Fourier transform is supported in the closed ball $\overline{B(0,t)}$ for some $0<t<\infty$. Assume that f lies in $L^p(\mathbf{R}^n)$ for some $0<p\leq\infty$. Then there is a constant $C(n,p)$ such that*

$$\big\|\{f(k)\}_{k\in\mathbf{Z}^n}\big\|_{\ell^p(\mathbf{Z}^n)}\leq C(n,p)\,t\,(1+t^{\frac{2n}{p}})\big\|f\big\|_{L^p(\mathbf{R}^n)}.$$

Proof. The proof is based on the following fact, whose proof can be found in [131] (Lemma 2.2.3). Let $0<r<\infty$. Then there exists a constant $C_2=C_2(n,r)$ such that for all $t>0$ and for all $\mathscr{C}^1$ functions u on $\mathbf{R}^n$ whose distributional Fourier transform is supported in the ball $|\xi|\leq t$ we have

$$\sup_{z\in\mathbf{R}^n}\frac{1}{t}\frac{|\nabla u(x-z)|}{(1+t|z|)^{\frac{n}{r}}}\leq C_2 M(|u|^r)(x)^{\frac{1}{r}}, \tag{6.6.22}$$

where M denotes the Hardy–Littlewood maximal operator.

Notice that f is a $\mathscr{C}^\infty$ function since its Fourier transform is compactly supported. Assuming (6.6.22), for each $k\in\mathbf{Z}^n$ and $x\in[0,1]^n$ we use the mean value theorem to obtain

$$\begin{aligned}|f(k)| &\leq |f(x+k)|+\sqrt{n}\sup_{z\in[0,1]^n}|\nabla f(z+k)|\\ &\leq |f(x+k)|+\sqrt{n}\sup_{z\in B(x+k,\sqrt{n})}|\nabla f(z)|.\end{aligned}$$

We raise this inequality to the power p, we integrate over the cube $[0,1]^n$, we sum over $k\in\mathbf{Z}^n$, and then we take the $1/p$ power. Let $c_p=\max(1,2^{1/p-1})$ and $c(n,r,t)=$

[1] A function is called tempered if there are constants C,M such that $|f(x)|\leq C(1+|x|)^M$ for all $x\in\mathbf{R}^n$. Tempered functions are tempered distributions.

$\sqrt{n}t(1+t\sqrt{n})^{n/r}$. The sum over k and the integral over $[0,1]^n$ yield an integral over $\mathbf{R}^n$ and thus we obtain

$$\begin{aligned}
\Big[\sum_{k\in\mathbf{Z}^n}|f(k)|^p\Big]^{\frac{1}{p}} &\le \Big[\int_{\mathbf{R}^n}|f(x)+\sqrt{n}\sup_{z\in B(x,\sqrt{n})}|\nabla f(z)|^p\,dx\Big]^{\frac{1}{p}}\\
&\le c_p\Big[\big\|f\big\|_{L^p}+\sqrt{n}\Big(\int_{\mathbf{R}^n}\sup_{z\in B(0,\sqrt{n})}|\nabla f(x-z)|^p\,dx\Big)^{\frac{1}{p}}\Big]\\
&\le c_p\Big[\big\|f\big\|_{L^p}+c(n,r,t)\Big(\int_{\mathbf{R}^n}\Big\{\sup_{z\in B(0,\sqrt{n})}\frac{|\nabla f(x-z)|}{t(1+t|z|)^{\frac{n}{r}}}\Big\}^p dx\Big)^{\frac{1}{p}}\Big]\\
&\le c_p\Big[\big\|f\big\|_{L^p}+c(n,r,t)\Big(\int_{\mathbf{R}^n}\Big\{\sup_{z\in\mathbf{R}^n}\frac{|\nabla f(x-z)|}{t(1+t|z|)^{\frac{n}{r}}}\Big\}^p dx\Big)^{\frac{1}{p}}\Big]\\
&\le c_p\Big[\big\|f\big\|_{L^p}+c(n,r,t)C_2\Big(\int_{\mathbf{R}^n}[M(|f|^r)(x)]^{\frac{p}{r}}\,dx\Big)^{\frac{1}{p}}\Big],
\end{aligned}$$

where the last step uses (6.6.22). We now select $r=p/2$ if $p<\infty$ and r to be any number if $p=\infty$. The required inequality follows from the boundedness of the Hardy-Littlewood maximal operator on L^2 if $p<\infty$ or on L^∞ if $p=\infty$. □

The next theorem could be considered a partial converse of Theorem 6.6.13

Theorem 6.6.13. *Suppose that an integrable function f has Fourier transform supported in the cube $[-(\frac{1}{2}-\varepsilon),\frac{1}{2}-\varepsilon]^n$ for some $0<\varepsilon<1/2$. Furthermore, suppose that the sequence of coefficients $\{f(k)\}_{k\in\mathbf{Z}^n}$ lies in $\ell^p(\mathbf{Z}^n)$ for some $0<p\le\infty$. Then f lies in $L^p(\mathbf{R}^n)$ and the following estimate is valid*

$$\big\|f\big\|_{L^p(\mathbf{R}^n)}\le C_{n,p,\varepsilon}\big\|\{f(k)\}_k\big\|_{\ell^p(\mathbf{Z}^n)}. \tag{6.6.23}$$

Proof. We fix a smooth function $\widehat{\Phi}$ supported in $[-\frac{1}{2},\frac{1}{2}]^n$ and equal to 1 on the smaller cube $[-(\frac{1}{2}-\varepsilon),\frac{1}{2}-\varepsilon]^n$. Then we may write $f=f*\Phi$, since $\widehat{\Phi}$ is equal to one on the support of $\widehat{f}$. Writing $\widehat{f}$ in terms of its Fourier series we have

$$\widehat{f}(\xi)=\sum_{k\in\mathbf{Z}^n}\widehat{\widehat{f}}(k)e^{2\pi ik\cdot\xi}\chi_{[-\frac{1}{2},\frac{1}{2}]^n}=\sum_{k\in\mathbf{Z}^n}f(-k)e^{2\pi ik\cdot\xi}\chi_{[-\frac{1}{2},\frac{1}{2}]^n} \tag{6.6.24}$$

Since f is integrable, $\widehat{f}$ is continuous and thus integrable over $[-\frac{1}{2},\frac{1}{2}]^n$. By Fourier inversion we have

$$f(x)=\int_{[-\frac{1}{2},\frac{1}{2}]^n}\widehat{f}(\xi)e^{2\pi ix\cdot\xi}d\xi=\int_{[-\frac{1}{2},\frac{1}{2}]^n}\widehat{f}(\xi)\widehat{\Phi}(\xi)e^{2\pi ix\cdot\xi}d\xi \tag{6.6.25}$$

for almost all $x\in\mathbf{R}^n$. Inserting (6.6.25) in (6.6.24) we obtain

$$f(x)=\int_{[-\frac{1}{2},\frac{1}{2}]^n}\sum_{k\in\mathbf{Z}^n}f(-k)e^{2\pi ik\cdot\xi}e^{2\pi ix\cdot\xi}\widehat{\Phi}(\xi)\,d\xi$$

$$= \sum_{k\in\mathbf{Z}^n} f(k) \int_{[-\frac{1}{2},\frac{1}{2}]^n} e^{-2\pi i k\cdot\xi} e^{2\pi i x\cdot\xi} \widehat{\Phi}(\xi)\,d\xi$$
$$= \sum_{k\in\mathbf{Z}^n} f(k)\Phi(x-k)\,.$$

This identity combined with the rapid decay of Φ yields (6.6.23) as follows. For $0<p\le 1$ we have

$$\big\|f\big\|_{L^p}^p \le \int_{\mathbf{R}^n} \sum_{k\in\mathbf{Z}^n} |f(k)|^p|\Phi(x-k)|^p = \big\|\{f(k)\}_k\big\|_{\ell^p(\mathbf{Z}^n)}^p \|\Phi\|_{L^p}^p$$

while for $1<p\le\infty$, setting $Q=[-\frac{1}{2},\frac{1}{2}]^n$ we write:

$$\begin{aligned}
\big\|f\big\|_{L^p(\mathbf{R}^n)} &\le \bigg[\sum_{l\in\mathbf{Z}^n}\int_{l+Q}\Big(\sum_{k\in\mathbf{Z}^n}|f(k)||\Phi(x-k)|\Big)^p dx\bigg]^{\frac{1}{p}} \\
&\le C_{n,N}\bigg[\sum_{l\in\mathbf{Z}^n}\int_{l+Q}\Big(\sum_{k\in\mathbf{Z}^n}|f(k)|\frac{1}{(2\sqrt{n}+|x-k|)^N}\Big)^p dx\bigg]^{\frac{1}{p}} \\
&\le C'_{n,N}\bigg[\sum_{l\in\mathbf{Z}^n}\int_{l+Q}\Big(\sum_{k\in\mathbf{Z}^n}|f(k)|\frac{1}{(\sqrt{n}+|l-k|)^N}\Big)^p dx\bigg]^{\frac{1}{p}} \\
&\le C'_{n,N}\bigg[\sum_{l\in\mathbf{Z}^n}\Big(\sum_{k\in\mathbf{Z}^n}|f(k)|\frac{1}{(\sqrt{n}+|l-k|)^N}\Big)^p\bigg]^{\frac{1}{p}}.
\end{aligned}$$

The preceding expression can be viewed as the ℓ^p norm of the discrete convolution of the sequences $\{f(k)\}_k$ and $\frac{1}{(\sqrt{n}+|k|)^N}$ and thus it is bounded by a constant multiple of $\big\|\{f(k)\}_k\big\|_{\ell^p(\mathbf{Z}^n)}$, since the sequence $\frac{1}{(\sqrt{n}+|k|)^N}$ is in $\ell^1(\mathbf{Z}^n)$ if N is large enough. This completes the proof. □

Exercise 6.6.6 gives examples of functions for which Theorem 6.6.13 fails if $\varepsilon=0$.

Exercises

6.6.1. (a) Let $A=[-1,-\frac{1}{2})\cup[\frac{1}{2},1)$. Show that the family $\{e^{2\pi i m x}\}_{m\in\mathbf{Z}}$ is an orthonormal basis of $L^2(A)$.
(b) Obtain the same conclusion for the family $\{e^{2\pi i m\cdot x}\}_{m\in\mathbf{Z}^n}$ in $L^2(A^n)$.
[*Hint:* To show completeness, given $f\in L^2(A)$, define h on $[0,1]$ by setting $h(x)=f(x-1)$ for $x\in[0,\frac{1}{2})$ and $h(x)=f(x)$ for $x\in[\frac{1}{2},1)$. Observe that $\widehat{h}(m)=\widehat{f}(m)$ for all $m\in\mathbf{Z}$ and expand h in Fourier series.]

6.6.2. Let g be an integrable function on $\mathbf{R}^n$.
(a) Suppose that g is supported in $[-b,b]^n$ for some $b>0$ and that the sequence $\{\widehat{g}(k/2b)\}_{k\in\mathbf{Z}^n}$ lies in $\ell^2(\mathbf{Z}^n)$. Show that

$$g(x) = (2b)^{-n} \sum_{k\in\mathbf{Z}^n} \widehat{g}(\tfrac{k}{2b})e^{2\pi i \frac{k}{2b}\cdot x}\chi_{[-b,b]^n},$$

where the series converges in $L^2(\mathbf{R}^n)$ and deduce that g is in $L^2(\mathbf{R}^n)$.
(b) Suppose that g is supported in $[0,b]^n$ for some $b>0$ and that the sequence $\{\widehat{g}(k/b)\}_{k\in\mathbf{Z}^n}$ lies in $\ell^2(\mathbf{Z}^n)$. Show that

$$g(x) = b^{-n} \sum_{k\in\mathbf{Z}^n} \widehat{g}(\tfrac{k}{b})e^{2\pi i \frac{k}{b}\cdot x}\chi_{[0,b]^n},$$

where the series converges in $L^2(\mathbf{R}^n)$ and deduce that g is in $L^2(\mathbf{R}^n)$.
(c) When $n=1$, obtain the same as the conclusion in part (b) for $x\in[-b,-\frac{b}{2})\cup[\frac{b}{2},b)$, provided g is supported in this set.
[*Hint:* Part (c): Use the result in Exercise 6.6.1.]

6.6.3. Show that the sequence of functions

$$H_k(x_1,\dots,x_n) = (2B)^{\frac{n}{2}} \prod_{j=1}^{n} \frac{\sin\big(\pi(2Bx_j-k_j)\big)}{\pi(2Bx_j-k_j)}, \qquad k\in\mathbf{Z}^n,$$

is orthonormal in $L^2(\mathbf{R}^n)$.
[*Hint:* Interpret the functions H_k as the Fourier transforms of known functions.]

6.6.4. Prove the following spherical multidimensional version of Theorem 6.6.9. Suppose that $\widehat{f}$ is supported in the ball $|\xi|\le R$. Show that

$$f(x) = \sum_{k\in\mathbf{Z}^n} f\big(-\tfrac{k}{2R}\big)\frac{1}{2^n}\frac{J_{\frac{n}{2}}(2\pi|Rx+\frac{k}{2}|)}{|Rx+\frac{k}{2}|^{\frac{n}{2}}},$$

where J_a is the Bessel function of order a.

6.6.5. Let $\{a_k\}_{k\in\mathbf{Z}^n}$ be in ℓ^p for some $1<p<\infty$. Show that the partial sums

$$\sum_{\substack{k\in\mathbf{Z}^n\\ |k|\le N}} a_k \prod_{j=1}^{n} \frac{\sin(2\pi Bx_j-\pi k_j)}{2\pi Bx_j-\pi k_j}$$

converge in $\mathscr{S}'(\mathbf{R}^n)$ as $N\to\infty$ to an L^p function on $\mathbf{R}^n$ whose Fourier transform is supported in $[-B,B]^n$. Here $k=(k_1,\dots,k_n)$. Moreover, the L^p norm of A is controlled by a constant multiple of the ℓ^p norm of $\{a_k\}_k$.

6.6.6. Consider the function $\prod_{j=1}^n \sin(\pi x_j)/(\pi x_j)$ on $\mathbf{R}^n$ to show that Theorem 6.6.13 fails when $\varepsilon=0$ and $p\le 1$. When $1<p\le\infty$ consider the function $x_1+\prod_{j=1}^n \sin(\pi x_j)/(\pi x_j)$.

6.6.7. (a) Let $\psi(x)$ be a nonzero continuous integrable function on $\mathbf{R}$ that satisfies $\int_{\mathbf{R}} \psi(x)\,dx = 0$ and

$$C_\psi = \int_{-\infty}^{+\infty} \frac{|\widehat{\psi}(t)|^2}{|t|}\,dt < \infty .$$

Define the *wavelet transform* of f in $L^2(\mathbf{R})$ by setting

$$W(f;a,b) = \frac{1}{\sqrt{|a|}} \int_{-\infty}^{+\infty} f(x)\overline{\psi\Big(\frac{x-b}{a}\Big)}\,dx$$

when $a \neq 0$ and $W(f;0,b) = 0$. Show that for any $f \in L^2(\mathbf{R})$ the following inversion formula holds:

$$f(x) = \frac{1}{C_\psi} \int_{-\infty}^{+\infty} \int_{-\infty}^{+\infty} \frac{1}{|a|^{\frac{1}{2}}} \psi\Big(\frac{x-b}{a}\Big) W(f;a,b)\,db\,\frac{da}{a^2}\,.$$

(b) State and prove an analogous wavelet transform inversion property on $\mathbf{R}^n$.
[*Hint:* Apply Theorem 2.2.14 (5) in the b-integral and use Fourier inversion.]

6.6.8. (*P. Casazza*) On $\mathbf{R}^n$ let e_j be the vector whose coordinates are zero everywhere except for the jth entry, which is 1. Set $q_j = e_j - \frac{1}{n}\sum_{k=1}^n e_k$ for $1 \le j \le n$ and also $q_{n+1} = \frac{1}{\sqrt{n}}\sum_{k=1}^n e_k$. Prove that

$$\sum_{j=1}^{n+1} |q_j \cdot x|^2 = |x|^2$$

for all $x \in \mathbf{R}^n$. This provides an example of a *tight frame* on $\mathbf{R}^n$.

HISTORICAL NOTES

An early account of square functions in the context of Fourier series appears in the work of Kolmogorov [196], who proved the almost everywhere convergence of lacunary partial sums of Fourier series of periodic square-integrable functions. This result was systematically studied and extended to L^p functions, $1 < p < \infty$, by Littlewood and Paley [227], [228], [229] using complex-analysis techniques. The real-variable treatment of the Littlewood and Paley theorem was pioneered by Stein [334] and allowed the higher-dimensional extension of the theory. The use of vector-valued inequalities in the proof of Theorem 6.1.2 is contained in Benedek, Calderón, and Panzone [22]. A Littlewood–Paley theorem for lacunary sectors in $\mathbf{R}^2$ was obtained by Nagel, Stein, and Wainger [264].

An interesting Littlewood–Paley estimate holds for $2 \le p < \infty$: There exists a constant C_p such that for all families of disjoint open intervals I_j in $\mathbf{R}$ the estimate $\|(\sum_j |(\widehat{f}\chi_{I_j})^\vee|^2)^{\frac{1}{2}}\|_{L^p} \le C_p \|f\|_{L^p}$ holds for all functions $f \in L^p(\mathbf{R})$. This was proved by Rubio de Francia [301], but the special case in which $I_j = (j, j+1)$ was previously obtained by Carleson [55]. An alternative proof of Rubio de Francia's theorem was obtained by Bourgain [34]. A higher-dimensional analogue of this estimate for arbitrary disjoint open rectangles in $\mathbf{R}^n$ with sides parallel to the axes was obtained by Journé [181]. Easier proofs of the higher-dimensional result were subsequently obtained by Sjölin [326], Soria [329], and Sato [311].

Part (a) of Theorem 6.2.7 is due to Mihlin [254] and the generalization in part (b) to Hörmander [159]. Theorem 6.2.2 can be found in Marcinkiewicz's article [241] in the context of one-dimensional Fourier series. Calderón and Torchinsky [45] have improved Theorem 6.2.7 in the following way: if for a suitable smooth bump η supported in an annulus the functions $m(2^k\xi)\eta(\xi)$ lie in the Sobolev space L^r_γ uniformly in $k \in \mathbf{Z}$, where $\gamma > n(\frac{1}{p} - \frac{1}{2})$, $1 < p < 2$, $\frac{1}{r} = \frac{1}{p} - \frac{1}{2}$, then m lies in $\mathscr{M}_p(\mathbf{R}^n)$. The power 6 in estimate (6.2.3) that appears in the statement of Theorem 6.2.2 is not optimal. Tao and Wright [357] proved that in dimension 1, the best power of $(p-1)^{-1}$ in this theorem is $\frac{3}{2}$ as $p \to 1$. An improvement of the Marcinkiewicz multiplier theorem in one dimension was obtained by Coifman, Rubio de Francia, and Semmes [69]. Weighted norm estimates for Hörmander–Mihlin multipliers were obtained by Kurtz and Wheeden [209] and for Marcinkiwiecz multipliers by Kurtz [208]. Heo, Nazarov, and Seeger [150] have obtained a very elegant characterization of radial L^p multipliers in large dimensions; precisely, they showed that for dimensions $n \ge 4$ and $1 < p < \frac{2n-2}{n+1}$, a radial function m on $\mathbf{R}^n$ is an L^p Fourier multiplier if and only if there exists a nonzero Schwartz function η such that $\sup_{t>0} t^{n/p} \|(m(\cdot)\eta(t\cdot))^\vee\|_{L^p} < \infty$. This characterization builds on and extends a previously obtained simple characterization by Garrigós and Seeger [124] of radial multipliers on the invariant subspace of radial L^p functions when $1 < p < \frac{2n}{n+1}$.

The method of proof of Theorem 6.3.4 is adapted from Duoandikoetxea and Rubio de Francia [102]. The method in this article is rather general and can be used to obtain L^p boundedness for a variety of rough singular integrals. A version of Theorem 6.3.6 was used by Christ [59] to obtain L^p smoothing estimates for Cantor–Lebesgue measures. When $p = q \neq 2$, Theorem 6.3.6 is false in general, but it is true for all r satisfying $|\frac{1}{r} - \frac{1}{2}| < |\frac{1}{p} - \frac{1}{2}|$ under the additional assumption that the m_j's are Lipschitz functions uniformly at all scales. This result was independently obtained by Carbery [52] and Seeger [316]. Miyachi [255] has obtained a complete characterization of the indices $a, b > 0$ such that the functions $|x|^{-b} e^{i|x|^a} \psi(x)$ are L^p Fourier multipliers; here ψ is a smooth function that is equal to 1 near infinity and vanishes near zero.

The probabilistic notions of conditional expectations and martingales have a strong connection with the Littlewood–Paley theory discussed in this chapter. For the purposes of this exposition we considered only the case of the sequence of σ-algebras generated by the dyadic cubes of side length 2^{-k} in $\mathbf{R}^n$. The L^p boundedness of the maximal conditional expectation (Doob [97]) is analogous to the L^p boundedness of the dyadic maximal function; likewise with the corresponding weak type $(1,1)$ estimate. The L^p boundedness of the dyadic martingale square function was obtained by Burkholder [39] and is analogous to Theorem 6.1.2. Moreover, the estimate $\big\|\sup_k |E_k(f)|\big\|_{L^p} \approx \big\|S(f)\big\|_{L^p}$, $0 < p < \infty$, obtained by Burkholder and Gundy [40] and also by Davis [90] is analogous to the square-function characterization of the Hardy space H^p norm. For an exposition on the different and unifying aspects of Littlewood–Paley theory we refer to Stein [337]. The proof of Theorem 6.4.8, which quantitatively expresses the almost orthogonality of the Littlewood–Paley and the dyadic martingale difference operators, is taken from Grafakos and Kalton [133].

The use of quadratic expressions in the study of certain maximal operators has a long history. We refer to the article of Stein [340] for a historical survey. Theorem 6.5.1 was first proved by Stein [339]. The proof in the text is taken from an article of Rubio de Francia [302]. Another proof when $n \ge 3$ is due to Cowling and Mauceri [76]. The more difficult case $n = 2$ was settled by Bourgain [36] about 10 years later. Alternative proofs when $n = 2$ were given by Mockenhaupt, Seeger, and Sogge [256] as well as Schlag [313]. The boundedness of maximal operators associated to more general smooth measures on compact surfaces of finite type were investigated by Iosevich and Sawyer [173]. The powerful machinery of Fourier integral operators was used by Sogge [328] to obtain the boundedness of spherical maximal operators on compact manifolds without boundary and positive injectivity radius; a simple proof for the boundedness of the spherical maximal function on the sphere was given by Nguyen [269]. Weighted norm inequalities for the spherical maximal operator were obtained by Duoandikoetxea and Vega [103]. The discrete spherical maximal function was studied by Magyar, Stein, and Wainger [237].

Much of the theory of square functions and the ideas associated with them has analogues in the dyadic setting. A dyadic analogue of the theory discussed here can be obtained. For an introduction to the area of dyadic harmonic analysis, we refer to Pereyra [276].

The idea of expressing (or reproducing) a signal as a weighted average of translations and dilations of a single function appeared in early work of Calderón [42]. This idea is in some sense a forerunner of wavelets. An early example of a wavelet was constructed by Strömberg [352] in his search for unconditional bases for Hardy spaces. Another example of a wavelet basis was obtained by Meyer [249]. The construction of an orthonormal wavelet presented in Theorem 6.6.7 is in Lemarié and Meyer [216]. A compactly supported wavelet was constructed by Daubechies [88]. Mallat [238] introduced the notion of multiresolution analysis, which led to a systematic production of wavelets. Theorem 6.6.9 is Shannon's [319] version of Nyquist's theorem [270] and is referred to as the Nyquist-Shannon sampling theorem. It is a fundamental result in telecommunications and signal processing, since it describes how to reconstruct a signal that contains no frequencies higher than B Hertz in terms of its values at a sequence of points spaced $1/(2B)$ seconds apart.

The area of wavelets has taken off significantly since its inception, spurred by these early results. A general theory of wavelets and its use in Fourier analysis was carefully developed in the two-volume monograph of Meyer [250], [251] and its successor Meyer and Coifman [253]. For further study and a deeper account of developments on the subject the reader may consult the books of Daubechies [89], Chui [64], Wickerhauser [374], Kaiser [184], Benedetto and Frazier [23], Hérnandez and Weiss [151], Wojtaszczyk [379], Mallat [239], Meyer [252], Frazier [120], Gröchenig [140], and the references therein. Theorems 6.6.12 and 6.6.13 first appeared in a combined form in the work of Plancherel and Pólya [285] for restrictions of entire functions of exponential type on the real line.

Chapter 7
Weighted Inequalities

Weighted inequalities arise naturally in Fourier analysis, but their use is best justified by the variety of applications in which they appear. For example, the theory of weights plays an important role in the study of boundary value problems for Laplace's equation on Lipschitz domains. Other applications of weighted inequalities include extrapolation theory, vector-valued inequalities, and estimates for certain classes of nonlinear partial differential equations.

The theory of weighted inequalities is a natural development of the principles and methods we have acquainted ourselves with in earlier chapters. Although a variety of ideas related to weighted inequalities appeared almost simultaneously with the birth of singular integrals, it was only in the 1970s that a better understanding of the subject was obtained. This was spurred by Muckenhoupt's characterization of positive functions w for which the Hardy–Littlewood maximal operator M maps $L^p(\mathbf{R}^n, w(x)\,dx)$ to itself. This characterization led to the introduction of the class A_p and the development of weighted inequalities. We pursue exactly this approach in the next section to motivate the introduction of the A_p classes.

7.1 The A_p Condition

A *weight* is a nonnegative locally integrable function on $\mathbf{R}^n$ that takes values in $(0,\infty)$ almost everywhere. Therefore, weights are allowed to be zero or infinite only on a set of Lebesgue measure zero. Hence, if w is a weight and $1/w$ is locally integrable, then $1/w$ is also a weight.

Given a weight w and a measurable set E, we use the notation

$$w(E) = \int_E w(x)\,dx$$

L. Grafakos, *Classical Fourier Analysis*, Graduate Texts in Mathematics 249, DOI 10.1007/978-1-4939-1194-3_7,

to denote the w-measure of the set E. Since weights are locally integrable functions, $w(E) < \infty$ for all sets E contained in some ball. The weighted L^p spaces are denoted by $L^p(\mathbf{R}^n, w)$ or simply $L^p(w)$. Recall the uncentered Hardy–Littlewood maximal operators on $\mathbf{R}^n$ over balls

$$M(f)(x) = \sup_{B \ni x} \operatorname{Avg}_B |f| = \sup_{B \ni x} \frac{1}{|B|} \int_B |f(y)|\, dy,$$

and over cubes

$$M_c(f)(x) = \sup_{Q \ni x} \operatorname{Avg}_Q |f| = \sup_{Q \ni x} \frac{1}{|Q|} \int_Q |f(y)|\, dy,$$

where the suprema are taken over all balls B and cubes Q (with sides parallel to the axes) that contain the given point x. A classical result (Theorem 2.1.6) states that for all $1 < p < \infty$ there is a constant $C_p(n) > 0$ such that

$$\int_{\mathbf{R}^n} M(f)(x)^p\, dx \le C_p(n)^p \int_{\mathbf{R}^n} |f(x)|^p\, dx \tag{7.1.1}$$

for all functions $f \in L^p(\mathbf{R}^n)$. We are concerned with the situation in which the measure dx in (7.1.1) is replaced by $w(x)\, dx$ for some weight $w(x)$.

7.1.1 Motivation for the A_p Condition

The question we raise is whether there is a characterization of all weights $w(x)$ such that the strong type (p,p) inequality

$$\int_{\mathbf{R}^n} M(f)(x)^p\, w(x)\, dx \le C_p^p \int_{\mathbf{R}^n} |f(x)|^p\, w(x)\, dx \tag{7.1.2}$$

is valid for all $f \in L^p(w)$.

Suppose that (7.1.2) is valid for some weight w and all $f \in L^p(w)$ for some $1 < p < \infty$. Apply (7.1.2) to the function $f\chi_B$ supported in a ball B and use that $\operatorname{Avg}_B |f| \le M(f\chi_B)(x)$ for all $x \in B$ to obtain

$$w(B)\big(\operatorname{Avg}_B |f|\big)^p \le \int_B M(f\chi_B)^p\, w\, dx \le C_p^p \int_B |f|^p\, w\, dx. \tag{7.1.3}$$

It follows that

$$\left(\frac{1}{|B|}\int_B |f(t)|\, dt\right)^p \le \frac{C_p^p}{w(B)} \int_B |f(x)|^p\, w(x)\, dx \tag{7.1.4}$$

for all balls B and all functions f. At this point, it is tempting to choose a function such that the two integrands are equal. We do so by setting $f = w^{-p'/p}$, which gives

$f^p w = w^{-p'/p}$. Under the assumption that $\inf_B w > 0$ for all balls B, it would follow from (7.1.4) that

$$\sup_{B \text{ balls}} \left(\frac{1}{|B|}\int_B w(x)\,dx\right)\left(\frac{1}{|B|}\int_B w(x)^{-\frac{1}{p-1}}\,dx\right)^{p-1} \le C_p^p. \tag{7.1.5}$$

If $\inf_B w = 0$ for some balls B, we take $f = (w+\varepsilon)^{-p'/p}$ to obtain

$$\left(\frac{1}{|B|}\int_B w(x)dx\right)\left(\frac{1}{|B|}\int_B (w(x)+\varepsilon)^{-\frac{p'}{p}}\,dx\right)^p\left(\frac{1}{|B|}\int_B \frac{w(x)\,dx}{(w(x)+\varepsilon)^{p'}}\right)^{-1} \le C_p^p \tag{7.1.6}$$

for all $\varepsilon > 0$. Replacing $w(x)\,dx$ by $(w(x)+\varepsilon)\,dx$ in the last integral in (7.1.6) we obtain a smaller expression, which is also bounded by C_p^p. Since $-p'/p = -p'+1$, (7.1.6) implies that

$$\left(\frac{1}{|B|}\int_B w(x)dx\right)\left(\frac{1}{|B|}\int_B (w(x)+\varepsilon)^{-\frac{p'}{p}}\,dx\right)^{p-1} \le C_p^p, \tag{7.1.7}$$

from which we can still deduce (7.1.5) via the Lebesgue monotone convergence theorem by letting $\varepsilon \to 0$. We have now obtained that every weight w that satisfies (7.1.2) must also satisfy the rather strange-looking condition (7.1.5), which we refer to in the sequel as the A_p condition. It is a remarkable fact, to be proved in this chapter, that the implication obtained can be reversed, that is, (7.1.2) is a consequence of (7.1.5). This is the first significant achievement of the theory of weights [i.e., a characterization of all functions w for which (7.1.2) holds]. This characterization is based on some deep principles discussed in the next section and provides a solid motivation for the introduction and careful examination of condition (7.1.5).

Before we study the converse statements, we consider the case $p = 1$. Assume that for some weight w the weak type $(1,1)$ inequality

$$w\big(\{x \in \mathbf{R}^n : M(f)(x) > \alpha\}\big) \le \frac{C_1}{\alpha}\int_{\mathbf{R}^n} |f(x)|w(x)\,dx \tag{7.1.8}$$

holds for all functions $f \in L^1(\mathbf{R}^n)$. Since $M(f)(x) \ge \operatorname{Avg}_B |f|$ for all $x \in B$, it follows from (7.1.8) that for all $\alpha < \operatorname{Avg}_B |f|$ we have

$$w(B) \le w\big(\{x \in \mathbf{R}^n : M(f)(x) > \alpha\}\big) \le \frac{C_1}{\alpha}\int_{\mathbf{R}^n} |f(x)|w(x)\,dx. \tag{7.1.9}$$

Taking $f\chi_B$ instead of f in (7.1.9), we deduce that

$$\operatorname*{Avg}_B |f| = \frac{1}{|B|}\int_B |f(t)|\,dt \le \frac{C_1}{w(B)}\int_B |f(x)|\,w(x)\,dx \tag{7.1.10}$$

for all functions f and balls B. Taking $f = \chi_S$, we obtain

$$\frac{|S|}{|B|} \le C_1 \frac{w(S)}{w(B)}, \tag{7.1.11}$$

where S is any measurable subset of the ball B.

Recall that the *essential infimum* of a function w over a set E is defined as

$$\operatorname*{ess.inf}_E(w) = \inf\big\{b>0 : \, |\{x \in E : \, w(x) < b\}| > 0\big\}.$$

Then for every $a > \text{ess.inf}_B(w)$ there exists a subset S_a of B with positive measure such that $w(x) < a$ for all $x \in S_a$. Applying (7.1.11) to the set S_a, we obtain

$$\frac{1}{|B|}\int_B w(t)\,dt \le \frac{C_1}{|S_a|}\int_{S_a} w(t)\,dt \le C_1 a, \tag{7.1.12}$$

which implies

$$\frac{1}{|B|}\int_B w(t)\,dt \le C_1 w(x) \qquad \text{for all balls } B \text{ and almost all } x \in B. \tag{7.1.13}$$

It remains to understand what condition (7.1.13) really means. For every ball B, there exists a null set $N(B)$ such that (7.1.13) holds for all x in $B \setminus N(B)$. Let N be the union of all the null sets $N(B)$ for all balls B with centers in $\mathbf{Q}^n$ and rational radii. Then N is a null set and for every x in $B \setminus N$, (7.1.13) holds for all balls B with centers in $\mathbf{Q}^n$ and rational radii. By density, (7.1.13) must also hold for all balls B that contain a fixed x in $\mathbf{R}^n \setminus N$. It follows that for $x \in \mathbf{R}^n \setminus N$ we have

$$M(w)(x) = \sup_{B \ni x} \frac{1}{|B|}\int_B w(t)\,dt \le C_1 w(x). \tag{7.1.14}$$

Therefore, assuming (7.1.8), we have arrived at the condition

$$M(w)(x) \le C_1 w(x) \qquad \text{for almost all } x \in \mathbf{R}^n, \tag{7.1.15}$$

where C_1 is the same constant as in (7.1.13).

We later see that this deduction can be reversed and we can obtain (7.1.8) as a consequence of (7.1.15). This motivates a careful study of condition (7.1.15), which we refer to as the A_1 condition. Since in all the previous arguments we could have replaced balls with cubes, we give the following definitions in terms of cubes.

Definition 7.1.1. A function $w(x) \ge 0$ is called an A_1 *weight* if

$$M(w)(x) \le C_1 w(x) \qquad \text{for almost all } x \in \mathbf{R}^n \tag{7.1.16}$$

for some constant C_1. If w is an A_1 weight, then the (finite) quantity

$$[w]_{A_1} = \sup_{Q \text{ cubes in } \mathbf{R}^n} \left(\frac{1}{|Q|}\int_Q w(t)\,dt\right) \|w^{-1}\|_{L^\infty(Q)} \tag{7.1.17}$$

is called the A_1 *Muckenhoupt characteristic constant* of w, or simply the A_1 *characteristic constant* of w. Note that A_1 weights w satisfy

$$\frac{1}{|Q|}\int_Q w(t)\,dt \le [w]_{A_1}\ \underset{y\in Q}{\text{ess.inf}}\, w(y) \tag{7.1.18}$$

for all cubes Q in $\mathbf{R}^n$.

Remark 7.1.2. We also define

$$[w]_{A_1}^{\text{balls}} = \sup_{B \text{ balls in } \mathbf{R}^n} \left(\frac{1}{|B|}\int_B w(t)\,dt\right) \left\|w^{-1}\right\|_{L^\infty(B)}. \tag{7.1.19}$$

Using (7.1.13), we see that the smallest constant C_1 that appears in (7.1.16) is equal to the A_1 characteristic constant of w as defined in (7.1.19). This is also equal to the smallest constant that appears in (7.1.13). All these constants are bounded above and below by dimensional multiples of $[w]_{A_1}$.

We now recall condition (7.1.5), which motivates the following definition of A_p weights for $1 < p < \infty$.

Definition 7.1.3. Let $1 < p < \infty$. A weight w is said to be *of class* A_p if

$$\sup_{Q \text{ cubes in } \mathbf{R}^n} \left(\frac{1}{|Q|}\int_Q w(x)\,dx\right)\left(\frac{1}{|Q|}\int_Q w(x)^{-\frac{1}{p-1}}\,dx\right)^{p-1} < \infty. \tag{7.1.20}$$

The expression in (7.1.20) is called the A_p *Muckenhoupt characteristic constant of* w (or simply the A_p characteristic constant of w) and is denoted by $[w]_{A_p}$.

Remark 7.1.4. Note that Definitions 7.1.1 and 7.1.3 could have been given with the set of all cubes in $\mathbf{R}^n$ replaced by the set of all balls in $\mathbf{R}^n$. Defining $[w]_{A_p}^{\text{balls}}$ as in (7.1.20) except that cubes are replaced by balls, we see that

$$\left(v_n 2^{-n}\right)^p \le \frac{[w]_{A_p}}{[w]_{A_p}^{\text{balls}}} \le \left(n^{n/2} v_n 2^{-n}\right)^p. \tag{7.1.21}$$

7.1.2 Properties of A_p Weights

It is straightforward that translations, isotropic dilations, and scalar multiples of A_p weights are also A_p weights with the same A_p characteristic. We summarize some basic properties of A_p weights in the following proposition.

Proposition 7.1.5. *Let* $w \in A_p$ *for some* $1 \le p < \infty$. *Then*

(1) $[\delta^\lambda(w)]_{A_p} = [w]_{A_p}$, *where* $\delta^\lambda(w)(x) = w(\lambda x_1, \dots, \lambda x_n)$.

(2) $[\tau^z(w)]_{A_p} = [w]_{A_p}$, *where* $\tau^z(w)(x) = w(x-z)$, $z \in \mathbf{R}^n$.

(3) $[\lambda w]_{A_p} = [w]_{A_p}$ *for all* $\lambda > 0$.

(4) When $1 < p < \infty$, *the function* $w^{-\frac{1}{p-1}}$ *is in* $A_{p'}$ *with characteristic constant*

$$\big[w^{-\frac{1}{p-1}}\big]_{A_{p'}} = [w]_{A_p}^{\frac{1}{p-1}}.$$

Therefore, $w \in A_2$ *if and only if* $w^{-1} \in A_2$ *and both weights have the same* A_2 *characteristic constant.*

(5) $[w]_{A_p} \geq 1$ *for all* $w \in A_p$. *Equality holds if and only if* w *is a constant.*

(6) The classes A_p *are increasing as* p *increases; precisely, for* $1 \leq p < q < \infty$ *we have*

$$[w]_{A_q} \leq [w]_{A_p}.$$

(7) $\lim_{q \to 1+} [w]_{A_q} = [w]_{A_1}$ *if* $w \in A_1$.

(8) The following is an equivalent characterization of the A_p *characteristic constant of* w:

$$[w]_{A_p} = \sup_{\substack{Q\, cubes \\ in\, \mathbf{R}^n}} \sup_{\substack{f \in L^p(Q, w\,dt) \\ \int_Q |f|^p w\,dt > 0}} \left\{ \frac{\left(\frac{1}{|Q|}\int_Q |f(t)|\,dt\right)^p}{\frac{1}{w(Q)}\int_Q |f(t)|^p w(t)\,dt} \right\}.$$

(9) The measure $w(x)\,dx$ *is doubling: precisely, for all* $\lambda > 1$ *and all cubes* Q *we have*

$$w(\lambda Q) \leq \lambda^{np} [w]_{A_p} w(Q).$$

(λQ *denotes the cube with the same center as* Q *and side length* λ *times the side length of* Q.*)*

Proof. The simple proofs of (1), (2), and (3) are left as an exercise. Property (4) is also easy to check and plays the role of duality in this context. To prove (5) we use Hölder's inequality with exponents p and p' to obtain

$$1 = \frac{1}{|Q|}\int_Q dx = \frac{1}{|Q|}\int_Q w(x)^{\frac{1}{p}} w(x)^{-\frac{1}{p}}\,dx \leq [w]_{A_p}^{\frac{1}{p}},$$

with equality holding only when $w(x)^{\frac{1}{p}} = c\,w(x)^{-\frac{1}{p}}$ for some $c > 0$ (i.e., when w is a constant). To prove (6), observe that $0 < q' - 1 < p' - 1 \leq \infty$ and that the statement

$$[w]_{A_q} \leq [w]_{A_p}$$

is equivalent to the fact

$$\big\|w^{-1}\big\|_{L^{q'-1}(Q, \frac{dx}{|Q|})} \leq \big\|w^{-1}\big\|_{L^{p'-1}(Q, \frac{dx}{|Q|})}.$$

Property (7) is a consequence of part (a) of Exercise 1.1.3.

To prove (8), apply Hölder's inequality with exponents p and p' to get

$$\begin{aligned}(\operatorname*{Avg}_Q |f|)^p &= \left(\frac{1}{|Q|}\int_Q |f(x)|\,dx\right)^p \\ &= \left(\frac{1}{|Q|}\int_Q |f(x)| w(x)^{\frac{1}{p}} w(x)^{-\frac{1}{p}}\,dx\right)^p \\ &\leq \frac{1}{|Q|^p}\left(\int_Q |f(x)|^p w(x)\,dx\right)\left(\int_Q w(x)^{-\frac{p'}{p}}\,dx\right)^{\frac{p}{p'}} \\ &= \left(\frac{1}{\omega(Q)}\int_Q |f(x)|^p w(x)\,dx\right)\left(\frac{1}{|Q|}\int_Q w(x)\,dx\right)\left(\frac{1}{|Q|}\int_Q w(x)^{-\frac{1}{p-1}}\,dx\right)^{p-1} \\ &\leq [w]_{A_p}\left(\frac{1}{\omega(Q)}\int_Q |f(x)|^p w(x)\,dx\right).\end{aligned}$$

This argument proves the inequality $\geq$ in (8) when $p > 1$. In the case $p = 1$ the obvious modification yields the same inequality. The reverse inequality follows by taking $f = (w+\varepsilon)^{-p'/p}$ as in (7.1.6) and letting $\varepsilon \to 0$.

Applying (8) to the function $f = \chi_Q$ and putting λQ in the place of Q in (8), we obtain

$$w(\lambda Q) \leq \lambda^{np}[w]_{A_p} w(Q),$$

which says that $w(x)\,dx$ is a doubling measure. This proves (9). □

Example 7.1.6. A positive measure $d\mu$ is called doubling if for some $C < \infty$,

$$\mu(2B) \leq C\mu(B) \tag{7.1.22}$$

for all balls B. We show that the measures $|x|^a\,dx$ are doubling when $a > -n$. We divide all balls $B(x_0,R)$ in $\mathbf{R}^n$ into two categories: balls of type I that satisfy $|x_0| \geq 3R$ and type II that satisfy $|x_0| < 3R$. For balls of type I we observe that

$$\int_{B(x_0,2R)} |x|^a\,dx \leq v_n(2R)^n \begin{cases} (|x_0|+2R)^a & \text{when } a \geq 0, \\ (|x_0|-2R)^a & \text{when } a < 0, \end{cases}$$

$$\int_{B(x_0,R)} |x|^a\,dx \geq v_n R^n \begin{cases} (|x_0|-R)^a & \text{when } a \geq 0, \\ (|x_0|+R)^a & \text{when } a < 0. \end{cases}$$

Since $|x_0| \geq 3R$, we have $|x_0|+2R \leq 4(|x_0|-R)$ and $|x_0|-2R \geq \frac{1}{4}(|x_0|+R)$, from which (7.1.22) follows with $C = 2^{3n}4^{|a|}$.

For balls of type II, we have $|x_0| \leq 3R$ and we note two things: first

$$\int_{B(x_0,2R)} |x|^a\,dx \leq \int_{|x|\leq 5R} |x|^a\,dx = c_n R^{n+a},$$

and second, since $|x|^a$ is radially decreasing for $a < 0$ and radially increasing for $a \geq 0$, we have

$$\int_{B(x_0,R)} |x|^a\,dx \geq \begin{cases} \displaystyle\int_{B(0,R)} |x|^a\,dx & \text{when } a \geq 0, \\ \displaystyle\int_{B(3R\frac{x_0}{|x_0|},R)} |x|^a\,dx & \text{when } a < 0. \end{cases}$$

For $x \in B(3R\frac{x_0}{|x_0|},R)$ we must have $|x| \geq 2R$, and hence both integrals on the right are at least a multiple of R^{n+a}. This establishes (7.1.22) for balls of type II.

Example 7.1.7. We investigate for which real numbers a the power function $|x|^a$ is an A_p weight on $\mathbf{R}^n$. For $1 < p < \infty$, we examine for which a the following expression is finite:

$$\sup_{B \text{ balls}} \left(\frac{1}{|B|}\int_B |x|^a\,dx\right)\left(\frac{1}{|B|}\int_B |x|^{-a\frac{p'}{p}}\,dx\right)^{\frac{p}{p'}}. \tag{7.1.23}$$

As in the previous example we split the balls in $\mathbf{R}^n$ into those of type I and those of type II. If $B = B(x_0,R)$ is of type I, then for x satisfying $|x - x_0| \leq R$ we must have

$$\frac{2}{3}|x_0| \leq |x_0| - R \leq |x| \leq |x_0| + R \leq \frac{4}{3}|x_0|,$$

thus the expression inside the supremum in (7.1.23) is comparable to

$$|x_0|^a\big(|x_0|^{-a\frac{p'}{p}}\big)^{\frac{p}{p'}} = 1.$$

If $B(x_0,R)$ is a ball of type II, then $B(0,5R)$ has size comparable to $B(x_0,R)$ and contains it. Since the measure $|x|^a\,dx$ is doubling, the integrals of the function $|x|^a$ over $B(x_0,R)$ and over $B(0,5R)$ are comparable. It suffices therefore to estimate the expression inside the supremum in (7.1.23), in which we have replaced $B(x_0,R)$ by $B(0,5R)$. But this is

$$\begin{aligned} &\left(\frac{1}{v_n(5R)^n}\int_{B(0,5R)} |x|^a\,dx\right)\left(\frac{1}{v_n(5R)^n}\int_{B(0,5R)} |x|^{-a\frac{p'}{p}}\,dx\right)^{\frac{p}{p'}} \\ &\qquad = \left(\frac{n}{(5R)^n}\int_0^{5R} r^{a+n-1}\,dr\right)\left(\frac{n}{(5R)^n}\int_0^{5R} r^{-a\frac{p'}{p}+n-1}\,dr\right)^{\frac{p}{p'}}, \end{aligned}$$

which is seen easily to be finite and independent of R exactly when $-n < a < n\frac{p}{p'}$. We conclude that $|x|^a$ is an A_p weight, $1 < p < \infty$, if and only if $-n < a < n(p-1)$.

The previous proof can be suitably modified to include the case $p = 1$. In this case we obtain that $|x|^a$ is an A_1 weight if and only if $-n < a \leq 0$. As we have seen, the measure $|x|^a\,dx$ is doubling on the larger range $-n < a < \infty$. Thus for $a > n(p-1)$, the function $|x|^a$ provides an example of a doubling measure that is not in A_p.

Example 7.1.8. On $\mathbf{R}^n$ the function

$$u(x) = \begin{cases} \log \frac{1}{|x|} & \text{when } |x| < \frac{1}{e}, \\ 1 & \text{otherwise,} \end{cases}$$

is an A_1 weight. Indeed, to check condition (7.1.19) it suffices to consider balls of type I and type II as defined in Example 7.1.6. In either case the required estimate follows easily.

We now return to a point alluded to earlier, that the A_p condition implies the boundedness of the Hardy–Littlewood maximal function M on the space $L^p(w)$. To this end we introduce four maximal functions acting on functions f that are locally integrable with respect to w:

$$M^w(f)(x) = \sup_{B \ni x} \frac{1}{w(B)} \int_B |f| \, w \, dy,$$

where the supremum is taken over open balls B that contain the point x and

$$\begin{aligned} \mathcal{M}^w(f)(x) &= \sup_{\delta>0} \frac{1}{w(B(x,\delta))} \int_{B(x,\delta)} |f| \, w \, dy, \\ M_c^w(f)(x) &= \sup_{Q \ni x} \frac{1}{w(Q)} \int_Q |f| \, w \, dy, \end{aligned}$$

where Q is an open cube containing the point x, and

$$\mathcal{M}_c^w(f)(x) = \sup_{\delta>0} \frac{1}{w(Q(x,\delta))} \int_{Q(x,\delta)} |f| \, w \, dy,$$

where $Q(x,\delta) = \prod_{j=1}^n (x_j - \delta, x_j + \delta)$ is a cube of side length 2δ centered at $x = (x_1, \dots, x_n)$. When $w = 1$, these maximal functions reduce to the standard ones $M(f)$, $\mathcal{M}(f)$, $M_c(f)$, and $\mathcal{M}_c(f)$, the uncentered and centered Hardy–Littlewood maximal functions with respect to balls and cubes, respectively.

Theorem 7.1.9. *(a) Let $w \in A_1$. Then we have*

$$\big\| \mathcal{M}_c \big\|_{L^1(w) \to L^{1,\infty}(w)} \le 3^n [w]_{A_1} . \tag{7.1.24}$$

(b) Let $w \in A_p(\mathbf{R}^n)$ for some $1 < p < \infty$. Then there is a constant $C_{n,p}$ such that

$$\big\| \mathcal{M}_c \big\|_{L^p(w) \to L^p(w)} \le C_{n,p} [w]_{A_p}^{\frac{1}{p-1}} . \tag{7.1.25}$$

Since the operators $\mathcal{M}_c$, M_c, $\mathcal{M}$, and M are pointwise comparable, a similar conclusions hold for the other three as well.

Proof. (a) Since $d\mu = w\,dx$ is a doubling measure and $d\mu(3Q) \le 3^n [w]_{A_1} \mu(Q)$, using Proposition 7.1.5 (9) and Exercise 2.1.1 we obtain that M_c^w maps $L^1(w)$ to $L^{1,\infty}(w)$ with norm at most $3^n [w]_{A_1}$. This proves (7.1.24).

(b) Fix a weight w in A_p and let $\sigma = w^{-\frac{1}{p-1}}$ be its dual weight. Fix an open cube $Q = Q(x_0, r)$ in $\mathbf{R}^n$ with center x_0 and side length $2r$ and write

$$\frac{1}{|Q|}\int_Q |f|\,dy = \frac{w(Q)^{\frac{1}{p-1}}\sigma(3Q)}{|Q|^{\frac{p}{p-1}}}\left\{\frac{|Q|}{w(Q)}\left(\frac{1}{\sigma(3Q)}\int_Q |f|\,dy\right)^{p-1}\right\}^{\frac{1}{p-1}}. \tag{7.1.26}$$

For any $x \in Q$, consider the cube $Q(x, 2r)$. Then $Q \subsetneqq Q(x,2r) \subsetneqq 3Q = Q(x_0, 3r)$ and thus

$$\frac{1}{\sigma(3Q)}\int_Q |f|\,dy \le \frac{1}{\sigma(Q(x,2r))}\int_{Q(x,2r)} |f|\,dy \le \mathcal{M}_c^{\sigma}(|f|\sigma^{-1})(x)$$

for any $x \in Q$. Inserting this expression in (7.1.26), we obtain

$$\frac{1}{|Q|}\int_Q |f|\,dy \le \frac{w(Q)^{\frac{1}{p-1}}\sigma(3Q)}{|Q|^{\frac{p}{p-1}}}\left\{\frac{1}{w(Q)}\int_Q \mathcal{M}_c^{\sigma}(|f|\sigma^{-1})^{p-1}\,dy\right\}^{\frac{1}{p-1}}. \tag{7.1.27}$$

Since one may easily verify that

$$\frac{w(Q)\sigma(3Q)^{p-1}}{|Q|^p} \le 3^{np}[w]_{A_p},$$

it follows that

$$\frac{1}{|Q|}\int_Q |f|\,dy \le 3^{\frac{np}{p-1}}[w]_{A_p}^{\frac{1}{p-1}}\left(\mathcal{M}_c^w\left[(\mathcal{M}_c^{\sigma}(|f|\sigma^{-1}))^{p-1}w^{-1}\right](x_0)\right)^{\frac{1}{p-1}},$$

since x_0 is the center of Q. Hence, we have

$$\mathcal{M}_c(f) \le 3^{\frac{np}{p-1}}[w]_{A_p}^{\frac{1}{p-1}}\left(\mathcal{M}_c^w\left[(\mathcal{M}_c^{\sigma}(|f|\sigma^{-1}))^{p-1}w^{-1}\right]\right)^{\frac{1}{p-1}}.$$

Applying $L^p(w)$ norms, we deduce

$$\begin{aligned}
\big\|\mathcal{M}_c(f)\big\|_{L^p(w)} &\le 3^{\frac{np}{p-1}}[w]_{A_p}^{\frac{1}{p-1}}\big\|\mathcal{M}_c^w\left[(\mathcal{M}_c^{\sigma}(|f|\sigma^{-1}))^{p-1}w^{-1}\right]\big\|_{L^{p'}(w)}^{\frac{1}{p-1}} \\
&\le 3^{\frac{np}{p-1}}[w]_{A_p}^{\frac{1}{p-1}}\big\|\mathcal{M}_c^w\big\|_{L^{p'}(w)\to L^{p'}(w)}^{\frac{1}{p-1}}\big\|(\mathcal{M}_c^{\sigma}(|f|\sigma^{-1}))^{p-1}w^{-1}\big\|_{L^{p'}(w)}^{\frac{1}{p-1}} \\
&= 3^{\frac{np}{p-1}}[w]_{A_p}^{\frac{1}{p-1}}\big\|\mathcal{M}_c^w\big\|_{L^{p'}(w)\to L^{p'}(w)}^{\frac{1}{p-1}}\big\|\mathcal{M}_c^{\sigma}(|f|\sigma^{-1})\big\|_{L^p(\sigma)} \\
&\le 3^{\frac{np}{p-1}}[w]_{A_p}^{\frac{1}{p-1}}\big\|\mathcal{M}_c^w\big\|_{L^{p'}(w)\to L^{p'}(w)}^{\frac{1}{p-1}}\big\|\mathcal{M}_c^{\sigma}\big\|_{L^p(\sigma)\to L^p(\sigma)}\big\|f\big\|_{L^p(w)},
\end{aligned}$$

and conclusion (7.1.25) follows, provided we show that

$$\big\|\mathcal{M}_c^w\big\|_{L^q(w)\to L^q(w)} \le C(q,n) < \infty \tag{7.1.28}$$

for any $1 < q < \infty$ and any weight w.

We obtain this estimate by interpolation. Obviously (7.1.28) is valid when $q=\infty$ with $C(\infty,n)=1$. If we prove that

$$\big\|\mathcal{M}_c^w\big\|_{L^1(w)\to L^{1,\infty}(w)} \le C(1,n)<\infty, \tag{7.1.29}$$

then (7.1.28) will follow from Theorem 1.3.2.

To prove (7.1.29) we fix $f\in L^1(\mathbf{R}^n,w\,dx)$. We first show that the set

$$E_\lambda=\{\mathcal{M}_c^w(f)>\lambda\}$$

is open. For any $r>0$, let $Q(x,r)$ denote an open cube of side length $2r$ with center $x\in\mathbf{R}^n$. If we show that for any $r>0$ and $x\in\mathbf{R}^n$ the function

$$x\mapsto \frac{1}{w(Q(x,r))}\int_{Q(x,r)}|f|\,w\,dy \tag{7.1.30}$$

is continuous, then $\mathcal{M}_c^w(f)$ is the supremum of continuous functions; hence it is lower semicontinuous and thus the set E_λ is open. But this is straightforward. If $x_n\to x_0$, then $w(Q(x_n,r))\to w(Q(x_0,r))$ and also $\int_{Q(x_n,r)}|f|\,w\,dy\to\int_{Q(x_0,r)}|f|\,w\,dy$ by the Lebesgue dominated convergence theorem. Since $w(Q(x_0,r))\neq 0$, it follows that the function in (7.1.30) is continuous.

Given K a compact subset of E_λ, for any $x\in K$ select an open cube Q_x centered at x such that

$$\frac{1}{w(Q_x)}\int_{Q_x}|f|\,w\,dy>\lambda\,.$$

Applying Lemma 7.1.10 (proved immediately afterward) we find a subfamily $\{Q_{x_j}\}_{j=1}^m$ of the family of the balls $\{Q_x:\ x\in K\}$ such that (7.1.31) and (7.1.32) hold. Then

$$w(K)\le\sum_{j=1}^m w(Q_{x_j})\le\sum_{j=1}^m\frac{1}{\lambda}\int_{Q_{x_j}}|f|\,w\,dy\le\frac{24^n}{\lambda}\int_{\mathbf{R}^n}|f|\,w\,dy,$$

where the last inequality follows by multiplying (7.1.32) by $|f|w$ and integrating over $\mathbf{R}^n$. Taking the supremum over all compact subsets K of E_λ and using the inner regularity of $w\,dx$, which is a consequence of the Lebesgue monotone convergence theorem, we deduce that $\mathcal{M}_c^w$ maps $L^1(w)$ to $L^{1,\infty}(w)$ with constant at most 24^n. Thus (7.1.29) holds with $C(1,n)=24^n$. □

Lemma 7.1.10. *Let K be a bounded set in $\mathbf{R}^n$ and for every $x\in K$, let Q_x be an open cube with center x and sides parallel to the axes. Then there are an $m\in\mathbf{Z}^+\cup\{\infty\}$ and a sequence of points $\{x_j\}_{j=1}^m$ in K such that*

$$K\subseteq\bigcup_{j=1}^m Q_{x_j} \tag{7.1.31}$$

and for almost all $y \in \mathbf{R}^n$ one has

$$\sum_{j=1}^{m} \chi_{Q_{x_j}}(y) \le 24^n. \tag{7.1.32}$$

Proof. Let $s_0 = \sup\{\ell(Q_x) : x \in K\}$. If $s_0 = \infty$, then there exists $x_1 \in K$ such that $\ell(Q_{x_1}) > 4L$, where $[-L,L]^n$ contains K. Then K is contained in Q_{x_1} and the statement of the lemma is valid with $m = 1$.

Suppose now that $s_0 < \infty$. Select $x_1 \in K$ such that $\ell(Q_{x_1}) > s_0/2$. Then define

$$K_1 = K \setminus Q_{x_1}, \qquad s_1 = \sup\{\ell(Q_x) : x \in K_1\},$$

and select $x_2 \in K_1$ such that $\ell(Q_{x_2}) > s_1/2$. Next define

$$K_2 = K \setminus (Q_{x_1} \cup Q_{x_2}), \qquad s_2 = \sup\{\ell(Q_x) : x \in K_2\},$$

and select $x_3 \in K_2$ such that $\ell(Q_{x_3}) > s_2/2$. Continue until the first integer m is found such that K_m is an empty set. If no such integer exists, continue this process indefinitely and set $m = \infty$.

We claim that for all $i \neq j$ we have $\frac{1}{3}Q_{x_i} \cap \frac{1}{3}Q_{x_j} = \emptyset$. Indeed, suppose that $i > j$. Then $x_i \in K_{i-1} = K \setminus (Q_{x_1} \cup \cdots \cup Q_{x_{i-1}})$; thus $x_i \notin Q_j$. Also $x_i \in K_{i-1} \subsetneqq K_{j-1}$, which implies that $\ell(Q_{x_i}) \le s_{j-1} < 2\ell(Q_{x_j})$. If $x_i \notin Q_j$ and $\ell(Q_{x_j}) > \frac{1}{2}\ell(Q_{x_i})$, it easily follows that $\frac{1}{3}Q_{x_i} \cap \frac{1}{3}Q_{x_j} = \emptyset$.

We now prove (7.1.31). If $m < \infty$, then $K_m = \emptyset$ and therefore $K \subsetneqq \bigcup_{j=1}^{m} Q_{x_j}$. If $m = \infty$, then there is an infinite number of selected cubes Q_{x_j}. Since the cubes $\frac{1}{3}Q_{x_j}$ are pairwise disjoint and have centers in a bounded set, it must be the case that some subsequence of the sequence of their lengths converges to zero. If there exists a $y \in K \setminus \bigcup_{j=1}^{\infty} Q_{x_j}$, this y would belong to all K_j, $j = 1, 2, \dots$, and then $s_j \ge \ell(Q_y)$ for all j. Since some subsequence of the s_j's tends to zero, it would follow that $\ell(Q_y) = 0$, which would force the open cube Q_y to be the empty set, a contradiction. Thus (7.1.31) holds.

Finally, we show that $\sum_{j=1}^{m} \chi_{Q_{x_j}}(y) \le 24^n$ for almost every point $y \in \mathbf{R}^n$. To prove this we consider the n hyperplanes H_i that are parallel to the coordinate hyperplanes and pass through the point y. Then we write $\mathbf{R}^n$ as a union of n hyperplanes H_i of n-dimensional Lebesgue measure zero and 2^n higher-dimensional open "octants" O_r, henceforth called orthants. We fix a $y \in \mathbf{R}^n$ and we show that there are only 12^n points x_j such that y lies in $O_r \cap Q_{x_j}$ for a given open orthant O_r. To prove this assertion, setting $|z|_{\ell^\infty} = \sup_{1 \le i \le n} |z_i|$ for points $z = (z_1, \dots, z_n)$ in $\mathbf{R}^n$, we pick an $x_{k_0} \in K \cap O_r$ such that $Q_{x_{k_0}}$ contains y and $|x_{k_0} - y|_{\ell^\infty}$ is the largest possible among all $|x_j - y|_{\ell^\infty}$. If x_j is another point in $K \cap O_r$ such that Q_{x_j} contains y, then we claim that $x_j \in Q_{x_{k_0}}$. Indeed, to show this we notice that for each $i \in \{1, \dots, n\}$ we have

$$\begin{aligned} |x_{j,i} - x_{k_0,i}| &= \big|x_{j,i} - y_i - (x_{k_0,i} - y_i)\big| \\ &= \big||x_{j,i} - y_i| - |x_{k_0,i} - y_i|\big| \end{aligned}$$

$$\begin{aligned}
&\leq \max\left(|x_{k_0,i}-y_i|,|x_{j,i}-y_i|\right)\\
&\leq \max\left(|x_{k_0}-y|_{\ell^\infty},|x_j-y|_{\ell^\infty}\right)\\
&= |x_{k_0}-y|_{\ell^\infty}\\
&< \tfrac{1}{2}\ell(Q_{x_{k_0}}),
\end{aligned}$$

where the second equality is due to the fact that x_j, x_{k_0} lie in the same orthant and the last inequality in the fact that $y \in Q_{x_{k_0}}$; it follows that x_j lies in $Q_{x_{k_0}}$.

We observed previously that $i > j$ implies $x_i \notin Q_j$. Since x_j lies in $Q_{x_{k_0}}$, one must then have $j \leq k_0$, which implies that $\frac{1}{2}\ell(Q_{x_{k_0}}) < \ell(Q_{x_j})$. Thus all cubes Q_{x_j} with centers in $K \cap O_r$ that contain the fixed point y have side lengths comparable to that of $Q_{x_{k_0}}$. A simple geometric argument now gives that there are at most finitely many cubes Q_{x_j} of side length between α and 2α that contain the given point y such that $\frac{1}{3}Q_{x_j}$ are pairwise disjoint. Indeed, let $\alpha = \frac{1}{2}\ell(Q_{x_{k_0}})$ and let $\{Q_{x_r}\}_{r\in I}$ be the cubes with these properties. Then we have

$$\frac{\alpha^n|I|}{3^n} \leq \sum_{r\in I}\left|\tfrac{1}{3}Q_{x_r}\right| = \Big|\bigcup_{r\in I}\tfrac{1}{3}Q_{x_r}\Big| \leq \Big|\bigcup_{r\in I}Q_{x_r}\Big| \leq (4\alpha)^n,$$

since all the cubes Q_{x_r} contain the point y and have length at most 2α and they must therefore be contained in a cube of side length 4α centered at y. This observation shows that $|I| \leq 12^n$, and since there are 2^n sets O_r, we conclude the proof of (7.1.32). □

Remark 7.1.11. Without use of the covering Lemma 7.1.10, (7.1.29) can be proved via the doubling property of w (cf. Exercise 2.1.1(a)), but then the resulting constant $C(q,n)$ would depend on the doubling constant of the measure $w\,dx$ and thus on $[w]_{A_p}$; this would yield a worse dependence on $[w]_{A_p}$ in the constant in (7.1.25).

Exercises

7.1.1. Let k be a nonnegative measurable function such that k, k^{-1} are in $L^\infty(\mathbf{R}^n)$. Prove that if w is an A_p weight for some $1 \leq p < \infty$, then so is kw.

7.1.2. Let w_1, w_2 be two A_1 weights and let $1 < p < \infty$. Prove that $w_1 w_2^{1-p}$ is an A_p weight by showing that

$$[w_1 w_2^{1-p}]_{A_p} \leq [w_1]_{A_1}[w_2]_{A_1}^{p-1}.$$

7.1.3. Suppose that $w \in A_p$ for some $p \in [1,\infty)$ and $0 < \delta < 1$. Prove that $w^\delta \in A_q$, where $q = \delta p + 1 - \delta$, by showing that

$$[w^\delta]_{A_q} \leq [w]_{A_p}^\delta.$$

7.1.4. Show that if the A_p characteristic constants of a weight w are uniformly bounded for all $p>1$, then $w\in A_1$.

7.1.5. Let $w_0\in A_{p_0}$ and $w_1\in A_{p_1}$ for some $1\le p_0,p_1<\infty$. Let $0\le\theta\le 1$ and define

$$\frac{1}{p}=\frac{1-\theta}{p_0}+\frac{\theta}{p_1} \qquad \text{and} \qquad w^{\frac{1}{p}}=w_0^{\frac{1-\theta}{p_0}}\,w_1^{\frac{\theta}{p_1}}.$$

Prove that

$$[w]_{A_p}\le [w_0]_{A_{p_0}}^{(1-\theta)\frac{p}{p_0}}[w_1]_{A_{p_1}}^{\theta\frac{p}{p_1}};$$

thus w is in A_p.

7.1.6. ([122]) Fix $1<p<\infty$. A pair of weights (u,w) that satisfies

$$[u,w]_{(A_p,A_p)}=\sup_{\substack{Q\text{ cubes}\\ \text{in }\mathbf{R}^n}}\bigg(\frac{1}{|Q|}\int_Q u\,dx\bigg)\bigg(\frac{1}{|Q|}\int_Q w^{-\frac{1}{p-1}}\,dx\bigg)^{p-1}<\infty$$

is said to be of class (A_p,A_p). The quantity $[u,w]_{(A_p,A_p)}$ is called the (A_p,A_p) *characteristic constant of the pair* (u,w).
(a) Suppose that pair of weights (u,w) is of class (A_p,A_p). Show that for all non-negative measurable functions f and all cubes Q' we have

$$\bigg(\frac{1}{|Q'|}\int_{Q'}|f|\,dx\bigg)^p u(Q')\le C_0\int_{Q'}|f|^p w\,dx,$$

where $C_0=[u,w]_{(A_p,A_p)}$.
(b) Suppose that a pair of weights (u,w) satisfies the inequality in part (a) for some constant C_0. Prove that M maps $L^p(w)$ to $L^{p,\infty}(u)$ with norm at most $C(n,p)C_0^{1/p}$, where $C(n,p)$ is a fixed constant.
(c) Suppose that for a pair of weights (u,w), M maps $L^p(w)$ to $L^{p,\infty}(u)$. Show that the pair (u,w) is of class (A_p,A_p).
[*Hint:* Part (b): Replacing f by $f\chi_Q$ in part (a), where $Q\subseteq Q'$, obtain that

$$u(Q')\le C_0|Q'|^p\frac{\int_Q|f|^p w\,dx}{(\int_Q|f|\,dx)^p}.$$

Then use Exercise 5.3.9 to find disjoint cubes Q_j such that the set $E_\alpha=\{x\in\mathbf{R}^n: M_c(f)(x)>\alpha\}$ is contained in the union of $3Q_j$ and $\frac{\alpha}{4^n}<\frac{1}{|Q_j|}\int_{Q_j}|f(t)|\,dt\le\frac{\alpha}{2^n}$. Then $u(E_\alpha)\le\sum_j u(3Q_j)$, and bound each $u(3Q_j)$ by taking $Q'=3Q_j$ and $Q=Q_j$ in the preceding estimate. Part (c): First prove the assertion in part (b) and then derive the inequality in part (a) by adapting the idea in the discussion in the beginning of Subsection 7.1.1.]

7.1.7. ([122]) Let $1 < p < \infty$ and let (u,w) be a pair of weights of class (A_p,A_p). Show that for any q with $p < q < \infty$ there is a constant $C_{p,q,n} < \infty$ such that for all $f \in L^q(w)$ we have

$$\left(\int_{\mathbf{R}^n} M(f)(x)^q u(x)\,dx\right)^{1/q} \le C_{p,q,n}\left(\int_{\mathbf{R}^n} f(x)^q w(x)\,dx\right)^{1/q}.$$

[*Hint:* Use Exercise 7.1.6 and interpolate between L^p and L^∞.]

7.1.8. Let $k > 0$. For an A_1 weight w show that $[\min(w,k)]_{A_1} \le [w]_{A_1}$. If $1 < p < \infty$ and $w \in A_p$, show that

$$[\min(w,k)]_{A_p} \le c_p [w]_{A_p},$$

where $c_p = 1$ if $1 < p \le 2$ and $c_p = 2^{p-1}$ if $2 < p < \infty$.
[*Hint:* Use the inequality $\frac{1}{|Q|}\int_Q \min(w,k)^{-\frac{1}{p-1}}dx \le \frac{1}{|Q|}\int_Q w^{-\frac{1}{p-1}}dx + k^{-\frac{1}{p-1}}$ and also $\frac{1}{|Q|}\int_Q \min(w,k)dx \le \min\left\{k, \frac{1}{|Q|}\int_Q w\,dx\right\}$.]

7.1.9. Suppose that $w_j \in A_{p_j}$ with $1 \le j \le m$ for some $1 \le p_1,\dots,p_m < \infty$ and let $0 < \theta_1,\dots,\theta_m < 1$ be such that $\theta_1 + \dots + \theta_m = 1$. Show that

$$w_1^{\theta_1}\cdots w_m^{\theta_m} \in A_{\max\{p_1,\dots,p_m\}}.$$

[*Hint:* First note that each weight w_j lies in $A_{\max\{p_1,\dots,p_m\}}$ and then apply Hölder's inequality.]

7.1.10. Let $w_1 \in A_{p_1}$ and $w_2 \in A_{p_2}$ for some $1 \le p_1, p_2 < \infty$. Prove that

$$[w_1 + w_2]_{A_p} \le [w_1]_{A_{p_1}} + [w_2]_{A_{p_2}},$$

where $p = \max(p_1, p_2)$.

7.1.11. Show that the function

$$u(x) = \begin{cases} \log\frac{1}{|x|} & \text{when } |x| < \frac{1}{e}, \\ 1 & \text{otherwise,} \end{cases}$$

in Example 7.1.8 is an A_1 weight on $\mathbf{R}^n$.
[*Hint:* Use $[u]_{A_1}^{\text{balls}}$ instead of $[u]_{A_1}$ and consider balls of type I and II as in Example 7.1.7.]

7.1.12. Let $1 < p < \infty$ and $w \in A_1$. Show that the uncentered Hardy-Littlewood maximal function M maps $L^{p,\infty}(w)$ to itself.
[*Hint:* Prove first the inequality

$$w(\{M(g) > \lambda\}) \le \frac{3^n([w]_{A_1}^{balls})^2}{\lambda}\int_{\{M(g)>\lambda\}} |g|\,w\,dx$$

and then use the characterization of $L^{p,\infty}$ given in Exercise 1.1.12.]

7.2 Reverse Hölder Inequality for A_p Weights and Consequences

An essential property of A_p weights is that they assign to subsets of balls mass proportional to the percentage of the Lebesgue measure of the subset within the ball. The following lemma provides a way to quantify this statement.

Lemma 7.2.1. *Let $w \in A_p$ for some $1 \le p < \infty$ and let $0 < \alpha < 1$. Then there exists $\beta < 1$ such that whenever S is a measurable subset of a cube Q that satisfies $|S| \le \alpha|Q|$, we have $w(S) \le \beta\, w(Q)$.*

Proof. Taking $f = \chi_A$ in property (8) of Proposition 7.1.5, we obtain

$$\left(\frac{|A|}{|Q|}\right)^p \le [w]_{A_p} \frac{w(A)}{w(Q)}. \tag{7.2.1}$$

We write $S = Q \setminus A$ to get

$$\left(1 - \frac{|S|}{|Q|}\right)^p \le [w]_{A_p} \left(1 - \frac{w(S)}{w(Q)}\right). \tag{7.2.2}$$

Given $0 < \alpha < 1$, set

$$\beta = 1 - \frac{(1-\alpha)^p}{[w]_{A_p}} \tag{7.2.3}$$

and use (7.2.2) to obtain the required conclusion. □

7.2.1 The Reverse Hölder Property of A_p Weights

We are now ready to state and prove one of the main results of the theory of weights, the reverse Hölder inequality for A_p weights.

Theorem 7.2.2. *Let $w \in A_p$ for some $1 \le p < \infty$. Then there exist constants C and $\gamma > 0$ that depend only on the dimension n, on p, and on $[w]_{A_p}$ such that for every cube Q we have*

$$\left(\frac{1}{|Q|}\int_Q w(t)^{1+\gamma}\,dt\right)^{\frac{1}{1+\gamma}} \le \frac{C}{|Q|}\int_Q w(t)\,dt. \tag{7.2.4}$$

Proof. Let us fix a cube Q and set

$$\alpha_0 = \frac{1}{|Q|}\int_Q w(x)\,dx.$$

We also fix $0 < \alpha < 1$. We define an increasing sequence of scalars

$$\alpha_0 < \alpha_1 < \alpha_2 < \cdots < \alpha_k < \cdots$$

for $k \geq 0$ by setting

$$\alpha_{k+1} = 2^n \alpha^{-1} \alpha_k \qquad \text{or} \qquad \alpha_k = (2^n \alpha^{-1})^k \alpha_0,$$

and for each $k \geq 1$ we apply a Calderón–Zygmund decomposition to w at height α_k. Precisely, for dyadic subcubes R of Q, we let

$$\frac{1}{|R|} \int_R w(x)\,dx > \alpha_k \tag{7.2.5}$$

be the selection criterion. Since Q does not satisfy the selection criterion, it is not selected. We divide the cube Q into a mesh of 2^n subcubes of equal side length, and among these cubes we select those that satisfy (7.2.5). We subdivide each unselected subcube into 2^n cubes of equal side length and we continue in this way indefinitely. We denote by $\{Q_{k,j}\}_j$ the collection of all selected subcubes of Q. We observe that the following properties are satisfied:

(1) $\alpha_k < \dfrac{1}{|Q_{k,j}|} \displaystyle\int_{Q_{k,j}} w(t)\,dt \leq 2^n \alpha_k$.

(2) For almost all $x \notin U_k$ we have $w(x) \leq \alpha_k$, where $U_k = \bigcup_j Q_{k,j}$.

(3) Each $Q_{k+1,j}$ is contained in some $Q_{k,l}$.

Property (1) is satisfied since the unique dyadic parent of $Q_{k,j}$ was not chosen in the selection procedure. Property (2) follows from the Lebesgue differentiation theorem using the fact that for almost all $x \notin U_k$ there exists a sequence of unselected cubes of decreasing lengths whose closures' intersection is the singleton $\{x\}$. Property (3) is satisfied since each $Q_{k,j}$ is the maximal subcube of Q satisfying (7.2.5). And since the average of w over $Q_{k+1,j}$ is also bigger than α_k, it follows that $Q_{k+1,j}$ must be contained in some maximal cube that possesses this property.

We now compute the portion of $Q_{k,l}$ that is covered by cubes of the form $Q_{k+1,j}$ for some j. We have

$$\begin{aligned}
2^n \alpha_k &\geq \frac{1}{|Q_{k,l}|} \int_{Q_{k,l} \cap U_{k+1}} w(t)\,dt \\
&= \frac{1}{|Q_{k,l}|} \sum_{j:\, Q_{k+1,j} \subseteq Q_{k,l}} |Q_{k+1,j}| \frac{1}{|Q_{k+1,j}|} \int_{Q_{k+1,j}} w(t)\,dt \\
&> \frac{\big|Q_{k,l} \cap U_{k+1}\big|}{|Q_{k,l}|} \alpha_{k+1} \\
&= \frac{\big|Q_{k,l} \cap U_{k+1}\big|}{|Q_{k,l}|} 2^n \alpha^{-1} \alpha_k\,.
\end{aligned}$$

It follows that $\big|Q_{k,l} \cap U_{k+1}\big| \leq \alpha |Q_{k,l}|$; thus, applying Lemma 7.2.1, we obtain

$$\frac{w(Q_{k,l} \cap U_{k+1})}{w(Q_{k,l})} < \beta = 1 - \frac{(1-\alpha)^p}{[w]_{A_p}},$$

from which, summing over all l, we obtain

$$w(U_{k+1}) \le \beta w(U_k).$$

The latter gives $w(U_k) \le \beta^k w(U_0)$. We also have $|U_{k+1}| \le \alpha |U_k|$; hence $|U_k| \to 0$ as $k \to \infty$. Therefore, the intersection of the U_k's is a set of Lebesgue measure zero. We can therefore write

$$Q = (Q \setminus U_0) \bigcup \Big(\bigcup_{k=0}^{\infty} U_k \setminus U_{k+1} \Big)$$

modulo a set of Lebesgue measure zero. Let us now find a $\gamma > 0$ such that the reverse Hölder inequality (7.2.4) holds. We have $w(x) \le \alpha_k$ for almost all x in $Q \setminus U_k$ and therefore

$$\begin{aligned}
\int_Q w(t)^{1+\gamma}\,dt &= \int_{Q\setminus U_0} w(t)^\gamma w(t)\,dt + \sum_{k=0}^{\infty} \int_{U_k\setminus U_{k+1}} w(t)^\gamma w(t)\,dt \\
&\le \alpha_0^\gamma w(Q\setminus U_0) + \sum_{k=0}^{\infty} \alpha_{k+1}^\gamma w(U_k) \\
&\le \alpha_0^\gamma w(Q\setminus U_0) + \sum_{k=0}^{\infty} ((2^n\alpha^{-1})^{k+1}\alpha_0)^\gamma \beta^k w(U_0) \\
&\le \alpha_0^\gamma \Big(1 + (2^n\alpha^{-1})^\gamma \sum_{k=0}^{\infty} (2^n\alpha^{-1})^{\gamma k}\beta^k \Big) w(Q) \\
&= \Big(\frac{1}{|Q|} \int_Q w(t)\,dt \Big)^\gamma \Big(1 + \frac{(2^n\alpha^{-1})^\gamma}{1-(2^n\alpha^{-1})^\gamma\beta} \Big) \int_Q w(t)\,dt,
\end{aligned}$$

provided $\gamma > 0$ is chosen small enough that $(2^n\alpha^{-1})^\gamma\beta < 1$. Keeping track of the constants, we conclude the proof of the theorem with

$$\gamma = \frac{1}{2}\frac{-\log\beta}{\log 2^n - \log\alpha} = \frac{\log([w]_{A_p}) - \log([w]_{A_p} - (1-\alpha)^p)}{2\log\frac{2^n}{\alpha}} \tag{7.2.6}$$

and

$$\begin{aligned}
C^{\gamma+1} &= 1 + \frac{(2^n\alpha^{-1})^\gamma}{1-(2^n\alpha^{-1})^\gamma\beta} \\
&= 1 + \frac{(2^n\alpha^{-1})^\gamma}{1-(2^n\alpha^{-1})^\gamma(1-\frac{(1-\alpha)^p}{[w]_{A_p}})} \\
&= 1 + \frac{1}{(2^n\alpha^{-1})^{-\gamma} - (1-\frac{(1-\alpha)^p}{[w]_{A_p}})},
\end{aligned}$$

which yields

$$C = \left[1 + \frac{1}{\left(1 - \frac{(1-\alpha)^p}{[w]_{A_p}}\right)^{\frac{1}{2}} - \left(1 - \frac{(1-\alpha)^p}{[w]_{A_p}}\right)}\right]^{\frac{2\log\frac{2^n}{\alpha}}{2\log\frac{2^n}{\alpha} - \log\left(1 - \frac{(1-\alpha)^p}{[w]_{A_p}}\right)}} . \tag{7.2.7}$$

Note that up to this point, α was an arbitrary number in $(0,1)$. □

Remark 7.2.3. It is worth observing that for α such that $(1-\alpha)^p = \frac{3}{4}$, the constant γ in (7.2.6) decreases as $[w]_{A_p}$ increases, while the constant C in (7.2.7) increases as $[w]_{A_p}$ increases. This is because $1 - \frac{3}{4}[w]_{A_p}^{-1} \geq \frac{1}{4}$ and for $t \in (\frac{1}{4}, 1)$ the function $\sqrt{t} - t$ is decreasing. This allows us to obtain the following stronger version of Theorem 7.2.2: For any $1 \leq p < \infty$ and $B > 1$, there exist positive constants $C = C(n,p,B)$ and $\gamma = \gamma(n,p,B)$ such that for all $w \in A_p$ satisfying $[w]_{A_p} \leq B$ the reverse Hölder condition (7.2.4) holds for every cube Q. See Exercise 7.2.4(a) for details.

Observe that in the proof of Theorem 7.2.2 it was crucial to know that for some $0 < \alpha, \beta < 1$ we have

$$|S| \leq \alpha\,|Q| \implies w(S) \leq \beta\, w(Q) \tag{7.2.8}$$

whenever S is a subset of the cube Q. No special property of Lebesgue measure was used in the proof of Theorem 7.2.2 other than its doubling property. Therefore, it is reasonable to ask whether Lebesgue measure in (7.2.8) can be replaced by a general measure μ satisfying the doubling property

$$\mu(3Q) \leq C_n\, \mu(Q) < \infty \tag{7.2.9}$$

for all cubes Q in $\mathbf{R}^n$. A straightforward adjustment of the proof of the previous theorem indicates that this is indeed the case.

Corollary 7.2.4. *Let w be a weight and let μ be a measure on $\mathbf{R}^n$ satisfying (7.2.9). Suppose that there exist $0 < \alpha, \beta < 1$, such that*

$$\mu(S) \leq \alpha\, \mu(Q) \implies \int_S w(t)\, d\mu(t) \leq \beta \int_Q w(t)\, d\mu(t)$$

whenever S is a μ-measurable subset of a cube Q. Then there exist $0 < C, \gamma < \infty$ [which depend only on the dimension n, the constant C_n in (7.2.9), α, and β] such that for every cube Q in $\mathbf{R}^n$ we have

$$\left(\frac{1}{\mu(Q)} \int_Q w(t)^{1+\gamma}\, d\mu(t)\right)^{\frac{1}{1+\gamma}} \leq \frac{C}{\mu(Q)} \int_Q w(t)\, d\mu(t). \tag{7.2.10}$$

Proof. The proof of the corollary can be obtained almost verbatim from that of Theorem 7.2.2 by replacing Lebesgue measure with the doubling measure $d\mu$ and the constant 2^n by C_n.

Precisely, we define $\alpha_k = (C_n\alpha^{-1})^k\alpha_0$, where α_0 is the μ-average of w over Q; then properties (1), (2), (3) concerning the selected cubes $\{Q_{k,j}\}_j$ are replaced by

(1_μ) $\alpha_k < \dfrac{1}{\mu(Q_{k,j})}\displaystyle\int_{Q_{k,j}} w(t)\,d\mu(t) \le C_n\,\alpha_k.$

(2_μ) On $Q \setminus U_k$ we have $w \le \alpha_k$ μ-almost everywhere, where $U_k = \bigcup_j Q_{k,j}$.

(3_μ) Each $Q_{k+1,j}$ is contained in some $Q_{k,l}$.

To prove the upper inequality in (1_μ) we use that the dyadic parent of each selected cube $Q_{k,j}$ was not selected and is contained in $3Q_{k,j}$. To prove (2_μ) we need a differentiation theorem for doubling measures, analogous to that in Corollary 2.1.16. This can be found in Exercise 2.1.1. The remaining details of the proof are trivially adapted to the new setting. The conclusion is that for

$$0 < \gamma < \frac{-\log\beta}{\log C_n - \log\alpha} \tag{7.2.11}$$

and

$$C = \left[1 + \frac{(C_n\alpha^{-1})^\gamma}{1-(C_n\alpha^{-1})^\gamma\beta}\right]^{\frac{1}{\gamma+1}}, \tag{7.2.12}$$

(7.2.10) is satisfied. Notice that the choice of the constants (7.2.6) and (7.2.7) is valid in this case with C_n in place of 2^n. □

7.2.2 Consequences of the Reverse Hölder Property

Having established the crucial reverse Hölder inequality for A_p weights, we now pass to some very important applications. Among them, the first result of this section yields that an A_p weight that lies a priori in $L^1_{\mathrm{loc}}(\mathbf{R}^n)$ must actually lie in the better space $L^{1+\sigma}_{\mathrm{loc}}(\mathbf{R}^n)$ for some $\sigma > 0$ depending on the weight.

Theorem 7.2.5. *If $w \in A_p$ for some $1 \le p < \infty$, then there exists a number $\gamma > 0$ (that depends on n, p, and $[w]_{A_p}$) such that $w^{1+\gamma} \in A_p$.*

Proof. Let C be the constant in the proof of Theorem 7.2.2. When $p = 1$, we apply the reverse Hölder inequality of Theorem 7.2.2 to the weight w to obtain

$$\frac{1}{|Q|}\int_Q w(t)^{1+\gamma}\,dt \le \left(\frac{C}{|Q|}\int_Q w(t)\,dt\right)^{1+\gamma} \le C^{1+\gamma}[w]_{A_1}^{1+\gamma} w(x)^{1+\gamma}$$

for almost all x in the cube Q. Therefore, $w^{1+\gamma}$ is an A_1 weight with characteristic constant at most $C^{1+\gamma}[w]_{A_1}^{1+\gamma}$. When $p > 1$, there exist $\gamma_1, \gamma_2 > 0$ and $C_1, C_2 > 0$ such that the reverse Hölder inequality of Theorem 7.2.2 holds for the weights $w \in A_p$ and $w^{-\frac{1}{p-1}} \in A_{p'}$, that is,

$$\left(\frac{1}{|Q|}\int_Q w(t)^{1+\gamma_1}\,dt\right)^{\frac{1}{1+\gamma_1}} \le \frac{C_1}{|Q|}\int_Q w(t)\,dt,$$

$$\left(\frac{1}{|Q|}\int_Q w(t)^{-\frac{1}{p-1}(1+\gamma_2)}\,dt\right)^{\frac{1}{1+\gamma_2}} \le \frac{C_2}{|Q|}\int_Q w(t)^{-\frac{1}{p-1}}\,dt.$$

Taking $\gamma=\min(\gamma_1,\gamma_2)$, both inequalities are satisfied with γ in the place of γ_1, γ_2. It follows that $w^{1+\gamma}$ is in A_p and satisfies

$$[w^{1+\gamma}]_{A_p} \le (C_1C_2^{p-1})^{1+\gamma}[w]_{A_p}^{1+\gamma}. \tag{7.2.13}$$

This concludes the proof of the theorem. □

Corollary 7.2.6. *For any $1<p<\infty$ and for every $w\in A_p$ there is a $q=q(n,p,[w]_{A_p})$ with $q<p$ such that $w\in A_q$. In other words, we have*

$$A_p=\bigcup_{q\in(1,p)} A_q\,.$$

Proof. Given $w\in A_p$, let γ, C_1, C_2 be as in the proof of Theorem 7.2.5. In view of the result in Exercise 7.1.3 with $\delta=1/(1+\gamma)$, if $w^{1+\gamma}\in A_p$ and

$$q=p\frac{1}{1+\gamma}+1-\frac{1}{1+\gamma}=\frac{p+\gamma}{1+\gamma},$$

then $w\in A_q$ and

$$[w]_{A_q}=[(w^{1+\gamma})^{\frac{1}{1+\gamma}}]_{A_q}\le [w^{1+\gamma}]_{A_p}^{\frac{1}{1+\gamma}}\le C_1C_2^{p-1}[w]_{A_p},$$

where the last estimate comes from (7.2.13). Since $1<q=\frac{p+\gamma}{1+\gamma}<p$, the required conclusion follows. Observe that the constants $C_1C_2^{p-1}$, q, and $\frac{1}{\gamma}$ increase as $[w]_{A_p}$ increases. □

Another powerful consequence of the reverse Hölder property of A_p weights is the following characterization of all A_1 weights.

Theorem 7.2.7. *Let w be an A_1 weight. Then there exist $0<\varepsilon<1$, a nonnegative function k such that $k,k^{-1}\in L^\infty$, and a nonnegative locally integrable function f that satisfies $M(f)<\infty$ a.e. such that*

$$w(x)=k(x)\,M(f)(x)^{\varepsilon}. \tag{7.2.14}$$

Conversely, given a nonnegative function k such that $k,k^{-1}\in L^\infty$ and given a nonnegative locally integrable function f that satisfies $M(f)<\infty$ a.e., define w via (7.2.14). *Then w is an A_1 weight that satisfies*

$$[w]_{A_1}\le\frac{C_n}{1-\varepsilon}\,\|k\|_{L^\infty}\|k^{-1}\|_{L^\infty}, \tag{7.2.15}$$

where C_n is a universal dimensional constant.

Proof. In view of Theorem 7.2.2, there exist $0 < \gamma, C < \infty$ such that the reverse Hölder condition

$$\left(\frac{1}{|Q|}\int_Q w(t)^{1+\gamma}\,dt\right)^{\frac{1}{1+\gamma}} \le \frac{C}{|Q|}\int_Q w(t)\,dt \le C\,[w]_{A_1} w(x) \tag{7.2.16}$$

holds for all cubes Q and for all x in $Q\setminus E_Q$, where E_Q is a null subset of Q. We set

$$\varepsilon = \frac{1}{1+\gamma} \qquad \text{and} \qquad f(x) = w(x)^{1+\gamma} = w(x)^{\frac{1}{\varepsilon}}.$$

Letting N be the union of E_Q over all Q with rational radii and centers in $\mathbf{Q}^n$, it follows from (7.2.16) that the uncentered Hardy–Littlewood maximal function $M_c(f)$ with respect to cubes satisfies

$$M_c(f)(x) \le C^{1+\gamma}[w]_{A_1}^{1+\gamma} f(x) \qquad \text{for } x \in \mathbf{R}^n \setminus N.$$

This implies that $M(f) \le C_n C^{1+\gamma}[w]_{A_1}^{1+\gamma} f$ a.e. for some constant C_n that depends only on the dimension. We now set

$$k(x) = \frac{f(x)^\varepsilon}{M(f)(x)^\varepsilon},$$

and we observe that $C^{-1}C_n^{-\varepsilon}[w]_{A_1}^{-1} \le k \le 1$ a.e.

It remains to prove the converse. Given a weight $w = kM(f)^\varepsilon$ in the form (7.2.14) and a cube Q, it suffices to show that

$$\frac{1}{|Q|}\int_Q M(f)(t)^\varepsilon\,dt \le \frac{C_n}{1-\varepsilon} M(f)^\varepsilon(x) \qquad \text{for almost all } x \in Q, \tag{7.2.17}$$

since then (7.2.15) follows trivially from (7.2.17) with $w = kM(f)^\varepsilon$ using that $k, k^{-1} \in L^\infty$. To prove (7.2.17), we write

$$f = f\chi_{3Q} + f\chi_{(3Q)^c}.$$

Then

$$\frac{1}{|Q|}\int_Q M(f\chi_{3Q})(t)^\varepsilon\,dt \le \frac{C_n'}{1-\varepsilon}\left(\frac{1}{|Q|}\int_{\mathbf{R}^n}(f\chi_{3Q})(t)\,dt\right)^\varepsilon \tag{7.2.18}$$

in view of Kolmogorov's inequality (Exercise 2.1.5). But the last expression in (7.2.18) is at most a dimensional multiple of $M(f)(x)^\varepsilon$ for almost all $x \in Q$, which proves (7.2.17) when f is replaced by $f\chi_{3Q}$ on the left-hand side of the inequality. And for $f\chi_{(3Q)^c}$ we only need to notice that

$$M(f\chi_{(3Q)^c})(t) \le 2^n \mathcal{M}(f\chi_{(3Q)^c})(t) \le 2^n n^{\frac{n}{2}} M(f)(x)$$

for all x, t in Q, since any ball B centered at t that gives a nonzero average for $f\chi_{(3Q)^c}$ must have radius at least the side length of Q, and thus $\sqrt{n}B$ must also contain x. (Here $\mathcal{M}$ is the centered Hardy–Littlewood maximal operator introduced in Definition 2.1.1.) Hence (7.2.17) also holds when f is replaced by $f\chi_{(3Q)^c}$ on the left-hand side. Combining these two estimates and using the subadditivity property $M(f_1+f_2)^\varepsilon \le M(f_1)^\varepsilon + M(f_2)^\varepsilon$, we obtain (7.2.17). □

We end this section with the following consequence of the reverse Hölder property of A_p weights which can be viewed as a reverse property to (7.2.1).

Proposition 7.2.8. *Let $1 \le p < \infty$ and $w \in A_p$. Then there exist $\delta \in (0,1)$ and $C > 0$ depending only on n, p, and $[w]_{A_p}$ such that for any cube Q and any measurable subset S of Q we have*

$$\frac{w(S)}{w(Q)} \le C\left(\frac{|S|}{|Q|}\right)^\delta .$$

Proof. Let C and γ be as in Theorem 7.2.2. We use Hölder's inequality to write

$$\begin{aligned}
\frac{w(S)}{w(Q)} &= \frac{1}{w(Q)}\int_Q w(x)\chi_S(x)\,dx \\
&\le \frac{1}{w(Q)}\left(\int_Q w(x)^{1+\gamma}dx\right)^{\frac{1}{1+\gamma}} |S|^{\frac{\gamma}{1+\gamma}} \\
&= \frac{1}{w(Q)}\left(\frac{1}{|Q|}\int_Q w(x)^{1+\gamma}dx\right)^{\frac{1}{1+\gamma}} |Q|^{\frac{1}{1+\gamma}}|S|^{\frac{\gamma}{1+\gamma}} \\
&= \frac{C}{w(Q)}\left(\int_Q w(x)\,dx\right)|Q|^{-\frac{\gamma}{1+\gamma}}|S|^{\frac{\gamma}{1+\gamma}} \\
&= C\left(\frac{|S|}{|Q|}\right)^\delta ,
\end{aligned}$$

where $\delta = \frac{\gamma}{1+\gamma}$. This proves the assertion. □

Exercises

7.2.1. Let $w \in A_p$ for some $1 < p < \infty$ and let $1 \le q < \infty$. Prove that the sublinear operator

$$S(f) = \left(M(|f|^q w)w^{-1}\right)^{\frac{1}{q}}$$

is bounded on $L^{p'q}(w)$.

7.2.2. Let v be a real-valued locally integrable function on $\mathbf{R}^n$ and let $1 < p < \infty$. For a cube Q, let v_Q be the average of v over Q.

(a) If e^{ν} is an A_p weight, show that

$$\sup_{Q \text{ cubes}} \frac{1}{|Q|}\int_Q e^{\nu(t)-\nu_Q}\,dt \le [e^{\nu}]_{A_p},$$

$$\sup_{Q \text{ cubes}} \frac{1}{|Q|}\int_Q e^{-(\nu(t)-\nu_Q)\frac{1}{p-1}}\,dt \le [e^{\nu}]_{A_p}.$$

(b) Conversely, if the preceding inequalities hold with some constant C in place of $[\nu]_{A_p}$, then ν lies in A_p with $[\nu]_{A_p} \le C$.
[*Hint:* Part (a): If $e^{\nu} \in A_p$, use that

$$\frac{1}{|Q|}\int_Q e^{\nu(t)-\nu_Q}\,dt \le \Big(\operatorname*{Avg}_Q e^{-\frac{\nu}{p-1}}\Big)^{p-1}\Big(\operatorname*{Avg}_Q e^{\nu}\Big)$$

and obtain a similar estimate for the second quantity.]

7.2.3. This exercise assumes familiarity with the space BMO.
(a) Show that if $\varphi \in A_2$, then $\log\varphi \in BMO$ and $\|\log\varphi\|_{BMO} \le [\varphi]_{A_2}$.
(b) Prove that every BMO function is equal to a constant multiple of the logarithm of an A_2 weight. Precisely, given $f \in BMO$ show that

$$\big[e^{cf}\big]_{A_2} \le 1+2e,$$

where $c = 1/(2^{n+1}\|f\|_{BMO})$.
(c) Prove that if φ is in A_p for some $1<p<\infty$, then $\log\varphi$ is in BMO by showing that

$$\|\log\varphi\|_{BMO} \le \begin{cases} [\varphi]_{A_p} & \text{when } 1<p\le 2,\\ (p-1)[\varphi]_{A_p}^{\frac{1}{p-1}} & \text{when } 2<p<\infty. \end{cases}$$

[*Hint:* Part (a): Use Exercise 7.2.2 with $p=2$. Part (b): Use Exercise 7.2.2 and Corollary 3.1.7 in [131]. Use Part (c): Use that $\varphi^{-\frac{1}{p-1}} \in A_{p'}$ when $p>2$.]

7.2.4. Prove the following quantitative versions of Theorem 7.2.2 and Corollary 7.2.6.
(a) For any $1\le p<\infty$ and $B>1$, there exists a positive constant $C_3(n,p,B)$ and $\gamma=\gamma(n,p,B)$ such that for all $w\in A_p$ satisfying $[w]_{A_p}\le B$, (7.2.4) holds for every cube Q with $C_3(n,p,B)$ in place of C.
(b) Given any $1<p<\infty$ and $B>1$ there exists a constant $C_4(n,p,B)$ and $\delta=\delta(n,p,B)$ such that for all $w\in A_p$ we have

$$[w]_{A_p}\le B \implies [w]_{A_{p-\delta}} \le C_4(n,p,B).$$

7.2.5. Given a positive doubling measure μ on $\mathbf{R}^n$, define the characteristic constant $[w]_{A_p(\mu)}$ and the class $A_p(\mu)$ for $1<p<\infty$.
(a) Show that statement (8) of Proposition 7.1.5 remains valid if Lebesgue measure is replaced by μ.

(b) Obtain as a consequence that if $w \in A_p(\mu)$, then for all cubes Q and all μ-measurable subsets A of Q we have

$$\left(\frac{\mu(A)}{\mu(Q)}\right)^p \le [w]_{A_p(\mu)} \frac{w(A)}{w(Q)}.$$

Conclude that if Lebesgue measure is replaced by μ in Lemma 7.2.1, then the lemma is valid for $w \in A_p(\mu)$.
(c) Use Corollary 7.2.4 to obtain that weights in $A_p(\mu)$ satisfy a reverse Hölder condition.
(d) Prove that given a weight $w \in A_p(\mu)$, there exists $1 < q < p$, which depends on $[w]_{A_p(\mu)}$, such that $w \in A_q(\mu)$.

7.2.6. Let $1 < q < \infty$ and μ a positive measure on $\mathbf{R}^n$. We say that a positive function K on $\mathbf{R}^n$ satisfies a *reverse Hölder condition* of order q with respect to μ, symbolically $K \in RH_q(\mu)$, if

$$[K]_{RH_q(\mu)} = \sup_{Q \text{cubes in } \mathbf{R}^n} \frac{\left(\frac{1}{\mu(Q)} \int_Q K^q \, d\mu\right)^{\frac{1}{q}}}{\frac{1}{\mu(Q)} \int_Q K \, d\mu} < \infty.$$

For positive functions u, v on $\mathbf{R}^n$ and $1 < p < \infty$, show that

$$[vu^{-1}]_{RH_{p'}(u\,dx)} = [uv^{-1}]_{A_p(v\,dx)}^{\frac{1}{p}},$$

that is, vu^{-1} satisfies a reverse Hölder condition of order p' with respect to $u\,dx$ if and only if uv^{-1} is in $A_p(v\,dx)$. Conclude that

$$w \in RH_{p'}(dx) \iff w^{-1} \in A_p(w\,dx),$$
$$w \in A_p(dx) \iff w^{-1} \in RH_{p'}(w\,dx).$$

7.2.7. ([125]) Suppose that a positive function K on $\mathbf{R}^n$ lies in $RH_p(dx)$ for some $1 < p < \infty$. Show that there exists a $\delta > 0$ such that K lies in $RH_{p+\delta}(dx)$.
[*Hint:* By Exercise 7.2.6, $K \in RH_p(dx)$ is equivalent to the fact that $K^{-1} \in A_{p'}(K\,dx)$, and the index p' can be improved by Exercise 7.2.5 (d).]

7.2.8. (a) Show that for any $w \in A_1$ and any cube Q in $\mathbf{R}^n$ and $a > 1$ we have

$$\operatorname*{ess.inf}_{Q} w \le a^n [w]_{A_1} \operatorname*{ess.inf}_{aQ} w.$$

(b) Prove that there is a constant C_n such that for all locally integrable functions f on $\mathbf{R}^n$ and all cubes Q in $\mathbf{R}^n$ we have

$$\operatorname*{ess.inf}_{Q} M(f) \le C_n \operatorname*{ess.inf}_{3Q} M(f),$$

and an analogous statement is valid for M_c.

[*Hint:* Part (a): Use (7.1.18). Part (b): Apply part (a) to $M(f)^{\frac{1}{2}}$, which is an A_1 weight in view of Theorem 7.2.7.]

7.2.9. ([223]) For a weight $w \in A_1(\mathbf{R}^n)$ define a quantity $r = 1 + \frac{1}{2^{n+1}[w]_{A_1}}$. Show that

$$M_c(w^r)^{\frac{1}{r}} \leq 2\,[w]_{A_1}\, w \qquad \text{a.e.}$$

[*Hint:* Fix a cube Q and consider the family $\mathscr{F}_Q$ of all cubes obtained by subdividing Q into a mesh of $(2^n)^m$ subcubes of side length $2^{-m}\ell(Q)$ for all $m = 1, 2, \dots$. Define $M_Q^d(f)(x) = \sup_{R \in \mathscr{F}_Q, R \ni x} |R|^{-1} \int_R |f|\, dy$. Using Corollary 2.1.21 obtain

$$\int_{Q \cap \{M_Q^d(w) > \lambda\}} w(x)\, dx \leq 2^n \lambda |\{x \in Q:\, M_Q^d(w)(x) > \lambda\}|$$

for $\lambda > w_Q = \frac{1}{|Q|}\int_Q w\, dt$. Multiply by $\lambda^{\delta - 1}$ and integrate to obtain

$$\int_Q M_Q^d(w)^\delta w\, dx \leq (w_Q)^\delta \int_Q w\, dx + \frac{2^n \delta}{\delta + 1} \int_Q M_Q^d(w)^{\delta+1} dx.$$

Replace w by $w_k = \min(k, w)$ and select $\delta = \frac{1}{2^{n+1}[w]_{A_1}}$ to deduce

$$\frac{1}{|Q|}\int_Q w_k^{\delta+1} dx \leq \frac{1}{|Q|}\int_Q M_Q^d(w_k)^\delta w_k\, dx \leq 2 (w_Q)^{\delta+1},$$

using $[w_k]_{A_1} \leq [w]_{A_1}$. Then let $k \to \infty$.]

7.2.10. Let $1 < p < \infty$. Recall that a pair of weights (u, w) that satisfies

$$[u, w]_{(A_p, A_p)} = \sup_{\substack{Q \text{ cubes} \\ \text{in } \mathbf{R}^n}} \left(\frac{1}{|Q|} \int_Q u\, dx \right) \left(\frac{1}{|Q|} \int_Q w^{-\frac{1}{p-1}}\, dx \right)^{p-1} < \infty$$

is said to be of *class* (A_p, A_p). The quantity $[u, w]_{(A_p, A_p)}$ is called the (A_p, A_p) *characteristic constant* of the pair (u, w).
(a) Show that for any $g \in L^1_{\text{loc}}(\mathbf{R}^n)$ with $0 < g < \infty$ a.e., the pair $(g, M(g))$ is of class (A_p, A_p) with characteristic constant independent of f.
(b) If (u, w) is of class (A_p, A_p), then the Hardy–Littlewood maximal operator M may not map $L^p(w)$ to $L^p(u)$.
(c) Given $g \in L^1_{\text{loc}}(\mathbf{R}^n)$ with $0 < g < \infty$ a.e., conclude that Hardy–Littlewood maximal operator M maps $L^p(M(g)dx)$ to $L^{p,\infty}(g dx)$ and also $L^q(M(g)dx)$ to $L^q(g dx)$ for any q with $p < q < \infty$.
[*Hint:* Part (a): Use Hölder's inequality and Theorem 7.2.7. Part (b): Try the pair $(M(g)^{1-p}, |g|^{1-p})$ for a suitable g. Part (c): Use Exercises 7.1.6 and 7.1.7.]

7.3 The A_∞ Condition

In this section we examine more closely the class of all A_p weights. It turns out that A_p weights possess properties that are p-independent but delicate enough to characterize them without reference to a specific value of p. The A_p classes increase as p increases, and it is only natural to consider their limit as $p \to \infty$. Not surprisingly, a condition obtained as a limit of the A_p conditions as $p \to \infty$ provides some unexpected but insightful characterizations of the class of all A_p weights.

7.3.1 The Class of A_∞ Weights

Let us start by recalling a simple consequence of Jensen's inequality:

$$\left(\int_X |h(t)|^q \, d\mu(t)\right)^{\frac{1}{q}} \geq \exp\left(\int_X \log|h(t)| \, d\mu(t)\right), \tag{7.3.1}$$

which holds for all measurable functions h on a probability space (X, μ) and all $0 < q < \infty$. See Exercise 1.1.3(b). Moreover, part (c) of the same exercise says that the limit of the expressions on the left in (7.3.1) as $q \to 0$ is equal to the expression on the right in (7.3.1).

We apply (7.3.1) to the function $h = w^{-1}$ for some weight w in A_p with $q = 1/(p-1)$. We obtain

$$\frac{w(Q)}{|Q|}\left(\frac{1}{|Q|}\int_Q w(t)^{-\frac{1}{p-1}} \, dt\right)^{p-1} \geq \frac{w(Q)}{|Q|} \exp\left(\frac{1}{|Q|}\int_Q \log w(t)^{-1} \, dt\right), \tag{7.3.2}$$

and the limit of the expressions on the left in (7.3.2) as $p \to \infty$ is equal to the expression on the right in (7.3.2). This observation provides the motivation for the following definition.

Definition 7.3.1. A weight w is called an A_∞ weight if

$$[w]_{A_\infty} = \sup_{Q \text{ cubes in } \mathbf{R}^n} \left\{\left(\frac{1}{|Q|}\int_Q w(t) \, dt\right) \exp\left(\frac{1}{|Q|}\int_Q \log w(t)^{-1} \, dt\right)\right\} < \infty.$$

The quantity $[w]_{A_\infty}$ is called the A_∞ *characteristic constant* of w.

It follows from the previous definition and (7.3.2) that for all $1 \leq p < \infty$ we have

$$[w]_{A_\infty} \leq [w]_{A_p}.$$

This means that

$$\bigcup_{1 \leq p < \infty} A_p \subseteq A_\infty, \tag{7.3.3}$$

but the remarkable thing is that equality actually holds in (7.3.3), a deep property that requires some work.

Before we examine this and other characterizations of A_∞ weights, we discuss some of their elementary properties.

Proposition 7.3.2. *Let $w \in A_\infty$. Then*

(1) $[\delta^\lambda(w)]_{A_\infty} = [w]_{A_\infty}$, *where* $\delta^\lambda(w)(x) = w(\lambda x_1, \dots, \lambda x_n)$ *and* $\lambda > 0$.

(2) $[\tau^z(w)]_{A_\infty} = [w]_{A_\infty}$, *where* $\tau^z(w)(x) = w(x-z)$, $z \in \mathbf{R}^n$.

(3) $[\lambda w]_{A_\infty} = [w]_{A_\infty}$ *for all* $\lambda > 0$.

(4) $[w]_{A_\infty} \ge 1$.

(5) The following is an equivalent characterization of the A_∞ characteristic constant of w:

$$[w]_{A_\infty} = \sup_{\substack{Q \text{cubes} \\ \text{in } \mathbf{R}^n}} \sup_{\substack{\log|f| \in L^1(Q) \\ \int_Q |f| w\, dt > 0}} \left\{ \frac{w(Q)}{\int_Q |f(t)| w(t)\, dt} \exp\left(\frac{1}{|Q|} \int_Q \log|f(t)|\, dt \right) \right\}.$$

(6) The measure $w(x)\,dx$ is doubling; precisely, for all $\lambda > 1$ and all cubes Q we have

$$w(\lambda Q) \le 2^{\lambda^n} [w]_{A_\infty}^{\lambda^n} w(Q).$$

As usual, λQ here denotes the cube with the same center as Q and side length λ times that of Q.

We note that estimate (6) is not as good as $\lambda \to \infty$ but it can be substantially improved using the case $\lambda = 2$. We refer to Exercise 7.3.1 for an improvement.

Proof. Properties (1)–(3) are elementary, while property (4) is a consequence of Exercise 1.1.3(b). To show (5), first observe that by taking $f = w^{-1}$, the expression on the right in (5) is at least as big as $[w]_{A_\infty}$. Conversely, (7.3.1) gives

$$\exp\left(\frac{1}{|Q|} \int_Q \log\left(|f(t)| w(t)\right) dt \right) \le \frac{1}{|Q|} \int_Q |f(t)| w(t)\, dt,$$

which, after a simple algebraic manipulation, can be written as

$$\frac{w(Q)}{\int_Q |f| w\, dt} \exp\left(\frac{1}{|Q|} \int_Q \log|f|\, dt \right) \le \frac{w(Q)}{|Q|} \exp\left(-\frac{1}{|Q|} \int_Q \log|w|\, dt \right),$$

whenever f does not vanish almost everywhere on Q. Taking the supremum over all such f and all cubes Q in $\mathbf{R}^n$, we obtain that the expression on the right in (5) is at most $[w]_{A_\infty}$.

To prove the doubling property for A_∞ weights, we fix $\lambda > 1$ and we apply property (5) to the cube λQ in place of Q and to the function

$$f = \begin{cases} c & \text{on } Q, \\ 1 & \text{on } \mathbf{R}^n \setminus Q, \end{cases} \tag{7.3.4}$$

where c is chosen so that $c^{1/\lambda^n} = 2[w]_{A_\infty}$. We obtain

$$\frac{w(\lambda Q)}{w(\lambda Q \setminus Q) + c\,w(Q)} \exp\left(\frac{\log c}{\lambda^n}\right) \le [w]_{A_\infty},$$

which implies (6) if we take into account the chosen value of c. □

7.3.2 Characterizations of A_∞ Weights

Having established some elementary properties of A_∞ weights, we now turn to some of their deeper properties, one of which is that every A_∞ weight lies in some A_p for $p < \infty$. It also turns out that A_∞ weights are characterized by the reverse Hölder property, which as we saw is a fundamental property of A_p weights. The following is the main theorem of this section.

Theorem 7.3.3. *Suppose that w is a weight. Then w is in A_∞ if and only if any one of the following conditions holds:*
(a) There exist $0 < \gamma, \delta < 1$ such that for all cubes Q in $\mathbf{R}^n$ we have

$$\left|\{x \in Q :\ w(x) \le \gamma \operatorname{Avg}_Q w\}\right| \le \delta\,|Q|.$$

(b) There exist $0 < \alpha, \beta < 1$ such that for all cubes Q and all measurable subsets A of Q we have

$$|A| \le \alpha\,|Q| \implies w(A) \le \beta\,w(Q).$$

(c) The reverse Hölder condition holds for w, that is, there exist $0 < C_1, \varepsilon < \infty$ such that for all cubes Q we have

$$\left(\frac{1}{|Q|}\int_Q w(t)^{1+\varepsilon}\,dt\right)^{\frac{1}{1+\varepsilon}} \le \frac{C_1}{|Q|}\int_Q w(t)\,dt.$$

(d) There exist $0 < C_2, \varepsilon_0 < \infty$ such that for all cubes Q and all measurable subsets A of Q we have

$$\frac{w(A)}{w(Q)} \le C_2\left(\frac{|A|}{|Q|}\right)^{\varepsilon_0}.$$

(e) There exist $0 < \alpha', \beta' < 1$ such that for all cubes Q and all measurable subsets A of Q we have

$$w(A) < \alpha'\,w(Q) \implies |A| < \beta'\,|Q|.$$

(f) There exist $p, C_3 < \infty$ *such that* $[w]_{A_p} \le C_3$. *In other words,* w *lies in* A_p *for some* $p \in [1,\infty)$.

All the constants $C_1, C_2, C_3, \alpha, \beta, \gamma, \delta, \alpha', \beta', \varepsilon, \varepsilon_0$, *and* p *in (a)–(f) depend only on the dimension* n *and on* $[w]_{A_\infty}$. *Moreover, if any of the statements in (a)–(f) is valid, then so is any other statement in (a)–(f) with constants that depend only on the dimension* n *and the constants that appear in the assumed statement.*

Proof. The proof follows from the sequence of implications

$$w \in A_\infty \Longrightarrow (a) \Longrightarrow (b) \Longrightarrow (c) \Longrightarrow (d) \Longrightarrow (e) \Longrightarrow (f) \Longrightarrow w \in A_\infty .$$

At each step we keep track of the way the constants depend on the constants of the previous step. This is needed to validate the last assertion of the theorem.

$w \in A_\infty \Longrightarrow (a)$

Fix a cube Q. Since multiplication of an A_∞ weight with a positive scalar does not alter its A_∞ characteristic, we may assume that $\int_Q \log w(t)\,dt = 0$. This implies that $\operatorname{Avg}_Q w \le [w]_{A_\infty}$. Then we have

$$\begin{aligned}
\big|\{x \in Q : w(x) \le \gamma \operatorname*{Avg}_Q w\}\big| &\le \big|\{x \in Q : w(x) \le \gamma [w]_{A_\infty}\}\big| \\
&= \big|\{x \in Q : \log(1 + w(x)^{-1}) \ge \log(1 + (\gamma [w]_{A_\infty})^{-1})\}\big| \\
&\le \frac{1}{\log(1 + (\gamma [w]_{A_\infty})^{-1})} \int_Q \log \frac{1 + w(t)}{w(t)}\,dt \\
&= \frac{1}{\log(1 + (\gamma [w]_{A_\infty})^{-1})} \int_Q \log(1 + w(t))\,dt \\
&\le \frac{1}{\log(1 + (\gamma [w]_{A_\infty})^{-1})} \int_Q w(t)\,dt \\
&\le \frac{[w]_{A_\infty} |Q|}{\log(1 + (\gamma [w]_{A_\infty})^{-1})} \\
&= \frac{1}{2} |Q| ,
\end{aligned}$$

which proves (a) with $\gamma = [w]_{A_\infty}^{-1} (e^{2[w]_{A_\infty}} - 1)^{-1}$ and $\delta = \frac{1}{2}$.

$(a) \Longrightarrow (b)$

Let Q be fixed and let A be a subset of Q with $w(A) > \beta w(Q)$ for some β to be chosen later. Setting $S = Q \setminus A$, we have $w(S) < (1 - \beta) w(Q)$. We write $S = S_1 \cup S_2$, where

$$S_1 = \{x \in S : w(x) > \gamma \operatorname{Avg}_Q w\} \quad \text{and} \quad S_2 = \{x \in S : w(x) \le \gamma \operatorname{Avg}_Q w\}.$$

For S_2 we have $|S_2| \le \delta |Q|$ by assumption (a). For S_1 we use Chebyshev's inequality to obtain

$$|S_1| \le \frac{1}{\gamma \operatorname*{Avg}_Q w} \int_S w(t)\,dt = \frac{|Q|}{\gamma} \frac{w(S)}{w(Q)} \le \frac{1 - \beta}{\gamma} |Q| .$$

Adding the estimates for $|S_1|$ and $|S_2|$, we obtain

$$|S| \le |S_1| + |S_2| \le \frac{1-\beta}{\gamma}|Q| + \delta|Q| = \Big(\delta + \frac{1-\beta}{\gamma}\Big)\,|Q|.$$

Choosing numbers α, β in $(0,1)$ such that $\delta + \frac{1-\beta}{\gamma} = 1-\alpha$, for example $\alpha = \frac{1-\delta}{2}$ and $\beta = 1 - \frac{(1-\delta)\gamma}{2}$, we obtain $|S| \le (1-\alpha)|Q|$, that is, $|A| > \alpha|Q|$.
$(b) \implies (c)$

This was proved in Corollary 7.2.4. To keep track of the constants, we note that the choices

$$\varepsilon = \frac{-\frac{1}{2}\log\beta}{\log 2^n - \log\alpha} \qquad \text{and} \qquad C_1 = 1 + \frac{(2^n\alpha^{-1})^\varepsilon}{1-(2^n\alpha^{-1})^\varepsilon\beta}$$

as given in (7.2.6) and (7.2.7) serve our purposes.
$(c) \implies (d)$

We apply first Hölder's inequality with exponents $1+\varepsilon$ and $(1+\varepsilon)/\varepsilon$ and then the reverse Hölder estimate to obtain

$$\begin{aligned}
\int_A w(x)\,dx &\le \Big(\int_A w(x)^{1+\varepsilon}\,dx\Big)^{\frac{1}{1+\varepsilon}} |A|^{\frac{\varepsilon}{1+\varepsilon}} \\
&\le \Big(\frac{1}{|Q|}\int_Q w(x)^{1+\varepsilon}\,dx\Big)^{\frac{1}{1+\varepsilon}} |Q|^{\frac{1}{1+\varepsilon}} |A|^{\frac{\varepsilon}{1+\varepsilon}} \\
&\le \frac{C_1}{|Q|}\int_Q w(x)\,dx\,|Q|^{\frac{1}{1+\varepsilon}} |A|^{\frac{\varepsilon}{1+\varepsilon}},
\end{aligned}$$

which gives

$$\frac{w(A)}{w(Q)} \le C_1\Big(\frac{|A|}{|Q|}\Big)^{\frac{\varepsilon}{1+\varepsilon}}.$$

This proves (d) with $\varepsilon_0 = \frac{\varepsilon}{1+\varepsilon}$ and $C_2 = C_1$.
$(d) \implies (e)$

Pick an $0 < \alpha'' < 1$ small enough that $\beta'' = C_2(\alpha'')^{\varepsilon_0} < 1$. It follows from (d) that

$$|A| < \alpha''|Q| \implies w(A) < \beta'' w(Q) \tag{7.3.5}$$

for all cubes Q and all A measurable subsets of Q. Replacing A by $Q \setminus A$, the implication in (7.3.5) can be equivalently written as

$$|A| \ge (1-\alpha'')|Q| \implies w(A) \ge (1-\beta'')w(Q).$$

In other words, for measurable subsets A of Q we have

$$w(A) < (1-\beta'')w(Q) \implies |A| < (1-\alpha'')|Q|, \tag{7.3.6}$$

which is the statement in (e) if we set $\alpha' = (1-\beta'')$ and $\beta' = 1-\alpha''$. Note that (7.3.5) and (7.3.6) are indeed equivalent.

$(e) \implies (f)$

We begin by examining condition (e), which can be written as

$$\int_A w(t)\,dt \le \alpha' \int_Q w(t)\,dt \implies \int_A w(t)^{-1} w(t)\,dt \le \beta' \int_Q w(t)^{-1} w(t)\,dt\,,$$

or, equivalently, as

$$\mu(A) \le \alpha' \mu(Q) \implies \int_A w(t)^{-1}\,d\mu(t) \le \beta' \int_Q w(t)^{-1}\,d\mu(t)$$

after defining the measure $d\mu(t) = w(t)\,dt$. As we have already seen, the assertions in (7.3.5) and (7.3.6) are equivalent. Therefore, we may use Exercise 7.3.2 to deduce that the measure μ is doubling, i.e., it satisfies property (7.2.9) for some constant $C_n = C_n(\alpha', \beta')$, and hence the hypotheses of Corollary 7.2.4 are satisfied. We conclude that the weight w^{-1} satisfies a reverse Hölder estimate with respect to the measure μ, that is, if γ, C are defined as in (7.2.11) and (7.2.12) [in which α is replaced by α', β by β', and C_n is the doubling constant of $w(x)\,dx$], then we have

$$\left(\frac{1}{\mu(Q)} \int_Q w(t)^{-1-\gamma}\,d\mu(t) \right)^{\frac{1}{1+\gamma}} \le \frac{C}{\mu(Q)} \int_Q w(t)^{-1}\,d\mu(t) \tag{7.3.7}$$

for all cubes Q in $\mathbf{R}^n$. Setting $p = 1 + \frac{1}{\gamma}$ and raising to the pth power, we can rewrite (7.3.7) as the A_p condition for w. We can therefore take $C_3 = C^p$ to conclude the proof of (f).

$(f) \implies w \in A_\infty$

This is trivial, since $[w]_{A_\infty} \le [w]_{A_p}$. □

An immediate consequence of the preceding theorem is the following result relating A_∞ to A_p.

Corollary 7.3.4. *The following equality is valid:*

$$A_\infty = \bigcup_{1 \le p < \infty} A_p.$$

Exercises

7.3.1. Let $\lambda > 0$, Q be a cube in $\mathbf{R}^n$, and $w \in A_\infty(\mathbf{R}^n)$.
(a) Show that property (6) in Proposition 7.3.2 can be improved to

$$w(\lambda Q) \le \min_{\varepsilon > 0} \frac{(1+\varepsilon)^{\lambda^n} [w]_{A_\infty}^{\lambda^n} - 1}{\varepsilon}\, w(Q)\,.$$

(b) Prove that

$$w(\lambda Q) \le (2\lambda)^{2^n(1+\log_2[w]_{A_\infty})} w(Q).$$

[*Hint:* Part (a): Take c in (7.3.4) such that $c^{1/\lambda^n} = (1+\varepsilon)[w]_{A_\infty}$. Part (b): Use the estimate in property (6) of Proposition 7.3.2 with $\lambda = 2$.]

7.3.2. Suppose that μ is a positive Borel measure on $\mathbf{R}^n$ with the property that for all cubes Q and all measurable subsets A of Q we have

$$|A| < \alpha|Q| \implies \mu(A) < \beta\mu(Q)$$

for some fixed $0 < \alpha, \beta < 1$. Show that μ is doubling [i.e., it satisfies (7.2.9)].
[*Hint:* Use that $|S| > (1-\alpha)|Q| \Rightarrow \mu(S) > (1-\beta)\mu(Q)$ when $S \subsetneq Q$.]

7.3.3. Prove that a weight w is in A_p if and only if both w and $w^{-\frac{1}{p-1}}$ are in A_∞.
[*Hint:* You may want to use the result of Exercise 7.2.2.]

7.3.4. ([33], [343]) Prove that if $P(x)$ is a polynomial of degree k in $\mathbf{R}^n$, then $\log|P(x)|$ is in *BMO* with norm depending only on k and n and not on the coefficients of the polynomial.
[*Hint:* Use that all norms on the finite-dimensional space of polynomials of degree at most k are equivalent to show that $|P(x)|$ satisfies a reverse Hölder inequality. Therefore, $|P(x)|$ is an A_∞ weight and thus Exercise 7.2.3 (c) is applicable.]

7.3.5. Show that the product of two A_1 weights may not be an A_∞ weight.

7.3.6. Let g be in $L^p(w)$ for some $1 \le p \le \infty$ and $w \in A_p$. Prove that $g \in L^1_{\text{loc}}(\mathbf{R}^n)$.
[*Hint:* Let B be a ball. In the case $p < \infty$, write $\int_B |g|\,dx = \int_B (|g|w^{-\frac{1}{p}})w^{\frac{1}{p}}\,dx$ and apply Hölder's inequality. In the case $p = \infty$, use that $w \in A_{p_0}$ for some $p_0 < \infty$.]

7.3.7. ([278]) Show that a weight w lies in A_∞ if and only if there exist $\gamma, C > 0$ such that for all cubes Q we have

$$w(\{x \in Q : w(x) > \lambda\}) \le C\lambda \big|\{x \in Q : w(x) > \gamma\lambda\}\big|$$

for all $\lambda > \operatorname{Avg}_Q w$.
[*Hint:* The displayed condition easily implies that

$$\frac{1}{|Q|}\int_Q w_k^{1+\varepsilon}\,dx \le \Big(\frac{w(Q)}{|Q|}\Big)^{\varepsilon+1} + \frac{C'\delta}{\gamma^{1+\varepsilon}}\frac{1}{|Q|}\int_Q w_k^{1+\varepsilon}\,dx,$$

where $k > 0$, $w_k = \min(w,k)$ and $\delta = \varepsilon/(1+\varepsilon)$. Take $\varepsilon > 0$ small enough to obtain the reverse Hölder condition (c) in Theorem 7.3.3 for w_k. Let $k \to \infty$ to obtain the same conclusion for w. Conversely, find constants $\gamma, \delta \in (0,1)$ as in condition (a) of Theorem 7.3.3 and for $\lambda > \operatorname{Avg}_Q w$ write the set $\{w > \lambda\} \cap Q$ as a union of maximal dyadic cubes Q_j such that $\lambda < \operatorname{Avg}_{Q_j} w \le 2^n\lambda$ for all j. Then $w(Q_j) \le 2^n\lambda|Q_j| \le \frac{2^n\lambda}{1-\delta}|Q_j \cap \{w > \gamma\lambda\}|$ and the required conclusion follows by summing on j.]

7.4 Weighted Norm Inequalities for Singular Integrals

We now address a topic of great interest in the theory of singular integrals, their boundedness properties on weighted L^p spaces. It turns out that a certain amount of regularity must be imposed on the kernels of these operators to obtain the aforementioned weighted estimates.

7.4.1 Singular Integrals of Non Convolution type

We introduce some definitions.

Definition 7.4.1. Let $0 < \delta, A < \infty$. A function $K(x,y)$ defined for $x,y \in \mathbf{R}^n$ with $x \neq y$ is called a *standard kernel* (with constants δ and A) if

$$|K(x,y)| \le \frac{A}{|x-y|^n}, \qquad x \neq y, \tag{7.4.1}$$

and whenever $|x-x'| \le \frac{1}{2}\max\big(|x-y|,|x'-y|\big)$ we have

$$|K(x,y)-K(x',y)| \le \frac{A|x-x'|^\delta}{(|x-y|+|x'-y|)^{n+\delta}} \tag{7.4.2}$$

and also when $|y-y'| \le \frac{1}{2}\max\big(|x-y|,|x-y'|\big)$ we have

$$|K(x,y)-K(x,y')| \le \frac{A|y-y'|^\delta}{(|x-y|+|x-y'|)^{n+\delta}}\,. \tag{7.4.3}$$

The class of all kernels that satisfy (7.4.1), (7.4.2), and (7.4.3) is denoted by $SK(\delta,A)$.

Definition 7.4.2. Let $0 < \delta, A < \infty$ and K in $SK(\delta,A)$. A *Calderón–Zygmund operator* associated with K is a linear operator T defined on $\mathscr{S}(\mathbf{R}^n)$ that admits a bounded extension on $L^2(\mathbf{R}^n)$,

$$\big\|T(f)\big\|_{L^2} \le B\big\|f\big\|_{L^2}, \tag{7.4.4}$$

and that satisfies

$$T(f)(x) = \int_{\mathbf{R}^n} K(x,y)f(y)\,dy \tag{7.4.5}$$

for all $f \in \mathscr{C}_0^\infty$ and x not in the support of f. The class of all Calderón–Zygmund operators associated with kernels in $SK(\delta,A)$ that are bounded on L^2 with norm at most B is denoted by $CZO(\delta,A,B)$. Note that there is no unique T associated with a given K. Given a Calderón–Zygmund operator T in $CZO(\delta,A,B)$, we define the truncated operator $T^{(\varepsilon)}$ as

$$T^{(\varepsilon)}(f)(x) = \int_{|x-y|>\varepsilon} K(x,y)\,f(y)\,dy$$

and the *maximal operator* associated with T as follows:

$$T^{(*)}(f)(x) = \sup_{\varepsilon>0} \left| T^{(\varepsilon)}(f)(x) \right| .$$

We note that if T is in $CZO(\delta,A,B)$, then $T^{(\varepsilon)}(f)$ and $T^{(*)}(f)$ are well defined for all f in $\bigcup_{1\le p<\infty} L^p(\mathbf{R}^n)$. It is also well defined whenever f is locally integrable and satisfies $\int_{|x-y|\ge\varepsilon} |f(y)|\,|x-y|^{-n}dy < \infty$ for all $x \in \mathbf{R}^n$ and $\varepsilon > 0$.

The class of kernels in $SK(\delta,A)$ extends the family of convolution kernels that satisfy conditions (5.3.10), (5.3.11), and (5.3.12). Obviously, the associated operators in $CZO(\delta,A,B)$ generalize the associated convolution operators.

A fundamental property of operators in $CZO(\delta,A,B)$ is that they have bounded extensions on all the $L^p(\mathbf{R}^n)$ spaces and also from $L^1(\mathbf{R}^n)$ to weak $L^1(\mathbf{R}^n)$. This is proved via an adaptation of Theorem 5.3.3; see Theorem 4.2.2 in [131]. There are analogous results for the maximal counterparts $T^{(*)}$ of elements of $CZO(\delta,A,B)$. In fact, an analogue of Theorem 5.3.5 yields that $T^{(*)}$ is L^p bounded for $1 < p < \infty$ and weak type $(1,1)$; this result is contained in Theorem 4.2.4 in [131].

We discuss weighted inequalities for singular integrals for general operators in $CZO(\delta,A,B)$. In Subsections 7.4.2 and 7.4.3, the reader may wish to replace kernels in $SK(\delta,A)$ by the more familiar functions $K(x)$ defined on $\mathbf{R}^n \setminus \{0\}$ that satisfy (5.3.10), (5.3.11), and (5.3.12).

7.4.2 A Good Lambda Estimate for Singular Integrals

The following theorem is the main result of this section.

Theorem 7.4.3. *Let $1 \le p \le \infty$, $w \in A_p$, and T in $CZO(\delta,A,B)$. Then there exist positive constants*[1] *$C_0 = C_0(n,p,[w]_{A_p})$, $\varepsilon_0 = \varepsilon_0(n,p,[w]_{A_p})$, and $c_0(n,\delta)$, such that if $\gamma_0 = c_0(n,\delta)/A$, then for all $0 < \gamma < \gamma_0$ we have*

$$w\big(\{T^{(*)}(f) > 3\lambda\} \cap \{M(f) \le \gamma\lambda\}\big) \le C_0 \gamma^{\varepsilon_0} (A+B)^{\varepsilon_0} w\big(\{T^{(*)}(f) > \lambda\}\big), \quad (7.4.6)$$

for all locally integrable functions f for which

$$\int_{|x-y|\ge\varepsilon} |f(y)|\,|x-y|^{-n}dy < \infty$$

for all $x \in \mathbf{R}^n$ and $\varepsilon > 0$. Here M denotes the Hardy–Littlewood maximal operator.

Proof. We write the open set

$$\Omega = \{T^{(*)}(f) > \lambda\} = \bigcup_j Q_j ,$$

[1] the dependence on p is relevant only when $p < \infty$

where Q_j are the Whitney cubes (see Appendix J). We set

$$Q_j^* = 10\sqrt{n}\,Q_j\,,$$
$$Q_j^{**} = 10\sqrt{n}\,Q_j^*\,,$$

where aQ denotes the cube with the same center as Q whose side length is $a\ell(Q)$, where $\ell(Q)$ is the side length of Q. We note that in view of the properties of the Whitney cubes, the distance from Q_j to Ω^c is at most $4\sqrt{n}\,\ell(Q_j)$. But the distance from Q_j to the boundary of Q_j^* is $(5\sqrt{n}-\frac{1}{2})\,\ell(Q_j)$, which is bigger than $4\sqrt{n}\,\ell(Q_j)$. Therefore, Q_j^* must meet Ω^c and for every cube Q_j we fix a point y_j in $\Omega^c \cap Q_j^*$. See Figure 7.1.

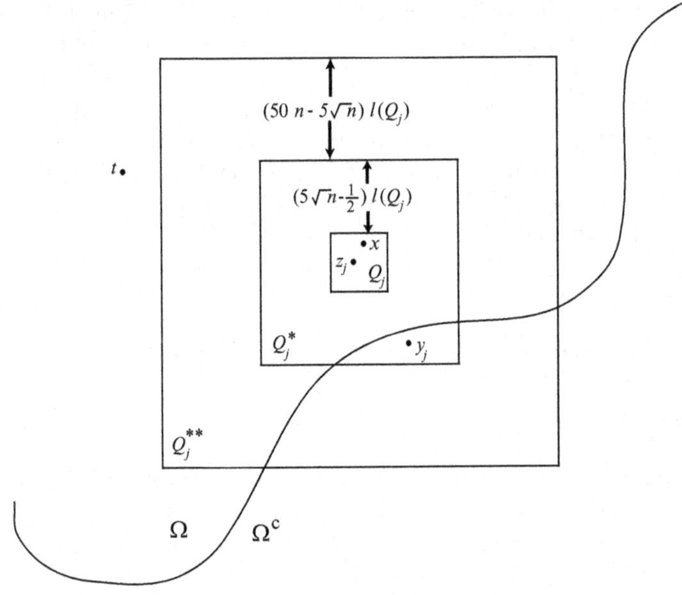

Fig. 7.1 A picture of the proof.

We also fix f in $\bigcup_{1\le p<\infty} L^p(\mathbf{R}^n)$, and for each j we write $f = f_0^j + f_\infty^j$, where $f_0^j = f\chi_{Q_j^{**}}$ is the part of f near Q_j and $f_\infty^j = f\chi_{(Q_j^{**})^c}$ is the part of f away from Q_j. We now claim that the following estimate is true:

$$\big|Q_j \cap \{T^{(*)}(f) > 3\lambda\} \cap \{M(f) \le \gamma\lambda\}\big| \le C_n\,\gamma(A+B)\big|Q_j\big|\,. \tag{7.4.7}$$

Once the validity of (7.4.7) is established, we apply Theorem 7.3.3 (d) when $p=\infty$ or Proposition 7.2.8 when $p<\infty$ to obtain constants $\varepsilon_0, C_2 > 0$, which depend on $[w]_{A_p}$, p, n when $p<\infty$ and on $[w]_{A_\infty}$ and n when $p=\infty$, such that

$$w\big(Q_j \cap \{T^{(*)}(f) > 3\lambda\} \cap \{M(f) \le \gamma\lambda\}\big) \le C_2\,(C_n)^{\varepsilon_0}\,\gamma^{\varepsilon_0}\,(A+B)^{\varepsilon_0}\,w(Q_j)\,.$$

Then a simple summation in j gives (7.4.6) with $C_0 = C_2(C_n)^{\varepsilon_0}$, and recall that C_2 and ε_0 depend on n and $[w]_{A_p}$ and on p if $p < \infty$.

In proving estimate (7.4.7), we may assume that for each cube Q_j there exists a $z_j \in Q_j$ such that $M(f)(z_j) \le \gamma\lambda$; otherwise, the set on the left in (7.4.7) is empty.

We now invoke Theorem 4.2.4 in [131], which states that $T^{(*)}$ maps $L^1(\mathbf{R}^n)$ to $L^{1,\infty}(\mathbf{R}^n)$ with norm at most $C(n)(A+B)$. We have the estimate

$$\big|Q_j \cap \{T^{(*)}(f) > 3\lambda\} \cap \{M(f) \le \gamma\lambda\}\big| \le I_0^\lambda + I_\infty^\lambda, \tag{7.4.8}$$

where

$$\begin{aligned} I_0^\lambda &= \big|Q_j \cap \{T^{(*)}(f_0^j) > \lambda\} \cap \{M(f) \le \gamma\lambda\}\big|, \\ I_\infty^\lambda &= \big|Q_j \cap \{T^{(*)}(f_\infty^j) > 2\lambda\} \cap \{M(f) \le \gamma\lambda\}\big|. \end{aligned}$$

To control I_0^λ we note that f_0^j is in $L^1(\mathbf{R}^n)$ and we argue as follows:

$$\begin{aligned} I_0^\lambda &\le \big|\{T^{(*)}(f_0^j) > \lambda\}\big| \\ &\le \frac{\|T^{(*)}\|_{L^1 \to L^{1,\infty}}}{\lambda} \int_{\mathbf{R}^n} |f_0^j(x)|\,dx \\ &\le C(n)\,(A+B)\,\frac{|Q_j^{**}|}{\lambda}\,\frac{1}{|Q_j^{**}|}\int_{Q_j^{**}} |f(x)|\,dx \\ &\le C(n)\,(A+B)\,\frac{|Q_j^{**}|}{\lambda} M_c(f)(z_j) \\ &\le \widetilde{C}(n)\,(A+B)\,\frac{|Q_j^{**}|}{\lambda} M(f)(z_j) \\ &\le \widetilde{C}(n)\,(A+B)\,\frac{|Q_j^{**}|}{\lambda}\,\lambda\,\gamma \\ &= C_n\,(A+B)\,\gamma\,|Q_j|. \end{aligned} \tag{7.4.9}$$

Next we claim that $I_\infty^\lambda = 0$ if we take γ sufficiently small. We first show that for all $x \in Q_j$ we have

$$\sup_{\varepsilon>0} \big|T^{(\varepsilon)}(f_\infty^j)(x) - T^{(\varepsilon)}(f_\infty^j)(y_j)\big| \le C_{n,\delta}^{(1)} A\, M(f)(z_j). \tag{7.4.10}$$

Indeed, let us fix an $\varepsilon > 0$. We have

$$\begin{aligned} \big|T^{(\varepsilon)}(f_\infty^j)(x) - T^{(\varepsilon)}(f_\infty^j)(y_j)\big| &= \Bigg| \int\limits_{|t-x|>\varepsilon} K(x,t) f_\infty^j(t)\,dt - \int\limits_{|t-y_j|>\varepsilon} K(y_j,t) f_\infty^j(t)\,dt \Bigg| \\ &\le L_1 + L_2 + L_3, \end{aligned}$$

where

$$L_1 = \left| \int\limits_{|t-y_j|>\varepsilon} \big[K(x,t) - K(y_j,t)\big] f_\infty^j(t)\,dt \right|,$$

$$L_2 = \left| \int\limits_{\substack{|t-x|>\varepsilon \\ |t-y_j|\le\varepsilon}} K(x,t) f_\infty^j(t)\,dt \right|,$$

$$L_3 = \left| \int\limits_{\substack{|t-x|\le\varepsilon \\ |t-y_j|>\varepsilon}} K(x,t) f_\infty^j(t)\,dt \right|,$$

in view of identity (5.4.7).

We now make a couple of observations. For $t \notin Q_j^{**}$, $x, z_j \in Q_j$, and $y_j \in Q_j^*$ we have

$$\frac{3}{4} \le \frac{|t-x|}{|t-y_j|} \le \frac{5}{4}, \qquad\qquad \frac{48}{49} \le \frac{|t-x|}{|t-z_j|} \le \frac{50}{49}. \tag{7.4.11}$$

Indeed,

$$|t-y_j| \ge (50n - 5\sqrt{n})\,\ell(Q_j) \ge 44\,n\,\ell(Q_j)$$

and

$$|x-y_j| \le \frac{1}{2}\sqrt{n}\,\ell(Q_j) + \sqrt{n}\,10\sqrt{n}\,\ell(Q_j) \le 11\,n\,\ell(Q_j) \le \frac{1}{4}|t-y_j|.$$

Using this estimate and the inequalities

$$\frac{3}{4}|t-y_j| \le |t-y_j| - |x-y_j| \le |t-x| \le |t-y_j| + |x-y_j| \le \frac{5}{4}|t-y_j|,$$

we obtain the first estimate in (7.4.11). Likewise, we have

$$|x-z_j| \le \sqrt{n}\,\ell(Q_j) \le n\,\ell(Q_j)$$

and

$$|t-z_j| \ge (50n - \tfrac{1}{2})\ell(Q_j) \ge 49\,n\,\ell(Q_j),$$

and these give

$$\frac{48}{49}|t-z_j| \le |t-z_j| - |x-z_j| \le |t-x| \le |t-z_j| + |x-z_j| \le \frac{50}{49}|t-z_j|,$$

yielding the second estimate in (7.4.11).

Since $|x-y_j| \le \frac{1}{2}|t-y_j| \le \frac{1}{2}\max\big(|t-x|, |t-y_j|\big)$, we have

$$|K(x,t) - K(y_j,t)| \le \frac{A|x-y_j|^\delta}{(|t-x| + |t-y_j|)^{n+\delta}} \le C'_{n,\delta} A \frac{\ell(Q_j)^\delta}{|t-z_j|^{n+\delta}};$$

hence, we obtain

$$L_1 \le \int_{|t-z_j|\ge 49n\ell(Q_j)} C'_{n,\delta} A \frac{\ell(Q_j)^\delta}{|t-z_j|^{n+\delta}} |f(t)|\,dt \le C''_{n,\delta} A M(f)(z_j)$$

using Theorem 2.1.10. Using (7.4.11) we deduce

$$L_2 \le \int_{|t-z_j|\le \frac{5}{4}\cdot\frac{49}{48}\varepsilon} \frac{A}{|x-t|^n} \chi_{|t-x|\ge\varepsilon} |f^j_\infty(t)|\,dt \le C'_n A M(f)(z_j).$$

Again using (7.4.11), we obtain

$$L_3 \le \int_{|t-z_j|\le \frac{49}{48}\varepsilon} \frac{A}{|x-t|^n} \chi_{|t-x|\ge\frac{3}{4}\varepsilon} |f^j_\infty(t)|\,dt \le C''_n A M(f)(z_j).$$

This proves (7.4.10) with constant $C^{(1)}_{n,\delta} = C''_{n,\delta} + C'_n + C''_n$.

Having established (7.4.10), we next claim that

$$\sup_{\varepsilon>0} \left| T^{(\varepsilon)}(f^j_\infty)(y_j) \right| \le T^{(*)}(f)(y_j) + C^{(2)}_n A M(f)(z_j). \tag{7.4.12}$$

To prove (7.4.12) we fix a cube Q_j and $\varepsilon > 0$. We let R_j be the smallest number such that

$$Q_j^{**} \subseteq B(y_j, R_j).$$

See Figure 7.2. We consider the following two cases.

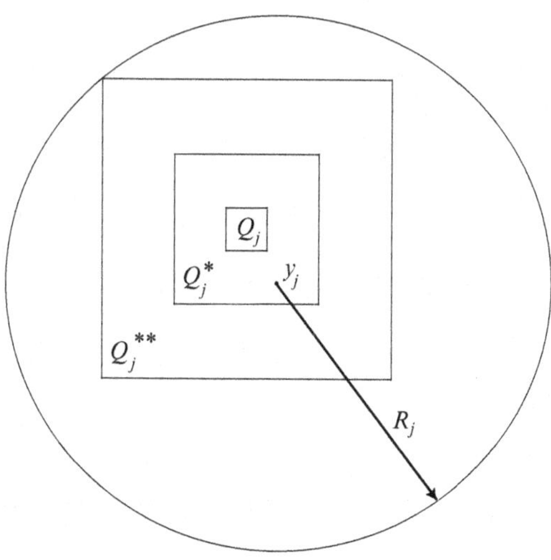

Fig. 7.2 The ball $B(y_j, R_j)$.

Case (1): $\varepsilon \geq R_j$. Since $Q_j^{**} \subseteq B(y_j,\varepsilon)$, we have $B(y_j,\varepsilon)^c \subseteq (Q_j^{**})^c$ and therefore

$$T^{(\varepsilon)}(f_\infty^j)(y_j) = T^{(\varepsilon)}(f)(y_j),$$

so (7.4.12) holds easily in this case.
Case (2): $0 < \varepsilon < R_j$. Note that if $t \in (Q_j^{**})^c$, then $|t-y_j| \geq 40n\ell(Q_j)$. On the other hand, $R_j \leq \operatorname{diam}(Q_j^{**}) = 100n^{\frac{3}{2}}\ell(Q_j)$. This implies that

$$R_j \leq \tfrac{5\sqrt{n}}{2}|t-y_j|, \qquad \text{when} \quad t \in (Q_j^{**})^c.$$

Notice also that in this case we have $B(y_j,R_j)^c \subseteq (Q_j^{**})^c$, hence

$$T^{(R_j)}(f_\infty^j)(y_j) = T^{(R_j)}(f)(y_j).$$

Therefore, we have

$$\begin{aligned}
\big|T^{(\varepsilon)}(f_\infty^j)(y_j)\big| &\leq \big|T^{(\varepsilon)}(f_\infty^j)(y_j) - T^{(R_j)}(f_\infty^j)(y_j)\big| + \big|T^{(R_j)}(f)(y_j)\big| \\
&\leq \int_{\varepsilon \leq |y_j-t| \leq R_j} |K(y_j,t)|\,|f_\infty^j(t)|\,dt + T^{(*)}(f)(y_j) \\
&\leq \int_{\frac{2}{5\sqrt{n}}R_j \leq |y_j-t| \leq R_j} |K(y_j,t)||f_\infty^j(t)|\,dt + T^{(*)}(f)(y_j) \\
&\leq \frac{A(\frac{2}{5\sqrt{n}})^{-n}}{R_j^n} \int_{|z_j-t| \leq \frac{5}{4}\cdot\frac{49}{48}R_j} |f(t)|\,dt + T^{(*)}(f)(y_j) \\
&\leq C_n^{(2)} A M(f)(z_j) + T^{(*)}(f)(y_j),
\end{aligned}$$

where in the penultimate estimate we used (7.4.11). The proof of (7.4.12) follows with the required bound $C_n^{(2)}A$.

Combining (7.4.10) and (7.4.12), we obtain

$$T^{(*)}(f_\infty^j)(x) \leq T^{(*)}(f)(y_j) + \big(C_{n,\delta}^{(1)} + C_n^{(2)}\big) A M(f)(z_j).$$

Recalling that $y_j \notin \Omega$ and that $M(f)(z_j) \leq \gamma\lambda$, we deduce

$$T^{(*)}(f_\infty^j)(x) \leq \lambda + \big(C_{n,\delta}^{(1)} + C_n^{(2)}\big) A \gamma\lambda.$$

Setting $\gamma_0 = \big(C_{n,\delta}^{(1)} + C_n^{(2)}\big)^{-1} A^{-1} = c_0(n,\delta)A^{-1}$, for $0 < \gamma < \gamma_0$, we have that the set

$$Q_j \cap \{T^{(*)}(f_\infty^j) > 2\lambda\} \cap \{M(f) \leq \gamma\lambda\}$$

is empty. This shows that the quantity I_∞^γ vanishes if γ is smaller than γ_0. Returning to (7.4.8) and using the estimate (7.4.9) proved earlier, we conclude the proof of (7.4.7), which, as indicated earlier, implies the theorem. □

Remark 7.4.4. We observe that for any $\delta > 0$, estimate (7.4.6) also holds for the operator

$$T_\delta^{(*)}(f)(x) = \sup_{\varepsilon \geq \delta} |T^{(\varepsilon)}(f)(x)| \tag{7.4.13}$$

with the same constant (which is independent of δ).

To see the validity of (7.4.6) for $T_\delta^{(*)}$, it suffices to prove

$$\left|T_\delta^{(*)}(f_\infty^j)(y_j)\right| \leq T_\delta^{(*)}(f)(y_j) + C_n^{(2)} A M(f)(z_j), \tag{7.4.14}$$

which is a version of (7.4.12) with $T^{(*)}$ replaced by $T_\delta^{(*)}$. The following cases arise: **Case (1′):** $R_j \leq \delta \leq \varepsilon$ or $\delta \leq R_j \leq \varepsilon$. Here, as in Case (1) we have

$$|T^{(\varepsilon)}(f_\infty^j)(y_j)| = |T^{(\varepsilon)}(f)(y_j)| \leq T_\delta^{(*)}(f)(y_j).$$

Case (2′): $\delta \leq \varepsilon < R_j$. As in Case (2) we have

$$T^{(R_j)}(f_\infty^j)(y_j) = T^{(R_j)}(f)(y_j),$$

thus

$$\left|T^{(\varepsilon)}(f_\infty^j)(y_j)\right| \leq \left|T^{(\varepsilon)}(f_\infty^j)(y_j) - T^{(R_j)}(f_\infty^j)(y_j)\right| + \left|T^{(R_j)}(f)(y_j)\right|.$$

As in the proof of Case (2), we bound the first term on the right of the last displayed expression by $C_n^{(2)} A M(f)(z_j)$ while the second term is at most $T_\delta^{(*)}(f)(y_j)$.

7.4.3 Consequences of the Good Lambda Estimate

Having obtained the important good lambda weighted estimate for singular integrals, we now pass to some of its consequences. We begin with the following lemma:

Lemma 7.4.5. *Let $1 \leq p < \infty$, $\varepsilon > 0$, $w \in A_p$, $x \in \mathbf{R}^n$, and $f \in L^p(w)$. Then we have*

$$\int_{|x-y|\geq \varepsilon} \frac{|f(y)|}{|x-y|^n}\, dy \leq C_{00}(w,n,p,x,\varepsilon) \left\|f\right\|_{L^p(w)}$$

for some constant C_{00} depending on the stated parameters. In particular, $T^{(\varepsilon)}(f)$ and $T^{(*)}(f)$ are defined for $f \in L^p(w)$.

Proof. For each $\varepsilon > 0$ and x pick a cube $Q_0 = Q_0(x,\varepsilon)$ of side length $c_n\varepsilon$ (for some constant c_n) such that $Q_0 \subsetneqq B(x,\varepsilon)$. Set $Q_j = 2^j Q_0$ for $j \geq 0$. We have

$$
\begin{aligned}
\int_{|y-x|\geq\varepsilon}\frac{|f(y)|}{|x-y|^n}\,dy &\leq C_n\sum_{j=0}^{\infty}(2^j\varepsilon)^{-n}\int_{Q_{j+1}\setminus Q_j}|f(y)|\,dy\\
&\leq C_n\sum_{j=1}^{\infty}\left(\frac{1}{|Q_j|}\int_{Q_j}|f(y)|^p w\,dy\right)^{\frac{1}{p}}\left(\frac{1}{|Q_j|}\int_{Q_j}w^{-\frac{p'}{p}}\,dy\right)^{\frac{1}{p'}}\\
&\leq C_n\,[w]_{A_p}^{\frac{1}{p}}\sum_{j=1}^{\infty}\left(\int_{Q_j}|f(y)|^p w\,dy\right)^{\frac{1}{p}}\left(\frac{1}{w(Q_j)}\right)^{\frac{1}{p}}\\
&\leq C_n\,[w]_{A_p}^{\frac{1}{p}}\,\big\|f\big\|_{L^p(w)}\sum_{j=1}^{\infty}\big(w(Q_j)\big)^{-\frac{1}{p}}.
\end{aligned}
$$

But Proposition 7.2.8 gives for some $\delta=\delta(n,p,[w]_{A_p})$ that

$$
\frac{w(Q_0)}{w(Q_j)}\leq C(n,p,[w]_{A_p})\frac{|Q_0|^\delta}{|Q_j|^\delta},
$$

from which it follows that

$$
w(Q_j)^{-\frac{1}{p}}\leq C'(n,p,[w]_{A_p})\,2^{-j\frac{n\delta}{p}}\,w(Q_0)^{-\frac{1}{p}}.
$$

In view of this estimate, the previous series converges. Note that C' and thus C_{00} depend on $[w]_{A_p},n,p,x,\varepsilon$, and $w(Q_0)$.

This argument is also valid in the case $p=1$ by an obvious modification. □

Theorem 7.4.6. *Let $A,B,\beta>0$ and let T be a $CZO(\beta,A,B)$. Then given $1<p<\infty$, there is a constant $C_p=C_p(n,\beta,[w]_{A_p})$ such that*

$$
\big\|T^{(*)}(f)\big\|_{L^p(w)}\leq C_p\,(A+B)\big\|f\big\|_{L^p(w)} \tag{7.4.15}
$$

for all $w\in A_p$ and $f\in L^p(w)$. There is also a constant $C_1=C_1(n,\beta,[w]_{A_1})$ such that

$$
\big\|T^{(*)}(f)\big\|_{L^{1,\infty}(w)}\leq C_1\,(A+B)\,\big\|f\big\|_{L^1(w)} \tag{7.4.16}
$$

for all $w\in A_1$ and $f\in L^1(w)$.

Proof. This theorem is a consequence of the estimate proved in the previous theorem. For technical reasons, it is useful to fix a $\delta>0$ and work with the auxiliary maximal operator $T_\delta^{(*)}$ defined in (7.4.13) instead of $T^{(*)}$. We begin by taking $1<p<\infty$ and $f\in L^p(w)$ for some $w\in A_p$. We write

$$
\begin{aligned}
\|T_\delta^{(*)}(f)\|_{L^p(w)}^p &= \int_0^\infty p\lambda^{p-1}w(\{T_\delta^{(*)}(f)>\lambda\})\,d\lambda\\
&= 3^p\int_0^\infty p\lambda^{p-1}w(\{T_\delta^{(*)}(f)>3\lambda\})\,d\lambda,
\end{aligned}
$$

which we control by

$$3^p \int_0^\infty p\lambda^{p-1} w(\{T_\delta^{(*)}(f) > 3\lambda\} \cap \{M(f) \le \gamma\lambda\})\, d\lambda \\ + 3^p \int_0^\infty p\lambda^{p-1} w(\{M(f) > \gamma\lambda\})\, d\lambda .$$

Using Theorem 7.4.3 (or rather Remark 7.4.4), there are $C_0 = C_0(n, [w]_{A_p})$, $\varepsilon_0 = \varepsilon_0(n, [w]_{A_p})$, and $\gamma_0 = c_0(n,\beta)A^{-1}$, such that the preceding displayed expression is bounded by

$$3^p C_0 \gamma^{\varepsilon_0} (A+B)^{\varepsilon_0} \int_0^\infty p\lambda^{p-1} w(\{T_\delta^{(*)}(f) > \lambda\})\, d\lambda \\ + \frac{3^p}{\gamma^p} \int_0^\infty p\lambda^{p-1} w(\{M(f) > \lambda\})\, d\lambda ,$$

which is equal to

$$3^p C_0 \gamma^{\varepsilon_0} (A+B)^{\varepsilon_0} \big\| T_\delta^{(*)}(f) \big\|_{L^p(w)}^p + \frac{3^p}{\gamma^p} \big\| M(f) \big\|_{L^p(w)}^p .$$

Taking $\gamma = \min\big(\frac{1}{2} c_0(n,\beta) A^{-1}, \frac{1}{2}(2C_0 3^p)^{-\frac{1}{\varepsilon_0}} (A+B)^{-1}\big) < \gamma_0$, we conclude that

$$\big\| T_\delta^{(*)}(f) \big\|_{L^p(w)}^p \\ \le \frac{1}{2} \big\| T_\delta^{(*)}(f) \big\|_{L^p(w)}^p + \widetilde{C}_p(n,\beta,[w]_{A_p})(A+B)^p \big\| M(f) \big\|_{L^p(w)}^p . \tag{7.4.17}$$

We now prove a similar estimate when $p = 1$. For $f \in L^1(w)$ and $w \in A_1$ we have

$$3\lambda w(\{T_\delta^{(*)}(f) > 3\lambda\}) \\ \le 3\lambda w(\{T_\delta^{(*)}(f) > 3\lambda\} \cap \{M(f) \le \gamma\lambda\}) + 3\lambda w(\{M(f) > \gamma\lambda\}) ,$$

and this expression is controlled by

$$3\lambda C_0 \gamma^{\varepsilon_0} (A+B)^{\varepsilon_0} w(\{T_\delta^{(*)}(f) > \lambda\}) + \frac{3}{\gamma} \big\| M(f) \big\|_{L^{1,\infty}(w)} .$$

Recalling that $\gamma_0 = c_0(n,\beta)A^{-1}$ and choosing $\gamma = \min\big(\frac{1}{2}\gamma_0, \frac{1}{2}(6C_0)^{-\frac{1}{\varepsilon_0}}(A+B)^{-1}\big)$, it follows that

$$\big\| T_\delta^{(*)}(f) \big\|_{L^{1,\infty}(w)} \\ \le \frac{1}{2} \big\| T_\delta^{(*)}(f) \big\|_{L^{1,\infty}(w)} + \widetilde{C}_1(n,\beta,[w]_{A_1})(A+B) \big\| M(f) \big\|_{L^{1,\infty}(w)} . \tag{7.4.18}$$

Estimate (7.4.15) would follow from (7.4.17) if we knew that $\|T_\delta^{(*)}(f)\|_{L^p(w)} < \infty$ whenever $1 < p < \infty$, $w \in A_p$ and $f \in L^p(w)$, while (7.4.16) would follow from (7.4.18) if we had $\|T_\delta^{(*)}(f)\|_{L^{1,\infty}(w)} < \infty$ whenever $w \in A_1$ and $f \in L^1(w)$. Since we do not know that these quantities are finite, a certain amount of work is needed.

To deal with this problem we momentarily restrict attention to a special class of functions on $\mathbf{R}^n$, the class of bounded functions with compact support. Such functions are dense in $L^p(w)$ when $w \in A_p$ and $1 \le p < \infty$; see Exercise 7.4.1. Let h be a bounded function with compact support on $\mathbf{R}^n$. Then $T_\delta^{(*)}(h) \le C_1 \delta^{-n} \|h\|_{L^1}$ and $T_\delta^{(*)}(h)(x) \le C_2(h)|x|^{-n}$ for x away from the support of h. It follows that

$$T_\delta^{(*)}(h)(x) \le C_3(h,\delta)(1+|x|)^{-n}$$

for all $x \in \mathbf{R}^n$. Furthermore, if h is nonzero, then

$$M(h)(x) \ge \frac{C_4(h)}{(1+|x|)^n},$$

and therefore for $w \in A_1$,

$$\big\|T_\delta^{(*)}(h)\big\|_{L^{1,\infty}(w\,dx)} \le C_5(h,\delta)\big\|M(h)\big\|_{L^{1,\infty}(w\,dx)} < \infty,$$

while for $1 < p < \infty$ and $w \in A_p$,

$$\int_{\mathbf{R}^n} (T_\delta^{(*)}(h)(x))^p w(x)\,dx \le C_5(h,p,\delta) \int_{\mathbf{R}^n} M(h)(x)^p w(x)\,dx < \infty$$

in view of Theorem 7.1.9. Using these facts, (7.4.17), (7.4.18), and Theorem 7.1.9 once more, we conclude that for all $\delta > 0$ and $1 < p < \infty$ we have

$$\begin{aligned} \big\|T_\delta^{(*)}(h)\big\|_{L^p(w)}^p &\le 2\widetilde{C}_p \big\|M(h)\big\|_{L^p(w)}^p \le \widetilde{C'}_p [w]_{A_p}^{\frac{p}{p-1}} \big\|h\big\|_{L^p(w)}^p = C_p^p \big\|h\big\|_{L^p(w)}^p, \\ \big\|T_\delta^{(*)}(h)\big\|_{L^{1,\infty}(w)} &\le 2\widetilde{C}_1 \big\|M(h)\big\|_{L^{1,\infty}(w)} \le \widetilde{C}_1 [w]_{A_1} \big\|h\big\|_{L^1(w)} = C_1 \big\|h\big\|_{L^1(w)}, \end{aligned} \tag{7.4.19}$$

whenever h a bounded function with compact support. The constants $\widetilde{C}_p$, $\widetilde{C'}_p$, and C_p depend only on the parameters n, β, p, and $[w]_{A_p}$.

We now extend estimates (7.4.16) and (7.4.15) to functions in $L^p(\mathbf{R}^n, w\,dx)$. Given $1 \le p < \infty$, $w \in A_p$, and $f \in L^p(w)$, let

$$f_N(x) = f(x)\chi_{|f| \le N}\chi_{|x| \le N}.$$

Then f_N is a bounded function with compact support that converges to f in $L^p(w)$ (i.e., $\|f_N - f\|_{L^p(w)} \to 0$ as $N \to \infty$) by the Lebesgue dominated convergence theorem. Also $|f_N| \le |f|$ for all N. Sublinearity and Lemma 7.4.5 give for all $x \in \mathbf{R}^n$,

$$\begin{aligned} |T_\delta^{(*)}(f_N)(x) - T_\delta^{(*)}(f)(x)| &\le T_\delta^{(*)}(f - f_N)(x) \\ &\le A\,C_{00}(w,n,p,x,\delta)\,\big\|f_N - f\big\|_{L^p(w)}, \end{aligned}$$

and this converges to zero as $N \to \infty$ since $C_{00}(w,n,p,x,\delta) < \infty$. Therefore

$$T_\delta^{(*)}(f) = \lim_{N\to\infty} T_\delta^{(*)}(f_N)$$

pointwise, and Fatou's lemma for weak type spaces [see Exercise 1.1.12 (d)] gives for $w \in A_1$ and $f \in L^1(w)$,

$$\begin{aligned}\big\|T_\delta^{(*)}(f)\big\|_{L^{1,\infty}(w)} &= \big\|\liminf_{N\to\infty} T_\delta^{(*)}(f_N)\big\|_{L^{1,\infty}(w)} \\ &\le \liminf_{N\to\infty} \big\|T_\delta^{(*)}(f_N)\big\|_{L^{1,\infty}(w)} \\ &\le C_1 \liminf_{N\to\infty} \big\|M(f_N)\big\|_{L^{1,\infty}(w)} \\ &\le C_1 \big\|M(f)\big\|_{L^{1,\infty}(w)},\end{aligned}$$

since $|f_N| \le |f|$ for all N. An analogous argument gives the estimate

$$\big\|T_\delta^{(*)}(f)\big\|_{L^p(w)} \le C_p \big\|f\big\|_{L^p(w)}$$

for $w \in A_p$ and $f \in L^p(w)$ when $1 < p < \infty$.

It remains to prove (7.4.15) and (7.4.16) for $T^{(*)}$. But this is also an easy consequence of Fatou's lemma, since the constants C_p and C_1 are independent of δ and

$$\lim_{\delta\to 0} T_\delta^{(*)}(f) = T^{(*)}(f)$$

for all $f \in L^p(w)$. □

We end this subsection by making the comment that if a given T in $CZO(\delta,A,B)$ is pointwise controlled by $T^{(*)}$, then the estimates of Theorem 7.4.6 also hold for it. This is the case for the Hilbert transform, the Riesz transforms, and other classical singular integral operators.

7.4.4 Necessity of the A_p Condition

We have established the main theorems relating Calderón–Zygmund operators and A_p weights, namely that such operators are bounded on $L^p(w)$ whenever w lies in A_p. It is natural to ask whether the A_p condition is necessary for the boundedness of singular integrals on L^p. We end this section by indicating the necessity of the A_p condition for the boundedness of the Riesz transforms on weighted L^p spaces.

Theorem 7.4.7. *Let w be a weight in $\mathbf{R}^n$ and let $1 \le p < \infty$. Suppose that each of the Riesz transforms R_j is of weak type (p,p) with respect to w. Then w must be an A_p weight. Similarly, let w be a weight in $\mathbf{R}$. If the Hilbert transform H is of weak type (p,p) with respect to w, then w must be an A_p weight.*

Proof. We prove the n-dimensional case, $n \geq 2$. The one-dimensional case is essentially contained in following argument, suitably adjusted.

Let Q be a cube and let f be a nonnegative function on $\mathbf{R}^n$ supported in Q that satisfies $\operatorname{Avg}_Q f > 0$. Let Q' be the cube that shares a corner with Q, has the same length as Q, and satisfies $x_j \geq y_j$ for all $1 \leq j \leq n$ whenever $x \in Q'$ and $y \in Q$. Then for $x \in Q'$ we have

$$\left|\sum_{j=1}^{n} R_j(f)(x)\right| = \frac{\Gamma(\frac{n+1}{2})}{\pi^{\frac{n+1}{2}}} \sum_{j=1}^{n} \int_Q \frac{x_j - y_j}{|x-y|^{n+1}} f(y)\,dy \geq \frac{\Gamma(\frac{n+1}{2})}{\pi^{\frac{n+1}{2}}} \int_Q \frac{f(y)}{|x-y|^n}\,dy.$$

But if $x \in Q'$ and $y \in Q$ we must have that $|x-y| \leq 2\sqrt{n}\,\ell(Q)$, which implies that $|x-y|^{-n} \geq (2\sqrt{n})^{-n}|Q|^{-1}$. Let $C_n = \Gamma(\frac{n+1}{2})(2\sqrt{n})^{-n}\pi^{-\frac{n+1}{2}}$. It follows that for all $0 < \alpha < C_n \operatorname{Avg}_Q f$ we have

$$Q' \subseteq \Big\{x \in \mathbf{R}^n : \big|\sum_{j=1}^{n} R_j(f)(x)\big| > \alpha\Big\}.$$

Since the operator $\sum_{j=1}^{n} R_j$ is of weak type (p,p) with respect to w (with constant C), we must have

$$w(Q') \leq \frac{C^p}{\alpha^p} \int_Q f(x)^p w(x)\,dx$$

for all $\alpha < C_n \operatorname{Avg}_Q f$, which implies that

$$\big(\operatorname{Avg}_Q f\big)^p \leq \frac{C_n^{-p}C^p}{w(Q')} \int_Q f(x)^p w(x)\,dx. \tag{7.4.20}$$

We observe that we can reverse the roles of Q and Q' and obtain

$$\big(\operatorname{Avg}_{Q'} g\big)^p \leq \frac{C_n^{-p}C^p}{w(Q)} \int_{Q'} g(x)^p w(x)\,dx \tag{7.4.21}$$

for all g supported in Q'. In particular, taking $g = \chi_{Q'}$ in (7.4.21) gives that

$$w(Q) \leq C_n^{-p}C^p w(Q').$$

Using this estimate and (7.4.20), we obtain

$$\big(\operatorname{Avg}_Q f\big)^p \leq \frac{(C_n^{-p}C^p)^2}{w(Q)} \int_Q f(x)^p w(x)\,dx. \tag{7.4.22}$$

Using the characterization of the A_p characteristic constant in Proposition 7.1.5 (8), it follows that

$$[w]_{A_p} \leq (C_n^{-p}C^p)^2 < \infty;$$

hence $w \in A_p$. □

Exercises

7.4.1. Let $1 \le p < \infty$ and let $w \in L^1_{\mathrm{loc}}(\mathbf{R}^n)$ satisfy $w > 0$ a.e. Show that $\mathscr{C}_0^\infty(\mathbf{R}^n)$ is dense in $L^p(w)$. In particular this assertion holds for any $w \in A_\infty$.

7.4.2. ([74]) Let T be in $CZO(\delta,A,B)$. Show that for all $\varepsilon > 0$ and all $1 < p < \infty$ there exists a constant $C_{n,p,\varepsilon,\delta}$ such that for all $f \in L^p(\mathbf{R}^n)$ and for all measurable nonnegative functions u with $u^{1+\varepsilon} \in L^1_{\mathrm{loc}}(\mathbf{R}^n)$ and $M(u^{1+\varepsilon}) < \infty$ a.e. we have

$$\int_{\mathbf{R}^n} |T^{(*)}(f)|^p u\,dx \le C_{n,p,\varepsilon,\delta}(A+B)^p \int_{\mathbf{R}^n} |f|^p M(u^{1+\varepsilon})^{\frac{1}{1+\varepsilon}}\,dx.$$

[*Hint:* Obtain this result as a consequence of Theorems 7.4.6 and 7.2.7.]

7.4.3. Use the idea of the proof of Theorem 7.4.6 to prove the following result. Suppose that for some fixed $A,B > 0$ the nonnegative μ-measurable functions F and G on a σ-finite measure space (X,μ) satisfy the distributional inequality

$$\mu(\{G > \alpha\} \cap \{F \le c\alpha\}) \le A\,\mu(\{G > B\alpha\})$$

for all $\alpha > 0$. Given $0 < p < \infty$, if $A < B^p$ and $\|G\|_{L^p(\mu)} < \infty$, show that

$$\|G\|_{L^p(\mu)} \le \frac{B}{(B^p - A)^{1/p}}\frac{1}{c}\,\|F\|_{L^p(\mu)}\,.$$

7.4.4. Let $\alpha > 0$, $w \in A_1$, and $f \in L^1(\mathbf{R}^n,w) \cap L^1(\mathbf{R}^n)$. Let $f = g + b$ be the Calderón–Zygmund decomposition of f at height $\alpha > 0$ given in Theorem 5.3.1, such that $b = \sum_j b_j$, where each b_j is supported in a dyadic cube Q_j, $\int_{Q_j} b_j(x)dx = 0$, and Q_j and Q_k have disjoint interiors when $j \ne k$. Prove that

(a) $\|g\|_{L^1(w)} \le [w]_{A_1}\|f\|_{L^1(w)}$ and $\|g\|_{L^\infty(w)} = \|g\|_{L^\infty} \le 2^n\alpha$,

(b) $\|b_j\|_{L^1(w)} \le (1+[w]_{A_1})\|f\|_{L^1(Q_j,w)}$ and $\|b\|_{L^1(w)} \le (1+[w]_{A_1})\|f\|_{L^1(w)}$,

(c) $\sum_j w(Q_j) \le \frac{[w]_{A_1}}{\alpha}\|f\|_{L^1(w)}$.

7.4.5. Assume that T is an operator associated with a kernel in $SK(\delta,A)$. Suppose that T maps $L^2(w)$ to $L^2(w)$ for all $w \in A_1$ with bound B_w. Prove that there is a constant $C_{n,\delta}$ such that

$$\|T\|_{L^1(w)\to L^{1,\infty}(w)} \le C_{n,\delta}(A+B_w)\,[w]_{A_1}^2$$

for all $w \in A_1$.
[*Hint:* Apply the idea of the proof of Theorem 5.3.3 using the Calderón-Zygmund decomposition $f = g + b$ of Exercise 7.4.4 at height $\gamma\alpha$ for a suitable γ. To estimate

$T(g)$ use an $L^2(w)$ estimate and Exercise 7.4.4. To estimate $T(b)$ use the mean value property, the fact that

$$\int_{\mathbf{R}^n \setminus Q_j^*} \frac{|y-c_j|^\delta}{|x-c_j|^{n+\delta}} w(x)\,dx \le C_{\delta,n} M(w)(y) \le C'_{\delta,n} [w]_{A_1} w(y),$$

and Exercise 7.4.4 to obtain the required estimate.]

7.4.6. Recall that the transpose T^t of a linear operator T is defined by

$$\langle T(f), g\rangle = \langle f, T^t(g)\rangle$$

for all suitable f and g. Suppose that T is a linear operator that maps $L^p(\mathbf{R}^n, v\,dx)$ to itself for some $1 < p < \infty$ and some $v \in A_p$. Show that the transpose operator T^t maps $L^{p'}(\mathbf{R}^n, w\,dx)$ to itself with the same norm, where $w = v^{1-p'} \in A_{p'}$.

7.4.7. Suppose that T is a linear operator that maps $L^2(\mathbf{R}^n, v\,dx)$ to itself for all v such that $v^{-1} \in A_1$. Show that the transpose operator T^t of T maps $L^2(\mathbf{R}^n, w\,dx)$ to itself for all $w \in A_1$.

7.4.8. Let $1 < p < \infty$. Suppose that T is a linear operator that maps $L^p(v)$ to itself for all v satisfying $v^{-1} \in A_p$. Show that the transpose operator T^t of T maps $L^{p'}(w)$ to itself for all w satisfying $w^{-1} \in A_{p'}$.

7.5 Further Properties of A_p Weights

In this section we discuss other properties of A_p weights. Many of these properties indicate deep connections with other branches of analysis. We focus attention on three such properties: factorization, extrapolation, and relations of weighted inequalities to vector-valued inequalities.

7.5.1 Factorization of Weights

Recall the simple fact that if w_1, w_2 are A_1 weights, then $w = w_1 w_2^{1-p}$ is an A_p weight (Exercise 7.1.2). The factorization theorem for weights says that the converse of this statement is true. This provides a surprising and striking representation of A_p weights.

Theorem 7.5.1. *Suppose that w is an A_p weight for some $1 < p < \infty$. Then there exist A_1 weights w_1 and w_2 such that*

$$w = w_1 w_2^{1-p}.$$

Proof. Let us fix a $p \ge 2$ and $w \in A_p$. We define an operator T as follows:

$$T(f) = \left(w^{-\frac{1}{p}} M(f^{p-1} w^{\frac{1}{p}})\right)^{\frac{1}{p-1}} + w^{\frac{1}{p}} M(f w^{-\frac{1}{p}}),$$

where M is the Hardy–Littlewood maximal operator. We observe that T is well defined and bounded on $L^p(\mathbf{R}^n)$. This is a consequence of the facts that $w^{-\frac{1}{p-1}}$ is an $A_{p'}$ weight and that M maps $L^{p'}(w^{-\frac{1}{p-1}})$ to itself and also $L^p(w)$ to itself. Thus the norm of T on L^p depends only on the A_p characteristic constant of w. Let $B(w) = \|T\|_{L^p \to L^p}$, the norm of T on L^p. Next, we observe that for $f, g \ge 0$ in $L^p(\mathbf{R}^n)$ and $\lambda \ge 0$ we have

$$T(f+g) \le T(f) + T(g), \quad T(\lambda f) = \lambda T(f). \tag{7.5.1}$$

To see the first assertion, we need only note that for every ball B, the operator

$$f \to \left(\frac{1}{|B|} \int_B |f|^{p-1} w^{\frac{1}{p}} \, dx \right)^{\frac{1}{p-1}}$$

is sublinear as a consequence of Minkowski's integral inequality, since $p-1 \ge 1$.

We now fix an L^p function f_0 with $\|f_0\|_{L^p} = 1$ and we define a function φ in $L^p(\mathbf{R}^n)$ as the sum of the L^p convergent series

$$\varphi = \sum_{j=1}^{\infty} (2B(w))^{-j} T^j(f_0). \tag{7.5.2}$$

We define

$$w_1 = w^{\frac{1}{p}} \varphi^{p-1}, \qquad w_2 = w^{-\frac{1}{p}} \varphi,$$

so that $w = w_1 w_2^{1-p}$. It remains to show that w_1, w_2 are A_1 weights. Applying T and using (7.5.1), we obtain

$$\begin{aligned} T(\varphi) &\le 2B(w) \sum_{j=1}^{\infty} (2B(w))^{-j-1} T^{j+1}(f_0) \\ &= 2B(w) \left(\varphi - \frac{T(f_0)}{2B(w)} \right) \\ &\le 2B(w) \, \varphi, \end{aligned}$$

that is,

$$\left(w^{-\frac{1}{p}} M(\varphi^{p-1} w^{\frac{1}{p}})\right)^{\frac{1}{p-1}} + w^{\frac{1}{p}} M(\varphi w^{-\frac{1}{p}}) \le 2B(w) \, \varphi .$$

Using that $\varphi = (w^{-\frac{1}{p}} w_1)^{\frac{1}{p-1}} = w^{\frac{1}{p}} w_2$, we obtain

$$M(w_1) \le (2B(w))^{p-1} w_1 \qquad \text{and} \qquad M(w_2) \le 2B(w) w_2 .$$

These show that w_1 and w_2 are A_1 weights whose characteristic constants depend on $[w]_{A_p}$ (and also the dimension n and p). This concludes the case $p \geq 2$.

We now turn to the case $p < 2$. Given a weight $w \in A_p$ for $1 < p < 2$, we consider the weight $w^{-1/(p-1)}$, which is in $A_{p'}$. Since $p' > 2$, using the result we obtained, we write $w^{-1/(p-1)} = v_1 v_2^{1-p'}$, where v_1, v_2 are A_1 weights. It follows that $w = v_1^{1-p} v_2$, and this completes the asserted factorization of A_p weights. □

Combining the result just obtained with Theorem 7.2.7, we obtain the following description of A_p weights.

Corollary 7.5.2. *Let w be an A_p weight for some $1 < p < \infty$. Then there exist locally integrable functions f_1 and f_2 with*

$$M(f_1) + M(f_2) < \infty \qquad a.e.,$$

constants $0 < \varepsilon_1, \varepsilon_2 < 1$, and a nonnegative function k satisfying $k, k^{-1} \in L^\infty$ such that

$$w = k M(f_1)^{\varepsilon_1} M(f_2)^{\varepsilon_2(1-p)}. \tag{7.5.3}$$

7.5.2 Extrapolation from Weighted Estimates on a Single L^{p_0}

Our next topic concerns a striking application of the class of A_p weights. It says that an estimate on $L^{p_0}(v)$ for a single p_0 and all A_{p_0} weights v implies a similar L^p estimate for all p in $(1,\infty)$. This property is referred to as extrapolation.

Surprisingly the operator T is not needed to be linear or sublinear in the following extrapolation theorem. The only condition required is that T be well defined on $\bigcup_{1\leq q<\infty} \bigcup_{w\in A_q} L^q(w)$. If T happens to be a linear operator, this condition can be relaxed to T being well defined on $\mathscr{C}_0^\infty(\mathbf{R}^n)$.

Theorem 7.5.3. *Suppose that T is defined on $\bigcup_{1\leq q<\infty} \bigcup_{w\in A_q} L^q(w)$ and takes values in the space of measurable complex-valued functions. Let $1 \leq p_0 < \infty$ and suppose that there exists a positive increasing function N on $[1,\infty)$ such that for all weights v in A_{p_0} we have*

$$\big\|T\big\|_{L^{p_0}(v)\to L^{p_0}(v)} \leq N\big([v]_{A_{p_0}}\big). \tag{7.5.4}$$

Then for any $1 < p < \infty$ and for all weights w in A_p we have

$$\big\|T\big\|_{L^p(w)\to L^p(w)} \leq K\big(n,p,p_0,[w]_{A_p}\big), \tag{7.5.5}$$

where

$$K\big(n,p,p_0,[w]_{A_p}\big) = \begin{cases} 2N\Big(\kappa_1(n,p,p_0)\,[w]_{A_p}^{\frac{p_0-1}{p-1}}\Big) & \text{when } p < p_0, \\[2ex] 2^{\frac{p-p_0}{p_0(p-1)}} N\big(\kappa_2(n,p,p_0)\,[w]_{A_p}\big) & \text{when } p > p_0, \end{cases}$$

and $\kappa_1(n,p,p_0)$ and $\kappa_2(n,p,p_0)$ are constants that depend on n, p, and p_0.

Proof. Let $1 < p < \infty$ and $w \in A_p$. We define an operator

$$M'(f) = \frac{M(fw)}{w},$$

where M is the Hardy–Littlewood maximal operator. We observe that since $w^{1-p'}$ is in $A_{p'}$, the operator M' maps $L^{p'}(w)$ to itself; indeed, we have

$$\begin{aligned} \big\|M'\big\|_{L^{p'}(w)\to L^{p'}(w)} &= \big\|M\big\|_{L^{p'}(w^{1-p'})\to L^{p'}(w^{1-p'})} \\ &\le C_{n,p}[w^{1-p'}]_{A_{p'}}^{\frac{1}{p'-1}} \\ &= C_{n,p}[w]_{A_p} \end{aligned} \tag{7.5.6}$$

in view of Theorem 7.1.9 and property (4) of Proposition 7.1.5.

We introduce operators $M^0(f) = |f|$ and $M^k = M \circ M \circ \cdots \circ M$, where M is the Hardy–Littlewood maximal function and the composition is taken k times. Likewise, we introduce powers $(M')^k$ of M' for $k \in \mathbf{Z}^+ \cup \{0\}$. The following lemma provides the main tool in the proof of Theorem 7.5.3. Its simple proof uses Theorem 7.1.9 and (7.5.6) and is omitted.

Lemma 7.5.4. *Let $1 < p < \infty$ and $w \in A_p$. Define operators R and R'*

$$R(f) = \sum_{k=0}^{\infty} \frac{M^k(f)}{\big(2\|M\|_{L^p(w)\to L^p(w)}\big)^k}$$

for functions f in $L^p(w)$ and also

$$R'(f) = \sum_{k=0}^{\infty} \frac{(M')^k(f)}{\big(2\big\|M'\big\|_{L^{p'}(w)\to L^{p'}(w)}\big)^k}$$

for functions f in $L^{p'}(w)$. Then there exist constants $C_1(n,p)$ and $C_2(n,p)$ that depend on n and p such that

$$|f| \le R(f), \tag{7.5.7}$$

$$\big\|R(f)\big\|_{L^p(w)} \le 2\big\|f\big\|_{L^p(w)}, \tag{7.5.8}$$

$$M(R(f)) \le C_1(n,p)\,[w]_{A_p}^{\frac{1}{p-1}}\,R(f), \tag{7.5.9}$$

for all functions f in $L^p(w)$ and such that

$$|h| \le R'(h), \tag{7.5.10}$$

$$\big\|R'(h)\big\|_{L^{p'}(w)} \le 2\big\|h\big\|_{L^{p'}(w)}, \tag{7.5.11}$$

$$M'(R'(h)) \le C_2(n,p)\,[w]_{A_p}\,R'(h), \tag{7.5.12}$$

for all functions h in $L^{p'}(w)$.

We now proceed with the proof of the theorem. It is natural to split the proof into the cases $p < p_0$ and $p > p_0$.

Case (1): $p < p_0$. Assume momentarily that $R(f)^{-\frac{p_0}{(p_0/p)'}}$ is an A_{p_0} weight. Then we have

$$\begin{aligned}
&\big\|T(f)\big\|_{L^p(w)}^p \\
&= \int_{\mathbf{R}^n} |T(f)|^p R(f)^{-\frac{p}{(p_0/p)'}} R(f)^{\frac{p}{(p_0/p)'}}\, w\,dx \\
&\le \bigg(\int_{\mathbf{R}^n} |T(f)|^{p_0} R(f)^{-\frac{p_0}{(p_0/p)'}}\, w\,dx\bigg)^{\frac{p}{p_0}} \bigg(\int_{\mathbf{R}^n} R(f)^p\, w\,dx\bigg)^{\frac{1}{(p_0/p)'}} \\
&\le N\Big(\big[R(f)^{-\frac{p_0}{(p_0/p)'}}\big]_{A_{p_0}}\Big)^p \bigg(\int_{\mathbf{R}^n} |f|^{p_0} R(f)^{-\frac{p_0}{(p_0/p)'}}\, w\,dx\bigg)^{\frac{p}{p_0}} \bigg(\int_{\mathbf{R}^n} R(f)^p\, w\,dx\bigg)^{\frac{1}{(p_0/p)'}} \\
&\le N\Big(\big[R(f)^{-\frac{p_0}{(p_0/p)'}}\big]_{A_{p_0}}\Big)^p \bigg(\int_{\mathbf{R}^n} R(f)^{p_0} R(f)^{-\frac{p_0}{(p_0/p)'}}\, w\,dx\bigg)^{\frac{p}{p_0}} \bigg(\int_{\mathbf{R}^n} R(f)^p\, w\,dx\bigg)^{\frac{1}{(p_0/p)'}} \\
&= N\Big(\big[R(f)^{-\frac{p_0}{(p_0/p)'}}\big]_{A_{p_0}}\Big)^p \bigg(\int_{\mathbf{R}^n} R(f)^p w\,dx\bigg)^{\frac{p}{p_0}} \bigg(\int_{\mathbf{R}^n} R(f)^p\, w\,dx\bigg)^{\frac{1}{(p_0/p)'}} \\
&\le N\Big(\big[R(f)^{-\frac{p_0}{(p_0/p)'}}\big]_{A_{p_0}}\Big)^p \big(2\big\|f\big\|_{L^p(w)}\big)^p,
\end{aligned}$$

where we used Hölder's inequality with exponents p_0/p and $(p_0/p)'$, the hypothesis of the theorem, (7.5.7), and (7.5.8). Thus, we have the estimate

$$\big\|T(f)\big\|_{L^p(w)} \le 2N\Big(\big[R(f)^{-\frac{p_0}{(p_0/p)'}}\big]_{A_{p_0}}\Big)\big\|f\big\|_{L^p(w)} \tag{7.5.13}$$

and it remains to obtain a bound for the A_{p_0} characteristic constant of $R(f)^{-\frac{p_0}{(p_0/p)'}}$. In view of (7.5.9), the function $R(f)$ is an A_1 weight with characteristic constant at most a constant multiple of $[w]_{A_p}^{\frac{1}{p-1}}$. Consequently, there is a constant C_1' such that

$$R(f)^{-1} \le C_1' \,[w]_{A_p}^{\frac{1}{p-1}} \bigg(\frac{1}{|Q|}\int_Q R(f)\,dx\bigg)^{-1}$$

for any cube Q in $\mathbf{R}^n$. Thus we have

$$\begin{aligned}
&\frac{1}{|Q|}\int_Q R(f)^{-\frac{p_0}{(p_0/p)'}}\, w\,dx \\
&\qquad \le \big(C_1' \,[w]_{A_p}^{\frac{1}{p-1}}\big)^{\frac{p_0}{(p_0/p)'}} \bigg(\frac{1}{|Q|}\int_Q R(f)\,dx\bigg)^{-\frac{p_0}{(p_0/p)'}} \bigg(\frac{1}{|Q|}\int_Q w\,dx\bigg).
\end{aligned} \tag{7.5.14}$$

Next we have

$$
\begin{aligned}
&\left(\frac{1}{|Q|}\int_Q \left(R(f)^{-\frac{p_0}{(p_0/p)'}}w\right)^{1-p_0'}dx\right)^{p_0-1}\\
&\quad=\left(\frac{1}{|Q|}\int_Q R(f)^{\frac{p_0(p_0'-1)}{(p_0/p)'}}w^{1-p_0'}\,dx\right)^{p_0-1} \qquad (7.5.15)\\
&\quad\le\left(\frac{1}{|Q|}\int_Q R(f)\,dx\right)^{\frac{p_0}{(p_0/p)'}}\left(\frac{1}{|Q|}\int_Q w^{1-p'}\right)^{p-1},
\end{aligned}
$$

where we applied Hölder's inequality with exponents

$$\left(\frac{p'-1}{p_0'-1}\right)' \qquad \text{and} \qquad \frac{p'-1}{p_0'-1},$$

and we used that

$$\frac{p_0(p_0'-1)}{(p_0/p)'}\left(\frac{p'-1}{p_0'-1}\right)'=1 \qquad \text{and} \qquad \frac{p_0-1}{\left(\frac{p'-1}{p_0'-1}\right)'}=\frac{p_0}{(p_0/p)'}.$$

Multiplying (7.5.14) by (7.5.15) and taking the supremum over all cubes Q in $\mathbf{R}^n$ we deduce that

$$\left[R(f)^{-\frac{p_0}{(p_0/p)'}}\right]_{A_{p_0}}\le\left(C_1'\,[w]_{A_p}^{\frac{1}{p-1}}\right)^{\frac{p_0}{(p_0/p)'}}[w]_{A_p}=\kappa_1(n,p,p_0)\,[w]_{A_p}^{\frac{p_0-1}{p-1}}.$$

Combining this estimate with (7.5.13) and using the fact that N is an increasing function, we obtain the validity of (7.5.5) in the case $p<p_0$.

Case (2): $p>p_0$. In this case we set $r=p/p_0>1$. Then we have

$$\big\|T(f)\big\|_{L^p(w)}^p=\big\|\,|T(f)|^{p_0}\big\|_{L^r(w)}^r=\left(\int_{\mathbf{R}^n}|T(f)|^{p_0}h\,w\,dx\right)^r \qquad (7.5.16)$$

for some nonnegative function h with $L^{r'}(w)$ norm equal to 1. We define a function

$$H=\left[R'\big(h^{\frac{r'}{p'}}\big)\right]^{\frac{p'}{r'}}.$$

Obviously, we have $0\le h\le H$ and thus

$$
\begin{aligned}
\int_{\mathbf{R}^n}|T(f)|^{p_0}h\,w\,dx &\le \int_{\mathbf{R}^n}|T(f)|^{p_0}H\,w\,dx\\
&\le N\big([H\,w]_{A_{p_0}}\big)^{p_0}\big\|f\big\|_{L^{p_0}(H\,w)}^{p_0}\\
&\le N\big([H\,w]_{A_{p_0}}\big)^{p_0}\big\|\,|f|^{p_0}\big\|_{L^r(w)}\big\|H\big\|_{L^{r'}(w)} \qquad (7.5.17)\\
&\le 2^{\frac{p'}{r'}}N\big([H\,w]_{A_{p_0}}\big)^{p_0}\big\|f\big\|_{L^p(w)}^{p_0},
\end{aligned}
$$

noting that

$$\|H\|_{L^{r'}(w)}^{r'} = \int_{\mathbf{R}^n} R'(h^{r'/p'})^{p'} w\,dx \le 2^{p'} \int_{\mathbf{R}^n} h^{r'} w\,dx = 2^{p'},$$

which is valid in view of (7.5.11). Moreover, this argument is based on the hypothesis of the theorem and requires that Hw be an A_{p_0} weight. To see this, we observe that condition (7.5.12) implies that $H^{r'/p'}w$ is an A_1 weight with characteristic constant at most a multiple of $[w]_{A_1}$. Thus, there is a constant C_2' that depends only on n and p such that

$$\frac{1}{|Q|}\int_Q H^{\frac{r'}{p'}} w\,dx \le C_2'\,[w]_{A_p} H^{\frac{r'}{p'}} w$$

for all cubes Q in $\mathbf{R}^n$. From this it follows that

$$(Hw)^{-1} \le \kappa_2(n,p,p_0)\,[w]_{A_p}^{\frac{p'}{r'}} \bigg(\frac{1}{|Q|}\int_Q H^{\frac{r'}{p'}} w\,dx\bigg)^{-\frac{p'}{r'}} w^{\frac{p'}{r'}-1},$$

where we set $\kappa_2(n,p,p_0) = (C_2')^{p'/r'}$. We raise the preceding displayed expression to the power $p_0'-1$, we average over the cube Q, and then we raise to the power p_0-1. We deduce the estimate

$$\begin{aligned}\bigg(\frac{1}{|Q|}\int_Q (Hw)^{1-p_0'}\,dx\bigg)^{p_0-1}&\\ \le \kappa_2(n,p,p_0)\,[w]_{A_p}^{\frac{p'}{r'}} \bigg(\frac{1}{|Q|}\int_Q H^{\frac{r'}{p'}} w\,dx\bigg)^{-\frac{p'}{r'}} &\bigg(\frac{1}{|Q|}\int_Q w^{1-p'}\,dx\bigg)^{p_0-1},\end{aligned} \tag{7.5.18}$$

where we use the fact that

$$\Big(\frac{p'}{r'}-1\Big)(p_0'-1) = 1-p'.$$

Note that $r'/p' \ge 1$, since $p_0 \ge 1$. Using Hölder's inequality with exponents r'/p' and $(r'/p')^{-1}$ we obtain that

$$\frac{1}{|Q|}\int_Q Hw\,dx \le \bigg(\frac{1}{|Q|}\int_Q H^{\frac{r'}{p'}} w\,dx\bigg)^{\frac{p'}{r'}} \bigg(\frac{1}{|Q|}\int_Q w\,dx\bigg)^{\frac{p_0-1}{p-1}}, \tag{7.5.19}$$

where we used that

$$\frac{1}{(\frac{r'}{p'})'} = \frac{p_0-1}{p-1}.$$

Multiplying (7.5.18) by (7.5.19), we deduce the estimate

$$\big[Hw\big]_{A_{p_0}} \le \kappa_2(n,p,p_0)\,[w]_{A_p}^{\frac{p'}{r'}}\,[w]_{A_p}^{\frac{p_0-1}{p-1}} = \kappa_2(n,p,p_0)\,[w]_{A_p}.$$

Inserting this estimate in (7.5.17) we obtain

$$\int_{\mathbf{R}^n} |T(f)|^{p_0} h\, w\, dx \le 2^{\frac{p'}{r'}} N\big(\kappa_2(n,p,p_0)\,[w]_{A_p}\big)^{p_0} \|f\|_{L^p(w)}^{p_0},$$

and combining this with (7.5.16) we conclude that

$$\|T(f)\|_{L^p(w)}^{p} \le 2^{\frac{p'r}{r'}} N\big(\kappa_2(n,p,p_0)\,[w]_{A_p}\big)^{p_0 r} \|f\|_{L^p(w)}^{p_0 r}.$$

This proves the required estimate (7.5.5) in the case $p > p_0$. □

There is a version of Theorem 7.5.3 in which the initial strong type assumption is replaced by a weak type estimate.

Theorem 7.5.5. *Suppose that T is a well defined operator on $\bigcup_{1<q<\infty}\bigcup_{w\in A_q} L^q(w)$ that takes values in the space of measurable complex-valued functions. Fix $1 \le p_0 < \infty$ and suppose that there is an increasing function N on $[1,\infty)$ such that for all weights v in A_{p_0} we have*

$$\|T\|_{L^{p_0}(v)\to L^{p_0,\infty}(v)} \le N([v]_{A_{p_0}}). \tag{7.5.20}$$

Then for any $1 < p < \infty$ and for all weights w in A_p we have

$$\|T\|_{L^p(w)\to L^{p,\infty}(w)} \le K\big(n,p,p_0,[w]_{A_p}\big), \tag{7.5.21}$$

where $K\big(n,p,p_0,[w]_{A_p}\big)$ is as in Theorem 7.5.3.

Proof. For every fixed $\lambda > 0$ we define

$$T_\lambda(f) = \lambda \chi_{|T(f)|>\lambda}.$$

The operator T_λ is not linear but is well defined on $\bigcup_{1<q<\infty}\bigcup_{w\in A_q} L^q(w)$, since T is well defined on this union. We show that T_λ maps $L^{p_0}(v)$ to $L^{p_0}(v)$ for every $v \in A_{p_0}$. Indeed, we have

$$\begin{aligned}
\left(\int_{\mathbf{R}^n} |T_\lambda(f)|^{p_0} v\, dx\right)^{\frac{1}{p_0}} &= \left(\int_{\mathbf{R}^n} \lambda^{p_0} \chi_{|T(f)|>\lambda}\, v\, dx\right)^{\frac{1}{p_0}} \\
&= \Big(\lambda^{p_0} v(\{|T(f)| > \lambda\})\Big)^{\frac{1}{p_0}} \\
&\le N([v]_{A_{p_0}}) \|f\|_{L^{p_0}(v)}
\end{aligned}$$

using the hypothesis on T. Applying Theorem 7.5.3, we obtain that T_λ maps $L^p(w)$ to itself for all $1 < p < \infty$ and all $w \in A_p$ with a constant independent of λ. Precisely, for any $w \in A_p$ and any $f \in L^p(w)$ we have

$$\|T_\lambda(f)\|_{L^p(w)} \le K\big(n,p,p_0,[w]_{A_p}\big) \|f\|_{L^p(w)}.$$

Since

$$\big\|T(f)\big\|_{L^{p,\infty}(w)} = \sup_{\lambda>0} \big\|T_\lambda(f)\big\|_{L^p(w)},$$

it follows that T maps $L^p(w)$ to $L^{p,\infty}(w)$ with the asserted norm. □

Assuming that the operator T in the preceding theorem is sublinear (or quasi-sublinear), we obtain the following result that contains a stronger conclusion.

Corollary 7.5.6. *Suppose that T is a sublinear operator on $\bigcup_{1<q<\infty}\bigcup_{w\in A_q} L^q(w)$ that takes values in the space of measurable complex-valued functions. Fix $1 \le p_0 < \infty$ and suppose that there is an increasing function N on $[1,\infty)$ such that for all weights v in A_{p_0} we have*

$$\big\|T\big\|_{L^{p_0}(v)\to L^{p_0,\infty}(v)} \le N([v]_{A_{p_0}}). \tag{7.5.22}$$

Then for any $1<p<\infty$ and any weight w in A_p there is a constant $K'(n,p,p_0,[w]_{A_p})$ such that

$$\big\|T(f)\big\|_{L^p(w)} \le K'(n,p,p_0,[w]_{A_p})\big\|f\big\|_{L^p(w)}.$$

Proof. The proof follows from Theorem 7.5.5 and the Marcinkiewicz interpolation theorem. □

We end this subsection by observing that the conclusion of the extrapolation Theorem 7.5.3 can be strengthened to yield vector-valued estimates. This strengthening may be achieved by a simple adaptation of the proof discussed.

Corollary 7.5.7. *Suppose that T is defined on $\bigcup_{1\le q<\infty}\bigcup_{w\in A_q} L^q(w)$ and takes values in the space of all measurable complex-valued functions. Fix $1 \le p_0 < \infty$ and suppose that there is an increasing function N on $[1,\infty)$ such that for all weights v in A_{p_0} we have*

$$\big\|T\big\|_{L^{p_0}(v)\to L^{p_0}(v)} \le N([v]_{A_{p_0}}).$$

Then for every $1<p<\infty$ and every weight $w\in A_p$ we have

$$\Big\|\Big(\sum_j |T(f_j)|^{p_0}\Big)^{\frac{1}{p_0}}\Big\|_{L^p(w)} \le K(n,p,p_0,[w]_{A_p})\Big\|\Big(\sum_j |f_j|^{p_0}\Big)^{\frac{1}{p_0}}\Big\|_{L^p(w)}$$

for all sequences of functions f_j in $L^p(w)$, where $K\big(n,p,p_0,[w]_{A_p}\big)$ is as in Theorem 7.5.3.

Proof. To derive the claimed vector-valued inequality follow the proof of Theorem 7.5.3 replacing the function f by $(\sum_j |f_j|^{p_0})^{\frac{1}{p_0}}$ and $T(f)$ by $(\sum_j |T(f_j)|^{p_0})^{\frac{1}{p_0}}$. □

7.5.3 Weighted Inequalities Versus Vector-Valued Inequalities

We now discuss connections between weighted inequalities and vector-valued inequalities. The next result provides strong evidence that there is a nontrivial connection of this sort. The following is a general theorem saying that any vector-valued inequality is equivalent to some weighted inequality. The proof of the theorem is based on a minimax lemma whose precise formulation and proof can be found in Appendix H.

Theorem 7.5.8. *(a) Let $0 < p < q, r < \infty$. Let $\{T_j\}_j$ be a sequence of sublinear operators that map $L^q(\mu)$ to $L^r(\nu)$, where μ and ν are arbitrary measures. Then the vector-valued inequality*

$$\Big\|\Big(\sum_j |T_j(f_j)|^p\Big)^{\frac{1}{p}}\Big\|_{L^r} \le C\Big\|\Big(\sum_j |f_j|^p\Big)^{\frac{1}{p}}\Big\|_{L^q} \tag{7.5.23}$$

holds for all $f_j \in L^q(\mu)$ if and only if for every $u \ge 0$ in $L^{\frac{r}{r-p}}(\nu)$ there exists $U \ge 0$ in $L^{\frac{q}{q-p}}(\mu)$ with

$$\begin{aligned} \|U\|_{L^{\frac{q}{q-p}}} &\le \|u\|_{L^{\frac{r}{r-p}}}, \\ \sup_j \int |T_j(f)|^p\, u\, d\nu &\le C^p \int |f|^p\, U\, d\mu. \end{aligned} \tag{7.5.24}$$

(b) Let $0 < q, r < p < \infty$. Let $\{T_j\}_j$ be as before. Then the vector-valued inequality (7.5.23) holds for all $f_j \in L^q(\mu)$ if and only if for every $u \ge 0$ in $L^{\frac{q}{p-q}}(\mu)$ there exists $U \ge 0$ in $L^{\frac{r}{p-r}}(\nu)$ with

$$\begin{aligned} \|U\|_{L^{\frac{r}{p-r}}} &\le \|u\|_{L^{\frac{q}{p-q}}}, \\ \sup_j \int |T_j(f)|^p\, U^{-1}\, d\nu &\le C^p \int |f|^p\, u^{-1}\, d\mu. \end{aligned} \tag{7.5.25}$$

Proof. We begin with part (a). Given $f_j \in L^q(\mathbf{R}^n, \mu)$, we use (7.5.24) to obtain

$$\begin{aligned} \Big\|\Big(\sum_j |T_j(f_j)|^p\Big)^{\frac{1}{p}}\Big\|_{L^r(\nu)} &= \Big\|\sum_j |T_j(f_j)|^p\Big\|_{L^{\frac{r}{p}}(\nu)}^{\frac{1}{p}} \\ &= \sup_{\|u\|_{L^{\frac{r}{r-p}}}\le 1} \Big(\int_{\mathbf{R}^n} \sum_j |T_j(f_j)|^p\, u\, d\nu\Big)^{\frac{1}{p}} \\ &\le \sup_{\|u\|_{L^{\frac{r}{r-p}}}\le 1} C\Big(\int_{\mathbf{R}^n} \sum_j |f_j|^p\, U\, d\mu\Big)^{\frac{1}{p}} \\ &\le \sup_{\|u\|_{L^{\frac{r}{r-p}}}\le 1} C\Big\|\sum_j |f_j|^p\Big\|_{L^{\frac{q}{p}}(\mu)}^{\frac{1}{p}} \|U\|_{L^{\frac{q}{q-p}}}^{\frac{1}{p}} \\ &\le C\Big\|\Big(\sum_j |f_j|^p\Big)^{\frac{1}{p}}\Big\|_{L^q(\mu)}, \end{aligned}$$

which proves (7.5.23) with the same constant C as in (7.5.24). To prove the converse, given a nonnegative $u \in L^{\frac{r}{r-p}}(v)$ with $\|u\|_{L^{\frac{r}{r-p}}} = 1$, we define

$$A = \{a = (a_0, a_1) : a_0 = \sum_j |f_j|^p, \quad a_1 = \sum_j |T_j(f_j)|^p, \quad f_j \in L^q(\mu)\}$$

and

$$B = \{b \in L^{\frac{q}{q-p}}(\mu) : b \geq 0, \quad \|b\|_{L^{\frac{q}{q-p}}} \leq 1 = \|u\|_{L^{\frac{r}{r-p}}}\}.$$

Notice that A and B are convex sets and B is weakly compact. (The sublinearity of each T_j is used here.) We define the function Φ on $A \times B$ by setting

$$\Phi(a,b) = \int a_1 u\, dv - C^p \int a_0 b\, d\mu = \sum_j \left(\int |T_j(f_j)|^p u\, dv - C^p \int |f_j|^p b\, d\mu \right).$$

Then Φ is concave on A and weakly continuous and convex on B. Thus the *minimax lemma* in Appendix H is applicable. This gives

$$\min_{b\in B} \sup_{a\in A} \Phi(a,b) = \sup_{a\in A} \min_{b\in B} \Phi(a,b). \tag{7.5.26}$$

At this point observe that for a fixed $a = \left(\sum_j |f_j|^p, \sum_j |T_j(f_j)|^p\right)$ in A we have

$$\begin{aligned}
\min_{b\in B} \Phi(a,b) &\leq \Big\|\sum_j |T_j(f_j)|^p\Big\|_{L^{\frac{r}{p}}(v)} \|u\|_{L^{\frac{r}{r-p}}} - C^p \max_{b\in B} \int \sum_j |f_j|^p b\, d\mu \\
&\leq \Big\|\sum_j |T_j(f_j)|^p\Big\|_{L^{\frac{r}{p}}(v)} - C^p \Big\|\sum_j |f_j|^p\Big\|_{L^{\frac{q}{p}}(\mu)} \leq 0
\end{aligned}$$

using the hypothesis (7.5.23). It follows that $\sup_{a\in A} \min_{b\in B} \Phi(a,b) \leq 0$ and hence (7.5.26) yields $\min_{b\in B} \sup_{a\in A} \Phi(a,b) \leq 0$. Thus there exists a $U \in B$ such that $\Phi(a,U) \leq 0$ for every $a \in A$. This completes the proof of part (a).

The proof of part (b) is similar. Using the result of Exercise 7.5.1 and (7.5.25), given $f_j \in L^q(\mathbf{R}^n, \mu)$ we have

$$\begin{aligned}
\Big\|\Big(\sum_j |f_j|^p\Big)^{\frac{1}{p}}\Big\|_{L^q(\mu)} &= \Big\|\sum_j |f_j|^p\Big\|_{L^{\frac{q}{p}}(\mu)}^{\frac{1}{p}} \\
&= \inf_{\|u\|_{L^{\frac{q}{p-q}}} \leq 1} \left(\int_{\mathbf{R}^n} \sum_j |f_j|^p u^{-1}\, d\mu \right)^{\frac{1}{p}} \\
&\geq \frac{1}{C} \inf_{\|U\|_{L^{\frac{r}{p-r}}} \leq 1} \left(\int_{\mathbf{R}^n} \sum_j |T_j(f_j)|^p U^{-1}\, dv \right)^{\frac{1}{p}}
\end{aligned}$$

$$= \frac{1}{C}\Big\|\sum_j |T_j(f_j)|^p\Big\|_{L^{\frac{r}{p}}(\nu)}^{\frac{1}{p}}$$

$$= \frac{1}{C}\Big\|\Big(\sum_j |T_j(f_j)|^p\Big)^{\frac{1}{p}}\Big\|_{L^r(\nu)}.$$

To prove the converse direction in part (b), given a fixed $u \geq 0$ in $L^{\frac{q}{p-q}}(\mu)$ with $\big\|u\big\|_{L^{\frac{q}{p-q}}} = 1$, we define A as in part (a) and

$$B = \big\{b \in L^{\frac{p}{p-r}}(\nu) : \, b \geq 0, \quad \|b\|_{L^{\frac{p}{p-r}}} \leq 1 = \|u\|_{L^{\frac{q}{p-q}}}\big\}.$$

We also define the function Φ on $A \times B$ by setting

$$\begin{aligned}\Phi(a,b) &= \int a_1 b^{-1} d\nu - C^p \int a_0 u^{-1}\, d\mu \\ &= \sum_j \Big(\int |T_j(f_j)|^p b^{-1}\, d\nu - C^p \int |f_j|^p u^{-1}\, d\mu\Big).\end{aligned}$$

Then Φ is concave on A and weakly continuous and convex on B. Also, using Exercise 7.5.1, for any $a = \big(\sum_j |f_j|^p, \sum_j |T_j(f_j)|^p\big)$ in A, we have

$$\min_{b\in B} \Phi(a,b) \leq \Big\|\sum_j |T_j(f_j)|^p\Big\|_{L^{\frac{r}{p}}(\nu)} - C^p\Big\|\sum_j |f_j|^p\Big\|_{L^{\frac{q}{p}}(\mu)} \leq 0.$$

Thus $\sup_{a\in A}\min_{b\in B}\Phi(a,b) \leq 0$. Using (7.5.26), yields $\min_{b\in B}\sup_{a\in A}\Phi(a,b) \leq 0$, and the latter implies the existence of a U in B such that $\Phi(a,U) \leq 0$ for all $a \in A$. This proves (7.5.25). □

Example 7.5.9. We use the previous theorem to obtain another proof of the vector-valued Hardy–Littlewood maximal inequality in Corollary 5.6.5. We take $T_j = M$ for all j. For given $1 < p < q < \infty$ and u in $L^{\frac{q}{q-p}}$ we set $s = \frac{q}{q-p}$ and $U = \|M\|_{L^s \to L^s}^{-1} M(u)$. In view of Exercise 7.1.7 we have

$$\|U\|_{L^s} \leq \|u\|_{L^s} \qquad \text{and} \qquad \int_{\mathbf{R}^n} M(f)^p\, u\, dx \leq C^p \int_{\mathbf{R}^n} |f|^p\, U\, dx.$$

Using Theorem 7.5.8, we obtain

$$\Big\|\Big(\sum_j |M(f_j)|^p\Big)^{\frac{1}{p}}\Big\|_{L^q} \leq C_{n,p,q}\Big\|\Big(\sum_j |f_j|^p\Big)^{\frac{1}{p}}\Big\|_{L^q} \tag{7.5.27}$$

whenever $1 < p < q < \infty$, an inequality obtained earlier in (5.6.25).

It turns out that no specific properties of the Hardy–Littlewood maximal function were used in the preceding inequality, and one could obtain a general result along these lines.

Exercises

7.5.1. Let (X,μ) be a measure space, $0<s<1$, and $f\in L^s(X,\mu)$. Show that

$$\|f\|_{L^s}=\inf\left\{\int_X |f|u^{-1}\,d\mu:\ \|u\|_{L^{\frac{s}{1-s}}}\le 1\right\}$$

and that the infimum is attained.
[*Hint:* Try $u=c|f|^{1-s}$ for a suitable constant c.]

7.5.2. (*K. Yabuta*) Let $w\in A_p$ for some $1<p<\infty$ and let f be in $L^p_{\text{loc}}(\mathbf{R}^n,w\,dx)$. Show that f lies in $L^1_{\text{loc}}(\mathbf{R}^n)$.
[*Hint:* Write $w=w_1/w_2^{p-1}$ via Theorem 7.5.1.]

7.5.3. Use the same idea of the proof of Theorem 7.5.1 to prove the following general result: Let μ be a positive measure on a measure space X and let T be a bounded sublinear operator on $L^p(X,\mu)$ for some $1\le p<\infty$. Suppose that $T(f)\ge 0$ for all f in $L^p(X,\mu)$. Prove that for all $f_0\in L^p(X,\mu)$, there exists an $f\in L^p(X,\mu)$ such that

(a) $f_0(x)\le f(x)$ for μ-almost all $x\in X$.

(b) $\|f\|_{L^p(X)}\le 2\,\|f_0\|_{L^p(X)}$.

(c) $T(f)(x)\le 2\,\|T\|_{L^p\to L^p}f(x)$ for μ-almost all $x\in X$.

[*Hint:* Try the expression in (7.5.2) starting the sum at $j=0$.]

7.5.4. ([100]) Suppose that T is an operator defined on $\bigcup_{1<q<\infty}\bigcup_{w\in A_q}L^q(w)$ that satisfies $\|T\|_{L^r(v)\to L^r(v)}\le N([v]_{A_r})$ for some increasing function $N:[1,\infty)\to\mathbf{R}^+$. Without using Theorem 7.5.3 prove that for $1<q<r$ and all $v\in A_1$, T maps $L^q(v)$ to $L^q(v)$ with constant depending on q,r,n, and $[v]_{A_1}$.
[*Hint:* Hölder's inequality gives that

$$\|T(f)\|_{L^q(v)}\le\left(\int_{\mathbf{R}^n}|T(f)(x)|^rM(f)(x)^{q-r}v(x)\,dx\right)^{\frac{1}{r}}\left(\int_{\mathbf{R}^n}M(f)(x)^q v(x)\,dx\right)^{\frac{r-q}{rq}}.$$

Then use the fact that the weight $M(f)^{\frac{r-q}{r-1}}$ is in A_1 and Exercise 7.1.2.]

7.5.5. Let T be a sublinear operator defined on $\bigcup_{2\le q<\infty}L^q$. Suppose that for all functions f and u we have

$$\int_{\mathbf{R}^n}|T(f)|^2u\,dx\le\int_{\mathbf{R}^n}|f|^2M(u)\,dx.$$

Prove that T maps $L^p(\mathbf{R}^n)$ to itself for all $2<p<\infty$.
[*Hint:* Use that

$$\|T(f)\|_{L^p}=\sup_{\|u\|_{L^{(p/2)'}}\le 1}\left(\int_{\mathbf{R}^n}|T(f)|^2u\,dx\right)^{\frac{1}{2}}$$

and Hölder's inequality.]

7.5.6. (*X. C. Li*) Let T be a sublinear operator defined on $\bigcup_{1<q\le 2}\bigcup_{w\in A_q} L^q(w)$. Suppose that T maps $L^2(w)$ to $L^2(w)$ for all weights w that satisfy $w^{-1}\in A_1$. Prove that T maps L^p to itself for all $1<p<2$.
[*Hint:* We have

$$\big\|T(f)\big\|_{L^p} \le \bigg(\int_{\mathbf{R}^n} |T(f)|^2 M(f)^{-(2-p)}\,dx\bigg)^{\frac{1}{2}} \bigg(\int_{\mathbf{R}^n} M(f)^p\,dx\bigg)^{\frac{2-p}{2p}}$$

by Hölder's inequality. Apply the hypothesis to the first term of the product.]

HISTORICAL NOTES

Weighted inequalities can probably be traced back to the beginning of integration, but the A_p condition first appeared in a paper of Rosenblum [298] in a somewhat different form. The characterization of A_p when $n=1$ in terms of the boundedness of the Hardy–Littlewood maximal operator was obtained by Muckenhoupt [260]. The estimate on the norm in (7.1.25) can also be reversed, as shown by Buckley [38]. The simple proof of Theorem 7.1.9 is contained in Lerner's article [218] and yields both the Muckenhoupt theorem and Buckley's optimal growth of the norm of the Hardy–Littlewood maximal operator in terms of the A_p characteristic constant of the weight. Another proof of this result is given by Christ and Fefferman [61]. Versions of Lemma 7.1.10 for balls were first obtained by Besicovitch [27] and independently by Morse [258]. The particular version of Lemma 7.1.10 that appears in the text is adapted from that in de Guzmán [93]. Another version of this lemma is contained in the book of Mattila [246]. The fact that A_∞ is the union of the A_p spaces was independently obtained by Muckenhoupt [261] and Coifman and Fefferman [66]. The latter paper also contains a proof that A_p weights satisfy the crucial reverse Hölder condition. This condition first appeared in the work of Gehring [125] in the following context: If F is a quasiconformal homeomorphism from $\mathbf{R}^n$ into itself, then $|\det(\nabla F)|$ satisfies a reverse Hölder inequality. The characterization of A_1 weights is due to Coifman and Rochberg [68]. The fact that $M(f)^\delta$ is in A_∞ when $\delta<1$ was previously obtained by Córdoba and Fefferman [74]. The different characterizations of A_∞ (Theorem 7.3.3) are implicit in [260] and [66]. Another characterization of A_∞ in terms of the Gurov-Reshetnyak condition $\sup_Q \frac{1}{|Q|}\int_Q |f-\mathrm{Avg}_Q f|\,dx \le \varepsilon\,\mathrm{Avg}_Q f$ for $f\ge 0$ and $0<\varepsilon<2$ was obtained by Korenovskyy, Lerner, and Stokolos [201]. The definition of A_∞ using the reverse Jensen inequality herein was obtained as an equivalent characterization of that space by García-Cuerva and Rubio de Francia [122] (p. 405) and independently by Hruščev [161]. The reverse Hölder condition was extensively studied by Cruz-Uribe and Neugebauer [82].

Weighted inequalities with weights of the form $|x|^a$ for the Hilbert transform were first obtained by Hardy and Littlewood [147] and later by Stein [332] for other singular integrals. The necessity and sufficiency of the A_p condition for the boundedness of the Hilbert transform on weighted L^p spaces was obtained by Hunt, Muckenhoupt, and Wheeden [167]. Historically, the first result relating A_p weights and the Hilbert transform is the Helson-Szegő theorem [149], which says that the Hilbert transform is bounded on $L^2(w)$ if and only if $\log w = u + Hv$, where $u,v\in L^\infty(\mathbf{R})$ and $\|v\|_{L^\infty}<\frac{\pi}{2}$. The Helson-Szegő condition easily implies the A_2 condition, but the only known direct proof for the converse gives $\|v\|_{L^\infty}<\pi$; see Coifman, Jones, and Rubio de Francia [67]. A related result in higher dimensions was obtained by Garnett and Jones [123]. Weighted L^p estimates controlling Calderón–Zygmund operators by the Hardy–Littlewood maximal operator were obtained by Coifman [65]. Coifman and Fefferman [66] extended one-dimensional weighted norm inequalities to higher dimensions and also obtained good lambda inequalities for A_∞ weights for more general singular integrals and maximal singular integrals (Theorem 7.4.3). Bagby and Kurtz [19], and later Alvarez and Pérez [4], gave a sharper version of Theorem 7.4.3, by replacing the good lambda inequality by a rearrangement inequality. See also the related work of Lerner [217]. The following relation $\|M_d(f)\|_{L^p(w)} \le C(p,n,[w]_{A_\infty})\|M^\#(f)\|_{L^p(w)}$ between the dyadic maximal function and the sharp maximal function is valid for any $w\in A_\infty$ under the condition $M(f)\in L^{p_0}$ but also under the weaker assumption that $w(\{|f|>t\})<\infty$ for every $t>0$; see Kurtz [208]. Using that $\min(M,w)$ is an A_∞ weight with constant independent of M and Fatou's lemma, this condition can be relaxed to $|\{|f|>t\}|<\infty$ for every $t>0$. A rearrangement inequality relating f and $M^\#(f)$ is given in Bagby and Kurtz [18].

The factorization of A_p weights was conjectured by Muckenhoupt and proved by Jones [179]. The simple proof given in the text can be found in [67]. Extrapolation of operators (Theorem 7.5.3) is due to Rubio de Francia [300]. An alternative proof of this theorem was given later by García-Cuerva [121]. The value of the constant $K(n,p,p_0,[w]_{A_p})$ first appeared in Dragičević, Grafakos, Pereyra, and Petermichl [98]. Another proof with sharp bounds (in terms of the characteristic constant of the weights) was given by Duoandikoetxea [101]. The present treatment of Theorem 7.5.3, based on crucial Lemma 7.5.4, was communicated to the author by J. M. Martell. One may also consult the related work of Cruz-Uribe, Martell, and Pérez [80]. The simple proof of Theorem 7.5.5 was conceived by J. M. Martell and first appeared in the treatment of extrapolation of operators of many variables; see Grafakos and Martell [135]. The idea of extrapolation can be carried to general pairs of functions, see Cruz-Uribe, Martell, and Pérez [78]. Estimates for the distribution function in extrapolation theory were obtained by Carro, Torres, and Soria [58]. The equivalence between vector-valued inequalities and weighted norm inequalities of Theorem 7.5.8 is also due to Rubio de Francia [299]. The difficult direction in this equivalence is obtained using a minimax principle (see Fan [111]). Alternatively, one can use the factorization theory of Maurey [247], which brings an interesting connection with Banach space theory. The book of García-Cuerva and Rubio de Francia [122] provides an excellent reference on this and other topics related to weighted norm inequalities.

A primordial double-weighted norm inequality is the observation of Fefferman and Stein [115] that the maximal function maps $L^p(M(w))$ to $L^p(w)$ for nonnegative measurable functions w (Exercise 7.1.7). Sawyer [312] obtained that the condition $\sup_Q\big(\int_Q v^{1-p'}dx\big)^{-1}\int_Q M(v^{1-p'}\chi_Q)^p w\,dx<\infty$ provides a characterization of all pairs of weights (v,w) for which the Hardy–Littlewood maximal operator M maps $L^p(v)$ to $L^p(w)$. Simpler proofs of this result were obtained by Cruz-Uribe [77] and Verbitsky [367]. The fact that Sawyer's condition reduces to the usual A_p condition when $v=w$ was shown by Hunt, Kurtz, and Neugebauer [166]. The two-weight problem for singular integrals is more delicate, since they are not necessarily bounded from $L^p(M(w))$ to $L^p(w)$. Known results in this direction are that singular integrals map $L^p(M^{[p]+1}(w))$ to $L^p(w)$, where M^r denotes the rth iterate of the maximal operator. See Wilson [377] (for $1<p<2$) and Pérez [277] for the remaining p's. A necessary condition for the boundedness of the Hilbert transform from $L^p(v)$ to $L^p(w)$ was obtained by Muckenhoupt and Wheeden [262].

For an approach to two-weighted inequalities using Bellman functions, we refer to the article of Nazarov, Treil, and Volberg [266]. The notion of Bellman functions originated in control theory; the article [267] of the previous authors analyzes the connections between optimal control and harmonic analysis. Bellman functions have been used to derive estimates for the norms of classical operators on weighted Lebesgue spaces; for instance, Petermichl [279] showed that for $w\in A_2(\mathbf{R})$, the norm of the Hilbert transform from $L^2(\mathbf{R},w)$ to $L^2(\mathbf{R},w)$ is bounded by a constant times the characteristic constant $[w]_{A_2}$.

The theory of A_p weights in this chapter carries through to the situation in which Lebesgue measure is replaced by a general doubling measure. This theory also has a substantial analogue when the underlying measure is nondoubling but satisfies $\mu(\partial Q)=0$ for all cubes Q in $\mathbf{R}^n$ with sides parallel to the axes; see Orobitg and Pérez [272]. A thorough account of weighted Littlewood–Paley theory and exponential-square function integrability is contained in the book of Wilson [378].

The conjecture whether $\|T\|_{L^1(M(w))\to L^{1,\infty}(w)}<\infty$ holds for a weight w was disproved by Reguera [287] when T is a Haar multiplier and then by Reguera and Thiele [288] for the Hilbert transform. However, the slightly weaker version of this inequality, in which $M(w)$ is replaced by the Orlicz maximal operator $M_{L(\log L)^\varepsilon}(w)$, holds for any $\varepsilon>0$ and any Calderón-Zygmund operator T, as shown by Pérez [277]. For A^1 weights w the aforementioned conjecture would imply $\|T\|_{L^1(w)\to L^{1,\infty}(w)}\le C[w]_{A_1}$. However, Nazarov, Reznikov, Vasyunin, and Volberg [265] disproved the weaker inequality $\|T\|_{L^1(w)\to L^{1,\infty}(w)}\le C[w]_{A_1}\big(\log(e+[w]_{A_1})\big)^\alpha$ for $\alpha<\frac{1}{5}$. Lerner, Ombrosi, and Pérez [223] had previously shown that the preceding inequality holds with $\alpha=1$ for any Calderón-Zygmund operator T.

Concerning the sharp weighted bound $\|T\|_{L^2(w)\to L^2(w)} \le c_T\,[w]_{A_2}$ for a Calderón-Zygmund operator T we have the work of Petermichl and Volberg [281] which answered a question by Astala, Iwaniecz and Saksman [14] on the regularity of solutions to the Beltrami equation. The proofs of this inequality for the Hilbert and Riesz transforms via the Bellman function technique were obtained soon afterwards by Petermichl [279], [280]. The use of Bellman functions was first avoided in the work of Lacey, Petermichl, and Reguera [210], whose proof recovered the already known cases and used Haar shift operators, the two-weight theory for them of Nazarov, Treil and Volberg [268], and corona decompositions. The simplest proof for these classical operators was obtained by Cruz-Uribe, Martell, and Pérez [79], [81] using a very powerful inequality due to Lerner [219]. The complete proof for a general Calderón-Zygmund operator was given by Hytönen [168]. A simplified proof was provided by Lerner [220], [221]. For other improvements and estimates involving A_p and A_∞ constants see the work of Lerner [222], Hytönen and Pérez [170], and Lacey, Hytönen, and Pérez [169].

Appendix A
Gamma and Beta Functions

A.1 A Useful Formula

The following formula is valid:

$$\int_{\mathbf{R}^n} e^{-|x|^2}\,dx = \left(\sqrt{\pi}\right)^n.$$

This is an immediate consequence of the corresponding one-dimensional identity

$$\int_{-\infty}^{+\infty} e^{-x^2}\,dx = \sqrt{\pi},$$

which is usually proved from its two-dimensional version by switching to polar coordinates:

$$I^2 = \int_{-\infty}^{+\infty}\int_{-\infty}^{+\infty} e^{-x^2}e^{-y^2}\,dy\,dx = 2\pi\int_0^{\infty} re^{-r^2}\,dr = \pi.$$

A.2 Definitions of $\Gamma(z)$ and $B(z,w)$

For a complex number z with $\operatorname{Re} z > 0$ define

$$\Gamma(z) = \int_0^{\infty} t^{z-1}e^{-t}\,dt.$$

$\Gamma(z)$ is called the gamma function. It follows from its definition that $\Gamma(z)$ is analytic on the right half-plane $\operatorname{Re} z > 0$.

Two fundamental properties of the gamma function are that

$$\Gamma(z+1) = z\Gamma(z) \qquad \text{and} \qquad \Gamma(n) = (n-1)!,$$

DOI 10.1007/978-1-4939-1194-3,

where z is a complex number with positive real part and $n \in \mathbf{Z}^+$. Indeed, integration by parts yields

$$\Gamma(z) = \int_0^\infty t^{z-1} e^{-t}\, dt = \left[\frac{t^z e^{-t}}{z}\right]_0^\infty + \frac{1}{z}\int_0^\infty t^z e^{-t}\, dt = \frac{1}{z}\Gamma(z+1).$$

Since $\Gamma(1) = 1$, the property $\Gamma(n) = (n-1)!$ for $n \in \mathbf{Z}^+$ follows by induction. Another important fact is that

$$\Gamma(\tfrac{1}{2}) = \sqrt{\pi}.$$

This follows easily from the identity

$$\Gamma(\tfrac{1}{2}) = \int_0^\infty t^{-\frac{1}{2}} e^{-t}\, dt = 2\int_0^\infty e^{-u^2}\, du = \sqrt{\pi}.$$

Next we define the beta function. Fix z and w complex numbers with positive real parts. We define

$$B(z,w) = \int_0^1 t^{z-1}(1-t)^{w-1}\, dt = \int_0^1 t^{w-1}(1-t)^{z-1}\, dt.$$

We have the following relationship between the gamma and the beta functions:

$$B(z,w) = \frac{\Gamma(z)\Gamma(w)}{\Gamma(z+w)},$$

when z and w have positive real parts.

The proof of this fact is as follows:

$$\begin{aligned}
\Gamma(z+w)B(z,w) &= \Gamma(z+w)\int_0^1 t^{w-1}(1-t)^{z-1}\, dt \\
&= \Gamma(z+w)\int_0^\infty u^{w-1}\left(\frac{1}{1+u}\right)^{z+w} du && t = u/(1+u) \\
&= \int_0^\infty \int_0^\infty u^{w-1}\left(\frac{1}{1+u}\right)^{z+w} v^{z+w-1}e^{-v}\, dv\, du \\
&= \int_0^\infty \int_0^\infty u^{w-1} s^{z+w-1} e^{-s(u+1)}\, ds\, du && s = v/(1+u) \\
&= \int_0^\infty s^z e^{-s} \int_0^\infty (us)^{w-1} e^{-su}\, du\, ds \\
&= \int_0^\infty s^{z-1} e^{-s}\Gamma(w) ds \\
&= \Gamma(z)\Gamma(w).
\end{aligned}$$

A.3 Volume of the Unit Ball and Surface of the Unit Sphere

We denote by v_n the volume of the unit ball in $\mathbf{R}^n$ and by ω_{n-1} the surface area of the unit sphere $\mathbf{S}^{n-1}$. We have the following:

$$\omega_{n-1} = \frac{2\pi^{\frac{n}{2}}}{\Gamma(\frac{n}{2})}$$

and

$$v_n = \frac{\omega_{n-1}}{n} = \frac{2\pi^{\frac{n}{2}}}{n\Gamma(\frac{n}{2})} = \frac{\pi^{\frac{n}{2}}}{\Gamma(\frac{n}{2}+1)}.$$

The easy proofs are based on the formula in Appendix A.1. We have

$$\left(\sqrt{\pi}\right)^n = \int_{\mathbf{R}^n} e^{-|x|^2}\,dx = \omega_{n-1}\int_0^\infty e^{-r^2} r^{n-1}\,dr,$$

by switching to polar coordinates. Now change variables $t = r^2$ to obtain that

$$\pi^{\frac{n}{2}} = \tfrac{\omega_{n-1}}{2}\int_0^\infty e^{-t} t^{\frac{n}{2}-1}\,dt = \tfrac{\omega_{n-1}}{2}\Gamma\left(\tfrac{n}{2}\right).$$

This proves the formula for the surface area of the unit sphere in $\mathbf{R}^n$.

To compute v_n, write again using polar coordinates

$$v_n = |B(0,1)| = \int_{|x|\le 1} 1\,dx = \int_{\mathbf{S}^{n-1}}\int_0^1 r^{n-1}\,dr\,d\theta = \frac{1}{n}\,\omega_{n-1}.$$

Here is another way to relate the volume to the surface area. Let $B(0,R)$ be the ball in $\mathbf{R}^n$ of radius $R > 0$ centered at the origin. Then the volume of the shell $B(0,R+h)\setminus B(0,R)$ divided by h tends to the surface area of $B(0,R)$ as $h \to 0$. In other words, the derivative of the volume of $B(0,R)$ with respect to the radius R is equal to the surface area of $B(0,R)$. Since the volume of $B(0,R)$ is $v_n R^n$, it follows that the surface area of $B(0,R)$ is $n\,v_n R^{n-1}$. Taking $R = 1$, we deduce $\omega_{n-1} = n\,v_n$.

A.4 Computation of Integrals Using Gamma Functions

Let $k_1,\dots,k_n$ be nonnegative even integers. The integral

$$\int_{\mathbf{R}^n} x_1^{k_1}\cdots x_n^{k_n} e^{-|x|^2}\,dx_1\cdots dx_n = \prod_{j=1}^n \int_{-\infty}^{+\infty} x_j^{k_j} e^{-x_j^2}\,dx_j = \prod_{j=1}^n \Gamma\left(\frac{k_j+1}{2}\right)$$

expressed in polar coordinates is equal to

$$\left(\int_{\mathbf{S}^{n-1}} \theta_1^{k_1}\cdots\theta_n^{k_n}\,d\theta\right)\int_0^\infty r^{k_1+\cdots+k_n} r^{n-1} e^{-r^2}\,dr,$$

where $\theta = (\theta_1, \dots, \theta_n)$. This leads to the identity

$$\int_{\mathbf{S}^{n-1}} \theta_1^{k_1} \cdots \theta_n^{k_n}\, d\theta = 2\Gamma\Big(\frac{k_1 + \cdots + k_n + n}{2}\Big)^{-1} \prod_{j=1}^{n} \Gamma\Big(\frac{k_j+1}{2}\Big).$$

Another classical integral that can be computed using gamma functions is the following:

$$\int_0^{\pi/2} (\sin\varphi)^a (\cos\varphi)^b\, d\varphi = \frac{1}{2}\frac{\Gamma(\frac{a+1}{2})\Gamma(\frac{b+1}{2})}{\Gamma(\frac{a+b+2}{2})},$$

whenever a and b are complex numbers with $\mathrm{Re}\, a > -1$ and $\mathrm{Re}\, b > -1$.

Indeed, change variables $u = (\sin\varphi)^2$; then $du = 2(\sin\varphi)(\cos\varphi)d\varphi$, and the preceding integral becomes

$$\frac{1}{2}\int_0^1 u^{\frac{a-1}{2}}(1-u)^{\frac{b-1}{2}}\, du = \frac{1}{2}B\Big(\frac{a+1}{2}, \frac{b+1}{2}\Big) = \frac{1}{2}\frac{\Gamma(\frac{a+1}{2})\Gamma(\frac{b+1}{2})}{\Gamma(\frac{a+b+2}{2})}.$$

A.5 Meromorphic Extensions of $B(z,w)$ and $\Gamma(z)$

Using the identity $\Gamma(z+1) = z\Gamma(z)$, we can easily define a meromorphic extension of the gamma function on the whole complex plane starting from its known values on the right half-plane. We give an explicit description of the meromorphic extension of $\Gamma(z)$ on the whole plane. First write

$$\Gamma(z) = \int_0^1 t^{z-1} e^{-t} dt + \int_1^\infty t^{z-1} e^{-t} dt$$

and observe that the second integral is an analytic function of z for all $z \in \mathbf{C}$. Write the first integral as

$$\int_0^1 t^{z-1} \Big\{ e^{-t} - \sum_{j=0}^{N} \frac{(-t)^j}{j!} \Big\} dt + \sum_{j=0}^{N} \frac{(-1)^j/j!}{z+j}.$$

The last integral converges when $\mathrm{Re}\, z > -N-1$, since the expression inside the curly brackets is $O(t^{N+1})$ as $t \to 0$. It follows that the gamma function can be defined to be an analytic function on $\mathrm{Re}\, z > -N-1$ except at the points $z = -j$, $j = 0, 1, \dots, N$, at which it has simple poles with residues $\frac{(-1)^j}{j!}$. Since N was arbitrary, it follows that the gamma function has a meromorphic extension on the whole plane.

In view of the identity

$$B(z,w) = \frac{\Gamma(z)\Gamma(w)}{\Gamma(z+w)},$$

the definition of $B(z,w)$ can be extended to $\mathbf{C} \times \mathbf{C}$. It follows that $B(z,w)$ is a meromorphic function in each argument.

A.6 Asymptotics of $\Gamma(x)$ as $x \to \infty$

We now derive *Stirling's formula*:

$$\lim_{x\to\infty}\frac{\Gamma(x+1)}{\left(\frac{x}{e}\right)^x\sqrt{2\pi x}}=1\,.$$

First change variables $t=x+sx\sqrt{\frac{2}{x}}$ to obtain

$$\Gamma(x+1)=\int_0^\infty e^{-t}t^x\,dt=\left(\frac{x}{e}\right)^x\sqrt{2x}\int_{-\sqrt{x/2}}^{+\infty}\frac{\left(1+s\sqrt{\frac{2}{x}}\right)^x}{e^{2s\sqrt{x/2}}}ds\,.$$

Setting $y=\sqrt{\frac{x}{2}}$, we obtain

$$\frac{\Gamma(x+1)}{\left(\frac{x}{e}\right)^x\sqrt{2x}}=\int_{-\infty}^{+\infty}\left(\frac{\left(1+\frac{s}{y}\right)^y}{e^s}\right)^{2y}\chi_{(-y,\infty)}(s)\,ds.$$

To show that the last integral converges to $\sqrt{\pi}$ as $y\to\infty$, we need the following:

(1) The fact that

$$\lim_{y\to\infty}\left(\frac{(1+s/y)^y}{e^s}\right)^{2y}\to e^{-s^2}\,,$$

which follows easily by taking logarithms and applying L'Hôpital's rule twice.

(2) The estimate, valid for $y\geq 1$,

$$\left(\frac{\left(1+\frac{s}{y}\right)^y}{e^s}\right)^{2y}\leq\begin{cases}\dfrac{(1+s)^2}{e^s} & \text{when } s\geq 0,\\[2ex] e^{-s^2} & \text{when } -y<s<0,\end{cases}$$

which can be easily checked using calculus. Using these facts, the Lebesgue dominated convergence theorem, the trivial fact that $\chi_{-y<s<\infty}\to 1$ as $y\to\infty$, and the identity in Appendix A.1, we obtain that

$$\begin{aligned}\lim_{x\to\infty}\frac{\Gamma(x+1)}{\left(\frac{x}{e}\right)^x\sqrt{2x}}&=\lim_{y\to\infty}\int_{-\infty}^{+\infty}\left(\frac{\left(1+\frac{s}{y}\right)^y}{e^s}\right)^{2y}\chi_{(-y,\infty)}(s)\,ds\\&=\int_{-\infty}^{+\infty}e^{-s^2}\,ds\\&=\sqrt{\pi}.\end{aligned}$$

As a consequence of Stirling's formula, for any $t > 0$, we obtain

$$\lim_{x\to\infty} \frac{\Gamma(x)}{\Gamma(x+t)} = 0.$$

A.7 Euler's Limit Formula for the Gamma Function

For n a positive integer and $\operatorname{Re} z > 0$ we consider the functions

$$\Gamma_n(z) = \int_0^n \left(1-\frac{t}{n}\right)^n t^{z-1}\,dt$$

We show that

$$\Gamma_n(z) = \frac{n!\,n^z}{z(z+1)\cdots(z+n)}$$

and we obtain *Euler's limit formula for the gamma function*

$$\lim_{n\to\infty} \Gamma_n(z) = \Gamma(z).$$

We write $\Gamma(z) - \Gamma_n(z) = I_1(z) + I_2(z) + I_3(z)$, where

$$\begin{aligned}
I_1(z) &= \int_n^\infty e^{-t} t^{z-1}\,dt,\\
I_2(z) &= \int_{n/2}^n \left(e^{-t} - \left(1-\frac{t}{n}\right)^n\right) t^{z-1}\,dt,\\
I_3(z) &= \int_0^{n/2} \left(e^{-t} - \left(1-\frac{t}{n}\right)^n\right) t^{z-1}\,dt.
\end{aligned}$$

Obviously $I_1(z)$ tends to zero as $n\to\infty$. For I_2 and I_3 we have that $0 \le t < n$, and by the Taylor expansion of the logarithm we obtain

$$\log\left(1-\frac{t}{n}\right)^n = n\log\left(1-\frac{t}{n}\right) = -t - L,$$

where

$$L = \frac{t^2}{n}\left(\frac{1}{2} + \frac{1}{3}\frac{t}{n} + \frac{1}{4}\frac{t^2}{n^2} + \cdots\right).$$

It follows that

$$0 < e^{-t} - \left(1-\frac{t}{n}\right)^n = e^{-t} - e^{-L}e^{-t} \le e^{-t},$$

and thus $I_2(z)$ tends to zero as $n\to\infty$. For I_3 we have $t/n \le 1/2$, which implies that

$$L \le \frac{t^2}{n}\sum_{k=0}^\infty \frac{1}{(k+1)2^{k-1}} = \frac{t^2}{n}c.$$

Consequently, for $t/n \le 1/2$ we have

$$0 \le e^{-t} - \left(1 - \frac{t}{n}\right)^n = e^{-t}(1 - e^{-L}) \le e^{-t} L \le e^{-t}\frac{ct^2}{n}.$$

Plugging this estimate into I_3, we deduce that

$$|I_3(z)| \le \frac{c}{n}\Gamma(\mathrm{Re}\, z + 2),$$

which certainly tends to zero as $n \to \infty$.

Next, n integrations by parts give

$$\Gamma_n(z) = \frac{n}{nz}\frac{n-1}{n(z+1)}\frac{n-2}{n(z+2)}\cdots\frac{1}{n(z+n-1)}\int_0^n t^{z+n-1}\,dt = \frac{n!\,n^z}{z(z+1)\cdots(z+n)}.$$

This can be written as

$$1 = \Gamma_n(z)\, z \exp\left\{z\left(1 + \frac{1}{2} + \frac{1}{3} + \cdots + \frac{1}{n} - \log n\right)\right\}\prod_{k=1}^{n}\left(1 + \frac{z}{k}\right)e^{-z/k}.$$

Taking limits as $n \to \infty$, we obtain an *infinite product form of Euler's limit formula*,

$$1 = \Gamma(z)\, z e^{\gamma z}\prod_{k=1}^{\infty}\left(1 + \frac{z}{k}\right)e^{-z/k},$$

where $\mathrm{Re}\, z > 0$ and γ is *Euler's constant*

$$\gamma = \lim_{n\to\infty} 1 + \frac{1}{2} + \frac{1}{3} + \cdots + \frac{1}{n} - \log n.$$

The infinite product converges uniformly on compact subsets of the complex plane that excludes $z = 0, -1, -2, \ldots$, and thus it represents a holomorphic function in this domain. This holomorphic function multiplied by $\Gamma(z)\, z e^{\gamma z}$ is equal to 1 on $\mathrm{Re}\, z > 0$ and by analytic continuation it must be equal to 1 on $\mathbf{C} \setminus \{0, -1, -2, \ldots\}$. But $\Gamma(z)$ has simple poles, while the infinite product vanishes to order one, at the nonpositive integers. We conclude that Euler's limit formula holds for all complex numbers z; consequently, $\Gamma(z)$ has no zeros and $\Gamma(z)^{-1}$ is entire.

An immediate consequence of Euler's limit formula is the identity

$$\frac{1}{|\Gamma(x+iy)|^2} = \frac{1}{|\Gamma(x)|^2}\prod_{k=0}^{\infty}\left(1 + \frac{y^2}{(k+x)^2}\right),$$

which holds for x and y real with $x \notin \{0, -1, -2, \ldots\}$. As a consequence we have that

$$|\Gamma(x+iy)| \le |\Gamma(x)|$$

and also that

$$\frac{1}{|\Gamma(x+iy)|} \le \frac{1}{|\Gamma(x)|} e^{C(x)|y|^2},$$

where

$$C(x) = \frac{1}{2}\sum_{k=0}^{\infty}\frac{1}{(k+x)^2},$$

whenever $y \in \mathbf{R}$ and $x \in \mathbf{R} \setminus \{0,-1,-2,\dots\}$. Before we find a similar estimate for $x \in \{0,-1,-2,\dots\}$ we provide a simpler expression for this estimate when $x > 0$.

When $x > 0$ we have

$$C(x) \le \frac{1}{2x^2} + \frac{1}{2}\sum_{k=1}^{\infty}\frac{1}{(k+x)^2} \le \frac{1}{2x^2} + \frac{1}{2}\int_0^{\infty}\frac{dt}{(t+x)^2} = \frac{1}{2x^2} + \frac{1}{2x}\int_1^{\infty}\frac{dt}{t^2} = \frac{1}{2x^2} + \frac{1}{2x}.$$

Thus we conclude that when $x > 0$ and $y \in \mathbf{R}$ we have

$$\frac{1}{|\Gamma(x+iy)|} \le \frac{1}{|\Gamma(x)|} e^{\max\{x^{-2},x^{-1}\}|y|^2}.$$

When $x = 0$ we write $\Gamma(iy)iy = \Gamma(1+iy)$ and use the preceding inequality to obtain

$$\frac{1}{|\Gamma(iy)|} \le \frac{|iy|}{|\Gamma(1)|} e^{|y|^2} = |y| e^{|y|^2}$$

and more generally for $x = -N \in \{-1,-2,\dots\}$ and $y \in \mathbf{R}$ we obtain by induction

$$\frac{1}{|\Gamma(-N+iy)|} \le |iy|\,|1+iy|\,|2+iy| \cdots |N+iy|\, e^{|y|^2}.$$

A.8 Reflection and Duplication Formulas for the Gamma Function

The *reflection formula* relates the values of the gamma function of a complex number z and its reflection about the point $1/2$ in the following way:

$$\frac{\sin(\pi z)}{\pi} = \frac{1}{\Gamma(z)}\frac{1}{\Gamma(1-z)}.$$

The *duplication formula* relates the entire functions $\Gamma(2z)^{-1}$ and $\Gamma(z)^{-1}$ as follows:

$$\frac{1}{\Gamma(z)\Gamma(z+\frac{1}{2})} = \frac{\pi^{-\frac{1}{2}}2^{2z-1}}{\Gamma(2z)}.$$

Both of these could be proved using Euler's limit formula. The reflection formula also uses the identity

$$\prod_{k=1}^{\infty}\left(1-\frac{z^2}{k^2}\right)=\frac{\sin(\pi z)}{\pi z},$$

while the duplication formula makes use of the fact that

$$\lim_{n\to\infty}\frac{(n!)^2\,2^{2n+1}}{(2n)!\,n^{1/2}}=2\,\pi^{1/2}.$$

These and other facts related to the gamma function can be found in Olver [271].

Appendix B
Bessel Functions

B.1 Definition

We survey some basics from the theory of Bessel functions J_ν of complex order ν with $\operatorname{Re}\nu > -1/2$. We define the Bessel function J_ν of order ν by its *Poisson representation formula*

$$J_\nu(t) = \frac{\left(\frac{t}{2}\right)^\nu}{\Gamma(\nu+\frac{1}{2})\Gamma(\frac{1}{2})} \int_{-1}^{+1} e^{its}(1-s^2)^\nu \frac{ds}{\sqrt{1-s^2}},$$

where $\operatorname{Re}\nu > -1/2$ and $t \ge 0$. Although this definition is also valid when t is a complex number, for the applications we have in mind, it suffices to consider the case that t is real and nonnegative; in this case $J_\nu(t)$ is also a real number.

B.2 Some Basic Properties

Let us summarize a few properties of Bessel functions. We take $t > 0$.
(1) We have the following recurrence formula:

$$\frac{d}{dt}\left(t^{-\nu}J_\nu(t)\right) = -t^{-\nu}J_{\nu+1}(t), \qquad \operatorname{Re}\nu > -1/2.$$

(2) We also have the companion recurrence formula:

$$\frac{d}{dt}\left(t^{\nu}J_\nu(t)\right) = t^{\nu}J_{\nu-1}(t), \qquad \operatorname{Re}\nu > 1/2.$$

L. Grafakos, *Classical Fourier Analysis*, Graduate Texts in Mathematics 249,
DOI 10.1007/978-1-4939-1194-3, © Springer Science+Business Media New York 2014

(3) $J_\nu(t)$ satisfies the differential equation:

$$t^2\frac{d^2}{dt^2}(J_\nu(t))+t\frac{d}{dt}(J_\nu(t))+(t^2-\nu^2)J_\nu(t)=0.$$

(4) If $\nu\in\mathbf{Z}^+$, then we have the following identity, which was taken by Bessel as the definition of J_ν for integer ν:

$$J_\nu(t)=\frac{1}{2\pi}\int_0^{2\pi}e^{it\sin\theta}e^{-i\nu\theta}\,d\theta=\frac{1}{2\pi}\int_0^{2\pi}\cos(t\sin\theta-\nu\theta)\,d\theta.$$

(5) For $\operatorname{Re}\nu>-1/2$ we have the following identity:

$$J_\nu(t)=\frac{1}{\Gamma(\frac{1}{2})}\left(\frac{t}{2}\right)^\nu\sum_{j=0}^{\infty}(-1)^j\frac{\Gamma(j+\frac{1}{2})}{\Gamma(j+\nu+1)}\frac{t^{2j}}{(2j)!},$$

which can also be written as

$$J_\nu(t)=\sum_{j=0}^{\infty}\frac{(-1)^j}{j!}\frac{1}{\Gamma(j+\nu+1)}\left(\frac{t}{2}\right)^{2j+\nu},$$

using the equality $(2j)!=2^{2j}j!(j-\frac{1}{2})(j-\frac{3}{2})\cdots\frac{1}{2}=2^{2j}j!\Gamma(j+\frac{1}{2})\Gamma(\frac{1}{2})^{-1}$.

(6) For $\operatorname{Re}\nu>1/2$ the identity below is valid:

$$\frac{d}{dt}(J_\nu(t))=\frac{1}{2}\big(J_{\nu-1}(t)-J_{\nu+1}(t)\big).$$

We first verify property (1). We have

$$\begin{aligned}\frac{d}{dt}(t^{-\nu}J_\nu(t)) &= \frac{i}{2^\nu\Gamma(\nu+\frac{1}{2})\Gamma(\frac{1}{2})}\int_{-1}^{1}se^{its}(1-s^2)^{\nu-\frac{1}{2}}\,ds\\ &= \frac{i}{2^\nu\Gamma(\nu+\frac{1}{2})\Gamma(\frac{1}{2})}\int_{-1}^{1}\frac{it}{2}e^{its}\frac{(1-s^2)^{\nu+\frac{1}{2}}}{\nu+\frac{1}{2}}\,ds\\ &= -t^{-\nu}J_{\nu+1}(t),\end{aligned}$$

where we integrated by parts and used the fact that $\Gamma(x+1)=x\Gamma(x)$.

We now prove Property (2) for $\operatorname{Re}\nu>1/2$:

$$\begin{aligned}&\frac{d}{dt}\big(t^\nu J_\nu(t)\big)\\ &=\frac{2\nu t^{2\nu-1}2^{-\nu}}{\Gamma(\nu+\frac{1}{2})\Gamma(\frac{1}{2})}\int_{-1}^{1}e^{its}(1-s^2)^{\nu-\frac{1}{2}}ds+\frac{t^{2\nu}2^{-\nu}}{\Gamma(\nu+\frac{1}{2})\Gamma(\frac{1}{2})}i\int_{-1}^{1}e^{its}is(1-s^2)^{\nu-\frac{1}{2}}ds\end{aligned}$$

$$
\begin{aligned}
&= \frac{2\nu t^{2\nu-1}2^{-\nu}}{\Gamma(\nu+\frac{1}{2})\Gamma(\frac{1}{2})}\int_{-1}^{1} e^{its}(1-s^2)^{\nu-\frac{1}{2}}\,ds + \frac{t^{2\nu}2^{-\nu}}{\Gamma(\nu+\frac{1}{2})\Gamma(\frac{1}{2})}\int_{-1}^{1}\Big(\frac{e^{its}}{t}\Big)'(1-s^2)^{\nu-\frac{1}{2}}s\,ds\\
&= \frac{2\nu t^{2\nu-1}2^{-\nu}}{\Gamma(\nu+\frac{1}{2})\Gamma(\frac{1}{2})}\int_{-1}^{1} e^{its}(1-s^2)^{\nu-\frac{1}{2}}\,ds - \frac{t^{2\nu}2^{-\nu}}{\Gamma(\nu+\frac{1}{2})\Gamma(\frac{1}{2})}\int_{-1}^{1}\frac{e^{its}}{t}\big((1-s^2)^{\nu-\frac{1}{2}}s\big)'\,ds\\
&= \frac{t^{2\nu-1}2^{-\nu}}{\Gamma(\nu+\frac{1}{2})\Gamma(\frac{1}{2})}\int_{-1}^{1} e^{its}\Big[2\nu(1-s^2)^{\nu-\frac{1}{2}} - \big((1-s^2)^{\nu-\frac{1}{2}}s\big)'\Big]\,ds\\
&= \frac{t^{2\nu-1}2^{-\nu}}{\Gamma(\nu+\frac{1}{2})\Gamma(\frac{1}{2})}\int_{-1}^{1} e^{its}(2\nu-1)(1-s^2)^{\nu-\frac{3}{2}}\,ds\\
&= \frac{t^{2\nu-1}2^{-(\nu-1)}}{\Gamma(\nu-\frac{1}{2})\Gamma(\frac{1}{2})}\int_{-1}^{1} e^{its}(1-s^2)^{\nu-\frac{1}{2}}\frac{ds}{\sqrt{1-s^2}}\\
&= t^{\nu}J_{\nu-1}(t).
\end{aligned}
$$

We proceed with the proof of property (3). A calculation using the definition of the Bessel function gives that the left-hand side of (3) is equal to

$$
\frac{2^{-\nu}t^{\nu+1}}{\Gamma(\nu+\frac{1}{2})\Gamma(\frac{1}{2})}\int_{-1}^{+1} e^{ist}\Big((1-s^2)t+2is(\nu+\tfrac{1}{2})\Big)(1-s^2)^{\nu-\frac{1}{2}}\,ds,
$$

which in turn is equal to

$$
-i\frac{2^{-\nu}t^{\nu+1}}{\Gamma(\nu+\frac{1}{2})\Gamma(\frac{1}{2})}\int_{-1}^{+1}\frac{d}{ds}\big(e^{ist}(1-s^2)^{\nu+\frac{1}{2}}\big)\,ds = 0.
$$

Property (4) can be derived directly from (1). Define

$$
G_\nu(t) = \frac{1}{2\pi}\int_0^{2\pi} e^{it\sin\theta}e^{-i\nu\theta}\,d\theta,
$$

for $\nu = 0,1,2,\dots$ and $t>0$. We can show easily that $G_0 = J_0$. If we had

$$
\frac{d}{dt}\big(t^{-\nu}G_\nu(t)\big) = -t^{-\nu}G_{\nu+1}(t), \qquad t>0,
$$

for $\nu\in\mathbf{Z}^+$, we would immediately conclude that $G_\nu = J_\nu$ for $\nu\in\mathbf{Z}^+$. We have

$$
\begin{aligned}
\frac{d}{dt}\big(t^{-\nu}G_\nu(t)\big) &= -t^{-\nu}\Big(\frac{\nu}{t}G_\nu(t) - \frac{dG_\nu}{dt}(t)\Big)\\
&= -t^{-\nu}\int_0^{2\pi}\frac{\nu}{2\pi t}e^{it\sin\theta}e^{-i\nu\theta} - \frac{1}{2\pi}\Big(\frac{d}{dt}e^{it\sin\theta}\Big)e^{-i\nu\theta}\,d\theta\\
&= -\frac{t^{-\nu}}{2\pi}\int_0^{2\pi} i\frac{d}{d\theta}\Big(\frac{e^{it\sin\theta-i\nu\theta}}{t}\Big) + (\cos\theta - i\sin\theta)e^{it\sin\theta}e^{-i\nu\theta}\,d\theta\\
&= -\frac{t^{-\nu}}{2\pi}\int_0^{2\pi} e^{it\sin\theta}e^{-i(\nu+1)\theta}\,d\theta\\
&= -t^{-\nu}G_{\nu+1}(t).
\end{aligned}
$$

For t real, the identity in (5) can be derived by inserting the expression

$$\sum_{j=0}^{\infty}(-1)^j\frac{(ts)^{2j}}{(2j)!}+i\sin(ts)$$

for e^{its} in the definition of the Bessel function $J_\nu(t)$ in Appendix B.1. Algebraic manipulations yield

$$\begin{aligned}J_\nu(t) &= \frac{(t/2)^\nu}{\Gamma(\frac{1}{2})}\sum_{j=0}^{\infty}(-1)^j\frac{1}{\Gamma(\nu+\frac{1}{2})}\frac{t^{2j}}{(2j)!}2\int_0^1 s^{2j-1}(1-s^2)^{\nu-\frac{1}{2}}s\,ds\\ &= \frac{(t/2)^\nu}{\Gamma(\frac{1}{2})}\sum_{j=0}^{\infty}(-1)^j\frac{1}{\Gamma(\nu+\frac{1}{2})}\frac{t^{2j}}{(2j)!}\frac{\Gamma(j+\frac{1}{2})\Gamma(\nu+\frac{1}{2})}{\Gamma(j+\nu+1)}\\ &= \frac{(t/2)^\nu}{\Gamma(\frac{1}{2})}\sum_{j=0}^{\infty}(-1)^j\frac{\Gamma(j+\frac{1}{2})}{\Gamma(j+\nu+1)}\frac{t^{2j}}{(2j)!}.\end{aligned}$$

To derive property (6) we first multiply (1) by t^ν and (2) by $t^{-\nu}$; then we use the product rule for differentiation and we add the resulting expressions.

For further identities on Bessel functions, one may consult Watson's monograph [371].

B.3 An Interesting Identity

Let $\operatorname{Re}\mu > -\frac{1}{2}$, $\operatorname{Re}\nu > -1$, and $t > 0$. Then the following identity is valid:

$$\int_0^1 J_\mu(ts)s^{\mu+1}(1-s^2)^\nu\,ds = \frac{\Gamma(\nu+1)2^\nu}{t^{\nu+1}}J_{\mu+\nu+1}(t).$$

To prove this identity we use formula (5) in Appendix B.2. We have

$$\begin{aligned}&\int_0^1 J_\mu(ts)s^{\mu+1}(1-s^2)^\nu\,ds\\ &\quad= \frac{(\frac{t}{2})^\mu}{\Gamma(\frac{1}{2})}\int_0^1\sum_{j=0}^{\infty}\frac{(-1)^j\Gamma(j+\frac{1}{2})t^{2j}}{\Gamma(j+\mu+1)(2j)!}s^{2j+\mu+\mu}(1-s^2)^\nu s\,ds\\ &\quad= \frac{1}{2}\frac{(\frac{t}{2})^\mu}{\Gamma(\frac{1}{2})}\sum_{j=0}^{\infty}\frac{(-1)^j\Gamma(j+\frac{1}{2})t^{2j}}{\Gamma(j+\mu+1)(2j)!}\int_0^1 u^{j+\mu}(1-u)^\nu\,du\\ &\quad= \frac{1}{2}\frac{(\frac{t}{2})^\mu}{\Gamma(\frac{1}{2})}\sum_{j=0}^{\infty}\frac{(-1)^j\Gamma(j+\frac{1}{2})t^{2j}}{\Gamma(j+\mu+1)(2j)!}\frac{\Gamma(\mu+j+1)\Gamma(\nu+1)}{\Gamma(\mu+\nu+j+2)}\\ &\quad= \frac{2^\nu\Gamma(\nu+1)}{t^{\nu+1}}\frac{(\frac{t}{2})^{\mu+\nu+1}}{\Gamma(\frac{1}{2})}\sum_{j=0}^{\infty}\frac{(-1)^j\Gamma(j+\frac{1}{2})t^{2j}}{\Gamma(j+\mu+\nu+2)(2j)!}\\ &\quad= \frac{\Gamma(\nu+1)2^\nu}{t^{\nu+1}}J_{\mu+\nu+1}(t).\end{aligned}$$

B.4 The Fourier Transform of Surface Measure on $\mathbf{S}^{n-1}$

Let $d\sigma$ denote surface measure on $\mathbf{S}^{n-1}$ for $n \geq 2$. Then the following is true:

$$\widehat{d\sigma}(\xi) = \int_{\mathbf{S}^{n-1}} e^{-2\pi i \xi\cdot\theta}\, d\theta = \frac{2\pi}{|\xi|^{\frac{n-2}{2}}} J_{\frac{n-2}{2}}(2\pi|\xi|).$$

To see this, use the result in Appendix D.3 to write

$$\begin{aligned}
\widehat{d\sigma}(\xi) &= \int_{\mathbf{S}^{n-1}} e^{-2\pi i \xi\cdot\theta}\, d\theta \\
&= \frac{2\pi^{\frac{n-1}{2}}}{\Gamma(\frac{n-1}{2})} \int_{-1}^{+1} e^{-2\pi i |\xi| s} (1-s^2)^{\frac{n-2}{2}} \frac{ds}{\sqrt{1-s^2}} \\
&= \frac{2\pi^{\frac{n-1}{2}}}{\Gamma(\frac{n-1}{2})} \frac{\Gamma(\frac{n-2}{2}+\frac{1}{2})\Gamma(\frac{1}{2})}{(\pi|\xi|)^{\frac{n-2}{2}}} J_{\frac{n-2}{2}}(2\pi|\xi|) \\
&= \frac{2\pi}{|\xi|^{\frac{n-2}{2}}} J_{\frac{n-2}{2}}(2\pi|\xi|).
\end{aligned}$$

B.5 The Fourier Transform of a Radial Function on $\mathbf{R}^n$

Let $f(x) = f_0(|x|)$ be a radial function defined on $\mathbf{R}^n$, where f_0 is defined on $[0,\infty)$. Then the Fourier transform of f is given by the formula

$$\widehat{f}(\xi) = \frac{2\pi}{|\xi|^{\frac{n-2}{2}}} \int_0^\infty f_0(r) J_{\frac{n}{2}-1}(2\pi r|\xi|) r^{\frac{n}{2}}\, dr.$$

To obtain this formula, use polar coordinates to write

$$\begin{aligned}
\widehat{f}(\xi) &= \int_{\mathbf{R}^n} f(x) e^{-2\pi i \xi\cdot x}\, dx \\
&= \int_0^\infty \int_{\mathbf{S}^{n-1}} f_0(r) e^{-2\pi i \xi\cdot r\theta}\, d\theta\, r^{n-1} dr \\
&= \int_0^\infty f_0(r) \widehat{d\sigma}(r\xi) r^{n-1} dr \\
&= \int_0^\infty f_0(r) \frac{2\pi}{(r|\xi|)^{\frac{n-2}{2}}} J_{\frac{n-2}{2}}(2\pi r|\xi|) r^{n-1} dr \\
&= \frac{2\pi}{|\xi|^{\frac{n-2}{2}}} \int_0^\infty f_0(r) J_{\frac{n}{2}-1}(2\pi r|\xi|) r^{\frac{n}{2}}\, dr.
\end{aligned}$$

As an application we take $f(x) = \chi_{B(0,1)}(x)$, where $B(0,1)$ is the unit ball in $\mathbf{R}^n$. We obtain

$$(\chi_{B(0,1)})\widehat{\ }(\xi) = \frac{2\pi}{|\xi|^{\frac{n-2}{2}}} \int_0^1 J_{\frac{n}{2}-1}(2\pi|\xi|r) r^{\frac{n}{2}}\, dr = \frac{J_{\frac{n}{2}}(2\pi|\xi|)}{|\xi|^{\frac{n}{2}}},$$

in view of the result in Appendix B.3. More generally, for $\operatorname{Re}\lambda > -1$, let

$$m_\lambda(\xi) = \begin{cases} (1-|\xi|^2)^\lambda & \text{for } |\xi| \le 1, \\ 0 & \text{for } |\xi| > 1. \end{cases}$$

Then

$$m_\lambda^\vee(x) = \frac{2\pi}{|x|^{\frac{n-2}{2}}} \int_0^1 J_{\frac{n}{2}-1}(2\pi|x|r) r^{\frac{n}{2}} (1-r^2)^\lambda\, dr = \frac{\Gamma(\lambda+1)}{\pi^\lambda} \frac{J_{\frac{n}{2}+\lambda}(2\pi|x|)}{|x|^{\frac{n}{2}+\lambda}},$$

using again the identity in Appendix B.3.

B.6 Bessel Functions of Small Arguments

We seek the behavior of $J_k(r)$ as $r \to 0+$. We fix $\nu \in \mathbf{C}$ with $\operatorname{Re}\nu > -\frac{1}{2}$. Then we have the identity

$$J_\nu(r) = \frac{r^\nu}{2^\nu \Gamma(\nu+1)} + S_\nu(r),$$

where

$$S_\nu(r) = \frac{(r/2)^\nu}{\Gamma(\nu+\frac{1}{2})\Gamma(\frac{1}{2})} \int_{-1}^{+1} (e^{irt}-1)(1-t^2)^{\nu-\frac{1}{2}}\, dt$$

and S_ν satisfies

$$|S_\nu(r)| \le \frac{2^{-\operatorname{Re}\nu} r^{\operatorname{Re}\nu+1}}{(\operatorname{Re}\nu+1)\,|\Gamma(\nu+\frac{1}{2})|\Gamma(\frac{1}{2})}.$$

To prove this estimate we note that

$$\begin{aligned} J_\nu(r) &= \frac{(r/2)^\nu}{\Gamma(\nu+\frac{1}{2})\Gamma(\frac{1}{2})} \int_{-1}^{+1} (1-t^2)^{\nu-\frac{1}{2}}\, dt + S_\nu(r) \\ &= \frac{(r/2)^\nu}{\Gamma(\nu+\frac{1}{2})\Gamma(\frac{1}{2})} \int_0^\pi (\sin^2\phi)^{\nu-\frac{1}{2}} (\sin\phi)\, d\phi + S_\nu(r) \\ &= \frac{(r/2)^\nu}{\Gamma(\nu+\frac{1}{2})\Gamma(\frac{1}{2})} \frac{\Gamma(\nu+\frac{1}{2})\Gamma(\frac{1}{2})}{\Gamma(\nu+1)} + S_\nu(r), \end{aligned}$$

where we evaluated the last integral using the result in Appendix A.4. Using that $|e^{irt}-1| \le r|t|$, we deduce the assertion regarding the size of $|S_\nu(r)|$.

It follows from these facts and the estimate in Appendix A.7 that for $0 < r \leq 1$ and $\operatorname{Re} \nu > -1/2$ we have

$$|J_\nu(r)| \leq C_0(\operatorname{Re}\nu)\, e^{\max\left(\frac{1}{(\operatorname{Re}\nu+\frac{1}{2})^2}, \frac{1}{\operatorname{Re}\nu+\frac{1}{2}}\right)|\operatorname{Im}\nu|^2}\, r^{\operatorname{Re}\nu},$$

where C_0 is a constant that depends smoothly on $\operatorname{Re}\nu \in (-1/2, \infty)$.

B.7 Bessel Functions of Large Arguments

For $r > 0$ and complex numbers ν with $\operatorname{Re}\nu > -1/2$ we prove the identity

$$J_\nu(r) = \frac{(r/2)^\nu}{\Gamma(\nu+\frac{1}{2})\Gamma(\frac{1}{2})}\left[ie^{-ir}\int_0^\infty e^{-rt}(t^2+2it)^{\nu-\frac{1}{2}}\,dt - ie^{ir}\int_0^\infty e^{-rt}(t^2-2it)^{\nu-\frac{1}{2}}\,dt\right].$$

Fix $0 < \delta < 1/10 < 10 < R < \infty$. We consider the region $\Omega_{\delta,R}$ in the complex plane whose boundary is the set consisting of the interval $[-1+\delta, 1-\delta]$ union a quarter circle centered at 1 of radius δ from $1-\delta$ to $1+i\delta$, union the line segments from $1+i\delta$ to $1+iR$, from $1+iR$ to $-1+iR$, and from $-1+iR$ to $-1+i\delta$, union a quarter circle centered at -1 of radius δ from $-1+i\delta$ to $-1+\delta$. This is a simply connected region on the interior of which the holomorphic function $(1-z^2)$ has no zeros. Since $\Omega_{\delta,R}$ is contained in the complement of the negative imaginary axis, there is a holomorphic branch of the logarithm such that $\log(t)$ is real, $\log(-t) = \log|t| + i\pi$, and $\log(it) = \log|t| + i\pi/2$ for $t > 0$. Since the function $\log(1-z^2)$ is well defined and holomorphic in $\Omega_{\delta,R}$, we may define the holomorphic function

$$(1-z^2)^{\nu-\frac{1}{2}} = e^{(\nu-\frac{1}{2})\log(1-z^2)}$$

for $z \in \Omega_{\delta,R}$. Since $e^{irz}(1-z^2)^{\nu-\frac{1}{2}}$ has no poles in $\Omega_{\delta,R}$, Cauchy's theorem yields

$$i\int_\delta^R e^{ir(1+it)}(t^2-2it)^{\nu-\frac{1}{2}}\,dt + \int_{-1+\delta}^{1-\delta} e^{irt}(1-t^2)^{\nu-\frac{1}{2}}\,dt$$
$$+\,i\int_R^\delta e^{ir(-1+it)}(t^2+2it)^{\nu-\frac{1}{2}}\,dt + E(\delta,R) = 0,$$

where $E(\delta,R)$ is the sum of the integrals over the two small quarter-circles of radius δ and the line segment from $1+iR$ to $-1+iR$. The first two of these integrals are bounded by constants times δ, the latter by a constant times $R^{2\operatorname{Re}\nu-1}e^{-rR}$; hence $E(\delta,R) \to 0$ as $\delta \to 0$ and $R \to \infty$. We deduce the identity

$$\int_{-1}^{+1} e^{irt}(1-t^2)^{\nu-\frac{1}{2}}\,dt = ie^{-ir}\int_0^\infty e^{-rt}(t^2+2it)^{\nu-\frac{1}{2}}\,dt - ie^{ir}\int_0^\infty e^{-rt}(t^2-2it)^{\nu-\frac{1}{2}}\,dt.$$

Estimating the two integrals on the right by putting absolute values inside and multiplying by the missing factor $r^\nu 2^{-\nu}(\Gamma(\nu+\frac{1}{2})\Gamma(\frac{1}{2}))^{-1}$, we obtain

$$|J_\nu(r)| \le 2\frac{(r/2)^{\mathrm{Re}\,\nu}\, e^{\frac{\pi}{2}|\mathrm{Im}\,\nu|}}{|\Gamma(\nu+\frac{1}{2})|\Gamma(\frac{1}{2})}\int_0^\infty e^{-rt}t^{\mathrm{Re}\,\nu-\frac{1}{2}}\big(\sqrt{t^2+4}\big)^{\mathrm{Re}\,\nu-\frac{1}{2}}\,dt\,,$$

since the absolute value of the argument of $t^2 \pm 2it$ is at most $\pi/2$. When $\mathrm{Re}\,\nu > 1/2$, we use the inequality $(\sqrt{t^2+4})^{\mathrm{Re}\,\nu-\frac{1}{2}} \le 2^{\mathrm{Re}\,\nu-\frac{3}{2}}\big(t^{\mathrm{Re}\,\nu-\frac{1}{2}}+2^{\mathrm{Re}\,\nu-\frac{1}{2}}\big)$ to get

$$|J_\nu(r)| \le 2\frac{(r/2)^{\mathrm{Re}\,\nu}\, e^{\frac{\pi}{2}|\mathrm{Im}\,\nu|}}{|\Gamma(\nu+\frac{1}{2})|\Gamma(\frac{1}{2})}2^{\mathrm{Re}\,\nu-\frac{3}{2}}\left[\frac{\Gamma(2\mathrm{Re}\,\nu)}{r^{2\mathrm{Re}\,\nu}}+2^{\mathrm{Re}\,\nu}\frac{\Gamma(\mathrm{Re}\,\nu+\frac{1}{2})}{r^{\mathrm{Re}\,\nu+\frac{1}{2}}}\right].$$

When $1/2 \ge \mathrm{Re}\,\nu > -1/2$ we use that $\big(\sqrt{t^2+4}\big)^{\mathrm{Re}\,\nu-\frac{1}{2}} \le 1$ to deduce that

$$|J_\nu(r)| \le 2\frac{(r/2)^{\mathrm{Re}\,\nu}\, e^{\frac{\pi}{2}|\mathrm{Im}\,\nu|}}{|\Gamma(\nu+\frac{1}{2})|\Gamma(\frac{1}{2})}\frac{\Gamma(\mathrm{Re}\,\nu+\frac{1}{2})}{r^{\mathrm{Re}\,\nu+\frac{1}{2}}}.$$

These estimates yield that for $\mathrm{Re}\,\nu > -1/2$ and $r \ge 1$ we have

$$|J_\nu(r)| \le C_1(\mathrm{Re}\,\nu)\, e^{\left(\max\left((\mathrm{Re}\,\nu+\frac{1}{2})^{-2},(\mathrm{Re}\,\nu+\frac{1}{2})^{-1}\right)+\frac{\pi}{2}\right)|\mathrm{Im}\,\nu|^2}\, r^{-1/2}$$

using the result in Appendix A.7, where C_1 is a constant that depends smoothly on $\mathrm{Re}\,\nu$ on the interval $(-1/2,\infty)$.

B.8 Asymptotics of Bessel Functions

We obtain asymptotics for $J_\nu(r)$ as $r \to \infty$ whenever $\mathrm{Re}\,\nu > -1/2$. We have the following identity for $r > 0$:

$$J_\nu(r) = \sqrt{\frac{2}{\pi r}}\cos\left(r-\frac{\pi\nu}{2}-\frac{\pi}{4}\right)+R_\nu(r)\,,$$

where R_ν is given by

$$\begin{aligned}R_\nu(r) &= \frac{(2\pi)^{-\frac{1}{2}}r^\nu}{\Gamma(\nu+\frac{1}{2})}e^{i(r-\frac{\pi\nu}{2}-\frac{\pi}{4})}\int_0^\infty e^{-rt}t^{\nu+\frac{1}{2}}\big[(1+\tfrac{it}{2})^{\nu-\frac{1}{2}}-1\big]\frac{dt}{t}\\ &\quad+\frac{(2\pi)^{-\frac{1}{2}}r^\nu}{\Gamma(\nu+\frac{1}{2})}e^{-i(r-\frac{\pi\nu}{2}-\frac{\pi}{4})}\int_0^\infty e^{-rt}t^{\nu+\frac{1}{2}}\big[(1-\tfrac{it}{2})^{\nu-\frac{1}{2}}-1\big]\frac{dt}{t}\end{aligned}$$

and satisfies $|R_\nu(r)| \le C_\nu r^{-3/2}$ whenever $r \ge 1$.

To see the validity of this identity we write

$$\begin{aligned} ie^{-ir}(t^2+2it)^{\nu-\frac{1}{2}} &= (2t)^{\nu-\frac{1}{2}}e^{-i(r-\frac{\nu\pi}{2}-\frac{\pi}{4})}(1-\tfrac{it}{2})^{\nu-\frac{1}{2}}, \\ -ie^{ir}(t^2-2it)^{\nu-\frac{1}{2}} &= (2t)^{\nu-\frac{1}{2}}e^{i(r-\frac{\nu\pi}{2}-\frac{\pi}{4})}(1+\tfrac{it}{2})^{\nu-\frac{1}{2}}. \end{aligned}$$

Inserting these expressions into the corresponding integrals in the formula proved in Appendix B.7, adding and subtracting 1 from each term $(1\pm\frac{it}{2})^{\nu-\frac{1}{2}}$, and multiplying by the missing factor $(r/2)^\nu/\Gamma(\nu+\frac{1}{2})\Gamma(\frac{1}{2})$, we obtain the claimed identity

$$J_\nu(r) = \sqrt{\frac{2}{\pi r}}\cos\left(r-\frac{\pi\nu}{2}-\frac{\pi}{4}\right)+R_\nu(r).$$

It remains to estimate $R_\nu(r)$. We begin by noting that for a,b real with $a>-1$ we have the pair of inequalities

$$\begin{aligned} &|(1\pm iy)^{a+ib}-1| \le 3\,(|a|+|b|)\,\big(2^{\frac{a+1}{2}}\,e^{\frac{\pi}{2}|b|}\big)\,y && \text{when } 0<y<1, \\ &|(1\pm iy)^{a+ib}-1| \le (1+y^2)^{\frac{a}{2}}e^{\frac{\pi}{2}|b|}+1 \le 2\,\big(2^{\frac{a+1}{2}}\,e^{\frac{\pi}{2}|b|}\big)\,y^a && \text{when } 1\le y<\infty. \end{aligned}$$

The first inequality is proved by splitting into real and imaginary parts and applying the mean value theorem in the real part. Taking $\nu-\frac{1}{2}=a+ib$, $y=t/2$, and inserting these estimates into the integrals appearing in R_ν, we obtain

$$|R_\nu(r)| \le \frac{2^{\frac{1}{2}\mathrm{Re}\,\nu+\frac{5}{4}}e^{\frac{\pi}{2}|\mathrm{Im}\,\nu|}r^{\mathrm{Re}\,\nu}}{(2\pi)^{1/2}|\Gamma(\nu+\frac{1}{2})|}\left[\frac{3\sqrt{2}|\nu|}{2}\int_0^2 e^{-rt}t^{\mathrm{Re}\,\nu+\frac{3}{2}}\frac{dt}{t}+\frac{2\sqrt{2}}{2^{\mathrm{Re}\,\nu}}\int_2^\infty e^{-rt}t^{2\mathrm{Re}\,\nu}\frac{dt}{t}\right].$$

It follows that for all $r>0$ we have

$$\begin{aligned} |R_\nu(r)| &\le 2\frac{2^{\frac{1}{2}\mathrm{Re}\,\nu}e^{\frac{\pi}{2}|\mathrm{Im}\,\nu|}}{|\Gamma(\nu+\frac{1}{2})|}\left[|\nu|\frac{\Gamma(\mathrm{Re}\,\nu+\frac{3}{2})}{r^{3/2}}+\frac{r^{-\mathrm{Re}\,\nu}}{2^{\mathrm{Re}\,\nu}}\int_{2r}^\infty e^{-t}t^{2\mathrm{Re}\,\nu}\frac{dt}{t}\right] \\ &\le 2\frac{2^{\frac{1}{2}\mathrm{Re}\,\nu}e^{\frac{\pi}{2}|\mathrm{Im}\,\nu|}}{|\Gamma(\nu+\frac{1}{2})|}\left[|\nu|\frac{\Gamma(\mathrm{Re}\,\nu+\frac{3}{2})}{r^{3/2}}+\frac{2^{\mathrm{Re}\,\nu}}{r^{\mathrm{Re}\,\nu}}\frac{\Gamma(2\mathrm{Re}\,\nu)}{e^r}\right], \end{aligned}$$

using that $e^{-t}\le e^{-t/2}e^{-r}$ for $t\ge 2r$. We conclude that for $r\ge 1$ and $\mathrm{Re}\,\nu>-1/2$ we have

$$\begin{aligned} |R_\nu(r)| &\le \frac{2^{\frac{1}{2}\mathrm{Re}\,\nu}e^{\frac{\pi}{2}|\mathrm{Im}\,\nu|}}{\Gamma(\mathrm{Re}\,\nu+\frac{1}{2})}e^{\max\left((\mathrm{Re}\,\nu+\frac{1}{2})^{-2},(\mathrm{Re}\,\nu+\frac{1}{2})^{-1}\right)|\mathrm{Im}\,\nu|^2} \\ &\qquad\qquad\left[|\nu|\frac{\Gamma(\mathrm{Re}\,\nu+\frac{3}{2})}{r^{3/2}}+\frac{2^{\mathrm{Re}\,\nu}}{r^{\mathrm{Re}\,\nu}}\frac{\Gamma(2\mathrm{Re}\,\nu)}{e^r}\right] \\ &\le C_2(\mathrm{Re}\,\nu)\,e^{\left(\max\left((\mathrm{Re}\,\nu+\frac{1}{2})^{-2},(\mathrm{Re}\,\nu+\frac{1}{2})^{-1}\right)+\frac{\pi}{2}+1\right)|\mathrm{Im}\,\nu|^2}\,r^{-3/2}, \end{aligned}$$

via the result in Appendix A.7, where C_2 is a constant that depends smoothly on $\mathrm{Re}\,\nu$ on the interval $(-1/2,\infty)$.

B.9 Bessel Functions of general complex indices

We now discuss how to extend $J_\nu(t)$ to complex values of ν when $t>0$. Identity (5)

$$J_\nu(t)=\sum_{j=0}^{\infty}\frac{(-1)^j}{j!}\frac{1}{\Gamma(j+\nu+1)}\left(\frac{t}{2}\right)^{2j+\nu}$$

of Appendix B.2 expresses $J_\nu(t)$ as a power series when ν is a complex number with $\operatorname{Re}\nu>-1/2$. But this power series converges absolutely for all complex values of ν by the ratio test. Indeed, for j sufficiently large we have

$$\left|\frac{\frac{(-1)^{j+1}}{(j+1)!}\frac{1}{\Gamma(j+1+\nu+1)}\left(\frac{t}{2}\right)^{2j+2+\nu}}{\frac{(-1)^j}{j!}\frac{1}{\Gamma(j+\nu+1)}\left(\frac{t}{2}\right)^{2j+\nu}}\right|=\left|\frac{\Gamma(j+\nu+1)\left(\frac{t}{2}\right)^2}{(j+1)\Gamma(j+\nu+2)}\right|=\left|\frac{\left(\frac{t}{2}\right)^2}{(j+1)(j+\nu+1)}\right|$$

which tends to zero as $j\to\infty$, thus the ratio test applies.

Therefore the power series defines an entire function of ν. Since this entire function coincides with J_ν on $(-\frac{1}{2},\infty)\times\mathbf{R}$, by analytic continuation, $J_\nu(t)$ has an entire extension for all $t>0$.

We obtain an integral representation of $J_z(t)$ when $-1/2\geq\operatorname{Re}z>-3/2$. Fix $t>0$. Then we write

$$\begin{aligned}J_z(t)&=\frac{\left(\frac{t}{2}\right)^z}{\Gamma(z+\frac{1}{2})\Gamma(\frac{1}{2})}\int_{-1}^{1}e^{its}(1-s^2)^z\frac{ds}{\sqrt{1-s^2}}\\&=\frac{2\left(\frac{t}{2}\right)^z}{\Gamma(z+\frac{1}{2})\Gamma(\frac{1}{2})}\int_0^1\cos(ts)(1-s^2)^{z-\frac{1}{2}}\,ds\\&=\frac{\left(\frac{t}{2}\right)^z}{\Gamma(z+\frac{1}{2})\Gamma(\frac{1}{2})}\int_0^1\frac{\cos(t\sqrt{1-u})}{\sqrt{1-u}}u^{z-\frac{1}{2}}\,du,\end{aligned}$$

where the last step follows by the change of variables $u=1-s^2$. By the mean value theorem, there is a constant c in $(0,1)$ such that

$$\begin{aligned}&\frac{\left(\frac{t}{2}\right)^z}{\Gamma(z+\frac{1}{2})\Gamma(\frac{1}{2})}\int_0^1\frac{\cos(t\sqrt{1-u})}{\sqrt{1-u}}u^{z-\frac{1}{2}}\,du\\&=\frac{\left(\frac{t}{2}\right)^z}{\Gamma(z+\frac{1}{2})\sqrt{\pi}}\int_0^1\frac{d}{du}\left[\frac{\cos(t\sqrt{1-u})}{\sqrt{1-u}}\right](c)\;u^{z+\frac{1}{2}}\,du+\frac{\left(\frac{t}{2}\right)^z\cos(t)}{\sqrt{\pi}}\frac{\int_0^1u^{z-\frac{1}{2}}\,du}{\Gamma(z+\frac{1}{2})}\\&=\frac{\left(\frac{t}{2}\right)^z}{\Gamma(z+\frac{1}{2})\sqrt{\pi}}\int_0^1\frac{d}{du}\left[\frac{\cos(t\sqrt{1-u})}{\sqrt{1-u}}\right](c)\;u^{z+\frac{1}{2}}\,du+\frac{\left(\frac{t}{2}\right)^z\cos(t)}{\sqrt{\pi}}\frac{1}{\Gamma(z+\frac{3}{2})},\end{aligned}$$

where the second identity is due to the fact that the integral $\int_0^1 u^{z-\frac{1}{2}}\,du$ is equal to $(z+\frac{1}{2})^{-1}$ when $\mathrm{Re}\,z > -1/2$ and thus when divided by $\Gamma(z+\frac{1}{2})$ can be analytically extended to an entire function. It follows from these calculations that

$$J_{-\frac{1}{2}}(t) = \sqrt{\frac{2}{\pi}}\,\frac{\cos(t)}{\sqrt{t}}.$$

We now estimate $J_z(t)$ when $-3/2 < \mathrm{Re}\,z \le -1/2$. Given $q \in (0,1)$ we write

$$J_z(t) = J_z^1(t;q) + J_z^2(t;q) + J_z^3(t;q),$$

where

$$\begin{aligned}
J_z^1(t;q) &= \frac{(\frac{t}{2})^z}{\Gamma(z+\frac{1}{2})\Gamma(\frac{1}{2})}\int_0^q \frac{d}{du}\left[\frac{\cos(t\sqrt{1-u})}{\sqrt{1-u}}\right](c)\; u^{z+\frac{1}{2}}\,du \\
J_z^2(t;q) &= \frac{(\frac{t}{2})^z\cos(t)}{\sqrt{\pi}}\,\frac{q^{z+\frac{1}{2}}}{\Gamma(z+\frac{3}{2})} \\
J_z^3(t;q) &= \frac{(\frac{t}{2})^z}{\Gamma(z+\frac{1}{2})\Gamma(\frac{1}{2})}\int_q^1 \frac{\cos(t\sqrt{1-u})}{\sqrt{1-u}}\,u^{z-\frac{1}{2}}\,du
\end{aligned}$$

for some c satisfying $0 < c < q$.

Suppose that $t > 2$. Then we pick $q = 1/t$. We have

$$|J_z^1(t;1/t)| \le \frac{(\frac{t}{2})^{\mathrm{Re}\,z}}{|\Gamma(z+\frac{1}{2})|\sqrt{\pi}}\,\frac{t}{(\sqrt{1-1/t})^3}\,\frac{1}{\mathrm{Re}\,z+\frac{3}{2}}\left(\frac{1}{t}\right)^{\mathrm{Re}\,z+\frac{3}{2}} \le \frac{C_1(\mathrm{Re}\,z)\,t^{-\frac{1}{2}}}{|\Gamma(z+\frac{1}{2})|},$$

for some constant C_1 depending on $\mathrm{Re}\,z$. The second term obviously satisfies

$$|J_z^2(t;1/t)| \le \frac{C_2(\mathrm{Re}\,z)\,t^{-\frac{1}{2}}}{|\Gamma(z+\frac{3}{2})|}.$$

In $J_z^3(t;1/t)$ we write $\frac{\cos(t\sqrt{1-u})}{\sqrt{1-u}} = \frac{-2}{t}\frac{d}{du}\left[\sin(t\sqrt{1-u})\right]$ and integrate by parts to obtain the estimate

$$|J_z^3(t;1/t)| \le \frac{(\frac{t}{2})^{\mathrm{Re}\,z}}{|\Gamma(z+\frac{1}{2})|\sqrt{\pi}}\left[\frac{2}{t}+\frac{2}{t^{\frac{1}{2}+\mathrm{Re}\,z}}\right] \le \frac{C_3(\mathrm{Re}\,z)\,t^{-\frac{1}{2}}}{|\Gamma(z+\frac{1}{2})|}$$

when $-1/2 \ge \mathrm{Re}\,z > -3/2$. Thus for these z's and $t > 2$ we have

$$|J_z(t)| \le C_0(\mathrm{Re}\,z)\,t^{-\frac{1}{2}}\left[\frac{1}{|\Gamma(z+\frac{1}{2})|}+\frac{1}{|\Gamma(z+\frac{3}{2})|}\right].$$

When $0 < t \leq 2$ we choose $q = 1/2$. Then each of $|J_z^k(t; 1/2)|$, $k = 1, 2, 3$ is bounded by a multiple of $t^{\operatorname{Re} z}$ and we obtain the estimate

$$|J_z(t)| \leq C_{00}(\operatorname{Re} z)\, t^{\operatorname{Re} z} \left[\frac{1}{|\Gamma(z+\frac{1}{2})|} + \frac{1}{|\Gamma(z+\frac{3}{2})|} \right].$$

when $-1/2 \geq \operatorname{Re} z > -3/2$ and $0 < t \leq 2$.

In the special case $\operatorname{Re} z = -1/2$ or $z = -1/2 + i\theta$, $\theta \in \mathbf{R}$, we deduce

$$|J_{-\frac{1}{2}+i\theta}(t)| \leq C t^{-\frac{1}{2}} \left[\frac{1}{|\Gamma(1+i\theta)|} + \frac{1}{|\Gamma(i\theta)|} \right] \leq C t^{-\frac{1}{2}} (1 + |\theta|)\, e^{|\theta|^2}$$

for some constant C independent of all parameters; see Appendix A.7 for the last inequality.

Appendix C
Rademacher Functions

C.1 Definition of the Rademacher Functions

The Rademacher functions are defined on $[0,1]$ as follows: $r_0(t)=1$; $r_1(t)=1$ for $0\le t\le 1/2$ and $r_1(t)=-1$ for $1/2<t\le 1$; $r_2(t)=1$ for $0\le t\le 1/4$, $r_2(t)=-1$ for $1/4<t\le 1/2$, $r_2(t)=1$ for $1/2<t\le 3/4$, and $r_2(t)=-1$ for $3/4<t\le 1$; and so on. According to this definition, we have that $r_j(t)=\operatorname{sgn}(\sin(2^j\pi t))$ for $j=0,1,2,\dots$. It is easy to check that the r_j's are mutually independent random variables on $[0,1]$. This means that for all functions f_j we have

$$\int_0^1 \prod_{j=0}^n f_j(r_j(t))\,dt = \prod_{j=0}^n \int_0^1 f_j(r_j(t))\,dt\,.$$

To see the validity of this identity, we write its right-hand side as

$$\begin{aligned} f_0(1)\prod_{j=1}^n \int_0^1 f_j(r_j(t))\,dt &= f_0(1)\prod_{j=1}^n \frac{f_j(1)+f_j(-1)}{2} \\ &= \frac{f_0(1)}{2^n}\sum_{S\subset\{1,2,\dots,n\}}\prod_{j\in S} f_j(1)\prod_{j\notin S} f_j(-1) \end{aligned}$$

and we observe that there is a one-to-one and onto correspondence between subsets S of $\{1,2,\dots,n\}$ and intervals $I_k=\left[\frac{k}{2^n},\frac{k+1}{2^n}\right]$, $k=0,1,\dots,2^n-1$, such that the restriction of the function $\prod_{j=1}^n f_j(r_j(t))$ on I_k is equal to

$$\prod_{j\in S} f_j(1)\prod_{j\notin S} f_j(-1)\,.$$

It follows that the last of the three equal displayed expressions is

$$f_0(1)\sum_{k=0}^{2^n-1}\int_{I_k}\prod_{j=1}^n f_j(r_j(t))\,dt = \int_0^1 \prod_{j=0}^n f_j(r_j(t))\,dt\,.$$

L. Grafakos, *Classical Fourier Analysis*, Graduate Texts in Mathematics 249, DOI 10.1007/978-1-4939-1194-3, © Springer Science+Business Media New York 2014

C.2 Khintchine's Inequalities

The following property of the Rademacher functions is of fundamental importance and with far-reaching consequences in analysis:

For any $0 < p < \infty$ and for any real-valued square summable sequences $\{a_j\}$ and $\{b_j\}$ we have

$$B_p\Big(\sum_j |a_j+ib_j|^2\Big)^{\frac{1}{2}} \le \Big\|\sum_j (a_j+ib_j)r_j\Big\|_{L^p([0,1])} \le A_p\Big(\sum_j |a_j+ib_j|^2\Big)^{\frac{1}{2}}$$

for some constants $0 < A_p, B_p < \infty$ that depend only on p.

These inequalities reflect the orthogonality of the Rademacher functions in L^p (especially when $p \neq 2$). Khintchine [193] was the first to prove a special form of this inequality, and he used it to estimate the asymptotic behavior of certain random walks. Later this inequality was systematically studied almost simultaneously by Littlewood [226] and by Paley and Zygmund [274], who proved the more general form stated previously. The foregoing inequalities are usually referred to by Khintchine's name.

C.3 Derivation of Khintchine's Inequalities

Both assertions in Appendix C.2 can be derived from an exponentially decaying distributional inequality for the function

$$F(t) = \sum_j (a_j+ib_j)r_j(t), \qquad t \in [0,1],$$

when a_j, b_j are square summable real numbers.

We first obtain a distributional inequality for the above function F under the following three assumptions:

(a) The sequence $\{b_j\}$ is identically zero.
(b) All but finitely many terms of the sequence $\{a_j\}$ are zero.
(c) The sequence $\{a_j\}$ satisfies $(\sum_j |a_j|^2)^{1/2} = 1$.

Let $\rho > 0$. Under assumptions (a), (b), and (c), independence gives

$$\begin{aligned}
\int_0^1 e^{\rho \sum a_j r_j(t)}\,dt &= \prod_j \int_0^1 e^{\rho a_j r_j(t)}\,dt \\
&= \prod_j \frac{e^{\rho a_j}+e^{-\rho a_j}}{2} \\
&\le \prod_j e^{\frac{1}{2}\rho^2 a_j^2} = e^{\frac{1}{2}\rho^2 \sum a_j^2} = e^{\frac{1}{2}\rho^2},
\end{aligned}$$

where we used the inequality $\frac{1}{2}(e^x+e^{-x}) \le e^{\frac{1}{2}x^2}$ for all real x, which can be checked using power series expansions. Since the same argument is also valid for $-\sum a_j r_j(t)$, we obtain that

$$\int_0^1 e^{\rho|F(t)|}\,dt \le 2e^{\frac{1}{2}\rho^2}.$$

From this it follows that

$$e^{\rho\alpha}\big|\{t\in[0,1]:\,|F(t)|>\alpha\}\big| \le \int_0^1 e^{\rho|F(t)|}\,dt \le 2e^{\frac{1}{2}\rho^2}$$

and hence we obtain the distributional inequality

$$d_F(\alpha) = \big|\{t\in[0,1]:\,|F(t)|>\alpha\}\big| \le 2e^{\frac{1}{2}\rho^2-\rho\alpha} = 2e^{-\frac{1}{2}\alpha^2},$$

by picking $\rho=\alpha$. The L^p norm of F can now be computed easily. Formula (1.1.6) gives

$$\big\|F\big\|_{L^p}^p = \int_0^\infty p\alpha^{p-1} d_F(\alpha)\,d\alpha \le \int_0^\infty p\alpha^{p-1} 2e^{-\frac{\alpha^2}{2}}\,d\alpha = 2^{\frac{p}{2}} p\Gamma(p/2).$$

We have now proved that

$$\big\|F\big\|_{L^p} \le \sqrt{2}\,(p\Gamma(p/2))^{\frac{1}{p}}\big\|F\big\|_{L^2}$$

under assumptions (a), (b), and (c).

We now dispose of assumptions (a), (b), and (c). Assumption (b) can be easily eliminated by a limiting argument and (c) by a scaling argument. To dispose of assumption (a), let a_j and b_j be real numbers. We have

$$\begin{aligned}
\Big\|\sum_j (a_j+ib_j)r_j\Big\|_{L^p} &\le \Big\|\Big|\sum_j a_j r_j\Big|+\Big|\sum_j b_j r_j\Big|\Big\|_{L^p}\\
&\le \Big\|\sum_j a_j r_j\Big\|_{L^p}+\Big\|\sum_j b_j r_j\Big\|_{L^p}\\
&\le \sqrt{2}\,(p\Gamma(p/2))^{\frac{1}{p}}\Big(\Big(\sum_j |a_j|^2\Big)^{\frac{1}{2}}+\Big(\sum_j |b_j|^2\Big)^{\frac{1}{2}}\Big)\\
&\le \sqrt{2}\,(p\Gamma(p/2))^{\frac{1}{p}}\sqrt{2}\Big(\sum_j |a_j+ib_j|^2\Big)^{\frac{1}{2}}.
\end{aligned}$$

Let us now set $A_p = 2\big(p\Gamma(p/2)\big)^{1/p}$ when $p>2$. Since we have the trivial estimate $\big\|F\big\|_{L^p} \le \big\|F\big\|_{L^2}$ when $0<p\le 2$, we obtain the required inequality $\big\|F\big\|_{L^p} \le A_p\big\|F\big\|_{L^2}$ with

$$A_p = \begin{cases} 1 & \text{when } 0<p\le 2,\\ 2p^{\frac{1}{p}}\Gamma(p/2)^{\frac{1}{p}} & \text{when } 2<p<\infty. \end{cases}$$

Using Sterling's formula in Appendix A.6, we see that A_p is asymptotic to $\sqrt{p}$ as $p \to \infty$.

We now discuss the converse inequality $B_p \|F\|_{L^2} \le \|F\|_{L^p}$. It is clear that $\|F\|_{L^2} \le \|F\|_{L^p}$ when $p \ge 2$ and we may therefore take $B_p = 1$ for $p \ge 2$. Let us now consider the case $0 < p < 2$. Pick an s such that $2 < s < \infty$. Find a $0 < \theta < 1$ such that

$$\frac{1}{2} = \frac{1-\theta}{p} + \frac{\theta}{s}.$$

Then

$$\|F\|_{L^2} \le \|F\|_{L^p}^{1-\theta} \|F\|_{L^s}^{\theta} \le \|F\|_{L^p}^{1-\theta} A_s^{\theta} \|F\|_{L^2}^{\theta}.$$

It follows that

$$\|F\|_{L^2} \le A_s^{\frac{\theta}{1-\theta}} \|F\|_{L^p}.$$

We have now proved the inequality $B_p \|F\|_{L^2} \le \|F\|_{L^p}$ with

$$B_p = \begin{cases} 1 & \text{when } 2 \le p < \infty, \\ \sup\limits_{s>2} A_s^{-\frac{\frac{1}{p}-\frac{1}{2}}{\frac{1}{2}-\frac{1}{s}}} & \text{when } 0 < p < 2. \end{cases}$$

Observe that the function $s \to A_s^{-\left(\frac{1}{p}-\frac{1}{2}\right)/\left(\frac{1}{2}-\frac{1}{s}\right)}$ tends to zero as $s \to 2+$ and as $s \to \infty$. Hence it must attain its maximum for some $s = s(p)$ in the interval $(2,\infty)$. We see that $B_p \ge 16 \cdot 256^{-1/p}$ when $p < 2$ by taking $s = 4$.

It is worthwhile to mention that the best possible values of the constants A_p and B_p in Khintchine's inequality are known when $b_j = 0$. In this case Szarek [353] showed that the best possible value of B_1 is $1/\sqrt{2}$, and later Haagerup [141] found that when $b_j = 0$ the best possible values of A_p and B_p are the numbers

$$A_p = \begin{cases} 1 & \text{when } 0 < p \le 2, \\ 2^{\frac{1}{2}} \pi^{-\frac{1}{2p}} \Gamma(\frac{p+1}{2}) & \text{when } 2 < p < \infty, \end{cases}$$

and

$$B_p = \begin{cases} 2^{\frac{1}{2}-\frac{1}{p}} & \text{when } 0 < p \le p_0, \\ 2^{\frac{1}{2}} \pi^{-\frac{1}{2p}} \Gamma(\frac{p+1}{2}) & \text{when } p_0 < p < 2, \\ 1 & \text{when } 2 < p < \infty, \end{cases}$$

where $p_0 = 1.84742\ldots$ is the unique solution of the equation $2\Gamma(\frac{p+1}{2}) = \sqrt{\pi}$ in the interval $(1,2)$.

C.4 Khintchine's Inequalities for Weak Type Spaces

We note that the following weak type estimates are valid:

$$4^{-\frac{1}{p}} B_{\frac{p}{2}} \Big(\sum_j |a_j + ib_j|^2 \Big)^{\frac{1}{2}} \le \Big\| \sum_j (a_j + ib_j) r_j \Big\|_{L^{p,\infty}} \le A_p \Big(\sum_j |a_j + ib_j|^2 \Big)^{\frac{1}{2}}$$

for all $0 < p < \infty$.

Indeed, the upper estimate is a simple consequence of the fact that L^p is a subspace of $L^{p,\infty}$. For the converse inequality we use the fact that $L^{p,\infty}([0,1])$ is contained in $L^{p/2}([0,1])$ and we have (see Exercise 1.1.11)

$$\big\| F \big\|_{L^{p/2}} \le 4^{\frac{1}{p}} \big\| F \big\|_{L^{p,\infty}}.$$

Since any Lorentz space $L^{p,q}([0,1])$ can be sandwiched between $L^{2p}([0,1])$ and $L^{p/2}([0,1])$, similar inequalities hold for all Lorentz spaces $L^{p,q}([0,1])$, $0 < p < \infty$, $0 < q \le \infty$.

C.5 Extension to Several Variables

We first extend the inequality on the right in Appendix C.2 to several variables. For a positive integer n we let

$$F_n(t_1, \dots, t_n) = \sum_{j_1} \cdots \sum_{j_n} c_{j_1,\dots,j_n} r_{j_1}(t_1) \cdots r_{j_n}(t_n),$$

for $t_j \in [0,1]$, where $c_{j_1,\dots,j_n}$ is a sequence of complex numbers and F_n is a function defined on $[0,1]^n$.

For any $0 < p < \infty$ and for any complex-valued square summable sequence of n variables $\{c_{j_1,\dots,j_n}\}_{j_1,\dots,j_n}$, we have the following inequalities for F_n:

$$B_p^n \Big(\sum_{j_1} \cdots \sum_{j_n} |c_{j_1,\dots,j_n}|^2 \Big)^{\frac{1}{2}} \le \big\| F_n \big\|_{L^p([0,1]^n)} \le A_p^n \Big(\sum_{j_1} \cdots \sum_{j_n} |c_{j_1,\dots,j_n}|^2 \Big)^{\frac{1}{2}},$$

where A_p, B_p are the constants in Appendix C.2. The norms are over $[0,1]^n$.

The case $n = 2$ is indicative of the general case. For $p \ge 2$ we have

$$\begin{aligned} \int_0^1 \int_0^1 |F_2(t_1,t_2)|^p \, dt_1 \, dt_2 &\le A_p^p \int_0^1 \Big(\sum_{j_1} \Big| \sum_{j_2} c_{j_1,j_2} r_{j_2}(t_2) \Big|^2 \Big)^{\frac{p}{2}} dt_2 \\ &\le A_p^p \Big(\sum_{j_1} \Big(\int_0^1 \Big| \sum_{j_2} c_{j_1,j_2} r_{j_2}(t_2) \Big|^p dt_2 \Big)^{\frac{2}{p}} \Big)^{\frac{p}{2}} \\ &\le A_p^{2p} \Big(\sum_{j_1} \sum_{j_2} |c_{j_1,j_n}|^2 \Big)^{\frac{p}{2}}, \end{aligned}$$

where we used Minkowski's integral inequality (with exponent $p/2 \geq 1$) in the second inequality and the result in the case $n = 1$ twice.

The case $p < 2$ follows trivially from Hölder's inequality with constant $A_p = 1$. The reverse inequalities follow exactly as in the case of one variable. Replacing A_p by A_p^n in the argument, giving the reverse inequality in the case $n = 1$, we obtain the constant B_p^n.

Likewise one may extend the weak type inequalities of Appendix C.3 in several variables.

Appendix D
Spherical Coordinates

D.1 Spherical Coordinate Formula

Switching integration from spherical coordinates to Cartesian is achieved via the following identity:

$$\int_{RS^{n-1}} f(x)\,d\sigma(x) = \int_{\varphi_1=0}^{\pi} \cdots \int_{\varphi_{n-2}=0}^{\pi} \int_{\varphi_{n-1}=0}^{2\pi} f(x(\varphi))J(n,R,\varphi)\,d\varphi_{n-1}\cdots d\varphi_1,$$

where

$$\begin{aligned}
x_1 &= R\cos\varphi_1\,,\\
x_2 &= R\sin\varphi_1\,\cos\varphi_2\,,\\
x_3 &= R\sin\varphi_1\,\sin\varphi_2\,\cos\varphi_3\,,\\
&\dots\\
x_{n-1} &= R\sin\varphi_1\,\sin\varphi_2\,\sin\varphi_3\cdots\sin\varphi_{n-2}\,\cos\varphi_{n-1}\,,\\
x_n &= R\sin\varphi_1\,\sin\varphi_2\,\sin\varphi_3\cdots\sin\varphi_{n-2}\,\sin\varphi_{n-1}\,,
\end{aligned}$$

and $0 \le \varphi_1,\dots,\varphi_{n-2} \le \pi$, $0 \le \varphi_{n-1} = \theta < 2\pi$,

$$x(\varphi) = (x_1(\varphi_1,\dots,\varphi_{n-1}),\dots,x_n(\varphi_1,\dots,\varphi_{n-1}))\,,$$

and

$$\begin{aligned}
J(n,R,\varphi) &= \left(\sum_{k=1}^{n}\left|\det\left(\frac{\partial x_i}{\partial \varphi_j}\right)_{\substack{1\le i\ne k\le n\\ 1\le j\le n-1}}\right|^2\right)^{\frac{1}{2}}\\
&= R^{n-1}(\sin\varphi_1)^{n-2}\cdots(\sin\varphi_{n-3})^2(\sin\varphi_{n-2})
\end{aligned}$$

is the Jacobian of the transformation.

L. Grafakos, *Classical Fourier Analysis*, Graduate Texts in Mathematics 249, DOI 10.1007/978-1-4939-1194-3, © Springer Science+Business Media New York 2014

D.2 A Useful Change of Variables Formula

The following formula is useful in computing integrals over the sphere $\mathbf{S}^{n-1}$ when $n \geq 2$. Let f be a function defined on $\mathbf{S}^{n-1}$. Then we have

$$\int_{R\mathbf{S}^{n-1}} f(x)\,d\sigma(x) = \int_{-R}^{+R} \int_{\sqrt{R^2-s^2}\,\mathbf{S}^{n-2}} f(s,\theta)\,d\theta\,\frac{R\,ds}{\sqrt{R^2-s^2}}\,.$$

To prove this formula, let $\varphi' = (\varphi_2,\dots,\varphi_{n-1})$ and

$$x' = x'(\varphi') = (\cos\varphi_2, \sin\varphi_2\cos\varphi_3,\dots,\sin\varphi_2\cdots\sin\varphi_{n-2}\sin\varphi_{n-1})\,.$$

Using the change of variables in Appendix D.1 we express

$$\int_{R\mathbf{S}^{n-1}} f(x)\,d\sigma(x)$$

as the iterated integral

$$\int_{\varphi_1=0}^{\pi}\left[\int_{\varphi_2=0}^{\pi}\cdots\int_{\varphi_{n-1}=0}^{2\pi} f(R\cos\varphi_1, R\sin\varphi_1\,x'(\varphi'))J(n-1,1,\varphi')\,d\varphi'\right]\frac{R\,d\varphi_1}{(R\sin\varphi_1)^{2-n}}\,,$$

and we can realize the expression inside the square brackets as

$$\int_{\mathbf{S}^{n-2}} f(R\cos\varphi_1, R\sin\varphi_1\,x')\,d\sigma(x')\,.$$

Consequently,

$$\int_{R\mathbf{S}^{n-1}} f(x)\,d\sigma(x) = \int_{\varphi_1=0}^{\pi}\int_{\mathbf{S}^{n-2}} f(R\cos\varphi_1, R\sin\varphi_1\,x')\,d\sigma(x')R^{n-1}(\sin\varphi_1)^{n-2}d\varphi_1\,,$$

and the change of variables

$$s = R\cos\varphi_1\,, \qquad \varphi_1 \in (0,\pi),$$

$$ds = -R\sin\varphi_1\,d\varphi_1\,, \qquad \sqrt{R^2-s^2} = R\sin\varphi_1\,,$$

yields

$$\int_{R\mathbf{S}^{n-1}} f(x)\,d\sigma(x) = \int_{-R}^{R}\left\{\int_{\mathbf{S}^{n-2}} f(s,\sqrt{R^2-s^2}\,\theta)\,d\theta\right\}\left(\sqrt{R^2-s^2}\right)^{n-2}\frac{R\,ds}{\sqrt{R^2-s^2}}\,.$$

Rescaling the sphere $\mathbf{S}^{n-2}$ to $\sqrt{R^2-s^2}\,\mathbf{S}^{n-2}$ yields the claimed identity.

D.3 Computation of an Integral over the Sphere

Let K be a function on the line. We use the result in Appendix D.2 to show that for $n \geq 2$ we have

$$\int_{\mathbf{S}^{n-1}} K(x\cdot\theta)\,d\theta = \frac{2\pi^{\frac{n-1}{2}}}{\Gamma\left(\frac{n-1}{2}\right)}\int_{-1}^{+1} K(s|x|)\left(\sqrt{1-s^2}\right)^{n-3}ds$$

when $x \in \mathbf{R}^n \setminus \{0\}$. Let $x' = x/|x|$ and pick a matrix $A \in O(n)$ such that $Ae_1 = x'$, where $e_1 = (1,0,\dots,0)$. We have

$$\begin{aligned}
\int_{\mathbf{S}^{n-1}} K(x\cdot\theta)\,d\theta &= \int_{\mathbf{S}^{n-1}} K(|x|(x'\cdot\theta))\,d\theta \\
&= \int_{\mathbf{S}^{n-1}} K(|x|(Ae_1\cdot\theta))\,d\theta \\
&= \int_{\mathbf{S}^{n-1}} K(|x|(e_1\cdot A^{-1}\theta))\,d\theta \\
&= \int_{\mathbf{S}^{n-1}} K(|x|\theta_1)\,d\theta \\
&= \int_{-1}^{+1} K(|x|s)\,\omega_{n-2}\left(\sqrt{1-s^2}\right)^{n-2}\frac{ds}{\sqrt{1-s^2}} \\
&= \omega_{n-2}\int_{-1}^{+1} K(s|x|)\left(\sqrt{1-s^2}\right)^{n-3}ds,
\end{aligned}$$

where $\omega_{n-2} = 2\pi^{\frac{n-1}{2}}\Gamma\left(\frac{n-1}{2}\right)^{-1}$ is the surface area of $\mathbf{S}^{n-2}$.

For example, we have

$$\int_{\mathbf{S}^{n-1}} \frac{d\theta}{|\xi\cdot\theta|^{\alpha}} = \omega_{n-2}\int_{-1}^{+1}\frac{1}{|s|^{\alpha}|\xi|^{\alpha}}(1-s^2)^{\frac{n-3}{2}}\,ds = \frac{1}{|\xi|^{\alpha}}\frac{2\pi^{\frac{n-1}{2}}\Gamma\left(\frac{1-\alpha}{2}\right)}{\Gamma\left(\frac{n-\alpha}{2}\right)},$$

and the integral converges only when $\operatorname{Re}\alpha < 1$.

D.4 The Computation of Another Integral over the Sphere

We compute the following integral for $n \geq 2$:

$$\int_{\mathbf{S}^{n-1}} \frac{d\theta}{|\theta - e_1|^{\alpha}},$$

where $e_1 = (1,0,\dots,0)$. Applying the formula in Appendix D.2, we obtain

$$\begin{aligned}
\int_{\mathbf{S}^{n-1}} \frac{d\theta}{|\theta - e_1|^\alpha} &= \int_{-1}^{+1} \int_{\theta \in \sqrt{1-s^2}\,\mathbf{S}^{n-2}} \frac{d\theta}{(|s-1|^2 + |\theta|^2)^{\frac{\alpha}{2}}} \frac{ds}{\sqrt{1-s^2}} \\
&= \int_{-1}^{+1} \omega_{n-2} \frac{(1-s^2)^{\frac{n-2}{2}}}{\big((1-s)^2 + 1 - s^2\big)^{\frac{\alpha}{2}}} \frac{ds}{\sqrt{1-s^2}} \\
&= \frac{\omega_{n-2}}{2^{\frac{\alpha}{2}}} \int_{-1}^{+1} \frac{(1-s^2)^{\frac{n-3}{2}}}{(1-s)^{\frac{\alpha}{2}}} ds \\
&= \frac{\omega_{n-2}}{2^{\frac{\alpha}{2}}} \int_{-1}^{+1} (1-s)^{\frac{n-3-\alpha}{2}} (1+s)^{\frac{n-3}{2}} ds,
\end{aligned}$$

which converges exactly when $\operatorname{Re}\alpha < n-1$.

D.5 Integration over a General Surface

Suppose that S is a hypersurface in $\mathbf{R}^n$ of the form $S = \{(u, \Phi(u)) : u \in D\}$, where D is an open subset of $\mathbf{R}^{n-1}$ and Φ is a continuously differentiable mapping from D to $\mathbf{R}$. Let σ be the canonical surface measure on S. If g is a function on S, then we have

$$\int_S g(y)\, d\sigma(y) = \int_D g(x, \Phi(x)) \Big(1 + \sum_{j=1}^{n-1} |\partial_j \Phi(x)|^2\Big)^{\frac{1}{2}} dx.$$

Specializing to the sphere, we obtain

$$\int_{\mathbf{S}^{n-1}} g(\theta)\, d\theta = \int_{\substack{\xi' \in \mathbf{R}^{n-1} \\ |\xi'|<1}} \Big[g(\xi', \sqrt{1-|\xi'|^2}) + g(\xi', -\sqrt{1-|\xi'|^2})\Big] \frac{d\xi'}{\sqrt{1-|\xi'|^2}}.$$

D.6 The Stereographic Projection

Define a map $\Pi : \mathbf{R}^n \to \mathbf{S}^n$ by the formula

$$\Pi(x_1, \dots, x_n) = \left(\frac{2x_1}{1+|x|^2}, \dots, \frac{2x_n}{1+|x|^2}, \frac{|x|^2 - 1}{1+|x|^2} \right).$$

It is easy to see that Π is a one-to-one map from $\mathbf{R}^n$ onto the sphere $\mathbf{S}^n$ minus the north pole $e_{n+1} = (0, \dots, 0, 1)$. Its inverse is given by the formula

$$\Pi^{-1}(\theta_1, \dots, \theta_{n+1}) = \left(\frac{\theta_1}{1-\theta_{n+1}}, \dots, \frac{\theta_n}{1-\theta_{n+1}} \right).$$

The Jacobian of the map is verified to be $J_\Pi(x) = \left(\frac{2}{1+|x|^2}\right)^{-n}$, and the following change of variables formulas are valid:

$$\int_{\mathbf{S}^n} F(\theta)\,d\theta = \int_{\mathbf{R}^n} F(\Pi(x))J_\Pi(x)\,dx$$

and

$$\int_{\mathbf{R}^n} F(x)\,dx = \int_{\mathbf{S}^n} F(\Pi^{-1}(\theta))\,J_{\Pi^{-1}}(\theta)\,d\theta\,,$$

where

$$J_{\Pi^{-1}}(\theta) = \frac{1}{J_\Pi(\Pi^{-1}(\theta))} = \left(\frac{|\theta_1|^2+\cdots+|\theta_n|^2+|1-\theta_{n+1}|^2}{2|1-\theta_{n+1}|^2}\right)^n.$$

Another interesting formula about the stereographic projection Π is

$$|\Pi(x)-\Pi(y)| = 2|x-y|(1+|x|^2)^{-1/2}(1+|y|^2)^{-1/2}, \qquad x,y\in\mathbf{R}^n.$$

Appendix E
Some Trigonometric Identities and Inequalities

The following inequalities are valid for t real:

$$
\begin{aligned}
0<t<\frac{\pi}{2} &\implies \sin(t)<t<\tan(t)\,,\\
0<|t|<\frac{\pi}{2} &\implies \frac{2}{\pi}<\frac{\sin(t)}{t}<1\,,\\
-\infty<t<+\infty &\implies |\sin(t)|\le|t|\,,\\
-\infty<t<+\infty &\implies |1-\cos(t)|\le\frac{|t|^2}{2}\,,\\
-\infty<t<+\infty &\implies |1-e^{it}|\le|t|\,,\\
|t|\le\frac{\pi}{2} &\implies |\sin(t)|\ge\frac{2|t|}{\pi}\,,\\
|t|\le\pi &\implies |1-\cos(t)|\ge\frac{2|t|^2}{\pi^2}\,,\\
|t|\le\pi &\implies |1-e^{it}|\ge\frac{2|t|}{\pi}\,.
\end{aligned}
$$

The following sum to product formulas are valid:

$$
\begin{aligned}
\sin(a)+\sin(b) &= 2\sin\left(\frac{a+b}{2}\right)\cos\left(\frac{a-b}{2}\right),\\
\sin(a)-\sin(b) &= 2\cos\left(\frac{a+b}{2}\right)\sin\left(\frac{a-b}{2}\right),\\
\cos(a)+\cos(b) &= 2\cos\left(\frac{a+b}{2}\right)\cos\left(\frac{a-b}{2}\right),\\
\cos(a)-\cos(b) &= -2\sin\left(\frac{a+b}{2}\right)\sin\left(\frac{a-b}{2}\right).
\end{aligned}
$$

L. Grafakos, *Classical Fourier Analysis*, Graduate Texts in Mathematics 249,
DOI 10.1007/978-1-4939-1194-3, © Springer Science+Business Media New York 2014

The following identities are also easily proved:

$$\sum_{k=1}^{N} \cos(kx) = -\frac{1}{2} + \frac{\sin((N+\frac{1}{2})x)}{2\sin(\frac{x}{2})},$$

$$\sum_{k=1}^{N} \sin(kx) = \frac{\cos(\frac{x}{2}) - \cos((N+\frac{1}{2})x)}{2\sin(\frac{x}{2})}.$$

Appendix F
Summation by Parts

Let $\{a_k\}_{k=0}^{\infty}$, $\{b_k\}_{k=0}^{\infty}$ be two sequences of complex numbers. Then for $N \geq 1$ we have

$$\sum_{k=0}^{N} a_k b_k = A_N b_N - \sum_{k=0}^{N-1} A_k (b_{k+1} - b_k),$$

where

$$A_k = \sum_{j=0}^{k} a_j .$$

More generally we have

$$\sum_{k=M}^{N} a_k b_k = A_N b_N - A_{M-1} b_M - \sum_{k=M}^{N-1} A_k (b_{k+1} - b_k),$$

whenever $0 \leq M \leq N$, where $A_{-1} = 0$ and

$$A_k = \sum_{j=0}^{k} a_j$$

for $k \geq 0$.

L. Grafakos, *Classical Fourier Analysis*, Graduate Texts in Mathematics 249,
DOI 10.1007/978-1-4939-1194-3, © Springer Science+Business Media New York 2014

Appendix G
Basic Functional Analysis

A quasi-norm is a nonnegative functional $\|\cdot\|$ on a vector space X that satisfies $\|x+y\|_X \le K(\|x\|_X + \|y\|_X)$ for some $K \ge 0$ and all $x,y \in X$ and also $\|\lambda x\|_X = |\lambda|\|x\|_X$ for all scalars λ. When $K=1$, the quasi-norm is called a norm. A quasi-Banach space is a quasi-normed space that is complete with respect to the topology generated by the quasi-norm. The proofs of the following theorems can be found in several books including Albiac and Kalton [1], Kalton Peck and Roberts [188], and Rudin [306].

The Hahn–Banach theorem. Let X be a normed vector space (over the real or complex numbers), let Y be a subspace of X, and let P be a positively homogeneous subadditive[1] functional on X. Then for every linear functional Λ on Y that satisfies

$$|\Lambda(y)| \le P(y)$$

for all $y \in Y$, there is a linear functional Λ' on X such that

$$\begin{aligned} \Lambda'(y) &= \Lambda(y) \qquad \text{for all } y \in Y, \\ |\Lambda'(x)| &\le P(x) \qquad \text{for all } x \in X. \end{aligned}$$

In particular, every bounded linear functional on a subspace has an extension on the entire space with the same norm.

Banach–Alaoglou theorem. Let X be a normed vector space and let X^* be the space of all bounded linear functionals on X. Then the closed unit ball of X^* is compact in the weak* topology.

A special case is the sequential version of this theorem, which asserts that the closed unit ball of the dual space of a separable normed vector space is sequentially compact in the weak* topology. Indeed, the weak* topology on the closed unit ball

[1] this means that $P(x) \ge 0$ for all $x \in X$, $P(\lambda x) = \lambda P(x)$ for all $\lambda > 0$ and all $x \in X$, and that $P(x+z) \le P(x) + P(z)$ for all $x,z \in X$.

L. Grafakos, *Classical Fourier Analysis*, Graduate Texts in Mathematics 249,
DOI 10.1007/978-1-4939-1194-3, © Springer Science+Business Media New York 2014

of the dual of a separable space is metrizable, and thus compactness and sequential compactness are equivalent.

Open mapping theorem. Suppose that X and Y are quasi-Banach spaces and T is a bounded surjective linear map from X onto Y. Then T maps open sets to open sets, i.e., it is an open mapping. Moreover, if T is injective, then there exists a constant $K < \infty$ such that for all $x \in X$ we have

$$\|x\|_X \leq K \|T(x)\|_Y .$$

Closed graph theorem. Suppose that X and Y are quasi-Banach spaces and T is a linear map from X to Y whose graph is a closed set, i.e., whenever $x_k, x \in X$ and $(x_k, T(x_k)) \mapsto (x, y)$ in $X \times Y$ for some $y \in Y$, then $T(x) = y$. Then T is a bounded linear map from X to Y.

Uniform boundedness principle. Suppose that X is a quasi-Banach space, Y is a quasi-normed space and $(T_\alpha)_{\alpha \in I}$ is a family of bounded linear maps from X to Y such that for all $x \in X$ there exists a $C_x < \infty$ such that

$$\sup_{\alpha \in I} \|T_\alpha(x)\|_Y \leq C_x .$$

Then there exists a constant $K < \infty$ such that

$$\sup_{\alpha \in I} \|T_\alpha\|_{X \to Y} \leq K .$$

Appendix H
The Minimax Lemma

Minimax type results are used in the theory of games and have their origin in the work of Von Neumann [369]. Much of the theory in this subject is based on convex analysis techniques. For instance, this is the case with the next proposition, which is needed in the "difficult" inequality in the proof of the minimax lemma. We refer to Fan [111] for a general account of minimax results. The following exposition is based on the simple presentation in Appendix A2 of [122].

Minimax Lemma. *Let A, B be convex subsets of certain vector spaces. Assume that a topology is defined in B for which it is a compact Hausdorff space and assume that there is a function $\Phi : A \times B \to \mathbf{R}\bigcup\{+\infty\}$ that satisfies the following:*
(a) $\Phi(\,.\,,b)$ is a concave function on A for each $b \in B$,
(b) $\Phi(a,\,.\,)$ is a convex function on B for each $a \in A$,
(c) $\Phi(a,\,.\,)$ is lower semicontinuous on B for each $a \in A$.
Then the following identity holds:

$$\min_{b\in B}\sup_{a\in A}\Phi(a,b) = \sup_{a\in A}\min_{b\in B}\Phi(a,b)\,.$$

To prove the lemma we need the following proposition:

Proposition. *Let B be a convex compact subset of a vector space and suppose that $g_j : B \to \mathbf{R}\bigcup\{+\infty\}$, $j = 1,2,\dots,n$, are convex and lower semicontinuous functions. If*

$$\max_{1\le j\le n} g_j(b) > 0 \quad \textit{for all} \quad b \in B,$$

then there exist nonnegative numbers $\lambda_1,\lambda_2,\dots,\lambda_n$ such that

$$\lambda_1 g_1(b) + \lambda_2 g_2(b) + \cdots + \lambda_n g_n(b) > 0 \quad \textit{for all} \quad b \in B\,.$$

Proof. We first consider the case $n = 2$. Define subsets of B

$$B_1 = \{b \in B:\ g_1(b) \le 0\}, \quad B_2 = \{b \in B:\ g_2(b) \le 0\}\,.$$

L. Grafakos, *Classical Fourier Analysis*, Graduate Texts in Mathematics 249,
DOI 10.1007/978-1-4939-1194-3, © Springer Science+Business Media New York 2014

If $B_1 = \emptyset$, we take $\lambda_1 = 1$ and $\lambda_2 = 0$, and we similarly deal with the case $B_2 = \emptyset$. If B_1 and B_2 are nonempty, then they are closed and thus compact. The hypothesis of the proposition implies that $g_2(b) > 0 \geq g_1(b)$ for all $b \in B_1$. Therefore, the function $-g_1(b)/g_2(b)$ is well defined and upper semicontinuous on B_1 and thus attains its maximum. The same is true for $-g_2(b)/g_1(b)$ defined on B_2. We set

$$\mu_1 = \max_{b \in B_1} \frac{-g_1(b)}{g_2(b)} \geq 0, \qquad \mu_2 = \max_{b \in B_2} \frac{-g_2(b)}{g_1(b)} \geq 0.$$

We need to find $\lambda > 0$ such that $\lambda g_1(b) + g_2(b) > 0$ for all $b \in B$. This is clearly satisfied if $b \notin B_1 \bigcup B_2$, while for $b_1 \in B_1$ and $b_2 \in B_2$ we have

$$\begin{aligned} \lambda g_1(b_1) + g_2(b_1) &\geq (1 - \lambda \mu_1) g_2(b_1), \\ \lambda g_1(b_2) + g_2(b_2) &\geq (\lambda - \mu_2) g_1(b_2). \end{aligned}$$

Therefore, it suffices to find a $\lambda > 0$ such that $1 - \lambda\mu_1 > 0$ and $\lambda - \mu_2 > 0$. Such a λ exists if and only if $\mu_1\mu_2 < 1$. To prove that $\mu_1\mu_2 < 1$, we can assume that $\mu_1 \neq 0$ and $\mu_2 \neq 0$. Then we take $b_1 \in B_1$ and $b_2 \in B_2$, for which the maxima μ_1 and μ_2 are attained, respectively. Then we have

$$\begin{aligned} g_1(b_1) + \mu_1 g_2(b_1) &= 0, \\ g_1(b_2) + \frac{1}{\mu_2} g_2(b_2) &= 0. \end{aligned}$$

But $g_1(b_1) < 0 < g_1(b_2)$; thus taking $b_\theta = \theta b_1 + (1-\theta) b_2$ for some θ in $(0,1)$, we have

$$g_1(b_\theta) \leq \theta g_1(b_1) + (1-\theta) g_1(b_2) = 0.$$

Considering the same convex combination of the last displayed equations and using this identity, we obtain that

$$\mu_1 \mu_2 \theta g_2(b_1) + (1-\theta) g_2(b_2) = 0.$$

The hypothesis of the proposition implies that $g_2(b_\theta) > 0$ and the convexity of g_2:

$$\theta g_2(b_1) + (1-\theta) g_2(b_2) > 0.$$

Since $g_2(b_1) > 0$, we must have $\mu_1\mu_2 g_2(b_1) < g_2(b_1)$, which gives $\mu_1\mu_2 < 1$. This proves the required claim and completes the case $n = 2$.

We now use induction to prove the proposition for arbitrary n. Assume that the result has been proved for $n-1$ functions. Consider the subset of B

$$B_n = \{ b \in B : \; g_n(b) \leq 0 \}.$$

If $B_n = \emptyset$, we choose $\lambda_1 = \lambda_2 = \cdots = \lambda_{n-1} = 0$ and $\lambda_n = 1$. If B_n is not empty, then it is compact and convex and we can restrict $g_1, g_2, \ldots, g_{n-1}$ to B_n. Using the induction hypothesis, we can find $\lambda_1, \lambda_2, \ldots, \lambda_{n-1} \geq 0$ such that

$$g_0(b) = \lambda_1 g_1(b) + \lambda_2 g_2(b) + \cdots + \lambda_{n-1} g_{n-1}(b) > 0$$

for all $b \in B_n$. Then g_0 and g_n are convex lower semicontinuous functions on B, and $\max(g_0(b), g_n(b)) > 0$ for all $b \in B$. Using the case $n = 2$, which was first proved, we can find $\lambda_0, \lambda_n \geq 0$ such that for all $b \in B$ we have

$$\begin{aligned} 0 &< \lambda_0 g_0(b) + \lambda_n g_n(b) \\ &= \lambda_0 \lambda_1 g_1(b) + \lambda_0 \lambda_2 g_2(b) + \cdots + \lambda_0 \lambda_{n-1} g_{n-1}(b) + \lambda_n g_n(b). \end{aligned}$$

This establishes the case of n functions and concludes the proof of the induction and hence of the proposition. □

We now turn to the proof of the minimax lemma.

Proof. The fact that the left-hand side in the required conclusion of the minimax lemma is at least as big as the right-hand side is obvious. We can therefore concentrate on the converse inequality. In doing this we may assume that the right-hand side is finite. Without loss of generality we can subtract a finite constant from $\Phi(a,b)$, and so we can also assume that

$$\sup_{a \in A} \min_{b \in B} \Phi(a,b) = 0.$$

Then, by hypothesis *(c)* of the minimax lemma, the subsets

$$B_a = \{b \in B : \ \Phi(a,b) \leq 0\}, \qquad a \in A$$

of B are closed and nonempty, and we show that they satisfy the finite intersection property. Indeed, suppose that

$$B_{a_1} \cap B_{a_2} \cap \cdots \cap B_{a_n} = \emptyset$$

for some $a_1, a_2, \ldots, a_n \in A$. We write $g_j(b) = \Phi(a_j, b)$, $j = 1, 2, \ldots, n$, and we observe that the conditions of the previous proposition are satisfied. Therefore we can find $\lambda_1, \lambda_2, \ldots, \lambda_n \geq 0$ such that for all $b \in B$ we have

$$\lambda_1 \Phi(a_1, b) + \lambda_2 \Phi(a_2, b) + \cdots + \lambda_n \Phi(a_n, b) > 0.$$

For simplicity we normalize the λ_j's by setting $\lambda_1 + \lambda_2 + \cdots + \lambda_n = 1$. If we set $a_0 = \lambda_1 a_1 + \lambda_2 a_2 + \cdots + \lambda_n a_n$, the concavity hypothesis (a) gives

$$\Phi(a_0, b) > 0$$

for all $b \in B$, contradicting the fact that $\sup_{a \in A} \min_{b \in B} \Phi(a,b) = 0$. Therefore, the family of closed subsets $\{B_a\}_{a \in A}$ of B satisfies the finite intersection property. The compactness of B now implies $\bigcap_{a \in A} B_a \neq \emptyset$. Take $b_0 \in \bigcap_{a \in A} B_a$. Then $\Phi(a, b_0) \leq 0$ for every $a \in A$, and therefore

$$\min_{b \in B} \sup_{a \in A} \Phi(a,b) \leq \sup_{a \in A} \Phi(a, b_0) \leq 0$$

as required. □

Appendix I
Taylor's and Mean Value Theorem in Several Variables

I.1 Mutlivariable Taylor's Theorem

For a multiindex $\alpha = (\alpha_1, \dots, \alpha_n) \in (\mathbf{Z}^+ \cup \{0\})^n$, we denote by $|\alpha| = \alpha_1 + \cdots + \alpha_n$ its size, we define $\alpha! = \alpha_1! \cdots \alpha_n!$ its factorial, and we set

$$h^\alpha = h_1^{\alpha_1} \cdots h_n^{\alpha_n},$$

where $h = (h_1, \dots, h_n)$; here $0^0 = 1$.

Let $k \in \mathbf{Z}^+ \cup \{0\}$. Suppose a real-valued $\mathscr{C}^{k+1}$ function f is defined on an open convex subset Ω of $\mathbf{R}^n$. Suppose that $x \in \Omega$ and $x + h \in \Omega$. Then we have the *Taylor expansion formula*

$$f(x+h) = \sum_{|\alpha| \le k} \frac{\partial^\alpha f(x)}{\alpha!} h^\alpha + R(h,x,k),$$

where the remainder $R(h,x,k)$ can be expressed either in Langrange's mean value form

$$R(h,x,k) = \sum_{|\alpha|=k+1} \frac{\partial^\alpha f(x+ch)}{\alpha!} h^\alpha$$

for some $c \in (0,1)$, or in integral form

$$R(h,x,k) = (k+1) \sum_{|\alpha|=k+1} \frac{h^\alpha}{\alpha!} \int_0^1 (1-t)^k \partial^\alpha f(x+th)\, dt.$$

L. Grafakos, *Classical Fourier Analysis*, Graduate Texts in Mathematics 249,
DOI 10.1007/978-1-4939-1194-3, © Springer Science+Business Media New York 2014

I.2 The Mean value Theorem

Suppose that f is as above and $k = 0$. Then for given $x, y \in \Omega$ we have

$$f(y) - f(x) = \int_0^1 \nabla f((1-t)x + ty) \cdot (y - x)\, dt = \nabla f((1-c)x + cy) \cdot (y - x)$$

for some $c \in (0,1)$. This is a special case of Taylor's formula when $k = 0$.

Appendix J
The Whitney Decomposition of Open Sets in $\mathbf{R}^n$

J.1 Decomposition of Open Sets

An arbitrary open set in $\mathbf{R}^n$ can be decomposed as a union of disjoint cubes whose lengths are proportional to their distance from the boundary of the open set. See, for instance, Figure J.1 when the open set is the unit disk in $\mathbf{R}^2$. For a given cube Q in $\mathbf{R}^n$, we denote by $\ell(Q)$ its length.

Proposition. *Let Ω be an open nonempty proper subset of $\mathbf{R}^n$. Then there exists a family of closed dyadic cubes $\{Q_j\}_j$ (called the Whitney cubes of Ω) such that*
(a) $\bigcup_j Q_j = \Omega$ and the Q_j's have disjoint interiors.
(b) $\sqrt{n}\,\ell(Q_j) \le \operatorname{dist}(Q_j,\Omega^c) \le 4\sqrt{n}\,\ell(Q_j)$. Thus $10\sqrt{n}\,Q_j$ meets Ω^c.
(c) If the boundaries of two cubes Q_j and Q_k touch, then

$$\frac{1}{4} \le \frac{\ell(Q_j)}{\ell(Q_k)} \le 4.$$

(d) For a given Q_j there exist at most $12^n - 4^n$ cubes Q_k that touch it.
(e) Let $0<\varepsilon<1/4$. If Q_j^* has the same center as Q_j and $\ell(Q_j^*) = (1+\varepsilon)\ell(Q_j)$ then

$$\chi_\Omega = \sum_j \chi_{Q_j^*} \le 12^n - 4^n + 1.$$

Proof. Let $\mathscr{D}_k$ be the collection of all dyadic cubes of the form

$$\{(x_1,\dots,x_n)\in\mathbf{R}^n:\ m_j 2^{-k} \le x_j < (m_j+1)2^{-k}\},$$

where $m_j \in \mathbf{Z}$. Observe that each cube in $\mathscr{D}_k$ gives rise to 2^n cubes in $\mathscr{D}_{k+1}$ by bisecting each side.

Write the set Ω as the union of the sets

$$\Omega_k = \{x\in\Omega:\ 2\sqrt{n}2^{-k} < \operatorname{dist}(x,\Omega^c) \le 4\sqrt{n}2^{-k}\}$$

L. Grafakos, *Classical Fourier Analysis*, Graduate Texts in Mathematics 249,
DOI 10.1007/978-1-4939-1194-3, © Springer Science+Business Media New York 2014

over all $k \in \mathbf{Z}$. Let $\mathscr{F}'$ be the set of all cubes Q in $\mathscr{D}_k$ for some $k \in \mathbf{Z}$ such that $Q \cap \Omega_k \neq \emptyset$. We show that the collection $\mathscr{F}'$ satisfies property (b). Let $Q \in \mathscr{F}'$ and pick $x \in \Omega_k \cap Q$ for some $k \in \mathbf{Z}$. Observe that

$$\sqrt{n}2^{-k} \leq \text{dist}(x, \Omega^c) - \sqrt{n}\ell(Q) \leq \text{dist}(Q, \Omega^c) \leq \text{dist}(x, \Omega^c) \leq 4\sqrt{n}2^{-k},$$

which proves (b).

Next we observe that

$$\bigcup_{Q \in \mathscr{F}'} Q = \Omega.$$

Indeed, every Q in $\mathscr{F}'$ is contained in Ω (since it has positive distance from its complement) and every $x \in \Omega$ lies in some Ω_k and in some dyadic cube in $\mathscr{D}_k$.

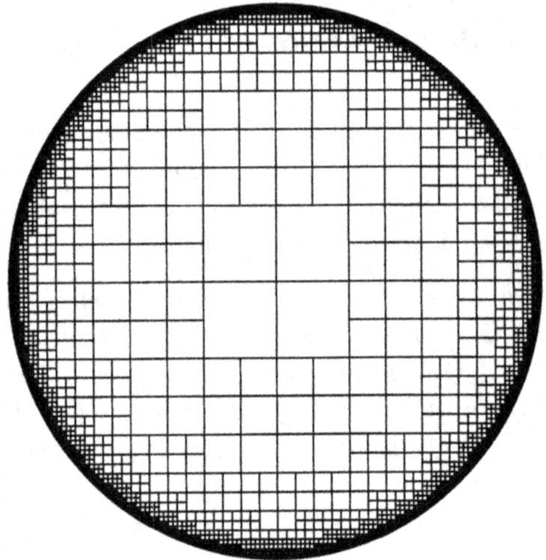

Fig. J.1 The Whitney decomposition of the unit disk.

The problem is that the cubes in the collection $\mathscr{F}'$ may not be disjoint. We have to refine the collection $\mathscr{F}'$ by eliminating those cubes that are contained in some other cubes in the collection. Recall that two dyadic cubes have disjoint interiors or else one contains the other. For every cube Q in $\mathscr{F}'$ we can therefore consider the unique *maximal* cube $Q^{\max}$ in $\mathscr{F}'$ that contains it. Two different such maximal cubes must have disjoint interiors by maximality. Now set $\mathscr{F} = \{Q^{\max} : Q \in \mathscr{F}'\}$.

The collection of cubes $\{Q_j\}_j = \mathscr{F}$ clearly satisfies (a) and (b), and we now turn our attention to the proof of (c). Observe that if Q_j and Q_k in $\mathscr{F}$ touch then

$$\sqrt{n}\ell(Q_j) \leq \text{dist}(Q_j, \Omega^c) \leq \text{dist}(Q_j, Q_k) + \text{dist}(Q_k, \Omega^c) \leq 0 + 4\sqrt{n}\ell(Q_k),$$

which proves (c). To prove (d), note that any cube Q in $\mathscr{D}_k$ is touched by exactly $3^n - 1$ other cubes in $\mathscr{D}_k$. But each cube Q in $\mathscr{D}_k$ can contain at most 4^n cubes of $\mathscr{F}$ of length at least one-quarter of the length of Q. This fact combined with (c) yields (d). To prove (e), notice that each Q_j^* is contained in Ω by part (b). If $x \in \Omega$, then

$x \in Q_{k_0}$ for some k_0. If Q_j does not touch Q_{k_0}, then Q_j^* does not touch Q_{k_0} as well. Consequently, the given $x \in \Omega$ may lie only in Q_k^* for these cubes Q_k that touch Q_{k_0} and there are $12^n - 4^n + 1$ such cubes including Q_{k_0}. □

J.2 Partition of Unity adapted to Whitney cubes

Let us fix an ε such that $0 < \varepsilon < 1/4$. For each cube Q we denote by Q^* the cube with the same center as Q and side length $(1+\varepsilon)\ell(Q)$, where $\ell(Q)$ is the side length of Q. Let us fix a nonnegative smooth function ϕ that is equal to 1 on the unit cube $Q_0 = [-1/2, 1/2]^n$ and equal to zero outside Q_0^*.

Let $\{Q_k\}_k$ be the family of Whitney cubes of Ω. We denote by c_k the center of Q_k and by ℓ_k its side length. Since $\operatorname{dist}(Q_k, \Omega^c) \geq \sqrt{n}\,\ell_k$, and $\operatorname{dist}((Q_k^*)^c, Q_k) \leq \varepsilon\sqrt{n}\,\ell_k$, it follows that

$$\operatorname{dist}(Q_k^*, \Omega^c) \geq (1-\varepsilon)\sqrt{n}\,\ell_k > 0,$$

hence Q_k^* is contained in Ω. Since the union of Q_k is Ω, then the union of Q_k^* is also Ω.

For each k we define

$$\phi_k(x) = \phi\left(\frac{x - c_k}{\ell_k}\right)$$

and we set

$$\Phi(x) = \sum_k \phi_k(x).$$

We notice that Φ is smooth and that $\Phi \geq 1$ on Ω. Then we define

$$\varphi_k(x) = \frac{\phi_k(x)}{\Phi(x)}.$$

Obviously φ_k are supported in Q_k^*, the union of Q_k^* is Ω, and we have

$$\sum_k \varphi_k = \chi_\Omega.$$

We would like to control $|\partial^\alpha \varphi_k(x)|$ in terms of ℓ_k. To achieve this we use the Leibniz rule of differentiation. For a fixed k we have that

$$\frac{\partial^\alpha}{\partial x^\alpha}\varphi_k(x) = \sum_{\beta \leq \alpha} \binom{\alpha_1}{\beta_1} \cdots \binom{\alpha_n}{\beta_n} \left(\frac{\partial^\beta}{\partial x^\beta}\frac{1}{\Phi(x)}\right) \frac{\partial^{\alpha-\beta}}{\partial x^{\alpha-\beta}}\phi_k(x)$$

and obviously $\left|\partial^{\alpha-\beta}\phi_k(x)\right| \leq \ell_k^{-|\alpha|+|\beta|}$. A simple inductive argument shows that

$$\left|\frac{\partial^\beta}{\partial x^\beta}\frac{1}{\Phi(x)}\right| \leq C \sum_{\substack{0 \leq q_j \leq |\beta| \\ q_1+\cdots+q_n = |\beta|}} \frac{\left|\partial_1^{q_1} \cdots \partial_n^{q_n}\Phi(x)\right|}{\Phi(x)^2},$$

for some constant C, and since for a given $x \in \Omega$, $\Phi(x)$ is the sum of at most 12^n functions with nonzero values, it follows that

$$\left|\partial_1^{q_1}\cdots\partial_n^{q_n}\Phi(x)\right| \le C_n\,\ell_k^{-(q_1+\cdots+q_n)}$$

when $x \in Q_k^*$ and thus

$$\left|\frac{\partial^\beta}{\partial x^\beta}\frac{1}{\Phi(x)}\right| \le C'_{n,\beta}\,\ell_k^{-|\beta|}$$

for $x \in Q_k^*$. We conclude that for every multiindex α there is a constant $C_{\alpha,n}$ such that

$$\left|\frac{\partial^\alpha}{\partial x^\alpha}\varphi_k(x)\right| \le C_{\alpha,n}\,\ell^{-|\alpha|}.$$

More on Whitney decompositions can be found in the article of Whitney [373] and the books of Stein [338], Krantz and Parks [204].

Glossary

$A \subseteq B$	A is a subset of B (also denoted by $A \subseteq B$)
$A \subsetneqq B$	A is a proper subset of B
$A \supset B$	B is a proper subset of A
A^c	the complement of a set A
χ_E	the characteristic function of the set E
d_f	the distribution function of a function f
f^*	the decreasing rearrangement of a function f
$f_n \uparrow f$	f_n increases monotonically to a function f
$\mathbf{Z}$	the set of all integers
$\mathbf{Z}^+$	the set of all positive integers $\{1,2,3,\dots\}$
$\mathbf{Z}^n$	the n-fold product of the integers
$\mathbf{R}$	the set of real numbers
$\mathbf{R}^+$	the set of positive real numbers
$\mathbf{R}^n$	the Euclidean n-space
$\mathbf{Q}$	the set of rationals
$\mathbf{Q}^n$	the set of n-tuples with rational coordinates
$\mathbf{C}$	the set of complex numbers
$\mathbf{C}^n$	the n-fold product of complex numbers
$\mathbf{T}$	the unit circle identified with the interval $[0,1]$
$\mathbf{T}^n$	the n-dimensional torus $[0,1]^n$
$\lvert x\rvert$	$\sqrt{\lvert x_1\rvert^2+\cdots+\lvert x_n\rvert^2}$ when $x=(x_1,\dots,x_n)\in\mathbf{R}^n$

L. Grafakos, *Classical Fourier Analysis*, Graduate Texts in Mathematics 249,
DOI 10.1007/978-1-4939-1194-3, © Springer Science+Business Media New York 2014

$\mathbf{S}^{n-1}$	the unit sphere $\{x \in \mathbf{R}^n : \ \lvert x\rvert = 1\}$
e_j	the vector $(0,\dots,0,1,0,\dots,0)$ with 1 in the jth entry and 0 elsewhere
$\log t$	the logarithm with base e of $t > 0$
$\log_a t$	the logarithm with base a of $t > 0$ $(1 \neq a > 0)$
$\log^+ t$	$\max(0, \log t)$ for $t > 0$
$[t]$	the integer part of the real number t
$x \cdot y$	the quantity $\sum_{j=1}^n x_j y_j$ when $x = (x_1,\dots,x_n)$ and $y = (y_1,\dots,y_n)$
$B(x,R)$	the ball of radius R centered at x in $\mathbf{R}^n$
ω_{n-1}	the surface area of the unit sphere $\mathbf{S}^{n-1}$
v_n	the volume of the unit ball $\{x \in \mathbf{R}^n : \ \lvert x\rvert < 1\}$
$\lvert A\rvert$	the Lebesgue measure of the set $A \subseteq \mathbf{R}^n$
dx	Lebesgue measure
$\mathrm{Avg}_B f$	the average $\frac{1}{\lvert B\rvert}\int_B f(x)\,dx$ of f over the set B
$\langle f, g\rangle$	the real inner product $\int_{\mathbf{R}^n} f(x)g(x)\,dx$
$\langle f \mid g\rangle$	the complex inner product $\int_{\mathbf{R}^n} f(x)\overline{g(x)}\,dx$
$\langle u, f\rangle$	the action of a distribution u on a function f
p'	the number $p/(p-1)$, whenever $0 < p \neq 1 < \infty$
$1'$	the number ∞
∞'	the number 1
$f = O(g)$	means $\lvert f(x)\rvert \leq M\lvert g(x)\rvert$ for some M for x near x_0
$f = o(g)$	means $\lvert f(x)\rvert\,\lvert g(x)\rvert^{-1} \to 0$ as $x \to x_0$
A^t	the transpose of the matrix A
A^*	the conjugate transpose of a complex matrix A
A^{-1}	the inverse of the matrix A
$O(n)$	the space of real matrices satisfying $A^{-1} = A^t$
$\lVert T\rVert_{X\to Y}$	the norm of the (bounded) operator $T : X \to Y$
$A \approx B$	means that there exists a $c > 0$ such that $c^{-1} \leq \frac{B}{A} \leq c$
$\lvert\alpha\rvert$	indicates the size $\lvert\alpha_1\rvert + \cdots + \lvert\alpha_n\rvert$ of a multi-index $\alpha = (\alpha_1,\dots,\alpha_n)$
$\partial_j^m f$	the mth partial derivative of $f(x_1,\dots,x_n)$ with respect to x_j
$\partial^\alpha f$	$\partial_1^{\alpha_1}\cdots\partial_n^{\alpha_n} f$

$\mathscr{C}^k$	the space of functions f with $\partial^\alpha f$ continuous for all $\lvert\alpha\rvert \le k$
$\mathscr{C}_0$	the space of continuous functions with compact support
$\mathscr{C}_{00}$	the space of continuous functions that vanish at infinity
$\mathscr{C}_0^\infty$	the space of smooth functions with compact support
$\mathscr{D}$	the space of smooth functions with compact support
$\mathscr{S}$	the space of Schwartz functions
$\mathscr{S}_0$	the space of Schwartz functions φ with the property $\int_{\mathbf{R}^n} x^\gamma \varphi(x)\,dx = 0$ for all multi-indices γ.
$\mathscr{C}^\infty$	the space of smooth functions $\bigcup_{k=1}^\infty \mathscr{C}^k$
$\mathscr{D}'(\mathbf{R}^n)$	the space of distributions on $\mathbf{R}^n$
$\mathscr{S}'(\mathbf{R}^n)$	the space of tempered distributions on $\mathbf{R}^n$
$\mathscr{E}'(\mathbf{R}^n)$	the space of distributions with compact support on $\mathbf{R}^n$
$\mathscr{P}$	the set of all complex-valued polynomials of n real variables
$\mathscr{S}'(\mathbf{R}^n)/\mathscr{P}$	the space of tempered distributions on $\mathbf{R}^n$ modulo polynomials
$\ell(Q)$	the side length of a cube Q in $\mathbf{R}^n$
∂Q	the boundary of a cube Q in $\mathbf{R}^n$
$L^p(X,\mu)$	the Lebesgue space over the measure space (X,μ)
$L^p(\mathbf{R}^n)$	the space $L^p(\mathbf{R}^n, \lvert\cdot\rvert)$
$L^{p,q}(X,\mu)$	the Lorentz space over the measure space (X,μ)
$L^p_{\text{loc}}(\mathbf{R}^n)$	the space of functions that lie in $L^p(K)$ for any compact set K in $\mathbf{R}^n$
$\lvert d\mu\rvert$	the total variation of a finite Borel measure μ on $\mathbf{R}^n$
$\mathscr{M}(\mathbf{R}^n)$	the space of all finite Borel measures on $\mathbf{R}^n$
$\mathscr{M}_p(\mathbf{R}^n)$	the space of L^p Fourier multipliers, $1 \le p \le \infty$
$\mathscr{M}^{p,q}(\mathbf{R}^n)$	the space of translation-invariant operators that map $L^p(\mathbf{R}^n)$ to $L^q(\mathbf{R}^n)$
$\lVert\mu\rVert_{\mathscr{M}}$	$\int_{\mathbf{R}^n} \lvert d\mu\rvert$ the norm of a finite Borel measure μ on $\mathbf{R}^n$
$\mathcal{M}$	the centered Hardy–Littlewood maximal operator with respect to balls
M	the uncentered Hardy–Littlewood maximal operator with respect to balls
$\mathcal{M}_c$	the centered Hardy–Littlewood maximal operator with respect to cubes
M_c	the uncentered Hardy–Littlewood maximal operator with respect to cubes

$\mathcal{M}_\mu$	the centered maximal operator with respect to a measure μ
M_μ	the uncentered maximal operator with respect to a measure μ
M_s	the strong maximal operator
M_d	the dyadic maximal operator

References

1. Albiac, F., Kalton, N. J., *Topics in Banach Space Theory*, Graduate Texts in Mathematics, 233, Springer, New York, 2006.
2. Aldaz, J. M., *The weak type (1,1) bounds for the maximal function associated to cubes grow to infinity with the dimension*, Ann. of Math. (2nd Ser.) **173** (2011), no. 2, 1013–1023.
3. Alimov, Sh. A., Ashurov, R. R., Pulatov, A. K., *Multiple Fourier series and Fourier integrals*, Proc. Commutative Harmonic Analysis, IV, pp. 1–95, Encyclopaedia Math. Sci. 42, Springer, Berlin, 1992.
4. Alvarez, J., Pérez, C., *Estimates with A_∞ weights for various singular integral operators*, Boll. Un. Mat. Ital. A (7) **8** (1994), no. 1, 123–133.
5. Antonov, N. Yu., *Convergence of Fourier series*, Proceedings of the XX Workshop on Function Theory (Moscow, 1995), East J. Approx. **2** (1996), no. 2, pp. 187–196.
6. Antonov, N. Yu., *The behavior of partial sums of trigonometric Fourier series* [Russian], Ph. D. Thesis, Institute of Mathematics and Mechanics, Ural Branch of the Russian Academy of Sciences, Ekaterinburg, 1998.
7. Aoki, T., *Locally bounded spaces*, Proc. Imp. Acad. Tokyo **18** (1942), no. 10, 585–634.
8. Arhipov, G. I., Karachuba, A. A., Čubarikov, V. N., *Trigonometric integrals*, Math. USSR Izvestija **15** (1980), 211–239.
9. Arias de Reyna, J., *Pointwise convergence of Fourier series*, J. London Math. Soc. (2) **65** (2002), no. 1, 139–153.
10. Arias de Reyna, J., *Pointwise Convergence of Fourier Series*, Lecture Notes in Mathematics, 1785, Springer-Verlag, Berlin, 2002.
11. Ash, J. M., *Multiple trigonometric series*, Studies in Harmonic Analysis (Proc. Conf., DePaul Univ., Chicago, IL, 1974), pp. 76–96, MAA Stud. Math., Vol. 13, Math. Assoc. Amer., Washington, D. C., 1976.
12. Asmar, N., Berkson, E., Gillespie, T. A., *Summability methods for transferring Fourier multipliers and transference of maximal inequalities*, Analysis and Partial Differential Equations, pp. 1–34, Lecture Notes in Pure and Appl. Math. 122, Dekker, New York, 1990.
13. Asmar, N., Berkson, E., Gillespie, T. A., *On Jodeit's multiplier extension theorems*, J. Anal. Math. **64** (1994), 337–345.
14. Astala, K., Iwaniec, T., Saksman, E., *Beltrami operators in the plane*, Duke Math. J. **107** (2001), no. 1, 27–56.
15. Aubrun, G., *Maximal inequality for high-dimensional cubes*, Confluentes Math. **1** (2009), no. 2, 169–179.
16. Babenko, K. I., *An inequality in the theory of Fourier integrals* [Russian], Izv. Akad. Nauk SSSR Ser. Mat. **25** (1961), 531–542.
17. Baernstein II, A., *Some sharp inequalities for conjugate functions*, Indiana Univ. Math. J. **27** (1978), 833–852.

L. Grafakos, *Classical Fourier Analysis*, Graduate Texts in Mathematics 249,
DOI 10.1007/978-1-4939-1194-3,

18. Bagby, R., Kurtz, D. S., *Covering lemmas and the sharp function,* Proc. Amer. Math. Soc. **93** (1985), no. 2, 291–296.
19. Bagby, R., Kurtz, D. S., *A rearranged good-λ inequality*, Trans. Amer. Math. Soc. **293** (1986), no. 1, 71–81.
20. Bary, N., *A Treatise on Trigonometric Series*, Vols. I, II, Authorized translation by Margaret F. Mullins, A Pergamon Press Book, The Macmillan Co., New York, 1964.
21. Beckner, W., *Inequalities in Fourier analysis*, Ann. of Math. (2nd Ser.) **102** (1975), no. 1, 159–182.
22. Benedek, A., Calderón, A.-P., Panzone, R., *Convolution operators on Banach-space valued functions*, Proc. Nat. Acad. Sci. U.S.A. **48** (1962), 356–365.
23. Benedetto, J., Frazier, M. (eds.), *Wavelets: Mathematics and Applications*, CRC Press, Boca Raton, FL, 1994.
24. Bennett, C., Sharpley, R., *Interpolation of Operators*, Pure and Applied Mathematics, 129 Academic Press, Inc., Boston, MA, 1988.
25. Bergh, J., Löfström, J., *Interpolation Spaces, An Introduction*, Grundlehren der Mathematischen Wissenschaften, No. 223, Springer-Verlag, Berlin-New York, 1976.
26. Bernstein, S., *Sur la convergence absolue des séries trigonométriques*, Comptes Rendus des Séances de l' Académie des Sciences, Paris, **158** (1914), 1661–1663.
27. Besicovitch, A., *A general form of the covering principle and relative differentiation of additive functions*, Proc. of Cambridge Philos. Soc. **41** (1945), 103–110.
28. Boas, R. P., Bochner, S., *On a theorem of M. Riesz for Fourier series*, J. London Math. Soc. **14** (1939), no. 1, 62–73.
29. Bôcher, M., *Introduction to the theory of Fourier's series*, Ann. of Math. (2nd Ser.) **7** (1906), no. 3, 81–152.
30. Bochner, S., *Summation of multiple Fourier series by spherical means*, Trans. Amer. Math. Soc. **40** (1936), no. 2, 175–207.
31. Bochner, S., *Harmonic Analysis and the Theory of Probability*, University of California Press, Berkeley and Los Angeles, 1955.
32. Bochner, S., *Lectures on Fourier Integrals, (with an author's supplement on monotonic functions, Stieltjes integrals, and harmonic analysis)*, Translated by Morris Tenenbaum and Harry Pollard. Annals of Mathematics Studies, No. 42, Princeton University Press, Princeton, NJ, 1959.
33. Bojarski, B., *Remarks on Markov's inequalities and some properties of polynomials* (Russian summary), Bull. Polish Acad. Sci. Math. **33** (1985), no. 7-8, 355–365.
34. Bourgain, J., *On square functions on the trigonometric system*, Bull. Soc. Math. Belg. Sér. B **37** (1985), no. 1, 20–26.
35. Bourgain, J., *Estimations de certaines fonctions maximales*, C. R. Acad. Sci. Paris Sér. I Math. **301** (1985), no. 10, 499–502.
36. Bourgain, J., *Averages in the plane over convex curves and maximal operators*, J. Analyse Math. **47** (1986), 69–85.
37. Bourgain, J., *On the Hardy-Littlewood maximal function for the cube*, Israel J. Math. **203** (2014), 275–293.
38. Buckley, S. M., *Estimates for operator norms on weighted spaces and reverse Jensen inequalities*, Trans. Amer. Math. Soc. **340** (1993), no. 1, 253–272.
39. Burkholder, D. L., *Martingale transforms*, Ann. Math. Statist. **37** (1966), 1494–1505.
40. Burkholder, D. L., Gundy, R. F., *Extrapolation and interpolation of quasilinear operators on martingales*, Acta Math. **124** (1970), no. 1, 249–304.
41. Calderón, A. P., *On the theorems of M. Riesz and Zygmund*, Proc. Amer. Math. Soc. **1** (1950), 533–535.
42. Calderón, A. P., *Intermediate spaces and interpolation, the complex method*, Studia Math. **24** (1964), 113–190.
43. Calderón, A. P., *Singular integrals*, Bull. Amer. Math. Soc. **72** (1966), 427–465.
44. Calderón, A. P., *Algebras of singular integral operators*, Singular Integrals (Proc. Sympos. Pure Math. (Chicago, IL, 1966), pp. 18–55, Amer. Math. Soc., Providence, RI, 1967.

45. Calderón, A. P., Torchinsky, A., *Parabolic maximal functions associated with a distribution, II*, Advances in Math. **24** (1977), no. 2, 101–171.
46. Calderón, A. P., Zygmund, A., *On the existence of certain singular integrals*, Acta Math. **88** (1952), no. 1, 85–139.
47. Calderón, A. P., Zygmund, A., *A note on interpolation of sublinear operators*, Amer. J. Math. **78** (1956), 282–288.
48. Calderón, A. P., Zygmund, A., *On singular integrals*, Amer. J. Math. **78** (1956), 289–309.
49. Calderón, A. P., Zygmund, A., *Algebras of certain singular integral operators*, Amer. J. Math. **78** (1956), 310–320.
50. Calderón, C., *Lacunary spherical means*, Illinois J. Math. **23** (1979), no. 3, 476–484.
51. Carbery, A., *An almost-orthogonality principle with applications to maximal functions associated to convex bodies*, Bull. Amer. Math. Soc. (N.S.) **14** (1986), no. 2, 269–273.
52. Carbery, A., *Variants of the Calderón–Zygmund theory for L^p-spaces*, Rev. Mat. Iberoamericana **2** (1986), no. 4, 381–396.
53. Carbery, A., Christ, M., Wright, J., *Multidimensional van der Corput and sublevel set estimates*, J. Amer. Math. Soc. **12** (1999), no. 4, 981–1015.
54. Carleson, L., *On convergence and growth of partial sums of Fourier series*, Acta Math. **116** (1966), no. 1, 135–157.
55. Carleson, L., *On the Littlewood–Paley Theorem*, Mittag-Leffler Institute Report, Djursholm, Sweden, 1967.
56. Carro, M. J., *New extrapolation estimates*, J. Funct. Anal. **174** (2000), no. 1, 155–166.
57. Carro, M. J., Raposo, J. A., Soria, J., *Recent developments in the theory of Lorentz spaces and weighted inequalities*, Mem. Amer. Math. Soc. **187** (2007), no. 877.
58. Carro, M. J., Torres, R. H., Soria, J., *Rubio de Francia's extrapolation theory: estimates for the distribution function*, J. Lond. Math. Soc. (2) **85** (2012), no. 2, 430–454.
59. Christ, M., *A convolution inequality for Cantor-Lebesgue measures*, Rev. Mat. Iberoamericana **1** (1985), no. 4, 79–83.
60. Christ, M., *Weak type (1,1) bounds for rough operators I*, Ann. of Math. (2nd Ser.) **128** (1988), no. 1, 19–42.
61. Christ, M., Fefferman, R., *A note on weighted norm inequalities for the Hardy–Littlewood maximal operator*, Proc. Amer. Math. Soc. **87** (1983), no. 3, 447–448.
62. Christ, M., Grafakos, L., *Best constants for two nonconvolution inequalities*, Proc. Amer. Math. Soc. **123** (1995), no. 6, 1687–1693.
63. Christ, M., Rubio de Francia, J.L.. *Weak type (1,1) bounds for rough operators II*, Invent. Math. **93** (1988), no. 1, 225–237.
64. Chui, C. (ed.), *Wavelets: A Tutorial in Theory and Applications*, Academic Press, San Diego, CA, 1992.
65. Coifman, R. R., *Distribution function inequalities for singular integrals*, Proc. Nat. Acad. Sci. U.S.A. **69** (1972), 2838–2839.
66. Coifman, R. R., Fefferman, C., *Weighted norm inequalities for maximal functions and singular integrals*, Studia Math. **51** (1974), 241–250.
67. Coifman, R. R., Jones, P., Rubio de Francia, J. L., *Constructive decomposition of BMO functions and factorization of A_p weights*, Proc. Amer. Math. Soc. **87** (1983), no. 4, 675–676.
68. Coifman, R. R., Rochberg, R., *Another characterization of BMO*, Proc. Amer. Math. Soc. **79** (1980), no. 2, 249–254.
69. Coifman, R. R., Rubio de Francia, J.-L., Semmes, S., *Multiplicateurs de Fourier de $L^p(\mathbb{R})$ et estimations quadratiques*, C. R. Acad. Sci. Paris Sér. I Math. **306** (1988), no. 8, 351–354.
70. Coifman, R. R., Weiss, G., *Transference Methods in Analysis*, Conference Board of the Mathematical Sciences Regional Conference Series in Mathematics, No. 31, American Mathematical Society, Providence, RI, 1976.
71. Coifman, R. R., Weiss, G., *Book review of "Littlewood–Paley and multiplier theory"*, Bull. Amer. Math. Soc. **84** (1978), no. 2, 242–250.
72. Connett, W. C., *Singular integrals near L^1*, Harmonic Analysis in Euclidean Spaces, (Williams Coll., Williamstown, MA, 1978), Part 1, pp. 163–165, Proc. Sympos. Pure Math., XXXV, Part, Amer. Math. Soc., Providence, RI, 1979.

73. Córdoba, A., Fefferman, R., *A geometric proof of the strong maximal theorem*, Ann. of Math. (2nd Ser.) **102** (1975), no. 1, 95–100.
74. Córdoba, A., Fefferman, C., *A weighted norm inequality for singular integrals*, Studia Math. **57** (1976), no. 1, 97–101.
75. Cotlar, M., *A unified theory of Hilbert transforms and ergodic theorems*, Rev. Mat. Cuyana, **1** (1955), 105–167.
76. Cowling, M., Mauceri, G., *On maximal functions*, Rend. Sem. Mat. Fis. Milano **49** (1979), 79–87.
77. Cruz-Uribe, D., *New proofs of two-weight norm inequalities for the maximal operator*, Georgian Math. J. **7** (2000), no. 1, 33–42.
78. Cruz-Uribe, D., Martell, J. M., Pérez, C., *Extrapolation results for A_∞ weights and applications*, J. Funct. Anal. **213** (2004), no. 2, 412–439.
79. Cruz-Uribe, D., Martell, J. M., Pérez, C., *Sharp weighted estimates for approximating dyadic operators*, Electron. Res. Announc. Math. Sci. **17** (2010), 12–19.
80. Cruz-Uribe, D., Martell, J. M., Pérez, C., *Weights, extrapolation and the theory of Rubio de Francia*, Operator Theory: Advances and Applications, 215, Birkhäuser/Springer Basel AG, Basel, 2011.
81. Cruz-Uribe, D., Martell, J. M., Pérez, C., *Sharp weighted estimates for classical operators*, Adv. Math. **229** (2012), no. 1, 408–441.
82. Cruz-Uribe, D., Neugebauer, C. J., *The structure of the reverse Hölder classes*, Trans. Amer. Math. Soc. **347** (1995), no. 8, 2941–2960.
83. Cwikel, M., *The dual of weak L^p*, Ann. Inst. Fourier (Grenoble) **25** (1975), no. 2, 81–126.
84. Cwikel, M., Fefferman, C., *Maximal seminorms on weak L^1*, Studia Math. **69** (1980/81), no. 2, 149–154.
85. Cwikel, M., Fefferman, C., *The canonical seminorm on Weak L^1*, Studia Math. **78** (1984), no. 3, 275–278.
86. Daly, J. E., *A necessary condition for Calderón–Zygmund singular integral operators*, J. Fourier Anal. Appl. **5** (1999), no. 4, 303–308.
87. Daly, J. E., Phillips, K., *On the classification of homogeneous multipliers bounded on $H^1(\mathbf{R}^2)$*, Proc. Amer. Math. Soc. **106** (1989), no. 3, 685–696.
88. Daubechies, I., *Orthonormal bases of compactly supported wavelets*, Comm. Pure Appl. Math. **41** (1988), no. 7, 909–996.
89. Daubechies, I., *Ten Lectures on Wavelets*, CBMS-NSF Regional Conference Series in Applied Math, 61, Society for Industrial and Applied Mathematics (SIAM), Philadelphia, PA, 1992.
90. Davis, B., *On the integrability of the martingale square function*, Israel J. Math. **8** (1970), 187–190.
91. Davis, B., *On the weak type (1,1) inequality for conjugate functions*, Proc. Amer. Math. Soc. **44** (1974), 307–311.
92. de Guzmán, M., *Real Variable Methods in Fourier Analysis*, North-Holland Mathematics Studies, 46, Notas de Matemática (75), North-Holland Publishing Co., Amsterdam-New York, 1981.
93. de Guzmán, M., *Differentiation of Integrals in $\mathbb{R}^n$*, Lecture Notes in Math. 481, Springer-Verlag, Berlin-New York, 1975.
94. de Leeuw, K., *On L_p multipliers*, Ann. of Math. (2nd Ser.) **81** (1965), no. 2, 364–379.
95. Diestel, J., Uhl, J. J. Jr., *Vector Measures*, With a foreword by B. J. Pettis, Mathematical Surveys, No. 15, American Mathematical Society, Providence, RI, 1977.
96. Dirichlet, P. G., *Sur la convergence des séries trigonométriques qui servent à représenter une fonction arbitraire entre des limites données*, J. Reine und Angew. Math. **4** (1829), 157–169.
97. Doob, J. L., *Stochastic Processes*, John Wiley & Sons, Inc., New York; Chapman & Hall, Limited, London, 1953.
98. Dragičević, O., Grafakos, L., C. Pereyra, and S. Petermichl, *Extrapolation and sharp norm estimates for classical operators on weighted Lebesgue spaces*, Publ. Mat. **49** (2005), no. 1, 73–91.

99. Dunford, N., Schwartz, J. T., *Linear Operators, I*, General Theory, With the assistance of W. G. Bade and R. G. Bartle, Pure and Applied Mathematics, Vol. 7, Interscience Publishers, Inc., New York; Interscience Publishers, Ltd., London, 1958.
100. Duoandikoetxea, J., *Fourier Analysis*, Grad. Studies in Math. 29, American Mathematical Society, Providence, RI, 2001.
101. Duoandikoetxea, J., *Extrapolation of weights revisited: new proofs and sharp bounds*, J. Funct. Anal. **260** (2011), no. 6, 1886–1901.
102. Duoandikoetxea, J., Rubio de Francia, J.-L., *Maximal and singular integral operators via Fourier transform estimates*, Invent. Math. **84** (1986), no. 3, 541–561.
103. Duoandikoetxea, J., Vega, L., *Spherical means and weighted inequalities*, J. London Math. Soc. (2) **53** (1996), no. 2, 343–353.
104. Duong, X. T., M[c]Intosh, A., *Singular integral operators with non-smooth kernels on irregular domains*, Rev. Mat. Iberoamericana **15** (1999), no. 2, 233–265.
105. Dym, H., McKean, H. P., *Fourier Series and Integrals*, Probability and Mathematical Statistics, No. 14, Academic Press, New York-London, 1972.
106. Edwards, R. E., *Fourier Series: A Modern Introduction. Vol. 1*, 2nd edition, Graduate Texts in Mathematics, 64, Springer-Verlag, New York-Berlin, 1979.
107. Erdélyi, A., *Asymptotic Expansions*, Dover Publications Inc., New York, 1956.
108. Essén, M., *A superharmonic proof of the M. Riesz conjugate function theorem*, Arkiv Math. **22** (1984), no. 2, 241–249.
109. Fan, D., Guo, K., Pan, Y., *A note of a rough singular integral operator*, Math. Inequal. and Appl. **2** (1999), no. 1, 73–81.
110. Fan, D., Pan, Y., *Singular integral operators with rough kernels supported by subvarieties*, Amer J. Math. **119** (1997), no. 4, 799–839.
111. Fan, K., *Minimax theorems*, Proc. Nat. Acad. Sci. U. S. A. **39** (1953), 42–47.
112. Fefferman, C., *On the convergence of multiple Fourier series*, Bull. Amer. Math. Soc. **77** (1971), 744–745.
113. Fefferman, C., *The multiplier problem for the ball*, Ann. of Math. (2nd Ser.) **94** (1971), no. 2, 330–336.
114. Fefferman, C., *The uncertainty principle*, Bull. Amer. Math. Soc. (N. S.) **9** (1983), no. 2, 129–206.
115. Fefferman, C., Stein, E. M., *Some maximal inequalities*, Amer. J. Math. **93** (1971), 107–115.
116. Fefferman, R., *A theory of entropy in Fourier analysis*, Adv. in Math. **30** (1978), no. 3, 171–201.
117. Fefferman, R., *Strong differentiation with respect to measures*, Amer J. Math. **103** (1981), no. 1, 33–40.
118. Folland, G. B., Sitaram, A., *The uncertainty principle: a mathematical survey*, J. Fourier Anal. Appl. **3** (1997), no. 3, 207–238.
119. Fourier, J., *Théorie Analytique de la Chaleur*, Institut de France, Paris, 1822.
120. Frazier, M., *An Introduction to Wavelets Through Linear Algebra*, Undergraduate Texts in Mathematics, Springer-Verlag, New York, 1999.
121. García-Cuerva, J., *An extrapolation theorem in the theory of A_p weights*, Proc. Amer. Math. Soc. **87** (1983), no. 3, 422–426.
122. García-Cuerva, J., Rubio de Francia, J.-L., *Weighted Norm Inequalities and Related Topics*, North-Holland Mathematics Studies, 116, Notas de Matemática (104), North-Holland Publishing Co., Amsterdam, 1985.
123. Garnett, J., Jones, P., *The distance in BMO to L^∞*, Ann. of Math. **108** (1978), no. 2, 373–393.
124. Garrigós, G., Seeger, A., *Characterizations of Hankel multipliers*, Math. Ann. **342** (2008), no. 1, 31–68.
125. Gehring, F. W., *The L^p-integrability of the partial derivatives of a quasiconformal mapping*, Acta Math. **130** (1973), no. 1, 265–277.
126. Gelfand, I. M., Šilov, G. E., *Generalized Functions, Vol. 1: Properties and Operations*, Academic Press, New York-London, 1964.
127. Gelfand, I. M., Šilov, G. E., *Generalized Functions, Vol. 2: Spaces of Fundamental and Generalized Functions*, Academic Press, New York-London, 1968.

128. Gibbs, J. W., *Fourier's Series*, Nature **59** (1899), 606.
129. Gohberg, I., Krupnik, N., *Norm of the Hilbert transformation in the L_p space*, Funct. Anal. Appl. **2** (1968), 180–181.
130. Grafakos, L., *Best bounds for the Hilbert transform on $L^p(\mathbb{R}^1)$*, Math. Res. Lett. **4** (1997), no. 4, 469–471.
131. Grafakos, L., *Modern Fourier Analysis*, 3rd edition, Graduate Texts in Math. 250, Springer, New York, 2014.
132. Grafakos, L., Honzík, P., Ryabogin, D., *On the p-independence property of Calderón–Zygmund theory*, J. Reine Angew. Math. **602** (2007), 227–234.
133. Grafakos, L., Kalton, N., *The Marcinkiewicz multiplier condition for bilinear operators*, Studia Math. **146** (2001), no. 2, 115–156.
134. Grafakos, L., Kinnunen, J., *Sharp inequalities for maximal functions associated to general measures*, Proc. Roy. Soc. Edinburgh Sect. A **128** (1998), no. 4, 717–723.
135. Grafakos, L., Martell, J. M., *Extrapolation of weighted norm inequalities for multivariable operators and applications* J. Geom. Anal. **14** (2004), no. 1, 19–46.
136. Grafakos, L., Montgomery-Smith, S., *Best constants for uncentred maximal functions*, Bull. London Math. Soc. **29** (1997), no. 1, 60–64.
137. Grafakos, L., Morpurgo, C., *A Selberg integral formula and applications*, Pacific J. Math. **191** (1999), no. 1, 85–94.
138. Grafakos, L., Stefanov, A., *L^p bounds for singular integrals and maximal singular integrals with rough kernels*, Indiana Univ. Math. J. **47** (1998), no. 2, 455–469.
139. Grafakos, L., Stefanov, A., *Convolution Calderón–Zygmund singular integral operators with rough kernels*, Analysis of Divergence (Orono, ME, 1997), pp. 119–143, Appl. Numer. Harmon. Anal., Birkhäuser Boston, Boston, MA, 1999.
140. Gröchenig, K., *Foundations of Time-Frequency Analysis*, Applied and Numerical Harmonic Analysis Birkhäuser Boston, Inc., Boston, MA, 2001.
141. Haagerup, U., *The best constants in Khintchine's inequality*, Studia Math. **70** (1981), no. 3, 231–283.
142. Hahn, L.-S., *On multipliers of p-integrable functions*, Trans. Amer. Math. Soc. **128** (1967), 321–335.
143. Hardy, G. H., *Theorems relating to the summability and convergence of slowly oscillating series*, Proc. London Math. Soc. (2) **8** (1910), no. 1, 301–320.
144. Hardy, G. H., *Note on a theorem of Hilbert*, Math. Z. **6** (1920), no. 3-4, 314–317.
145. Hardy, G. H., *Note on some points in the integral calculus, LX. An inequality between integrals*, Messenger of Math. **54** (1925), 150–156.
146. Hardy, G. H., Littlewood, J. E., *A maximal theorem with function-theoretic applications*, Acta Math. **54** (1930), no. 1, 81–116.
147. Hardy, G. H., Littlewood, J. E., *Some more theorems concerning Fourier series and Fourier power series*, Duke Math. J. **2** (1936), no. 2, 354–381.
148. Hardy, G. H., Littlewood, J. E., Pólya, G., *Inequalities*, 2nd edition, Cambridge University Press, Cambridge, 1952.
149. Helson, H., Szegő, G., *A problem in prediction theory*, Ann. Mat. Pura Appl. (4) **51** (1960), 107–138.
150. Heo, Y., Nazarov, F., Seeger, A., *Radial Fourier multipliers in high dimensions*, Acta Math. **206** (2011), no. 1, 55–92.
151. Hérnandez, E., Weiss, G., *A First Course on Wavelets (Studies in Advanced Mathematics)*, CRC Press, Boca Raton, FL, 1996.
152. Hewitt, E., Ross, K., *Abstract Harmonic Analysis, Vol I,* 2nd edition, Grundlehren der mathematischen Wissenschaften, 115, Springer-Verlag, Berlin, New York, 1979.
153. Hewitt, E., Stromberg, K., *Real and Abstract Analysis*, Graduate Texts in Math. 25, Springer-Verlag, New York, 1965.
154. Hirschman, I. I. Jr., *A convexity theorem for certain groups of transformations*, J. Analyse Math. **2** (1953), 209–218.
155. Hofmann, S., *Weak (1,1) boundedness of singular integrals with nonsmooth kernel*, Proc. Amer. Math. Soc. **103** (1988), no. 1, 260–264.

156. Hollenbeck, B., Verbitsky, I. E., *Best constants for the Riesz projection*, J. Funct. Anal. **175** (2000), no. 2, 370–392.
157. Honzík, P., *An example of an unbounded maximal singular operator*, J. Geom. Anal. **20** (2010), 153–167.
158. Honzík, P., *On p dependent boundedness of singular integral operators*, Math. Z. **267** (2011), no. 3–4, 931–937.
159. Hörmander, L., *Estimates for translation invariant operators in L^p spaces*, Acta Math. **104** (1960), no. 1–2, 93–140.
160. Hörmander, L., *The Analysis of Linear Partial Differential Operators I*, 2nd edition, Springer-Verlag, Berlin, 1990.
161. Hruščev, S. V., *A description of weights satisfying the A_∞ condition of Muckenhoupt*, Proc. Amer. Math. Soc. **90** (1984), no. 2, 253–257.
162. Hudson, S., *A covering lemma for maximal operators with unbounded kernels*, Michigan Math. J. **34** (1987), no. 1, 147–151.
163. Hunt, R., *An extension of the Marcinkiewicz interpolation theorem to Lorentz spaces*, Bull. Amer. Math. Soc. **70** (1964), 803–807.
164. Hunt, R., *On $L(p,q)$ spaces*, Einseignement Math. (2) **12** (1966), 249–276.
165. Hunt, R., *On the convergence of Fourier series*, Orthogonal Expansions and Their Continuous Analogues (Proc. Conf., Edwardsville, IL, 1967), pp. 235–255, Southern Illinois Univ. Press, Carbondale IL, 1968.
166. Hunt, R., Kurtz, D., Neugebauer, C. J., *A note on the equivalence of A_p and Sawyer's condition for equal weights*, Conference on Harmonic Analysis in Honor of Antoni Zygmund, Vol. I, II (Chicago, IL, 1981), pp. 156–158, Wadsworth Math. Ser., Wadsworth, Belmont, CA, 1983.
167. Hunt, R., Muckenhoupt, B., Wheeden, R., *Weighted norm inequalities for the conjugate function and the Hilbert transform*, Trans. Amer. Math. Soc. **176** (1973), 227–251.
168. Hytönen, T. P., *The sharp weighted bound for general Calderón-Zygmund operators*, Ann. of Math. (2nd Ser.) **175** (2012), no.3, 1473–1506.
169. Hytönen, T. P., Lacey, M. T., Pérez, C., *Sharp weighted bounds for the q-variation of singular integrals*, Bull. Lond. Math. Soc. **45** (2013), no. 3, 529–540.
170. Hytönen, T. P., Pérez, C., *Sharp weighted bounds involving A_∞*, Anal. PDE **6** (2013), no. 4, 777–818.
171. Igari, S., *Lectures on Fourier Series of Several Variables*, Univ. Wisconsin Lecture Notes, Madison, WI, 1968.
172. Ionescu, A. D., *An endpoint estimate for the Kunze-Stein phenomenon and related maximal operators*, Ann. Math. (2) **152** (2000), no. 1, 259–275.
173. Iosevich, A., Sawyer, E., *Maximal averages over surfaces*, Adv. Math. **132** (1997), no. 1, 46–119.
174. Ismagilov, R. S., *On the Pauli problem*, Funct. Anal. Appl. **30** (1996), no. 2, 138–140.
175. Iwaniec, T., Martin, G., *Riesz transforms and related singular integrals*, J. Reine Angew. Math. **473** (1996), 25–57.
176. Jessen, B., Marcinkiewicz, J., Zygmund, A., *Note on the differentiability of multiple integrals*, Fund. Math. **25** (1935), 217–234.
177. Jodeit, M. Jr., *Restrictions and extensions of Fourier multipliers*, Studia Math. **34** (1970), 215–226.
178. Jodeit, M., *A note on Fourier multipliers*, Proc. Amer. Math. Soc. **27** (1971), 423–424.
179. Jones, P., *Factorization of A_p weights*, Ann. of Math. (2nd Ser.) **111** (1980), no. 3, 511–530.
180. Jordan, C., *Sur la série de Fourier*, Comptes Rendus Hebdomadaires des Séances de l' Académie des Sciences **92** (1881), 228–230.
181. Journé, J.- L., *Calderón–Zygmund operators on product spaces*, Rev. Mat. Iberoamericana **1** (1985), no. 3, 55–91.
182. Kahane, J.-P., *The heritage of Fourier*, Perspectives in Analysis, 83–95, Math. Phys. Stud., **27**, Springer, Berlin, 2005.
183. Kahane, J.-P., Katznelson, Y., *Sur les ensembles de divergence des séries trigonométriques*, Studia Math. **26** (1966), 305–306.

184. Kaiser, G., *A Friendly Guide to Wavelets*, Birkhäuser, Boston, MA, 1994.
185. Kalton, N. J., *Convexity, type, and the three space problem*, Studia Math. **69** (1980/81), no. 3, 247–287.
186. Kalton, N. J., *Plurisubharmonic functions on quasi-Banach spaces*, Studia Math. **84** (1986), no. 3, 297–323.
187. Kalton, N. J., *Analytic functions in non-locally convex spaces and applications*, Studia Math. **83** (1986), no. 3, 275–303.
188. Kalton, N. J., Peck, N. T., Roberts, J. W., *An F-space Sampler*, London Mathematical Society Lecture Notes Series, 89, Cambridge University Press, Cambridge, 1984.
189. Katznelson, Y., *Sur les ensembles de divergence des séries trigonométriques*, Studia Math. **26** (1966), 301–304.
190. Katznelson, Y., *An Introduction to Harmonic Analysis*, 2nd corrected edition, Dover Publications, Inc., New York, 1976.
191. Kendall, D. G., *On the number of lattice points inside a random oval*, Quart. J. Math., Oxford Ser. **19** (1948), 1–26.
192. Kenig, C., Tomas, P. A., *Maximal operators defined by Fourier multipliers*, Studia Math. **68** (1980), no. 1, 79–83.
193. Khintchine, A., *Über dyadische Brüche*, Math. Z. **18** (1923), no. 1, 109–116.
194. Kislyakov, S., Kruglyak, N., *Extremal Problems in Interpolation Theory, Whitney–Besicovitch Coverings, and Singular Integrals*, Birkhüser/Springer Basel AG, Basel, 2013.
195. Kolmogorov, A. N., *Une série de Fourier–Lebesgue divergente presque partout*, Fund. Math. **4** (1923), 324–328.
196. Kolmogorov, A. N., *Une contribution à l' étude de la convergence des séries de Fourier*, Fund. Math. **5** (1924), 96–97.
197. Kolmogorov, A. N., *Sur les fonctions harmoniques conjuguées et les séries de Fourier*, Fund. Math. **7** (1925), 23–28.
198. Kolmogorov, A. N., *Une série de Fourier–Lebesgue divergente partout*, C. R. Acad. Sci. Paris **183** (1926), 1327–1328.
199. Kolmogorov, A. N., *Zur Normierbarkeit eines topologischen Raumes*, Studia Math. **5** (1934), 29–33.
200. Konyagin, S. V., *On the divergence everywhere of trigonometric Fourier series*, Sb. Math. **191** (2000), no. 1–2, 97–120.
201. Korenovskyy, A. A., Lerner, A. K., Stokolos, A. M., *A note on the Gurov–Reshetnyak condition*, Math. Res. Lett. **9** (2002), no. 5–6, 579–583.
202. Körner, T. W., *Fourier Analysis*, Cambridge University Press, Cambridge, 1988.
203. Krantz, S. G., *A panorama of Harmonic Analysis*, Carus Mathematical Monographs, 27, Mathematical Association of America, Washington, DC, 1999.
204. Krantz, S. G., Parks, H. R., *The Geometry of Domains in Space*, Birkhäuser Advanced Texts: Basler Lehrbücher, Birkhäuser Boston, Inc., Boston, MA, 1999.
205. Krantz, S. G., Parks, H. R., *A Primer of Real Analytic Functions*, 2nd edition, Birkhäuser Advanced Texts: Basler Lehrbücher, Birkhäuser Boston, Inc., Boston, MA, 2002.
206. Krivine, J. L., *Théorèmes de factorisation dans les espaces réticulés*, Séminaire Maurey-Schwartz 1973–74: Espaces L^p, applications radonifiantes et géométrie des espaces de Banach, Exp. Nos. 22 et 23, Centre de Math., École Polytech., Paris, 1974.
207. Kruglyak, N., *An elementary proof of the real version of the Riesz-Thorin theorem*, Interpolation Theory and Applications, pp. 179–182, Contemp. Math. 445, Amer. Math. Soc., Providence, RI, 2007.
208. Kurtz, D., *Operator estimates using the sharp function*, Pacific J. Math. **139** (1989), no. 2, 267–277.
209. Kurtz, D., Wheeden, R., *Results on weighted norm inequalities for multipliers*, Trans. Amer. Math. Soc. **255** (1979), 343–362.
210. Lacey, M. T., Petermichl, S., Reguera, M. C., *Sharp A_2 inequality for Haar shift operators*, Math. Ann. **348** (2010), no. 1, 127–141.
211. Laeng, E., *On the L^p norms of the Hilbert transform of a characteristic function*, J. Funct. Anal. **262** (2012), no. 10, 4534–4539.

212. Landau, E., *Zur analytischen Zahlentheorie der definiten quadratischen Formen (Über die Gitterpunkte in einem mehrdimensionalen Ellipsoid)*, Berl. Sitzungsber. **31** (1915), 458–476; reprinted in E. Landau "Collected Works," Vol. 6, pp. 200–218, Thales-Verlag, Essen, 1986.
213. Lang, S., *Real Analysis*, 2nd edition, Addison-Wesley Publishing Company, Advanced Book Program, Reading, MA, 1983.
214. Lebesgue, H., *Intégrale, longeur, aire*, Annali Mat. Pura Appl. **7** (1902), 231–359.
215. Lebesgue, H., *Oeuvres Scientifiques (en cinq volumes)*, Vol. I, (French) Sous la rédaction de F. Châtelet et G. Choquet, Institut de Mathématiques de l' Université de Genève, Geneva, 1972.
216. Lemarié, P., Meyer, Y., *Ondelettes et bases hilbertiennes*, Rev. Mat. Iberoamericana **2** (1986), no. 1–2, 1–18.
217. Lerner, A. K., *On pointwise estimates for maximal and singular integral operators*, Studia Math. **138** (2000), no. 3, 285–291.
218. Lerner, A. K., *An elementary approach to several results on the Hardy–Littlewood maximal operator*, Proc. Amer. Math. Soc., **136** (2008), no. 8, 2829–2833.
219. Lerner, A. K., *A pointwise estimate for the local sharp maximal function with applications to singular integrals*, Bull. London Math. Soc. **42** (2010), no. 5, 843–856.
220. Lerner, A. K., *On an estimate of Calderón-Zygmund operators by dyadic positive operators*, J. Anal. Math. **121** (2013), 141–161.
221. Lerner, A. K., *A simple proof of the A_2 conjecture*, Int. Math. Res. Not. IMRN **14** (2013), 3159–3170.
222. Lerner, A. K., *Mixed A_p-A_r inequalities for classical singular integrals and Littlewood-Paley operators*, J. Geom. Anal. **23** (2013), no. 3, 1343–1354.
223. Lerner, A. K., Ombrosi, S., Pérez, C., *A_1 bounds for Calderón–Zygmund operators related to a problem of Muckenhoupt and Wheeden*, Math. Res. Lett. **16** (2009), no. 1, 149–156.
224. Liang, Y. Y., Liu, L. G., Yang, D. C., *An Off-diagonal Marcinkiewicz interpolation theorem on Lorentz spaces*, Acta Math. Sin. (Engl. Ser.) **27** (2011), no. 8, 1477–1488.
225. Lions, J.-L., Peetre, J., *Sur une classe d'espaces d' interpolation*, Inst. Hautes Études Sci. Publ. Math. No. **19** (1964), 5–68.
226. Littlewood, J. E., *On a certain bilinear form*, Quart. J. Math. Oxford Ser. **1** (1930), 164–174.
227. Littlewood, J. E., Paley, R. E. A. C., *Theorems on Fourier series and power series*, J. London Math. Soc. **6** (1931), no. 3, 230–233.
228. Littlewood, J. E., Paley, R. E. A. C., *Theorems on Fourier series and power series (II)*, Proc. London Math. Soc. **42** (1936), no. 1, 52–89.
229. Littlewood, J. E., Paley, R. E. A. C., *Theorems on Fourier series and power series (III)*, Proc. London Math. Soc. **43** (1937), no. 2, 105–126.
230. Loomis, L. H., *A note on the Hilbert transform*, Bull. Amer. Math. Soc. **52** (1946), 1082–1086.
231. Loomis, L. H., Whitney, H., *An inequality related to the isoperimetric inequality*, Bull. Amer. Math. Soc. **55** (1949), 961–962.
232. Lorentz, G. G., *Some new function spaces*, Ann. of Math. (2nd Ser.) **51** (1950), no. 1, 37–55.
233. Lorentz, G. G., *On the theory of spaces Λ*, Pacific. J. Math. **1** (1951), 411–429.
234. Lu, S., Ding, Y., Yan, D., *Singular Integrals and Related Topics*, World Scientific Publishing Co. Pte. Ltd., Hackensack, NJ, 2007.
235. Luzin, N., *Sur la convergence des séries trigonométriques de Fourier*, C. R. Acad. Sci. Paris **156** (1913), 1655–1658.
236. Lyapunov, A., *Sur les fonctions-vecteurs complètement additives* [Russian], Izv. Akad. Nauk SSSR Ser. Mat. **4** (1940), 465–478.
237. Magyar, A., Stein, E. M., Wainger, S., *Discrete analogues in harmonic analysis: spherical averages*, Ann. of Math. (2nd Ser.) **155** (2002), no. 1, 189–208.
238. Mallat, S., *Multiresolution approximations and wavelet orthonormal bases of $L^2(R)$*, Trans. Amer. Math. Soc. **315** (1989), no. 1, 69–87.
239. Mallat, S., *A Wavelet Tour of Signal Processing*, Academic Press, Inc., San Diego, CA, 1998.
240. Marcinkiewicz, J., *Sur l'interpolation d'operations*, C. R. Acad. Sci. Paris **208** (1939), 1272–1273.

241. Marcinkiewicz, J., *Sur les multiplicateurs des séries de Fourier*, Studia Math. **8** (1939), 78–91.
242. Marcinkiewicz, J., Zygmund, A., *Quelques inégalités pour les opérations linéaires*, Fund. Math. **32** (1939), 112–121.
243. Marcinkiewicz, J., Zygmund, A., *On the summability of double Fourier series*, Fund. Math. **32** (1939), 122–132.
244. Mateu, J., Orobitg, J., Verdera, J., *Estimates for the maximal singular integral in terms of the singular integral: the case of even kernels*, Ann. of Math. (2nd Ser.) **174** (2011), no. 3, 1429–1483.
245. Mateu, J., Verdera, J., *L^p and weak L^1 estimates for the maximal Riesz transform and the maximal Beurling transform*, Math. Res. Lett. **13** (2006), no. 5–6, 957–966.
246. Mattila, P., *Geometry of sets and measures in Euclidean spaces. Fractals and rectifiability*, Cambridge Studies in Advanced Mathematics, 44, Cambridge University Press, Cambridge, 1995.
247. Maurey, B., *Théorèmes de factorization pour les opérateurs linéaires à valeurs dans les espace L^p*, Astérisque, No. 11, Société Mathématique de France, Paris, 1974.
248. Melas, A., *The best constant for the centered Hardy–Littlewood maximal inequality*, Ann. of Math. (2nd Ser.) **157** (2003), no. 2, 647–688.
249. Meyer, Y., *Principe d' incertitude, bases hilbertiennes et algèbres d' opérateurs*, Séminaire Bourbaki, Vol. 1985/86, Astérisque No. 145-146 (1987) **4**, 209–223.
250. Meyer, Y., *Ondelettes et Opérateurs, I, Ondelletes*, Hermann, Paris, 1990.
251. Meyer, Y., *Ondelettes et Opérateurs, II, Opérateurs de Calderón-Zygmund*, Hermann, Paris, 1990.
252. Meyer, Y., *Wavelets, Vibrations, and Scalings*, CRM Monograph Series, 9, American Mathematical Society, Providence, RI, 1998.
253. Meyer, Y., Coifman, R. R., *Ondelettes et Opérateurs, III, Opérateurs multilinéaires* , Hermann, Paris, 1991.
254. Mihlin, S. G., *On the multipliers of Fourier integrals* [Russian], Dokl. Akad. Nauk. SSSR (N.S.) **109** (1956), 701–703.
255. Miyachi, A., *On some singular Fourier multipliers*, J. Fac. Sci. Univ. Tokyo Sect. IA Math. **28** (1981), no. 2, 267–315.
256. Mockenhaupt, G., Seeger, A., Sogge, C., *Wave front sets, local smoothing and Bourgain's circular maximal theorem*, Ann. of Math. (2nd Ser.) **136** (1992), no. 1, 207–218.
257. Moon, K. H., *On restricted weak type (1,1)*, Proc. Amer. Math. Soc. **42** (1974), 148–152.
258. Morse, A. P., *Perfect blankets*, Trans. Amer. Math. Soc. **69** (1947), 418–442.
259. Muckenhoupt, B., *On certain singular integrals*, Pacific J. Math. **10** (1960), 239–261.
260. Muckenhoupt, B., *Weighted norm inequalities for the Hardy maximal function*, Trans. Amer. Math. Soc. **165** (1972), 207–226.
261. Muckenhoupt, B., *The equivalence of two conditions for weight functions*, Studia Math. **49** (1973/1974), 101–106.
262. Muckenhoupt, B., Wheeden, R. L., *Two weight function norm inequalities for the Hardy–Littlewood maximal function and the Hilbert transform*, Studia Math. **55** (1976), no. 3, 279–294.
263. Müller, D., *A geometric bound for maximal functions associated to convex bodies*, Pacific J. Math. **142** (1990), no. 2, 297–312.
264. Nagel, A., Stein, E. M., Wainger, S., *Differentiation in lacunary directions*, Proc. Nat. Acad. Sci. U.S.A. **75** (1978), no. 3, 1060–1062.
265. Nazarov, F., Reznikov, A., Vasyunin, V., Volberg, A., *A_1 conjecture: weak norm estimates of weighted singular integrals and Bellman functions*, http://sashavolberg.wordpress.com
266. Nazarov, F., Treil, S., Volberg, A., *The Bellman functions and two-weight inequalities for Haar multipliers*, J. Amer. Math. Soc. **12** (1999), no. 4, 909–928.
267. Nazarov, F., Treil, S., Volberg, A., *Bellman function in stochastic control and harmonic analysis,* Systems, Approximation, Singular Integral Operators, and Related Topics (Bordeaux, 2000), pp. 393–423, Oper. Theory Adv. Appl. 129, Birkhäuser, Basel, 2001.

268. Nazarov, F., Treil, S., Volberg, A., *Two weight inequalities for individual Haar multipliers and other well localized operators*, Math. Res. Lett. **15** (2008), no. 3, 583–597.
269. Nguyen, H. V., *A maximal Fourier integral operator and an application*, J. of Pseudodiffential Operators **4** (2013), 443–456.
270. Nyquist, H., *Certain topics in telegraph transmission theory*, Trans. AIEE **47** (Apr. 1928), 617–64.
271. Olver, F. W. J., *Asymptotics and Special Functions*, Academic Press, New York, London, 1974.
272. Orobitg, J., Pérez, C., *A_p weights for nondoubling measures in R^n and applications*, Trans. Amer. Math. Soc. **354** (2002), no. 5, 2013–2033 (electronic).
273. Paley, R. E. A. C., *A remarkable series of orthogonal functions (II)*, Proc. London Math. Soc. **34** (1932), 241–264.
274. Paley, R. E. A. C., Zygmund, A., *On some series of functions*, Proc. Cambridge Phil. Soc. **26** (1930), 337–357.
275. Peetre, J., *Nouvelles propriétés d' espaces d' interpolation*, C. R. Acad. Sci. Paris **256** (1963), 1424–1426.
276. Pereyra, M. C., *Lecture Notes on Dyadic Harmonic Analysis*, Second Summer School in Analysis and Mathematical Physics (Cuernavaca, 2000), pp. 1–60, Contemp. Math., 289, American Mathematical Society, Providence, RI, 2001.
277. Pérez, C., *Weighted norm inequalities for singular integral operators*, J. London Math. Soc. (2) **49** (1994), no. 2, 296–308.
278. Pérez, C., *Some topics from Calderón–Zygmund theory related to Poincaré-Sobolev inequalities, fractional integrals and singular integral operators*, Function Spaces Lectures, Spring School in Analysis, pp. 31–94, Jaroslav Lukeš and Luboš Pick (eds.), Paseky, 1999.
279. Petermichl, S., *The sharp bound for the Hilbert transform on weighted Lebesgue spaces in terms of the classical A_p characteristic*, Amer. J. Math. **129** (2007), no. 5, 1355–1375.
280. Petermichl, S., *The sharp weighted bound for the Riesz transforms*, Proc. Amer. Math. Soc. **136** (2008), no. 4, 1237–1249.
281. Petermichl, S., Volberg, A., *Heating of the Ahlfors-Beurling operator: weakly quasiregular maps on the plane are quasiregular*, Duke Math. J. **112** (2002), no. 2, 281–305.
282. Pichorides, S., *On the best values of the constants in the theorems of M. Riesz, Zygmund and Kolmogorov*, Studia Math. **44** (1972), 165–179.
283. Pinsky, M., *Introduction to Fourier Analysis and Wavelets*, Graduate Studies in Mathematics 102, American Mathematical Society, Providence, RI, 2009.
284. Pinsky, M., Stanton, N. K., Trapa, P. E., *Fourier series of radial functions in several variables*, J. Funct. Anal. **116** (1993), no. 1, 111–132.
285. Plancherel, M., Pólya, G., *Fonctions entières et intégrales de fourier multiples*, Comment. Math. Helv. **10** (1937), no. 1, 110–163.
286. Reed, M., Simon, B., *Methods of Modern Mathematical Physics, II, Fourier analysis, self-adjointness*, Academic Press, New York, London, 1975.
287. Reguera, M. C., *On Muckenhoupt-Wheeden conjecture*, Adv. Math. **227** (2011), no. 4, 1436–1450.
288. Reguera, M., C., Thiele, C., *The Hilbert transform does not map $L^1(Mw)$ to $L^{1,\infty}(w)$*, Math. Res. Lett. **19** (2012), no. 1, 1–7.
289. Ricci, F., Weiss, G., *A characterization of $H^1(\Sigma_{n-1})$*, Harmonic Analysis in Euclidean Spaces Proc. Sympos. Pure Math. (Williams Coll., Williamstown, MA, 1978), Part 1, pp. 289–294, Amer. Math. Soc., Providence, RI, 1979.
290. Riesz, F., *Untersuchungen über Systeme integrierbarer Funktionen*, Math. Ann. **69** (1910), no. 4, 449–497.
291. Riesz, F., *Sur un théorème du maximum de Mm. Hardy et Littlewood*, J. London Math. Soc. **7** (1932), 10–13.
292. Riesz, M., *Les fonctions conjuguées et les séries de Fourier*, C. R. Acad. Sci. Paris **178** (1924), 1464–1467.
293. Riesz, M., *Sur les maxima des formes bilinéaires et sur les fonctionnelles linéaires*, Acta Math. **49** (1927), no. 3-4, 465–497.

294. Riesz, M., *Sur les fonctions conjuguées*, Math. Z. **27** (1928), no. 1, 218–244.
295. Riesz, M., *L'intégrale de Riemann–Liouville et le problème de Cauchy*, Acta Math. **81** (1949), no. 1, 1–223.
296. Riviere, N., *Singular integrals and multiplier operators*, Ark. Mat. **9** (1971), 243–278.
297. Rolewicz, S., *Metric Linear Spaces*, 2nd edition, Mathematics and its Applications (East European Series), 20. D. Reidel Publishing Co., Dordrecht; PWN–Polish Scientific Publishers, Warsaw, 1985.
298. Rosenblum, M., *Summability of Fourier series in $L^p(d\mu)$*, Trans. Amer. Math. Soc. **105** (1962), 32–42.
299. Rubio de Francia, J.-L., *Weighted norm inequalities and vector valued inequalities*, Harmonic Analysis (Minneapolis, MN, 1981), pp. 86–101, Lect. Notes in Math. 908, Springer, Berlin, New York, 1982.
300. Rubio de Francia, J.-L., *Factorization theory and A_p weights*, Amer. J. Math. **106** (1984), no. 3, 533–547.
301. Rubio de Francia, J.-L., *A Littlewood–Paley inequality for arbitrary intervals*, Rev. Mat. Iberoamericana **1** (1985), no. 2, 1–14.
302. Rubio de Francia, J.-L., *Maximal functions and Fourier transforms*, Duke Math. J. **53** (1986), no. 2, 395–404.
303. Rubio de Francia, J.-L., Ruiz, F. J., Torrea, J. L., *Calderón–Zygmund theory for operator-valued kernels*, Adv. in Math. **62** (1986), no. 1, 7–48.
304. Rubio de Francia, J.-L., Torrea, J. L., *Vector extensions of operators in L^p spaces*, Pacific J. Math. **105** (1983), no. 1, 227–235.
305. Rudin, W., *Trigonometric series with gaps*, J. Math. Mech. **9** (1960), 203–227.
306. Rudin, W., *Functional Analysis*, Tata McGraw-Hill Publishing Company, New Delhi, 1973.
307. Rudin, W., *Real and Complex Analysis*, 2nd edition, Tata McGraw-Hill Publishing Company, New Delhi, 1974.
308. Ryabogin, D., Rubin, B., *Singular integrals generated by zonal measures*, Proc. Amer. Math. Soc. **130** (2002), no. 3, 745–751.
309. Sadosky, C., *Interpolation of Operators and Singular Integrals*, Marcel Dekker Inc., New York, 1979.
310. Saks, S., *Theory of the Integral*, Hafner Publ. Co, New York, 1938.
311. Sato, S., *Note on a Littlewood–Paley operator in higher dimensions*, J. London Math. Soc. (2) **42** (1990), no. 3, 527–534.
312. Sawyer, E., *A characterization of a two-weight norm inequality for maximal operators*, Studia Math. **75** (1982), no. 1, 1–11.
313. Schlag, W., *A geometric proof of the circular maximal theorem*, Duke Math. J. **93** (1998), no. 3, 505–533.
314. Schwartz, L., *Théorie de Distributions. Tome I*, Hermann & Cie, Paris, 1950.
315. Schwartz, L., *Théorie de Distributions. Tome II*, Hermann & Cie, Paris, 1951.
316. Seeger, A., *Some inequalities for singular convolution operators in L^p-spaces*, Trans. Amer. Math. Soc. **308** (1988), no. 1, 259–272.
317. Seeger, A., *Singular integral operators with rough convolution kernels*, J. Amer. Math. Soc. **9** (1996), no. 1, 95–105.
318. Seeger, A., Tao, T., *Sharp Lorentz estimates for rough operators*, Math. Ann. **320** (2001), no. 2, 381–415.
319. Shannon, C. E., *Communication in the presence of noise*, Proc. I.R.E. **37** (1949), 10–21.
320. Shapiro, V. L., *Fourier series in several variables*, Bull. Amer. Math. Soc. **70** (1964), 48–93.
321. Shapiro, V. L., *Fourier series in several variables with applications to partial differential equations*, Chapman & Hall/CRC Applied Mathematics and Nonlinear Science Series, CRC Press, Boca Raton, FL, 2011.
322. Siegel, A., *Historical review of the isoperimetric theorem in 2-D, and its place in elementary plane geometry*, http://www.cs.nyu.edu/faculty/siegel/SCIAM.pdf
323. Sjögren, P., *A remark on the maximal function for measures in $\mathbb{R}^n$*, Amer. J. Math. **105** (1983), no. 5, 1231–1233

324. Sjögren, P., Soria, F., *Rough maximal operators and rough singular integral operators applied to integrable radial functions*, Rev. Math. Iberoamericana **13** (1997), no. 1, 1–18.
325. Sjölin, P., *Convergence almost everywhere of certain singular integrals and multiple Fourier series*, Ark. Math. **9** (1971), 65–90.
326. Sjölin, P., *A note on Littlewood–Paley decompositions with arbitrary intervals*, J. Approx. Theory **48** (1986), no. 3, 328–334.
327. Sjölin, P., Soria, F., *Remarks on a theorem by N. Yu. Antonov*, Studia Math. **158** (2003), no. 1, 79–97.
328. Sogge, C., *Fourier Integrals in Classical Analysis*, Cambridge Tracts in Mathematics, 105, Cambridge University Press, Cambridge, 1993.
329. Soria, F., *A note on a Littlewood–Paley inequality for arbitrary intervals in* $\mathbb{R}^2$, J. London Math. Soc. (2) **36** (1987), no. 1, 137–142.
330. Soria, F., *On an extrapolation theorem of Carleson-Sjölin with applications to a.e. convergence of Fourier series*, Studia Math. **94** (1989), no. 3, 235–244.
331. Stein, E. M., *Interpolation of linear operators*, Trans. Amer. Math. Soc. **83** (1956), 482–492.
332. Stein, E. M., *Note on singular integrals*, Proc. Amer. Math. Soc. **8** (1957), 250–254.
333. Stein, E. M., *Localization and summability of multiple Fourier series*, Acta Math. **100** (1958), no. 1–2, 93–147.
334. Stein, E. M., *On the functions of Littlewood–Paley, Lusin, and Marcinkiewicz*, Trans. Amer. Math. Soc. **88** (1958), 430–466.
335. Stein, E. M., *On limits of sequences of operators*, Ann. of Math. (2nd Ser.) **74** (1961), no. 1, 140–170.
336. Stein, E. M., *Note on the class* $L\log L$, Studia Math. **32** (1969), 305–310.
337. Stein, E. M., *Topics in Harmonic Analysis Related to the Littlewood–Paley Theory*, Annals of Mathematics Studies, No. 63, Princeton University Press, Princeton, NJ, 1970.
338. Stein, E. M., *Singular Integrals and Differentiability Properties of Functions*, Princeton Mathematical Series, No. 30, Princeton University Press, Princeton, NJ, 1970.
339. Stein, E. M., *Maximal functions I: Spherical means*, Proc. Nat. Acad. Sci. U.S.A. **73** (1976), no. 7, 2174–2175.
340. Stein, E. M., *The development of square functions in the work of A. Zygmund*, Bull. Amer. Math. Soc. (N.S.) **7** (1982), no. 2, 359–376.
341. Stein, E. M., *Some results in harmonic analysis in* $\mathbf{R}^n$, *for* $n \to \infty$, Bull. Amer. Math. Soc. (N.S.) **9** (1983), no. 1, 71–73.
342. Stein, E. M., *Boundary behavior of harmonic functions on symmetric spaces: Maximal estimates for Poisson integrals*, Invent. Math. **74** (1983), no. 1, 63–83.
343. Stein, E. M., *Oscillatory integrals in Fourier analysis*, Beijing Lectures in Harmonic Analysis (Beijing, 1984), pp. 307–355, Annals of Math. Studies 112, Princeton Univ. Press, Princeton, NJ, 1986.
344. Stein, E. M., *Harmonic Analysis, Real Variable Methods, Orthogonality, and Oscillatory Integrals*, Princeton Mathematical Series, 43, Monographs in Harmonic Analysis, III, Princeton University Press, Princeton, NJ, 1993.
345. Stein, E. M., Strömberg, J.-O., *Behavior of maximal functions in* $\mathbf{R}^n$ *for large* n, Ark. Math. **21** (1983), no. 2, 259–269.
346. Stein, E. M., Weiss, G., *Interpolation of operators with change of measures*, Trans. Amer. Math. Soc. **87** (1958), 159–172.
347. Stein, E. M., Weiss, G., *An extension of theorem of Marcinkiewicz and some of its applications*, J. Math. Mech. **8** (1959), 263–284.
348. Stein, E. M., Weiss, G., *Introduction to Fourier Analysis on Euclidean Spaces*, Princeton Mathematical Series, No. 32, Princeton University Press, Princeton, NJ, 1971.
349. Stein, E. M., Weiss, N. J., *On the convergence of Poisson integrals*, Trans. Amer. Math. Soc. **140** (1969), 34–54.
350. Stein, P., *On a theorem of M. Riesz*, J. London Math. Soc. **8** (1933), no. 4, 242–247.
351. Stepanov, V. D., *Convolution integral operators [Russian]*, Dokl. Akad. Nauk SSSR **243** (1978), no. 1, 45–48.

352. Strömberg, J.-O., *A modified Franklin system and higher-order spline systems on* $\mathbb{R}^n$ *as unconditional bases for Hardy spaces*, Conference on Harmonic Analysis in Honor of Antoni Zygmund, Vol. I, II (Chicago, IL, 1981), pp. 475–494, Wadsworth Math. Ser., Wadsworth, Belmont, CA, 1983.
353. Szarek, S. J., *On the best constant in the Khinchin inequality*, Studia Math. **58** (1976), no. 2, 197–208.
354. Tamarkin, J. D., Zygmund, A., *Proof of a theorem of Thorin*, Bull. Amer. Math. Soc. **50** (1944), 279–282.
355. Tao, T., *The weak type (1,1) of* $L\log L$ *homogeneous convolution operators*, Indiana Univ. Math. J. **48** (1999), no. 4, 1547–1584.
356. Tao, T., *A converse extrapolation theorem for translation-invariant operators*, J. Funct. Anal. **180** (2001), no. 1, 1–10.
357. Tao, T., Wright, J., *Endpoint multiplier theorems of Marcinkiewicz type*, Rev. Mat. Iberoamericana **17** (2001), no. 3, 521–558.
358. Tauber, A., *Ein Satz aus der Theorie der unendlichen Reihen* Monatsh. Math. Phys. **8** (1897), no. 1, 273–277.
359. Tevzadze, N. R., *The convergence of the double Fourier series of quadratic at a square summable function*, Sakharth. SSR Mecn. Akad. Moambe **58** (1970), 277–279.
360. Thorin, G. O., *An extension of a convexity theorem due to M. Riesz*, Fys. Säellsk. Förh. **8** (1938), No. 14.
361. Thorin, G. O., *Convexity theorems generalizing those of M. Riesz and Hadamard with some applications*, Comm. Sem. Math. Univ. Lund [Medd. Lunds Univ. Mat. Sem.] **9** (1948), 1–58.
362. Titchmarsh, E. C., *The Theory of the Riemann Zeta Function*, Oxford, at the Clarendon Press, 1951.
363. Torchinsky, A., *Real-Variable Methods in Harmonic Analysis*, Pure and Applied Mathematics, 123, Academic Press, Inc., Orlando, FL, 1986.
364. van der Corput, J. G., *Zahlentheoretische Abschätzungen*, Math. Ann. **84** (1921), no. 1–2, 53–79.
365. Vargas, A., *On the maximal function for rotation invariant measures in* $\mathbb{R}^n$, Studia Math. **110** (1994), no. 1, 9–17.
366. Verbitsky, I. E., *Estimate of the norm of a function in a Hardy space in terms of the norms of its real and imaginary part*, Amer. Math. Soc. Transl. **124** (1984), 11–15.
367. Verbitsky, I. E., *Weighted norm inequalities for maximal operators and Pisier's theorem on factorization through* $L^{p\infty}$, Integral Equations Operator Theory **15** (1992), no. 1, 124–153.
368. Vitali, G., *Sui gruppi di punti e sulle funzioni di variabili reali*, Atti Accad. Sci. Torino **43** (1908), 229–246.
369. Von Neuman, J., *Zur Theorie des Gesellschaftsspiele*, Math. Ann. **100** (1928), 295–320.
370. Wainger, S., *Special trigonometric series in k dimensions*, Mem. Amer. Math. Soc. 59, 1965.
371. Watson, G. N., *A Treatise on the Theory of Bessel Functions*, Cambridge University Press, Cambridge, England, The Macmillan Company, New York, 1944.
372. Weiss, M., Zygmund, A., *An example in the theory of singular integrals*, Studia Math. **26** (1965), 101–111.
373. Whitney, H., *Analytic extensions of differentiable functions defined in closed sets*, Trans. Amer. Math. Soc. **36** (1934), no. 1, 63–89.
374. Wickerhauser, M. V., *Adapted Wavelet Analysis from Theory to Software*, A. K. Peters, Ltd., Wellesley, MA, 1994.
375. Wiener, N., *The ergodic theorem*, Duke Math. J. **5** (1939), no. 1, 1–18.
376. Wilbraham, H., *On a certain periodic function*, The Cambridge and Dublin Mathematical Journal, **3** (1848), 198–201.
377. Wilson, J. M., *Weighted norm inequalities for the continuous square function*, Trans. Amer. Math. Soc. **314** (1989), no. 2, 661–692.
378. Wilson, M., *Weighted Littlewood–Paley Theory and Exponential-Square Integrability*, Lecture Notes in Mathematics, 1924 Springer, Berlin, 2008.
379. Wojtaszczyk, P., *A Mathematical Introduction to Wavelets*, London Mathematical Society Student Texts 37, Cambridge University Press, Cambridge, 1997.

380. Yano, S., *Notes on Fourier analysis, XXIX, An extrapolation theorem*, J. Math. Soc. Japan **3** (1951), 296–305.
381. Yanushauskas, A. I., *Multiple trigonometric series* [Russian], "Nauka" Sibirsk. Otdel., Novosibirsk, 1986.
382. Yosida, K., *Functional Analysis*, Reprint of the 6th edition (1980), Classics in Mathematics Springer-Verlag, Berlin, 1995.
383. Zhizhiashvili, L. V., *Some problems in the theory of simple and multiple trigonometric and orthogonal series*, Russian Math. Surveys, **28** (1973), 65–127.
384. Zhizhiashvili, L. V., *Trigonometric Fourier Series and Their Conjugates*, Mathematics and Its Applications, 372, Kluwer Academic Publishers Group, Dordrecht,1996.
385. Ziemer, W. P., *Weakly Differentiable Functions, Sobolev Spaces and Functions of Bounded Variation*, Graduate Texts in Mathematics, 120, Springer-Verlag, New York, 1989.
386. Zo, F., *A note on approximation of the identity*, Studia Math. **55** (1976), no. 2, 111–122.
387. Zygmund, A., *On a theorem of Marcinkiewicz concerning interpolation of operators*, J. Math. Pures Appl. (9) **35** (1956), 223–248.
388. Zygmund, A., *Trigonometric Series*, Vol. I, 2nd edition, Cambridge University Press, New York, 1959.
389. Zygmund, A., *Trigonometric Series*, Vol. II, 2nd edition, Cambridge University Press, New York, 1959.
390. Zygmund, A., *Notes on the history of Fourier series,* Studies in harmonic analysis, (Proc. Conf. DePaul Univ., Chicago, IL, 1974) pp. 1–19, MAA Stud. Math., Vol. 13, Math. Assoc. Amer., Washington, DC, 1976.

Index

L. Grafakos, *Classical Fourier Analysis*, Graduate Texts in Mathematics 249,
DOI 10.1007/978-1-4939-1194-3,

Printed in the United States
By Bookmasters